The Field Theoretic Renormalization Group in Critical Behavior Theory and Stochastic Dynamics

T0239620

The Field Theoretic Renormalization Group in Critical Behavior Theory and Stochastic Dynamics

A. N. Vasil'ev

St. Petersburg State University
Russia

Translated by

Patricia A. de Forcrand-Millard

CRC Press
Taylor & Francis Group
Boca Raton London New York

CRC Press is an imprint of the
Taylor & Francis Group, an **informa** business
A CHAPMAN & HALL BOOK

Originally published in Russian in 1998 as
Квантовополевая ренормгруппа в теории критического поведения и стохастической динамике by St. Petersburg Institute of Nuclear Physics Press, St. Petersburg

Published 2004 by CRC Press
Taylor & Francis Group
6000 Broken Sound Parkway NW, Suite 300
Boca Raton, FL 33487-2742

© 1998 A.N. Vasil'ev
CRC Press is an imprint of Taylor & Francis Group, an Informa business

No claim to original U.S. Government works

ISBN 13: 978-0-367-57837-4 (pbk)
ISBN 13: 978-0-415-31002-4 (hbk)

Visit the Taylor & Francis Web site at
http://www.taylorandfrancis.com

and the CRC Press Web site at
http://www.crcpress.com

Library of Congress Cataloging-in-Publication Data

Vasil'ev, A.N. (Aleksandr Nikolaevich)
The field theoretic renormalization group in critical behavior theory and stochastic dynamics / by A.N. Vasil'ev.
p. cm.
Includes bibliographical references and index.
ISBN 0-415-31002-4 (alk. paper)
1. Renormalization group. 2. Critical phenomena (Physics). 3. Stochastic processes. 4. Statistical physics. I. Title.

QC20.7.R43V37 2004
530.13'3—dc22   2004043573

Library of Congress Card Number 2004043573

Contents

CHAPTER 6 Stochastic Theory of Turbulence   **581**

Preface

The goal of this book is to present a detailed exposition of the quantum field renormalization-group (RG) technique and its applications to various problems in the classical theory of critical behavior and stochastic dynamics. The common feature of these problems is the presence of a nontrivial scale invariance, "scaling" (critical scaling in the theory of critical phenomena, Kolmogorov scaling in the theory of turbulence, and so on). The RG is the simplest and most effective method of establishing scaling in specific models and calculating the corresponding critical exponents.

There are many different formulations of the RG technique. While they are all conceptually equivalent, they differ considerably in their technical aspects. In this book we present the quantum field version of the RG, which is the most formal and technically most effective, especially for calculations in higher orders of perturbation theory. It is based on the techniques of ultraviolet (UV) renormalization, which is well developed in quantum field theory and provides a reliable foundation for the quantum field version of the RG.

The fundamental principles of UV renormalization and the associated RG technique were developed in quantum field theory in the 1950s. The formalism of quantum field theory was initially created as a mathematical tool for the quantum physics of elementary particles. However, it subsequently became clear that the formalism itself is not related to the quantum nature of the problem, and that it is equally applicable (in its Euclidean version) to classical problems involving random fields. This makes it possible to apply the techniques developed in quantum field theory (graph representations, functional integration, and so on) to problems in the theory of critical phenomena involving a classical random field (the order parameter), and also to various problems in stochastic dynamics with time-dependent classical random fields.

All these problems involve an infinite number of degrees of freedom, and complete information about the system is contained in the infinite family of Green functions: the correlation functions of the random field and various response functions in dynamical models. As a rule, here one does not try to obtain an exact solution. From the practical point of view, it is usually the asymptote of the solution in some region that is of interest. In the problems that will be considered in this book, this will always be the infrared (IR) asymptote of the Green functions, corresponding to small momenta (or large relative separations) and parameters close to the critical point. For any particular quantum field model, the RG technique can be used to prove (or disprove) the existence of the corresponding IR scaling and, if scaling does occur, to obtain explicit expressions for calculating the critical exponents in the form of expansions in the parameter ε characterizing the deviation from logarithmic behavior (in the original scheme of Wilson, for the classical φ^4 model $\varepsilon \equiv 4 - d$, where d is the variable spatial dimension). This technique, first applied to problems in critical statics (thermodynamics and the equilibrium statistical physics of the order-parameter field) by Wilson and coworkers (awarded the Nobel Prize in 1982), was later generalized to critical dynamics (analysis of the behavior of the relaxation time and the various kinetic coefficients in the neighborhood of a critical point), and then to many other related problems, in particular, the theory of polymers, the stochastic theory of turbulence and magnetic hydrodynamics, and various problems in random diffusion. The list of such problems continues to grow, and so acquaintance with the RG technique should be considered one of the essential elements in the education of a modern theoretical physicist.

This book is based on a series of lectures on the quantum field RG technique given by the author at St. Petersburg State University for graduate students specializing in theoretical physics, and is intended as a general textbook and also a reference book on this topic. No prior acquaintance with the techniques of quantum field theory is required of the reader, as all the necessary background material on the functional and graph technique and the theory of UV renormalization is given in Chs. 2 and 3.

There are a number of other books (see, for example, [44], [45], [48], and [49]) dealing with the ideas and techniques of the RG approach in the theory of critical phenomena. The present book is primarily devoted to the detailed description of the computational techniques. Therefore, the general ideas are always explained using examples of particular calculations, giving as many details as possible (for example, in Ch. 3 we present the complete calculation of the renormalization constants of the φ^4 model in the three-loop approximation). The present book can therefore be regarded as a textbook on "how to calculate." The problems of justifying the various models and comparing the results with experiment lie outside its scope, and, as a rule, are not discussed.

This text can also be regarded as a reference book, as all the final results of the calculations are presented in universal notation with the highest level of accuracy attained at present and with all misprints in earlier studies corrected. This should help the reader obtain information without resorting to the original studies.

Another distinguishing feature of this text is its discussion of problems in stochastic dynamics (Chs. 5 and 6), which is much more detailed than elsewhere.

A few words about the organization of the text. It consists of six chapters, divided into sections. The first chapter presents the general scheme of the RG approach in the theory of critical phenomena (statics) without discussing the actual techniques for calculating the renormalization constants. It includes a historical review, the phenomenology of critical scaling, the general principles of constructing fluctuation models and their multiplicative renormalization, a derivation of the RG equations and analysis of their general solutions, the justification of critical scaling using the RG equations, and a description of the technique for calculating the critical dimensions and scaling functions in the form of ε expansions.

In the next two chapters we present background information on the techniques of quantum field theory used to calculate the renormalization constants and RG functions. In Ch. 2 we formulate the basic rules of the functional and graph technique of quantum field theory, derive the Schwinger equations and the Ward identities, and explain the technique of constructing large-n expansions. Chapter 3 is devoted to the theory of UV renormalization, with special focus on modern technical tools (dimensional regularization, minimal subtraction, the passage to massless graphs), which significantly simplify the computational procedure. We discuss in detail the combinatorics of the R operation, the technique for calculating counterterms, the matrix multiplicative renormalization of composite operators, and the Wilson operator expansion (the short-distance expansion).

The longest chapter, Ch. 4, contains material that substantially amplifies and particularizes the general scheme described in Ch. 1. Here we study various specific models of critical statics, explaining the details of the calculations for different variants of the asymptotic expansions of the exponents, in particular, the $4-\varepsilon$, $2+\varepsilon$, and large-n expansions in the standard $O_n \; \varphi^4$ model. We give the generalization of the RG equations to the case of functionals involving vacuum loops, and explain in detail and illustrate by examples the technique of calculating the critical dimensions for families of composite operators that mix under renormalization. In this chapter we also present an alternative to the RG technique for calculating the critical dimensions using various self-consistency equations, including the conformal bootstrap method. In connection with this, we discuss the problem of justifying the critical conformal invariance of the Green functions in various models, and also the criterion for the critical conformal invariance of composite operators. The description of a number of special technical tricks for calculating the contributions of massless graphs in arbitrary spatial dimension is of independent interest.

Chapter 5 is devoted to critical dynamics. First we discuss the general form of the equations of stochastic dynamics and the rules for going from them to the equivalent quantum field models. Problems in critical dynamics involve equations of a special form, namely, stochastic Langevin equations with possible contributions from the intermode interaction. We formulate the general principles for constructing such models, present the standard list [153] of specific dynamical A–J models with their description and interpretation, and then give the RG analysis and the results of calculating the dynamical exponents for each model. In the last section of this chapter we study the interaction of a sound wave with the critical fluctuations near the critical point of the liquid–gas transition and derive the corresponding dispersion law (the speed of sound and damping).

Finally, in Ch. 6 we discuss dynamical problems of the non-Langevin type, namely, the stochastic theory of turbulence and its generalizations: stochastic magnetic hydrodynamics, the turbulent mixing of a passive admixture, and the Langmuir turbulence of a plasma.

The small Addendum at the end of the book contains a brief summary of information about some of the new results on topics discussed earlier.

This book is organized such that a general idea of the subject can be obtained by reading only the first section in Ch. 1, the historical review. Careful reading of the entire Ch. 1 is sufficient for basic acquaintance with the general principles of the RG technique and its application to typical problems in critical statics. The reader interested in the details and techniques of specific calculations must turn to Ch. 4. The reader well acquainted with the techniques of quantum field theory can skip Chs. 2 and 3, referring to them only when necessary. The final Chs. 5 and 6 are for those interested in applications of the RG technique to dynamical problems.

Finally, the reader should be warned that Sec. 3.15 introduces nonstandard compact notation for the gamma functions and products of them, and this notation is used frequently in the later discussion without explanation.

I would like to express my gratitude to all the students and coworkers at the Theoretical Physics Department of St. Petersburg State University with whom I have worked and discussed many of the problems touched upon in this book. I am also grateful to J. Zinn-Justin for many useful discussions and for providing the corrected data for the tables in his book [45]. I am especially grateful to A. S. Stepanenko, N. A. Kivel', and F. V. Andreev, who as students willingly and unselfishly devoted a large amount of effort to typesetting the Russian text on computer. At a time when it appeared that there was no hope that such a large and commercially unviable book could be produced in Russia, the book has been published owing to the financial support of the Russian Foundation for Basic Research, for which I am deeply grateful.

The English version of the book includes many improvements and corrections made during the translation of the Russian version. In addition, the updated English Addendum and Bibliography include recent results and references which appeared after the publication of the Russian edition.

Chapter 1

Foundations of the Theory of Critical Phenomena

In this chapter we give a general overview of the modern theory of critical phenomena. In the first section we review the history of the field, and in the sections that follow we discuss the basic concepts and ideas.

1.1 Historical review

The existence of a critical point was discovered in 1869 by Andrews [1] in experiments involving CO_2 ($T_c = 304$ K). Earlier it had been assumed that some gases do not become liquids under pressure at all. Four years later, in his doctoral dissertation Van der Waals [2] proposed the now well-known equation of state of a real gas, thereby constructing what was historically the first theory of a self-consistent field. The analogous theory of ferromagnetism was constructed in 1907 by Weiss [3]. Subsequently, more and more phase transitions were discovered and models describing them were constructed. It naturally became necessary to organize the accumulated results in a systematic form. In 1933 Ehrenfest proposed a classification of phase transitions. It is now known to be inaccurate, and in practice is used only to make the basic distinction between first-order transitions with a finite jump of the order parameter (the magnetization of a ferromagnet, the density in the liquid–gas transition, and so on) and second-order transitions, where the order parameter is continuous. In the space D of all the external parameters specifying the state of a system, the singular points at which a first-order transition occurs form a manifold M of lower dimension (for example, if D is two-dimensional, then M is the one-dimensional phase-coexistence line). When the external parameters are varied smoothly, the point representing the state of the system moves along some trajectory in D. If the trajectory intersects the manifold M at some point x, a first-order transition with finite jump of the order parameter occurs at the intersection point. The magnitude of the jump depends on the location of x inside M and tends to zero when x tends to boundary points of the manifold M. Such points are called *critical points*. If the trajectory arriving from the region of nonsingular points reaches a critical point x_c and then passes through the manifold M, it is said that a second-order phase transition occurs in the system at the point of passage through x_c. This is usually observed as the temperature is decreased, and so in what follows we shall use the abbreviated expressions "above T_c," "below T_c," and "at T_c" to mean "before the transition," "after the transition," and "at the transition" or "at the critical point."

If M is a one-dimensional phase-coexistence line, the endpoint of this line is usually a critical point. However, this is not the general case. For example, for a Bose gas the space D is four-dimensional (the dimensions being the temperature, the pressure, and the complex external field thermodynamically conjugate to the order parameter). The manifold M is a two-dimensional region in the temperature–pressure plane at zero external field, and its boundary forms an entire line of critical points. Therefore, as the temperature is decreased, a second-order phase transition to the superconducting state of liquid helium does not occur at a definite value of the pressure, as in an ordinary liquid–gas transition, but at any value within some range.

It is known from experiments that in approaching a critical point the susceptibility of the system diverges, and anomalies also arise in the specific heat and other quantities. The theory of critical phenomena is devoted to describing such critical behavior.

One of the most important features of this theory is universality: different systems manifest roughly identical critical behavior. For example, the susceptibility of a magnet diverges as $T \to T_c$, roughly like the compressibility of a gas, the spontaneous magnetization behaves like the difference of the liquid and gas densities on the coexistence curve, and so on. This has led to the concept of *universality class* to describe systems with identical critical behavior. According to current ideas, this behavior is determined only by the general properties of the system: the spatial dimension, the nature (number of components, tensor properties, etc.) of the order parameter, the symmetry of the problem, and the general nature of the interaction (long- or short-range), but is independent of the details of the system.

A unified theory of critical phenomena that naturally explains universality was developed in 1937 by Landau [4]. There it is postulated that the equilibrium value α_0 of the order parameter α can be found from the condition that the free energy F, treated as a functional of α and the other variables specifying the state of the system, is a minimum. The second postulate is that F is analytic in all variables in the neighborhood of a critical point. Since the deviations of these variables from the critical values must always be assumed small (continuity), only the first few terms (sufficient to explain the existence of the transition itself) need to be kept in the Taylor expansion of F in the deviations. In the end, the explicit form of F for a given set of variables and a given symmetry is determined practically uniquely, which explains universality.

The Landau theory gives completely defined predictions regarding the singularities of various quantities at the critical point. For the simplest second-order transition in a magnet or a liquid–gas system, it predicts a finite jump of the specific heat and a singularity of the form $|T - T_c|^{-1}$

for the susceptibility (compressibility) and $(T_c - T)^{1/2}$ for the spontaneous magnetization (difference of the densities $\rho_l - \rho_g \equiv \Delta\rho$ on the coexistence curve). However, it later became clear that these predictions are inaccurate. The exact calculation of the partition function of the two-dimensional Ising model for zero external field performed by Onsager in 1942 (and published [5] in 1944) showed that the specific heat in this model has a logarithmic singularity instead of a finite jump. In 1943 Guggenheim [6], who carried out experiments involving several gases, confirmed the power law $\Delta\rho \sim (T_c - T)^\beta$, but with exponent $\beta = \frac{1}{3}$ instead of the $\frac{1}{2}$ given by the Landau theory.

Intensive experimental and theoretical study of critical singularities began in the 1960s. It was decided fairly early on that in the overwhelming majority of cases the singularities are actually power-law singularities, but the exponents, now called the *critical exponents*, differ from the canonical values predicted by the Landau theory.

The Ising model has played a very important role in the construction of the modern theory of critical phenomena, and so we shall briefly describe its curious history [7]. The Ising model is a very simple model of a uniaxial classical ferromagnet with exchange interaction of adjacent spins located at the sites of a given spatial lattice. It is also used to describe other systems such as a lattice gas [8] and binary alloys [9]. It would be more correct to refer to it as the Lentz–Ising model, as it was proposed in 1920 by Lentz [10], and in 1925 his student Ising succeeded in finding an exact solution for a one-dimensional spin chain [11]. He was disappointed at not finding a phase transition at a finite temperature in the solution. Ising considered this to be a fundamental defect of the model, as did other scientists initially (for example, when constructing his quantum model of ferromagnetism, Heisenberg [12] noted that the classical Ising model does not explain the phase transition). Only later did it become clear that the flaw did not lie within the model itself, but was due to the one-dimensional nature of the problem solved by Ising. In 1936 Peierls presented the now well-known arguments showing that it is impossible to have a phase transition in one-dimensional systems with short-range forces [13].

The first accurate information on the phase transition in the Ising model was obtained in 1941 by Kramers and Wannier [14], who succeeded in finding the exact value of T_c for the two-dimensional square Ising lattice from symmetry considerations. On February 28, 1942, at a session of the New York Academy of Sciences, Onsager presented his remarkable exact solution of the two-dimensional model: the partition function for zero external field [15]. The details of the calculation were only published two years later [5]. He later calculated the spontaneous magnetization in the same model and presented the result during a discussion at a conference on phase transitions at Cornell University in August 1948. He then published the formula for the spontaneous magnetization without derivation in [16] (see [17] for the history of these developments). Onsager never published the derivation of this expression, and only four years later did Yang succeed in finding his own solution of the problem and published a complete calculation of the spontaneous magnetization [18]. Onsager [5] used the matrix method proposed earlier in [19], [20]. Kaufman improved this method [21], which allowed her and Onsager to also calculate the spin pair correlation function in zero field [22].

These are all the exact results available for the Ising model, and they all pertain to the two-dimensional model in zero external field. They were later reproduced many times using various techniques (see [23] for references), but no exact results have been obtained for the three-dimensional model or even for the two-dimensional model with an external field. Nevertheless, the critical behavior of the three-dimensional Ising model is now viewed as being fairly well-known owing to the development of effective methods for extrapolating the high- and low-temperature expansions. The technique of constructing such expansions has long been known (see, for example, [24]), but their effective use for analyzing critical singularities requires knowledge of fairly long segments of the series and became possible only in the late 1950s with the appearance of powerful computers. This method has made it possible to obtain fairly reliable and accurate estimates of the critical temperatures and exponents for various spatial lattices [24]. In some sense these results are even more valuable than experimental ones, because they pertain to an exactly defined model, while in actual experiments any deviations from the theoretical predictions can always be attributed to the effect of impurities, unaccounted-for interactions, and so on. Monte Carlo calculations also provide important information about critical behavior.

The results of numerical calculations in the three-dimensional Ising model have proved very important. First of all, they have clearly confirmed the idea of universality: the critical temperatures obtained for various lattices (simple cubic, volume-centered, face-centered, and so on) have turned out to be quite different from each other, while the critical exponents are identical, but different from those in the two-dimensional model. Second, analysis of the exponents [25] has led to the idea of *critical scaling* (or the *scaling hypothesis*), on which the modern theory is based.

The *thermodynamical scaling hypothesis* was formulated practically simultaneously and independently by Domb and Hunter [25] for the Ising model, by Widom [26] for the liquid–gas transition, and by Patashinskiĭ and Pokrovskiĭ [27] for ferromagnetic systems. In magnetic terminology the hypothesis essentially amounts to the statement that the part of the free energy per unit volume responsible for the critical singularities can be assumed to be a generalized homogeneous function of the variable $\tau = T - T_c$ and the external field h, and so all the exponents are uniquely expressed in terms of two parameters: the critical dimensions of the variables τ and h. Since there are many exponents, well-defined relations between them must exist. Such relations do actually hold for the known (see above) exponents of the Ising model, and this observation is the leading argument in favor of the thermodynamical scaling hypothesis.

Kadanoff [28] generalized the thermodynamical scaling hypothesis to the correlation functions and presented heuristic arguments explaining the mechanism that gives rise to critical scaling. The Kadanoff block construction later played an important role in the conceptual foundation of the renormalization-group method. Mention should also be made of the earlier study by Patashinskiĭ and Pokrovskiĭ [29] that was devoted to the phase transition in liquid helium (superfluidity). Here there is no clear, general formulation of the scaling hypothesis, although the authors came very close, even pointing out the analogy with Kolmogorov scaling in the theory of turbulence. That study also contains other important ideas on which the theory was subsequently based: the possibility of substituting the exact microscopic model by a fluctuation model and the idea that the system can "forget" the bare charges in the critical region.

After all these (and many more) studies, the idea of critical scaling became generally accepted in the second half of the 1960s. The classical Landau theory fits within the thermodynamical scaling hypothesis, but gives incorrect predictions for the critical dimensions. The reason is clear: it is a typical theory of a self-consistent field neglecting fluctuations of the order parameter, the magnitude of which grows as T_c is approached, as is clearly manifested in critical opalescence. This implies that near T_c the order parameter should be treated as a random quantity, or, more precisely, as a random field, since the fluctuations will in general be spatially nonuniform. From the viewpoint of the mathematical apparatus, the theory of a classical random field is identical to Euclidean quantum field theory, and in the language of field theory the Landau variational functional is the action functional of the model. According to current ideas, the *fluctuation field model* thus defined is completely equivalent to the original exact microscopic model as far (and only as far) as the critical behavior is concerned. The possibility of substituting a fluctuation model for the exact one is one of the most important postulates of the modern theory of critical phenomena. It is just a postulate, because the validity of the substitution (as far as the author knows) has not been proved completely rigorously in any nontrivial case. The value of such a substitution is obvious: the fluctuation model already contains the idea of universality (the Landau functional), and at the same time includes fluctuations, because it involves not the average value of the order parameter, but the corresponding random field. The need to include fluctuations by going from the simple Landau theory to the corresponding fluctuation field model was realized long ago (see, for example, [29]). However, this could not inspire optimism, as the fact that fluctuations are important near T_c implies that we are dealing with a theory of a strongly interacting field. The difficulties that arise are obvious, and in elementary particle physics, where an analogous technique is used, they have not yet been overcome. The next step, taken by Wilson in the early 1970s, was therefore crucial. He noticed that even for a strong interaction it is possible to obtain specific information about the critical behavior using the *renormalization-group* (RG) *method*.

The RG method was developed in relativistic quantum field theory in the mid-1950s. The existence of a group of renormalization transformations (the renormalization group) describing the arbitrariness of the procedure of ultraviolet renormalization was first pointed out by Stueckelberg and Petermann [30] in 1953. In 1954, Gell–Mann and Low [31] independently proposed a simple method of using a differential equation to obtain the leading logarithmic asymptote in quantum electrodynamics, which had been calculated earlier by Landau, Abrikosov, and Khalatnikov [32] by direct summation of the leading singularities of the graphs of perturbation theory. Right after this, Bogolyubov and Shirkov [33], [34] showed that the approach of [31] is closely related to the renormalization group [30]. They constructed a complete, general quantum field theory of the renormalization group as it is understood today, and a separate chapter was devoted to this topic in the first edition (1957) of their monograph [35].

From the practical point of view, the RG technique is an effective method of calculating a nontrivial asymptote of Green functions at large [ultraviolet (UV)] or small [infrared (IR)] momenta. An asymptote is nontrivial if the individual terms of the perturbation series contain singularities in the momenta, for example, large logarithms that cancel the smallness of the coupling constant (charge) g. In such a situation, in obtaining the desired asymptote it is impossible to deal with only a finite order of perturbation theory; the entire series must be summed. It should be noted that only in models involving the interaction of massless fields can the Green functions have singularities at small momenta. There were none among the relativistic field models discussed in the 1950s. Therefore, in those days people were always dealing with a nontrivial ultraviolet asymptote, and the RG method was originally developed to analyze it.

Let us describe the problem in more detail for the typical example of a propagator (the field analog of a correlator) in relativistic quantum electrodynamics. It depends on the single momentum $k \equiv \sqrt{k^2}$, and the general term of the perturbation series behaves at large momentum as $g^n P_n(z)$, where $g = \frac{1}{137}$ is the coupling constant and P_n is a polynomial of order n in the "large logarithm" $z \equiv \ln(k/k_0)$ with constant k_0 to make the momentum k dimensionless. Clearly, at sufficiently large momentum, when $gz \simeq 1$, it is necessary to sum the entire series. A natural first approximation is to sum the "leading logarithms" $(gz)^n$, keeping only the contribution of the highest power z^n in each polynomial P_n (this was done in [32]). One can then include the first correction corresponding to the sum of all terms of the type $g(gz)^n$ missing one large logarithm, then the next largest correction missing two logarithms, and so on.

Such summations are easily performed using the RG technique. It allows a linear partial differential equation to be obtained for the Green functions. The coefficients of the corresponding differential *RG operator*, called the *RG functions*, can be calculated from perturbation theory, but if the equation itself is then solved exactly, the final results for the Green functions will correspond to the result of the desired infinite summation. For example, by calculating the RG functions in the equation for the propagator in the lowest nontrivial order of perturbation theory and then solving the equation with these RG functions exactly, we obtain the result of [32] for the propagator, which corresponds to the sum of the contributions of all the leading logarithms. If the RG functions are calculated in the next highest order, it becomes possible to find the first correction to the leading-log approximation, and so on.

Therefore, the RG technique is very effective for summing the leading logarithms, as shown in [31], [33]. However, it was noticed almost immediately [35] that this summation is sometimes meaningless, or, more precisely, it gives the correct answer only for one of the two asymptotes (if they are both nontrivial), either the infrared or the ultraviolet. Let us explain in more detail. If the electron in quantum electrodynamics were massless, the representation $g^n P_n(z)$ for the general term of the perturbation series would hold at all momenta, not only large ones, and we would have a large logarithm not only for $k \to \infty$, but also for $k \to 0$. The statement is that the result of summing the leading logarithms will in this case correctly give only one of the two asymptotes: either the one at $z \to -\infty$ or the one at $z \to +\infty$. When the formal result is known, it is easy to determine which one: in the incorrect region the answer necessarily contains some pathology, usually a singularity from a zero in the denominator at finite gz, causing the answer to have the incorrect sign. Such a pathology also occurs in the result of [32] for large momenta. This "Moscow zero" (to use the old term) was once actively discussed in the literature, until it became clear that there was nothing to discuss, because the answer does not at all have the meaning of an ultraviolet asymptote.

In the RG scheme there is a simple way of determining the region of applicability of the obtained asymptote without calculating it explicitly. One of the RG functions is the β *function*, the coefficient of the derivative ∂_g in the RG operator. This is a function of the charge (coupling constant) g, and is actually calculated in perturbation theory as a series in g. For all the standard relativistic models, including quantum electrodynamics, this series begins at g^2: $\beta(g) = \beta_2 g^2 + \beta_3 g^3 + \ldots$. In the general RG theory it is shown that nontrivial asymptotes of the Green functions are associated with *fixed points* g_*, zeros of the β function [i.e., roots of the equation $\beta(g) = 0$], one of which is always the trivial point $g_* = 0$. A fixed point can be IR-attractive or UV-attractive, depending on the behavior of the β function near g_*. A given point g_* determines only its own (according to its type) asymptote of the Green function. The large logarithms mentioned above are associated with the trivial fixed point $g_* = 0$, and the type of fixed point is determined by the sign of the first coefficient β_2 in the expansion of the β function. For quantum electrodynamics it turns out to be IR-attractive ($\beta_2 > 0$) and therefore does not determine the UV asymptote. Calculations performed right after the development of the RG method showed that all the other field models known at that time (φ^4, various Yukawa-type interactions, etc.) possess the same property. If they were massless, the problem of calculating the nontrivial IR asymptote would arise, and the RG technique would be the ideal method for doing this. However, this problem did not arise in elementary particle physics, because real particles and the corresponding fields are massive, and the two exceptions, the photon and the neutrino, interact only with massive particles. For such models the Green functions are analytic at small momenta, and so there is no problem with IR singularities and their summation. Therefore, the RG method was developed in the hope that it could be used to study the nontrivial UV asymptote of the Green functions. However, it proved to be inapplicable for this, and in the end it turned out that for about 15 years the powerful RG technique had essentially no applications.

The situation changed dramatically in the early 1970s owing to the nearly simultaneous appearance of two broad areas in which the RG technique is applicable. One is high-energy relativistic physics, where the ultraviolet asymptotic freedom of non-Abelian gauge theories was discovered. The models of this new class, unknown in the 1950s, are now considered the leading candidate for the theory of the strong interaction in elementary particle physics. It was shown in [36] that for these models the first coefficient β_2 in the expansion of the β function is negative, i.e., it has a sign opposite to that of all the relativistic models previously known. The point $g_* = 0$ is therefore UV-attractive, i.e., it determines the UV asymptote of the Green functions. This corresponds to asymptotic freedom. It thereby became possible to use the RG method to study various high-energy processes, because high energies are associated with the UV asymptote.

The second area where the RG technique is applicable is the theory of critical phenomena. This was Wilson's idea (1971) [37]. It proved to be extremely fruitful, and over a period of several years led to a complete revolution in this area of physics.

As mentioned above, the fluctuation theory of critical behavior is equivalent to Euclidean quantum field theory. The critical point corresponds directly to massless fields, and we deal with a nontrivial IR asymptote of the Green functions (in statistical physics the latter have the meaning of correlation functions, and we shall use the two terms synonymously).

The typical models of statistical physics belong to the same class as the relativistic field models studied using the RG when it was first developed. However, there is one important difference: owing to the absence of time in equilibrium statistical physics, the natural spatial dimension is $d = 3$ instead of the $d = 4$ (space + time) of relativistic physics. If we wanted to study the critical behavior for such models in the abstract dimension $d = 4$ rather than the real dimension $d = 3$, we would not encounter any problems: the Green functions would then contain the standard logarithmic singularities, which are easily summed by the RG method. In fact, this was done by Larkin and Khmel'nitskiĭ in 1969 [38], in their study of the critical behavior of a uniaxial ferroelectric. That model is exceptional in that for $d = 3$ it has the same IR behavior as ordinary field models in $d = 4$, and so there the quantum field RG technique is applicable directly to the real three-dimensional problem.

The study of [38] is historically the first actual example of the use of the RG technique in the theory of critical phenomena. However, it did not significantly affect the further development of this theory, because the model studied in [38] is exceptional, and the situation is different in the standard models mentioned above. These models are logarithmic for $d = 4$, and as the dimension decreases the IR singularities get stronger, becoming power-law singularities. On the other hand, the very origin of the quantum field RG technique is related to the procedure of ultraviolet renormalization for removing UV divergences. These are important in ordinary models at $d = 4$, but practically absent for $d = 3$. It therefore appears at first glance that the quantum field RG technique associated with UV renormalization does not have any connection with real three-dimensional problems.

In Wilson's first study (1971) [37], the derivation of the RG differential equations for the theory of critical phenomena was based on considerations completely different from those in quantum field theory, although Wilson noted the existence of a formal analogy (hence the term "renormalization group"). He began with the Kadanoff block construction [28] for the Ising model, in which the original interaction of individual spins is replaced by the interaction of large-scale objects: blocks of spins. The Kadanoff hypothesis [28], proposed to explain critical scaling, consists of two assumptions: (1) near T_c the interaction Hamiltonian of the blocks can be assumed to have the same form as the original microscopic interaction but with different parameters; (2) the dependence of the parameters on the block size L has a scaling form for $L \to \infty$. Wilson [37] introduced the RG differential equations as the infinitesimal form of the Kadanoff transformations and showed that for sufficiently general assumptions the second part of the Kadanoff hypothesis is an automatic consequence of these equations, i.e., a general property of their solutions. He then proposed [39] a specific computational procedure, the method of recursion relations, and demonstrated its technical possibilities by numerically calculating the exponents for the Ising model.

An obvious weakness of the proposed scheme was the absence of any parameter that was even formally small. The next step taken by Wilson and Fisher in [40] was therefore very important. They showed that such a parameter arises if the problem is studied in variable spatial dimension $d = 4 - \varepsilon$, where the exponents can be calculated as series in ε. The introduction of the formal parameter ε ensures the internal self-consistency of the entire scheme; in particular, it allows the classification of interactions as IR-relevant or IR-irrelevant. All the former must be taken into account when analyzing the critical behavior, and all the latter can be neglected, since, in the language of the Kadanoff procedure, their contributions fall off as a power for $L \to \infty$. The first part of the Kadanoff hypothesis is also justified in the $(4 - \varepsilon)$ scheme: the discarded terms of the interaction Hamiltonian of the blocks are actually IR-irrelevant at small ε.

In a later study [41] Wilson improved the computational procedure, showing how the standard Feynman diagram technique can be used to calculate the ε expansions of the exponents. It was quickly realized that these studies were important, and a variety of applications of their ideas and techniques were promptly found. A detailed discussion and review of the results during this first stage of the development of the RG theory of critical phenomena can be found in [42].

The development of the general ideas of the RG approach was actually completed during this stage, but the technique continued to be developed in its more formal aspects and made closer to the traditional methodology of quantum field theory. In the end it was realized that all the results of the Wilson approach could also be obtained within the earlier quantum field RG technique adapted to dimension $d = 4 - \varepsilon$, and that this is the technically simplest method of calculation, especially in higher orders in ε. The work of French physicists at Saclay was very important for the development and popularization of the "quantum field" version of the Wilson approach (see their large review [43]). For a long time this review and the book based on it [44] were the main sources of information on this topic, and only more recently has the much more extensive monograph [45] appeared.

Let us briefly describe how the exponents are calculated using the quantum field RG technique. The RG differential equation obtained for the correlation functions (Green functions) W in the models discussed above can be written as

$$[\mathcal{D}_{\mathrm{RG}} + \gamma_W(g)]W = 0, \ \mathcal{D}_{\mathrm{RG}} = \mu\frac{\partial}{\partial\mu} + \beta(g)\frac{\partial}{\partial g} - \sum_i \gamma_i(g)e_i\frac{\partial}{\partial e_i}, \quad (1.1)$$

where g is the charge (coupling constant), μ is an auxiliary dimensional parameter called the renormalization mass, and e_i are other parameters characterizing the deviation from the critical point ($T - T_{\mathrm{c}}$, an external field, and so on); γ_W depends on the function W studied, while the operator $\mathcal{D}_{\mathrm{RG}}$ is the same for all. The coefficient of $\frac{\partial}{\partial g}$ in $\mathcal{D}_{\mathrm{RG}}$ is called the β function, and γ is the anomalous dimension (γ_i for the variable e_i and γ_W for the function W). These coefficients in general are referred to as RG functions. In practice, they are calculated from graphs as series in g. The asymptotic regimes are associated with fixed points g_* at which $\beta(g_*) = 0$. The type of point (IR or UV) is determined by the behavior of $\beta(g)$ in its vicinity, and for a first-order zero by the sign of the derivative $\omega = \beta'(g_*)$ ($\omega > 0$ corresponds to an IR fixed point, and $\omega < 0$ to a UV fixed point). As already mentioned, in dimension $d = 4$ the expansion of the β function takes the form $\beta(g) = \beta_2 g^2 + \beta_3 g^3 + ...$. For our models $\beta_2 > 0$, which corresponds to trivial fixed point $g_* = 0$ of the IR type. It turns out that for $d = 4 - \varepsilon$ a linear term appears in the expansion of the β function: $\beta(g) = -\varepsilon g + \beta_2 g^2 + ...,$ while the signs of the coefficients β_n with $n \geq 2$ remain as before (in some calculational schemes these coefficients do not depend at all on ε). Therefore, the β function at small ε behaves as shown in Fig. 1.1, and has two fixed points: $g_* = 0$ and $g_* \sim \varepsilon$. We see from the graph that the first of these

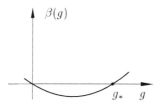

Figure 1.1 Behavior of the β function for $d = 4 - \varepsilon$.

is of the UV type, while the second is of the IR type and therefore determines the desired IR asymptote of the correlation functions W in this case. If the β function behaves as in Fig. 1.1, then independently of the smallness of ε it can be shown by rigorous mathematics that the leading term W_{IR} of the asymptote of W in the critical region (all momenta and parameters e_i are small) satisfies Eq. (1.1) with all the RG functions replaced by their values at the IR-attractive fixed point $g_* \sim \varepsilon$ [we recall that $\beta(g_*) = 0$]:

$$\left[\mu \frac{\partial}{\partial \mu} - \sum_i \gamma_i(g_*) e_i \frac{\partial}{\partial e_i} + \gamma_W(g_*) \right] W_{\mathrm{IR}} = 0. \tag{1.2}$$

This equation is the differential form of the condition for generalized homogeneity of the function W_{IR}. Taking into account ordinary (trivial) scale invariance, (1.2) leads to the existence of critical scaling with dimensions

$$\Delta[e_i] = d[e_i] + \gamma_i(g_*), \quad \Delta[W] = d[W] + \gamma_W(g_*), \tag{1.3}$$

where $\Delta[F]$ is the desired critical dimension and $d[F]$ the canonical dimension (i.e., that obtained in the Landau theory) of F. For the momentum both are taken to be unity. The canonical dimensions are determined uniquely from the form of the Landau functional (i.e., the action in the field formulation), and the additions $\gamma(g_*)$ in (1.3) are calculated as series in ε: knowing all the RG functions as series in g, from the equation $\beta(g_*) = 0$ it is possible to obtain the ε expansion for g_*, and its substitution into $\gamma(g)$ gives the desired series for $\gamma(g_*)$, which will be the Wilson ε expansions. Of course, in reality it is possible to calculate only the first few terms of the series, as the complexity of the calculation grows rapidly with increasing order in ε. For the standard φ^4 model, the Wilson technique has been used to calculate [42] only the order-ε and ε^2 terms for the principal exponents. The highest order that has been reached at present is ε^5. The calculation of higher orders primarily involves purely technical problems, and here the advantages of the quantum field formulation become especially apparent. One is that in this approach all the RG functions can be calculated in the form of series in g in terms of so-called renormalization constants, which do not depend on variables of the coordinate or momentum type and are therefore objects much simpler than the correlation functions used in the Wilson technique.

Other exponent expansions have been obtained for certain models by analogy with the $4-\varepsilon$ expansion. For example, for the O_n-symmetric φ^4 model describing the classical isotropic Heisenberg ferromagnet with n-component spin, in addition to the standard Wilson expansion in the parameter $\varepsilon = 4-d$ it is now possible to also construct the $2 + \varepsilon$ expansion in the parameter $\varepsilon = d - 2$ and the $1/n$ expansion in the inverse number of spin components.

Critical scaling is a consequence of the behavior of the β function shown in Fig. 1.1. For asymptotically small ε this behavior is guaranteed, but in order to be able to use the results at the real value $\varepsilon = 1$ one must believe that the picture in Fig. 1.1 is distorted but not qualitatively changed with increasing ε, i.e., that the IR point g_* does not disappear. It is impossible to prove this knowing only the first few terms of the series in g for the β function, and so this is essentially the second fundamental postulate of the modern theory of critical phenomena (we recall that the first is the possibility of changing over to a fluctuation model). The main justification of these postulates is provided by the results, primarily the qualitative ones. As far as the numerical values of the exponents are concerned, it cannot really be hoped that good agreement with an experiment (either a real experiment or a numerical one) will be obtained for the real value $\varepsilon = 1$. Therefore, it is surprising that rather often this agreement is not bad. For example, the inclusion of the first two corrections to the Landau exponents leads to fairly good agreement with the results of a numerical calculation of the exponents for the three-dimensional Ising model (in particular, the susceptibility exponent γ in the Landau theory is $\gamma = 1$; inclusion of the ε and ε^2 corrections gives $\gamma = 1.244$, and the exact numerical result is $\gamma = 1.250 \pm 0.003$). At some point the inclusion of higher orders in ε makes the result worse, i.e., the series are clearly asymptotic. This problem has now been overcome fairly successfully by using methods (Padé–Borel) for summing divergent series, which became possible owing to the development of a technique for calculating higher-order asymptotes [46], [47].

However, the main achievement of the Wilson scheme is that it gives not new numbers, but a new view of the problem: the RG approach explains the mechanism by which critical scaling and corrections to the canonical Landau exponents appear, and it allows study of the approach to the limiting scaling asymptote and the corrections to it. For complex multicharge models it allows explanation of the sometimes highly nontrivial qualitative picture of the critical behavior, the effect of random impurities on this behavior, and so on. It admits natural generalization to other types of critical behavior, for example, tricritical behavior. First developed in equilibrium statistical physics ("critical statics"), it was then generalized to critical dynamics, the study of the singularities in the kinetic coefficients and the relaxation times at the critical point, and other related problems: the theory of polymers, various types of percolation and random walks, the theory of developed turbulence, and the magnetic hydrodynamics of a turbulent medium (references will be given later). The list of such problems is expected to grow in the future.

It can be stated that the RG approach represents a qualitatively new stage in the development of the theory of critical phenomena and related problems. It has led to the construction of a new language no less universal than that of the old Landau theory, but possessing much greater flexibility and predictive power. The 1970s witnessed the rapid development of this area. By the early 1980s the construction of this approach was essentially complete and the RG method became the common language of theoreticians and experimentalists for discussing critical behavior. In 1982 Wilson was awarded the Nobel Prize for his contributions in this area.

The theory continued to develop in subsequent years, although it passed through a quiet phase during which the earlier results were refined, new models were studied, and entire new areas of applicability of the RG technique were discovered, in particular, the theory of developed turbulence. It can be stated that the RG technique has become one of the most universal and powerful methods for studying a wide range of problems.

There are many different ways of explaining the RG method. The ideas are the same, but there are technical differences. In this book we shall use the quantum field formulation, as it is the clearest and most formally developed. It makes use of the nontrivial techniques of ultraviolet renormalization well developed in relativistic field theory, which is important for calculating higher orders in ε. The main technical tool is the calculation of the RG functions in terms of the UV-renormalization constants using dimensional regularization and the minimal subtraction scheme, which is most convenient for practical calculations. The entire discussion is rigorous within the framework of the ε expansion with formal small parameter ε, and $\varepsilon = 1$ is taken only in the final expressions (as for other expansion variants). The quantum field formulation of the RG is at present the most powerful technical tool for studying critical behavior. However, as it is the final result of a long evolution, it is not as clear conceptually as the original Wilson scheme. The ideas of the RG approach were explained briefly above. More details can be found in [42], [48], [49], where an excellent description is given. In the present book we shall focus our attention on the computational technique, which itself is nontrivial in higher orders.

Let us now turn from the historical review to the systematic description of the basics of the theory.

1.2   Generalized homogeneity

Since the concept of generalized homogeneity is used to formulate the similarity hypothesis, we shall give the necessary definitions here. A function $W(e)$ of numerical variables $e \equiv \{e_1...e_n\}$ is termed generally homogeneous or simply dimensional if

$$W(\lambda^{\Delta_1} e_1, ..., \lambda^{\Delta_n} e_n) = \lambda^{\Delta_W} W(e_1, ..., e_n)$$

for some set of numbers Δ and any $\lambda > 0$. In abbreviated form,

$$W(e_\lambda) = \lambda^{\Delta_W} W(e), \quad e_{i\lambda} = \lambda^{\Delta_i} e_i \quad \text{for all } i = 1, ..., n. \tag{1.4}$$

The parameters $\Delta_i \equiv \Delta_{e_i} \equiv \Delta[e_i]$ (these are different variants of the notation) are called the dimensions of the corresponding variables e_i, and $\Delta_W \equiv \Delta[W]$ is the dimension of the function W. For $\Delta_W = 0$ the function W is termed scale-invariant or simply dimensionless. A dimensional function $W(e)$ of a single variable e is a multiple of a simple power $|e|^\beta$ with known exponent $\beta = \Delta_W / \Delta_e$, and when W is a function of n variables it can be represented as the product of a power and an arbitrary *scaling function* f of the $n - 1$ dimensionless combinations of arguments, for example,

$$W(e_1, ..., e_n) = |e_n|^\beta f\left(\frac{e_1}{|e_n|^{\beta_1}}, ..., \frac{e_{n-1}}{|e_n|^{\beta_{n-1}}}\right), \tag{1.5}$$

where $\beta = \Delta_W / \Delta_n$, $\beta_i = \Delta_i / \Delta_n$ for all $i = 1, 2, ..., n - 1$. Differentiating (1.4) with respect to λ and then setting $\lambda = 1$, we obtain a differential equation equivalent to (1.4):

$$\left[\sum_e \Delta_e \mathcal{D}_e - \Delta_W\right] W(e) = 0, \quad \mathcal{D}_e \equiv e\frac{\partial}{\partial e} \equiv e\partial_e \tag{1.6}$$

with summation over all the variables e.

In thermodynamics, one often uses the Legendre transformation of a function with respect to all or some of its arguments. Let $e = \{e', e''\}$, where e' are the variables with respect to which the transformation is done, and e'' are the other variables, which act as parameters. The Legendre transform of the function $W(e) = W(e', e'')$ with respect to the variables e' is the function $\Gamma(\alpha', e'')$ defined as

$$\Gamma(\alpha', e'') = W(e) - \sum_i e_i' \frac{\partial W(e)}{\partial e_i'}, \quad \frac{\partial W(e)}{\partial e_i'} = \alpha_i', \tag{1.7}$$

where $e' = e'(\alpha', e'')$ in the expression for Γ, this dependence being determined implicitly by the second expression in (1.7) (on the right-hand side it is possible to change the signs and introduce additional coefficients, thereby changing the definition of α'). In thermodynamics, pairs of variables e_i', α_i' are called conjugate variables (temperature–entropy, pressure–volume, external field–magnetization, chemical potential–density, and so on), and only one of the two variables of a pair is independent. From the definition (1.7) we easily derive the relations

$$\frac{\partial \Gamma(\alpha', e'')}{\partial \alpha_i'} = -e_i', \quad \frac{\partial \Gamma(\alpha', e'')}{\partial e_k''} = \frac{\partial W(e)}{\partial e_k''}, \tag{1.8}$$

and from (1.7), (1.8), and the expression

$$\delta_{ik} = \frac{\partial e_i'}{\partial e_k'} = \sum_s \frac{\partial e_i'}{\partial \alpha_s'} \frac{\partial \alpha_s'}{\partial e_k'},$$

we find

$$-\sum_s \frac{\partial^2 \Gamma(\alpha', e'')}{\partial \alpha_i' \partial \alpha_s'} \frac{\partial^2 W(e)}{\partial e_s' \partial e_k'} = \delta_{ik}, \tag{1.9}$$

i.e., the matrices of second derivatives of W with respect to e' and of Γ with respect to α' are the inverses of each other up to a sign. We shall need all of this later on.

Derivatives of a dimensional function with respect to dimensional variables are also dimensional, so that if W and e' are dimensional, Γ and α' will also be

$$\Delta[\alpha_i'] = \Delta \left[\frac{\partial W}{\partial e_i'} \right] = \Delta[W] - \Delta[e_i'], \quad \Delta[\Gamma] = \Delta[W]. \tag{1.10}$$

The replacement $\lambda \to \lambda^a$ in (1.4) is equivalent to multiplication of all the dimensions Δ by a, so that one of them can be fixed arbitrarily. Using this fact, it is always possible to take the dimension of the momentum to be unity,[*] and then the dimension of the coordinate (a length) will be the minus of unity:

$$\Delta[\text{momentum}] = 1, \quad \Delta[\text{coordinate}] = -1. \tag{1.11}$$

We shall use this convention throughout the book.

We conclude by noting that the property of generalized homogeneity is sensitive to the choice of variables and is lost even in a simple linear change of variables if the latter mixes variables of different dimension.

1.3 The scaling hypothesis (critical scaling) in thermodynamics

In statistical physics, the thermodynamics of a system is completely determined by the partition function Z, treated as a function of the thermodynamical variables e specifying the state of the system. The latter will always be assumed to be chosen such that all the $e_i = 0$ at the critical point. The scaling hypothesis is, crudely speaking, the statement that $Z(e)$ is dimensionless for the correct choice of variables e and their dimensions $\Delta[e]$.

A more accurate statement is needed. First, it must be stressed that we are speaking not of the ordinary dimensions (like centimeters, grams, and so on), but of certain new *critical dimensions*. The partition function is always dimensionless in the usual sense, but this has nothing to do with its e dependence, because Z also involves other model parameters that are dimensional in the usual sense and taken to be fixed when studying the thermodynamics.

---

[*]The terms "momentum" and "wave vector" will always be used synonymously.

If desired, all the variables e can be chosen to be dimensionless in the usual sense, but this is not necessary.

Second, when studying phase transitions it is necessary to consider the system in the thermodynamical limit of infinite volume, because phase transitions cannot occur in a finite volume V. For spatially uniform systems, the quantity that is well defined in the limit $V \to \infty$ is not the partition function, but the specific value of its logarithm:

$$W = \lim V^{-1} \ln[Z_V] \quad \text{for } V \to \infty. \tag{1.12}$$

If we understand the scaling hypothesis as the statement that the critical dimension of the partition function is zero, we can conclude from (1.11) and (1.12) that for a model in d-dimensional space W is a dimensional quantity with dimension $\Delta[W] = d$, because $\Delta[V] = -d$ according to (1.11). For lattice models the volume V is the total number of sites, which is assigned the same critical dimension as the volume, $-d$.

However, this formulation is still imprecise, because the statement of the scaling hypothesis actually pertains not to the entire function W, but only to its singular part $W^{(s)} = W - W^{(r)}$ responsible for the critical behavior. We shall define the regular part $W^{(r)}$ as the sum of the contributions of all the existing first terms of the Taylor expansion of $W(e)$ in e, i.e., the terms for which the corresponding partial derivatives of W are finite at $e = 0$.

A simplified statement of the scaling hypothesis is the following. For the correct choice of the variables e and their dimensions $\Delta[e]$, the singular part $W^{(s)}(e)$ is a function of dimension d possessing the property of generalized homogeneity [see (1.4)]:

$$W^{(s)}(e_\lambda) = \lambda^d W^{(s)}(e). \tag{1.13}$$

This statement is accurate when applied to only the leading critical singularities. However, if the corrections to these are taken into account, the scaling hypothesis must be understood as the following more complicated statement regarding the asymptote of $W^{(s)}(e_\lambda)$ for $\lambda \to 0$:

$$W^{(s)}(e_\lambda) \underset{\lambda \to 0}{=} \lambda^d \left[W_0^{(s)}(e) + \lambda^{\omega_1} W_1^{(s)}(e) + \lambda^{\omega_2} W_2^{(s)}(e) + ... \right] \tag{1.14}$$

with certain unknown *correction exponents* $0 < \omega_1 < \omega_2 < ....$ This implies that $W^{(s)}(e)$ is the sum of a series of functions possessing generalized homogeneity $W_n^{(s)}(e)$ with growing dimensions $\Delta[W_n^{(s)}] = d + \omega_n$, $\omega_0 = 0$. The statement (1.13) pertains only to the leading contribution $W_0^{(s)}$, which can be isolated from $W^{(s)}$ as follows: $W_0^{(s)}(e) = \lim[\lambda^{-d} W^{(s)}(e_\lambda)]$ for $\lambda \to 0$. The scaling hypothesis states that the limit exists, and its generalized homogeneity with dimension d is an automatic consequence of the definition.

For our choice of variables ($e = 0$ at the critical point) their dimensions must obviously be positive, in order for the asymptote (1.14) to correspond to approach the critical point. Therefore, the partial derivatives of W

with respect to e of sufficiently high order certainly diverge for $e \to 0$, so that the above-defined regular part $W^{(\mathrm{r})}$ is a polynomial in e with $\Delta[\text{monomials}] < d = \Delta[W_0^{(\mathrm{s})}]$. Sometimes $W^{(\mathrm{r})}$ is defined as the sum of all the analytic and $W^{(\mathrm{s})}$ as the sum of all the nonanalytic contributions to W. This is possible if the explicit form of the nonanalytic contributions is known (and naturally in the interpolation formulas used to analyze an experiment). However, in general the splitting into analytic and nonanalytic parts is nonunique, so that we prefer the formally unique definition of $W^{(\mathrm{r})}$ as the sum of all the analytic contributions with small ($< d$) dimensions. The analytic contributions of higher ($> d$) dimension are included in the singular part and there play the role of the analytic corrections $W_n^{(\mathrm{s})}$ with known ω_n in (1.14).

In thermodynamics it is often more convenient to deal not with the function (1.12), but with the specific (per unit volume) thermodynamical potential Ω:

$$W = -\beta\Omega, \qquad \beta = 1/kT, \tag{1.15}$$

where k is the Boltzmann constant and T is the temperature. For the leading singular contributions we have $W_0^{(\mathrm{s})} = -\beta_{\mathrm{c}}\Omega_0^{(\mathrm{s})}$, where $\beta_{\mathrm{c}} = 1/kT_{\mathrm{c}}$ (the inclusion of $\beta - \beta_{\mathrm{c}}$ represents one of the corrections to scaling), so that from the viewpoint of the simple scaling hypothesis (1.13) the quantities W and Ω are completely equivalent.

In conclusion, we note that the scaling hypothesis (1.13) pertains only to the most commonly encountered case of power-law critical singularities. It is supported (see Sec. 1.1) by the RG method, and the critical dimensions $\Delta[e]$ and correction exponents ω are calculated using the renormalization group. However, this is not the most general case. Sometimes (for example, for a tricritical point or a uniaxial ferroelectric) the RG analysis shows that the corrections to the Landau theory are not power-law corrections, but logarithmic ones. Of course, the scaling hypothesis (1.13) does not apply to such systems.

1.4 The Ising model and thermodynamics of a ferromagnet

In the Ising model describing a uniaxial magnet, a classical random spin variable $\widehat{s}_i$ taking "up" and "down" values $s_i = \pm 1$ with equal probability is associated with each site i of a given spatial lattice.* The Hamiltonian is the sum of the exchange interaction and the interaction with an external field h:

$$\widehat{H} = -J \sum_{(ik)} \widehat{s}_i \widehat{s}_k - h \sum_i \widehat{s}_i \equiv \widehat{H}_{\mathrm{exc}} - h\widehat{S}, \tag{1.16}$$

---

*A hat will always be used to denote random (fluctuating) quantities, which should be distinguished from their specific realizations (configurations).

where the positive (negative for an antiferromagnet) constant J is the "exchange integral," and the sum in the first term runs over all pairs (ik) of nearest neighbors on the lattice.

The partition function Z and the average value of an arbitrary random quantity $\widehat{a}$ are given by the general expressions

$$Z = \operatorname{tr}\left[\exp(-\beta\widehat{H})\right], \quad \langle\widehat{a}\rangle = Z^{-1}\operatorname{tr}\left[\widehat{a}\exp(-\beta\widehat{H})\right], \qquad (1.17)$$

where $\beta = 1/kT$ and tr denotes summation over all configurations of independent random variables, in this case, lattice spins. The meaning of the partial derivatives of the specific logarithm of the partition function $W(T,h)$ is easily found from the definitions [in (1.12) the volume V is the total number of lattice sites]: $\partial_h W = M/kT$ and $\partial_T W = U/kT^2$, where $M \equiv V^{-1}\langle\widehat{S}\rangle$ is the magnetization and $U \equiv V^{-1}\langle\widehat{H}\rangle$ is the internal energy. All extensive quantities are understood to be calculated per site, and $\partial_x \equiv \partial/\partial x$ for all x.

The definition of U as the specific average of $\widehat{H}$ connects statistical physics with thermodynamics. The thermodynamics of a magnet is based on the equation $dU = TdS - Mdh$, where S is the entropy, $TdS = dQ$ is the heat gained by the system, and Mdh is the work done by the system. Then for the free energy $F = U - TS$ and the function $\Gamma = F + Mh$ we obtain $dF = -SdT - Mdh$ and $d\Gamma = -SdT + hdM$. This determines the meaning of the partial derivatives of the functions U, F, and Γ in their natural variables: $U = U(S,h)$, $F = F(T,h)$, and $\Gamma = \Gamma(T,M)$. The partition function is used to calculate the function Ω directly in the variables T, h natural for the free energy F. We shall show that in this case $F = \Omega(T,h) = -kTW(T,h)$. From the definition of Ω in (1.15) and the expressions given above for the partial derivatives of W, we easily obtain the equation $U = \Omega - T\partial_T\Omega$, and from the equations of thermodynamics we find $U = F + TS = F - T\partial_T F$, so that Ω and F satisfy the same equation. However, it does not follow that $\Omega = F$, because the equation $U = F - T\partial_T F$ does not determine $F(T,h)$ in terms of $U(T,h)$ uniquely, but only up to a term aT with arbitrary coefficient a independent of T (for the difference $\Delta F = F_1 - F_2$ of two solutions we have $\Delta F - T\partial_T\Delta F = 0$, from which $\Delta F = aT$). The thermodynamical definition of F in terms of U becomes unique only when an additional requirement, for example, $S = -\partial_T F = 0$ at $T = 0$ (the Nernst theorem), is imposed. It is easily verified from the definition of Ω in terms of the partition function that Ω satisfies this requirement and also the equation (see above), so that we can take $F(T,h) = \Omega(T,h)$ in agreement with the Nernst theorem.

Other important thermodynamical quantities are the susceptibility $\chi = dM/dh$ (the isothermal susceptibility χ_T at constant temperature and the adiabatic susceptibility χ_S at constant entropy) and the specific heat $C = dQ/dT = TdS/dT$ (C_h at constant field and C_M at constant magnetization). From the known function $F = \Omega(T,h)$ it is easy to find $M = -\partial_h F$, $S = -\partial_T F$, $\chi_T = -\partial_h^2 F$, and $C_h = -T\partial_T^2 F$. In terms of the function $\Gamma(T,M)$ (the Legendre transform of F with respect to h) we have $h = \partial_M\Gamma$,

$S = -\partial_T \Gamma$, $\chi_T = [\partial_M^2 \Gamma]^{-1}$, and $C_M = -T\partial_T^2 \Gamma$. Finally, from the function $U(S, h)$ we easily find $\chi_S = -\partial_h^2 U$. The equation $h = h(T, M)$ is called the *equation of state* of the magnet.

1.5 The scaling hypothesis for the uniaxial ferromagnet

Let us now turn to the description of the critical behavior of a magnet using the scaling hypothesis. The phase-separation line in the T, h plane is the line segment $0 \leq T \leq T_c$, $h = 0$ ending at the Curie point $T = T_c$, $h = 0$. The value of a quantity X at the critical point will be denoted X_c. Owing to the symmetry of the problem, the correct natural variables are simply $e_1 = h$ and $e_2 = \tau = T - T_c$, and their dimensions are positive (Sec. 1.3). Here and below when speaking of the dimension of any quantity we mean the critical dimension of its leading singular contribution. The free energy has regular part $F^{(r)}(e) = F_c - \tau S_c$, and its singular part $F^{(s)} = \Omega^{(s)}$ is, according to the scaling hypothesis (1.13) and (1.15), a generally homogeneous function with dimension d. From this, using the equations of Sec. 1.4 and (1.10), we easily find the dimensions of all the fundamental quantities, which are given in Table 1.1.

Table 1.1   Critical dimensions of the leading singular contributions of various thermodynamical quantities for a ferromagnet.

Quantity	$h \equiv e_1$	$\tau \equiv e_2$	F, Γ	U, S	M	χ_T, χ_S	C_h, C_M
Dimension	$\Delta_h \equiv \Delta_1$	$\Delta_\tau \equiv \Delta_2$	d	$d - \Delta_\tau$	$d - \Delta_h$	$d - 2\Delta_h$	$d - 2\Delta_\tau$

Although U is the Legendre transform of F, their dimensions are not the same, in spite of (1.10), because the transform is not done with respect to a dimensional variable. In fact, isolating the singular contributions in the equation $U = F + TS$, we obtain

$$
\begin{aligned}
U &= F^{(r)} + F^{(s)} + (T_c + \tau)(S_c + S^{(s)}) \\
&= F^{(r)} + (T_c + \tau)S_c + T_c S^{(s)} + F^{(s)} + \tau S^{(s)}.
\end{aligned}
$$

The leading term is the singular term $T_c S^{(s)}$ with dimension $d - \Delta_\tau < d$. However, this term does not contribute to $\chi_S = -\partial_h^2 U(S, h)$, because in the variables S, h it is independent of h, so that $\Delta[\chi_S] = \Delta[\chi_T]$.

All the thermodynamical quantities are functions of a single variable on the lines $h = 0$ or $\tau = 0$. Even before the scaling hypothesis was first stated, it was assumed that the thermodynamical quantities have a simple power-law asymptote when a critical point is approached along these lines:

$$M \sim |\tau|^\beta \sim |h|^{1/\delta}, \quad \chi \sim |\tau|^{-\gamma}, \quad C \sim |\tau|^{-\alpha}. \tag{1.18}$$

The notation for the critical exponents is standard. The power of $|h|$ corresponds to the line $\tau = 0$, and the power of $|\tau|$ corresponds to the line $h = 0$ (for M this is only for $\tau < 0$, and then M is the spontaneous magnetization). The scaling hypothesis justifies the asymptotes (1.18), and, with the data of Table 1.1, it can be used to express all the critical exponents in terms of d, Δ_h, and Δ_τ (for a single variable the exponent is the quotient of the dimensions of the function and the argument; see Sec. 1.2):

$$\alpha = \frac{2\Delta_\tau - d}{\Delta_\tau}, \quad \beta = \frac{d - \Delta_h}{\Delta_\tau}, \quad \gamma = \frac{2\Delta_h - d}{\Delta_\tau}, \quad \delta = \frac{\Delta_h}{d - \Delta_h}. \quad (1.19)$$

There are two relations between these: $\alpha + 2\beta + \gamma = 2$ and $\beta\delta = \beta + \gamma$. Satisfation of these relations for independently determined (measured or calculated) exponents provides direct verification of the thermodynamical scaling hypothesis. As an illustration, in Table 1.2 we give the values of the critical exponents (including the indices η and ν, which will be defined in Sec. 1.9) for the classical Landau theory and the Ising model. All the exponents are known exactly for the Landau theory (the φ^4 model), and all except δ are known for the two-dimensional Ising model (the value $\alpha = 0$ corresponds to a logarithmic singularity of the specific heat). The exponent δ of the two-dimensional model and all the exponents of the three-dimensional model have been obtained by extrapolation of high-temperature expansions (Sec. 1.1). No generally accepted values of the critical exponents of the three-dimensional Ising model exist yet, and the data given in different studies differ slightly. However, the differences are not very important, and an idea of them can be obtained from Table 1.2, where we give data from various sources, including the review of Wilson and Kogut [42] and the later monograph [45].

Table 1.2 Values of the critical exponents in the Landau theory and the Ising model.

Exponent	α	β	γ	δ	η	ν
Landau theory	0	1/2	1	3	0	1/2
Ising model $(d = 2)$	0	1/8	7/4	$\cong 15$	1/4	1
Ising model $(d = 3)$ [42]	0.125 ±0.015	0.312 ±0.003	1.250 ±0.003	5.15 ±0.02	0.055 ±0.010	0.642 ±0.003
Same, other sources	0.105 ±0.010 [45]	0.312 ±0.005 [45]	1.2385 ±0.0025 [45]	5.00 ±0.05 [48]	0.041 ±0.010 [44]	0.6305 ±0.0015 [45]

Using the data of Table 1.2, it can be checked that the relations given above between the exponents (1.19) are satisfied exactly in the Landau theory and the two-dimensional Ising model with $\delta = 15$, while in the three-dimensional Ising model they are satisfied fairly well. Here $\Delta_h = 15/8$ and $\Delta_\tau = 1$ for $d = 2$ and $\Delta_h \cong 2.48$ and $\Delta_\tau \cong 1.57$ for $d = 3$.

Away from the special lines discussed above, all quantities are functions of the two variables h and τ. The scaling hypothesis (1.13) predicts that their leading singular contributions have a scaling form of the type (1.5), for example,

$$F^{(s)}(h, \tau) = |\tau|^* B_\pm(h|\tau|^*) = |h|^* A(\tau|h|^*). \tag{1.20}$$

Here and below the stars denote any known (from the dimensions) exponents, and they are all expressed in terms of the exponents (1.19). In general, the scaling functions (for which there is no standard notation) are different for $\tau > 0$ and $\tau < 0$. For A these regions differ by the sign of the argument, and for B it is necessary to introduce two different functions $B_\pm$. Another example is the equation of state [$\text{sgn}(h) = \pm 1$ is the sign of h]:

$$h(M, \tau) = |\tau|^* g_\pm(M|\tau|^*) = \text{sgn}(h)|M|^* f(\tau|M|^*). \tag{1.21}$$

The scaling hypothesis does not determine the form of the scaling functions, but the scaling form itself is a nontrivial and experimentally verifiable statement. For example, it follows from (1.21) that in the "reduced" variables $h' = h|\tau|^*$, $M' = M|\tau|^*$ the equation of state takes the form $h' = g_\pm(M')$, i.e., the points h' and M' obtained at different temperatures must lie on the same curve (for a given sign of τ). This does occur (see Ch. 2, Sec. 4 in [50] and Ch. 2, Sec. 6 in [49]), and the curves for different signs of τ are different, i.e., the functions g_+ and g_- in (1.21) actually do not coincide.

Let us briefly discuss the main features of the scaling functions [26], [49]. First, it is clear that a scaling function of any quantity can be expressed (by change of variables, differentiation, and Legendre transformation) in terms of the functions A or $B_\pm$ in (1.20). Second, owing to the symmetry of the problem a scaling function always has definite parity in h and M, in particular, $B_\pm(x)$ in (1.20) are even and $g_\pm(x)$ in (1.21) are odd. Third, it can usually be assumed that the forms of the asymptotic expansions at the origin and at infinity are known. This information is obtained by applying to representations of the type (1.20) and (1.21) the following statements, which have been proved rigorously for the Ising model. For $h \neq 0$ the free energy is analytic in τ near $\tau = 0$, and for $\tau > 0$ it is analytic in h and M [8]. Assuming that this is also true for the representations (1.20) and (1.21), from the analyticity in τ for $h \neq 0$ we conclude that the functions $A(x)$ and $f(x)$ are analytic in x near $x = 0$, so that their asymptote for $x \to 0$ is a series in integer powers of x. On the other hand, the analyticity in h and M for $\tau > 0$ uniquely determines the form of the asymptotic expansions of $A(x)$ and $f(x)$ for $x \to +\infty$. They are series of the type $x^* \sum c_n(x^*)^n$, and substitution of them into (1.20) gives a series in even integer powers of h, while substitution into (1.21) gives a series in odd integer powers of M. This uniquely determines the exponents denoted by the stars in the asymptotic expansions of the exponent. We shall not give them here, because it is simpler to remember the basic ideas: analyticity in τ for $h \neq 0$ and in h and M for $\tau > 0$. From this it is easy to find the form of the needed asymptotic expansions in any particular case. In the region $\tau < 0$ there is no simple analyticity in h at the origin, and for M the region of definition does not contain the origin at all, because $|M|$ cannot be smaller than the spontaneous magnetization $M_0(\tau)$. Therefore, no exact statements about the corresponding asymptotes can be made, although some ideas are under discussion (see, for example, [26] and [49]).

Scaling functions of the type $A(x)$ in (1.20) are not yet universal, because they depend on the choice of units in which h and τ are measured. Arbitrary rescalings of these variables correspond to transformations $A(x) \rightarrow c_1 A(c_2 x)$ with arbitrary constants $c_{1,2}$.

Therefore, only the *normalized scaling functions* are universal characteristics. For them this arbitrariness is eliminated by a suitable normalization condition, for example, $A(x) = 1 + x + O(x^2)$. Universality is understood as systems of the same class having identical critical behavior (neglecting corrections). That is, for such systems not only the critical exponents, but also the normalized scaling functions must coincide [we recall that they can all be expressed in terms of $A(x)$ in (1.20)]. A particular manifestation of the universality of the scaling functions is the universality of the amplitude ratios $\mathcal{A}_+/\mathcal{A}_-$ in the laws

$$X(\tau, h = 0) \cong \mathcal{A}_\pm |\tau|^{\Delta_X/\Delta_\tau} \quad \text{for} \quad \tau \rightarrow \pm 0. \tag{1.22}$$

Here X is any quantity for which both asymptotes at $\tau \rightarrow \pm 0$ are meaningful [for example, χ and C in (1.18)]. We note that before the scaling hypothesis was formulated, it was thought that the exponents in (1.22) could differ when T_c was approached from different sides. The scaling hypothesis excludes this, and experiment and computer calculations have confirmed that the exponents are the same.

1.6 The O_n-symmetric classical Heisenberg ferromagnet

The Hamiltonian of this model has the same form as (1.16), but now the external field h and the spins $\widehat{s}_i$ at the sites are assumed to be n-component vectors, and $\widehat{s}_i \widehat{s}_k$ and $h\widehat{s}_i$ are understood as scalar products of vectors. By condition, the realizations of the fluctuating variable $\widehat{s}_i$ are classical vectors s_i of unit length, all of whose directions are equally probable. Now the trace in (1.17) denotes the integral over the directions of all the s_i. The Ising model is obtained in the special case $n = 1$; when speaking of the "Heisenberg model" without further specifics one usually has in mind the case $n = 3$ for $d = 3$, i.e., the ordinary spin vector (only for $n = d$ can the spin actually be assumed to be a vector under spatial rotations).

All the general thermodynamical relations of Sec. 1.4 and the general conclusions following from the scaling hypothesis remain valid for this model (of course, the exponents and scaling functions depend on n), with the obvious difference that h and M become O_n vectors and the susceptibility becomes a tensor $\chi_{ab} = dM_a/dh_b$. Owing to the O_n symmetry, the free energy $F = F(T, |h|)$ is independent of the direction $n_a = h_a/|h|$ of the vector h, so that $M_a = -\partial F/\partial h_a = -F'n_a$, and for the isothermal susceptibility we have $\chi_{ab} = \partial M_a/\partial h_b = -F'(\delta_{ab} - n_a n_b)/|h| - F''n_a n_b$, where the primes everywhere denote derivatives of F with respect to $|h|$. The equation $\chi_{ab} = \chi_\perp(\delta_{ab} - n_a n_b) + \chi_\parallel n_a n_b$ determines the longitudinal ($\chi_\parallel$) and transverse ($\chi_\perp$) susceptibilities. It follows from these expressions that $\chi_\perp = -F'/|h| = |M|/|h|$ and $\chi_\parallel = -F''$. The first of these shows that at fixed $\tau \equiv T - T_c < 0$ the transverse susceptibility $\chi_\perp$ diverges as $|h|^{-1}$ for $h \to 0$ because M has a finite limit, the spontaneous magnetization. In contrast to $\chi_\perp$, the behavior of $\chi_\parallel$ for $h \to 0$ is not uniquely determined by symmetry considerations; the answer depends on the model. In the simple Landau theory $\chi_\parallel$ is finite for $h = 0$ and $\tau < 0$, but the more accurate fluctuation theory of critical phenomena for $2 < d < 4$ predicts that $\chi_\parallel$ diverges, though more weakly than $\chi_\perp$: $\chi_\parallel \sim |h|^{d/2-2}$. This will be discussed in detail in Ch. 4.

As far as the nature of the variables is concerned, the susceptibilities $\chi_\perp$ and $\chi_\parallel$ are the same as χ in the uniaxial model, and the scaling hypothesis allows the standard scaling representations (Sec. 1.5) to be written down for them. As before, there is analyticity in τ for $h \neq 0$ and in h for $\tau > 0$. For $h = 0$, $\tau > 0$ the system must be O_n-isotropic, i.e., $\chi_{ab} \sim \delta_{ab}$, which implies that the susceptibilities are equal: $\chi_\perp(0, \tau) = \chi_\parallel(0, \tau)$ for $\tau > 0$. The main difference from the uniaxial model is the singular behavior of the susceptibilities for $h \to 0$, $\tau < 0$ discussed above.

1.7 The classical nonideal gas: the model and thermodynamics

The Hamiltonian in (1.17) for a canonical ensemble of N classical identical structureless particles of mass m is

$$\widehat{H} = \sum_{i=1}^{N} \frac{\widehat{p}_i^2}{2m} + v(\widehat{x}_1, ..., \widehat{x}_N), \tag{1.23}$$

where $\widehat{p}_i$ and $\widehat{x}_i$ are the random momenta and coordinates of the particles and v is the potential energy. The trace in (1.17) is now understood as the integral over N-particle phase space, i.e., over all p_i, x_i (realizations of $\widehat{p}_i$, $\widehat{x}_i$) with measure $dpdx/2\pi\hbar$ for each one-dimensional pair and overall coefficient $1/N!$. The momenta are integrated over all space, and the coordinates are integrated over a given region of finite volume V. The coefficient $1/N!$ is needed to ensure finiteness of the limit in (1.12), and Planck's constant $\hbar$ ensures that the partition function has zero canonical dimension. The observables are functions of the random variables $\hat{p}$, $\hat{x}$ and their realizations are ordinary functions p, x in the phase space.

The analogy with magnetic systems is seen more clearly when we consider the grand canonical ensemble. Then N becomes an additional variable of the "configurations" and observables, and the trace in (1.17) includes summation over N. Moreover, the replacement $\hat{H} \to \hat{H} - \mu\hat{N}$ in (1.17) introduces a new independent variable, the chemical potential μ. In the N-particle sector the random quantity $\hat{N}$ coincides with N. Later on we will use the fact that the observable $\hat{n}(x)$ represented by the function $n(x; x_1 \ldots x_N) = \sum_{i=1}^{N} \delta(x - x_i)$ in the phase space of the N-particle sector has the meaning of the particle number density at the point x. This random quantity is the analog of the spin variable $\hat{s}_i$ in magnets.

The definitions (1.17) can be used to express logarithmic derivatives of the partition function with respect to parameters in terms of the averages of known observables. The quantity $\langle\hat{H}\rangle$ is the internal energy, thus establishing the correspondence with thermodynamics. For the canonical ensemble we obtain $\partial_T \ln Z = \langle\hat{H}\rangle/kT^2$, and for the grand canonical ensemble $\partial_\mu \ln Z = \langle\hat{N}\rangle/kT$ and $\partial_T \ln Z = \langle\hat{H} - \mu\hat{N}\rangle/kT^2$.

When studying phase transitions we are interested in spatially uniform systems in the thermodynamical limit $V \to \infty$ with T and μ fixed for the grand canonical ensemble or $V \to \infty$, $N \to \infty$ with T and $\rho = N/V$ fixed for the canonical ensemble. In both cases $\ln Z \sim V$ up to corrections that are unimportant for $V \to \infty$. Therefore, the V dependence is always easy to isolate, and the final expressions can always be written as specific (per unit volume) quantities.

In statistical physics it is also possible to define the pressure p independently of thermodynamics, expressing it in terms of the equilibrium expectation values of the components of the microscopic strength tensor. However, there is no unique definition of the entropy in statistical physics. It is defined only up to an additive constant proportional to N, i.e., up to a constant that adds to the "entropy per particle." An addition of the type cN to the total entropy of a system corresponds to an addition $-cTN$ to the free energy and $-cT$ to the chemical potential (see below), so that none of these quantities can be considered uniquely defined. No quantity that can actually be measured depends on the choice of normalization of the entropy or related quantities, and so the choice is a matter of convenience. Here there is an obvious analogy with potentials in classical electrodynamics: the choice of a specific normalization (= gauge) of the potentials is also a matter of convenience, and no observables depend on it. Therefore, potentials in a particular gauge can be calculated theoretically, but cannot be measured directly. The same is true for objects like the entropy and related quantities in statistical physics.

Let us give the basic thermodynamical relations related to the statistical physics of these two ensembles (the canonical ensemble is closer to experiment, while the grand canonical ensemble is more convenient for theory). We shall use primed letters to denote extensive quantities (except for V and N) that pertain to the entire volume, and letters without a prime to denote specific quantities. The heat capacity C' and the compressibility K are given by

$$C'_x = dQ'/dT|_x = TdS'/dT|_x, \quad K_x = \rho^{-1}d\rho/dp|_x, \qquad (1.24)$$

where $dQ' = TdS'$ is the heat added to the system, S' is the entropy, p is the pressure, and x indicates what is held fixed in the differentiation. The condition $N = \text{const}$ is always assumed.

Let us begin with the canonical ensemble: N is a fixed parameter, and the thermodynamical variables are the pairs of conjugate variables T, S' and p, V, only one of the variables in each pair is independent. The basic thermodynamical functions are the internal energy U' with $dU' = TdS' - pdV$ (the heat added minus the work done) and its Legendre transforms. We shall study only one of the latter, namely, the free energy $F' = U' - TS'$ with $dF' = -S'dT - pdV$. The form of the differential determines the "natural variables" of the function and the meaning of its partial derivatives with respect to these variables, for example, $F' = F'(T, V)$, $\partial_T F' = -S'$, $\partial_V F' = -p$, and so on.

In statistical physics the function $\ln Z = -\Omega'/kT$ is calculated directly in the variables T, V [see (1.12) and (1.15)]. Differentiating both sides with respect to T and using the expression $\partial_T \ln Z = \langle \widehat{H} \rangle / kT^2$ given above, we obtain $U' = \Omega' - T\partial_T\Omega'$. From this it follows, as for the magnet in Sec. 1.4, that Ω' coincides with the free energy F' up to a term linear in T: $F' = \Omega' + Ta(V; N)$, with coefficient a linear in V at the asymptote $V \to \infty$, $N/V = \rho = \text{const}$. The simplest choice $\Omega' = F'$ leads to the correct (according to statistical physics; see above) expression for the pressure p, and so the addition Ta should not contribute to the pressure $p = -\partial_V F'$, i.e., the coefficient $a(V; N)$ should not depend on V at the asymptote $V \to \infty$. Taking into account its linearity in $V \sim N$ for $V \to \infty$, we find the equation $F' = \Omega' - cTN$ with some constant c independent of the parameters, which corresponds to the arbitrariness $S' \to S' + cN$ in the definition of the entropy mentioned above. We can therefore postulate the equation $F' = \Omega'$ as the definition of the free energy F', thereby fixing a particular normalization of the entropy $S' = -\partial_T F'(T, V; N)$. This definition of the free energy and the entropy following from it for the canonical ensemble is commonly used and will be taken below. The quantity $F' = \Omega'$ is calculated directly from the partition function in its natural variables T, V, and so the entropy and pressure can be calculated as the corresponding partial derivatives of $F' = \Omega'$.

At the asymptote $V \sim N \to \infty$ it can be assumed that $\Omega' = F' = VF(T, \rho)$, where $\rho = N/V$. Equating the expressions $dF' = -S'dT - pdV$ and $d(VF) = FdV + VdF$ and using $d\rho/\rho = -dV/V$ and $p = -\partial_V[VF(T, \rho)]$, for the function $F(T, \rho)$ we obtain

$$dF = -SdT + \mu d\rho, \quad F - \rho\partial_\rho F = -p, \qquad (1.25)$$

where $S = S'/V$ is the specific entropy and $\mu \equiv \partial_\rho F$ by definition.

Using Eqs. (1.24) and (1.25), the specific heats $C = C'/V$ and the compressibility can be expressed in terms of the function $F(T, \rho)$ known from the partition function:

$$C_\rho = -T\partial_T^2 F, \quad K_T = (\rho^2\partial_\rho^2 F)^{-1},$$

$$C_p = C_\rho + TK_T R^2, \quad K_{S'} = K_T C_\rho C_p^{-1},$$

where $R \equiv \rho\partial_\rho\partial_T F - \partial_T F = \partial_T p(T, \rho)$. The second equation follows from the first and from (1.25). For the canonical ensemble $C_\rho = C_V$, because N is fixed. We note that when calculating dS'/dT in (1.24), for $S' = VS(T, \rho)$ it is necessary to also know dV/dT if $V \neq$ const. The definition $\rho = N/V$ for $N =$ const allows dV/dT to be expressed in terms of $d\rho/dT$, and with the given condition this derivative is easily found by differentiating the second equation in (1.25).

Let us now briefly discuss the analogous relations for the grand canonical ensemble: the functions $\ln Z = -\Omega'(T, V, \mu)/kT \cong -V\Omega(T, \mu)/kT$ are known for $V \to \infty$ from the partition function. By definition, $\langle\hat{H}\rangle = U'$ and $\langle\hat{N}\rangle = N$, and the logarithmic derivatives of the partition function with respect to T and μ are expressed in terms of these averages (see above), which leads to the relations $U' = \Omega' - T\partial_T\Omega' - \mu\partial_\mu\Omega'$ and $\partial_\mu\Omega' = -N$. The first of these shows that U' is the Legendre transform of Ω' with respect to T and μ. Defining S' and p by $d\Omega'(T, V, \mu) = -S'dT - pdV - Nd\mu$, we obtain $dU' = TdS' - pdV + \mu dN$, in agreement with thermodynamics (Sec. 24 in [51]). This proves the correctness of defining p and S' in terms of Ω' already with some particular normalization of the entropy S'. In this case the normalization of the entropy and the free energy is fixed by the actual choice of the form of the statistical ensemble, i.e., by (1.17) with $\hat{H} \to \hat{H} - \mu\hat{N}$. In fact, the arbitrariness in the free energy discussed above, $F' \to F' - cTN$ (i.e., $S' \to S' + cN$ for the entropy), corresponds to replacement of the chemical potential $\mu = \partial_N F' \to \mu - cT$. This replacement in the usual relations like (1.17) for the grand canonical ensemble does not reduce to a simple change of the numerical values of the parameters T and μ, and so it must be interpreted as a transformation to a different (physically equivalent) statistical ensemble. By choosing the standard notation for relations like (1.17) for the grand canonical ensemble and identifying the parameter μ appearing in these relations as the chemical potential $\mu = \partial_N F'(T, V; N)$, we have thereby fixed the normalization of the free energy F' and the entropy S'. In this respect the grand canonical ensemble differs from the canonical ensemble, for which the normalization of F' is fixed by choosing the constant c in the definition $F'(T, V; N) = \Omega'(T, V; N) - cTN$ (the standard choice is $c = 0$).

From the equation $\Omega' = V\Omega(T,\mu)$ and the expression given above for $d\Omega'$, we obtain the following for the specific thermodynamical potential $\Omega(T,\mu)$:

$$\Omega(T,\mu) = -p, \quad dp = SdT + \rho d\mu. \tag{1.26}$$

Therefore, the quantity Ω defined in (1.15) for the grand canonical ensemble is minus the pressure, whereas for the canonical ensemble it was the specific free energy F (see above). The equivalence of the ensembles in the thermodynamical limit implies, in particular, that p in (1.25) and (1.26) is the same quantity. It therefore follows from (1.25) that F and $-p$ are related by a Legendre transform. In natural variables, $F = F(T,\rho)$ and $p = p(T,\mu)$. For the grand canonical ensemble the function $p(T,\mu)$ is calculated directly from the partition function, and the other quantities are expressed in terms of it:

$$C_\rho/T = \partial_T^2 p - (\partial_T \partial_\mu p)^2/\partial_\mu^2 p,$$

$$C_p/T = \partial_T^2 p + \rho^{-2}\partial_T p(\partial_T p \partial_\mu^2 p - 2\rho \partial_T \partial_\mu p), \tag{1.27}$$

$$K_T = \rho^{-2}\partial_\mu^2 p, \quad K_{S'} = K_T C_\rho C_p^{-1}.$$

The quantities T, ρ, and p are directly measurable, and among the objects containing second derivatives of F or p there are only three independent measurable quantities, for example, K_T, C_ρ $(= C_V)$, and C_p. It is easily checked that the arbitrariness $F' \to F' - cNT$ in the total free energy or $F \to F - c\rho T$ in the specific free energy does not affect any of these quantities, but changes only the normalization of the entropy and the chemical potential. For the ensembles we have considered with fixed (commonly used) normalization of the free energy and entropy, these quantities can be calculated theoretically but cannot be measured, just like potentials with fixed gauge in electrodynamics.

1.8 The thermodynamical scaling hypothesis for the critical point of the liquid–gas transition

The quantities μ, ρ, and $-p(T,\mu)$ for a gas are the formal analogs of h, M, and the specific free energy $F(T,h)$ for a magnet. However, from the viewpoint of the theory of critical phenomena the analogy between h and μ is not exact, because the line of coexistence of the liquid and gas phases in the T, μ plane, $\mu = \mu_0(T)$, $T \leq T_c$, has finite slope near the endpoint T_c, $\mu_c = \mu_0(T_c)$. Therefore, the exact analogs of the variables $e_1 = h$ and $e_2 = \tau = T - T_c$ of a magnet will now be certain linear combinations (see, for example, [49]):

$$e_1 = \Delta\mu + a\Delta T, \quad e_2 = \Delta T + b\Delta\mu. \tag{1.28}$$

Here and below $\Delta F \equiv F - F_c$ for any quantity F (not to be confused with the dimension $\Delta_F = \Delta[F]$), and a and b in (1.28) are numerical coefficients. We assume that the phase-coexistence line near T_c is $e_1 = 0$ [the inclusion of curvature and possible nonlinearities in (1.28) are corrections to scaling, which will not be discussed or included here]. The parameter a is determined by its slope. The coefficients a and b in (1.28) can be calculated theoretically from the partition function of the grand canonical ensemble, but cannot be measured directly (see the end of this section for more details).

According to the current ideas, this system and the uniaxial ferromagnet belong to the same universality class, i.e., their critical behavior is identical when corrections to scaling are neglected. Not only the critical exponents, but also the normalized scaling functions (Sec. 1.5) coincide. It then follows from the correspondence h, τ, $F \leftrightarrow e_1$, e_2, $-p$ that the singular part of the pressure, defined as

$$p(T, \mu) = p_c + S_c \Delta T + \rho_c \Delta\mu + p^{(s)}, \qquad (1.29)$$

satisfies a relation analogous to (1.20)

$$-p^{(s)} = |e_1|^{d/\Delta_1} A(x), \quad x \equiv e_2 |e_1|^{-\Delta_2/\Delta_1}, \qquad (1.30)$$

with the same (up to the normalization) scaling function $A(x)$ and the same values of the critical dimensions $\Delta_i \equiv \Delta[e_i]$. In (1.30) we have made the exponents denoted by stars in (1.20) explicit. We again remind the reader that when speaking of equivalence we always are referring only to the leading singular contributions.

From (1.28) we have $\partial_\mu = \partial_1 + b\partial_2$, $\partial_T = a\partial_1 + \partial_2$, and $\partial_i \equiv \partial/\partial e_i$, so that from (1.26) and (1.29) it follows that

$$\Delta\rho = (\partial_1 + b\partial_2)p^{(s)}, \quad \Delta S = (a\partial_1 + \partial_2)p^{(s)}. \qquad (1.31)$$

The quantities $p^{(s)}$ and e have definite dimensions, and (1.27) and (1.31) are the sums of terms of different dimension. For such quantities, Δ_F will always be understood as the dimension of the most singular term (the term with the smallest Δ). Since $\Delta_1 > \Delta_2$ (Sec. 1.5), the operator ∂_1 is "more singular" than ∂_2, and it determines the leading contributions. Therefore, $\Delta\rho \equiv \rho - \rho_c$ and $\Delta S \equiv S - S_c$ in (1.31) have the same dimension $d - \Delta_1$ in the sense discussed above, as for the magnetization in the Ising model. In (1.27) any of the second derivatives of p contains the leading singular contribution $\partial_1^2 p^{(s)}$ with the dimension of the magnetic susceptibility $d - 2\Delta_1$. This will also be the dimension of C_p and K_T in (1.27), but not of C_ρ or $K_{S'} \sim C_\rho$, because in these quantities the most singular contributions $\sim \partial_1^2 p^{(s)}$ cancel out. In fact, we see from (1.27) that C_ρ is proportional to the combination $\partial_T^2 p \cdot \partial_\mu^2 p - (\partial_T \partial_\mu p)^2$, which reduces to the form $(1 - ab)^2[\partial_1^2 p \cdot \partial_2^2 p - (\partial_1 \partial_2 p)^2]$ after changing to the derivatives ∂_i (see above), and so it has definite dimension $2d - 2\Delta_1 - 2\Delta_2$. In the expression for C_ρ this combination is divided by $\partial_\mu^2 p$ with dimension (of the leading term) $d - 2\Delta_1$, and the dimension of the ratio is $d - 2\Delta_2$. It follows from (1.27) for the adiabatic compressibility $K_{S'}$, taking into account the equality of the dimensions of K_T and C_p, that this quantity has the same dimension as C_ρ. Therefore, C_ρ ($= C_V$) and $K_{S'}$ have the dimension of the specific heat in the Ising model ($d-2\Delta_2$), while C_p and K_T have the dimension of the susceptibility ($d - 2\Delta_1$).

Let us now discuss the behavior of various quantities in approaching the critical point along different straight lines in the T, μ plane. We shall always speak of the leading term of the asymptote, neglecting corrections, so that all the quantities discussed above $F \equiv \{\Delta\rho \sim \Delta S,\ C_\rho \sim K_{S'}, K_T \sim C_p\}$ can be assumed to be dimensional quantities with $\Delta_F = \{d-\Delta_1,\ d-2\Delta_2,\ d-2\Delta_1\}$, respectively. First excluding the special case of the line $e_1 = 0$, let us consider any other straight line passing through the critical point. On such a line, all the nonzero quantities e_1, e_2, ΔT, $\Delta\mu$, and Δp are proportional to each other [the contribution $p^{(s)}$ in (1.29) is an unimportant correction]. By condition, $e_1 \neq 0$ and $e_2 \sim e_1$, so that on the line the argument x of the scaling function in representations like (1.30) reduces to a power $|e_1|^*$ with positive exponent $* = (\Delta_1 - \Delta_2)/\Delta_1$, so that $x \to 0$ for $e_1 \to 0$ and $A(x) \to A(0)$ owing to the analyticity of $A(x)$ near zero (Sec. 1.5). This implies that on any line except $e_1 = 0$ the asymptote is exactly the same as on the line $e_2 = 0$, $e_1 \to 0$, and any dimensional function F behaves as a simple power of any of the nonzero variables $e_1 \sim \Delta T \sim \Delta\mu \sim \Delta p$ with the known exponent $\lambda_F = \Delta_F/\Delta_1$. In the special case of the lines $\Delta T = 0$ (the critical isotherm) and $\Delta p = 0$ (the critical isobar) it is usual to write $F \sim |\Delta p|^{\lambda_F}$ for the isotherm and $F \sim |\Delta T|^{\lambda_F}$ for the isobar. In both cases it would also be possible to write $F \sim |\Delta\mu|^{\lambda_F}$. The exponents $\lambda_F = \Delta_F/\Delta_1$ for the quantities $F = \{\Delta\rho,\ C_\rho,\ K_T\}$ can be expressed in terms of the traditional exponents (1.19). In particular, for $\Delta\rho \equiv \rho - \rho_c$ we have $\lambda_F = (d - \Delta_1)/\Delta_1 = 1/\delta$, so that on an isobar $\Delta\rho \sim |\Delta T|^{1/\delta}$ and on an isotherm $\Delta\rho \sim |\Delta p|^{1/\delta}$. Comparing the last expression with (1.18), it is sometimes said that Δp is the analog of the magnetic field h, but such terminology can lead to confusion. The exact analogs of h and τ of the magnet are e_1 and e_2 in (1.28). On the critical isotherm $e_1 \sim \Delta p \sim \Delta\mu$, and on the isobar $e_1 \sim \Delta T \sim \Delta\mu$; in both cases $F \sim |e_1|^{\lambda_F}$. This formulation is exact and complete.

Let us now discuss the special case of the line $e_1 = 0$, which so far we have not considered. This line is the analog of the zero-field line in a magnet. For $\Delta T < 0$ it is the phase-coexistence line, and for $\Delta T > 0$ it is its extension. All the quantities e_2, ΔT, Δp, and $\Delta\mu$ are proportional to each other on this line and are nonzero (for Δp this follows from the finiteness of the slope of the phase-coexistence line in the p, T plane near the endpoint). Any single-valued quantity F behaves as $A_{\pm}|e_2|^{\lambda'_F}$ on the line $e_1 = 0$, and the exponent $\lambda'_F = \Delta_F/\Delta_2$ is the same on both sides of T_c (since they are proportional to each other, ΔT and e_2 change sign simultaneously). The amplitude factors $A_{\pm}$ can differ. However, now for $\Delta T < 0$ we must take into account the possibility that quantities on the line $e_1 = 0$ are nonunique, i.e., the $e_1 \to \pm 0$ limits can differ (discontinuities on the line of first-order transitions). The formal rule is simple: quantities even in e_1 are single-valued on the line $e_1 = 0$, i.e., their $e_1 \to \pm 0$ limits coincide, while for odd quantities at $\Delta T < 0$ these limits have different signs. For $\Delta T > 0$ all quantities are single-valued, and so those that are odd on the line $e_1 = 0$, $\Delta T > 0$ vanish.

We see from (1.30) that $p^{(s)}$ is even, and the parity of any derivative of it is determined by the number of odd operators ∂_1. Accordingly, the leading contributions of all the quantities in (1.27) are even, i.e., when corrections are neglected the quantities (1.27) are single-valued on the line $e_1 = 0$ and have asymptote $F \cong A_\pm |e_2|^{\lambda_F'}$, $e_2 \sim \Delta T \sim \Delta p \sim \Delta \mu$, with the known exponent $\lambda_F' = \Delta_F / \Delta_2$. In general, these differ on different sides of T_c by the amplitude factors $A_\pm$.

The quantities in (1.31) have leading odd contribution $\sim \partial_1 p^{(s)}$ with dimension $d - \Delta_1$ and even correction term $\sim \partial_2 p^{(s)}$ with dimension $d - \Delta_2$. The corresponding exponents λ_F' are $(d - \Delta_1)/\Delta_2 = \beta$ and $(d - \Delta_2)/\Delta_2 = 1 - \alpha$ in the notation of (1.19). Therefore, on the line $e_1 = 0$ the quantities $\Delta \rho$ and ΔS are single-valued for $\Delta T > 0$ and have asymptotes $\Delta \rho \equiv \rho - \rho_c \cong bA_+|e_2|^{1-\alpha}$ and $\Delta S \equiv S - S_c \cong A_+|e_2|^{1-\alpha}$. For $\Delta T < 0$ they are double-valued, and for the two phases (liquid and gas) we have

$$
\begin{aligned}
\rho_l - \rho_c &\cong B|e_2|^\beta + bA_-|e_2|^{1-\alpha}, \\
\rho_g - \rho_c &\cong -B|e_2|^\beta + bA_-|e_2|^{1-\alpha}, \\
S_l - S_c &\cong aB|e_2|^\beta + A_-|e_2|^{1-\alpha}, \\
S_g - S_c &\cong -aB|e_2|^\beta + A_-|e_2|^{1-\alpha},
\end{aligned}
\tag{1.32}
$$

where $A_\pm$ are the amplitude factors in $\partial_2 p^{(s)}$ and B is the amplitude factor in $\partial_1 p^{(s)}$ for $\Delta T < 0$. In all of the expressions e_2 can be replaced by any of the quantities ΔT, Δp, or $\Delta \mu$, which leads only to a change of the amplitudes [in particular, $e_2 = (1 - ab)\Delta T$]. If we completely neglect the small even correction, for $\Delta T > 0$ we obtain $\Delta \rho = 0$, so that in a real experiment with $N = \mathrm{const}$ the line $e_1 = 0$ above T_c can approximately be considered a critical isochore. An analogous error is also present in the definition of the linear isobar owing to neglect of the contribution $p^{(s)}$ in (1.29).

The small even correction is important when studying the problem of the *diameter of the phase-coexistence curve* in the T, ρ plane, constructed from the graphs of $\rho_l(T)$ and $\rho_g(T)$ obtained by substituting $e_2 = (1 - ab)\Delta T$ into (1.32). The diameter is defined as $\rho_0 \equiv (\rho_l + \rho_g - 2\rho_c)/2$. From (1.32) we have $\rho_0 = bA_-|(1 - ab)\Delta T|^{1-\alpha}$. This and other predictions of the scaling hypothesis have been confirmed experimentally [52] (the diameter is discussed in Sec. 5.2 of that study). However, to obtain quantitative agreement it is necessary to include both analytic corrections (for example, ones $\sim \Delta T$ in ρ_0) and the nonanalytic corrections to the leading singular contributions associated with the exponents ω (Sec. 1.3). Since $\rho_0 \sim b$ (see above), the experimental fact $\rho_0 \neq 0$ indicates that we must have $b \neq 0$ in (1.28).

Let us conclude by discussing the measurability of the coefficients a and b in (1.28). The quantities ρ_c and $\Delta \rho$ are measurable and S_c and ΔS are not, because the arbitrariness in the total entropy $S' \to S' + cN$

corresponds to $S \to S + c\rho$ for the specific entropy, and the undefined parameter c does not cancel, even in the difference $\Delta S = S - S_c$. However, if we consider the entropy per particle $\overline{S} = S/\rho$, we would have $\overline{S} \to \overline{S} + c$, and so the difference $\Delta \overline{S} = \overline{S} - \overline{S}_c$ is measurable. In the linear approximation $\Delta \overline{S} = \rho_c^{-2}[\rho_c \Delta S - S_c \Delta \rho]$, and so from (1.32) we can obtain analogous expressions for $\Delta \overline{S}$. Comparing the amplitudes of contributions of the same type in the measurable quantities $\Delta \rho$ and $\Delta \overline{S}$, we find $a\rho_c - S_c$ and $b^{-1}\rho_c - S_c$, i.e., these two combinations are measurable [we note that the ratio $(1 - ab)/b$ can be found from their difference]. It is easily checked that the measurability of (1.27) does not contribute any new information about the parameters a, b, and S_c.

Therefore, only two relations for these three parameters can be obtained from the experimental data. At the level of phenomenology one of them (a or S_c) can be fixed arbitrarily, taking, for example, $a = 0$. We have not done this, because we are taking the original microscopic model to be the grand canonical ensemble, for which all three parameters a, b, and S_c are theoretically calculable.

1.9 The scaling hypothesis for the correlation functions

Let $\widehat{\varphi}(x)$ be the random field of the order parameter. In different problems it will mean different things, but it will always be defined such that at the critical point $\langle \widehat{\varphi}(x) \rangle = 0$. For a gas $\widehat{\varphi}(x) = \widehat{n}(x) - \rho_c$, where $\widehat{n}(x)$ is the fluctuating particle number density at the point x (Sec. 1.7). For lattice spin models $\widehat{\varphi}$ is the spin field on the lattice. It is convenient to picture it geometrically, by specifying the sites i by their spatial coordinates x_i, which assumes the introduction of a dimensional lattice constant (the spacing) playing the role of the *minimum length*. Then $\widehat{\varphi}(x) = \widehat{s}(x)$, where x runs through the discrete set of site coordinates x_i and $\widehat{s}(x_i) \equiv \widehat{s}_i$ is the spin at site i. For a gas and the Ising model $\widehat{\varphi}$ is a one-component field. In other problems it can have indices, and there can be many fields that are also distinguished by additional indices. We shall adopt the following *convention* in writing general definitions and expressions. The complete set of fields of a model will be denoted by the single symbol $\widehat{\varphi}(x)$, where the argument x includes all continuous and discrete variables (indices) on which the field depends. The symbol $\int dx...$ will be understood as the integral over all continuous and the sum over all discrete components of x, and $\delta(x)$ will denote the product of δ functions for all continuous and the product of Kronecker deltas for all discrete components of x. We shall refer to this as *universal notation*.

The *full correlation functions* G_n of the field $\widehat{\varphi}$ (the *normalized full Green functions* in field theory) written in universal notation are

$$G_n(x_1...x_n) = \langle \widehat{\varphi}(x_1)...\widehat{\varphi}(x_n) \rangle \quad \text{for all } n \geq 1, \quad G_0 \equiv 1. \tag{1.33}$$

For a gas the averaging is always done over the grand canonical ensemble with the replacement $\widehat{H} \to \widehat{H} - \mu\widehat{N}$ in (1.17). The functions (1.33) are symmetric under permutations of the x_i and also depend on all the parameters of the model. If necessary, this will be indicated explicitly. The functions (1.33) are multivalued on the phase-coexistence line and single-valued off it. For spatially uniform systems, including lattices with periodic boundary conditions, the functions (1.33) are translationally invariant, i.e., $\langle\widehat{\varphi}(x)\rangle$ is independent of the spatial components of the argument x, and the higher functions depend only on their differences. In this case $\langle\widehat{\varphi}(x)\rangle$ is the same as its spatial average and has a simple thermodynamical interpretation. For example, for a magnet $\langle\widehat{\varphi}(x)\rangle = M$ is the magnetization. For a gas, $\langle\widehat{\varphi}(x)\rangle = \rho - \rho_c$. The quantity $\widehat{\varphi}(x) - \langle\widehat{\varphi}(x)\rangle$ is the random fluctuation field of the order parameter. Its average is zero by definition. The average of the product of two such fields, given by

$$D(x, x') \equiv \langle\widehat{\varphi}(x)\widehat{\varphi}(x')\rangle - \langle\widehat{\varphi}(x)\rangle\langle\widehat{\varphi}(x')\rangle, \qquad (1.34)$$

is called the pair correlation function of the fluctuation field or simply the *correlator* (the *propagator* in quantum field theory). For a uniform system it depends only on the coordinate difference and then is simply related to the thermodynamical susceptibility. Let us demonstrate this for the example of the Ising model. From the definitions of Sec. 1.4 we have $\partial_h \ln Z = \langle\beta\widehat{S}\rangle = \beta V M$. Differentiating this expression with respect to h, on the right-hand side we obtain the susceptibility $dM/dh = \chi_T$, and on the left-hand side the derivative of the average $\langle\beta\widehat{S}\rangle$ written as in (1.17) reduces to the variance of the same quantity: $\beta^2[\langle\widehat{S}^2\rangle - \langle\widehat{S}\rangle^2] = \beta V \chi_T$. The observable $\widehat{S} = \int dx\widehat{\varphi}(x)$ is the total spin of the system (for discrete x the integral is understood as a sum), so that the expression in square brackets is the double integral of the correlator (1.34) over x and x'. Owing to its translational invariance, after the first integration over x we obtain a constant independent of x', and then integration over x' gives a factor of V, which cancels in the above equation, finally giving $\int dx D(x, x') = \beta^{-1}\chi_T$. The same arguments for the grand canonical ensemble of a gas lead to the equation $\int dx D(x, x') = \beta^{-1}\rho^2 K_T$.

The Fourier transforms of translationally invariant functions like (1.34) depending only on the difference $x - x'$ of spatial coordinates in a space of arbitrary dimension d will always be defined as

$$F(x, x') = (2\pi)^{-d}\int dk F(k)\exp[ik(x - x')], \qquad (1.35a)$$

$$F(k) = \int d(x - x')F(x, x')\exp[ik(x' - x)], \qquad (1.35b)$$

with the coordinate and momentum representations distinguished only by the arguments. In (1.35) x and x' are only d-dimensional spatial coordinates, k is the d-dimensional momentum (wave vector), and $k(x - x')$ is the scalar product of vectors. When the fields have discrete indices all the F are matrices in these indices. The integration over the difference $x - x'$ in (1.35b) can be replaced by integration over either of the two variables x or x'; the result will not depend on the second variable owing to the assumed translational invariance of $F(x, x')$. For a periodic lattice the integration in (1.35a) runs over a restricted region, the first Brillouin zone, and in (1.35b) it is understood as summation over the discrete coordinates of the lattice sites.

The positive definiteness of the correlator (1.34), which is obvious from its physical meaning, is equivalent to positivity of its Fourier transform $D(k)$ (a matrix when there are discrete indices), and the susceptibility (the compressibility for a gas) is expressed in terms of its value at the origin:

$$D(k) \geq 0, \quad D(k)|_{k=0} = \beta^{-1} \chi_T. \tag{1.36}$$

Let us now generalize the thermodynamical scaling hypothesis (1.13). The field $\widehat{\varphi}(x)$ and, consequently, all its correlation functions (1.33) are recognized as dimensional quantities and are included in the general scheme of the scaling hypothesis on the same footing. It should be noted immediately that this does not lead to the appearance of new critical dimensions Δ, because $\Delta_x = -1$ according to (1.11), and the dimension of the field $\Delta_\varphi \equiv \Delta[\widehat{\varphi}(x)]$ must naturally coincide with the dimension of its average, which is already known from thermodynamics; the dimension of the function (1.33) is the sum of the dimensions of the fields entering into it. Henceforth, for simplicity we shall assume that $\widehat{\varphi}(x)$ is a one-component field, with x being only a spatial coordinate, and we shall explicitly indicate the dependence of the functions (1.33) on the thermodynamical variables e. The analog of (1.13) is expressed in the following statement. For $\lambda \to 0$ and e_λ from (1.13),

$$G_n(\lambda^{-1} x_1 ... \lambda^{-1} x_n; e_\lambda) = \lambda^{n\Delta_\varphi} G_n(x_1 ... x_n; e). \tag{1.37}$$

Remarks. (1) The critical region corresponds to the $\lambda \to 0$ asymptote of the left-hand side of (1.37), i.e., this is the region of small e (closeness to the critical point) and, at the same time, relative distances $|x_i - x_k|$ that are large compared to the minimum length of the microscopic model (the lattice spacing, the interatomic separation in the case of a gas, and so on), or small momenta in momentum space.

(2) As in (1.13), the equality in (1.37) is exact only for the leading singular contribution in the critical region.

(3) For a gas, restriction to the leading term of the asymptote implies the neglect of all effects related to the concept of diameter (Sec. 1.8). To take them into account, the field $\widehat{\varphi}$ must be assumed to be the sum of terms of different dimension, but such corrections to the correlation functions are not observable at the current level of experimental accuracy [52].

(4) The discrete variables x in lattice models can be replaced by continuous variables, because in the critical region the functions (1.33) practically do not vary over distances of the order of the lattice spacing owing to the smoothness of the asymptotes. In momentum space the argument goes as follows. We are interested in the small-momentum asymptote, and the lattice is manifested only in the cutoff of the momentum integrals at the upper limit corresponding to the first Brillouin zone.

(5) For a gas, when there is a uniform external field the potential energy in (1.23) is spatially isotropic, i.e., invariant under a simultaneous, identical rotation of all the coordinate vectors. Then all the functions in (1.33) possess the same property, in particular, the correlator (1.34) depends only on the relative separation $r = |x - x'|$. This is also true for lattice spin models in the critical region for the correct choice of coordinates x (see Sec. 2.26 for more details).

It follows from this discussion than when corrections are neglected, the correlators for the gas and the uniaxial magnet have the same critical behavior. Henceforth, for definiteness we take the example of a magnet and assume that its correlator is uniform and isotropic: $D = D(r, \tau, h)$. Even before the formulation of the scaling hypothesis it was assumed that for $h = 0$ and $\tau \equiv T - T_c > 0$ the critical asymptote (small τ, large r) of the correlator has the scaling form

$$D(r, \tau, h = 0) = Ar^{2-d-\eta} f(r/r_c), \quad r_c = r_0[(T - T_c)/T_c]^{-\nu} \sim \tau^{-\nu}, \quad (1.38)$$

where d is the spatial dimension and η and ν are new critical exponents. The coefficients r_0 and $A = A(r_0)$ have zero critical dimension. The length r_0 is the characteristic scale of the fluctuations, and the coefficient A is needed to ensure the correct canonical dimension of the entire expression. In momentum space we have

$$D(k, \tau, h = 0) = Ak^{-2+\eta} g(kr_c) \quad (1.39)$$

with r_c from (1.38) and a different scaling function.

Since the correlator must be defined even right at the critical point, the scaling functions in (1.38) and (1.39) must have a finite limit for $r_c \to \infty$. We note that the correlator (1.39) for $\tau > 0$ is usually expanded in a series in integer powers of k^2, but in general the expansion in τ for $k \neq 0$ also contains fractional powers (see Ch. 4).

The parameter r_c is called the *correlation length* (or correlation radius), η is the Fisher exponent, and ν is the critical exponent of the correlation length. The latter characterizes the rate of increase of the correlation length as $T \to T_c$. We see from (1.38) that r_c has the meaning of the characteristic scale of variation of the function f. Experiment and calculations show that for $h = 0$, $T > T_c$ the correlator falls off exponentially with distance, i.e., $D \sim \exp(-r/r_c)$ for the appropriate choice of the parameter r_0 in the definition of r_c. This means that r_c is not only the characteristic scale, but also the actual distance up to which correlations are important. At the critical point, f in (1.38) becomes a constant, and the exponential falloff of the correlator is replaced by power-law falloff.

As already mentioned, the asymptote (1.38) was proposed before the formulation of the scaling hypothesis, and the exponents η and ν were at that time considered to be independent critical exponents. The scaling hypothesis (1.37) justifies the asymptote (1.38) and its generalizations for $h \neq 0$ and $\tau < 0$; moreover, it allows η and ν to be expressed in terms of the dimensions Δ_τ and Δ_h. In fact, from (1.5) and (1.11) it follows that the argument of f in (1.38) must be a multiple of the dimensionless combination $r\tau^{1/\Delta_\tau}$, so that $1/\nu = \Delta_\tau$. On the other hand, it follows from (1.38) that $2 - d - \eta = -2\Delta_\varphi$, because the dimension of the correlator (1.34) is $2\Delta_\varphi$. Then

$$\Delta_\varphi = d/2 - 1 + \eta/2 = d - \Delta_h, \quad \Delta_\tau = 1/\nu. \tag{1.40}$$

We recall that $\Delta_\varphi = \Delta_M = d - \Delta_h$, because the dimension of the field coincides with the dimension of its average value, the magnetization M.

Equations (1.40) and (1.19) lead to the relations $d\nu = 2 - \alpha$ and $\gamma = \nu(2 - \eta)$. From the data in Table 1.2 of Sec. 1.5 it can be verified that they are satisfied for the Ising model, exactly in the two-dimensional case and approximately in the three-dimensional one. The Landau theory corresponds to the Ornstein–Zernike correlator $D(k) = (k^2 + \tau)^{-1}$, for which $\eta = 0$ and $\nu = 1/2$ according to (1.39). For this value of ν and $\alpha = 0$, the relation $d\nu = 2 - \alpha$ is satisfied only in four-dimensional space. Therefore, for $d \neq 4$ the Landau theory does not agree with the predictions of the scaling hypothesis for the correlation functions, although it is always contained within the framework of the thermodynamical scaling hypothesis (see Sec. 1.15 for the reason).

1.10 The functional formulation

The fundamental tool in what follows will be the functional and diagrammatic technique of quantum field theory, and so we shall often use its terminology. In field theory the correlation functions are called the *Green functions* (of a given field), and in what follows these terms will be used synonymously. In field theory one distinguishes between full (normalized and unnormalized), connected, 1-irreducible, and other types of Green function. The functions in (1.33) are the *normalized full functions*, and the others will be introduced later. The adjective "normalized" will often be omitted.

The infinite set of functions (1.33) is conveniently written as a single object, the generating functional (all general expressions are given in universal notation; see Sec. 1.9):

$$G(A) = \sum_{n=0}^{\infty} \frac{1}{n!} \int \cdots \int dx_1 ... dx_n G_n(x_1...x_n) A(x_1)...A(x_n). \tag{1.41}$$

The argument of the functional $A(x)$ is an arbitrary function with x the same as for the field $\widehat{\varphi}(x)$. Equation (1.41) is a functional Taylor expansion. The functions G_n are its coefficients:

$$G_n(x_1...x_n) = \frac{\delta^n G(A)}{\delta A(x_1)...\delta A(x_n)}\bigg|_{A=0}, \tag{1.42}$$

and they are always symmetric with respect to permutations of $x_1...x_n$.

We introduce new quantities, the *connected Green functions* $W_n(x_1...x_n)$ of the field $\widehat{\varphi}$. By definition, they are specified by the generating functional

$$W(A) = \ln G(A) = \sum_n \frac{1}{n!} W_n A^n \tag{1.43}$$

as the coefficients of the Taylor expansion of $W(A)$ analogous to (1.41), which is written symbolically in (1.43). We rewrite (1.43) as $G(A) = \exp W(A)$, expand both sides in A, and equate the coefficients of identical powers of A (symmetrization with respect to permutations of the arguments x is always understood). As a result, we obtain $1 = G_0 = \exp W_0$, and

$$W_0 = 0; \quad G_1(x) = W_1(x); \quad G_2(x, x') = W_2(x, x') + W_1(x)W_1(x');$$

$$G_3(x, x', x'') = W_3(x, x', x'') + W_1(x)W_2(x', x'') + W_1(x')W_2(x, x'')$$
$$+ W_1(x'')W_2(x, x') + W_1(x)W_1(x')W_1(x''),$$

and so on. Comparing these expressions with (1.33) and (1.34), we see that

$$W_1(x) = \langle \widehat{\varphi}(x) \rangle, \quad W_2(x, x') = D(x, x'). \tag{1.44}$$

It is easily checked that the scaling hypothesis (1.37) is equivalent to the following invariance property of the generating functionals:

$$G(A_\lambda, e_\lambda) = G(A, e), \quad W(A_\lambda, e_\lambda) = W(A, e), \tag{1.45}$$

$$A_\lambda(x) = \lambda^{\Delta_A} A(\lambda x), \quad \Delta_A = d - \Delta_\varphi, \tag{1.46}$$

where $\Delta_A = \Delta[A(x)]$ is the critical dimension of $A(x)$. The second equation in (1.45) follows from the first and from (1.43).

Let us conclude this section with an explanation of the physical meaning of the functional (1.41). Substituting (1.33) in the form (1.17) into it, we can form the series in n, which gives

$$G(A) = \langle \exp(\widehat{\varphi}A) \rangle = Z^{-1} \, \mathrm{tr} \, \exp[-\beta\widehat{H} + \widehat{\varphi}A], \tag{1.47}$$

$$\widehat{\varphi}A \equiv \int dx \, \widehat{\varphi}(x)A(x). \tag{1.48}$$

For the Ising model $\widehat{\varphi}(x) = \widehat{s}(x)$, the spin field on the lattice (where x is discrete). Comparing with (1.16), we see that the addition $\widehat{\varphi}A$ in the exponent in (1.47) can be interpreted as the interaction of the spins with a nonuniform external field $h(x) = \beta^{-1}A(x)$. For a gas $\widehat{\varphi}(x) = \widehat{n}(x) - \rho_c$ (Sec. 1.9), with the replacement $\widehat{H} \to \widehat{H} - \mu\widehat{N}$ understood in (1.17) and (1.47). It follows from the definition of $\widehat{n}(x)$ in Sec. 1.7 that inside the trace in the N-particle sector we have $\widehat{\varphi}A = \sum_{i=1}^N A(x_i) - A\rho_c$. The sum can be interpreted as the interaction of N particles with an external potential field $v(x) = -\beta^{-1}A(x)$ [see (1.23)], and the contribution $A\rho_c \equiv \int dx\, A(x)\rho_c$ is taken outside the trace in (1.47) as a trivial overall factor $\exp(-A\rho_c)$.

Therefore, in both cases up to the normalization $G(A)$ is the partition function in a nonuniform external field proportional to $A(x)$. The normalization is determined by the condition $G(0) = 1$, i.e., $G(A)$ reproduces only the A-dependent factor in the partition function. Henceforth we shall refer to $A(x)$ as the *source* conjugate to the field of the order parameter $\widehat{\varphi}(x)$, independently of the actual physical meaning of these quantities. The equation $\Delta_\varphi + \Delta_A = d$ following from (1.46) and expressing the condition that the scalar product $\widehat{\varphi}A$ be dimensionless is sometimes referred to as the *shadow relation*: the critical dimensions of a mutually conjugate field and source add up to the spatial dimension d.

1.11 Exact variational principle for the mean field

As pointed out in Sec. 1.9, below T_c on the phase-coexistence line the correlation functions are multivalued. A typical example is the spontaneous magnetization of a magnet $\langle\widehat{\varphi}(x)\rangle = W_1(x)$ in zero field, which has arbitrary direction. We shall work with connected correlation functions and their generating functional $W(A)$, and for definiteness we shall consider a magnet, though all the general expressions in universal notation (Sec. 1.9) are true for any system. It follows from the discussion in Sec. 1.10 that the quantities

$$W_n(x_1...x_n; A) = \frac{\delta^n W(A)}{\delta A(x_1)...\delta A(x_n)} \tag{1.49}$$

have the meaning of the connected Green functions for the system with an auxiliary external field A. In the special case $A(x) = $ const the source reproduces a uniform external field. We shall therefore assume that it is absent in the original Hamiltonian, and then (1.49) is the connected functions for a system with a given external field A. For $T < T_c$ and some special value of A [for a magnet $A = 0$, for a gas some $A(x) = $ const $\neq 0$] the functions (1.49) are multivalued. This multivaluedness can be eliminated by transforming to an object with behavior more regular than that of $W(A)$ by a Legendre transform $\Gamma(\alpha)$ of the functional W with respect to A:

$$\Gamma(\alpha) = W(A) - \alpha A, \quad \alpha(x) = \frac{\delta W(A)}{\delta A(x)}, \tag{1.50}$$

where αA is a scalar product like (1.48). The functional variables A and α are conjugates of each other, and either can be taken as the independent variable. The second equation in (1.50) determines $\alpha(A)$ explicitly and $A(\alpha)$ implicitly, and the substitution $A = A(\alpha)$ is understood in the definition of Γ. These are all obvious generalizations of the equations of Sec. 1.2. Here $\alpha(x) = \langle\widehat{\varphi}(x)\rangle$ (the magnetization) for a system with external field A. From (1.50) we find [see (1.8)] that

$$\frac{\delta\Gamma(\alpha)}{\delta\alpha(x)} = -A(x). \tag{1.51}$$

The analog of (1.9) is

$$-\int dx'' \frac{\delta^2\Gamma(\alpha)}{\delta\alpha(x)\delta\alpha(x'')} \frac{\delta^2 W(A)}{\delta A(x'')\delta A(x')} = \delta(x-x'), \qquad (1.52)$$

and the analog of the second equation in (1.8) is

$$\frac{\delta W(A,a)}{\delta a} = \frac{\delta\Gamma(\alpha,a)}{\delta a}, \qquad (1.53)$$

where a is any auxiliary numerical or functional parameter on which the functionals in question depend. The quantities

$$\Gamma_n(x_1...x_n;\alpha) = \frac{\delta^n\Gamma(\alpha)}{\delta\alpha(x_1)...\delta\alpha(x_n)} \qquad (1.54)$$

are, by definition, the *1-irreducible Green functions* for the system with given value of α or A, if in (1.54) we take $\alpha = \alpha(A)$. In the notation of (1.44), (1.49), and (1.54), Eq. (1.52) can be written compactly as

$$-\Gamma_2 W_2 = 1, \quad D = W_2 = -\Gamma_2^{-1}, \qquad (1.55)$$

understood as equality of linear operators [the right-hand side of (1.52) is the kernel of the unit operator and $D = W_2$ is the correlator (1.34) for the system with external field A]. The second equation in (1.55) expresses the connected function W_2 in terms of the 1-irreducible function Γ_2. These definitions and expressions can also be used to express all the higher-order connected functions W_n with $n \geq 3$ in terms of the Γ_n. The explicit expressions will not be needed in this chapter and will be given in Sec. 2.5.

The problem of the theory is to calculate the average $\alpha = \langle\hat\varphi\rangle = W_1$ and all the other connected Green functions W_n with $n \geq 2$ for a given external field A. The full functions (1.33) are found trivially from the connected ones. If the functional $\Gamma(\alpha)$ for the model with zero external field is known [in Ch. 2 we describe a technique for calculating $\Gamma(\alpha)$ directly for a given quantum field model], the problem stated above is solved as follows. First, the desired $\alpha = \alpha(A)$ is found from the given $\Gamma(\alpha)$ and A using (1.51). Then (1.54) is used to determine all the 1-irreducible Green functions — the coefficients of the expansion of $\Gamma(\alpha)$ at the point $\alpha = \alpha(A)$. These are then used to find the corresponding connected Green functions W_n with $n \geq 2$: for the correlator $W_2 = D$ using (1.55), and for the higher-order functions W_n with $n \geq 3$ using the expressions of Sec. 2.5.

We introduce the quantity

$$\Phi(\alpha;A) = \Gamma(\alpha) + \alpha A, \qquad (1.56)$$

understanding it to be a functional of independent α and A. From (1.51) and (1.50) we have

$$\frac{\delta\Phi(\alpha,A)}{\delta\alpha(x)}\bigg|_{\alpha=\alpha(A)} = 0, \quad \Phi(\alpha,A)|_{\alpha=\alpha(A)} = W(A), \qquad (1.57)$$

i.e., the desired $\alpha(A)$ is a stationarity point of Φ with respect to α at fixed A, and the value of Φ at this point coincides with $W(A)$. If there are many stationarity points for a given A, then $\alpha(A)$ is multivalued, but the value of Φ at all points $\alpha(A)$ is the same (see below).

Multivaluedness is possible only in the limit of infinite volume V, because for finite V the functionals W and Γ are well defined and strictly convex [convexity refers to the sign of the second variation. The fact that the sign is positive for $W(A)$ is easily checked from (1.43) and (1.47); for $W(A)$ convex downward, the convexity upward of $\Gamma(\alpha)$ follows from (1.52)]. For strictly convex functionals there is a one-to-one correspondence between A and α. For $T < T_c$ and $V \to \infty$ kinks appear on the limiting surface of $W(A)$ and flat segments on $\Gamma(\alpha)$, i.e., the convexity of Γ is preserved, but becomes nonrigorous. Then the functional (1.56) for some particular A (at $A = 0$ for a magnet) has many stationarity points corresponding to the maximum of Φ with respect to α. Owing to the convexity (even though it is nonrigorous) of the functional Φ, the set of its stationarity points is necessarily convex, i.e., together with any two points it contains a line segment completely joining them. The extreme points of this set correspond to *pure phases* (a liquid or a gas, states with definite direction of the spontaneous magnetization of a ferromagnet, and so on), and the interior points to statistical mixtures of phases. Experimentally, mixtures are realized only for the liquid–gas transition as a two-phase system (part of the volume filled with gas and part with liquid). In what follows we shall discuss only pure phases, taking $A = 0$ (a magnet). The Green functions constructed from the single-valued functional $\Gamma(\alpha)$ (see above) are in general nonunique and depend on the choice of phase, because they are determined not only by the functional $\Gamma(\alpha)$ itself, but also by the choice of specific stationarity point $\alpha_0 = \alpha(A = 0)$.

The functionals W and Γ, and therefore also the set $\mathcal{O}$ of all stationarity points of Γ, always possess the symmetry of the problem in zero field, i.e., Γ and $\mathcal{O}$ are invariant under transformations $\alpha \to g\alpha$ with arbitrary g from the symmetry group. If $\mathcal{O}$ consists of a single point $(T \geq T_c)$ it is invariant, while if $\mathcal{O}$ is a nontrivial convex set $(T < T_c)$ its extreme points $\alpha \in \mathcal{O}$ are noninvariant, and it is said that *spontaneous symmetry breaking* occurs in the corresponding pure phase. If the broken symmetry group is continuous (for example, the rotation group for the Heisenberg ferromagnet), then in any pure phase $\alpha_0 \in \mathcal{O}$ the susceptibility of the system in the direction of a group shift $\alpha_0 \to g\alpha_0$ becomes infinite (the Goldstone theorem). In fact, in this case $g\alpha_0 \in \mathcal{O}$ for any g, i.e., $\delta\Gamma(\alpha)/\delta\alpha(x)|_{\alpha=g\alpha_0} = 0 \ \forall g$. Using $\delta_g...$ to denote the first variation with respect to g near $g = 1$, from the above discussion we conclude that

$$0 = \delta_g \left\{ \left. \frac{\delta\Gamma(\alpha)}{\delta\alpha(x)} \right|_{\alpha=g\alpha_0} \right\} = \int dx' \, \Gamma_2(x, x'; \alpha_0)\delta_g\alpha_0(x')$$

with Γ_2 from (1.54). We write the resulting equation compactly as $\Gamma_2\delta\alpha = 0$, understanding the left-hand side as the action of the linear operator Γ_2 on the "vector" $\delta\alpha = \delta_g\alpha_0$. The equation $\Gamma_2\delta\alpha = 0$ implies that $\delta\alpha$ is an eigenvector of the operator Γ_2 with zero eigenvalue; owing to (1.55), it will be an eigenvector of the correlator $D = W_2$ (in the phase α_0) with infinite eigenvalue. This also implies that the "susceptibility in the direction $\delta\alpha$" is infinite. For uniform systems and global (independent of x) transformations g, α_0 and its variation $\delta\alpha = \delta_g\alpha_0$ are also usually uniform, i.e., independent of x. It then follows that $D(k)|_{k=0}\delta\alpha = \infty$, where $D(k)$ is the correlator in momentum space. The Goldstone theorem is therefore usually formulated as the statement that there exist gapless excitations (massless particles in field theory) in a system with spontaneously broken continuous symmetry. The theorem is not applicable to the Ising model with discrete symmetry.

In practice, the functionals W and Γ can be calculated only in some sort of perturbation theory (for example, as high-temperature expansions in the Ising model), which means they can be obtained only to some finite order. The W and Γ thus obtained can always be expanded about zero in their arguments A and α. In this language it is clear why Γ is better than W: the expandability of $W(A)$ in A certainly excludes the possibility that its derivatives with respect to A are multivalued at $A = 0$, whereas the expandability of $\Gamma(\alpha)$ in α does not prevent it from having many stationarity points. This leads only to a loss of convexity: on the surface $\Gamma(\alpha)$ instead of a flat segment there appears a well, the existence of which is forbidden in the rigorous theory. However, this defect is not very important.

The variational formulation in the language of Γ is therefore natural for phase transitions. It is also efficient in practice. For example, for the Ising model, when Γ is calculated in the lowest nontrivial order of the high-temperature expansion (the contributions of zeroth and first order in the expansion parameter $\beta = 1/kT$), from (1.51) we obtain the Weiss self-consistent field equation. Adding higher-order terms of the β expansion in Γ allows systematic improvement of the Weiss approximation, refining the value of T_c [53]. Therefore, even in a finite order of perturbation theory the variational formulation in the language of Γ allows the qualitative description of phase transitions, and it can be used to find the solution in an approximation of the self-consistent field type with further improvement possible. However, this scheme has a fundamental flaw. Owing to the expandability of Γ in α, in any finite order of perturbation theory the critical exponents turn out to be exactly the same as in the Landau theory. This scheme is therefore inapplicable for calculating exponents, although theoretically it allows arbitrarily high accuracy to be obtained away from T_c.

1.12   The Landau theory

The Landau theory postulates the existence of a "free-energy functional" $F(\alpha)$, the minimization condition for which gives the equilibrium average of the order parameter $\alpha(x) = \langle\hat{\varphi}(x)\rangle$. It is assumed that this functional (1) possesses the symmetry of the problem; (2) is, as a rule, local, i.e., represented by a single integral, over d-dimensional x, of the energy density, which is a function of the field and its derivatives at the point x (although sometimes the nonlocal contribution of dipole forces is also included); (3) is analytic in all the thermodynamical variables, including α, near the critical point. The theory is designed only for describing critical behavior, and so the deviations of all variables from their values at the critical point are assumed to be small. Higher powers of the deviations are neglected, and only the contributions absolutely necessary for describing the transition are kept. These considerations and the symmetry determine the explicit form of F practically uniquely up to constants.

The equation $\Phi = -\beta \widetilde{F}$ with $\beta = 1/kT$ and Φ from (1.56) determines the exact functional $\widetilde{F}$, the minimization condition for which gives the desired α for given A and other parameters. This $\widetilde{F}$ is calculated using a microscopic model (actually, only using perturbation theory for Γ), whereas in the Landau theory the form of F is simply postulated. However, there is practically no difference between $\widetilde{F}$ and F for describing critical behavior, because in any finite order of perturbation theory $\widetilde{F}$ gives the same exponents as F (Sec. 1.11). The calculation of $\widetilde{F}$ using a microscopic model can only improve the values of constants like T_{c}, but the Landau theory (and the theory of critical behavior in general) does not claim to define such constants. Instead, it focuses on the calculation of critical exponents and other universal characteristics of critical behavior. Any analytic functional gives the same exponents as the Landau functional, and so in order to obtain nontrivial exponents using the original microscopic model it is necessary to go beyond the simple perturbation theory for Γ. Attempts to do this have not met with great success, and so the theory has evolved along a different path: replacement of the exact microscopic model by a fluctuation model and then application of the renormalization-group method to the latter.

1.13 The fluctuation theory of critical behavior

The Landau theory deals with only the average value $\alpha(x) = \langle \widehat{\varphi}(x) \rangle$ of the random field $\widehat{\varphi}(x)$. However, this does not amount to neglecting fluctuations. If the functional $F(\alpha)$ were exact, from the variational principle we would obtain the exact results of the original microscopic model fully including the fluctuations of $\widehat{\varphi}$, even though we would only be working with its average value. Fluctuations apparently must lead to the appearance of nonanalyticity in the exact F. Therefore, the defect of the Landau theory is not the fact that it deals with only the average value $\langle \widehat{\varphi} \rangle = \alpha$, but the assumption that $F(\alpha)$ is analytic. It is the latter that amounts to neglecting fluctuations.

The main idea of the fluctuation theory is that fluctuations are taken into account by using not the average α, but the random field $\widehat{\varphi}$ itself, taking the simple Landau functional as the Hamiltonian instead of the exact microscopic model. Then the averaging must be done over all configurations $\varphi(x)$ of the random field $\widehat{\varphi}(x)$ near its equilibrium average. The probability density in configuration space is determined by the weight

$$\rho(\varphi) = \exp S(\varphi), \quad S(\varphi) = -\beta_{\mathrm{c}} F(\varphi), \tag{1.58}$$

where $F(\varphi)$ is the Landau functional and $\beta_{\mathrm{c}} \equiv 1/kT_{\mathrm{c}}$. We recall that the Landau functional, in contrast to the Hamiltonian of the microscopic model, is assumed to be explicitly (and analytically) dependent on all the important thermodynamical variables, including the temperature. The theory is designed only for describing critical behavior, and so (1.58) involves β_{c} rather than β (the correction terms are always discarded in the Landau functional). For the same reason, lattice microscopic systems are always replaced by continuous ones in (1.58).

By analogy with field theory, we shall refer to $S(\varphi)$ in the exponential (1.58) as the *action functional* or simply the action of a given model and write it as

$$S(\varphi) = S_0(\varphi) + V(\varphi), \quad S_0(\varphi) = -\tfrac{1}{2}\varphi K\varphi, \tag{1.59}$$

where S_0 is the *free action*, a quadratic form on the space of fields [it is written symbolically in (1.59)], and the term $V(\varphi)$ is the *interaction functional* (or simply the interaction).

Therefore, the fluctuation theory of critical behavior is the theory of a classical random field or system of fields (the convention of Sec. 1.9) $\widehat{\varphi}(x)$ with a given distribution function (1.58). The partition function Z and the expectation value of an arbitrary observable (random quantity) $Q(\widehat{\varphi})$ are determined by the usual expressions of the type (1.17):

$$Z = \int D\varphi \exp S(\varphi), \quad \langle Q(\widehat{\varphi}) \rangle = Z^{-1} \int D\varphi \, Q(\varphi) \exp S(\varphi). \tag{1.60}$$

In particular, for the functions (1.33) we have

$$G_n(x_1...x_n) = Z^{-1} \int D\varphi \, \varphi(x_1)...\varphi(x_n) \exp S(\varphi), \tag{1.61}$$

and their generating functional is

$$G(A) = Z^{-1} \int D\varphi \exp[S(\varphi) + A\varphi]. \tag{1.62}$$

The symbol $\int D\varphi...$ in these expressions stands for the functional (path) integral over the infinite-dimensional manifold E_{int} of configurations $\varphi(x)$, which will be defined below.

Convention. We shall use E to denote the linear space of functions $\varphi(x)$ rapidly decreasing for $|x| \to \infty$, and $\psi + E$ to denote the manifold of functions of the form $\psi(x) + \varphi(x)$ with fixed ψ and arbitrary $\varphi \in E$ (for $\psi \notin E$ the manifold $\psi + E$ is flat, but not linear). Functions from E will be termed localized, and "rapidly decreasing" will mean exponential: $\varphi \in E$ if a number $\alpha > 0$ can be found such that $\varphi(x)$ for $|x| \to \infty$ falls off no more slowly than $\exp(-\alpha|x|)$. We note that then any function from E has a Fourier transform analytic in some circle $|k| < \alpha$. The formal source in generating functionals like (1.41) will always be assumed to be an arbitrary function $A \in E$. The location of the maximum (minimum) of a functional will be understood to be the value of φ at which the functional reaches a maximum (minimum); such a point is a stationarity point. According to (1.58), a maximum of the action corresponds to a minimum of the energy in the Landau theory.

The integration region E_{int} is defined by the following rules:

Rule 1. In any integral $E_{\text{int}} = \psi + \overline{E}$, where ψ is the point of the absolute maximum of the argument of the integrated exponential, and $\overline{E} \supseteq E$ is a linear space of configurations describing fluctuations near zero.

Let $\psi(x; A)$ be the point of maximum of the argument of the exponential in (1.62), and $\psi(x; 0) \equiv \varphi_0$ be the point of maximum of the action S (we assume that it is translationally invariant and that φ_0 is uniform). For (1.61) and (1.62) to be consistent with each other it is necessary that the integration region E_{int} in (1.62) be independent of A, i.e., $\psi(x; A) + \overline{E} = \varphi_0 + \overline{E}$, which, owing to the assumed linearity of $\overline{E}$, implies that $\psi(x; A) - \varphi_0 \in \overline{E}$. However, the difference $\psi(x; A) - \varphi_0 \equiv v(x; A)$ has the meaning of a perturbation introduced at the location of the maximum by a localized source A. This explains the next rule.

Rule 2. $\overline{E}$ is the minimal linear space containing all the localized functions and also all the perturbations $v(x; A)$ introduced at the location of the maximum of the action by an arbitrary localized source.

The class of functions $v(x; A)$ depends on the model, i.e., on the form of S. For any reasonable model a perturbation $v(x; A)$ is damped out at infinity for any $A \in E$. The difference between models is manifested only in the damping rate: usually it is fast, and then $\overline{E} = E$ (models with massive fields). Sometimes it is only a power law, and then $\overline{E}$ is larger than E (models with massless fields). In any case, the next rule is valid.

Rule 3. For all φ from the integration region, $\varphi(\infty) = \varphi_0$.

It is this simple rule that is usually used to define the integration region $E_{\text{int}} = E_{\text{int}}(\varphi_0)$. Rules 1 and 2 stated above actually only clarify and refine this definition, giving the rate of falloff of the difference $\varphi(x) - \varphi_0$ for $|x| \to \infty$, i.e., the rate at which configurations $\varphi \in E_{\text{int}}(\varphi_0)$ reach the limiting uniform asymptote φ_0.

When spontaneous symmetry breaking occurs there are many different, completely equivalent maxima of the action (minima of the energy) φ_0, corresponding to individual pure phases. Equations (1.60)–(1.62) are meaningful for each of them; only $E_{\text{int}} = E_{\text{int}}(\varphi_0)$ contains a dependence on the choice of phase. Transformations from the spontaneously broken symmetry group take different φ_0 [and thereby different $E_{\text{int}} = E_{\text{int}}(\varphi_0)$] into each other, and so the partition function Z and the averages of any invariants (under the broken group) do not depend on the choice of phase, while the averages of noninvariant quantities like (1.61) and (1.62) do depend on it (via E_{int}).

The connected and 1-irreducible Green functions are determined from the generating functional of the full Green functions (1.62) by the standard rules of Secs. 1.10 and 1.11. When spontaneous symmetry breaking occurs the functional $W(A)$ depends on the choice of phase, while the corresponding $\Gamma(\alpha)$ is independent of it; in the language of Γ, different phases correspond to choosing different solutions of (1.51) for $A = 0$, $T < T_c$ (see Ch. 2 for more details).

Mathematically, the theory of a classical random field is identical to (Euclidean) quantum field theory, for which the functional integration technique was actually developed. This will be discussed in detail in Ch. 2. Here we only note that for the free theory with quadratic action, the Gaussian functional integrals can be calculated exactly, and the interaction V in (1.59) can be taken into account using perturbation theory. The convenient technique of Feynman diagrams has been developed to describe the terms of the perturbation series.

In this formalism critical behavior is not described correctly in any finite order of perturbation theory, as was also the case for the original microscopic model. It is therefore natural to ask, what is the advantage of changing from the exact microscopic model to a fluctuation model? The answer is that ordinary Landau functionals are polynomials in φ, and the corresponding field theories are renormalizable theories in some space dimension. This makes it possible to use the quantum field RG technique for analyzing critical behavior, which is not possible for the original microscopic model.

It should be noted that the substitution of the exact model by a fluctuation model is the most vulnerable feature of the modern theory of critical phenomena. The arguments justifying this substitution always rely on plausibility considerations rather than proofs. The strongest argument is that the results obtained in this manner are reasonable and nontrivial, and the scheme is universal.

In conclusion, we remark that a return to the simple Landau theory corresponds to the saddle-point approximation in the calculation of the integrals (1.60)–(1.62), which is equivalent to the use of the *loopless approximation* $\Gamma(\alpha) = S(\alpha)$ for the functional Γ in the equations of Sec. 1.11 [$S(\alpha)$ is the action functional of the model under study with $\varphi(x) \to \alpha(x)$].

1.14 Examples of specific models

1. The simplest Ginzburg–Landau *standard* φ^4 *model* describing a second-order phase transition in any system with one-component order parameter $\varphi = \varphi(x)$ and $\varphi \to -\varphi$ symmetry for zero external field h has action

$$S(\varphi) = \int dx \, \left[-\tfrac{1}{2}(\partial\varphi)^2 - \tfrac{1}{2}\tau\varphi^2 - \tfrac{1}{24}g\varphi^4 + h\varphi \right]. \qquad (1.63)$$

Here $\tau = T - T_c$, the parameter g characterizes the strength of the fluctuation interaction (it is called the *charge* or *coupling constant* in field theory),

$(\partial\varphi)^2 \equiv \partial_i\varphi \cdot \partial_i\varphi$ is the squared gradient, $\partial_i \equiv \partial/\partial x_i$, and repeated indices are always summed over. Additional coefficients, including the factor β_c from (1.58), are not written out in (1.63), because they can always be eliminated by appropriate rescaling of φ, τ, and h, i.e., by choice of the measurement units. The coefficients $\frac{1}{2}$ and $\frac{1}{24}$ are left in (1.63) for convenience.

2. *Tricritical behavior.* In constructing the action (1.63) it is assumed that the coefficients $a_{2,4}$ of φ^2 and φ^4 in the Landau functional depend only on T [inclusion of their h dependence would only lead to corrections that are negligible compared to $h\varphi$ in (1.63)]. The critical temperature is then determined by the condition $a_2 = 0$. In general, $a_4 \neq 0$, and owing to the assumed analyticity (see Sec. 1.12) we have $a_2 \sim \tau$ and $a_4 = \text{const}$ up to unimportant corrections. However, in some problems the coefficients a do actually depend on two parameters. Then the equation $a_2 = 0$ for $a_4 > 0$ (see Sec. 1.16 for more details) determines an entire line of critical points on the phase plane of the parameters. A tricritical point, at which simultaneously $a_2 = a_4 = 0$, can exist on this line. Near it the φ^6 contribution becomes important, and is then added to (1.63). Tricritical behavior, i.e., anomalous smallness of a_4, is observed experimentally in many systems [54], [55].

3. *The O_n φ^4 model.* This is an O_n-symmetric generalization of the model (1.63) with n-component field $\varphi = \{\varphi_a(x), a = 1, ..., n\}$ and similarly for h. The form of the action (1.63) is preserved if the quantities entering into it are defined as follows: $(\partial\varphi)^2 \equiv \partial_i\varphi_a \cdot \partial_i\varphi_a$, $\varphi^2 \equiv \varphi_a\varphi_a$, $\varphi^4 \equiv (\varphi^2)^2$, $h\varphi \equiv h_a\varphi_a$. The field φ is sometimes referred to as an n-component spin vector, and the model as an n-component isotropic ferromagnet. However, this terminology can lead to confusion, because real spin is always associated with spatial rotations, whereas here we have isotropy in the abstract space of the internal degrees of freedom, which in field theory is usually referred to as isospin. Under the group of spatial rotations the field $\{\varphi_a\}$ is a set of n scalars.

4. *Cubic symmetry.* The field is the same as in Example 3, but now it is not complete O_n but only cubic symmetry that is required. Cubic symmetry is invariance under permutations of the components of φ and reflections $\varphi_a \to -\varphi_a$ of each component separately. All the O_n-symmetric forms given above also possess cubic symmetry, there are no new invariants quadratic in φ, and for the φ^4 interaction a second independent, cubically symmetric structure $\sum \varphi_a^4$ is possible.

5. *Spatial (rotational or cubic) symmetry.* In this case the number n of field components coincides with the spatial dimension d, and φ and h are assumed to be vectors under spatial rotations. It is this case that corresponds to a real isotropic magnet and a ferroelectric. For local forms of the type φ^{2n} the spatial symmetry is equivalent to the isospin symmetry studied above, and so the invariant forms are the same as before. However, new invariant structures appear for $(\partial\varphi)^2$, namely, the isotropic structure $(\partial_i\varphi_i)^2 = \partial_i\varphi_i \cdot \partial_k\varphi_k$ and the cubic structure $(\partial_1\varphi_1)^2 + (\partial_2\varphi_2)^2 + ... + (\partial_n\varphi_n)^2$. They are included in the action when studying real systems [56].

6. *Dipole forces.* For a d-dimensional vector field φ interpreted as the dipole (magnetic for magnets, electric for ferroelectrics) moment density, in real systems one often includes a nonlocal dipole interaction energy of the form

$$\text{const} \iint dx dx' \; \varphi_i(x) R^{-d} [\delta_{is} - dn_i n_s] \varphi_s(x'),$$

where d is the spatial dimension, $R_i \equiv x_i - x'_i$, $R \equiv |R|$, $n_i \equiv R_i/R$, and const is a positive coefficient. Taking the Fourier transform

$$\varphi(x) = (2\pi)^{-d/2} \int dk \; \widetilde{\varphi}(k) \exp(ikx) \tag{1.64}$$

(the indices on φ have been omitted), the dipole contribution to the action can be written as a single integral:

$$S_{\mathrm{d}}(\varphi) = \text{const} \int dk \; \widetilde{\varphi}_i^*(k) \left[\delta_{is} - dk_i k_s/k^2 \right] \widetilde{\varphi}_s(k), \tag{1.65}$$

where the constant is positive and $\widetilde{\varphi}^*(k) = \widetilde{\varphi}(-k)$ for real $\varphi(x)$. The nonlocality is manifested in (1.65) by the presence of k^2 in the denominator. When (1.65) is added to the action (1.63), the local contribution involving δ_{is} in (1.65), grouped together with $\tau \varphi^2 \equiv (T - T_c)\varphi_i \delta_{is} \varphi_s$ in (1.63), leads only to a shift of the critical temperature $T_c \to T'_c$. Therefore, the general form of the spatially isotropic, free (quadratic in φ) action taking into account dipole forces is

$$S_0(\varphi) = -\tfrac{1}{2} \int dk \; \widetilde{\varphi}_i^*(k) \left[(k^2 + \tau)\delta_{is} + \nu^2 k_i k_s/k^2 \right] \widetilde{\varphi}_s(k), \tag{1.66}$$

where $\tau = T - T'_c$ and ν is a constant parameter associated with the dipole gap in the longitudinal fluctuation spectrum. If only cubic symmetry rather than isotropy is required, a structure $\sim k_i^2 \delta_{is}$ (no summation over i) can be added to (1.66), as mentioned in Example 5 above. The second structure mentioned there $\sim k_i k_s$ is unimportant compared to $k_i k_s/k^2$ in (1.66). The contributions to the action invariant only under combined rotation of the vectors k and φ are sometimes termed anisotropic, by analogy with spin anisotropy.

We shall limit ourselves to these examples for now, as others will be introduced later on. When necessary, the dimension d in models like (1.63) will be made explicit by using standard notation of the type φ_d^4, for example, φ_4^4 for $d = 4$.

The O_n φ^4 model with various n has many applications in the theory of critical phenomena. For $n = 1$ it coincides with (1.63) and describes the critical behavior of the Ising model and other systems in the same equivalence class (the liquid–gas transition point, the demixing point in binary mixtures, and so on). For $n = 3$ it describes the isotropic, and for $n = 2$ the planar, Heisenberg magnet (both ferro- and antiferromagnets). For $n = 2$ it also describes the transition to the superfluid phase of liquid He$_4$. Finally, the formal limit $n \to 0$ is used to describe the statistics of long polymer chains [45], [57].

1.15 Canonical dimensions and canonical scale invariance

All the quantities in functionals like (1.63) can be assigned *canonical dimensions*, defined from the requirement that each term be dimensionless. The canonical dimension of an arbitrary quantity F will be denoted d_F, or $d[F]$ for complicated F. The dimensions d_F should not be confused with the critical dimensions Δ_F introduced earlier in Secs. 1.2 and 1.3. The latter are unknown and will be calculated by the RG method, whereas the d_F are simply determined from the form of the action up to an overall factor. The latter will be fixed by a normalization condition like (1.11): $d_p = -d_x = 1$. Then in d-dimensional space the quantity dx in (1.63) has dimension $-d$, and ∂ has the momentum dimension $+1$. The condition that the contribution of $(\partial\varphi)^2$ in (1.63) be dimensionless takes the form $2d_\varphi + 2 - d = 0$, from which we find d_φ and then the dimensions of all the other parameters in (1.63). The results are given in Table 1.3.

Table 1.3   Canonical dimensions d_F of quantities F in the action $S(\varphi) = -\int dx \left[\frac{1}{2}(\partial\varphi)^2 + \sum_n \frac{1}{n!}g_n\varphi^n\right]$ in d-dimensional space.

F	x	∂, p	$\varphi(x)$	$h = -g_1$	$\tau = g_2$	g_4	g_6	g_n
d_F	-1	1	$\frac{1}{2}d - 1$	$\frac{1}{2}d + 1$	2	$4 - d$	$6 - 2d$	$n - \frac{1}{2}d(n-2)$

If for an action S it is not possible to make all the terms dimensionless by some choice of d_F, this means that S contains too few parameters and some more must be introduced. On the other hand, if the dimensions d_F are not determined uniquely, this means that S contains too many parameters, and some must be eliminated by appropriate rescaling of the fields and redefinition of the constants.

The action $S(\varphi, e)$ (where φ is the set of all fields and e is the set of all parameters) being dimensionless is equivalent to scale invariance:

$$S(\varphi_\lambda, e_\lambda) = S(\varphi, e), \quad \varphi_\lambda(x) = \lambda^{d_\varphi}\varphi(\lambda x), \quad e_{i\lambda} = \lambda^{d_i}e_i \tag{1.67}$$

with $d_i \equiv d[e_i]$. A consequence of (1.62) and (1.67) is the invariance of the functional G and its logarithm W [the Jacobian $D\varphi_\lambda/D\varphi$ is independent of φ and cancels out in (1.62)]:

$$W(A_\lambda, e_\lambda) = W(A, e), \quad A_\lambda(x) = \lambda^{d_A}A(\lambda x), \quad d_A = d - d_\varphi. \tag{1.68}$$

These expressions extend the concept of generalized homogeneity introduced in Sec. 1.2 to the case of functionals. Here d_φ and d_A are the canonical dimensions of the functional arguments $\varphi(x)$ and $A(x)$. The rule (1.10) is generalized to functionals as

$$d\left[\frac{\delta\Phi}{\delta\psi(x)}\right] = d[\Phi] + d - d_\psi, \quad d_\psi \equiv d[\psi(x)], \tag{1.69}$$

i.e., each variational differentiation with respect to the "dimensional argument" $\psi(x)$ increases the dimension of the functional by $d - d_\psi$. Equations (1.67) and (1.68) indicate that the corresponding functionals are dimensionless, and the fact that $W(A, e)$ is dimensionless and the definitions of Sec. 1.11 imply that the Legendre transform $\Gamma(\alpha, e)$ is dimensionless under scale transformations with $d_\alpha \equiv d[\alpha(x)] = d - d_A = d_\varphi$. These considerations determine the canonical dimensions of all the Green functions — the variational derivatives of the corresponding dimensionless functionals:

$$d[W_n(x_1...x_n)] = d\left[\frac{\delta^n W}{(\delta A)^n}\right] = n(d - d_A) = nd_\varphi,$$

$$d[\Gamma_n(x_1...x_n)] = d\left[\frac{\delta^n \Gamma}{(\delta\alpha)^n}\right] = n(d - d_\alpha) = n(d - d_\varphi). \tag{1.70}$$

Equation (1.45), the analog of (1.68), leads to analogs of (1.70) with all the canonical dimensions d_F replaced by the critical dimensions Δ_F. Equations (1.45) and (1.68) differ in that the variables e in (1.68) denote all the parameters of the model, while in (1.45) they denote only the IR-relevant external parameters characterizing the deviation from the critical point. Denoting the latter by e' and the others by e'', altogether $e = (e', e'')$ [for example, in (1.63) $e' = (\tau, h)$ and $e'' = g$], it can be said that all the parameters e in (1.68) are transformed, while in (1.45) only the e' are transformed for fixed e''. Therefore, (1.68) does not imply (1.45). If all the parameters e'' had zero canonical dimension, then critical scaling (1.45) with $\Delta[e'] = d[e']$ would follow from (1.68), just as for the simple Landau theory, as a consequence of the invariance (1.67) for the functional $\Gamma(\alpha) = S(\alpha)$ (see the discussion at the end of Sec. 1.13). The presence of additional parameters e'' with $d[e''] \neq 0$ is the reason that critical scaling (the scaling hypothesis) is violated for the correlation functions in the Landau theory with the action (1.63) (see the discussion at the end of Sec. 1.19). In this model $e'' = g$ and $d_g = 4 - d$ (see Table 1.3), and so scaling occurs in the Landau theory only for $d = 4$, when $d_g = 0$. Thermodynamical scaling exists for any d, because the transition to thermodynamics is effected by restricting $S(\varphi, e)$ to the set of uniform fields φ, so that the contribution $(\partial\varphi)^2$ in (1.63) vanishes, and the parameter g becomes redundant and is eliminated by rescaling φ. In going beyond the Landau theory, which corresponds to the stationary-phase approximation in the integral (1.60) with the action (1.63), critical scaling does not follow from canonical scale invariance even for $d = 4$, because the presence of ultraviolet divergences in the graphs (discussed in detail below) requires the introduction of an additional dimensional cutoff parameter Λ, which is included in the e''.

1.16 Relevant and irrelevant interactions. The logarithmic dimension

The free part of the action (1.59) is assumed to be a quadratic form $S_0(\varphi) = -\frac{1}{2}\varphi K\varphi$ with some linear operator K [for example, in (1.63) $K = -\partial^2 + \tau$ or $p^2 + \tau$ in momentum space]. Above T_c all the independent terms of K are positive and become small in the critical region (divergence of the susceptibility). We shall assume that S_0 contains only IR-relevant parameters like e' with $d[e'] > 0$ that vanish at the critical point (Sec. 1.15). This does not lead to loss of generality, because parameters like e'', for example, the coefficient of ∂^2 in K, can always be eliminated from S_0 by redefinition of the fields and parameters e'.

Let us add an interaction $V(\varphi) = gv(\varphi)$ to S_0 and ask: how important is this interaction compared to S_0 in the critical region? The answer is simple. In lowest order, the relative magnitude of the correction is proportional to the coupling constant g, which is made dimensionless by some parameter from S_0 or momentum e, so that the combination ge^{-d_g/d_e} is a dimensionless characteristic of the interaction strength, where d_F are the canonical dimensions.* We know that $d_e > 0$ and that $e \to 0$ at the critical asymptote (here it is important that S_0 by condition does not contain any dimensional parameters like e'', which remain finite at the critical asymptote), and so for $d_g > 0$ the relative magnitude of the correction grows as $e \to 0$, while for $d_g < 0$ it falls. An interaction with $d_g \geq 0$ is termed *IR-relevant*, and one with $d_g < 0$ is termed *IR-irrelevant*. For the former the dimensionless parameter of the perturbation series is not small at the critical asymptote, and so the entire series must somehow be summed (this will be done by the RG method). An irrelevant interaction can be neglected if small corrections are not of interest. When there are two or more competing interactions with $d_g \geq 0$ it is necessary to include only the most important one with the maximum magnitude of d_g, and all the others contribute only small corrections. If several interactions simultaneously have the same maximum $d_g > 0$, they must all be taken into account on an equal footing in the absence of any symmetry constraints.

The value of d_g is determined from the form of the interaction according to the rules of Sec. 1.15 and depends on the spatial dimension d; usually d_g decreases with increasing d. *Definition*: the value of $d = d^{**}$ at which $d_g = 0$ is referred to as the *upper critical dimension* for a given model.

Let $v(\varphi) = \int dx F(x; \varphi)$, where F is a local monomial constructed from the field $\varphi(x)$ and (in general) its derivatives, without any additional dimensional factors. It has definite canonical dimension d_F. *Definition*: the value of $d = d^*$ at which $d_v = d_F - d = 0$ is termed *logarithmic* for the given interaction. For example, for S_0 from (1.63) and the interaction $v(\varphi) = \varphi^n$ [here and below $\varphi^n \equiv \int dx\varphi^n(x)$], from Table 1.3 in Sec. 1.15 we obtain $d^* = 2n/(n-2)$, in

---

*The correction may also contain the UV-cutoff parameter Λ, but it can be ignored. In Sec. 1.19 it will be shown that all contributions containing Λ either redefine parameters like T_c, or are IR-irrelevant compared to the others.

particular, $d^* = 6, 4, 3$ for the φ^3, φ^4, and φ^6 interactions, respectively. The term "logarithmic" is also used for the interaction itself or the model as a whole, for example, one says that the φ^4 interaction is logarithmic at $d = 4$.

As a rule, the values of d^* and d^{**} coincide. They differ only in rather exotic cases where, owing to physical considerations, the coefficient of the interaction contains an additional power of an IR-relevant parameter. A simple example is the interaction $V(\varphi) = g\tau^\alpha \varphi^4$ with given exponent α. The usual value $d^* = 4$ is logarithmic for the interaction $\sim \varphi^4$, but the upper critical dimension is $d^{**} = 4 - 2\alpha$, since it is determined by the dimension $d_g = 4 - d - 2\alpha$ of the parameter g, and not by the coefficient $g\tau^\alpha$ as a whole.

As an example, let us consider the model

$$S(\varphi) = -\int dx \left[\tfrac{1}{2}(\partial\varphi)^2 + a_2\varphi^2 + a_4\varphi^4 + a_6\varphi^6 \right], \qquad (1.71)$$

assuming that the coefficients a are regular functions of two variables $z \equiv \{z', z''\}$, where z' is the temperature and z'' is an auxiliary parameter (cases often encountered experimentally are the pressure in liquid crystals, the concentration in solutions, and so on [52], [54], [55]). For stability it is necessary that $a_6 > 0$, and the line of critical points in the space of the parameters z is given by the conditions $a_2 = 0$ and $a_4 > 0$ (for $a_2 = 0$ and $a_4 < 0$ the situation is not critical, because for small a_2 the energy minimum with respect to φ is determined by the competition of the contributions with $a_4 < 0$ and $a_6 > 0$, and so it does not "feel" the change of sign of a_2). Let us assume that at the end of this line there is a tricritical point z_t at which simultaneously $a_2 = a_4 = 0$, and that we are interested in the behavior of the system near z_t. The experimental conditions specify a trajectory $z(\tau)$ in the plane of the variables z, where τ is some parameter (not necessarily $T - T_c$) along the trajectory. We assume that the trajectory $z(\tau)$ passes through the tricritical point z_t, and that the parameter τ is chosen such that $z(0) = z_t$. Then τ characterizes the degree of nearness to z_t. By condition, $a_2 = a_4 = 0$ at $\tau = 0$, and the relative magnitude of a_2 and a_4 at small τ depends on the shape of the trajectory $z(\tau)$. If it is a straight line, then $a_2 \sim a_4 \sim \tau$ owing to the analyticity of $a_{2,4}$ in the variables z assumed in the Landau theory. However, the trajectory $z(\tau)$ can also be chosen such that (at least in theory; the experimental possibilities are another matter) a_2 and a_4 will be of different orders of smallness on it, for example, $a_4 \sim a_2^\alpha$ with given α. Let us assume that we approach from the side $a_{2,4} > 0$ and choose the coefficient a_2 itself as τ. Then $a_2 = \tau$, $a_4 = g_4\tau^\alpha$, and $a_6 = g_6$, where $g_{4,6}$ are positive constants independent of τ.

We want to understand which of the two interactions in (1.71) is more important for this trajectory. According to the general rule (see above), this is determined by the canonical dimensions $d[g_6] = 6 - 2d$ and $d[g_4] = 4 - d - 2\alpha$ (we recall that $d_\tau = 2$). We see from the expressions for d_g that in the actual dimension $d = 3$, for $\alpha > \tfrac{1}{2}$ (in particular, for straight-line trajectories with $\alpha = 1$) it is necessary to keep only φ^6, which corresponds to *tricritical behavior.* For $\alpha < \frac{1}{2}$ it is necessary to keep only $\tau^\alpha \varphi^4$, while for $\alpha = \frac{1}{2}$ it is necessary to keep both interactions.

This is all true, of course, only at the asymptote, i.e., at sufficiently small τ. In a real situation the numerical values of the parameters $g_{4,6}$ are also important, and an interaction asymptotically irrelevant for $\tau \to 0$ can play an important role for a long time. This is also true of trajectories that intersect the line of critical points not at the tricritical point z_t itself, but somewhere in its vicinity. The limiting behavior in that case is simply critical (φ^6 is irrelevant), but in practice owing to the smallness of the coefficient of φ^4 the behavior of the system can appear to be tricritical for a rather long time.

Above we have used the common terminology, referring to behavior as tricritical or critical if it is controlled by the φ^6 or φ^4 interaction, respectively. It follows from this analysis that two other types of behavior are also possible in the class of nontrivial trajectories approaching a tricritical point $a_2 \equiv \tau \to 0$, $a_4 \sim \tau^\alpha \to 0$: *combined tricriticality*, when both interactions φ^6 and $\tau^\alpha \varphi^4$ are important [this is observed at some fixed value $\alpha = \alpha_0$, with $\alpha_0 = \frac{1}{2}$ for $d = 3$], and *modified criticality*, which is controlled only by the interaction $\tau^\alpha \varphi^4$ and is observed for $\alpha < \alpha_0$. These will be studied in detail using the RG method in Ch. 4. The RG analysis confirms the general conclusions made above, and leads only to refinement of the value of α_0 owing to the inclusion of the anomalous dimensions.

Another important refinement should be added to the above discussion. The general conclusions of this section were based on analysis of the canonical dimensions of the contributions in first-order perturbation theory. In the critical region the perturbation series must be summed, which for the given model will be done using the RG method and leads to replacement of the canonical dimensions by the critical ones (Sec. 1.1). Therefore, accurate estimates of the relative importance of the various contributions must be based not on the canonical, but on the critical dimensions. They differ [see Eq. (1.3)] only by corrections of order ε, where ε is a formal small parameter of the type $d^* - d$ characterizing the deviation of the main interaction(s) from logarithmicity, for example, $\varepsilon = 4 - d$ in the φ^4 model. Since these corrections are unknown *a priori* (i.e., before the construction of the model and its RG analysis), contributions of the same order $\sim \varepsilon$ should not be included in the canonical dimensions. This means that arguments about the relative importance of the two interactions based on the inequality $d_{g_1} < d_{g_2}$ are definitive and reliable (at small ε) only when this inequality is satisfied already in zeroth order in ε. However, if the values of d_g differ by only quantities of order ε, then in constructing the model the two interactions must be considered equally relevant.

The situation is delicate because ε itself can be exactly determined only for a particular model, and a comparative analysis of the relevance of the interactions is needed already at the construction stage. In practice, the accuracy in ε can be checked only in hindsight: first, it is necessary to choose a model after discarding all contributions irrelevant according to the canonical dimensions, then the logarithmic dimension and (thereby) the corresponding parameter ε must be found, and only then can it be checked whether or not the discarded terms of the original interaction contain any that differ in dimension from the ones included only by contributions of order ε or higher. If there are such terms, they must be added in, and the final judgment of the relative importance of such contributions with "nearly identical dimension" can be made only after RG analysis of the model. This judgment will be reliable only within the ε expansion, i.e., for small ε. The extrapolation to finite ε without knowing the exact critical dimensions will always be a matter of faith and is justified only by comparing the results with experiment.

Let us briefly summarize. In studying critical behavior, the initial model is always maximally simplified by discarding all contributions that are certainly (i.e., in zeroth order in ε) IR-irrelevant compared to the leading contributions, as they give only corrections to scaling and shift (renormalize) the coefficients of the IR-relevant terms. A shift of parameters such as T_c and g in (1.63) is meaningless, because these parameters are not at all controlled by the theory of critical behavior and do not affect its universal characteristics such as the critical indices and the normalized scaling functions. Shifts of parameters of the IR-relevant part of the action by IR-irrelevant contributions can, in principle, violate the original symmetry of the IR-relevant part (and then one sometimes speaks of "dangerous" IR-irrelevant operators). However, this cannot happen if from the very start the IR-relevant part includes all the contributions allowed by symmetry, as required by the general ideas of the theory of critical behavior.

It should also be noted that discarding the IR-irrelevant contributions is not only desirable, because it simplifies the model, but also necessary. It is only in such a correctly constructed, simplified model that the correlation between IR and UV behavior appears, which makes it possible to study the IR asymptote by the RG method.

Therefore, the RG technique is always used just for models simplified in this way, and only then, if desired, does one study the effect of various correction terms irrelevant at the level of the canonical dimensions.

1.17   An example of a two-scale model: the uniaxial ferroelectric

In a uniaxial ferroelectric (or magnet) the d-dimensional polarization vector φ_i is always assumed to point along the distinguished axis, i.e., $\varphi_i = \varphi n_i$, where φ is the one-component order parameter and n_i is a fixed unit vector. A realistic model [38] is the standard φ^4 model of the type (1.63) taking into account dipole forces. Its free action is obtained by substituting $\varphi_i = \varphi n_i$

into (1.66):

$$S_0(\varphi) = -\frac{1}{2} \int dk \; \tilde{\varphi}^*(k) \left[k^2 + \tau + \nu^2 \frac{k_\parallel^2}{k^2} \right] \tilde{\varphi}(k), \qquad (1.72)$$

where $k^2 = k_\parallel^2 + k_\perp^2$, the longitudinal momentum $k_\parallel$ is the projection of the vector k along n, and $k_\perp$ is the projection on the space orthogonal to n. The dipole gap parameter ν in (1.72) is assumed to be fixed, in contrast to the variable τ.

The critical asymptote corresponds to the vanishing of all the independent terms of the operator K in the free action $S_0(\varphi) = -\frac{1}{2}\varphi K \varphi$ (Sec. 1.16). For the functional (1.63) this implies $k^2 \to 0$ and $\tau \to 0$, and for (1.72) the condition $k_\parallel^2/k^2 \to 0$ is added, from which it follows that $k_\parallel^2/k_\perp^2 \to 0$. This means that in the critical region the momenta $k_\parallel$ and $k_\perp$ must be assumed to be of different orders of smallness. Therefore, in the form $k^2 = k_\perp^2 + k_\parallel^2$ the second term is an IR-irrelevant correction that can be discarded in analyzing the leading singularities. The complete action functional finally obtained can be written symbolically as

$$S(\varphi) = -\frac{1}{2}\varphi \left[-\partial_\perp^2 + \tau + \nu^2 \frac{\partial_\parallel^2}{\partial_\perp^2} \right] \varphi - \frac{1}{24} g \varphi^4 + h\varphi, \qquad (1.73)$$

where $\partial_\parallel$ is the derivative with respect to the longitudinal coordinate, $\partial_\perp^2$ is the transverse part of the Laplacian $\partial^2 = \partial_\perp^2 + \partial_\parallel^2$, and the needed integrations over the coordinates x (single integration in local terms and double in nonlocal ones) are understood. At the critical asymptote $k \to 0$, $\tau \to 0$, $\nu = \text{const}$ all the contributions in the square brackets in (1.73) must be of the same order of smallness (at the level of the canonical dimensions). It then follows that $k_\parallel \propto k_\perp^2$ in order of magnitude.

The model (1.73) is a two-scale model: for any F it is possible to introduce two independent canonical dimensions into it, longitudinal $d_F^\parallel$ and transverse $d_F^\perp$, and the total dimension is defined as $d_F = d_F^\perp + 2d_F^\parallel$ (this form of d_F follows from $k_\parallel \propto k_\perp^2$). The two-scale nature is a consequence of the invariance of the action (1.73) under two independent scale transformations. In the first all quantities F and the longitudinal coordinates are scaled for fixed transverse coordinates, and in the second the F and the transverse coordinates are scaled for fixed longitudinal coordinates. From these we can construct the total transformation with scaling $x_\perp, x_\parallel \to \lambda x_\perp, \lambda^2 x_\parallel$. In d-dimensional space one coordinate is longitudinal and $d-1$ are transverse. The dimensions of all the F in (1.73) are determined by the general rules of Sec. 1.15 from the requirement that each term be dimensionless with respect to longitudinal and transverse transformations separately with normalization of the type (1.11) in the two cases. The results are given in Table 1.4.

The relevance of the interaction (Sec. 1.16) in this case can be judged from the total canonical dimension $d_g = 3 - d$ (see Table 1.4), i.e., the

Table 1.4 Canonical dimensions of the quantities F in the action (1.73).

F	$k_\perp$	$k_\parallel$	$x_\perp$	$x_\parallel$	$\varphi(x)$	h	τ	ν	g
$d_F^\perp$	1	0	-1	0	$\frac{1}{2}(d-3)$	$\frac{1}{2}(d+1)$	2	2	$5-d$
$d_F^\parallel$	0	1	0	-1	$\frac{1}{2}$	$\frac{1}{2}$	0	-1	-1
$d_F = d_F^\perp + 2d_F^\parallel$	1	2	-1	-2	$\frac{1}{2}(d-1)$	$\frac{1}{2}(d+3)$	2	0	$3-d$

model is logarithmic in real three-dimensional space. This allowed Larkin and Khmel'nitskiĭ [38] (1969) to calculate the leading logarithmic correction to the Landau theory by the RG method. After the development of the technique of the Wilson ε expansion, the critical indices in dimension $d = 3 - \varepsilon$ were calculated in first order in ε (in [58]) and in second order in ε (in [59]); see Ch. 4 for more details.

1.18 Ultraviolet multiplicative renormalization

Now let us turn to the quantum field RG technique, limiting ourselves to a description of the general scheme. Details of the calculations and explanations will be given in subsequent chapters.

A particular model is specified by the action functional S, and its Green functions are calculated as infinite graph expansions in perturbation theory (Ch. 2). The graphs correspond to integrals over momenta. If they diverge at large momenta (as will always be the case in any model studied in its logarithmic dimension), it is said that the model contains ultraviolet (UV) divergences. This problem was first encountered in quantum field theory, where, as a rule, one deals with models that are logarithmic in dimension $d = 4$. It was there that the renormalization technique of eliminating UV divergences was developed in the 1950s [35] (details are given in Ch. 3). The first step is *regularization* of the model using any particular procedure that gives meaningful results. The simplest regularization is cutoff $|k| \leq \Lambda$ of the integration over momenta in (1.64). When the cutoff is removed ($\Lambda \to \infty$) the answers diverge. The problem is how to rearrange the theory without distorting its physical meaning and at the same time ensure finite limits for $\Lambda \to \infty$. Models for which this can be done are termed *renormalizable*, and the rearrangement procedure is referred to as *renormalization*.

It should immediately be stated that the situation is different in the theory of critical phenomena, because here we are dealing with a completely real, not artificially introduced, cutoff Λ, for example, the size of the first Brillouin zone of a lattice. In general, in the order of magnitude $\Lambda \sim r_{\min}^{-1}$, where $r_{\min}$ is the characteristic minimum size of the inhomogeneities of the fluctuation field (the lattice spacing, the interatomic separation in a gas, and so on). If distances are measured in units of $r_{\min}$, then Λ is a fixed parameter of order unity. Here it is not necessary to take the limit $\Lambda \to \infty$ and ensure that it exists, and so there is no direct need to renormalize the model. However, it turns out that the possibility itself of doing this implicitly contains valuable information about the original model with finite Λ — it is this possibility that ensures that the RG technique is applicable.

In renormalization theory, it is proved that for some class of models (including the φ_4^4 model) the UV divergences can be eliminated completely by redefinition of the parameters and multiplicative renormalization of the field $\widehat{\varphi}$:

$$\widehat{\varphi}(x) \to \widehat{\varphi}_{\mathrm{R}}(x) = Z_\varphi^{-1}\widehat{\varphi}(x), \tag{1.74}$$

where Z_φ is a dimensionless field renormalization constant.[*]

Warning! In (1.74) it is traditional to write $Z_\varphi^{-1/2}$ instead of Z_φ^{-1}. However, this is inconvenient for many reasons, in particular, because then Z_φ is interpreted as the renormalization constant not of the field, but of its correlator. Therefore, we prefer to violate tradition and will always use the notation of (1.74), in which Z_φ is the field renormalization constant in the exact sense.

From the relation (1.74) between the fields $\widehat{\varphi}$ and $\widehat{\varphi}_{\mathrm{R}}$ we find the relation

$$G_{n\mathrm{R}}(x_1...x_n) = Z_\varphi^{-n}G_n(x_1...x_n) \tag{1.75}$$

between the corresponding Green functions (1.33).

Let us explain the above statement in more detail. The original action S and the corresponding Green functions are termed *unrenormalized*, and the parameters in S are termed *bare*. The complete set of bare parameters will be denoted e_0. The unrenormalized functions on the right-hand side of (1.75) depend on the parameters e_0 and the cutoff Λ. The above-mentioned redefinition of the parameters consists of going from the bare parameters e_0 to some other *renormalized* parameters e that are functions of e_0 and Λ. The correspondence between e_0 and e is assumed to be one-to-one, i.e., $e = e(e_0, \Lambda)$ and $e_0 = e_0(e, \Lambda)$. Either the parameters e_0 or e can be taken as independent. We shall take e to be the independent parameters in the renormalized Green functions $G_{n\mathrm{R}}$, and e_0 and Z_φ in (1.75) to be certain functions of e and Λ. With this convention, (1.75) can be written in detail as

$$G_{n\mathrm{R}}(x_1...x_n; e, \Lambda) = Z_\varphi^{-n}G_n(x_1...x_n; e_0, \Lambda), \tag{1.76}$$

---

[*]In some models with multicomponent field, Z_φ may carry matrix indices. However, this situation is relatively rare (the components of φ must mix in the renormalization), and for simplicity in this chapter we shall always assume that Z_φ is a constant. The generalization to the case of matrix Z_φ is trivial and requires only care in the notation. In addition, in models of the φ^3 type without the symmetry $\varphi \to -\varphi$ a shift of $\widehat{\varphi}$ is also sometimes added to the rescaling (1.74). In the renormalization a shift of $\widehat{\varphi}$ can always be replaced by a shift of the external field h (see Sec. 4.24 for more details), and so without loss of generality the renormalization of $\widehat{\varphi}$ can always be assumed multiplicative.

$$e_0 = e_0(e, \Lambda), \quad Z_\varphi = Z_\varphi(e, \Lambda). \tag{1.77}$$

Definition: A model is termed *multiplicatively renormalizable* if UV finiteness of the renormalized functions (1.76) can be ensured by a suitable choice of the functions (1.77). Here and below by UV finiteness we mean finiteness of the limit when the UV cutoff is removed, here, $\Lambda \to \infty$, for fixed renormalized parameters e.

All the usual models of the theory of critical behavior, including the φ^4 model (1.63), are multiplicatively renormalizable in their logarithmic dimension. It is this property that will subsequently ensure the possibility of using the standard RG technique for them.

The renormalization procedure described above is not unique. The regularization, the parameters e, and the form of the functions (1.77) can be chosen differently, leading to different UV-finite sets of functions G_{nR}. They are all physically equivalent, because the following statement [35] is valid: any two UV-finite sets of Green functions $G_{nR}(...; e, \Lambda)$ and $\widetilde{G}_{nR}(...; \widetilde{e}, \Lambda)$ (where the ellipsis denotes the arguments $x_1...x_n$) obtained by different renormalization prescriptions from the same set of unrenormalized functions are related to each other by a *finite renormalization transformation*:

$$\widetilde{e} = \widetilde{e}(e, \Lambda), \quad G_{nR}(...; e, \Lambda) = \widetilde{Z}_\varphi^n \widetilde{G}_{nR}(...; \widetilde{e}, \Lambda) \tag{1.78}$$

with some UV-finite functions $\widetilde{e} = \widetilde{e}(e, \Lambda)$ and $\widetilde{Z}_\varphi = \widetilde{Z}_\varphi(e, \Lambda)$. The transformations (1.78) form a group called the *renormalization group* (RG).

The UV finiteness of all the quantities in (1.78) allows us to take the limit $\Lambda \to \infty$ in them. The physical content of the theory (in particular, the universal characteristics of critical behavior) obviously cannot depend on the choice of variables and normalization of the field. Therefore, any two sets of Green functions related by the transformation (1.78) are physically equivalent. The unrenormalized and renormalized functions related by (1.76) at finite fixed Λ are equivalent in the same sense. Therefore, the renormalized functions can be used just as well as the original unrenormalized ones for analyzing the critical behavior.

The renormalization expression (1.75) can be translated into the language of generating functionals (1.41), (1.43), and (1.50):

$$G_R(A) = G(Z_\varphi^{-1}A), \quad W_R(A) = W(Z_\varphi^{-1}A), \quad \Gamma_R(\alpha) = \Gamma(Z_\varphi \alpha). \tag{1.79}$$

From this for the connected and 1-irreducible Green functions we obtain

$$W_{nR}(x_1...x_n) = Z_\varphi^{-n} W_n(x_1...x_n), \tag{1.80}$$

$$\Gamma_{nR}(x_1...x_n) = Z_\varphi^n \Gamma_n(x_1...x_n). \tag{1.81}$$

Let us also introduce the *renormalized action* S_R, defining it from the known functional $G_R(A)$ by the general expressions (1.60) and (1.62):

$$G_R(A) = \frac{\int D\varphi \, \exp\left[S_R(\varphi) + A\varphi\right]}{\int D\varphi \, \exp S_R(\varphi)}. \tag{1.82}$$

On the other hand, from Eqs. (1.60) and (1.62) for the unrenormalized model and the first equation in (1.79) we have

$$G_{\mathrm{R}}(A) = \frac{\int D\varphi \, \exp\left[S(\varphi) + Z_\varphi^{-1} A\varphi\right]}{\int D\varphi \, \exp S(\varphi)}. \tag{1.83}$$

Replacing the integration variable $\varphi \to Z_\varphi \varphi$ [the Jacobian is a constant independent of φ and cancels in (1.83)] and equating the resulting equation to (1.82), we conclude that $S_{\mathrm{R}}(\varphi) = S(Z_\varphi \varphi)$, or, in detail,

$$S_{\mathrm{R}}(\varphi; e, ...) = S(Z_\varphi(e, ...)\varphi; e_0(e, ...)). \tag{1.84}$$

The ellipsis denotes additional arguments, here, Λ. According to the definition (1.82), the renormalized Green functions for the model with action S are the ordinary Green functions for the model with the different (renormalized) action S_{R}.[*]

In general, $S(\varphi)$ contains a contribution linear in φ, $h_0\varphi$ [a form of the type (1.48)], interpreted as an interaction with the bare external field h_0. Let $S(\varphi) = S'(\varphi) + h_0\varphi$, where S' is the unrenormalized action for the model without an external field. If the functional $S'(\varphi)$ has some symmetry guaranteeing that $\langle\widehat{\varphi}\rangle = 0$ in the absence of spontaneous breakdown (for example, φ parity), then the renormalization of S' will not require a shift of $\widehat{\varphi}$, or, equivalently [see the footnote following (1.74)], of the external field. In other words, in this (quite common) situation the external-field contribution absent from S' will not appear in S_{R}', i.e., the renormalization of S' will also be multiplicative. In this case all the renormalization problems for the full model $S(\varphi) = S'(\varphi) + h_0\varphi$ are solved within the model without an external field. If for it we define the functional $S_{\mathrm{R}}'(\varphi) = S'(Z_\varphi \varphi)$, then for the full model

$$S_{\mathrm{R}}(\varphi) = S_{\mathrm{R}}'(\varphi) + h\varphi, \quad h_0 = hZ_h, \quad Z_h = Z_\varphi^{-1}, \tag{1.85}$$

where h is the renormalized external field, assumed to be an independent, UV-finite parameter in S_{R}, and Z_φ is the renormalization constant of $\widehat{\varphi}$ in (1.74), which is identical for S' and S. The equations (1.85) for the renormalization of h then follow from the general rule (1.84) and the first expression in (1.85).

To justify the above arguments, it is sufficient to note that when S_{R} from (1.85) is substituted into (1.82), the contribution $h\varphi$ is grouped with $A\varphi$, i.e., it leads only to a UV-finite shift $A \to A+h$ of the argument of the functional G_{R}' of the model without a field. When the change of normalization [i.e., the denominator in (1.82)] is taken into account, the exact relation has

---

[*]Equation (1.82) determines S_{R} from the Green functions only up to an arbitrary additive constant, which cancels in (1.82). The addition of a suitable UV-divergent constant to the right-hand side of (1.84) is required when we want to renormalize not only the Green functions, but also the partition function (1.60). This will be discussed in detail in Ch. 3. Here we shall limit ourselves to renormalization of only the Green functions, for which the multiplicative renormalization of the action (1.84) is sufficient.

the form $G_R(A) = G'_R(A+h)/G'_R(h)$ and guarantees the UV finiteness of G_R owing to the UV finiteness of G'_R.

For models like the φ^4 model with zero external field there is a difference between the symmetric $(T \geq T_c)$ and nonsymmetric $(T < T_c)$ phases. Later on (Sec. 1.28) it will be shown that all the renormalization problems can be solved for any sign of $T - T_c$ just by learning how to renormalize the model in the symmetric phase. Then what was said above can be stated as the following rule: when studying renormalization in models like the φ^4 model, it is sufficient to study the model in the symmetric phase, i.e., for $h = 0$ and $T > T_c$. The renormalization of the external field is then determined by the general rule (1.85) with the constant Z_φ determined for the symmetric phase.

This rule is inapplicable to models of the φ^3 type, in which h in (1.85) undergoes a shift in addition to rescaling [if we want to preserve (1.74)]. The renormalization of such models should be treated as for the problem with an external field. The φ^3 model will be discussed in detail in Ch. 4.

1.19 Dimensional regularization. Relation between the exact and formal expressions for one-loop integrals[*]

UV divergences are fully manifested in the logarithmic theory, i.e., when the model is studied directly in its logarithmic dimension $d = d^*$ (Sec. 1.16). The dimensional regularization is effected by an ε-shift:

$$d = d^* - 2\varepsilon, \quad \varepsilon > 0, \tag{1.86}$$

(this notation is more convenient than $d^* - \varepsilon$ for actual calculations), and removal of the regularization corresponds to $\varepsilon \to 0$. This all presupposes the ability to calculate the integrals in arbitrary spatial dimension.

The technique for these calculations is discussed in detail in the following chapters. Here we limit ourselves to the general information needed to understand renormalization in dimensional regularization. For clarity, as an example let us give several simple expressions containing integrals over the d-dimensional vector k; $n \equiv k/|k|$ is its direction, and everywhere $dk \equiv d^d k = |k|^{d-1}d|k|dn$ and $\alpha \equiv d/2$:

$$\int dn = \frac{2\pi^\alpha}{\Gamma(\alpha)}, \tag{1.87a}$$

$$\int dk f(|k|) = \frac{2\pi^\alpha}{\Gamma(\alpha)} \int_0^\infty ds\, s^{d-1} f(s), \tag{1.87b}$$

$$\int dk\, f(|k|) \exp(ikx) = (2\pi)^\alpha r^{1-\alpha} \int_0^\infty ds\, s^\alpha J_{\alpha-1}(sr) f(s), \tag{1.87c}$$

---

[*]The material in this section is purely technical and may appear complicated to the reader without experience in calculating integrals in arbitrary dimension. Such a reader should just focus on the general ideas and conclusions in this section, since they will be needed for understanding later material.

where kx is the scalar product of d-dimensional vectors, $s \equiv |k|$, $r \equiv |x|$, $\Gamma(z)$ is the gamma function, $J_\nu(z)$ is the Bessel function, and Eq. (1.87a) is the angular integral in d-dimensional space. All the equations in (1.87) are valid for any integer $d \geq 1$. Their right-hand sides can usually be analytically continued in d to meromorphic functions defined in the entire complex d plane (see below for more details). These continuations, where they exist, i.e., away from poles, are taken by definition to be the values of the d-dimensional integrals without a cutoff for noninteger or complex d.

When dealing with integrals with cutoff Λ in the dimension (1.86), for accuracy we shall use the terms Λ convergence (divergence) to characterize the behavior for $\Lambda \to \infty$, and ε convergence (divergence) for $\varepsilon \to 0$. Right now we are interested only in the UV divergences from large integration momenta k, and so we shall always assume that there are no IR singularities in the integrands at zero or finite k.

Let us consider in more detail the integral with cutoff Λ in a single d-dimensional momentum k:

$$I \equiv I(e, d, \Lambda) = \int_{|k| \leq \Lambda} dk \, f(k, e), \quad d_I = d + d_f, \tag{1.88}$$

where f is a function with definite canonical dimension d_f depending on some set $e \equiv \{e_i\}$ of IR-relevant parameters, i.e., parameters tending to zero at the critical asymptote, with $d_e > 0$ for each one (τ, the external momenta, and so on). We shall assume that $f(k, e)$ is analytic for $k \neq 0$ in the set of parameters e (in the neighborhood of $e = 0$, which henceforth will always be understood), that f is analytic in k for all or almost all $e \neq 0$ [which ensures the absence of IR singularities in d in the integral (1.88)], and that $f(k, 0) \neq 0$ [which excludes functions of the type $f = e_i \tilde{f}_i(k, e)$, for which the UV convergence is determined by the properties of $\tilde{f}$ rather than f]. All these conditions will usually be satisfied for integrals corresponding to one-loop graphs for $T > T_c$ in any quantum field model. Typical examples of "good" functions are $f = (k^2 + \tau)^\alpha$ or $f = (k^2 + \tau_1)^\alpha [(k - p)^2 + \tau_2]^\beta$. For the first $d_f = 2\alpha$ and $e = \tau$, and for the second $d_f = 2\alpha + 2\beta$ and $e = \{p, \tau_1, \tau_2\}$, where the vector p is the external momentum, α and β are arbitrary, and all the τ necessarily are strictly positive, because factors like $k^{2\alpha}$ even with a positive exponent generate additional IR singularities in d, which we have agreed not to consider at present.

Expressions like (1.87) can be used to define integrals (1.88) in the half-plane Re $d > 0$ (this restriction is related to the divergence at $|k| = 0$ in the integral over $|k|$ owing to the factor $|k|^{d-1}$ with Re $d \leq 0$), and then to analytically continue to the region Re $d < 0$. This procedure is explained in detail in [60] (Sec. 4.2), and we shall not dwell on it here, since we are mainly interested in the neighborhood of $d^* > 0$ in (1.86). We only note that in continuing (1.87b) to negative d, poles appear at the points $d = 0, -2, -4, \ldots$ in the one-dimensional integral over $|k| \equiv s$, which cancel the poles of $\Gamma(d/2)$ in the denominator of the coefficient of the integral (the gamma function has no zeros).

The following statement given without proof characterizes the analytic properties of integrals like (1.88) corresponding to graphs without IR singularities.

Statement 1. The integral (1.88) with cutoff Λ is well defined and analytic in d in the half-plane $\mathrm{Re}\ d > 0$ and can be continued to a function $I(e, d, \Lambda)$ analytic in the entire d plane. For this function the limit $\lim_{\Lambda \to \infty} I(e, d, \Lambda) \equiv I(e, d, \infty)$ exists and is analytic in d in the region $\mathrm{Re}\ d < -d_f$, i.e., for $\mathrm{Re}\ d_I < 0$. This limit can be analytically continued to a meromorphic function $I_{\mathrm{form}}(e, d)$ defined in the entire complex d plane and having poles at the points $d = -d_f + d_M$, where $d_M \geq 0$ are the canonical dimensions of all possible monomials M, constructed from the parameters e, in which I is expanded, including $M = 1$ with $d_M = 0$. *Definition:* We shall call the meromorphic function $I_{\mathrm{form}}(e, d)$ the *formal result* for an integral in dimensional regularization.

The formal result I_{form} is independent of Λ and exists at all values of d except for poles located at the points for which $d_I = d + d_f = d_M$ for any monomial M.

For what follows it will be necessary to clearly characterize the relation between the formal result and the exact integral (1.88). We restrict ourselves to real values of d except for the points $d = -d_f + d_M$, at which I_{form} does not exist. The latter condition implies that $d_I \neq d_M$ for all M, in particular, $d_I \neq 0$, and allows for a given d the unique splitting of the set E of all monomials M into a finite subset $E_1 \equiv \{M : d_M < d_I\}$ and an infinite subset $E_2 \equiv \{M : d_M > d_I\}$. The projectors onto these sets will be denoted $P_{1,2}$, and $P_1 + P_2 = 1$. For a Λ-convergent integral ($d_I < 0$) the set E_1 is empty, and so $E_2 = E$, $P_1 = 0$, and $P_2 = 1$.

By condition, the function f in (1.88) for $k \neq 0$ can be expanded in e, i.e., $f(k, e) = \sum_{M \in E} M \alpha_M(k)$ with coefficients $\alpha_M(k) \propto |k|^{d_f - d_M} = |k|^{d_I - d_M - d}$. By condition, only Λ divergences from large k are possible in the integral (1.88), and in the term-by-term integration of the series for f the region of small $k \to 0$ also becomes dangerous. It is clear from dimensional considerations that the coefficients $\alpha_M(k)$ with $M \in E_1$ are integrable for $k = 0$, while those with $M \in E_2$ are integrable for $k \to \infty$. From this it follows that for the integral (1.88) with $d_I > 0$ there exists an initial (and only an initial) segment of the Taylor expansion in e obtained by the term-by-term integration of the corresponding segment of the expansion of f and consisting of all monomials $M \in E_1$. We then have the following definitions:

$$I^{(\mathrm{r})}(e, d, \Lambda) \equiv \int_{|k| \leq \Lambda} dk\ P_1 f(k, e) = \sum_{M \in E_1} M \Lambda^{d_I - d_M} c_M, \qquad (1.89a)$$

$$I^{(\mathrm{s})}(e, d, \Lambda) \equiv I - I^{(\mathrm{r})} = \int_{|k| \leq \Lambda} dk\ (1 - P_1) f(k, e), \qquad (1.89b)$$

$$\Delta I^{(\mathrm{s})}(e, d, \Lambda) \equiv \int_{|k| \geq \Lambda} dk\ (1 - P_1) f(k, e) = \sum_{M \in E_2} M \Lambda^{d_I - d_M} c_M \qquad (1.89c)$$

with dimensionless coefficients $c_M = c_M(d)$ independent of e and Λ.

We shall call $I^{(\mathrm{r})}$ the *regular* and $I^{(\mathrm{s})}$ the *singular* parts of I (analogous to the definitions in Sec. 1.3). The right-hand side of (1.89b) will be called the *integral with subtractions*, and (1.89c) the *remainder* of $I^{(\mathrm{s})}$. It follows from the definitions of $E_{1,2}$ that all the exponents $d_I - d_M$ in (1.89a) are positive, while all those in (1.89c) are negative, and that the regular part is nonzero only for Λ-divergent integrals with $d_I > 0$. From this follows the next statement.

Statement 2. Λ divergences occur in the integral (1.88) only for $d_I > 0$ and are concentrated in its regular part (1.89a), which is the initial segment of the Taylor expansion of $I(e, d, \Lambda)$ in e and consists of monomials M with $d_M < d_I$. For any $d_I \neq d_M$ for all M the integral with subtractions (1.89b) is Λ-convergent, because its remainder vanishes for $\Lambda \to \infty$. Therefore, $I^{(\mathrm{s})}(e, d, \infty)$, the integral with subtractions without a cutoff, exists. The remainder $\Delta I^{(\mathrm{s})}(e, d, \Lambda)$ is analytic in e and contains only monomials M with $d_M > d_I$, which are IR-irrelevant at the critical asymptote $e \to 0$, $\Lambda = \mathrm{const}$, compared to $I^{(\mathrm{s})}(e, d, \infty)$ of dimension d_I.

The relation between the exact and formal results is given by the next statement.

Statement 3. For any $d > 0$ away from the singular points $d = -d_f + d_M$,

$$I_{\mathrm{form}}(e, d) = I^{(\mathrm{s})}(e, d, \infty) = \int dk \, (1 - P_1) f(k, e), \tag{1.90a}$$

$$I(e, d, \Lambda) = \begin{cases} I_{\mathrm{form}}(e, d) - \Delta I^{(\mathrm{s})}(e, d, \Lambda) & \text{for } d_I < 0, \\ I_{\mathrm{form}}(e, d) + I^{(\mathrm{r})}(e, d, \Lambda) - \Delta I^{(\mathrm{s})}(e, d, \Lambda) & \text{for } d_I > 0. \end{cases} \tag{1.90b}$$

Let us explain this without rigorous proof, which can always be obtained for specific f. The powers of Λ in (1.89) are analytic functions of d, and the coefficients c_M easily calculated for specific integrals are meromorphic functions of d with a pole at the corresponding point $d = -d_f + d_M$. Depending on the value of d, a given monomial M can enter either (1.89a) or (1.89c). In the first case the coefficient c_M is determined (see above) by the integral over the region $|k| \leq \Lambda$, and in the second by that over the region $|k| \geq \Lambda$. It is important that these seemingly different definitions lead to the same meromorphic function $c_M(d)$ up to a sign, and this forms the basis of the proof of Statement 3. In fact, in the Λ-convergence region $d < -d_f$ of the integral I we have $I(e, d, \Lambda) = I_{\mathrm{form}}(e, d) - \sum_{M \in E} M \Lambda^{d_I - d_M} c_M(d)$, where the first term is the integral without a cutoff, which coincides with I_{form} by definition, and the second is the remainder (1.89c) with $E_2 = E$. This equation can be continued in d directly to all $d \neq -d_f + d_M$ for all M. For $d > -d_f$ the initial segment of the series becomes the sum (1.89a) and the remainder becomes (1.89c), as follows from the above-mentioned property of the coefficients $c_M(d)$. Equation (1.90b) is thereby proved, and it and the definitions (1.89) lead to (1.90a).

Therefore, for a Λ-convergent integral the exact result (1.88) differs from the formal result only by the IR-irrelevant remainder $\Delta I^{(\mathrm{s})}$, while for a Λ-divergent integral it differs also by the regular contribution $I^{(\mathrm{r})}$. In any case, the formal result coincides with the integral without a cutoff with subtractions ensuring Λ-convergence in the integrand (for $d_I < 0$ these are not needed) and is the leading (for $e \to 0$) singular contribution of the exact result for the integral (1.88). We note that the absence of poles in d on the left-hand side of (1.90b) (Statement 1) indicates that the poles cancel on the right-hand side: the poles of $I_{\mathrm{form}}(e, d)$ at the points $d = -d_f + d_M$, $M \in E$, are cancelled by the contributions of the analogous poles of the coefficients $c_M(d)$ of the regular part $(M \in E_1)$ and the remainder $(M \in E_2)$.

In general, a graph corresponds to an l-fold integral over d-dimensional momenta, where l is the number of loops in the graph. Above we considered only one-loop integrals (1.88), but all the conclusions we obtained regarding the structure of the divergences and the analyticity in d remain valid for any l-loop graph, as long as all subdivergences (divergences of subgraphs) are removed from it first by an auxiliary R' operation. The divergences remaining after this procedure are called *superficial divergences* , and the statements given above apply to them. All these topics will be dealt with in detail in Ch. 3 in our discussion of renormalization theory, where we will also give a more precise discussion of IR singularities.

1.20   The renormalization problem in dimensional regularization

As already noted, the UV divergences of a given model are fully manifested in its logarithmic dimension d^*. In the theory of critical behavior we are interested in models with given "real" dimension d_{r} with specific UV cutoff $\Lambda \sim r_{\min}^{-1}$ (Sec. 1.18). Usually $d_{\mathrm{r}} \leq d^*$, and two variants are possible: for example, in three-dimensional problems $d_{\mathrm{r}} = 3$, while for the φ^4 model $d^* = 4$, and for the φ^6 interaction (Sec. 1.16) or the unaxial ferroelectric (Sec. 1.17) $d^* = 3$. In any case, the model is studied in variable dimension (1.86), at the end taking $\varepsilon \to 0$ if $d_{\mathrm{r}} = d^*$ or $\varepsilon = \varepsilon_{\mathrm{r}}$ if $d_{\mathrm{r}} < d^*$.

When studying divergences and renormalization one usually considers graphs of 1-irreducible functions (1.54) in momentum space for a model without an external field (see the end of Sec. 1.18) for $T \geq T_{\mathrm{c}}$, i.e., in the symmetric phase. In this case in (1.54) $\alpha = 0$ and Γ_n are simply the coefficients of the expansion of the functional (1.50) about $\alpha = 0$. Parameters like τ and the external momenta p [variables conjugate to the coordinates x in (1.54)] play the role of e in the integrals corresponding to the graphs. In this chapter we shall study only Green functions like (1.33) and (1.49) or (1.54) with $n \geq 1$, and in Ch. 3 we shall include vacuum loops, i.e., the graphs of the free energy Γ_0, which have no effect on the general ideas discussed below. Generalization of the renormalization procedure to the case $T < T_{\mathrm{c}}$, i.e., to the phase with spontaneous symmetry breaking, also usually does not present any difficulties (Sec. 1.28), and everywhere below in discussing renormalization, unless stated otherwise, the symmetric phase in zero field will always be understood.

For $d = d^*$ the integrals I corresponding to the graphs of Γ_n usually have integer dimension d_I, the same for all the graphs of a given function Γ_n. Graphs for which $d_I \geq 0$ for $d = d^*$ are termed *superficially divergent*. All the graphs of the first few functions Γ_n are usually superficially divergent. Graphs with $d_I = 0$ are termed logarithmically divergent, graphs with $d_I = 1$ are linearly divergent, graphs with $d_I = 2$ are quadratically divergent, and so on. A typical example is the φ^4 model for $d = d^* = 4$. In the symmetric phase only even functions Γ_n are nonzero, all the graphs of Γ_4 diverge logarithmically, and those of Γ_2 diverge quadratically.

The small ε shift (1.86) decreases the value of d_I for an l-loop graph by $2\varepsilon l$ (2ε for each d-dimensional integration momentum). This is sufficient for ensuring the Λ convergence of graphs that diverge logarithmically at $\varepsilon = 0$ (we recall that for multiloop graphs these are superficial divergences; see the discussion at the end of Sec. 1.19). However, the small ε shift does not entirely eliminate the Λ divergences. They remain in graphs that diverge linearly or more strongly at $\varepsilon = 0$. These remaining Λ divergences are concentrated in the regular contributions of the integrals corresponding to the graphs and are polynomials in the IR-relevant parameters e (Sec. 1.19). It should be stressed that there are no ε divergences in a model with finite cutoff Λ, since all the results are analytic in d.

The shift (1.86) ensures that the condition $d_I \neq d_M$ is satisfied for all M (Sec. 1.19) and that representations like (1.90b) are valid. Their left-hand sides and the contributions $I^{(\mathrm{r})}$ on the right-hand sides do not have poles in $\varepsilon \sim d - d^*$ [$I^{(\mathrm{r})}$ contains singularities in d, but not at $d = d^*$]. However, I_{form} and $\Delta I^{(\mathrm{s})}$ do contain such poles, and they cancel each other out. Poles in ε appear in the results when the contributions $\Delta I^{(\mathrm{s})}$ are discarded, which is always done in calculations using dimensional regularization. This is justified by the fact that, owing to their analyticity in e and their IR-irrelevance (Statement 2 in Sec. 1.19), the contributions $\Delta I^{(\mathrm{s})}$ represent only trivial corrections at the critical asymptote.

Let us explain this in more detail. According to the general ideas of the theory of critical phenomena (Sec. 1.3), we are interested only in the leading singular contributions in the critical region and the nontrivial corrections to them characterized by the correction critical exponents ω. In this sense the corrections analytic in e are not interesting, because they are always present in the general phenomenology and the corresponding indices are uniquely determined by the critical dimensions Δ_e of the parameters e. The trivial corrections $\Delta I^{(\mathrm{s})}$ are not needed for calculating the leading critical singularities and the nontrivial correction terms.

Let us now discuss regular contributions of the type $I^{(\mathrm{r})}$ in (1.90b). It should be stated immediately that the regular contributions are not controlled by the fluctuation models and have nothing to do with the adequacy of the theory of critical behavior (see Secs. 1.3 and 1.12). It is therefore meaningful to discuss them only from the viewpoint of the internal self-consistency of the model under study: the model must contain parameters whose renormalization can absorb all the regular contributions arising in the model. Let us explain for the example of the φ^4 model in dimension $d = 4 - 2\varepsilon$. The only type of regular term in the graphs of the Green functions Γ_{2n} with $n \geq 1$ is the constant in the graphs of Γ_2 independent of the external momentum and τ. The inclusion of this constant leads only to a shift of T_c (see Sec. 1.21 for more details). This is the case in all reasonable field models: Λ-divergent regular contributions renormalize parameters like T_c that are irrelevant from the viewpoint of the theory of critical behavior (it should be recalled that T_c is an irrelevant parameter, while $\tau = T - T_c$ is relevant).

The regular terms in (1.90b) always have positive dimensions and therefore cannot contribute to the dimensionless field renormalization constant in (1.74). They are involved only in the renormalization of bare parameters: $e_0 \to e_0' = e_0 + \Delta e_0$, where Δe_0 are the Λ-divergent corrections from the regular terms. This step corresponds to the Λ *renormalization* of the model: if e_0' instead of e_0 are used as the independent variables, setting $e_0 = e_0(e_0', \Lambda)$, the Green functions expressed in terms of e_0' will no longer contain any Λ divergences. They will all be concentrated in the expressions relating e_0 and e_0'. If these relations affect only irrelevant parameters like T_c and are therefore not of interest, this renormalization step corresponds to the simple replacement $e_0 \to e_0'$ in the Green functions with the simultaneous omission of all the regular contributions in the calculation of the integrals corresponding to the graphs, which is equivalent to making subtractions which ensure Λ convergence in the integrands [see (1.89b)].

At this stage there are no ε divergences in the theory for finite cutoff Λ. However, if we stop here we will not have used some essential information: the multiplicative renormalizability of the model. One therefore always takes the next step, the $\Lambda \to \infty$ limit in Λ-convergent integrals with subtractions, which is equivalent to discarding contributions like $\Delta I^{(\mathrm{s})}$ in (1.90b) (see above for the justification). This means that in the graphs of unrenormalized Green functions one uses the formal results I_{form} for all the integrals, simultaneously with the replacement $e_0 \to e_0'$ corresponding to the inclusion of all the regular contributions. The formal results contain poles in ε [which are exactly cancelled by the analogous poles in $\Delta I^{(\mathrm{s})}$], the elimination of which is the goal of the next step: the nontrivial multiplicative ε *renormalization* of the field and parameters.

The full renormalization procedure can be written symbolically as

$$\{e_0, \widehat{\varphi}\} \xrightarrow{\Lambda\text{-ren}} \{e_0', \widehat{\varphi}\} \xrightarrow{\varepsilon\text{-ren}} \{e, \widehat{\varphi}_{\mathrm{R}} = Z_\varphi^{-1}\widehat{\varphi}\}. \tag{1.91}$$

The first step is the trivial Λ renormalization:

$$G_n(...; e_0, \Lambda) \equiv G_n'(...; e_0'(e_0, \Lambda), \Lambda), \tag{1.92}$$

where here and below the ellipsis denotes the coordinate or momentum arguments of the Green functions (1.33). The dependence on the parameter ε from (1.86) is not explicitly indicated, but is always understood. The explicit Λ dependence in G'_n corresponds to the cutoff of the Λ-convergent integrals, and all the Λ divergences are concentrated in the functions $e'_0 = e'_0(e_0, \Lambda)$. The second step in (1.91), the nontrivial ε renormalization, plays the main role in what follows. The ε renormalization eliminates the poles in ε in the quantities

$$G_n(...; e'_0) \equiv G'_n(...; e'_0, \Lambda = \infty), \quad e'_0 = e'_0(e_0, \Lambda), \qquad (1.93)$$

which are taken by definition to be the unrenormalized Green functions in the ε-renormalization procedure. The functions (1.93) depend on Λ only implicitly through $e'_0 = e'_0(e_0, \Lambda)$, and if the e'_0 are chosen as the independent variables, the Λ dependence in (1.93) completely vanishes. The parameters e'_0, like e_0, will be called bare parameters, and which is which will be clear from the context. They are different from the new renormalized parameters e introduced in the second step of the procedure (1.91). This amounts to going from the unrenormalized functions (1.93) to the new renormalized functions

$$G_{n\mathrm{R}}(...; e) = Z_\varphi^{-n} G_n(...; e'_0) \qquad (1.94)$$

with certain functions $Z_\varphi = Z_\varphi(e)$, $e'_0 = e'_0(e)$ (ε is understood), selected from the condition that the poles in ε cancel in (1.94).

We immediately note that the formulation (1.94) is suitable only when the model contains dimensional parameters like $T - T_{\mathrm{c}}$. In general, to correctly determine the renormalized Green functions it is necessary to introduce an auxiliary dimensional parameter, the *renormalization mass* μ with $d_\mu = 1$. Instead of (1.94) we then write

$$G_{n\mathrm{R}}(...; e, \mu) = Z_\varphi^{-n} G_n(...; e'_0), \qquad (1.95a)$$

$$Z_\varphi = Z_\varphi(e, \mu), \quad e'_0 = e'_0(e, \mu). \qquad (1.95b)$$

This formulation is also suitable in the absence of parameters like $T - T_{\mathrm{c}}$, for example, for the φ^4 model studied right at the critical point.

In all the following discussion we shall use the term renormalization to refer to the second step of the procedure (1.91): multiplicative renormalization in the form (1.94) or (1.95). All the general statements and expressions of Sec. 1.18 are valid here if the UV divergences are understood to be poles in ε, and the ellipsis in (1.84) represents dependence on ε and μ. In the RG technique, the critical behavior is always determined from the asymptote of the renormalized Green functions, since they, on the one hand, are just as good for this purpose as the original unrenormalized functions, while, on the other, they contain important information about the renormalizability of the model in a simple, explicit form (the absence of poles in ε). During the calculations the parameter ε in the renormalized functions is taken to be arbitrary, and it is assigned the real value $\varepsilon = \varepsilon_r$ only in the final expressions. For $\varepsilon_r > 0$ there is no direct need for eliminating the poles by ε renormalization. Nevertheless, such renormalization is done even in this case, because important information about the renormalizability of the model is introduced in this manner.

If it is known *a priori* that the first step in (1.91) leads only to a change of quantities like T_c that are irrelevant for the theory of critical behavior, then this step is usually completely omitted in actual calculations, with the following *convention*: the Λ-independent formal results obtained by analytic continuation in d or similar parameters from the Λ-convergence region of the integral are used to calculate all the integrals in variable dimension d corresponding to the graphs. We shall refer to this calculational method as the formal scheme of dimensional regularization or simply the *formal scheme*.

The use of this scheme is equivalent to discarding all the regular contributions and remainders $\Delta I^{(s)}$ in equations like (1.90b). The parameters e_0 and e'_0 are then identical, the Λ dependence completely vanishes, and the UV divergences are assumed to be only poles in ε. The information about the Λ-renormalization $e_0 \to e'_0$, which is lost as a result is of no real value, since the fluctuation models alone, as explained in Sec. 1.12, are useless for determining parameters like T_c.

1.21 Explicit renormalization formulas

We shall explain the general ideas for the example of the φ^4 model (1.63) with zero external field and $T \geq T_c$ in dimension $d = 4 - 2\varepsilon$. The original action

$$S(\varphi) = -\int dx [\tfrac{1}{2}(\partial\varphi)^2 + \tfrac{1}{2}\tau_0\varphi^2 + \tfrac{1}{24}g_0\varphi^4] \tag{1.96}$$

is assumed to be unrenormalized (Sec. 1.18), and its parameters $e_0 \equiv \{\tau_0, g_0\}$ are the bare parameters, with $\tau_0 = T - T_{c0}$, where T_{c0} is the critical temperature in the Landau approximation. All the canonical dimensions are known (Sec. 1.15):

$$d = 4 - 2\varepsilon, \quad d[g_0] = 2\varepsilon, \quad d[\tau_0] = 2, \quad d_\varphi = 1 - \varepsilon. \tag{1.97}$$

In studying the renormalization we consider the 1-irreducible Green functions (1.54) in momentum space, where in the symmetric phase only even functions Γ_{2n} are nonzero. The function Γ_2 is related to the correlator D in momentum space by (1.55):

$$\Gamma_2(p, ...) = -D^{-1}(p, ...), \tag{1.98}$$

where p is the external momentum (denoted as k in Sec. 1.9), and the ellipsis denotes other arguments. Owing to isotropy, the scalar functions in (1.98) depend only on p^2. The role of K in (1.59) for the action (1.96) is played by the operator $-\partial^2 + \tau_0$, i.e., $p^2 + \tau_0$ in momentum space, and the unperturbed correlator is the inverse $D_0 = K^{-1} = (p^2 + \tau_0)^{-1}$.

Only the graphs for the Green functions Γ_4 and Γ_2 have superficial divergences (Sec. 1.20). At $\varepsilon = 0$ the former diverge logarithmically ($d_I = 0$) and the latter quadratically ($d_I = 2$). For finite ε the value of d_I is decreased by $2\varepsilon l$, where l is the number of loops in the graph, so that Λ-divergent regular contributions occur only in the graphs for Γ_2. They depend on the parameters $e \equiv \{p^2, \tau_0\}$ (using the notation of Sec. 1.19), where both these parameters have dimension 2, exceeding $d_I = 2 - 2\varepsilon l$. Therefore, the regular contribution determined from the rule (1.89a) for any graph of Γ_2 is a constant $\sim \Lambda^{2-2\varepsilon l}$, independent of the parameters e (contributions $\sim |p|$ of suitable dimension are not allowed owing to their nonanalyticity in e). The inclusion of these regular contributions of the graphs of Γ_2 leads only to a shift of the bare critical temperature T_{c0}. This will be shown rigorously in Ch. 3, and here we limit ourselves to the qualitative aspects. The exact (within the model) value of T_c is by definition the value of T at which a divergence appears at zero momentum in the full correlator $D = D(p, g_0, \tau_0(T), \Lambda, \varepsilon)$. Owing to (1.98), this is equivalent to the condition $\Gamma_2(p = 0, g_0, \tau_0(T_c), \Lambda, \varepsilon) = 0$. This equation implicitly defines $\tau_0(T_c) = T_c - T_{c0} \equiv \Delta T_c$ as a function of Λ, g_0, and ε, in perturbation theory as a series in g_0:

$$\Delta T_c \equiv T_c - T_{c0} = \sum_{l=1}^{\infty} c_l(\varepsilon) g_0^l \Lambda^{2-2\varepsilon l} \tag{1.99}$$

with dimensionless coefficients c_l. Clearly, all the terms of the series (1.99) have the structure of the regular contributions of the graphs of Γ_2 and can be generated only by these contributions. The nontrivial statement is that the regular contributions have no other effect except for shifting T_c. This is easily proved using the general theory of renormalization (Ch. 3). We note that the coefficients c_l in (1.99) do not have poles in ε, i.e., singularities in d at $d = d^*$. According to the general rules of Sec. 1.19, the coefficient c_l has a singularity only at the value of d for which $d_I = d_M$; in this case, $d_I = 2 - 2\varepsilon l$ and $d_M = 0$.

Therefore, the first step of the procedure (1.91) in this model reduces to the replacement $e_0 \equiv \{\tau_0, g_0\} \to e_0' \equiv \{\tau_0', g_0' = g_0\}$, i.e.,

$$\tau_0 \equiv T - T_{c0} \to \tau_0' \equiv T - T_c = \tau_0 - \Delta T_c \tag{1.100}$$

with ΔT_c from (1.99).

Let us now go to the second step of the procedure (1.91): the nontrivial multiplicative ε renormalization. For the choice of renormalized parameters $e = \{\tau, g\}$ and constant Z_φ in (1.74), we shall limit the arbitrariness by adopting the following *convention*: the renormalized charge g is always chosen to be dimensionless, while τ has the dimension of τ_0, and $\tau = 0$ corresponds to the critical point:

$$d_g = 0, \quad d_\tau = 2, \quad \tau = 0 \Longleftrightarrow T = T_c. \tag{1.101}$$

Moreover, we shall always require that in lowest-order perturbation theory the equations $Z_\varphi = 1$ and $\tau = \tau_0$ be satisfied, and the parameters g and g_0 differ by only a UV-finite factor that cancels the difference of their canonical dimensions.

Assuming that all these conditions are satisfied, the general renormalization rule (1.94) can be made specific for our model:

$$\widehat{\varphi}(x) = Z_\varphi \widehat{\varphi}_{\mathrm{R}}(x), \quad g_0 = g\tau^\varepsilon Z_g, \quad \tau_0' = \tau Z_\tau \qquad (1.102)$$

with τ_0' from (1.100). The dimensionless factors Z, called *renormalization constants*, can depend only on the dimensionless parameters g and ε. In perturbation theory all the Z are series in g beginning at unity with ε-dependent coefficients containing poles in ε.

Equations (1.102) presuppose the presence of a dimensional parameter τ and are therefore inapplicable right at the critical point $\tau = 0$. In this case we resort to the formulation (1.95) with the additional parameter μ, the renormalization mass, and instead of the second equation in (1.102) we write $g_0 = g\mu^{2\varepsilon} Z_g$. To ensure that the renormalized Green functions are continuous for $\tau \to 0$, it is convenient to retain the parameter μ also for $\tau > 0$. Then instead of (1.102) we have

$$\widehat{\varphi}(x) = Z_\varphi \widehat{\varphi}_{\mathrm{R}}(x), \quad g_0 = g\mu^{2\varepsilon} Z_g, \quad \tau_0' = \tau Z_\tau, \qquad (1.103)$$

and now, in general, $Z = Z(g, \lambda, \varepsilon)$ with $\lambda \equiv \tau\mu^{-2}$.

1.22   The constants Z in the minimal subtraction scheme

Dimensional considerations determine the general structure of (1.103), but they do not fix the explicit form of the functions $Z(g, \lambda, \varepsilon)$. According to the main statement of renormalization theory (Sec. 1.18), these can be chosen so as to eliminate all the UV divergences in the Green functions, in this case, poles in ε. This is the main requirement on the functions Z, but it does not determine them uniquely. The remaining arbitrariness is fixed by imposing some auxiliary conditions. This is referred to as the choice of *subtraction scheme*. There are various schemes (see Ch. 3 for details), for which the renormalized Green functions differ only by a UV-finite renormalization (1.78) and from the viewpoint of physics are equivalent (Sec. 1.18). Therefore, the scheme is chosen on the basis of convenience.

The most convenient scheme for calculations is the minimal subtraction (MS) scheme proposed in [61], since in it even for $\tau \neq 0$ all the constants Z in (1.103) are independent of λ and have the following form in perturbation theory:

$$Z = Z(g, \varepsilon) = 1 + \sum_{n=1}^{\infty} g^n \sum_{k=1}^{n} \varepsilon^{-k} c_{nk}, \qquad (1.104)$$

where c_{nk} are numbers independent of any parameters. The term "minimum" reflects the fact that ε enters into (1.104) only in the form of poles. The restriction $k \leq n$ on their order is a consequence of the general structure of the divergences of the graphs, which will be discussed in detail in Ch. 3.

The fact that the constants Z are independent of λ in the MS scheme greatly simplifies the calculations using the RG technique and allows all the constructions to be easily generalized to the case $T < T_c$, i.e., to the phase with spontaneous symmetry breakdown (Sec. 1.28). It is also important that in this scheme the parameter τ has a simple physical interpretation: it is proportional to $\tau_0' = T - T_c$, whereas in the general expression (1.103) the relation between τ and the temperature is complicated when Z_τ depends on $\lambda = \tau \mu^{-2}$.

Since the choice of subtraction scheme is only a matter of convenience, everywhere in this chapter we shall assume, unless stated otherwise, that the MS scheme is used for the renormalization, so that the constants Z have the form (1.104).

1.23 The relation between the IR and UV problems

The main problem of the theory of critical behavior is the determination of the IR asymptote (where the momenta and τ tend to zero, and the parameters g and μ are fixed) of the Green functions. In solving this problem it is impossible to restrict ourselves to any finite order of perturbation theory even for a very weak interaction. Let us consider, for example, the unrenormalized correlator D for $T = T_c$. The parameter Λ does not enter into this quantity (since not τ_0 but $\tau_0' = T - T_c = 0$ is fixed; see Sec. 1.20), and so it depends only on ε, g_0, and the momentum p. In perturbation theory in g_0 using (1.97) we have

$$D\,|_{T=T_c} = D_0(p)\left[1 + \sum_{n=1}^{\infty}(g_0 p^{-2\varepsilon})^n c_n(\varepsilon)\right], \qquad (1.105)$$

where $D_0(p) = p^{-2}$ is the zeroth-order approximation and c_n are dimensionless coefficients. The expansion parameter in (1.105) is actually not g_0, but the dimensionless combination $g_0 p^{-2\varepsilon}$, so that for arbitrarily small g_0 the entire series (1.105) must be summed in order to determine the $p \to 0$ asymptote. This is the IR problem, i.e., the problem of IR singularities (here for $p \to 0$, $T = T_c$, and in general for $p \to 0$, $T \to T_c$). It is much more complicated than the problem of UV divergences, since there is no analog of the theory of UV renormalization allowing the explicit isolation and summation of the infrared singularities in the general case.

However, for small ε these two problems are related. We see from (1.105) that the IR singularities for $p \to 0$ weaken with decreasing ε, and would disappear completely for $\varepsilon = 0$ if we could simply set $\varepsilon = 0$ in (1.105). However, it is impossible to do this owing to the UV divergences — the poles in ε in the coefficients c_n. Therefore, the problem of IR singularities at $p \to 0$ for asymptotically small ε is somehow related to the UV problem, the presence of poles in ε. If the latter did not occur, neither would the former.

This to some degree explains why the UV-renormalization technique for eliminating poles in ε and the associated RG technique can somehow be related to the solution of the nontrivial IR problem.

However, this is true only for "correctly constructed" models with the IR-irrelevant contributions discarded (Sec. 1.16), in any case, in the free part of the action. Let us explain this by an example. If we add an explicitly IR-irrelevant contribution $\sim \varphi \partial^4 \varphi$ to the action (1.96) (which is equivalent to adding a term $\sim k^4$ to the denominator $k^2 + \tau_0$ of the bare correlator), the leading IR singularities of the graphs of perturbation theory would obviously not be affected, but their UV behavior would be completely changed. For such a model the needed correlation between the IR and UV behavior is violated, which makes it impossible to use the standard RG technique.

1.24 The differential RG equations

The quantum field renormalization group is, by definition, the set of all the UV-finite renormalization transformations (1.78). The RG transformations relate the different sets of Green functions obtained from the same unrenormalized functions by different renormalization procedures. In dimensional regularization, renormalization will always be understood to be the procedure discussed in Sec. 1.20. A particular manifestation of the nonuniqueness of the renormalization is the arbitrariness of the parameter μ in (1.95), which is used for the simplest derivation of the RG equations.

If in (1.95) we vary μ keeping the bare parameters e_0 and the cutoff Λ [and thereby $e_0' = e_0'(e_0, \Lambda)$] fixed, some quantities will change and others will remain unchanged. Quantities of the second type are termed *renormalization-invariant*. Examples of these are the parameters Λ, e_0, e_0', and anything else expressed in terms of them [for example, the shift of T_c in (1.99)], and also the original unrenormalized field $\widehat{\varphi}$ and any (full, connected, 1-irreducible) of its Green functions and their generating functionals. The noninvariant quantities will be the renormalized parameters and the field $\widehat{\varphi}_R$ and all its Green functions. We immediately note that the condition $\tau = 0$, i.e., $T = T_c$, is renormalization-invariant, although the parameter τ itself is noninvariant.

The renormalization invariance of F can be expressed as

$$\widetilde{\mathcal{D}}_\mu F = 0, \quad \widetilde{\mathcal{D}}_\mu \equiv \mu \partial_\mu |_{\Lambda, e_0} = \mu \partial_\mu |_{e_0'} . \tag{1.106}$$

By definition, the differentiation with respect to μ in the operator $\widetilde{\mathcal{D}}_\mu$ is done for fixed e_0 and Λ, i.e., for fixed parameters e_0' in the notation of (1.95). The notation $\widetilde{\mathcal{D}}_\mu$ will be used for the operator (1.106) throughout this book, as will the notation $\mathcal{D}_u \equiv u \partial_u$ for any operator involving partial derivatives in renormalized variables.

The differential RG equations are written for the renormalized Green functions, which are usually connected or 1-irreducible, and express the renormalization invariance of the corresponding unrenormalized functions. For example, in (1.106) we can take F to be the unrenormalized full function G_n from (1.95a) or the corresponding (Sec. 1.10) connected function $W_n(...; e_0')$, where here and below the ellipsis denotes coordinate or momentum arguments. Then, using (1.80) to write this as $W_n = Z_\varphi^n W_{nR}$ and contracting the factor Z_φ^n after differentiation, we obtain the desired RG equation for the renormalized connected function W_{nR}:

$$[\mathcal{D}_{RG} + n\gamma_\varphi] W_{nR}(...; e, \mu) = 0, \quad \gamma_\varphi \equiv \widetilde{D}_\mu \ln Z_\varphi, \tag{1.107}$$

where $\mathcal{D}_{RG}$ is the operator $\widetilde{\mathcal{D}}_\mu$ from (1.106) in the variables e and μ:

$$\mathcal{D}_{RG} = \widetilde{\mathcal{D}}_\mu = \mathcal{D}_\mu + \sum_e (\widetilde{\mathcal{D}}_\mu e)\partial_e, \quad \mathcal{D}_\mu \equiv \mu\partial_\mu, \tag{1.108}$$

with summation over all the renormalized parameters e.

Similarly, from (1.81) and (1.106) for the 1-irreducible functions we obtain

$$[\mathcal{D}_{RG} - n\gamma_\varphi] \Gamma_{nR}(...; e, \mu) = 0. \tag{1.109}$$

Equations (1.107) and (1.109) are valid for any $n \geq 1$. Each of these infinite systems of equations can be written as a single equation for the corresponding generating functional (1.43) or (1.50):

$$[\mathcal{D}_{RG} + \gamma_\varphi \mathcal{D}_A] W_R(A; e, \mu) = 0, \tag{1.110}$$

$$[\mathcal{D}_{RG} - \gamma_\varphi \mathcal{D}_\alpha] \Gamma_R(\alpha; e, \mu) = 0, \tag{1.111}$$

where $\mathcal{D}_A$ and $\mathcal{D}_\alpha$ are functional operators:

$$\mathcal{D}_A \equiv \int dx\, A(x)\frac{\delta}{\delta A(x)}, \quad \mathcal{D}_\alpha \equiv \int dx\, \alpha(x)\frac{\delta}{\delta\alpha(x)}. \tag{1.112}$$

The coefficients of the derivatives in the operator (1.108) and γ_φ from (1.107) expressed in terms of the variables e and μ [ε is always understood, and the cutoff Λ does not enter at all into the expressions in our notation of (1.95)] are called the *RG functions*. Their definition involves the bare parameters and renormalization constants, and so it can be expected that UV divergences — poles in ε — will also appear. However, it is easily shown that there are no UV divergences in the RG functions. To prove this it is sufficient to note that (1.107) with different n can be viewed as linear, inhomogeneous equations for determining the unknown RG functions from the given functions W_{nR} and their derivatives with respect to e and μ, which are UV-finite by definition. Isolating the needed finite number of independent equations from the infinite system (1.107), all the RG functions can be expressed in terms of known UV-finite quantities (there is no reason for the determinant of this system to vanish at $\varepsilon = 0$), and so all the RG functions must be UV-finite. This conclusion is, of course, confirmed by explicit calculation of the RG functions (see below).

There are other notations for and derivations of the RG equations. The elementary derivation given above was suggested in [62]. All variants of the RG equations give the same final results.

1.25 The RG functions in terms of the renormalization constants

The possibility of expressing the RG functions in terms of the renormalized Green functions demonstrates the UV finiteness of the RG functions, but in practice it is more convenient to calculate them in terms of simpler quantities: the renormalization constants Z. The scheme we shall describe here is general, but for definiteness we shall consider the φ^4 model in dimension $d = 4 - 2\varepsilon$ (Sec. 1.21). For it $e = \{\tau, g\}$ and the renormalization is accomplished by (1.103). From this using the operator $\widetilde{\mathcal{D}}_\mu$ from (1.106) we define the quantities

$$\beta_g \equiv \beta \equiv \widetilde{\mathcal{D}}_\mu g, \quad \gamma_a \equiv \widetilde{\mathcal{D}}_\mu \ln Z_a. \tag{1.113}$$

The first of these is conventionally called the *beta function* (of a given charge g), and γ_a is the *anomalous dimension* of the quantity a (here $a = g, \tau, \varphi$). The general term is RG functions, as for the coefficients of the operator (1.108). The latter are simply related to the functions (1.113): if we take the logarithm of (1.103) for g_0 and τ_0' and then operate on both sides with $\widetilde{\mathcal{D}}_\mu$, the left-hand sides vanish owing to the definition of $\widetilde{\mathcal{D}}_\mu$ in (1.106), while the right-hand sides are expressed in terms of (1.113), which gives

$$\widetilde{\mathcal{D}}_\mu g \equiv \beta = -g(2\varepsilon + \gamma_g), \quad \widetilde{\mathcal{D}}_\mu \tau = -\tau\gamma_\tau. \tag{1.114}$$

From this it follows, in particular, that all the RG functions (1.113) are also UV-finite.

The definitions (1.113) and (1.114) can be used to express all the RG functions in terms of the renormalization constants Z. We shall assume that the MS scheme (Sec. 1.22) is used for the renormalization, and so the constants Z have the form (1.104). It follows from (1.114) that for any function F depending only on g (dependence on ε is always understood if not indicated explicitly) we have

$$\widetilde{\mathcal{D}}_\mu F(g) = \beta\partial_g F(g) = -(2\varepsilon + \gamma_g)\mathcal{D}_g F(g). \tag{1.115}$$

Taking $F = \ln Z_g$ and using the definition of γ_g in (1.113), we have

$$\gamma_g = -(2\varepsilon + \gamma_g)\mathcal{D}_g \ln Z_g. \tag{1.116}$$

From this we find γ_g, and then from it the β function using (1.114):

$$\gamma_g = -\frac{2\varepsilon\mathcal{D}_g \ln Z_g}{1 + \mathcal{D}_g \ln Z_g}; \quad \beta = -\frac{2\varepsilon g}{1 + \mathcal{D}_g \ln Z_g}. \tag{1.117}$$

Assuming that these quantities are known, the other anomalous dimensions in (1.113) are easily expressed in terms of the corresponding constants Z using (1.115):

$$\gamma_a = \beta\partial_g \ln Z_a = -(2\varepsilon + \gamma_g)\mathcal{D}_g \ln Z_a. \tag{1.118}$$

In general, the RG functions depend on the same variables as the renormalization constants Z, which here are g and ε. However, in the MS scheme using the explicit form (1.104) of the constants Z, it can be shown that all the anomalous dimensions γ_a are actually independent of ε. To prove this, we rewrite (1.104) for the three constants Z_a as

$$Z_a(g, \varepsilon) = 1 + \sum_{n=1}^{\infty} A_{na}(g)\varepsilon^{-n}. \tag{1.119}$$

The coefficients A_{na} are independent of ε, and their expansion in g begins with g^n. Substituting (1.119) into (1.117) and (1.118), it is easily shown that for all three anomalous dimensions we obtain the representation $\gamma_a = -2\mathcal{D}_g A_{1a}(g)+$ contributions containing only poles in ε. The latter must cancel owing to the known UV-finiteness of the RG functions γ_a, and so

$$\gamma_a = -2\mathcal{D}_g A_{1a}. \tag{1.120}$$

We have thereby shown that in the MS scheme all the anomalous dimensions γ_a are independent of ε and are completely determined by the residues $A_{1a}(g)$ at the first-order pole in ε of the corresponding constants Z_a. In perturbation theory

$$\gamma_a(g) = \sum_{n=1}^{\infty} \gamma_{na} g^n, \quad \beta(g) = -2\varepsilon g + \sum_{n=2}^{\infty} \beta_n g^n \tag{1.121}$$

with numerical coefficients γ_{na} independent of any parameters, and from (1.114) we find $\beta_{n+1} = -\gamma_{ng}$.

It follows from (1.114) that the operator (1.108) in our case has the form

$$\mathcal{D}_{\text{RG}} = \widetilde{\mathcal{D}}_\mu = \mathcal{D}_\mu + \beta(g)\partial_g - \gamma_\tau(g)\mathcal{D}_\tau \tag{1.122}$$

with the RG functions (1.121). The explicit ε dependence is contained only in the first term of the series (1.121) for the β function.

Above we assumed that the constants Z depend only on g and ε. Expressions for the RG functions in terms of the renormalization constants can also be written down in the general case, when they also depend on the variable $\lambda = \tau\mu^{-2}$ (Sec. 1.21). However, these expressions will, of course, be more awkward, and so in practical calculations it is always more convenient to use the MS scheme. Here we shall first use the graphs directly to calculate the renormalization constants Z, and then we shall use the latter to calculate the RG functions from (1.117) and (1.118). All objects are constructed as series in g, and in reality it is, of course, possible to find only the first few terms of the series. The technique of calculating the constants Z from graphs is rather complicated if we do not limit ourselves to the lowest nontrivial order of perturbation theory. This will be studied in detail in the following chapters, and here we shall just discuss the consequences of the RG equations, assuming that the RG functions entering into them are known.

1.26   Relations between the residues of poles in Z of various order in ε. Representation of Z in terms of RG functions

The UV finiteness of all the RG functions γ_a is ensured by the mutual cancellation of all the poles in ε on the right-hand side of (1.118), which requires a well defined correlation between the residues A_{na} in the expansion (1.119). Taking the logarithm, we write $\ln Z_a = \sum_{n=1}^{\infty} B_{na} \varepsilon^{-n}$. The coefficients B_{na} are uniquely expressed in terms of the A_{na} from (1.119): $B_{1a} = A_{1a}$, $B_{2a} = A_{2a} - \frac{1}{2} A_{1a}^2$, $B_{3a} = A_{3a} - A_{1a} A_{2a} + \frac{1}{3} A_{1a}^3$, and so on. On the other hand, from (1.118) we have

$$\mathcal{D}_g \ln Z_a = -\frac{\gamma_a}{2\varepsilon + \gamma_g} = -\frac{\gamma_a}{2\varepsilon} \sum_{n=0}^{\infty} \left[-\frac{\gamma_g}{2\varepsilon} \right]^n. \qquad (1.123)$$

Substituting the above expression for $\ln Z_a$ into the left-hand side and equating the coefficients of ε^{-n} on both sides of (1.123), from the known (Sec. 1.25) ε independence of all the γ_a we obtain

$$\mathcal{D}_g B_{na} = (-\tfrac{1}{2})^n \gamma_a \gamma_g^{n-1}, \qquad n \geq 1, \qquad a = g, \tau, \varphi. \qquad (1.124)$$

For $n = 1$ these relations coincide with (1.120) and express γ_a in terms of $B_{1a} = A_{1a}$, and for $n \geq 2$ using the relation between B_{na} and A_{na} (see above) from (1.124) we obtain

$$\mathcal{D}_g A_{2a} = \mathcal{D}_g (\tfrac{1}{2} A_{1a}^2) + \tfrac{1}{4} \gamma_a \gamma_g,$$

$$\mathcal{D}_g A_{3a} = \mathcal{D}_g [A_{1a} A_{2a} - \tfrac{1}{3} A_{1a}^3] - \tfrac{1}{8} \gamma_a \gamma_g^2, \qquad (1.125)$$

and so on. Knowledge of $\mathcal{D}_g A_{na}$ is equivalent to knowledge of the A_{na} themselves, because the action of the operator $\mathcal{D}_g = g \partial_g$ on the desired series in g for A_{na} gives only an additional factor of m in the coefficient of g^m. [It is important that the g series for A_{na} do not contain constants, since according to (1.104) they begin at g^n.] Therefore, taking into account (1.120) it can be stated that the equations (1.125) systematically determine the residues A_{na} of all the higher-order poles in ε in the renormalization constants (1.119) in terms of the residues A_{1a} at the first-order poles of the three constants Z_a. This property follows from the UV finiteness of the RG functions (1.118), which in the MS scheme reduces to the complete independence of these functions from ε and is, in turn, a consequence of the renormalizability of the model we are studying.

From (1.118) taking into account the normalization $Z_a = 1$ at $g = 0$ we find

$$Z_a(g, \varepsilon) = \exp \left\{ \int_0^g dg' \frac{\gamma_a(g')}{\beta(g')} \right\}. \qquad (1.126)$$

This expression with $a = g$ and (1.114) give

$$Z_g(g, \varepsilon) = \exp\left\{-\int_0^g dg' \left[\frac{2\varepsilon}{\beta(g')} + \frac{1}{g'}\right]\right\}. \tag{1.127}$$

Therefore, the constant Z_g is uniquely determined by the β function, and the other Z_a by the corresponding γ_a and the β functions.

1.27 Relation between the renormalized and bare charges

In the unrenormalized theory the natural scale parameter for all the variables is the UV cutoff Λ. In particular, it is usually assumed that $g_0 = u\Lambda^{2\varepsilon}$ with dimensionless coefficient u of order unity, if there is no special reason to assume otherwise. In the renormalized theory g is taken to be the formal independent parameter, and the renormalization mass μ, introduced in (1.103) as an arbitrary parameter, is taken as the scaling variable. The second equation in (1.103) for known function $Z_g(g)$ (ε understood) determines $g_0 = g_0(g, \mu)$ explicitly and $g = g(g_0, \mu)$ implicitly. Actually, in the theory of critical behavior the known initial parameters of the model are the bare variables and Λ, and all the renormalized parameters, in particular, $g = g(g_0, \mu)$, are calculated in terms of the bare ones. If we use perturbation theory (series in g or g_0) everywhere, the relations between the bare and renormalized parameters do not impose any constraints on their values, and so the choice of which are independent is arbitrary (Sec. 1.18).

However, in going beyond perturbation theory the "symmetry" between the bare and renormalized parameters is lost, and then the bare parameters must be taken as the initial ones (this is the essential difference from the formulation of the problem in relativistic quantum field theory). The departure from perturbation theory is ensured, in particular, by (1.126) and (1.127): if the RG functions in some finite order of perturbation theory are substituted into them, the resulting expressions for Z_a will be infinite series in g, which must be understood as the result of a partial summation of the entire perturbation series.

In this way we can obtain information about the behavior of the function $Z_g(g)$, assuming that the behavior of the β function in (1.127) is known (Sec. 1.1). In the models of interest to us, the first nontrivial coefficient β_2 in the expansion of the β function (1.121) is positive. Therefore, the function $\beta(g)$ for small ε must behave as shown in Fig. 1.1 of Sec. 1.1, vanishing at some point $g_* \sim \varepsilon$. One of the main postulates of the modern theory of critical behavior (Sec. 1.1) is that the behavior of the β function shown in Fig. 1.1 is preserved also at large ε, including the real value $\varepsilon_r = \frac{1}{2}$ in (1.86) for the φ^4 model. If this is so, which is always assumed in what follows, from (1.127) we can then make definite conclusions about the behavior of the function $gZ_g(g)$ in the interval $g \in [0, g_*]$, where $\beta < 0$. Actually, the second equation in (1.117), rewritten as $\partial_g \ln(g Z_g) = -2\varepsilon/\beta$, demonstrates the monotonic increase of the function $g Z_g(g)$ in this interval, and from the nonintegrability of the singularity at $g' = g_*$ in (1.127) it follows that $g Z_g \to \infty$ for $g \to g_* - 0$. The behavior of the function $g Z_g(g)$ is shown in Fig. 1.2, where we also give the graphical solution of the equation $g_0 \mu^{-2\varepsilon} = g Z_g(g)$ for g following from (1.103).

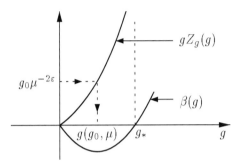

Figure 1.2 Behavior of the functions $\beta(g)$ and $g Z_g(g)$ in the φ^4 model and graphical solution of the equation $g_0 \mu^{-2\varepsilon} = g Z_g(g)$ for g.

We see from this figure that for any g_0 and μ a solution $g = g(g_0, \mu)$ with the required properties (expandability of g in g_0 and vice versa) exists, is unique, and is always located in the interval $[0, g_*]$. For clarity, let us give the explicit expression for Z_g, obtained from (1.127), in the simplest *one-loop approximation* for the β function, keeping only the first two terms of the series (1.121):

$$\beta(g)\,|_{1-\text{loop}} = -2\varepsilon g + \beta_2 g^2, \tag{1.128a}$$

$$Z_g(g)\,|_{1-\text{loop}} = \frac{2\varepsilon}{2\varepsilon - \beta_2 g} = \frac{g_*}{g_* - g}, \tag{1.128b}$$

where $g_* = 2\varepsilon/\beta_2$ in this approximation (we assume that $\beta_2 > 0$). In the exact expression (1.127) the singularity for $g \to g_* - 0$ will not be a simple pole as in (1.128b), but a fractional power $(g_* - g)^\lambda$ with exponent $\lambda = -2\varepsilon/\beta'(g_*)$.

Therefore, even though the formal perturbation series make sense for any g, the natural region of definition for the renormalized φ^4 model is the interval $0 \le g < g_*(\varepsilon)$.

For $\varepsilon \to 0$ this interval vanishes, and so the question arises of the meaning of the renormalized theory at $\varepsilon = 0$, i.e., directly in the logarithmic dimension. According to current ideas (see, for example, [60], Sec. 7.11), the renormalized φ^4 model with cutoffs completely removed ($\Lambda = \infty$, $\varepsilon = 0$) does not exist. Only the model with the cutoff Λ is physically meaningful at $\varepsilon = 0$, and the finiteness of the cutoff ensures finiteness of the interval of possible values of the renormalized charge g. In practice, the RG equations of Sec. 1.24

are also used at $\varepsilon = 0$, arguing that all the quantities entering into them are continuous in ε and that different renormalization schemes are physically equivalent (Sec. 1.18), so that finiteness of Λ can be replaced by finiteness of ε. The latter is used only when really necessary, in particular, in the expressions relating g_0 and g (which ensures the existence of the interval of allowed g), while $\varepsilon = 0$ is taken directly in the RG functions and the renormalized Green functions, and thereby also in the RG equations. The logarithmic corrections to the Landau theory are obtained in this manner from RG equations like (1.107) (see Secs. 1.43 and 1.44 below). Here the parameter g enters only into the nonuniversal amplitude factors, and so the specific value of g is not important. Only the fact that some region of applicability in g exists is important.

1.28 Renormalization and RG equations for $T < T_c$

For quantum field models like (1.96) there are known methods (Ch. 2) of directly calculating the functional (1.50) as a whole in the form of graphs defined for any sign of $T - T_c$ and expandable in α at the origin for $T \geq T_c$. These expansions generate the graphical representations of 1-irreducible Green functions Γ_n, which are studied in the renormalization procedure: the superficial divergences of the graphs of Γ_n are determined by power counting, and then it is checked that they can all be eliminated by introducing the needed constants Z into expressions like (1.103), thus proving the multiplicative renormalizability of the model (Ch. 3). If the constants Z are independent of $\tau \sim T - T_c$ (for example, in the MS scheme), then expressions like (1.103) are meaningful also for $\tau < 0$. In this case the following *statement* is valid (see Ch. 3 for the proof): the renormalization constants Z calculated from the Green functions Γ_n in the symmetric phase with $\tau \geq 0$ automatically eliminate all the UV divergences from the graphs of the functional $\Gamma(\alpha)$ as a whole for any sign of τ, i.e., they ensure the UV finiteness of the renormalized functional

$$\Gamma_R(\alpha; e, \mu) = \Gamma(Z_\varphi \alpha; e_0'), \tag{1.129}$$

with all quantities having the same meaning as in (1.95).

This can be explained as follows. We go from the graphs of Γ_n to those of $\Gamma(\alpha)$ by summing the functional Taylor series in α. This operation and the subsequent continuation of the graphs of $\Gamma(\alpha)$ to the region $\tau < 0$ do not introduce new UV divergences, and so the UV finiteness of the functions Γ_n for $\tau \geq 0$ guarantees the UV finiteness of all the graphs of the functional $\Gamma(\alpha)$ for any sign of τ.

The functional $\Gamma(\alpha)$ is always constructed using the model without an external field, and all the Green functions are determined from it: the spontaneous magnetization α_0 is the saddle point of $\Gamma(\alpha)$, the 1-irreducible Green functions Γ_n are the coefficients of its expansion in α at this point, and the connected functions W_n and the full functions G_n are expressed in terms of the Γ_n in the usual way (Sec. 1.11). The functional $\Gamma(\alpha)$ is always unique, and the Green functions are single-valued only for $T \geq T_c$, when $\alpha_0 = 0$. For $T < T_c$ they are not single-valued and depend on the choice of "pure phase," i.e., the specific point α_0 (Sec. 1.11). All these general definitions and relations are assumed to hold in both the unrenormalized and the renormalized theory. New UV divergences do not arise in the process of constructing the Green functions from the functional $\Gamma(\alpha)$, and so the UV finiteness of the functional $\Gamma_R(\alpha)$ in (1.129) guarantees the UV finiteness of all the corresponding Green functions, i.e., of the renormalized theory as a whole. We note that the IR convergence of the graphs of $\Gamma(\alpha)$ for $T < T_c$ is guaranteed only in the "physical" region $|\alpha| \geq |\alpha_0|$. Inside the sphere $|\alpha| \leq |\alpha_0|$ the surface Γ has a flat segment, i.e., $\Gamma(\alpha) = \text{const}$ (Sec. 1.11). The graphical representation of $\Gamma(\alpha)$ is inapplicable (and unnecessary) in this region.

From the relation between the functionals (1.129) and the definitions, it is easily shown that for $T < T_c$ for any pure phase,

$$\alpha_0 = Z_\varphi \alpha_{0R}, \quad \Gamma_{nR} = Z_\varphi^n \Gamma_n \quad \text{for all } n \geq 1, \tag{1.130}$$

i.e., the functions Γ_n are renormalized according to the same rule (1.81) as in the case $T \geq T_c$. It follows from the explicit expressions for the functions W_n and G_n in terms of Γ_n (Sec. 1.11) that the same is true for the connected and full Green functions, i.e., the renormalization rules (1.75) and (1.80) remain valid also for $T < T_c$ for each of the pure phases. Therefore, in schemes like the MS scheme for $T < T_c$ all the renormalized Green functions and the corresponding generating functionals for each of the pure phases satisfy the same RG equations as for $T \geq T_c$.

So far we have discussed the problem with zero external field, taking $\Gamma(\alpha)$ to be the functional of the model without a field. In the language of Γ the field is introduced into the problem as an independent functional argument $A(x)$ in (1.51), from which we find $\alpha = \alpha(A)$. In accordance with the terminology of Sec. 1.18, the quantity A in (1.51) for the unrenormalized functional $\Gamma(\alpha)$ in (1.129) should be called the bare external field, and that for $\Gamma_R(\alpha)$ the renormalized external field, denoting the first by $A_0(x)$ and the second by $A(x)$. Then from the equations $\delta\Gamma(\alpha)/\delta\alpha = -A_0$ and $\delta\Gamma_R(\alpha)/\delta\alpha = -A$ and (1.129) we find $A_0(x) = Z_\varphi^{-1} A(x)$, in complete analogy with (1.85). This makes sense, as in the formalism of Sec. 1.11 the quantity $A(x)$ in the final expressions plays the role of the external field h.

1.29 Solution of the linear partial differential equations

In actual problems the RG equation always can be written as ($\mathcal{D}_x \equiv x\partial_x$ for all x)

$$LF(u) = \gamma(u), \quad L \equiv -\mathcal{D}_s + \sum_{i=2}^{n} Q_i(u)\partial/\partial e_i, \tag{1.131}$$

where $s \equiv u_1$ is a particular scaling variable, $e_i \equiv u_i$ with $i = 2, ..., n$ are the other variables, $u \equiv s, e$ is compact notation for the entire set of variables, all the Q_i and γ are given ($Q_1 = -s$), and F is the desired function.

The general solution of the inhomogeneous equation (1.131) is the sum of its particular solution and the general solution of the homogeneous equation. The latter is an arbitrary function of any complete set of independent first integrals. A first integral is any solution of the homogeneous equation, and every function of first integrals is also a first integral; only one which is not a function of the other first integrals is treated as independent. The total number of independent first integrals is one less than the number of variables of F, i.e., there are as many as there are variables e. The first integrals can be chosen differently. In our problems it will be convenient to use the integrals $\bar{e}_i(u) = \bar{e}_i(s, e)$, which are defined as

$$L\bar{e}_i(s, e) = 0, \quad \bar{e}_i(s, e) \mid_{s=1} = e_i. \tag{1.132}$$

We shall call these the *invariant variables* e (invariant charge, temperature, and so on). Finding the first integrals directly from the definition (1.132) is just as difficult as the original problem. However, it reduces to the simpler problem of solving a system of ordinary differential characteristic equations.

The characteristic curves of (1.131) are the lines $\bar{\bar{u}}(t)$ in the space of the variables u on which $d\bar{\bar{u}}_i \sim Q_i(\bar{\bar{u}})dt$ for all $i = 1, ..., n$. The proportionality factor is arbitrary; it depends on the choice of parameter t on the curve and will be fixed below.

The following statement is valid:

$$LF(u) = 0 \Leftrightarrow F(\bar{\bar{u}}(t)) = \text{const}, \tag{1.133}$$

i.e., every solution of a homogeneous equation is a constant (independent of t) on any characteristic curve and vice versa.

Let us define the canonical family of characteristic curves $\bar{\bar{u}}(t) = \bar{\bar{s}}(t), \bar{\bar{e}}(t)$ as the solution of the following Cauchy problem ($\mathcal{D}_t \equiv t\partial_t$):

$$\mathcal{D}_t\bar{\bar{s}}(t) = -\bar{\bar{s}}(t), \quad \bar{\bar{s}}(t) \mid_{t=1} = s, \tag{1.134a}$$

$$\mathcal{D}_t\bar{\bar{e}}_i(t) = Q_i(\bar{\bar{s}}(t), \bar{\bar{e}}(t)), \quad \bar{\bar{e}}_i(t) \mid_{t=1} = e_i. \tag{1.134b}$$

Here the proportionality factor (see above) is already fixed. The Cauchy problem (1.134a) is easily solved explicitly ($\bar{\bar{s}} = s/t$); the solutions of (1.134b) are functions of t depending parametrically on the initial data $u = s, e$. Written out in detail,

$$\bar{\bar{s}} = \bar{\bar{s}}(t; s) = s/t, \quad \bar{\bar{e}}_i = \bar{\bar{e}}_i(t; s, e). \tag{1.135}$$

On the curves (1.135), every solution $F(u)$ of the inhomogeneous equation (1.131) satisfies the equation $\mathcal{D}_t F(\bar{\bar{u}}(t)) = \gamma(\bar{\bar{u}}(t))$ with the condition $F(\bar{u}) = F(u)$ at $t = 1$, which is equivalent to the equation

$$F(u) = F(\bar{\bar{u}}(t)) - \int_1^t \frac{dt'}{t'}\gamma(\bar{\bar{u}}(t')). \tag{1.136}$$

The following statement relating the first integrals (1.132) and the solutions of the characteristic system (1.134) is fundamental for what follows:

$$\bar{\bar{s}}(t;s)\mid_{t=s}=1, \qquad \bar{\bar{e}}_i(t;s,e)\mid_{t=s}=\bar{e}_i(s,e). \qquad (1.137)$$

The first equation is obvious from (1.135), and the others are easily proved. From (1.132) and (1.133) we find that the quantities $\bar{e}_i(\bar{\bar{s}}(t),\bar{\bar{e}}(t))$ are independent of t. Equating their values at $t=1$ and $t=s$ taking into account the initial data for $\bar{\bar{u}}$ in (1.134) and the first equation in (1.137), we find

$$\bar{e}_i(s,e) = \bar{e}_i(1,\bar{\bar{e}}\mid_{t=s}) = \bar{\bar{e}}_i\mid_{t=s}.$$

The second equation is a consequence of the normalization of $\bar{e}_i$ in (1.132).

Statement 1. Let $F(s,e)$ be a solution of the homogeneous equation $LF=0$ with operator L from (1.131). Then

$$F(s,e) = F(1,\bar{e}(s,e)). \qquad (1.138)$$

According to (1.133), the quantity $F(\bar{\bar{u}}(t))$ is independent of t; its value at $t=1$ is the left-hand side of (1.138), and its value at $t=s$ taking into account (1.137) is the right-hand side.

Statement 2. Let $F(s,e)$ be a solution of the inhomogeneous equation $LF(s,e)=\gamma(s,e)$. Then

$$F(s,e) = F(1,\bar{e}(s,e)) - \int_1^s \frac{dt'}{t'}\gamma(\bar{\bar{s}}(t';s),\bar{\bar{e}}(t';s,e)). \qquad (1.139)$$

To prove this we must set $t=s$ in (1.136) and use (1.137).

Statement 3. If all the functions Q_i in (1.131) with $i \geq 2$ are independent of s, the first integrals in (1.132) are the solutions of the following Cauchy problem for a system of ordinary differential equations in the variable s:

$$\mathcal{D}_s\bar{e}_i(s,e) = Q_i(\bar{e}(s,e)), \qquad \bar{e}_i\mid_{s=1}=e_i. \qquad (1.140)$$

If γ in (1.131) is also independent of s, then

$$F(s,e) = F(1,\bar{e}(s,e)) - \int_1^s \frac{ds'}{s'}\gamma(\bar{e}(s',e)). \qquad (1.141)$$

In fact, with our assumptions the variable s does not enter into the Cauchy problem (1.134b) (neither the equation nor the initial data), and so it also does not enter into its solutions: $\bar{\bar{e}}_i = \bar{\bar{e}}_i(t;e)$. Then in (1.134b) we can make the replacement $t \to s$, $\mathcal{D}_t \to \mathcal{D}_s$ (it is important that s does not enter implicitly via the initial data), and when (1.137) is taken into account this transforms (1.134b) into (1.140). With the assumptions we have made, (1.141) follows from (1.139) and (1.137).

An important special case is the equation

$$\left[-\mathcal{D}_s + \beta(g)\partial_g - \sum_a \Delta_a(g)\mathcal{D}_a + \gamma(g)\right]\Phi(s,g,a) = 0, \qquad (1.142)$$

which the substitution $\Phi = \exp F$ reduces to (1.131). Here $e \equiv \{g, a\}$, where g is the charge and a are the other variables, $Q_g = \beta(g)$ and $Q_a = -a\Delta_a(g)$ for all a. These functions Q and γ in (1.142) are independent of s, and so (1.140) and (1.141) are valid. The variables a do not enter into either the equation or the intial data for the Cauchy problem (1.140) for the invariant charge $\bar{g}$, and so

$$\bar{g} = \bar{g}(s, g), \qquad \mathcal{D}_s \bar{g} = \beta(\bar{g}), \qquad \bar{g}\,|_{s=1} = g. \qquad (1.143)$$

This equation is easily integrated:

$$\ln s = \int_g^{\bar{g}} \frac{dx}{\beta(x)}, \qquad (1.144)$$

and thereby implicitly determines the function $\bar{g} = \bar{g}(s, g)$.

For any of the variables a from (1.142) and (1.140) we have

$$\mathcal{D}_s \bar{a} = -\bar{a} \Delta_a(\bar{g}), \qquad \bar{a}\,|_{s=1} = a. \qquad (1.145)$$

Here it is convenient to use (1.143) to make the variable substitution $s \to \bar{g}$:

$$\mathcal{D}_s \equiv s\frac{d}{ds} = \beta(\bar{g})\frac{d}{d\bar{g}}, \qquad \beta(\bar{g})\frac{d\ln\bar{a}}{d\bar{g}} = -\Delta_a(\bar{g}), \qquad (1.146)$$

from which, taking into account the initial condition (1.145) for $\bar{a}$, we find

$$\bar{a} = \bar{a}(s, g, a) = a \exp\left[-\int_g^{\bar{g}} dx \frac{\Delta_a(x)}{\beta(x)}\right]. \qquad (1.147)$$

By a similar substitution $s' \to \bar{g}(s')$ in (1.141) for $\Phi = \exp F$ we obtain

$$\Phi(s, g, a) = \Phi(1, \bar{g}, \bar{a}) \exp\left[\int_g^{\bar{g}} dx \frac{\gamma(x)}{\beta(x)}\right]. \qquad (1.148)$$

This expression with an arbitrary function of the first integrals $\Phi(1, \bar{g}, \bar{a})$ on the right-hand side is the general solution of (1.142).

In our problems the function $\Phi(s, g, a)$ in (1.142) usually also has a definition independent of (1.142), for example, as the sum of the contributions of all the graphs of renormalized perturbation theory. Then (1.148) must be interpreted as a property of the function $\Phi(s, g, a)$ following from (1.142).

1.30 The RG equation for the correlator of the φ^4 model in zero field

From dimensional considerations, the renormalized correlator D_R of the model (1.96) in momentum space has the form (see Sec. 1.21 for the notation)

$$D_R(p, g, \tau, \mu) = p^{-2}\Phi(s, g, z), \qquad s \equiv p/\mu, \qquad z \equiv \tau/\mu^2, \qquad (1.149)$$

where p is the momentum (always understood as the modulus of a vector), and Φ is a function of three canonically dimensionless arguments. From (1.44) and (1.107) we have

$$[\mathcal{D}_{\mathrm{RG}} + 2\gamma_\varphi]D_{\mathrm{R}} = 0. \tag{1.150}$$

We shall assume that the RG functions have been calculated in the MS scheme (Sec. 1.25). Then the RG operator has the form (1.122), and (1.150) is valid for either sign of τ (Sec. 1.28). Substitution of (1.149) into (1.150) leads to an equation for Φ, where in the RG operator (1.122) $\mathcal{D}_\mu \to -\mathcal{D}_s - 2\mathcal{D}_z$ and $\mathcal{D}_\tau \to \mathcal{D}_z$. In the end we obtain

$$[-\mathcal{D}_s + \beta\partial_g - (2+\gamma_\tau)\mathcal{D}_z + 2\gamma_\varphi]\Phi(s, g, z) = 0. \tag{1.151}$$

This is an equation like (1.142), since all the RG functions depend only on g (Sec. 1.25). Therefore, the solution of (1.151) is given by (1.148) with $\gamma = 2\gamma_\varphi$:

$$\Phi(s, g, z) = \Phi(1, \overline{g}, \overline{z}) \exp\left[2\int_g^{\overline{g}} dx \, \frac{\gamma_\varphi(x)}{\beta(x)}\right]. \tag{1.152}$$

The invariant charge $\overline{g}(s, g)$ is given by (1.144), and the invariant variable $\overline{z}$ by (1.147) with $a = z$ and $\Delta_a = 2 + \gamma_\tau$. The contribution of the two to Δ_a taking into account (1.144) reduces to a factor of s^{-2} in $\overline{z}$:

$$\overline{z} = \overline{z}(s, g, z) = zs^{-2} \exp\left[-\int_g^{\overline{g}} dx \, \frac{\gamma_\tau(x)}{\beta(x)}\right]. \tag{1.153}$$

The representation (1.152) for the function Φ from (1.149) is a property, following from the RG equation (1.150), of the renormalized correlator determined by the summed contributions of the graphs of perturbation theory.

1.31   Fixed points and their classification

Fixed points g_* (the traditional notation) are zeros of the β function, i.e., roots of the equation $\beta(g_*) = 0$. One of them is always the trivial fixed point $g_* = 0$ (often called the Gaussian point). In general, there are also others, for example, $g_* \sim \varepsilon$ in Fig. 1.2 for the β function of the φ^4 model. Actually, the β function is calculated as the first few terms of a series in g, and its behavior for $g \simeq 1$ is unknown. Therefore, in what follows we shall just discuss the general ideas about possible situations. When analyzing fixed points in this section we will treat g as an arbitrary independent parameter of the renormalized theory, and will not use the estimate $g < g_*$ obtained for it in Sec. 1.27.

The requirement that the system be stable usually imposes some bound on g, for example, $g \geq 0$ for the φ^4 model, and the problem is studied only in this *stability region*. The fixed points g_* divide this region into intervals $\mathcal{O}$ in each of which the β function has fixed sign. In general, for a finite number of fixed points there are several finite segments $\mathcal{O}$ and one infinite one (g_*, ∞).

Assuming that $\beta(g)$ is analytic in the neighborhood of g_*, we shall say that a given point g_* is of the type $[n, +]$ if in a neighborhood of it $\beta(g) \cong \omega(g - g_*)^n$ with $n \geq 1$ and constant $\omega > 0$, and of the type $[n, -]$ if $\omega < 0$.

Let $\mathcal{O}$ be the interval in which the value of g given in (1.143) falls. For g and $\overline{g}$ from $\mathcal{O}$ the integral of (1.144) exists and for fixed g determines $\ln s$ as a single-valued, monotonic (owing to the fact that β has fixed sign in $\mathcal{O}$) function of the variable $\overline{g} \in \mathcal{O}$. If $\mathcal{O}$ is a finite segment bounded by two fixed points g_* (for example, $[0, g_*]$ in Fig. 1.2), then $\ln s(\overline{g})$ monotonically runs over the entire axis $(-\infty, \infty)$ when $\overline{g}$ runs over $\mathcal{O}$ in the appropriate direction. In fact, from (1.144), using the known (see above) behavior of the β function near g_*, it is easy to find the asymptote of $\ln s(\overline{g})$ for $\overline{g} \to g_*$, namely, $\ln s(\overline{g}) \cong \omega^{-1} \ln |\overline{g} - g_*|$ for $n = 1$ and $\ln s(\overline{g}) \cong (\overline{g} - g_*)^{1-n}/\omega(1 - n)$ for $n \geq 2$. It follows from these expressions that for $\overline{g} \to g_* \pm 0$ the quantity $\ln s(\overline{g})$ tends to $-\infty$ for a point of the type $[2k + 1, +]$, to $+\infty$ for a point of the type $[2k + 1, -]$, to $\mp\infty$ for a point of the type $[2k, +]$, and to $\pm\infty$ for a point of the type $[2k, -]$. If we return to the original formulation, in which $\overline{g}$ is assumed a function of the independent variable $s = p/\mu$, it then follows that for one of the two asymptotes (which one is known) $\ln s \to \pm\infty$ the corresponding value of $\overline{g}(s, g)$ tends to the given g_*. The behavior of $\overline{g}$ at the IR asymptote $s \to 0$ for points of various types is shown in Fig. 1.3.

1	2	3	4

Figure 1.3 Behavior of the invariant charge $\overline{g}(s, g)$ for $s = p/\mu \to 0$ for fixed points of various types. The curves show the behavior of the β function near g_*, and the arrow indicates the direction of motion of $\overline{g}$ along the g axis for $s \to 0$. For $s \to \infty$ (the UV asymptote) the direction of all the arrows must be reversed.

Points of type 1 in Fig. 1.3 are called *IR-stable* (or IR-attractive, UV-unstable, UV-repulsive), points of type 2 are called *UV-stable* (UV-attractive, IR-unstable, IR-repulsive), points of type 3 are called IR-attractive from the right and UV-attractive from the left, and vice versa for points of type 4. The meaning of this terminology is clear from Fig. 1.3. It follows from the monotonicity of the function $\ln s(\overline{g})$ in $\mathcal{O}$ that for a finite segment $\mathcal{O}$ one of its endpoints will always be an IR-attractive fixed point for $\overline{g} \in \mathcal{O}$ while the other will be UV-attractive. Here $\overline{g}(s, g)$ always remains inside the interval $\mathcal{O}$ in which the value of g specified in (1.143) falls, tending to its endpoints for $\ln s \to \pm\infty$.

The unbounded interval $\mathcal{O} = (g_*, \infty)$ is a special case. If g and therefore also $\overline{g}(s, g)$ fall inside this interval, the behavior of $\ln s(\overline{g})$ for $\overline{g} \to \infty$ depends on the behavior of $\beta(g)$ for $g \to \infty$. If it is such that the integral of the function $1/\beta(x)$ in (1.144) diverges at infinity, the value of $\overline{g}(s, g)$ will be uniquely determined by (1.144) for all $s > 0$, as for a finite interval $\mathcal{O}$. However, if this integral converges [which will always be the case when $\beta(g)$ is approximated by a finite segment of the series in g], the region of values of the function $\ln s(\bar{g})$, i.e., the region of definition in s of the function $\bar{g}(s, g)$, is naturally bounded. The formal solution of the differential equation (1.143) can usually be obtained for all s, but at the boundary of the natural region of definition in s it has a discontinuity, and beyond the boundary it acquires the wrong sign [see, for example, the s behavior of the function (1.154) for $g > g_*$]. This formal solution is devoid of physical meaning in the "forbidden region," because, owing to the discontinuity, it cannot be considered a solution of the Cauchy problem (1.143) in the exact sense of the term, as such a solution is smooth by definition.

In conclusion, we recall that the actual behavior of the β function at $g \cong 1$ is usually unknown, and given this situation the penetration of the solution $\bar{g}$ into the region $\bar{g} \cong 1$ actually just implies going beyond the limit where the approximation is valid.

1.32   Invariant charge of the RG equation for the correlator

Unless further qualified, the β function of the charge g is always understood to be the quantity defined in (1.113), and the invariant charge for any equation of the type (1.142) is expressed by (1.144) in terms of the β *function of the given RG equation*, defined as the coefficient of ∂_g in the operator (1.142). These two quantities are usually the same, but not always: if the scaling argument s in (1.142) is a variable like $s = \tau\mu^{-2}$ with nontrivial critical dimension, the β function of (1.142) can differ from β in (1.113) by an additional factor (see Sec. 1.39 for an example). This factor is always positive, and therefore does not affect the location and type of fixed points (Sec. 1.31), and so their classification is uniquely determined by the β function (1.113). However, it should be recalled that the invariant charge $\bar{g}(s, g)$ and its argument s can mean different things in different RG equations.

Let us discuss in more detail the invariant charge $\bar{g}(s, g)$ for (1.151), in which $s = p/\mu$ and the β function coincides with that defined in (1.113) and is given by the series (1.121) with $\beta_2 > 0$. Its behavior [known at small ε and assumed at larger ε up to the real value $\varepsilon_r = \frac{1}{2}$ in (1.86)] is shown in Figs. 1.1 and 1.2. In calculating $\bar{g}$ using (1.144) we shall assume that g [and therefore also $\bar{g}(s, g)$] is always located inside the first interval $\mathcal{O} = (0, g_*)$, the left-hand end of which is a UV-attractive fixed point, and the right-hand end is an IR-attractive fixed point (Secs. 1.27 and 1.31), so that $\bar{g}(s, g) \to g_* \sim \varepsilon$ at the IR asymptote $s \to 0$. To make this clear, we give the explicit expression, obtained easily from (1.144), for $\bar{g}$ in the one-loop approximation (1.128a) for the β function, $\beta(g) = -2\varepsilon g + \beta_2 g^2$:

$$\bar{g}(s,g) = \frac{2\varepsilon g}{2\varepsilon s^{2\varepsilon} + \beta_2 g(1 - s^{2\varepsilon})} = \frac{gg_*}{g_* s^{2\varepsilon} + g(1 - s^{2\varepsilon})}, \tag{1.154}$$

where $g_* = 2\varepsilon/\beta_2$ in this approximation. It is easily checked that all the required general features (Sec. 1.31) are preserved in the approximation (1.154): $\bar{g} \in \mathcal{O} = (0, g_*)$ for all $g \in \mathcal{O}$, $\bar{g} \to 0$ for $s \to \infty$, and $\bar{g} \to g_*$ for $s \to 0$.

Let us now use the example (1.154) to discuss the condition for $\bar{g}$ to enter the critical regime $|\bar{g} - g_*| \ll g_*$, which in (1.154) is reached for

$$(g_* - g)s^{2\varepsilon} \ll g, \quad s \equiv p/\mu. \tag{1.155}$$

From this we see that the critical regime for $\bar{g}$ is certainly reached at sufficiently small momentum p. It may appear from (1.155) that reaching the critical regime can be ensured for any p by choosing a sufficiently large value of the renormalization mass μ, which is an arbitrary parameter (Sec. 1.21). This is true if p, g, and μ are assumed to be arbitrary independent parameters. However, the actual statement of the problem is different (Sec. 1.27): the bare parameters and μ must be assumed fixed, while g is calculated from the given g_0 and μ using the second equation in (1.103). Substituting Z_g from (1.128b) into it, we can find the relation between g_0 and g in this approximation. Then, using this to express g in terms of g_0 in (1.155), it is easily verified that this inequality takes the "renormalization-invariant" form $g_* \ll g_0 p^{-2\varepsilon}$, i.e., μ drops out.

This is not a consequence of the approximation, but is a general property. It can be shown that the invariant charge $\bar{g}(s,g)$ defined by (1.144) with the β function from (1.113) and $s = p/\mu$ is actually a function of the single renormalization-invariant (Sec. 1.24) variable $g_0 p^{-2\varepsilon} \equiv \xi$. In fact, from (1.103) we have $\xi = gZ_g(g)s^{-2\varepsilon}$, i.e., ξ is a function of the same variables s and g as $\bar{g}$. On such functions the operator (1.122) takes the form $\mathcal{D}_{\mathrm{RG}} = -\mathcal{D}_s + \beta\partial_g \equiv L$. It follows from the obvious renormalization invariance of $\xi = g_0 p^{-2\varepsilon}$ (Sec. 1.24) that $L\xi = 0$, i.e., ξ is the first integral of the equation $LF(s,g) = 0$ (Sec. 1.29). The same is also true for $\bar{g}(s,g)$ owing to the general definition (1.132), and since this equation has only a single independent first integral (F has only two variables), either of the two first integrals is a function of the other, i.e., we have $\bar{g} = f(\xi)$ and vice versa [for example, in the one-loop approximation it is easily shown from (1.128) and (1.154) that $\bar{g} = \xi g_*/(\xi + g_*)$].

Similar arguments are also valid for the other invariant variables, for example, for $\bar{z}$ in (1.152). Generalizing these arguments, it can be shown that all the invariant variables of (1.132) with $L = \mathcal{D}_{\mathrm{RG}}$ are functions of the bare parameters and the momentum, i.e., of only renormalization-invariant quantities (Sec. 1.24).

1.33   Critical scaling, anomalous critical dimensions, scaling function of the correlator

Let us return to the solution (1.152) of the RG equation for the correlator and study its IR asymptote $p \to 0$, $\tau \to 0$ at fixed g and μ. The behavior of the β function and the invariant charge was studied in detail in Sec. 1.32. In this case the β function has a first-order zero at the IR-attractive point $g_* \sim \varepsilon$, near which

$$\beta(g) \cong \omega(g - g_*), \qquad \omega \equiv \beta'(g_*) > 0. \tag{1.156}$$

Here ω is the slope of the β function at the point g_*. It will be shown in Sec. 1.34 that this quantity is the critical exponent determining the corrections to scaling (Sec. 1.3).

For $s = p/\mu \to 0$ we have $\overline{g}(s, g) \to g_*$ (Sec. 1.32), and the asymptote of the integrals entering into (1.152) and (1.153) is easily found by substituting $\gamma(x) = \gamma(g_*) + [\gamma(x) - \gamma(g_*)]$ into them. We assume that all the RG functions are regular near g_* (in actual calculations they are polynomials), and so in integrals involving $\gamma(x) - \gamma(g_*)$ the zero of the β function in the denominator is cancelled by a zero in the numerator. These integrals are regular in the upper limit $\overline{g}$, which can be replaced by g_* with an error of order $\overline{g} - g_*$. The individual contributions involving $\gamma(g_*)$ are easily calculated using (1.144), and we find

$$\int_g^{\overline{g}} dx \frac{\gamma_i(x)}{\beta(x)} = \gamma_i^* \ln s + \widetilde{c}_i(g) + ..., \qquad i = \tau, \varphi, \tag{1.157}$$

where the ellipsis denotes corrections $\sim \overline{g} - g_*$ unimportant for $s \to 0$ and

$$\gamma_i^* \equiv \gamma_i(g_*), \qquad \widetilde{c}_i(g) \equiv \int_g^{g_*} dx \frac{\gamma_i(x) - \gamma_i(g_*)}{\beta(x)}. \tag{1.158}$$

Substitution of (1.157) into (1.152) and (1.153) neglecting corrections gives

$$\Phi(s, g, z) \cong C_\Phi \Phi(1, g_*, \overline{z}) s^{2\gamma_\varphi^*}, \tag{1.159a}$$

$$\overline{z} \cong C_z \overline{z}_{\text{scal}}, \qquad \overline{z}_{\text{scal}} \equiv z s^{-2 - \gamma_\tau^*} = \tau p^{-2} (\mu/p)^{\gamma_\tau^*}, \tag{1.159b}$$

where C are nonuniversal (g-dependent) normalization factors:

$$C_\Phi = \exp[2\widetilde{c}_\varphi(g)], \qquad C_z = \exp[-\widetilde{c}_\tau(g)]. \tag{1.160}$$

Substitution of (1.159) into (1.149) leads to a representation like (1.39) for the renormalized correlator:

$$D_{\text{R}}(p, g, \tau, \mu) \cong p^{-2} (p/\mu)^\eta C_\Phi f(C_z \overline{z}_{\text{scal}}) \tag{1.161}$$

with critical exponents

$$\eta = 2\gamma_\varphi^*, \qquad 1/\nu = 2 + \gamma_\tau^* \tag{1.162}$$

and scaling function

$$f(z) \equiv \Phi(1, g_*, z). \tag{1.163}$$

Therefore, the RG equation for the correlator implies the existence of critical scaling with exponents (1.162) calculable from the RG functions. They are related to the critical dimensions $\Delta_{\varphi,\tau}$ by (1.40), which can be rewritten using (1.158) and (1.162) as ($i = \varphi, \tau$):

$$\Delta_i = d_i + \gamma_i^*, \quad \gamma_i^* \equiv \gamma_i(g_*), \tag{1.164}$$

where d_i are the known (Sec. 1.15) canonical dimensions $d_\varphi = (d-2)/2$ and $d_\tau = 2$. The additions γ_i^* are called the *critical anomalous dimensions*, often just anomalous dimensions, although the same term is also used for the RG functions $\gamma_i(g)$. What is meant will usually be clear from the context.

The two independent critical dimensions can be used with (1.19) and (1.40) or their consequences

$$2\alpha + \beta + \gamma = 2, \quad \beta\delta = \beta + \gamma, \quad d\nu = 2 - \alpha, \quad \gamma = \nu(2 - \eta) \tag{1.165}$$

to find all the other traditional critical indices of the model.

Knowledge of all the needed critical dimensions and the normalized scaling function (Sec. 1.5) corresponds to complete information about the critical behavior of a given quantity. The normalized scaling function is, by definition, independent of the normalization, in particular, of the factors C in (1.161), so that in our case it can also be determined from the function (1.163), which we shall call the *reduced scaling function*, in contrast to the full function with coefficients $C_{\Phi,z}$ in (1.161). The RG equation leads to the representation (1.163), which allows the reduced scaling function f to be expressed in terms of the initial quantity, in this case, the correlator (1.149), so that f can be calculated as an ε expansion (Sec. 1.36). It must be emphasized that (1.163) allows f to be found only with the use of auxiliary information — perturbation theory for the function on the right-hand side of (1.163). The RG equation itself never determines the explicit form of the scaling function, because in the general solution (1.148) of (1.142) the function of the first integrals on the right-hand side remains arbitrary. Actually, in the RG technique one actually always calculates the reduced scaling functions like (1.163), and they can always be used to find the universal normalized scaling functions for any particular normalization conditions.

It is useful to note that all the important information about the critical behavior of the correlator can be obtained simply by setting $g = g_*$ in the RG equation (1.151). Then the contribution involving $\beta(g)$ vanishes in it, all the RG functions $\gamma_i(g)$ are transformed into constants $\gamma_i(g_*) = \gamma_i^*$, and we obtain the equation of generalized scaling of the type (1.6), whose general solution is given by (1.159) with $C_\Phi = C_z = 1$. The loss of these normalization factors is not important (see above), and so the above procedure for constructing and analyzing the first integrals is actually necessary only in order to justify the following general *statement*: up to unimportant normalization factors, the IR asymptote of the renormalized Green functions for any value of g from the basin of attraction of an IR-stable fixed point g_* of the type (1.156) is exactly the same as for $g = g_*$. The RG equation written in canonically dimensionless variables becomes an equation of critical scaling of the type (1.6) for $g = g_*$. Its solution leads to a representation like (1.163) for the reduced scaling function, and the coefficients of the RG operator determine the critical dimensions of all quantities according to the general rule (1.6).

Let us explain the practical application of this statement in more detail. Let F be some multiplicatively renormalizable quantity $F = Z_F F_R$, where F is an unrenormalized object, F_R is the renormalized version, and Z_F is a dimensionless renormalization constant. The renormalization invariance of F in terms of F_R is expressed as in (1.107) by the RG equation:

$$[\mathcal{D}_{\mathrm{RG}} + \gamma_F] F_{\mathrm{R}} = 0, \qquad \gamma_F \equiv \widetilde{\mathcal{D}}_\mu \ln Z_F, \tag{1.166}$$

where the RG function γ_F is the anomalous dimension of F. Let $\{e_i\}$ be the set of all variables on which F_R depends. The canonical scale invariance is then expressed by an equation like (1.6):

$$[\sum_i d_i \mathcal{D}_i - d_F] F_{\mathrm{R}} = 0, \tag{1.167}$$

in which d_i are the canonical dimensions of the variables e_i, and d_F is the dimension of F (it is the same for F and F_R owing to the fact that Z_F is dimensionless).

If we now set $g = g_*$ everywhere, then (1.166) is also transformed to the form (1.6). Equations of this type describe scaling with the rescaling of variables, the derivatives with respect to which are contained in the operator (1.6). If we are interested in scale transformations in which a given variable e_i is not rescaled, then to obtain the required scaling equation the contribution with the given $\mathcal{D}_i$ must be eliminated by combining the available equations. For critical scaling the eliminated variable is μ. From (1.166) and (1.167) with $g = g_*$ we then obtain a single equation of the type (1.6). The coefficients of the $\mathcal{D}_i$ remaining in it will be the desired critical dimensions of the variables e_i, and the quantity analogous to Δ_W in (1.6) will be the critical dimension of F.

As an example, let us consider the correlator (1.149). For $g = g_*$ the corresponding RG equation (1.150) with the operator (1.122) and the canonical scaling equation of the type (1.167) have the form

$$[\mathcal{D}_\mu - \gamma_\tau^* \mathcal{D}_\tau + 2\gamma_\varphi^*] D_{\mathrm{R}}(p, g_*, \tau, \mu) = 0, \tag{1.168a}$$

$$[\mathcal{D}_\mu + \mathcal{D}_p + 2\mathcal{D}_\tau + 2] D_{\mathrm{R}}(p, g_*, \tau, \mu) = 0, \tag{1.168b}$$

where we have used the fact that the canonical dimension of (1.149) is -2. Eliminating $\mathcal{D}_\mu$ from the system (1.168), we obtain

$$[\mathcal{D}_p + (2 + \gamma_\tau^*)\mathcal{D}_\tau + 2 - 2\gamma_\varphi^*] D_{\mathrm{R}}(p, g_*, \tau, \mu) = 0. \tag{1.169}$$

This is the desired equation of critical scaling. The coefficients of $\mathcal{D}_p$ and $\mathcal{D}_\tau$ are the critical dimensions of the corresponding variables, and $-2 + 2\gamma_\varphi^*$ is the critical dimension of D_R. The general solution of the system (1.168) is given by (1.161)–(1.163) with $C_\Phi = C_z = 1$.

1.34 Conditions for reaching the critical regime. Corrections to scaling

In obtaining the asymptote (1.159) we discarded the corrections of order $\bar{g} - g_*$ denoted by the ellipsis in (1.157). Their asymptote for $s = p/\mu \to 0$ can easily be found from (1.144) and (1.156):

$$\bar{g}(s, g) - g_* \cong \text{const} \cdot s^\omega. \tag{1.170}$$

These quantities, which are small for $s \to 0$, enter into the exponents (Secs. 1.30 and 1.33), and so they have the meaning of "relative" corrections, i.e., they correspond to correction factors $1 + \text{const} \cdot s^\omega$ in the leading contributions.

Smallness of the relative corrections (1.170) is ensured only by smallness of the momentum and does not require smallness of τ. However, to actually reach the critical regime the corrections must be not relatively, but absolutely small. If there is no reason to assume that the leading contribution is large, as, for example, for Φ in (1.159a), then it can be assumed that relative smallness of the corrections corresponds to absolute smallness. However, the situation regarding the invariant variable $\bar{z}$ in (1.152) is different. Its leading contribution (1.159b) is proportional to the combination $\tau p^{-\Delta_\tau}$, which for $p \to 0$ and fixed τ diverges. If we did not know the asymptotes of the function $\Phi(1, \bar{g}, \bar{z})$ at $\bar{z} \to \infty$, we would not even be justified in replacing the argument $\bar{z}$ by its asymptotic expression (1.159b). [A simple example: $\sin(1000)$ and $\sin(1001)$ differ by one hundred percent, although the relative difference in the arguments is only 0.001.] In our case there is no such problem, because the asymptote of the correlator for $p \to 0$, $\tau = \text{const}$ is known and trivial: it is completely characterized as "analytic in p^2 at fixed τ."

Nontrivial critical scaling corresponds to an asymptote at which the argument of the scaling function (1.161) is a quantity of order (or smaller than) unity, which implies the simultaneous and mutually consistent smallness of p and τ: $p \to 0$, $\tau \sim p^{\Delta_\tau} \to 0$. Clearly, this completely corresponds to the general ideas of critical scaling (Sec. 1.3).

The following *statement* represents the generalization of these ideas to an arbitrary model. Nontrivial critical scaling corresponds to the region of parameter values in which the invariant charge is sufficiently close to a fixed point, while all the other invariant variables are quantities of order (or smaller than) unity. The first is ensured by sufficiently small scaling variable, and the second by consistent (via the critical dimensions) smallness of the other variables.

We note that in some models, variables like $\bar{z}$ are generated not only by the mass parameters, but also by additional dimensional coupling constants. For such $\bar{z}$ the asymptote of Φ for $\bar{z} \to \infty$ is not trivial, as in the case studied above, but, on the contrary, is nontrivial and unknown. Actually, $\bar{z} \to \infty$ implies that this interaction is more important than the one treated as the fundamental interaction, since the fundamental invariant charge $\bar{g} \leq g_*$ remains finite. A detailed analysis of such a situation for the example of the model (1.71) is performed in Ch. 4.

It should also be noted that the criterion for reaching the critical regime was formulated above completely in terms of the invariant variables. It is therefore renormalization-invariant (i.e., insensitive to the choice of the arbitrary parameter μ) owing to the statement at the end of Sec. 1.32.

Let us return to the corrections to scaling. The relative corrections of order (1.170) are related to the nonexactness of reaching the critical asymptote in the model in question. They are characterized by the correction critical exponent ω (Sec. 1.3) defined in (1.156), which is easily calculated as an ε expansion in terms of the known β function.

We can also consider corrections of another type, related to the inaccuracy of the model itself owing to all the additional interactions that are IR-irrelevant compared to the fundamental one (Sec. 1.16) and that were discarded when constructing the model. The new exponents ω corresponding to these corrections (Sec. 1.3) can also be calculated as ε expansions using the technique of renormalization of composite operators, which will be discussed in later chapters.

1.35 What is summed in the solution of the RG equation?

In renormalized perturbation theory the object studied, for example, the correlator (1.149), is constructed as a series in g, and the representation (1.152), which in the critical region implies (1.161), is the result of an infinite resummation of the contributions of this series. At first glance it appears that at small ε and $g \sim \varepsilon$ (Sec. 1.27) such resummation is not necessary owing to the smallness of the expansion parameter $g \sim \varepsilon$ and the absence of poles in ε in the coefficients of the series (because the object is renormalized). However, in the critical region this is actually not true. Calculations show that the coefficients of the series contain factors like $(s^{-2\varepsilon} - 1)/\varepsilon$ [an example is the series expansion of (1.154)], which are UV-finite (i.e., finite in the limit $\varepsilon \to 0$, $s = \mathrm{const}$) and of order unity for $s \cong 1$, but become of order $1/\varepsilon$ or larger for $|\ln s| \gtrsim 1/\varepsilon$, i.e., for sufficiently small s. The maximum number of such "large" factors $\sim 1/\varepsilon$ in the terms of the perturbation series never exceeds the number of "small" factors $g \sim \varepsilon$, but can be equal to it or differ from it only by a fixed finite number.

This means that at small s we actually have a new parameter

$$\xi \equiv g(s^{-2\varepsilon} - 1)/\varepsilon \gtrsim 1, \qquad (1.171)$$

all powers of which must be summed when contributions of the same order in ε are collected. This summation is accomplished by the representation (1.152), i.e., the solution of the RG equation allowing the construction of the answer as a series in ε even when the factors (1.171) are present. To avoid misunderstandings, we note that here we are dealing not with universal (g-independent) ε expansions, as for exponents, but only with the classification of the orders in ε for $g \sim \varepsilon$.

To justify these statements, let us first consider the integrals in (1.152) and (1.153). The RG functions entering into them are series (1.121) with coefficients $\gamma_{na} \cong 1$, since they do not depend on any of the parameters. Therefore, for $g \sim \varepsilon$ these series are actually (nonuniversal) ε expansions, which allows the analogous expansions to be made for all the quantities constructed from them. For the β function, for example, for $g \sim \varepsilon$ the one-loop approximation (1.128a) is of lowest order in ε. It corresponds to the approximation (1.154) for the invariant charge. Expanding (1.154) in a perturbation series in g, we obtain a geometric progression containing powers of ξ, all the terms of which are of the same order in ε for $\xi \cong 1$. To obtain the first correction in ε to the approximation (1.154) it is necessary to include as a perturbation in (1.144) the first correction $\beta_3 g^3 \sim \varepsilon^3$ in the expansion of the β function (1.121), and so on.

It is necessary to sum all powers of the parameter (1.171) already in the precritical region $|\ln s| \cong 1/\varepsilon$, where $\xi \cong 1$. This is not yet the critical region, since for $\xi \cong 1$ the deviation of the invariant charge (1.154) from its maximum value g_* is a quantity of order g_*. In the critical region we must have $|\bar{g} - g_*| \ll g_*$, which is ensured by the condition $|\ln s| \gg 1/\varepsilon$, from which it follows that $|\xi| \gg 1$, so that summation of powers of ξ is essential.

Let us now consider the scaling function $\Phi(1, \bar{g}, \bar{z})$ in (1.152). In it $s = 1$, and so the factors (1.171) are absent, and only the analogous large factors from possible singularities in the argument $\bar{z}$ are dangerous. For all ordinary models like the φ^4 model such singularities are present only for $\bar{z} \to 0$ or $\bar{z} \to \infty$ and are absent for $\bar{z} \cong 1$. Therefore, in the critical region $s \ll 1$, $\bar{z} \cong 1$, i.e., for simultaneously small p and $\tau \sim p^{\Delta_\tau}$ (Sec. 1.34) the usual perturbation series in powers of g for Φ, which becomes the universal ε expansion when the substitutions $g \to \bar{g} \to g_*$ are made in (1.152) and (1.163), will not have dangerous factors of order $1/\varepsilon$ in the series coefficients.

In summary, the solution of the RG equation allows all dangerous singularities like (1.171) in the scaling variable to be grouped together in invariant variables like $\bar{z}$ and $\bar{g}$ with known simple behavior for $s \to 0$. The RG equation does not determine the dependence on the invariant variables, and so when singularities in invariant variables like $\bar{z}$ appear in scaling functions analogous to (1.171), their summation lies beyond the scope of the RG method.

In some cases the summation can be accomplished by other methods. For example, singularities like (1.171) in $\bar{z}$ for $\bar{z} \to 0$ in the scaling function in (1.152) can be summed by the Wilson operator expansion, which will be discussed in Ch. 3.

1.36 Algorithm for calculating the coefficients of ε expansions of critical exponents and scaling functions

These topics have already been discussed, and so here we limit ourselves to a brief description of the general scheme for the example of the φ^4 model.

The exponents for a given multiplicatively renormalizable model are calculated as follows. First, all the needed renormalization constants Z are calculated directly from the graphs as series in g. They are then used in the same form with the expressions of Sec. 1.25 to calculate all the RG functions. Critical scaling with critical exponents admitting ε expansion occurs only when the β function has an IR-stable fixed point $g_* \sim \varepsilon$. Its coordinate $g_*(\varepsilon)$ is then found as a series in ε from the known series (1.121) for the β function. Substitution of the ε expansion of g_* into the series (1.121) for $\gamma_a(g)$ gives the ε expansions of the anomalous critical dimensions $\gamma_a^* = \gamma_a(g_*)$. These are then used with expressions like (1.164) to find the total critical dimensions Δ_a of all the fundamental quantities a (for example, $a = \varphi, \tau$ for the φ^4 model). The other critical exponents (Sec. 1.5) are found from Δ_a using the general relations of the scaling hypothesis, the validity of which is guaranteed by the presence of the critical scaling following from the RG equation. The leading correction critical exponent ω (Sec. 1.3) is calculated as an ε expansion from the slope of the β function at the fixed point [see (1.156)]. In practice, it is of course possible to calculate only the first few terms of the series, as the calculations rapidly get more complicated with increasing order.

For reduced scaling functions like (1.163) the ε expansion is constructed directly from the definitions. In perturbation theory the original function Φ in (1.149) is calculated as a series (for clarity, we shall now explicitly indicate the ε dependence):

$$\Phi(s, g, z, \varepsilon) = \sum_{n=0}^{\infty} g^n \Phi_n(s, z, \varepsilon). \tag{1.172}$$

After expanding everything (i.e., g_* and Φ_n) in ε and grouping together contributions of the same order, this series with the replacements $s \to 1$ and $g \to g_*$ indicated in (1.163) becomes the desired ε expansion:

$$f(z, \varepsilon) = \Phi(1, g_*(\varepsilon), z, \varepsilon) = \sum_{n=0}^{\infty} \varepsilon^n f_n(z). \tag{1.173}$$

It is important that this object is renormalized, as this guarantees the absence in the coefficients Φ_n in (1.172) of poles in ε, which could cancel out the small factors $g_* \sim \varepsilon$. Therefore, to calculate a finite number of terms of the ε expansion (1.173) it is sufficient to keep the same number of first terms of the series (1.172), i.e., no infinite summations are necessary.

This scheme is universal and is based on two assumptions: (1) the multiplicative renormalizabiliy of the model, and (2) the existence of an IR-stable fixed point $g_* \sim \varepsilon$. In models like the φ^4 model with a single charge g, there can be only a single fixed point $g_* \sim \varepsilon$, and passage to the corresponding critical regime for sufficiently small s is always guaranteed by the condition $g < g_*$ (Sec. 1.27). In multicharge models with several g there are many more possibilities. We shall discuss them in detail in Sec. 1.42.

The results of calculating the ε expansions of the exponents of the φ^4 model are given in the following section, and the scaling functions are given in Ch. 4.

1.37 Results of calculating the ε expansions of the exponents of the O_n φ^4 model in dimension $d = 4 - 2\varepsilon$

The actual computational technique will be discussed in detail in the following chapters. Here as an illustration we present the final results, that is, all the presently known coefficients of the ε expansions of the critical exponents of the O_n φ^4 model. In our notation (1.86) $4-d \equiv 2\varepsilon$, but often $4-d \equiv \varepsilon$ is used; here, to facilitate comparisons, all the ε expansions will be given as series in powers of $4 - d \equiv 2\varepsilon$. In the following expressions $\zeta(z)$ is the Riemann zeta function; numerically, $\zeta(3) \cong 1.20205$, $\zeta(4) \cong 1.08232$, and $\zeta(5) \cong 1.03693$. The calculations give

$$\eta = 2\gamma_\varphi^* = \frac{(2\varepsilon)^2(n+2)}{2(n+8)^2} \left\{ 1 + \frac{(2\varepsilon)}{4(n+8)^2}[-n^2 + 56n + 272] \right.$$

$$+ \frac{(2\varepsilon)^2}{16(n+8)^4}[-5n^4 - 230n^3 + 1124n^2 + 17\,920n + 46\,144$$

$$- 384(n+8)(5n+22)\zeta(3)]$$

$$- \frac{(2\varepsilon)^3}{64(n+8)^6}[13n^6 + 946n^5 + 27\,620n^4 + 121\,472n^3 - 262\,528n^2$$

$$- 2\,912\,768n - 5\,655\,552 - \zeta(3)(n+8)16(n^5 + 10n^4 + 1220n^3$$

$$- 1136n^2 - 68\,672n - 171\,264) + \zeta(4)(n+8)^3 1152(5n+22)$$

$$\left. - \zeta(5)(n+8)^2 5120(2n^2 + 55n + 186)] \right\} + O(\varepsilon^6),$$

$$1/\nu = \Delta_\tau = 2 + \gamma_\tau^* = 2 - \frac{(2\varepsilon)(n+2)}{(n+8)}\left\{1 + \frac{(2\varepsilon)}{2(n+8)^2}[13n+44]\right.$$

$$+\frac{(2\varepsilon)^2}{8(n+8)^4}[-3n^3 + 452n^2 + 2672n + 5312$$

$$-96(n+8)(5n+22)\zeta(3)]$$

$$+\frac{(2\varepsilon)^3}{32(n+8)^6}[-3n^5 - 398n^4 + 12\,900n^3 + 81\,552n^2 + 219\,968n$$

$$+357\,120 + 1280(n+8)^2(2n^2+55n+186)\zeta(5)$$

$$-288(n+8)^3(5n+22)\zeta(4) - 16(n+8)(3n^4 - 194n^3$$

$$+148n^2 + 9472n + 19\,488)\zeta(3)]$$

$$-\frac{(2\varepsilon)^4}{128(n+8)^8}[3n^7 - 1198n^6 - 27\,484n^5 - 1\,055\,344n^4$$

$$-5\,242\,112n^3 - 5\,256\,704n^2 + 6\,999\,040n - 626\,688$$

$$-\zeta(3)(n+8)16(13n^6 - 310n^5 + 19\,004n^4 + 102\,400n^3$$

$$-381\,536n^2 - 2\,792\,576n - 4\,240\,640)$$

$$-\zeta^2(3)(n+8)^21024(2n^4 + 18n^3 + 981n^2 + 6994n + 11\,688)$$

$$+\zeta(4)(n+8)^348(3n^4 - 194n^3 + 148n^2 + 9472n + 19\,488)$$

$$+\zeta(5)(n+8)^2256(155n^4 + 3026n^3 + 989n^2 - 66\,018n - 130\,608)$$

$$-\zeta(6)(n+8)^46400(2n^2+55n+186) + \zeta(7)(n+8)^356\,448(14n^2$$

$$\left.+189n+526)]\right\} + O(\varepsilon^6),$$

$$\omega = (2\varepsilon) - \frac{(2\varepsilon)^2}{(n+8)^2}[9n+42] + \frac{(2\varepsilon)^3}{4(n+8)^4}[33n^3 + 538n^2 + 4288n$$

$$+9568 + 96(n+8)(5n+22)\zeta(3)]$$

$$-\frac{(2\varepsilon)^4}{16(n+8)^6}[-5n^5 + 1488n^4 + 46\,616n^3 + 419\,528n^2 + 1\,750\,080n$$

$$+2\,599\,552 + 96(n+8)(63n^3 + 548n^2 + 1916n + 3872)\zeta(3)$$

$$-288(n+8)^3(5n+22)\zeta(4) + 1920(n+8)^2(2n^2+55n+186)\zeta(5)]$$

$$+\frac{(2\varepsilon)^5}{64(n+8)^8}[13n^7 + 7196n^6 + 240\,328n^5 + 3\,760\,776n^4$$

$$+38\,877\,056n^3 + 223\,778\,048n^2 + 660\,389\,888n + 752\,420\,864$$

$$-\zeta(3)(n+8)16(9n^6 - 1104n^5 - 11\,648n^4 - 243\,864n^3$$

$$-2\,413\,248n^2 - 9\,603\,328n - 14\,734\,080)$$

$$-\zeta^2(3)(n+8)^2768(6n^4 + 107n^3 + 1826n^2 + 9008n + 8736)$$

$$-\zeta(4)(n+8)^3288(63n^3 + 548n^2 + 1916n + 3872)$$

$$+\zeta(5)(n+8)^2256(305n^4 + 7386n^3 + 45\,654n^2 + 143\,212n$$

$$+226\,992) - \zeta(6)(n+8)^49600(2n^2 + 55n + 186)$$

$$+\zeta(7)(n+8)^3112\,896(14n^2 + 189n + 526)] + O(\varepsilon^6).$$

The two anomalous dimensions γ_φ^* and γ_τ^* can be used with (1.165), (1.19), and (1.40) to find all the traditional critical exponents. They can be obtained from the above expansions through order ε^5. However, here we shall give them only through order ε^3:

$$\alpha = \frac{(2\varepsilon)(4-n)}{2(n+8)} - \frac{(2\varepsilon)^2(n+28)(n+2)^2}{4(n+8)^3} - \frac{(2\varepsilon)^3(n+2)}{8(n+8)^5}[n^4 + 50n^3$$

$$+920n^2 + 3472n + 4800 - 96(n+8)(5n+22)\zeta(3)] + O(\varepsilon^4),$$

$$\beta = \frac{1}{2} - \frac{3(2\varepsilon)}{2(n+8)} + \frac{(2\varepsilon)^2(n+2)(2n+1)}{2(n+8)^3} + \frac{(2\varepsilon)^3(n+2)}{8(n+8)^5}[3n^3$$

$$+128n^2 + 488n + 848 - 24(n+8)(5n+22)\zeta(3)] + O(\varepsilon^4),$$

$$\gamma = 1 + \frac{(2\varepsilon)(n+2)}{2(n+8)} + \frac{(2\varepsilon)^2(n+2)}{4(n+8)^3}[n^2 + 22n + 52]$$

$$+\frac{(2\varepsilon)^3(n+2)}{8(n+8)^5}[n^4 + 44n^3 + 664n^2 + 2496n + 3104$$

$$-48(n+8)(5n+22)\zeta(3)] + O(\varepsilon^4),$$

$$\delta = 3 + (2\varepsilon) + \frac{(2\varepsilon)^2}{2(n+8)^2}[n^2 + 14n + 60]$$

$$+\frac{(2\varepsilon)^3}{4(n+8)^4}[n^4 + 30n^3 + 276n^2 + 1376n + 3168] + O(\varepsilon^4),$$

$$\nu = \frac{1}{2} + \frac{(2\varepsilon)(n+2)}{4(n+8)} + \frac{(2\varepsilon)^2(n+2)}{8(n+8)^3}[n^2 + 23n + 60]$$

$$+\frac{(2\varepsilon)^3(n+2)}{32(n+8)^5}[2n^4 + 89n^3 + 1412n^2 + 5904n + 8640$$

$$-96(n+8)(5n+22)\zeta(3)] + O(\varepsilon^4).$$

The contributions of order ε and ε^2 were obtained already in Wilson's first study (1972) [41], and the contributions of order ε^3 in all the exponents and ε^4 in η were obtained in [63] (1973). The ε^4 contributions in the other exponents were found in [64] (1979), and the ε^5 contribution to η in [65] (1981). Later on, in [66] (1983) the ε^5 contributions to the other exponents were calculated. Some of the five-loop graphs were evaluated numerically in those studies. It was later shown [67] that they can all in fact be calculated exactly as zeta functions with rational coefficients (but this is not guaranteed for six loops). It was then shown [68] that there are errors in [66]. The correct answers were found, and these are given above.

1.38   Summation of the ε expansions. Results

For a clear evaluation of the convergence of the ε expansions, in Table 1.5 we give the coefficients of the powers of $2\varepsilon = 4 - d$ for several exponents of the $O_n \, \varphi^4$ model with $n = 1$.

Table 1.5   Numerical values of the coefficients of powers of $2\varepsilon = 4 - d$ for several exponents of the $O_n \, \varphi^4$ model for $n = 1$.

	1	2ε	$(2\varepsilon)^2$	$(2\varepsilon)^3$	$(2\varepsilon)^4$
η	0	0	0.0185	0.0187	-0.0083
$1/\nu$	2	-0.3333	-0.1173	0.1245	-0.3069
γ	1	0.1667	0.0772	-0.0490	0.1436
ω	0	1	-0.63	1.62	-5.24

If these numbers are compared with the results (assumed reliable) of calculating the exponents by the method of high-temperature expansions for the three-dimensional Ising model (see Table 1.2 in Sec. 1.5), we arrive at the following conclusions about the applicability of the ε expansions for the real problem with $2\varepsilon = 1$: the first corrections to the canonical dimensions have the required sign and reasonable order of magnitude (for example, the susceptibility exponent γ including ε and ε^2 contributions for $2\varepsilon = 1$ coincides almost exactly with the known value $\gamma \cong 1.25$). However, then the contributions of the terms of the ε expansion begin to grow and the agreement deteriorates. Therefore, the direct summation of more and more terms of the ε expansions is meaningless.

This behavior is characteristic of divergent asymptotic expansions. It is easy to understand qualitatively: ε expansions are constructed from ordinary perturbation series, and the divergence of the latter for interactions of the type $g\varphi^{2m}$ is a well-known fact. The standard argument is that the behavior of the system, and, consequently, the results, must qualitatively change when the sign of g is changed, which makes it impossible to have absolute convergence of the series guaranteeing analyticity in some circle in the complex g plane.

Viewing the perturbation series as an invalid expansion of some well de-fined function, we can seek ways of summing the formal series that lead to the exact answer. The technique of Borel summation, first used to calculate the critical exponents in [69], has become quite popular. To use this procedure it is necessary to know the first few terms (the more the better) of the expansion of the desired quantity and the asymptote of the higher orders for the general term of the series. A technique for calculating this asymptote was proposed in [46]. It has been used for the ε expansions $f(\varepsilon) = \sum_k f_k(2\varepsilon)^k$ of the critical exponents of the $O_n\,\varphi^4$ model to obtain the following asymptote [47]:

$$f_k \underset{k\to\infty}{\cong} \text{const} \cdot k!(-a)^k k^b, \qquad a = \tfrac{3}{n+8}, \tag{1.174}$$

where f are the various exponents, n is the number of field components, and the constant b is $3+n/2$ for the exponent η, $4+n/2$ for $1/\nu$, and $5+n/2$ for ω.

The procedure of Borel summation is equivalent to a rearrangement of the original divergent series to form a convergent one. Estimates like (1.174) are necessary to establish the possibility of such a rearrangement and to fix the auxiliary parameters used in it. The more terms of the series that are known, the more accurate is the calculation of the sum of the rearranged convergent series. The coefficients of the latter are expressed in terms of the coefficients of the original series and auxiliary fitted parameters, which are usually chosen from certain optimization conditions when several series are dealt with simultaneously. The presence of fitted parameters and the arbi-trariness in their optimization make the Borel summation procedure rather delicate. Reliability of the results is ensured by application of the procedure to exactly solvable problems with known answers, various stability tests, and, most important, the simultaneous processing of several series.

Table 1.6 Critical exponents of the $O_n\,\varphi^4$ model ($d = 3$), obtained by Borel summation of the 5-loop ε expansions.

n	0	1	2	3
γ	1.157 ± 0.003	1.2390 ± 0.0025	1.315 ± 0.007	1.390 ± 0.010
ν	0.5880 ± 0.0015	0.6310 ± 0.0015	0.671 ± 0.005	0.710 ± 0.007
η	0.0320 ± 0.0025	0.0375 ± 0.0025	0.040 ± 0.003	0.040 ± 0.003
β	0.3035 ± 0.0020	0.3270 ± 0.0015	0.3485 ± 0.0035	0.368 ± 0.004
ω	0.82 ± 0.04	0.81 ± 0.04	0.80 ± 0.04	0.79 ± 0.04

We shall not discuss the summation technique in more detail, as this topic is covered in the original studies and in the book by Zinn-Justin [45]. Instead, we shall compare the results obtained by this method with the other available data. In the original study [69] the series were summed directly in dimension $d = 3$ (see Sec. 1.41). The ε expansions were summed in [64], [66], [70]. All the data given in this section were taken from [45] with that author's corrections of misprints in Table 1.8.

The data given in Table 1.6 can be compared with the results obtained for three-dimensional lattices using high-temperature expansions (Sec. 1.1) and other numerical methods of the same type (Table 1.7), and also with the experimental data (Table 1.8).

Table 1.7   Critical exponents of three-dimensional O_n-symmetric lattice models (high-temperature expansions, etc.).

n	0	1	2	3
γ	1.161 ± 0.002	1.2385 ± 0.0025	1.33 ± 0.02	1.38 ± 0.02
ν	0.592 ± 0.003	0.6305 ± 0.0015	0.672 ± 0.007	0.715 ± 0.020
α	0.25	0.105 ± 0.010	0.00 ± 0.02	-0.08 ± 0.04
β		0.312 ± 0.005		0.38 ± 0.03
$\omega\nu$	0.465	0.52 ± 0.07	0.60 ± 0.08	0.54 ± 0.10

Table 1.8   Experimental values of the critical exponents for several three-dimensional systems: binary mixtures (1), the liquid–gas transition (2), the isotropic ferromagnet (3), He$_4$ (4), polymers (5).

	γ	ν	β	α	$\omega\nu$
1	1.236 ± 0.008	0.625 ± 0.010	0.325 ± 0.005	0.110 ± 0.005	0.50 ± 0.03
2	$1.23 - 1.25$	0.625 ± 0.006	$0.316 - 0.327$	$0.101 - 0.116$	0.50 ± 0.03
3	1.40 ± 0.03	$0.700 - 0.725$	0.35 ± 0.03	$-0.09 - -0.12$	0.54 ± 0.10
4		0.672 ± 0.001		-0.013 ± 0.003	
5		0.586 ± 0.004			

In Table 1.8 we give the averaged experimental values of the exponents for several specific systems ($d = 3$) whose critical behavior should be described by the O_n φ^4 model: the critical demixing point in binary mixtures and the liquid–gas transition ($n = 1$), the isotropic ferromagnet ($n = 3$), the transition to the superfluid phase of He$_4$ ($n = 2$), and the analog of the exponent ν for polymers ($n = 0$). We give only those exponents that can be measured directly in an experiment. The first two systems listed belong to the same universality class as the Ising model. Therefore, their exponents should be identical, which is confirmed by experiment.

Borel summation of the ε expansions of the exponents of the O_n φ^4 model has also been done for two-dimensional systems ($d = 2$). Here only the cases $n = 0, 1$ are of interest, since for $n \geq 2$ the exponents must be canonical (on the basis of the $2 + \varepsilon$ and $1/n$ expansions; see Ch. 4). For two-dimensional systems with continuous symmetry group, in particular, for the O_n φ^4 model with $n \geq 2$, the general Mermin–Wagner theorem [71] forbids a transition to the phase with $\langle \widehat{\varphi} \rangle \neq 0$. The case $n = 0$ in this model corresponds to polymers, and $n = 1$ corresponds to the two-dimensional Ising model. For the latter all the fundamental exponents are known exactly (the Onsager solution). However, no nonanalytic corrections to scaling have been found there at all (in any case, in the objects studied), and so the traditional methods cannot be used to determine the correction exponent ω. The hypothetical value $\omega = 4/3$ for this model is based on the new technique of two-dimensional conformal theories, where the value $\omega = 4/(m + 1)$ is predicted for models of the φ^{2m} family ($d = 2$, $n = 1$, $m \geq 2$) [45]. The same technique has been used to exactly determine all the fundamental exponents of the two-dimensional polymer problem (i.e., the O_n φ^4 model with $n = 0$), but the exponent ω remains unknown (the other exponents were found earlier in [72] by reduction of the problem to a Coulomb gas). In Table 1.9 we give the results for these two systems.

Table 1.9 Critical exponents of the two-dimensional O_n φ^4 model with $n = 0, 1$, obtained by Borel summation of the 5-loop ε expansions, and the corresponding exact values.

	γ	ν	η	β	ω
$n = 0$	1.39 ± 0.04	0.76 ± 0.03	0.21 ± 0.05	0.065 ± 0.01	1.7 ± 0.2
exact	$43/32 = 1.34375$	$3/4 = 0.75$	$5/24 \simeq 0.2083$	$5/64 \simeq 0.0781$	?
$n = 1$	1.73 ± 0.06	0.99 ± 0.04	0.26 ± 0.05	0.120 ± 0.015	1.6 ± 0.2
exact	$7/4 = 1.75$	1	$1/4 = 0.25$	$1/8 = 0.125$	$4/3 \simeq 1.33$ (?)

Analysis of the full set of data presented here leads to the conclusion that the theoretical predictions obtained by the various methods agree fairly well with each other and with experiment. This agreement is all the more impressive when we consider the results, systematized in [45], of similar calculations of various universal ratios of amplitudes (Sec. 1.5) and the critical exponents calculated in the "real" dimension (Sec. 1.41). This all indicates that the RG approach is effective on the whole and that the Borel technique for summing series is fairly reliable when properly used.

1.39 The RG equation for $\Gamma(\alpha)$ (the equation of state)

We have studied the RG equation for the correlator in zero field as a basic example. The RG analysis can also be applied to other objects, such as quantities entering into the equation of state. We immediately note that the critical dimensions of fundamental quantities obtained from the RG analysis of any particular object characterize the model as a whole, i.e., they do not change in going from one object to another in the same model — only the corresponding scaling functions will be new. This is explained by the fact that all the RG equations for specific objects like (1.150) follow from the single RG equation (1.110) or its equivalent (1.111) for functionals containing complete information about all the Green functions.

Let $W_R(A)$ and $\Gamma_R(\alpha)$ be the generating functionals of the renormalized connected and 1-irreducible Green functions of the model defined in the usual manner (Sec. 1.11). When they are restricted to the set of homogeneous (x-independent) arguments A and α, a factor corresponding to the infinite volume of the system $V = \int dx$ is isolated from each functional; the coefficients of this factor, i.e., the specific quantities, will be denoted as $\overline{W}$ and $\overline{\Gamma}$. In magnet terminology, the numerical parameter A is the uniform external field, α is the magnetization, $\overline{W}_R(A)$ is the specific logarithm of the partition function (1.12), $\overline{\Gamma}_R(\alpha)$ is its Legendre transform with respect to A, and the equation

$$\partial \overline{\Gamma}_R(\alpha)/\partial \alpha = -A, \tag{1.175}$$

following from (1.51), is called the equation of state (Sec. 1.4). In the O_n-symmetric model, A and α carry the appropriate indices.

The form of the general equations (1.110) and (1.111) for the functionals remains the same also for the specific values with the natural replacement of the functional operators (1.112) by the ordinary operators $\mathcal{D}_A = A\partial_A$ and $\mathcal{D}_\alpha = \alpha\partial_\alpha$. In particular, from (1.111) we find that

$$[\mathcal{D}_{RG} - \gamma_\varphi \mathcal{D}_\alpha]\overline{\Gamma}_R(\alpha, e, \mu) = 0. \tag{1.176}$$

This is all true for any multiplicatively renormalizable model. Let us write out (1.176) specifically for the φ^4 model, in which $e = \tau$, g and the operator $\mathcal{D}_{RG}$ has the form (1.122). The canonical dimensions of all the quantities are known: $d[\overline{W}_R] = d[\overline{\Gamma}_R] = d$ by definition, $d_\alpha = d_\varphi = \frac{d}{2} - 1$, and $d_A = d - d_\varphi$ (Sec. 1.15). Therefore,

$$\overline{\Gamma}_R(\alpha, g, \tau, \mu) = \mu^d \Phi(s, g, z), \quad s \equiv \tau\mu^{-2}, \quad z \equiv \alpha\mu^{-d_\varphi}. \tag{1.177}$$

We use the same letters as in (1.149) to denote the dimensionless variables, but now they have a different meaning. The fundamental scaling variable is now the temperature τ rather than the momentum. Upon substituting (1.177) into (1.176), we obtain the equation for Φ with $\mathcal{D}_\alpha \rightarrow \mathcal{D}_z$, and in the operator (1.122) we make the replacements $\mathcal{D}_\mu \rightarrow d - 2\mathcal{D}_s - d_\varphi\mathcal{D}_z$

and $\mathcal{D}_\tau \to \mathcal{D}_s$. Finally, for Φ we find the expression

$$[\beta \partial_g - (2 + \gamma_\tau)\mathcal{D}_s - (d_\varphi + \gamma_\varphi)\mathcal{D}_z + d]\Phi(s, g, z) = 0. \tag{1.178}$$

Dividing by $2 + \gamma_\tau$, we arrive at an equation like (1.142), in which the β function (see Sec. 1.32) is the quantity $\beta/(2 + \gamma_\tau)$ in the notation of (1.178). Therefore, the corresponding invariant charge (1.143) is now given by

$$\ln s = \int_g^{\bar{g}} dx \frac{2 + \gamma_\tau(x)}{\beta(x)}, \quad \bar{g} = \bar{g}(s, g), \quad s = \tau \mu^{-2}. \tag{1.179}$$

The general solution of (1.178) is obtained from the expressions of Sec. 1.30 by simple replacements of the RG functions. Its substitution into (1.177) and (1.175) gives the RG form of the equation of state.

If we are interested only in the critical regime, all the essential information can, as explained in Sec. 1.33, be obtained by the simple replacement $g \to g_*$ in (1.176). For the φ^4 model this gives

$$[\mathcal{D}_\mu - \gamma_\tau^* \mathcal{D}_\tau - \gamma_\varphi^* \mathcal{D}_\alpha]\overline{\Gamma}_{\mathrm{R}}(\alpha, g_*, \tau, \mu) = 0. \tag{1.180}$$

Adding an equation like (1.6) for the canonical dimensions

$$[\mathcal{D}_\mu + 2\mathcal{D}_\tau + d_\varphi \mathcal{D}_\alpha - d]\overline{\Gamma}_{\mathrm{R}}(\alpha, g_*, \tau, \mu) = 0 \tag{1.181}$$

and then eliminating $\mathcal{D}_\mu$, we obtain the desired equation of critical scaling:

$$[\Delta_\tau \mathcal{D}_\tau + \Delta_\varphi \mathcal{D}_\alpha - d]\overline{\Gamma}_{\mathrm{R}}(\alpha, g_*, \tau, \mu) = 0 \tag{1.182}$$

with the former critical dimensions (1.164). The only fundamentally new quantity is the reduced scaling function for Φ from (1.177) analogous to (1.163).

When obtaining the critical asymptote for the integrals in (1.147) and (1.148) using (1.144) and (1.157), it should be remembered that β and γ were used to denote the RG functions of the canonical equation (1.142). Therefore, in γ and γ^* in (1.157) we must include the factors appearing when the original RG equation is reduced to the canonical form (1.142), for example, the division by $2 + \gamma_\tau$ in (1.178).

In conclusion, we note that for small ε in the region $g \sim \varepsilon$ the quantity $2 + \gamma_\tau(g)$ is certainly positive, and so dividing (1.178) by it does not change the coordinate of the fixed point $g_* \sim \varepsilon$ or the signs of all quantities.

1.40 Subtraction-scheme independence of the critical exponents and normalized scaling functions

As noted in Sec. 1.18, the renormalized theory is not determined uniquely from the original theory. In particular, the use of different subtraction schemes

(Sec. 1.22) to calculate the renormalization constants leads to different renormalized theories. However, they are not independent, but are related to each other by UV-finite renormalization transformations like (1.78). In the formal scheme of dimensional regularization without Λ (Sec. 1.20) these will be the transformations

$$e \to e' = e'(e,\mu), \quad \widehat{\varphi}_R \to \widehat{\varphi}'_R = \widehat{\varphi}_R Z'_\varphi(e,\mu) \tag{1.183}$$

(possible ε dependence is understood everywhere) of the renormalized parameters and field with UV-finite, i.e., not containing poles in ε, functions $e'(e,\mu)$ and $Z'_\varphi(e,\mu)$. In the language of generating functionals, the transformations (1.183) correspond to

$$W_R(A,e,\mu) \to W'_R(A,e',\mu) = W_R(Z'_\varphi A, e, \mu). \tag{1.184}$$

For given functions e' and Z'_φ the RG equation for W'_R can be obtained by replacement of the variables from the RG equation (1.110) for W_R, and vice versa. In the broad sense of the word, (1.110) is a linear, homogeneous, first-order partial differential equation in all the variables, including the functional argument A, and it remains such for any change of variables. For nonsingular changes of variable the new equation is equivalent to the old one, i.e., the amount of information it contains remains the same. However, the explicit form of the operator in (1.110) depends on the choice of subtraction scheme: it is simpler in some than in others. The MS scheme (Sec. 1.22) is convenient precisely because the operator in (1.110) has a very simple form: all the RG functions (1.113) depend only on g, and so for $g \to g_*$ in the critical regime (Sec. 1.33) (1.110) is immediately transformed into an equation like (1.6). The form of the RG equation can become considerably more complicated after the replacement (1.184), which makes its analysis difficult. If the arbitrariness of the replacement is not restricted, it can, of course, even affect the critical dimensions. For example, if in the φ^4 model we set $\widehat{\varphi}'_R = (\tau\mu^{-2})^\alpha \widehat{\varphi}_R$, the critical dimension of the field changes: $\Delta_{\varphi'} = \Delta_\varphi + \alpha\Delta_\tau$. Obviously, such manipulations do not affect the essence of the problem, because the change of critical dimension is generated simply by the singularity of the coefficient. In general, analysis of the complicated RG equations obtained in non-MS schemes is meaningless: once they are reduced by changes of variable to the simple RG equation of the MS scheme, the information on the critical behavior obtained from the latter is objective and definitive up to possible changes of variables in the expressions for critical scaling. The variables which have definite critical dimensions are "correct" in the sense of the definitions of Sec. 1.3. Therefore, when no dimensions coincide accidentally, the correct variables are determined by this property almost uniquely and independently of which variables were used originally. The arbitrariness reduces only to normalization factors like C in (1.161) and is not reflected in the exponents and normalized scaling functions (Sec. 1.5).

Therefore, if in calculations using the MS scheme it is shown that a given physical quantity has a definite critical dimension, the same result (existence of a definite dimension and its specific value) must be obtained for this quantity when any other subtraction scheme is used for the calculations. However, it should also be remembered that sometimes the same letter denotes different physical quantities in different schemes (for example, the quantity traditionally called τ in the MS scheme stands for $T - T_c$, while in another popular scheme, that of subtractions at zero momenta, it stands for the inverse susceptibility; see Ch. 3 for more details). Of course, these different quantities have different critical dimensions.

Let us explain these general statements in more detail using two simple examples. First, we consider a UV-finite change of the charge $g \to g'$, i.e., $g' = g'(g)$ and vice versa $g = g(g')$, with possible ε dependence understood. For this change, in the RG equation we have $\gamma(g) \to \gamma(g(g')) \equiv \gamma'(g')$ and $\beta(g)\partial_g \to \beta(g(g'))(\partial g'/\partial g)\partial_{g'} \equiv \beta'(g')\partial_{g'}$. Monotonicity is required for the change to be correct, and so the derivative $\partial g'/\partial g$ will be assumed to be strictly positive. It is then obvious that the fixed points in the variables g and g' are in one-to-one correspondence with each other, and the critical anomalous dimensions are invariant: $g_* = g(g'_*)$ and $\gamma'(g'_*) = \gamma(g_*)$. The latter is also true for the exponent ω: using $\beta(g_*) = 0$ and the expression obtained above for $\beta'(g')$, we have

$$\omega' = \left.\frac{\partial\beta'(g')}{\partial g'}\right|_{g'_*} = \left.\frac{\partial}{\partial g'}\left[\beta(g(g'))\frac{\partial g'}{\partial g}\right]\right|_{g'_*} = \left.\frac{\partial\beta(g)}{\partial g}\right|_{g_*} = \omega. \tag{1.185}$$

As the second example, we consider an arbitrary UV-finite (i.e., with UV-finite Z) renormalization:

$$\tau' = \tau Z'_\tau(g), \qquad \widehat{\varphi}'_R = \widehat{\varphi}_R Z'_\varphi(g) \tag{1.186}$$

with the old charge g left unchanged. It generates additions to the anomalous dimensions (1.113): $\gamma_a \to \gamma_a + \gamma'_a$, where $a = \tau, \varphi$ and

$$\gamma'_a(g) = \beta(g)\partial_g \ln Z'_a(g) \tag{1.187}$$

according to (1.115). At the fixed point $g = g_*$ the quantities (1.187) vanish because $\beta(g_*) = 0$, and the coefficient of the β function in (1.187) does not have a pole that cancels the zero of the β function owing to the assumed UV-finiteness of the constants Z'_a. In fact, without loss of generality these constants can be assumed to be series in g beginning at unity whose coefficients do not contain poles in ε. For $g = g_* \sim \varepsilon$ these series are transformed into ε expansions beginning at unity without any infinite summations [owing to the absence of the combination g/ε characteristic of the constants (1.104)], and so at the point $g = g_* \sim \varepsilon$ the constants $Z'_a(g)$ cannot have either zeros (since there are no contributions competing with unity) or poles (since there are no infinite summations). The latter is true also for the derivatives $\partial_g \ln Z'_a(g)$.

This is the principal difference between the UV-finite constants Z'_a and the UV-divergent constants (1.104), whose logarithmic derivatives must have a pole at $g = g_*$ which cancels the zero of the β function in (1.118). It is easiest to verify the presence of the pole from the representation (1.123). The denominator is, according to (1.114), a multiple of the β function, and so it has a zero at the point $g = g_*$, while the numerator γ_a cannot contain a zero at this point, because g_* depends on ε, while the RG functions γ_a do not (Sec. 1.25). It is the absence of the zero in the numerator, and not the representation (1.123) itself, which plays the decisive role: (1.187) leads to representations analogous to (1.123) also for the constants (1.186), but in them a zero of the denominator will necessarily be canceled by a zero of the numerator.

Here we should stress the role of the renormalizability of the model. In proving the existence of the needed pole in (1.123), we used the fact that the RG functions γ_a are independent of ε, but the real reason is that they are UV-finite, while the independence of ε is only a consequence of the UV finiteness in the MS scheme (Sec. 1.25). In turn, the UV finiteness of the RG functions is a consequence of the UV finiteness of the renormalized Green functions (Sec. 1.24), in other words, of the renormalizability of the model.

Returning to the UV-finite transformations (1.186), we conclude that they generate nontrivial additions (1.187) to the RG functions of the anomalous dimensions, but for $g = g_*$ these additions vanish, i.e., the critical exponents are invariant. This is true for any value of ε and, consequently, for any particular order of the ε expansion of the exponents.

The above examples illustrate the general statement that the exact critical dimensions of specific physical quantities and the corresponding normalized scaling functions (Sec. 1.5) are objective characteristics of the model studied, independent of the choice of subtraction scheme used for the renormalization. Therefore, the coefficients of the expansions of these quantities in any parameter, for example, the coefficients of the ε expansion, are also objective. These coefficients must be the same no matter what calculational scheme is used. This is one of the main advantages of any regular expansion in a parameter.

1.41 The renormalization group in real dimension

This frequently encountered term has many different meanings. In what follows, we shall use it to mean the following: all calculations using the RG technique are performed directly in real space (for example, $d_{\mathrm{r}} = 3$ for the φ^4 model) without resorting to the logarithmic theory and ε expansions. The constants Z are then determined not from the requirement of eliminating UV divergences for $\varepsilon \to 0$, but from the normalization conditions for the renormalized Green functions. An arbitrary parameter μ can be introduced into these conditions, and the RG equations for the renormalized functions can be obtained in the standard manner (Sec. 1.24). The RG functions are determined by the usual expressions (1.113) and are calculated as series in g. The coordinate of the fixed point g_* and the values of $\gamma^* = \gamma(g_*)$ are found numerically from the known first few terms of these series or, more precisely, from their Borel sums (Sec. 1.38). Formally, this is analogous to the subtraction scheme described above, with the difference that now ε is treated not as an arbitrary parameter, but as a fixed finite number. An important difference arises only in the final stage: in the ε scheme, the equation $\beta(g_*) = 0$ for g_* is solved and $\gamma(g_*) \equiv \gamma^*$ is calculated by iterating in ε, and the real finite value ε_r is used only in the final expressions for γ^*. In the new, "real" scheme the substitution $\varepsilon = \varepsilon_r$ is made directly in the RG functions, and the equation $\beta(g_*) = 0$ is solved numerically. Since the RG functions are not known exactly, but only as the first few terms of series in g, these two procedures are obviously inequivalent.

Actual calculations performed using the real scheme can give good quantitative results. The critical exponents have been calculated in this manner for the $O_n \ \varphi^4$ model using Borel summation of the series for the RG functions, which for $d = 3$ are known with six-loop accuracy, and for $d = 2$ are known through four loops [73]. The results of the calculation are presented in Tables 1.10 and 1.11 taken from [45].

Table 1.10 Critical exponents of the φ^4 model ($n = 1$), obtained by Borel summation through 4 loops directly in dimension $d = 2$.

	γ	ν	η	ω
φ^4 model	1.79 ± 0.04	0.96 ± 0.04	0.18 ± 0.04	1.3 ± 0.2
Ising model ($d = 2$)	$7/4 = 1.75$	1	$1/4 = 0.25$	$4/3$?

Table 1.11 Critical exponents of the $O_n \ \varphi^4$ model, obtained by Borel summation through 6 loops directly in dimension $d = 3$.

n	0	1	2	3
γ	1.1615 ± 0.0020	1.2405 ± 0.0015	1.316 ± 0.0025	1.386 ± 0.0040
ν	0.5880 ± 0.0015	0.6300 ± 0.0015	0.6695 ± 0.0020	0.705 ± 0.0030
η	0.027 ± 0.004	0.032 ± 0.003	0.033 ± 0.004	0.033 ± 0.004
β	0.302 ± 0.0015	0.325 ± 0.0015	0.3455 ± 0.0020	0.3645 ± 0.0025
α	0.236 ± 0.0045	0.110 ± 0.0045	-0.007 ± 0.006	-0.115 ± 0.009
ω	0.80 ± 0.04	0.79 ± 0.03	0.78 ± 0.025	0.78 ± 0.02
$\omega\nu$	0.470 ± 0.025	0.498 ± 0.020	0.522 ± 0.0018	0.550 ± 0.0016

The results in these tables are in good agreement with the analogous data in the tables of Sec. 1.38, indicating that the scheme works well in practice. However, we cannot rely on it completely, as is sometimes done, arguing that there is no need to go to the logarithmic dimension and check the multiplicative renormalizability of the model for $\varepsilon \to 0$. This viewpoint cannot withstand serious criticism. Its main defect is the loss in specific calculations of the difference between UV-finite and infinite renormalization. We have seen (Sec. 1.40) that in the usual ε scheme, nontrivial critical anomalies $\gamma(g_*)$ can be generated only by UV-divergent (for $\varepsilon \to 0$) renormalization constants and do not change in any finite renormalization, because the pole cancelling the zero of the β function at the fixed point occurs only in the logarithmic derivatives of UV-divergent constants. However, these derivatives, like the β function, are in practice calculated only as the first few terms of series in g, which are then multiplied term by term in (1.118) and (1.187). In this approximation the pole is, of course, not overlooked, and also information about the presence of a zero in one of the cofactors is lost if the contributions of higher order in g which "exceed the accuracy" are dropped in a product of polynomials. In the ε scheme this procedure is consistent: errors of order g^n and higher for $g \to g_* \sim \varepsilon$ generate errors of order ε^n and higher in the results, and the coefficients of all the lower powers of ε are reliably determined. This scheme pretends to do no more. However, in the real scheme with $\varepsilon_r \simeq 1$ this procedure makes it possible, in principle, to obtain any *a priori* specified numerical values of the anomalous critical dimensions. For example, if two polynomials are multiplied without discarding contributions in the product which formally exceed the accuracy in g (for $g \to g_* \simeq 1$ there is no criterion), then we always obtain $\gamma(g_*) = 0$. If we proceed, as usual, using perturbation theory, discarding contributions of higher order (compared with the cofactors) in g, then in general we will find $\gamma(g_*) \neq 0$, but the numerical values of these quantities can be changed at will using the arbitrariness of the finite renormalization of the type (1.186). Actually, for fixed $\varepsilon = \varepsilon_r \cong 1$ when only the first few terms of the series are used, the additions (1.187) are the same as the basic expressions (1.118), which actually makes the second cofactor of the product of polynomials completely arbitrary. The completely satisfactory quantitative results for the exponents in real dimension given above are actually obtained only because the calculations are always performed using some fixed subtraction scheme (Sec. 1.22) with a definite set of natural normalization conditions for determining the constants Z. The errors given in Tables 1.10 and 1.11 are the computational errors in the fixed subtraction scheme, and do not include possible changes of the results when the scheme itself is changed (the "naturalness" of certain normalization conditions does not at all imply their necessity). As explained above, these changes can be arbitrarily large, and so a given technique with natural normalization conditions for determining the constants Z may be applicable as a computational method, but that is all it is.

In conclusion, we again note that the internal consistency of the RG technique is ensured only by the introduction of the formal small parameter ε. The passage to the real value $\varepsilon_r \simeq 1$ in the final results should be understood as an extrapolation. The multiplicative renormalizability of the model plays a key role here: it is this that guarantees the presence of the correlations between the contributions of different order in $1/\varepsilon$ in the renormalization constants Z (Sec. 1.26) that lead to the appearance of poles at $g = g_*$ in their logarithmic derivatives and, consequently, to nontrivial anomalies $\gamma(g_*)$. From the theoretical viewpoint, only quantities independent of the arbitrariness of the finite renormalization are of objective value, in particular, the coefficients of the ε expansions (or analogous expansions in other parameters) of the critical exponents, and the results of summing such series. The "objectivity" of the coefficients of the ε expansions allows the results obtained for them by different methods to be compared and systematized, so that information is accumulated as the accuracy of the calculations is improved by including higher and higher orders in ε. In contrast, in calculations performed in the real dimension it is impossible in principle to eliminate the effect of the arbitrariness of the finite renormalization, because the choice of definite subtraction scheme is only a convention. In this regard, the ε and other expansions possess definite advantages.

1.42 Multicharge theories

The RG technique described above for the example of the φ^4 model is applicable to any multiplicatively renormalizable (in its logarithmic dimension) model. In general, the renormalization constants Z and, consequently, the RG functions, depend on several charges $g \equiv \{g_1...g_n\}$. According to the general rule (1.113), each of them is associated with its own β function $\widetilde{\mathcal{D}}_\mu g_i \equiv \beta_i(g)$, and at a fixed point $g_* \equiv \{g_{1*}...g_{n*}\}$ they all vanish. The evolution equations (1.140) for the invariant charges $\bar{g}_i(s, g)$ take the form

$$\mathcal{D}_s \bar{g}_i = \beta_i(\bar{g}), \qquad \bar{g}_i|_{s=1} = g_i, \tag{1.188}$$

($s \to 0$ at the IR asymptote). The solutions of (1.188) are called the *phase trajectories* in n-dimensional charge space, and now there are many more possibilities than in single-charge models. The analog of the exponent ω in (1.156) will now be the set of n eigenvalues ω_α of the matrix

$$\omega_{ik} = \frac{\partial \beta_i(g)}{\partial g_k}\bigg|_{g=g_*}. \tag{1.189}$$

This matrix is real but, in general, nonsymmetric, and so some of its eigenvalues can be complex-conjugate pairs. The fixed point g_* is *IR-stable* (Sec. 1.31) if for it the real parts of all the ω_α are strictly positive, and *UV-stable* if all are strictly negative. If some are positive and the others negative, g_* is a *saddle point*.

In multicharge theories there can be many fixed points $g_* \sim \varepsilon$, and in general they are of different types. Critical scaling will be observed only when some of these fixed points g_* are IR-stable, and the phase trajectory is determined from (1.188) with given initial data g runs into the basin of attraction of one of them. The behavior of $\overline{g}$ near g_* is determined by the linearized equations (1.188) $\mathcal{D}_s \Delta \overline{g}_i = \omega_{ik} \Delta \overline{g}_k$ for deviations $\Delta \overline{g}_i \equiv \overline{g}_i - g_{i*}$. It is clear from analysis of their solutions for IR-stable points that for $\omega_\alpha > 0$ for all α, the trajectory $\overline{g}(s, g)$ approaches g_* smoothly for $s \to 0$, and if there are complex-conjugate pairs among the ω_α, then $\overline{g}$ approaches g_* along a shrinking spiral. An IR-stable point g_* of this type is called an *attractive focus*, and it gives rise to characteristic oscillations in correction terms like (1.170) owing to the complex values of the exponent. For substitutions $g = g(g')$, the analog of (1.185) for the matrix (1.189) will be the similarity transformation $\omega' = V^{-1} \omega V$ with matrix $V_{ik} = \partial g_i / \partial g'_k$ at the fixed point. The eigenvalues ω_α are invariants under these transformations.

It should be stressed that in a multicharge theory, even when IR-stable fixed points are present a trajectory determined by (1.188) with given initial data g may not reach any of these points at the asymptote $s \to 0$. In this case the trajectory can behave in various possible ways. The following two are encountered most often: (1) passage to the boundary of the stability region, and (2) passage to infinity inside the stability region. Let us explain. In the n-dimensional space of the charges g there is usually a natural *stability region* outside which the system is unstable (like $g > 0$ for the φ^4 model). Passage of the trajectory $\overline{g}(s, g)$ outside the stability region for some finite value s_0 is usually interpreted as indicating the existence of a first-order phase transition at s_0. This interpretation corresponds to the Landau theory: when stability is lost, the energy minimum of the system suddenly collapses to $-\infty$ (but in reality it must be stabilized by the higher-order contributions discarded as IR-irrelevant in constructing the model). The question can arise of whether this interpretation is valid for a scale variable like $p/\mu \equiv s$, since it is not clear how to interpret a phase transition in momentum. However, there is no problem, because the critical regime actually corresponds to the simultaneous and consistent vanishing of all the IR-relevant variables, for example, p and τ in the φ^4 model. It is therefore not important which of them is chosen as the scale variable. The boundary value s_0 for $s = p/\mu$ corresponds to a definite value of τ_0, so that a transition is always associated with a change of the thermodynamical parameters.

Regarding the second type of behavior of the trajectory $\overline{g}$ (passage to infinity), here it should be remembered that the behavior of the trajectory is usually determined on the basis of numerical integration of the equations (1.188), with some approximation for the β functions that becomes meaningless at large $\overline{g}$. In this situation the arguments about the behavior of the trajectory at large $\overline{g}$ become unreliable, and it would be more correct to say that we have simply gone beyond the region of applicability of the approximation. In both cases the only reliable statement that can be made is that critical scaling will not occur for such behavior of the trajectory.

We conclude by noting that the behavior of the solutions of multicharge equations like (1.188) can, in principle, be quite exotic. It is sufficient to recall *limit cycles* (trajectories that follow a closed curve and "circle" around it without approaching a fixed point) and *strange attractors* (trajectories which are attracted not to a point, but to a region inside which they behave chaotically). The system of Lorenz equations, in which a strange attractor was first discovered, is a special case of (1.188) for three charges $g = x, y, z$ with $\beta_x = \sigma(x - y)$, $\beta_y = y - rx + xz$, and $\beta_z = bz - xy$, where $r > 1$, $b > 0$, and $\sigma > b + 1$ are parameters (for the Lorenz system, $\sigma = 10$, $b = 8/3$, and $r = 28$). There are three fixed points in this system: $(0, 0, 0)$, $(\alpha, \alpha, r-1)$, and $(-\alpha, -\alpha, r-1)$, where $\alpha = \sqrt{b(r - 1)}$. The first of these is always IR-unstable (a saddle point), and the other two for $1 < r < r_c = \sigma(\sigma + b + 3)/(\sigma - b - 1)$ (in the Lorenz system, $r_c \cong 24.74$) are IR-attractive foci and become unstable for $r > r_c$. Then a strange attractor in the exact sense of the term appears, but the behavior of the trajectories for $t = -\ln s \to +\infty$ becomes nontrivial already in approaching r_c [74]. Let us also give an example of a limit cycle: the system (1.188) for two charges $g = x, y$ with $\beta_x = -x - \alpha y + x(x^2 + y^2)$ and $\beta_y = -y + \alpha x + y(x^2 + y^2)$. It has only one IR-unstable fixed point $(0, 0)$ and an IR-stable limit cycle $x^2 + y^2 = 1$, and the parameter α is interpreted as the angular velocity of rotation about the cycle.

So far, these two types of exotic behavior of the phase trajectories have not been observed in realistic models.

1.43 Logarithmic corrections for $\varepsilon = 0$

Critical scaling with nontrivial exponents follows from the RG equations only for $d < d^*$, while in the logarithmic dimension $d = d^*$ (Sec. 1.16), i.e., for $\varepsilon = 0$, the corrections to the Landau theory will no longer be power-law corrections, but only logarithmic ones. As usual, we shall discuss them for the example of the φ^4 model, assuming that it is defined for any value of g (see the discussion in Sec. 1.27).

All the quantities in the RG equations of Sec. 1.24 are UV-finite, i.e., they do not have poles in ε, and so they can be continued directly to the point $\varepsilon = 0$. For $\varepsilon = 0$ the linear term in the expansion of the β function (1.121) vanishes [we recall that in the MS scheme all the other coefficients of the series (1.121) are independent of ε]. The IR-stable fixed point is shifted to zero and becomes a point of type 3 in the classification of Sec. 1.31, i.e., IR-attractive from the right. The one-loop approximation (1.154) for the invariant charge $\bar{g}$ of the RG equation (1.151) for the correlator at $\varepsilon = 0$ becomes

$$\bar{g}(s, g) = \frac{g}{1 - g\beta_2 \ln s}, \qquad s \equiv p/\mu. \qquad (1.190)$$

From this we see that $\bar{g}$ falls off as $1/\ln s$ for $s \to 0$, tending to the fixed point $g_* = 0$. This conclusion is independent of the specific approximation (1.128a), because the inclusion of higher-order terms in the expansion of the β function (1.121) leads only to corrections to (1.190) that are relatively small for $s \to 0$, and these if desired can be found explicitly from the definition (1.144).

Naturally, a change of the type of fixed point leads to a change of the critical asymptotes. As an example, let us consider the RG equation (1.151) for the renormalized correlator (1.149) of the φ^4 model. All the general expressions of Sec. 1.30 for $\varepsilon = 0$ remain the same. Only the $s \to 0$ asymptote of the integrals appearing in (1.152) and (1.153) changes. In our model for $\varepsilon = 0$ the expansion (1.121) of the RG function γ_τ begins at g, and the series for γ_φ and the β function begin at g^2 (these will be calculated in later chapters), and so the asymptotes of the integrals in (1.152) and (1.153) are different: for $\bar g \to 0$ the former has a finite limit, while the second contains a logarithmic singularity. Isolating it, we obtain

$$\int_g^{\bar g} dx \frac{\gamma_\tau(x)}{\beta(x)} = \frac{\gamma_{1\tau}}{\beta_2} \ln(\bar g/g) + \tilde c_\tau(g) + ..., \qquad (1.191)$$

where $\gamma_{1\tau}$ and β_2 are the first coefficients of the corresponding series (1.121),

$$\tilde c_\tau(g) \equiv \int_g^0 dx \left[\frac{\gamma_\tau(x)}{\beta(x)} - \frac{\gamma_{1\tau}}{\beta_2 x} \right], \qquad (1.192)$$

and the ellipsis in (1.191) and below denotes corrections of order $\bar g \sim 1/\ln s$ unimportant for $s \to 0$. The singularity at $x = 0$ in the integrand of (1.192) is eliminated, which allowed us to replace the upper limit $\bar g$ by zero with an error of order $\bar g$. By the same arguments,

$$\int_g^{\bar g} dx \frac{\gamma_\varphi(x)}{\beta(x)} = \tilde c_\varphi(g) + ..., \qquad \tilde c_\varphi \equiv \int_g^0 dx \frac{\gamma_\varphi(x)}{\beta(x)}. \qquad (1.193)$$

Using the known series (1.172) for the function Φ in the renormalized correlator (1.149) at $\varepsilon = 0$, the quantity $\Phi(1, \bar g, \bar z)$ on the right-hand side of (1.152) is determined as a series in the small parameter $\bar g \sim 1/\ln s$. With our accuracy,

$$\Phi(1, \bar g, \bar z) = \Phi_0(\bar z) + ..., \qquad (1.194)$$

where $\Phi_0(z) = (1 + z)^{-1}$ is the zeroth-order approximation, corresponding to the unperturbed correlator $D_{0R} = (p^2 + \tau)^{-1}$. The desired $s \to 0$ asymptote of the renormalized correlator is obtained by substituting the above expressions into the general equations of Sec. 1.30.

Let us consider in more detail the behavior of the correlator in the special cases $\tau = 0$ and $p = 0$. In the first case the answer is obtained directly from the expressions discussed above, since we can simply set $\bar z \sim z \sim \tau = 0$ in them. It then follows from (1.149), (1.152), (1.193), and (1.194) that up to the normalization, the leading term of the $\tau = 0$, $p \to 0$ asymptote is exactly the same as for the unperturbed correlator with $\tau = 0$:

$$D_R \Big|_{\substack{\tau=0 \\ p\to 0}} \cong p^{-2} C_\Phi (1 + ...), \qquad C_\Phi \equiv \exp[2\tilde c_\varphi(g)], \qquad (1.195)$$

where the ellipsis denotes small corrections of order $\bar{g} \sim 1/\ln s$, analogous to the corrections (1.170) for $\varepsilon > 0$. If necessary, these can be calculated by raising the accuracy of the estimates in (1.193) and (1.194).

Now we consider the $\tau \to 0$ asymptote of the correlator with zero external momentum, which, according to (1.36), is proportional to the susceptibility χ. The equations of Sec. 1.30 do not admit direct passage to the case $p = 0$, because we have isolated the factor $1/p^2$ from (1.149). Instead of (1.149) we now take

$$D_{\mathrm{R}}(p = 0, g, \tau, \mu) = \tau^{-1}\Phi(s, g), \qquad s \equiv \tau/\mu^2, \tag{1.196}$$

and then after substitution into (1.150) we find that

$$[-(2 + \gamma_\tau)\mathcal{D}_s + \beta\partial_g + \gamma_\tau + 2\gamma_\varphi]\Phi(s, g) = 0. \tag{1.197}$$

Division of this equation by $2 + \gamma_\tau$ leads to the form (1.142). The corresponding invariant charge is then determined by (1.179) and in the one-loop approximation differs from (1.190) by only the replacement $s \to s^{1/2}$ (equivalent to $\beta_2 \to \beta_2/2$), and so it also falls off as $1/\ln s$ for $s = \tau\mu^{-2} \to 0$. The general solution of (1.197) is given by (1.148):

$$\Phi(s, g) = \Phi(1, \bar{g})\exp\left\{\int_g^{\bar{g}} dx \frac{\gamma_\tau(x) + 2\gamma_\varphi(x)}{\beta(x)}\right\}. \tag{1.198}$$

The asymptote of the integral in the exponential resembles (1.191), and the expansion of the function $\Phi(1, \bar{g})$ in (1.198) analogous to (1.194) in this case begins simply at unity. Therefore, up to the normalization, for the asymptote of the susceptibility we finally obtain

$$D_{\mathrm{R}}\Big|_{\substack{p=0 \\ \tau\to 0}} \sim \chi(\tau) \sim \frac{1}{\tau}(\bar{g}/g)^{\gamma_{1\tau}/\beta_2}(1 + ...). \tag{1.199}$$

We recall that $\bar{g}(s, g) \sim 1/\ln s$ for $s \equiv \tau\mu^{-2} \to 0$.

Therefore, for $\varepsilon = 0$ the canonical law $\chi(\tau) \sim \tau^{-1}$ is only corrected by a fractional power of $\ln s = \ln(\tau/\mu^2)$, whereas for $\varepsilon > 0$ the exponent of τ itself is changed. At the asymptote (1.199) the exponent of the correction logarithm is determined by the first coefficients of the expansions (1.121) of the RG functions γ_τ and β. For the O_n φ^4 model $\gamma_{1\tau}/\beta_2 = -(n+2)/(n+8)$ (the calculations are given later in Chs. 3 and 4). The logarithmic corrections to the equation of state (1.175) for $\varepsilon = 0$ are found just as simply from the RG equation (1.178).

Finally, we note that in some models[*] the expansion of the RG function γ_φ begins not at g^2, but at g, and then logarithmic contributions also appear in integrals like (1.193) and the corresponding asymptotes (1.195).

---

[*]For example, in relativistic electrodynamics or chromodynamics, where the square of the elementary charge plays the role of g.

1.44 Summation of the $g \ln s$ contributions at $\varepsilon = 0$ using the RG equations

As an example, let us consider (1.197), rewriting it as

$$2\mathcal{D}_s F = [\beta \partial_g - \gamma_\tau \mathcal{D}_s] F + \gamma_\tau + 2\gamma_\varphi, \quad F \equiv \ln \Phi. \quad (1.200)$$

We shall seek a solution in the form of a perturbation series: $F(s, g) = g F_1(s) + g^2 F_2(s) + \ldots$ (the series for Φ begins at unity, and so that for F begins at zero). Writing all quantities in (1.200) as series in g, in particular, the series (1.121) for the RG functions with $\varepsilon = 0$, and grouping together contributions of the same order in g, we obtain an infinite system of equations: $2\mathcal{D}_s F_1 = \gamma_{1\tau}$, $2\mathcal{D}_s F_2 = \beta_2 F_1 - \gamma_{1\tau} \mathcal{D}_s F_1 + \gamma_{2\tau} + 2\gamma_{2\varphi}$, and so on. They are easily solved:

$$F_1(s) = \tfrac{1}{2}\gamma_{1\tau} \ln s + c_1,$$
$$F_2(s) = \tfrac{1}{8}\beta_2 \gamma_{1\tau} \ln^2 s + \tfrac{1}{2}[\gamma_{2\tau} + 2\gamma_{2\varphi} - \tfrac{1}{2}\gamma_{1\tau}^2 + \beta_2 c_1] \ln s + c_2, \quad (1.201)$$

and so on, where the parameters c_i are arbitrary integration constants. Analyzing this solution, we see that

$$F(s, g) = \sum_{n=1}^{\infty} g^n P_n(\ln s), \quad (1.202)$$

where P_n are polynomials of order n. The contributions $(g \ln s)^m$ with any $m \geq 0$ are called *leading logarithms*, the contributions $g(g \ln s)^m$ are the first correction, $g^2(g \ln s)^m$ are the second correction, and so on. If the parameter g is small and $\ln s$ is large, so that $u \equiv g \ln s \lesssim 1$, then in the solution (1.202) it is natural to first sum the contributions of all the leading logarithms, and then calculate the corrections to them. For this, we shall seek a solution of (1.200) in the form

$$F(s, g) = \sum_{n=0}^{\infty} g^n f_n(u), \quad u \equiv g \ln s. \quad (1.203)$$

We see from the procedure of obtaining (1.201) that the leading logarithms in all orders in g are uniquely determined by the first coefficients β_2 and $\gamma_{1\tau}$ in the expansions (1.121) of the corresponding RG functions ($\gamma_{1\varphi} = 0$). To determine the first correction we need the next coefficients β_3, $\gamma_{2\tau}$, and $\gamma_{2\varphi}$, and also the constant c_1 from (1.201). In the next order we need the next coefficients of the RG functions and c_2, and so on. We therefore see that the sum of all the leading logarithms, i.e., the function $f_0(u)$ in (1.203), can be found from the simplified equation (1.200), taking into account only the needed leading terms in the expansions of the RG functions: $2\mathcal{D}_s f_0 = \beta_2 g^2 \partial_g f_0 + g\gamma_{1\tau}$, with the condition $f_0(0) = 0$ following from (1.201). In acting on a function of a single variable $u = g \ln s$ we have $\mathcal{D}_s = g\partial_u$ and $g^2 \partial_g = gu\partial_u$, and so the above equation is rewritten as $(2 - \beta_2 u)f_0'(u) = \gamma_{1\tau}$. Taking into account the condition at the origin, it is easily solved:

$$f_0(u) = \frac{\gamma_{1\tau}}{\beta_2} \ln \left| \frac{2}{2 - \beta_2 u} \right|. \qquad (1.204)$$

To determine the corrections, it is necessary to make the change of variable $s, g \to u, g$ in (1.200) and then seek a solution in the form (1.203), systematically selecting the contributions of the required order in g. It should be emphasized that already the first correction $g f_1(u)$, in contrast to $f_0(u)$, is not determined uniquely by the coefficients of the RG functions, because it also depends on the constant c_1 from (1.201). To determine $f_2(u)$ the constant c_2 also must be known, and so on. In fact, for a function $f_n(u)$ from (1.200) we obtain an equation of the form $(2 - \beta_2 u) f'_n(u) =$ a known quantity, and the constant $c_n = f_n(0)$ is needed to make the solution unique. The constant c_n is interpreted as the contribution of the nonleading logarithms studied in the lowest order of simple perturbation theory in which they first appear. From this contribution and the RG functions, the RG equation reconstructs the nonleading logarithms of the given type in all higher orders of simple perturbation theory. This is the essence of the RG equation from this viewpoint.

All the results can also be obtained from the general solution (1.198) of (1.197), which is equivalent to (1.200). All the leading logarithms are contained in the one-loop approximation for the corresponding invariant charge (1.179), differing from (1.190) by only the replacement $\beta_2 \to \beta_2/2$. Therefore, in this approximation $\bar{g}/g = 2/(2 - \beta_2 u)$, which coincides with the argument of the logarithm in (1.204). All the leading logarithms are contained in the fractional power of $\bar{g}/g$ explicitly isolated in (1.199), which up to notation coincides with (1.204). However, it is difficult to find even the first correction $g f_1(u)$ to $F = \ln \Phi$ from (1.198), because it comes from several sources: the correction for β_3 in the ratio $\bar{g}/g$, the contribution of order $\bar{g}$ in $\Phi(1, \bar{g})$, and also the contribution of order g in the constant of the type (1.192).

In the example considered above, the expansion of the inhomogeneous term in the equation for $F = \ln \Phi$ began at g. If it began at g^2, the solution would not contain any leading logarithms, and the expansion (1.203) would then start with $g f_1(u)$. This will happen, for example, in (1.151) for the correlator at $\tau = 0$ (then $s = p/\mu$) owing to the absence of a contribution to $\gamma_\varphi(g)$ linear in g. Again in this case the first term $g f_1(u)$ is determined only by the RG functions, since for $\tau = 0$ the expansion of the function $F = \ln \Phi$ in the simple perturbation theory of the φ^4 model also begins at g^2. This gives the condition $f_1(0) = 0$, allowing the unique determination of the function $f_1(u)$ from the coefficients β_2 and $\gamma_{2\varphi}$ in the expansions of the RG functions.

With this we end our general description of the ideas and methods of the renormalization group. We shall return to it only in Ch. 4 after developing the techniques needed for specific calculations in Chs. 2 and 3.

Chapter 2

Functional and Diagrammatic Technique of Quantum Field Theory

In this chapter we give a brief but sufficiently complete review of the technical tools of quantum field theory. The use of the term "quantum" is traditional: the techniques described below were first developed for quantum (pseudo-Euclidean) field theory. However, the techniques themselves are not related at all to quantum physics, and can equally well be applied to the theory of a classical random field, in particular, the fluctuation theory of critical behavior. The term "operator" in quantum field theory corresponds to the term "random quantity" in the problems we shall discuss.

2.1 Basic formulas

As explained in Sec. 1.13, the fluctuation theory of critical behavior is the theory of a classical random field (or set of fields) $\widehat{\varphi}(x)$. The weight of a "configuration" $\varphi(x)$ is given by $\exp S(\varphi)$, where $S(\varphi)$ is the action functional specifying the model. We immediately note that all the general formulas in the present chapter written in universal notation (Sec. 1.9) are valid for any field or set of fields.

The fundamental definitions of Sec. 1.13 involve functional (path) integrals, and so we need a precise formulation of the rules for calculating such integrals. These objects are traditionally defined using various interpolation procedures (see, for example, [75] and [76]), but careful analysis reveals that there are many problems associated with dependence of the results on the interpolation method. Technically, it is simpler to take the following formal rule for calculating Gaussian integrals as a fundamental postulate:

$$\int D\varphi \exp[-\tfrac{1}{2}\varphi K\varphi + A\varphi] = \det(K/2\pi)^{-1/2} \exp[\tfrac{1}{2}AK^{-1}A], \qquad (2.1)$$

where $A\varphi$ is a general linear form of the type (1.48), $\varphi K\varphi$ is a quadratic form specified by a linear symmetric operator K acting on the fields φ, and K^{-1} on the right-hand side of (2.1) is the inverse operator. Without loss of generality (see below), all linear operators can be written as integral operators, i.e.,

$$[K\varphi](x) \equiv (K\varphi)_x = \int dx' K(x, x')\varphi(x'), \qquad (2.2)$$

$$\varphi K\varphi = \int\int dx dx' \varphi(x) K(x, x')\varphi(x'). \qquad (2.3)$$

The function $K(x, x')$ is called the *kernel of the operator* K, and the unit operator corresponds to kernel $\delta(x - x')$. *Symmetry* of K is expressed as $K = K^\top$, where $\top$ denotes the transpose of the operator. For kernels $K^\top(x, x') = K(x', x)$, and so symmetry of the operator K is equivalent to symmetry of its kernel. Symmetry of K implies symmetry of the inverse operator K^{-1} and its kernel.

Let us make some explanatory remarks regarding (2.1).

1. If the argument x is a discrete index taking a finite number of values, then (2.2) and (2.3) are the usual constructions of linear algebra, and (2.1) is the usual expression for a Gaussian integral in a finite-dimensional space. It is written such that the spatial dimension does not enter explicitly into the result, and so it can be used without modification for infinite-dimensional functional spaces. This is the general principle used to obtain such expressions.

2. Convergence of the integral in (2.1) is ensured by positive-definiteness of K, but (2.1) is often used also in other situations, where the answer is understood as an analytic continuation from the region of convergence.

3. The notation in (2.2) and (2.3) remains meaningful also for differential operators K. Then $K(x, x') = K_x \delta(x - x')$, where K_x is a given symmetric (for derivatives $\partial^\top = -\partial$) differential operator with respect to the argument x, the integration in (2.2) is removed, and that in (2.3) becomes single integration. In practice, the kernel K^{-1} usually has the meaning of the *Green function of the linear problem* $K\varphi = A$, $A \in E$ (E is the space of rapidly decreasing functions; see Sec. 1.13), where for $K \geq 0$ the additional condition $\varphi(\infty) = 0$ will always be imposed. This condition fixes the choice of Green function in the general case where the latter is nonunique.

4. In calculating the integral (2.1) we make the shift $\varphi \to \varphi + K^{-1}A$, which for any $A \in E$ should not change the integration region E_{int} in (2.1). Therefore, according to the general rules of Sec. 1.13 (see also Sec. 6 of [53]), it should be assumed that $E_{\text{int}} = K^{-1}E$, i.e., that E_{int} is the set of all functions of the form $\varphi = K^{-1}A$, $A \in E$. For differential operators K with nonunique Green function K^{-1}, in defining E_{int} we use those functions that we want to obtain on the right-hand side of (2.1). In our problems for $K \geq 0$ we shall always make the choice in remark 3 above. Then for strictly positive operators like $K = -\partial^2 + \tau$ with $\tau > 0$, all functions of the form $K^{-1}A$, $A \in E$, are contained in E, and in this case $E_{\text{int}} = E$. However, for massless operators like $K = -\partial^2$, the falloff of functions $\varphi = K^{-1}A$, $A \in E$, at infinity will in general not be fast, but only power-law falloff. These details are important only in rare cases (see, for example, Sec. 2.16). It is usually sufficient to know that $E_{\text{int}} \supseteq E$ and consists of functions that fall off at infinity. In any case, the form $\varphi K \varphi$ is well defined (finite) for any $\varphi \in E_{\text{int}} = K^{-1}E$.

5. By definition, $\varphi K = K^\top \varphi$, the *trace of an operator* is defined as $\text{tr} K = \int dx\, K(x, x)$, and a product of operators corresponds to convolution of their kernels: $(LM)(x, x') = \int dx''\, L(x, x'') M(x'', x')$. All these definitions are analogous to the usual expressions of linear algebra.

6. For translationally invariant operators (including differential operators with constant coefficients), the kernel $K(x, x')$ depends only on the difference of spatial coordinates $x - x'$ (of course, this does not pertain to discrete indices if they are included in x). If the Fourier transform $K(k)$ is defined by Eq. (1.35b), then for a differential operator it will be a simple polynomial in the momenta. A convolution of kernels then corresponds to a product of Fourier transforms without an additional coefficient, and so the inverse operator corresponds simply to $K^{-1}(k)$. All operators remain matrices in their discrete indices, if there are any.

7. To calculate the determinants of arbitrary (not necessarily symmetric) operators, we can use the expressions

$$\det(LM) = \det L \cdot \det M, \quad \det(K^\alpha) = (\det K)^\alpha,$$

$$\det L / \det M = \det(LM^{-1}) = \det(M^{-1}L) \equiv \det(L/M), \quad (2.4)$$

$$\det(K^\top) = \det K, \quad \det K = \exp \operatorname{tr} \ln K.$$

For a translationally invariant operator M,

$$\operatorname{tr} M = V(2\pi)^{-d} \int dk \ \operatorname{tr}_{\mathrm{ind}} M(k), \quad V \equiv \int dx,$$

$$\operatorname{tr} \ln M = V(2\pi)^{-d} \int dk \ \ln \det_{\mathrm{ind}} M(k), \quad (2.5)$$

where x now denotes only the d-dimensional spatial coordinates, V is the infinite volume of the system, $M(k)$ is the Fourier transform (1.35b) of the kernel M, which is a matrix in the discrete indices if there are any, and $\operatorname{tr}_{\mathrm{ind}}$ and $\det_{\mathrm{ind}}$ are the corresponding operators for these indices.

8. Derivatives of determinants with respect to parameters (numerical or functional) are calculated using the equation

$$\delta \ \ln \det K = \delta \operatorname{tr} \ln K = \operatorname{tr}(K^{-1} \delta K), \quad (2.6)$$

where δK is an arbitrary variation of K.

9. A change of variable $\varphi \to \varphi'$ in the functional integral corresponds to the *Jacobian* $D\varphi / D\varphi' = \det[\delta\varphi/\delta\varphi']$, i.e., the determinant of the linear operator with kernel $\delta\varphi(x)/\delta\varphi'(x')$.

10. Expressions in field theory will always involve ratios of Gaussian integrals. The factors of 2π from (2.1) cancel, and constructions of the type $\det(K - L)/\det K = \det[K^{-1}(K - L)] = \det(1 - M)$ appear, where $M \equiv K^{-1}L$. Such quantities can be calculated as series in M using the last expression in (2.4):

$$\ln \det(1 - M) = \operatorname{tr} \ln(1 - M) = -\operatorname{tr}[M + \tfrac{1}{2}M^2 + \tfrac{1}{3}M^3 + ...]. \quad (2.7)$$

11. In universal notation (Sec. 1.9), the general expression (2.1) is valid for any fields or sets of fields, in particular, for complex fields ψ, ψ^+ (we can introduce an index to distinguish these: $\psi \equiv \varphi_1$, $\psi^+ \equiv \varphi_2$, and then include

it in x). For completeness, let us give the complex analog of (2.1) in ordinary notation including all normalizations:

$$\int D\psi D\psi^+ \exp[-\psi^+ L\psi + \psi^+ A + A^+\psi] = \det[iL/2\pi]^{-1} \exp[A^+ L^{-1} A].$$

Here the operator L, in contrast to K in (2.1), need no longer be symmetric. The symmetry is manifested only in going to universal notation with $\varphi = (\psi, \psi^+)$, i.e., $\varphi_1 = \psi$, $\varphi_2 = \psi^+$:

$$\psi^+ L\psi = \frac{1}{2} \begin{pmatrix} \psi \\ \psi^+ \end{pmatrix} \begin{pmatrix} 0 & L^\top \\ L & 0 \end{pmatrix} \begin{pmatrix} \psi \\ \psi^+ \end{pmatrix} \equiv \frac{1}{2}\varphi_i K_{is}\varphi_s \equiv \frac{1}{2}\varphi K\varphi \qquad (2.8)$$

(unless stated otherwise, repeated indices are always summed over). The 2×2 matrix operator K in the form (2.8) is symmetric for any L.

Let us give several useful formulas involving variational derivatives. *Variational differentiation* is performed according to the same rules as ordinary differentiation, using the basic definition $\delta\varphi(x)/\delta\varphi(x') = \delta(x - x')$ in universal notation (Sec. 1.9). The following relations are valid:

$$F(\delta/\delta\varphi)\exp(A\varphi)... = \exp(A\varphi)F(A + \delta/\delta\varphi)..., \qquad (2.9)$$

$$F(\delta/\delta\varphi)\exp(A\varphi) = F(A)\exp(A\varphi), \qquad (2.10)$$

$$\exp(A\delta/\delta\varphi)F(\varphi) = F(\varphi + A), \qquad (2.11)$$

$$F(\delta/\delta\varphi)\prod_i F_i(\varphi) = F\left(\sum_i \delta/\delta\varphi_i\right)\prod_i F_i(\varphi_i)\,|_{\varphi_1=\varphi_2=...=\varphi}, \qquad (2.12)$$

in which $A\varphi$ and $A\delta/\delta\varphi$ are linear forms like (1.48), and all F are arbitrary functionals. The dots in (2.9) denote an arbitrary expression on which the differential operator acts; if it is independent of φ, then (2.9) becomes (2.10). Equation (2.11) is compact notation for the functional Taylor expansion in A.

When (2.1) is used in field theory, the free action in (1.59) will play the role of the quadratic form. The symmetric operator kernel $\Delta \equiv K^{-1}$, called the *free (bare) propagator* or correlator of the field, will be represented as a line in the graphs. Then,

$$S(\varphi) = S_0(\varphi) + V(\varphi), \quad S_0(\varphi) = -\frac{1}{2}\varphi K\varphi, \quad \Delta = \Delta^\top = K^{-1}. \qquad (2.13)$$

Another important construction is the differential *reduction operator* P_φ, defined in terms of $S_0(\varphi)$ as

$$P_\varphi \equiv \exp\left[\frac{1}{2}\frac{\delta}{\delta\varphi}\Delta\frac{\delta}{\delta\varphi}\right] \equiv \exp\left[\frac{1}{2}D_\varphi\right], \qquad (2.14a)$$

$$D_\varphi \equiv \int\int dx dx' [\delta/\delta\varphi(x)]\Delta(x, x')[\delta/\delta\varphi(x')], \qquad (2.14b)$$

where the subscript φ specifies the functional argument with respect to which the differentiation is done. The formal substitution $A(x) \to \delta/\delta\psi(x)$ from (2.1) can be used to obtain the expression [with the same notation as in (2.13) and (2.14)]

$$c\int D\varphi \exp[S_0(\varphi) + \varphi(\delta/\delta\psi)] = P_\psi, \qquad (2.15)$$

in which c is a normalization constant defined as

$$c^{-1} \equiv \int D\varphi \exp S_0(\varphi) = \det(K/2\pi)^{-1/2}. \qquad (2.16)$$

From (2.11) and (2.15) for an arbitrary functional $F(\psi)$ we have

$$P_\psi F(\psi) = c\int D\varphi F(\varphi + \psi) \exp S_0(\varphi), \qquad (2.17)$$

and for $\psi = 0$ from (2.17) we find that

$$P_\varphi F(\varphi) \,|_{\varphi=0} = c\int D\varphi F(\varphi) \exp S_0(\varphi). \qquad (2.18)$$

These expressions are fundamental for calculating non-Gaussian functional integrals. We note that (2.17) can be used to rewrite any expression involving the reduction operator P as a functional integral.

2.2 The universal graph technique

From a given action we define the functional [in the notation of (2.13) and (2.14)]

$$H(\varphi) = P \exp V(\varphi), \quad P = \exp(\tfrac{1}{2}\mathrm{D}), \quad P \equiv P_\varphi, \quad \mathrm{D} \equiv \mathrm{D}_\varphi. \qquad (2.19)$$

Using quantum field terminology, we shall call $H(\varphi)$ the *S-matrix functional*,[*] and the coefficients H_n of its expansion in φ the *S-matrix Green functions* of the field $\widehat{\varphi}$. We shall first describe the graph representation of H, and then we shall show that the results apply directly to functionals of the type (1.62) that we are interested in. We note that the partition functions of the classical Heisenberg magnet (and also of the Ising magnet as a special case; see Sec. 2.26) and of a gas with two-body forces (the grand canonical ensemble) in an arbitrary nonuniform external field, which corresponds to φ in (2.19), also reduce to the form (2.19) [53].

Expanding the exponentials in (2.19) in series, we obtain

$$H = \sum_{n=0}^{\infty} \frac{PV^n}{n!} = \sum_{n,m=0}^{\infty} \frac{\mathrm{D}^m V^n}{2^m m! n!}. \qquad (2.20)$$

---

[*]In quantum field theory, the functional $H(\varphi)$ represents the N-form of the S matrix: the S-matrix operator is $NH(\widehat{\varphi})$, where $\widehat{\varphi}$ is the free field operator and N denotes normal ordering.

This expression can be written as a sum of contributions of various graphs with definite "symmetry" coefficients.

A *graph* or *diagram* is a figure consisting of some number of points (vertices) and lines connecting them. In a *labeled graph* the vertices are labeled 1, 2, ... A graph without labels is termed *free*. A labeled graph is uniquely determined by its *adjacency matrix* π; by definition, the matrix element $\pi_{ik} = \pi_{ki}$ is equal to the number of lines (directly) joining the vertices i and k. The diagonal elements correspond to *contracted* lines, both ends of which are joined to the same vertex. Labeled graphs are considered equal if their adjacency matrices are equal; two labeled graphs differing only by a permutation of the vertex labels are termed equivalent and correspond a single free graph. Two free graphs are equal if for some method of labeling the corresponding labeled graphs are equal (the concept of "different" free graphs is defined accordingly). The *symmetry group* of a labeled graph is the subgroup of permutations of vertex labels which take the graph into itself, i.e., which do not change the adjacency matrix; the order (number of elements) of this group is referred to as the *symmetry number* s of the graph. The symmetry groups of equivalent labeled graphs are isomorphic, and so they are of the same order. It then follows that the symmetry number s is actually a characteristic of a free graph independent of how its vertices are labeled. The total number of ways of labeling the vertices in a graph with n vertices is $n!$, and the corresponding labeled graphs are divided into classes, each consisting of s identical labeled graphs. Therefore, the number of different labeled graphs corresponding to a given free graph is $n!/s$. A *connected graph* is a graph in which one can go from one vertex to any other by following lines. These very simple concepts from graph theory will suffice for what follows.

We associate V^n in (2.20) with a graph consisting of n isolated points (vertices), each corresponding to a single factor of V. In acting on V^n, the operator D joins pairs of points by a line Δ in all possible ways, because each of the two derivatives $\delta/\delta\varphi$ in D can act on any of the factors V. Therefore, the graph representation of $D^m V^n$ in (2.20) contains all free graphs with n vertices and m lines, and the full series (2.20) is the sum of unity and all the different free graphs with any number of vertices and lines. To make these statements precise, we need to formulate the rule for the correspondence between a graph and an analytic expression, and also find the symmetry coefficient of each graph.

Let us start with the first problem. It is clear from the representation (2.20) and the definition of the operator D that the joining of another line to a given vertex is accompanied by differentiation of the vertex with respect to φ. Therefore, the vertex of a graph to which $m \geq 0$ lines are joined is associated with the *vertex factor*

$$V_m(x_1...x_m; \varphi) = \delta^m V(\varphi)/\delta\varphi(x_1)...\delta\varphi(x_m), \qquad V_0 \equiv V, \qquad (2.21)$$

and its arguments x are contracted with the arguments of the lines Δ connected to the given vertex.

This completely defines the correspondence rule. Let us explain it for the example of graph (2.22a):

$$(a) \equiv \left\{ \begin{array}{c} \cdot \quad \bigcirc \\ \bullet\!\!-\!\!\bigcirc \end{array} \right\} \; ; \; (b) \equiv \left\{ \begin{array}{c} \cdot \quad \overset{1}{\bigcirc}\!\!\overset{2}{} \\ \underset{8}{\overset{3}{\bullet}}\!\!\overset{4}{\underset{7}{\bigcirc}}\!\!\overset{5\ 6}{} \end{array} \right\} . \tag{2.22}$$

The ends of the lines Δ correspond to independent arguments x, and so altogether there are twice as many of them as there are lines; in our example, eight. By choosing any labeling, for example, that shown in (2.22b), we thereby define the arguments x of all the lines Δ and all the vertex factors (2.21), which allows the graph (2.22a) to be associated with the analytic expression

$$V\int...\int dx_1...dx_8\Delta(12)\Delta(34)\Delta(56)\Delta(78)V_2(12)V_1(3)V_3(458)V_2(67).$$

For brevity, we have used the notation $\Delta(ik) \equiv \Delta(x_i, x_k)$ and similarly for the vertex factors (2.21); the order of the arguments in them is not important since they are symmetric. This example clearly illustrates the general correspondence rule. From it we also see that the contribution of a disconnected graph is always equal to the product of the contributions of all its connected components; in our example there are three of these.

Turning now to the second problem, we can give the answer immediately: each of the free graphs enters into (2.20) with coefficient

$$C = \left[s \cdot 2^{\mathrm{tr}\,\pi} \prod_i \pi_{ii}! \prod_{i<k} \pi_{ik}! \right]^{-1}, \quad \mathrm{tr}\,\pi \equiv \sum_i \pi_{ii}, \tag{2.23}$$

where s is the symmetry number of a given graph and π_{ik} are the elements of the adjacency matrix (see above) of the labeled graph obtained from a given free graph by the introduction of arbitrary labeling of the vertices. The quantity in (2.23) is independent of the choice of labeling, because it involves only permutation invariants.

Let us give the derivation of (2.23). For PV^n in (2.20) we use (2.12):

$$PV^n = \left\{ \exp\left[\sum_i \tfrac{1}{2}\mathrm{D}_{ii} + \sum_{i<k} \mathrm{D}_{ik} \right] \right\} V_1...V_n \mid_*, \tag{2.24}$$

where we have introduced n independent arguments φ_i, $V_i \equiv V(\varphi_i)$, $\mathrm{D}_{ik} = \mathrm{D}_{ki} = (\delta/\delta\varphi_i)\Delta(\delta/\delta\varphi_k)$ (the symmetry of Δ has been taken into account), and the symbol $\mid_*$ in (2.24) denotes $\varphi_1 = \varphi_2 = ... = \varphi_n = \varphi$ after the differentiations are performed. The representation (2.24) introduces labeling of the vertices, and D_{ik} is the operator that attaches a line to the pair of vertices specified by the subscripts $i \leq k$. Writing the exponential of the sum in (2.24) as a product of exponentials and expanding each factor in a series, we can write the exponential (2.24) as a multiple sum:

$$\sum_{\pi} \prod_{i} \left[(\tfrac{1}{2}\mathrm{D}_{ii})^{\pi_{ii}} / \pi_{ii}! \right] \prod_{i<k} \left[\mathrm{D}_{ik}^{\pi_{ik}} / \pi_{ik}! \right]. \qquad (2.25)$$

The summation runs over all independent sets $\pi \equiv \{\pi_{ik}, i \leq k\}$ of non-negative integers π_{ik}, each of which takes all values from zero to infinity. Clearly, each term of the sum (2.25) is in one-to-one correspondence with a labeled graph, and the set π specifies the adjacency matrix of the graph. It is also clear that (2.25) involves all possible labeled graphs, and that no two of them are identical, because each independent set π enters only one time. Therefore, the quantity (2.24) is the sum of the contributions of all possible different labeled graphs, whose coefficients are determined by the factorials and factors of two in (2.25), which combine to give the expression (2.23) without the $1/s$. Since that expression involves only permutation invariants, the coefficients of equivalent labeled graphs are identical. By setting all the fields in (2.24) equal after differentiation, we erase the vertex label and go over to the language of free graphs. When the different free graphs are summed, their coefficients in (2.20) are obtained by multiplying the coefficients of the labeled graphs by the total number of different but equivalent labeled graphs corresponding to a given free graph. This gives $n!/s$ (see above), and when the coefficient $1/n!$ in (2.20) is included we obtain (2.23).

The equation $2\delta H/\delta \Delta(x, x') = \delta^2 H/\delta\varphi(x)\delta\varphi(x')$ following from the definitions (2.14) and (2.19) can be used to prove [53] the following useful recursion relation:

$$C_2/C_1 = \varepsilon N[D_1 \to D_2]/N[D_2 \to D_1]. \qquad (2.26)$$

Here $C_{1,2}$ are the symmetry coefficients (2.23) of the graphs $D_{1,2}$ such that D_2 is obtained by joining a single line to D_1, $N[D_1 \to D_2]$ is the number of equivalent methods of changing D_1 into D_2 by adding a line, $N[D_2 \to D_1]$ is the number of equivalent ways of changing D_2 into D_1 by removing a line, and $\varepsilon = \tfrac{1}{2}$ if the added line is contracted and $\varepsilon = 1$ otherwise. Let us give some examples. Let D_1 be a chain with $m - 1$ links and D_2 the m-hedron obtained by joining the ends of the chain by a line. Then $N[D_1 \to D_2] = 1$ and $N[D_2 \to D_1] = m$. Another example: let D_1 be a triangle and D_2 be a triangle, one of whose sides is doubled. Then $N[D_1 \to D_2] = 3$ and $N[D_2 \to D_1] = 2$. In both examples $\varepsilon = 1$.

The graphs in (2.20) are connected and disconnected. The *connected part* of a functional having a graph representation is the sum of the contributions of all its connected graphs with their coefficients, i.e., when the connected part is selected, all the disconnected graphs and contributions like unity are discarded. The following statement is valid for the functional (2.19):

$$\ln H(\varphi) = \text{connected part of } H(\varphi). \qquad (2.27)$$

This is actually a statement about the coefficients of the disconnected graphs: each is specified by enumerating the set of independent connected components D_a and the multiplicity m_a of each, while (2.27) implies that the coefficient of a disconnected graph is equal to the product of the coefficients C_a of all its connected components [which gives $\prod_a (C_a)^{m_a}$] divided by the product of factors $m_a!$. This statement, and along with it (2.27), can be proved using (2.23) by analyzing the symmetry group of a disconnected graph. A simpler proof not requiring information about the symmetry coefficients will be given in Sec. 2.12.

Some additional remarks:

1. We see from (2.24) that all the graphs with contracted lines in (2.20) can be excluded by replacing the original interaction functional $V(\varphi)$ in the vertex factors (2.21) of the other graphs by the *reduced interaction functional* $V_*(\varphi) = PV(\varphi)$.[*]

2. If V is a polynomial in φ, then only the first few factors in (2.21) are nonzero, and the number of graphs is reduced accordingly.

3. In this technique, a vertex to which m lines are joined is associated with a factor (2.21) with m independent arguments x. However, if V is local, i.e., if it can be represented as a single integral over x of some function of the field and its derivatives at x (which is the usual case), then the factors (2.21) will contain δ functions making all its arguments x coincide, and one argument x will actually be associated with the vertex.

Let us conclude by briefly stating the basic rules of the universal graph technique.

Rule 1. Any functional of the form $P \exp V(\varphi)$ with $P \equiv P_\varphi$ from (2.14) is represented as the sum of unity and all free graphs with symmetry coefficients (2.23), lines Δ, and vertex factors (2.21).

Rule 2. Let $v_n(x_1...x_n)$ be a given set of symmetric functions independent of φ, $v_n \varphi^n \equiv \int ... \int dx_1...dx_n v_n(x_1...x_n)\varphi(x_1)...\varphi(x_n)$. The integral G defined as

$$G = c \int D\varphi \exp\left[\sum_{n=0}^\infty \tfrac{1}{n!} v_n \varphi^n\right], \qquad c^{-1} = \int D\varphi \exp\left[\tfrac{1}{2}v_2\varphi^2\right] \qquad (2.28)$$

is written as the sum of unity and all free graphs with symmetry coefficients (2.23), lines $\Delta = -v_2^{-1}$, and vertex factors v_n with $n \neq 2$. This follows from Rule 1 and the representation

$$G = P \exp\left[\sum_{n \neq 2} \tfrac{1}{n!} v_n \varphi^n\right]\Big|_{\varphi=0}, \qquad (2.29)$$

---

[*]In quantum field theory, $V(\varphi)$ is the Sym-form of the quantum interaction operator $\widehat{V}$, and $V_*(\varphi)$ is its N-form, i.e., $\widehat{V} = \mathrm{Sym}V(\widehat{\varphi}) = NV_*(\widehat{\varphi})$, where $\widehat{\varphi}$ is the free field operator, Sym denotes the symmetrized product, and N is the normal-ordered product of such fields. In the discussion of [35] it is assumed that the interaction is always written in N-form, i.e., the primary object is assumed to be $V_*(\varphi)$, not $V(\varphi)$. This makes it possible to ignore graphs with contracted lines.

obtained from (2.28) using (2.18). We note that a part of the quadratic contribution $v_2\varphi^2$ can, if desired, be included in the interaction, changing the line Δ and the normalization factor c in (2.28) accordingly.

Rule 3. Taking the logarithm of these objects is equivalent to selecting their connected parts.

2.3   Graph representations of Green functions

The generating functionals of the normalized full $[G(A)]$ and corresponding connected $[W(A)]$ Green functions introduced in Sec. 1.10 are given by the expressions of Sec. 1.13 in the language of functional integrals. It is more convenient to work with unnormalized Green functions, defining their generating functionals (retaining the same notation G and W) in the absence of spontaneous symmetry breaking (discussed in Sec. 2.4) as

$$G(A) = \exp W(A) = c \int D\varphi \exp[S(\varphi) + A\varphi], \qquad (2.30)$$

with arbitrary action functional (2.13) and constant c from (2.16). The integration region in (2.30) is defined according to the action functional by the general rules of Sec. 1.13. The functionals (2.30) and (1.62) differ only in their normalization: in (1.62) the integral was normalized by dividing by the exact partition function Z, while in (2.30) the unperturbed (free) one Z_0 was used. Therefore, the functional $G(A)$ from (2.30) differs from (1.62) only by the additional numerical factor $G_0 = G(0) = Z/Z_0$, and its logarithm $W(A)$ contains the additional term $W_0 = W(0) = \ln[Z/Z_0]$, which was not present in (1.43). All the other coefficients W_n, $n \geq 1$, of the Taylor expansion in A of the functionals $W(A)$ from (2.30) and (1.43) coincide and have the meaning of the connected Green functions of the model (Sec. 1.10). The *unnormalized full Green functions* determined by the generating functional (2.30) differ from the normalized full functions (1.33) only by the additional factor $G_0 = Z/Z_0$.

In practice, the integral in (2.30) can be calculated only as some perturbation theory; in the simplest case, in powers of the interaction V from (2.13). Strictly speaking, any perturbation theory can be assumed good only when the perturbation is insignificant for the integration region in (2.30) defined according to the rules of Sec. 1.13. In practice, this is often not true. For example, the presence in V of a contribution from a uniform external field h leads to a shift of the equilibrium average φ_0 determining the integration region [by the condition $\varphi(\infty) = \varphi_0$]. In such cases, to ensure self-consistency when perturbation theory is used it must be assumed that the integration region is defined by the unperturbed rather than the exact problem, in particular, by the free action S_0 for perturbation theory in V. We shall have this in mind, so that the use of equations like (2.17) and (2.18) is justified.

The functionals (2.19) and (2.30) defined using the action (2.13) are related simply to each other. From (2.13), (2.18), and (2.30) we find

$$G(A) = P \exp[V(\varphi) + A\varphi]|_{\varphi=0}. \tag{2.31}$$

Substituting $P \equiv P_\varphi$ in the form (2.14) and pulling $\exp A\varphi$ through P using the rule (2.9), taking into account the symmetry of Δ we find

$$G(A) = \left\{\exp[\tfrac{1}{2}A\Delta A + A\Delta(\delta/\delta\varphi)]\right\} P \exp V(\varphi)|_{\varphi=0}.$$

From this, using (2.11) and (2.19) we obtain

$$G(A) = H(\Delta A) \exp[\tfrac{1}{2}A\Delta A], \tag{2.32}$$

where ΔA in the argument of H is a convolution of the type (2.2).

The graph representations of the functionals (2.30) and (2.31) are determined by the general rules of Sec. 2.2. It follows from comparison of (2.31) and (2.29) that for arbitrary action (2.13), $\Delta = K^{-1}$ plays the role of a line in the graphs, while the vertex factors in the notation of (2.21) and (2.28) are

$$v_1 = A + V_1|_{\varphi=0} \equiv A + h, \quad v_n = V_n|_{\varphi=0} \text{ for } n \neq 1. \tag{2.33}$$

We have omitted the arguments x and have used h to denote the coefficient of φ in $V(\varphi)$.

As an example, let us consider the model (1.63) in more detail. Writing its action functional in the form (2.13), we have

$$\Delta(k) = (k^2 + \tau)^{-1}, \quad V(\varphi) = \int dx[-\tfrac{1}{24}g\varphi^4(x) + h\varphi(x)]. \tag{2.34}$$

The propagator Δ is taken in momentum space, and the coordinate kernel $\Delta(x, x')$ is defined in terms of $\Delta(k)$ using (1.35a). In the graphs of the S-matrix functional (2.19) the vertex factors (2.21) for the interaction (2.34) are the following:

$$V_1(x; \varphi) = h - \tfrac{1}{6}g\varphi^3(x),$$

$$V_2(x_1, x_2; \varphi) = -\tfrac{1}{2}g\varphi^2(x_1)\delta(x_1 - x_2),$$

$$V_3(x_1, x_2, x_3; \varphi) = -g\varphi(x_1)\delta(x_1 - x_2)\delta(x_2 - x_3),$$

$$V_4(x_1, x_2, x_3, x_4; \varphi) = -g\delta(x_1 - x_2)\delta(x_2 - x_3)\delta(x_3 - x_4),$$

and $V_n = 0$ for all $n > 4$. The graphs of the functionals (2.30) are determined by Rules 2 and 3 in Sec. 2.2, the line $\Delta(x, x')$ in them will be the same as in the graphs of H, and the only nonzero vertex factors in (2.33) are

$$v_1(x) = h + A(x), \quad v_4(x_1...x_4) = -g\delta(x_1 - x_2)\delta(x_2 - x_3)\delta(x_3 - x_4), \quad (2.35)$$

i.e., the graphs will contain only 1-leg and 4-leg vertices. Owing to the locality of the interaction, the latter actually corresponds to a single argument x and factor $-g$, and the number of such factors is the order of the perturbation theory. According to Rule 3 of Sec. 2.2, going from G to W in (2.30) corresponds to selecting connected graphs. As an illustration, we give all the graphs of $W(A)$ of order 1, g, and g^2:

$$\tag{2.36}$$

The coefficients are determined by (2.23), and the symmetry numbers s of all the graphs in (2.36) in the order in which they appear are $s = 2$, 2, 1, 24, 2, 2, 2, 2, 8, 2, 6, 72. Using the general rules of Sec. 2.2, for each graph we can easily write down the corresponding analytic expression; for example, for the second graph in (2.36) without the coefficient we have

$$-g\iiint dx dx' dx'' v_1(x)\Delta(x, x')\Delta(x', x')\Delta(x', x'')v_1(x'')$$

with v_1 from (2.35).

In going from the functional $W(A)$ to the individual connected functions $W_n(x_1...x_n) = \delta^n W(A)/\delta A(x_1)...\delta A(x_n)|_{A=0}$, the variational differentiation with respect to $A(x)$ corresponds graphically to removing the vertex factor $v_1(x)$ in all possible ways. Here the end of the line Δ attached to the vertex v_1 becomes free, and its coordinate x becomes one of the arguments of the function W_n. In this manner we can obtain the expansion of any of the functions W_n in the graphs (2.36) of the functional $W(A)$. This is done most simply for zero external field h, when $v_1(x) = A(x)$ and the functions W_n correspond to the graphs (2.36) with n 1-leg vertices. Observing the expansion of $W(A)$ analogous to (1.43), we see that in going to W_n the coefficients in (2.36) are multiplied by $n!$. Therefore, for $h = 0$ we obtain from (2.36)

$$W_0 = \ln\left[\frac{Z}{Z_0}\right] = \tfrac{1}{8}\; \text{(graph)} \;+\; \tfrac{1}{48}\; \text{(graph)} \;+\; \tfrac{1}{16}\; \text{(graph)} \;+\; ...,$$

$$W_2 = 2!\left[\tfrac{1}{2}\; \text{------} \;+\; \tfrac{1}{4}\; \text{(graph)} \;+\; \tfrac{1}{12}\; \text{(graph)} \;+ \right.$$

$$\left. +\; \tfrac{1}{8}\; \text{(graph)} \;+\; \tfrac{1}{8}\; \text{(graph)} \;+\; ... \right],$$

$$\tag{2.37}$$

$$W_4 = 4!\left[\tfrac{1}{24}\; \text{(graph)} \;+\; \tfrac{1}{16}\; \text{(graph)} \;+\; \tfrac{1}{12}\; \text{(graph)} \;+\; ... \right]$$

and so on. The outer ends of the legs now correspond to free arguments x of the functions $W_n(x_1...x_n)$, and they are always understood to be completely symmetrized. Let us explain by an example. The contribution to the function $W_4(x_1...x_4)$ from the second graph in (2.37) without the coefficient is

$$(-g)^2\mathrm{Sym}\int\!\int dx\,dx'\,\Delta(x_1,x)\Delta(x_2,x)\Delta^2(x,x')\Delta(x',x_3)\Delta(x',x_4),$$

where Sym denotes complete symmetrization with respect to permutations of $x_1...x_4$.

The graphs of W_0 not containing free arguments x are called *vacuum loops* (and also for $h \neq 0$). The passage to the normalized functionals (1.41) and (1.43) corresponds in (2.30) to discarding the contribution of W_0 in $W(A)$ and division by $G_0 = G(0)$ for $G(A)$. In graph language this is equivalent to discarding all the vacuum loops of $W(A)$ and all the graphs of $G(A)$ having at least one such loop as a connected component.

2.4 Graph technique for spontaneous symmetry breaking ($\tau < 0$)

The graph technique described in Sec. 2.3 for the model (1.63) corresponds to simple perturbation theory in powers of g and makes sense only for $\tau \geq 0$, when the unperturbed system with $g = 0$ is stable, and the propagator (2.34) is positive-definite. These properties are violated for $\tau < 0$, because the maximum of the action (1.63), corresponding to the energy minimum in the

Landau theory (Sec. 1.13), for $h = 0$ and $\tau < 0$ is reached not at $\varphi(x) = 0$, as in the case $\tau \geq 0$, but at $\varphi(x) = \kappa u$, where $u = \sqrt{-6\tau/g}$ and $\kappa = \pm 1$. The double-valuedness is a consequence of the $\varphi \to -\varphi$ symmetry at $h = 0$. The parameter κ labels the energy minima equivalent for $h = 0$, i.e., the "pure phases" in thermodynamics (Sec. 1.11). The perturbation theory can be constructed for any of these, using the general rules of Secs. 2.3 and 1.13 to define the functionals analogous to (2.30) for the phase κ as

$$G_\kappa(A) = \exp W_\kappa(A) = \bar{c} \int_{\varphi(\infty)=\kappa u} D\varphi \exp[S(\varphi) + A\varphi] \qquad (2.38)$$

with the new normalization constant $\bar{c}$ [the definition (2.16) becomes meaningless for $\tau < 0$ in S_0], which is found below. We are considering the simple φ^4 model and are defining the integration space in (2.38) using the action with $h = 0$ [see the discussion following (2.30)], which is indicated explicitly by the condition on φ in the integral.

Making the replacement $\varphi(x) = \overline{\varphi}(x) + \kappa u$ in the integral (2.38), we find

$$G_\kappa(A) = \bar{c} \int_{\overline{\varphi}(\infty)=0} D\overline{\varphi} \exp[\overline{S}_\kappa(\overline{\varphi}) + A(\overline{\varphi} + \kappa u)], \qquad (2.39)$$

$$\overline{S}_\kappa(\overline{\varphi}) \equiv S(\overline{\varphi} + \kappa u) = \overline{S}_0(\overline{\varphi}) + \overline{V}_\kappa(\overline{\varphi}), \qquad (2.40)$$

where $\overline{S}_0$ contains all contributions quadratic in $\overline{\varphi}$ (owing to symmetry, $\overline{S}_0$ is independent of κ), and $\overline{V}_\kappa$ contains all the rest. In contrast to the original S_0, the new functional $\overline{S}_0$ has definite sign (since it corresponds to the energy minimum), and it is used to define the constant $\bar{c}$ in (2.38) by the usual expression (2.16).

The contributions $S(\kappa u) + A\kappa u$ singular for $g \to 0$ in the argument of the exponential in (2.39) are taken outside the integral as coefficients. This leaves an interaction of the usual type with coupling constants that are formally small in g (of order g for $\overline{\varphi}^4$ and $gu \sim \sqrt{g}$ for $\overline{\varphi}^3$), to which the standard technique of Sec. 2.3 is applicable.

As a general scheme, the formalism described above is applicable to any system with spontaneous symmetry breaking. The meaning of the parameter κ labeling the pure phases depends on the problem. For the simple φ^4 model (1.63) $\kappa = \pm 1$, for its O_n generalization $\kappa = e \equiv \{e_a\}$, the unit vector directed along the spontaneous magnetization, and so on.

An important feature of models with the spontaneous breakdown of a continuous symmetry group is the presence for $h = 0$ and any $\tau \leq 0$ of massless Goldstone modes (Sec. 1.11), which for the O_n φ^4 model correspond to components of the field $\varphi \equiv \{\varphi_a\}$ transverse to the vector e. If the shift parameter u is defined, as above, using the action with $h = 0$ and the graphs of perturbation theory with $h \neq 0$ are studied, then, owing to the fact that the Goldstone modes are massless, IR divergences appear in the graphs at small momenta (this problem does not occur in the simple φ^4 model with $n = 1$).

To eliminate these divergences, the shift parameter in $\varphi_a = ue_a + \overline{\varphi}_a$ in the presence of an external field $h_a = he_a$ must be found from the condition that the total action including h be stationary. For the O_n φ^4 model the parameter u and the new τ after the shift for the single longitudinal and $n-1$ transverse components of φ will then be determined by the equations

$$\tau u + \tfrac{1}{6}gu^3 = h, \quad \tau_\| = -2\tau + 3h/u, \quad \tau_\perp = h/u, \tag{2.41}$$

where $u > 0$ is the root of the first equation in (2.41) with the original $\tau < 0$, corresponding to the absolute minimum of the energy including h. For this definition of the shift parameter, the transverse modes for $h \neq 0$ turn out to be massive ($\tau_\perp = h/u > 0$), and so the IR divergences of the graphs vanish. However, they leave a trace in the form of Goldstone singularities for $h \to 0$, which will be discussed in detail in Ch. 4.

For $h = 0$ all the phases labeled by the vector $\kappa = e$ become completely equivalent. The symmetry of the problem is preserved only in the form $W_{\kappa'}(A') = W_\kappa(A)$ and is spontaneously broken in each particular solution with a given κ. We shall return to these topics in Sec. 2.12.

2.5 One-irreducible Green functions

Let $W(A, h)$ be the functional W from (2.30) for the problem with an external field h, which enters into the action such that $W(A, h) = W(A + h, 0)$ (the usual case). Then the theory with an external field can be studied using the functional $W(A, 0) \equiv W(A)$ of the theory without a field, simply assuming in the final expressions that the role of the external field is played by the source A itself (after the differentiations with respect to it have been performed). In particular, the quantities

$$W_n = D_A^n W(A), \quad D_A \equiv \delta/\delta A \tag{2.42}$$

(the needed arguments x are always understood) for $n > 0$ then have the meaning of the connected Green functions of the theory with an external field A, and the quantity $W(A)$ itself is $\ln[Z/Z_0]$ (see Sec. 2.3).

Let $\Gamma(\alpha)$ be the Legendre transform of $W(A)$ with respect to A, defined in Sec. 1.11. We recall that α and A are mutually conjugate variables like temperature–entropy, pressure–volume, etc. in thermodynamics; $\alpha(x) = \langle \widehat{\varphi}(x) \rangle = W_1(x; A)$ is the first connected function (2.42) (the magnetization for a magnet). For Γ the independent variable is taken to be α, and the quantities

$$\Gamma_n = D_\alpha^n \Gamma(\alpha), \quad D_\alpha \equiv \delta/\delta\alpha \tag{2.43}$$

are by definition the 1-irreducible Green functions (sometimes they are also called the vertex or strongly-coupled Green functions) of the theory with a given value of α. We stress the fact that in (2.42) and (2.43) we are speaking of the Green functions of the same theory, even though they are expressed in different variables: for W_n it is natural to use the variable A, while for Γ_n it is natural to use α, where $\alpha = \alpha(A)$ and vice versa. Either variable can be taken to be the independent one.

The fundamental expressions of Sec. 1.11 can be written in the notation of (2.42) and (2.43) as

$$\Gamma(\alpha) = W(A) - \alpha A, \quad W_1 = \alpha, \quad \Gamma_1 = -A, \quad \Gamma_2 W_2 = -1, \qquad (2.44)$$

where the last equation is a matrix equation: its left-hand side is the product of linear operators with kernels Γ_2 and W_2, and the right-hand side is a multiple of the unit operator [see (1.52)]. Equations (2.44) can be used to express the connected functions (2.42) in terms of the 1-irreducible functions (2.43) and vice versa. Say, for example, that we want to express $W_n(A)$ (the arguments $x_1...x_n$ are understood) in terms of α and $\Gamma(\alpha)$. For the first two connected functions the result is contained in (2.44): $W_1 = \alpha$, $W_2 = -\Gamma_2^{-1}$, where $W_2 = D$ has the meaning of the *dressed propagator* (the quantum field synonym of the term "correlator") according to (1.44). The next W_n are obtained by the recursion relation $W_{n+1} = D_A W_n$ following from (2.42), using (2.44) to express the operator $D_A \equiv \delta/\delta A$ in the variables α: $D_A = [D_A \alpha] D_\alpha = W_2 D_\alpha = -\Gamma_2^{-1} D_\alpha$. Following (2.2), the result is understood as the action of the linear operator Γ_2^{-1} on the vector D_α. From these equations we find that

$$W_3 = D_A W_2 = [-\Gamma_2^{-1} D_\alpha][-\Gamma_2^{-1}] = [-\Gamma_2^{-1}]^3 \Gamma_3.$$

In the derivation we have used the definition (2.43) and the rule for differentiating an inverse matrix (operator): $D_\alpha[M^{-1}] = -M^{-1}[D_\alpha M]M^{-1}$, which is a matrix equation. Acting again with the operator $D_A = -\Gamma_2^{-1} D_\alpha$ on the expression obtained for W_3, we find the desired representation for W_4 and so on. The results are conveniently represented graphically, which also allows the location of the arguments x dropped above to be specified:

$$(2.45)$$

and so on. The heavy lines correspond to the dressed propagator (correlator) $D = W_2 = -\Gamma_2^{-1}$, the circles to the 1-irreducible functions (2.43) with $n \geq 3$, the points where circles and lines are joined correspond to the integration variables x, and the outer ends of the legs are the free arguments x of the functions W_n. Let us explain by an example: $W_4(1,2,3,4) = $ Sym $\int ... \int d1'd2'd3'd4'D(1,1') \cdot D(2,2') \cdot D(3,3') \cdot D(4,4') \cdot [\Gamma_4(1',2',3',4')$

$+3 \int\int d5' d6' \Gamma_3(1',2',5') \cdot D(5',6') \cdot \Gamma_3(6',3',4')]$, where $D = W_2$, the arguments x_i, x_i' are denoted i, i' for simplicity, and Sym denotes symmetrization over permutations of 1 to 4. We note that the coefficients 3, 10, and 15 in (2.45) coincide with the number of permutations needed for complete symmetrization of the corresponding contribution in (2.45) taking into account the symmetry of the functions Γ_n. In addition, in field theory one also uses the concept of *amputated Green functions* : the quantities obtained from the connected functions (2.45) by removing the legs D.

The problem of the theory is to calculate the partition function and the functions W_n for a given action S and external field A. For known $\Gamma(\alpha)$ the solution is constructed (Sec. 1.11) as follows. The desired field average $\alpha = \langle \hat{\varphi} \rangle = W_1$ is found from the given A using the equation $\Gamma_1(\alpha) = -A$, equivalent to the stationarity condition for the functional (1.56). Its value at the saddle point $\alpha(A)$ determines $W(A) = \ln[Z/Z_0]$. The higher derivatives of Γ with respect to α at this point are the functions (2.43) with $n \geq 2$, and they are used to find the dressed propagator $D = W_2 = -\Gamma_2^{-1}$ from (2.44) and all the higher functions W_n from (2.45). This procedure is also applicable in the case of spontaneous symmetry breaking, where the solution of the variational problem is nonunique (Sec. 1.11).

2.6   Graph representations of $\Gamma(\alpha)$ and the functions Γ_n

Using the known (Sec. 2.3) graph representation of $W(A)$ and the definitions of Sec. 2.5, for any model we can explicitly construct the corresponding graph expansion of $\Gamma(\alpha)$. However, this direct procedure is complicated, because it requires the graphical, iteration solution of the equation $\alpha = W_1(A) \equiv \delta W(A)/\delta A$ for A. There is a much simpler way of constructing the graphs of Γ from the graphs of W, as we now explain.

Definition: A connected graph that remains connected when any one line is removed from it is termed *1-irreducible*, and the *1-irreducible part of a functional* is the sum of the contributions of all its 1-irreducible graphs with their coefficients, discarding the reducible graphs. Earlier we simply called the functions (2.43) 1-irreducible, and now we shall verify that their graphs do actually possess this property.

Let us consider an arbitrary model (2.13) without the contribution linear in φ in the interaction ($h = 0$). First, we calculate the functionals W and Γ in zeroth order in V. The integral (2.30) for $S = S_0 = -\frac{1}{2}\varphi K \varphi$ is Gaussian, and its calculation using (2.1) gives $W(A) = \frac{1}{2}A\Delta A$ with $\Delta = K^{-1}$, for which from the definitions of Sec. 2.5 we easily find the corresponding functional $\Gamma(\alpha) = -\frac{1}{2}\alpha\Delta^{-1}\alpha = S_0(\alpha)$, where everything is a quadratic form of the type (2.3). Isolating these zeroth-order approximations in W and Γ, we write

$$W(A) = \tfrac{1}{2}A\Delta A + \widetilde{W}(A), \qquad \Gamma(\alpha) = -\tfrac{1}{2}\alpha\Delta^{-1}\alpha + \widetilde{\Gamma}(\alpha), \qquad (2.46)$$

where the terms $\widetilde{W}$ and $\widetilde{\Gamma}$ contain only nontrivial graphs with interaction vertices. The basic statement completely characterizing the graphs of Γ is

$$\widetilde{\Gamma}(\alpha) = \text{1-ir part of } \widetilde{W}(A = \Delta^{-1}\alpha). \tag{2.47}$$

Let us explain. According to (2.33), for $h = 0$ the argument A plays the role of the vertex factor v_1 of one-leg vertices in the graphs of $W(A)$, and so it always enters in the combination ΔA [a convolution of the type (2.2)], where Δ is the line attached to the one-leg vertex. Let us represent the independent argument α of Γ by a wavy external line (a tail). For the substitution indicated in (2.47), the combination ΔA becomes α, i.e., graphically, lines with a point at the end in graphs like those in (2.36) become tails α attached directly to the interaction vertices. [The exception is the first graph in (2.36), which was therefore isolated in (2.46), with the others forming $\widetilde{W}(A)$.] Among the graphs of $\widetilde{W}(A = \Delta^{-1}\alpha)$ obtained in this way there will be both 1-irreducible and reducible graphs. According to (2.47), the latter must be discarded, leaving the former with their coefficients, which corresponds to $\widetilde{\Gamma}$. For example, for the φ^4 model, from (2.36), (2.46), and (2.47) we obtain

$$\Gamma(\alpha) = -\tfrac{1}{2}\alpha\Delta^{-1}\alpha + \tfrac{1}{4}\;\raisebox{-0.5ex}{\includegraphics{}}\; + \tfrac{1}{8}\;\raisebox{-0.5ex}{\includegraphics{}}\; + \tfrac{1}{24}\;\raisebox{-0.5ex}{\includegraphics{}}\; + \tfrac{1}{48}\;\raisebox{-0.5ex}{\includegraphics{}}\; +$$

$$\tag{2.48}$$

$$+ \tfrac{1}{12}\;\raisebox{-0.5ex}{\includegraphics{}}\; + \tfrac{1}{16}\;\raisebox{-0.5ex}{\includegraphics{}}\; + \tfrac{1}{8}\;\raisebox{-0.5ex}{\includegraphics{}}\; + \tfrac{1}{16}\;\raisebox{-0.5ex}{\includegraphics{}}\; + \cdots \; .$$

After the replacement $\Delta A \rightarrow \alpha$, only the last three graphs in (2.36) generate 1-reducible contributions. Let us explain the meaning of the contributions (2.48) by two examples:

$$\raisebox{-0.5ex}{\includegraphics{}} = (-g)\int dx\,\alpha^4(x), \quad \raisebox{-0.5ex}{\includegraphics{}} = (-g)^2 \int\int dx\,dx'\,\alpha(x)\Delta^3(x,x')\alpha(x').$$

Let us give a brief proof of the statement (2.47) [53]. Assuming that α is an independent variable, from (2.44) and (2.46) we have $A = -\Gamma_1 \equiv -\delta\Gamma/\delta\alpha = \Delta^{-1}\alpha - \widetilde{\Gamma}_1$, where $\widetilde{\Gamma}_1 \equiv \delta\widetilde{\Gamma}/\delta\alpha$ consists only of nontrivial graphs with interaction vertices [since the trivial term in (2.46) has been isolated]. Expressing A in terms of α in this manner and substituting this expression along with (2.46) into the basic definition $\Gamma = W - \alpha A$, after cancellations we obtain the equation $\widetilde{\Gamma} = \widetilde{W} + \tfrac{1}{2}\widetilde{\Gamma}_1\Delta\widetilde{\Gamma}_1$. Owing to the nontriviality of the blocks $\widetilde{\Gamma}_1$, all the graphs of the second term on the right-hand side are certainly 1-reducible, and so by selecting the 1-irreducible part in the equation obtained above, we find: 1-ir part of $\widetilde{\Gamma}(\alpha)$ = 1-ir part of $\widetilde{W}(A(\alpha))$, where $A(\alpha) = \Delta^{-1}\alpha - \widetilde{\Gamma}_1$. The argument A enters into the graphs of $\widetilde{W}(A)$ only in the combination ΔA (see above), and with the replacement $A = A(\alpha)$ we obtain $\Delta A = \Delta[\Delta^{-1}\alpha - \widetilde{\Gamma}_1] = \alpha - \Delta\widetilde{\Gamma}_1$. Inclusion of the contribution $\Delta\widetilde{\Gamma}_1$

in the graphs of $\widetilde{W}(A(\alpha))$ leads to certainly 1-reducible graphs with the nontrivial block $\widetilde{\Gamma}_1$ connected by a single line Δ to the rest of the graph, and so this contribution can be dropped when the 1-irreducible part is selected: 1-ir part of $\widetilde{\Gamma}(\alpha)$ = 1-ir part of $\widetilde{W}(A = \Delta^{-1}\alpha)$.

This is "half" of the statement (2.47). The second half is obtained from the equations of motion for Γ, which will be studied in Secs. 2.11 and 2.12. It will be shown (Sec. 2.12) that all the graphs of $\widetilde{\Gamma}(\alpha)$ can be constructed by iterating these equations, and then the 1-irreducibility of all the graphs obtained in the iterations becomes obvious, which is equivalent to the equation: 1-ir part of $\widetilde{\Gamma}(\alpha) = \widetilde{\Gamma}(\alpha)$. This provides the proof of the statement (2.47).

From (2.32), (2.27), (2.46), and (2.47) we find that

$$\widetilde{W}(A) = \ln \, H(\Delta A) = \text{conn. part of } H(\Delta A),$$
$$\widetilde{\Gamma}(\alpha) = \text{1-ir part of } H(\alpha),$$

$$(2.49)$$

where H is the S-matrix functional (2.19). We note that the selection of the 1-ir part in the second equation in (2.49) automatically (according to the definition of 1-irreducibility) includes *a priori* selection of the connected part.

When the functions (2.43) are used, the operator D_α corresponds graphically to removing a tail α in all possible ways. For a theory with a given A, the functions (2.43) will be taken at the saddle point $\alpha(A)$. For a functional Γ even in α, in the absence of spontaneous symmetry breaking $\alpha(0) = 0$, and then the Γ_n are simply the coefficients of the expansion of $\Gamma(\alpha)$ in α at the origin. In particular, for the φ^4 model with zero external field ($A = 0$) and $\tau \geq 0$, from (2.48) we obtain

$$\Gamma_2 = -\Delta^{-1} + \tfrac{1}{2} \, \vcenter{\hbox{⊙}} + \tfrac{1}{4} \, \vcenter{\hbox{⊗}} + \tfrac{1}{6} \, \vcenter{\hbox{⬭}} + \cdots,$$

$$(2.50)$$

$$\Gamma_4 = \times + \tfrac{3}{2} \, \vcenter{\hbox{⋈}} + \cdots.$$

In contrast to the graphs of (2.37), the legs in the graphs of (2.50) are not associated with propagators Δ. These lines simply indicate the points at which the removed tails α were attached, and the coordinates x of the corresponding vertices are the arguments of the functions Γ_n. Removing two tails from a single vertex leads to the appearance of a δ function making the corresponding arguments x coincide. In detailed notation, the contributions of the three graphs in (2.50) without their coefficients to the function $\Gamma_2(x, x')$ have the form $-g\delta(x - x')\Delta(x, x)$, $(-g)^2\delta(x - x') \int dx'' \Delta^2(x, x'')\Delta(x'', x'')$, and $(-g)^2\Delta^3(x, x')$, and for $\Gamma_4(x_1...x_4)$ the first contribution is v_4 from (2.35), while the second is $(-g)^2\text{Sym}[\delta(x_1 - x_2)\delta(x_3 - x_4)\Delta^2(x_1, x_3)]$.

The equation $D^{-1} = \Delta^{-1} - \Sigma$ for the dressed propagator $D = W_2$ determines the quantity Σ, called the *self-energy*. From (2.44) and (2.46) we find

$$D^{-1} = -\Gamma_2 = \Delta^{-1} - \Sigma, \qquad \Sigma = \widetilde{\Gamma}_2 \equiv \delta^2\widetilde{\Gamma}(\alpha)/\delta\alpha\delta\alpha, \qquad (2.51)$$

which proves the 1-irreducibility of the graphs of $\Sigma(\alpha)$ (however, if α is expressed in terms of A, then for $A \neq 0$ the irreducibility is spoiled). From (2.51) we have $D = [\Delta^{-1} - \Sigma]^{-1} = \Delta + \Delta\Sigma\Delta + \Delta\Sigma\Delta\Sigma\Delta + ...$, or, graphically,

$$D = \quad - \quad + \quad -\!\!\blacksquare\!\!- \quad + \quad -\!\!\blacksquare\!\!-\!\!\blacksquare\!\!- \quad + \ldots , \qquad (2.52)$$

where a line denotes Δ and a block Σ. Equation (2.51) and its equivalent $D = \Delta + \Delta\Sigma D$ are usually called the *Dyson equations* .

The technique we have described corresponds to simple perturbation theory, which is applicable only for $\tau \geq 0$. For spontaneous symmetry breaking ($\tau < 0$), we can start from the modified graph technique for the functionals $W_\kappa(A)$ from (2.38). According to the general ideas of the variational principle (Sec. 1.11), the Legendre transform Γ will be the same for all $W_\kappa(A)$, i.e., Γ will not depend on κ: in the language of Γ, κ labels different solutions of the saddle-point equation, but the functional $\Gamma(\alpha)$ itself is unique. In practice, the most convenient graph technique for Γ in the case of spontaneous symmetry breakdown is the loop expansion, which will be discussed in Sec. 2.9.

Let us conclude by giving a useful simple rule for calculating the symmetry coefficients of graphs like (2.50) for an arbitrary monomial interaction $V = -g\varphi^N/N!$. Each vertex in these graphs is associated with a factor $-g$, and the coefficient C for any graph Γ_n can be found from

$$C = n!C_H\prod_a(1/n_a!). \qquad (2.53)$$

The index a labels the external vertices of the graph, n_a is the number of external lines leaving the vertex a, and C_H is the symmetry coefficient (2.23) of the "reduced graph" of the functional $H(\varphi)$, obtained by imagining the removal of all the external lines from the original graph Γ_n. The rule (2.53) follows from (2.49) taking into account the explicit form of the vertex factors (2.21) for the interaction in question [it is these which give rise to the factors $1/n_a!$ in (2.53); $\sum n_a = n$].

The rule (2.53) has an obvious generalization to monomial interactions of the type $V = -g(\varphi_1^{N_1}/N_1!)(\varphi_2^{N_2}/N_2!)...$ for a multicomponent field $\varphi = \{\varphi_1, \varphi_2...\}$. For polynomial interactions involving several monomials (for example, $g_4\varphi^4/4! + g_6\varphi^6/6!$), the rule (2.53) must be modified by introducing an additional combinatorial factor to take into account crossings. In the example given above, this factor is $(v_4+v_6)!/v_4!v_6!$, where $v_{4,6}$ are the numbers of vertices $g_{4,6}$ in the graph in question.

2.7   Passage to momentum space

Graphs are usually calculated in momentum space, and so we shall briefly review the commonly used rules for passing from coordinate to momentum space. In this section we shall take x to mean only d-dimensional spatial coordinates, and it is always understood that there are discrete indices on all Green functions when φ carries them.

In a uniform (in particular, zero) external field, all the Green functions are translationally invariant (we shall not consider problems like the liquid–solid phase transition with spontaneous breakdown of this symmetry), i.e., they depend only on coordinate (x) differences. For objects like Δ, $D = W_2$, Γ_2 the Fourier transform is then given by (1.35). In general, the Fourier transforms of such functions

$$\widetilde{F}_n(p_1...p_n) = \int...\int dx_1...dx_n F_n(x_1...x_n) \exp[-i\textstyle\sum_s p_s x_s] \tag{2.54}$$

contain a d-dimensional δ function ensuring momentum conservation:

$$\widetilde{F}_n(p_1...p_n) = (2\pi)^d \delta[\textstyle\sum_s p_s] F_{np}(p_1...p_n). \tag{2.55}$$

Here F_{np} is a function on the manifold $\sum p_s = 0$, where we use the additional subscript p to distinguish it from the coordinate function F_n in (2.54). The factor $(2\pi)^d$ has been isolated in (2.55) for convenience: with this normalization, the Fourier transform (1.35b) of the function $F_2(x, x')$ coincides with $F_{2p}(k, -k)$, and the vertex factor v_4 in (2.35) corresponds to $v_{4p} = -g$ without additional coefficients. It follows from (2.55) and (2.54) that

$$F_{np}(p_1...p_n) = \int...\int \underbrace{dx_1...dx_n}_{\text{one excluded}} F_n(x_1...x_n) \exp[-i\textstyle\sum_s p_s x_s] \tag{2.56}$$

with integration over all coordinates except any particular one (for $\sum p_s = 0$ the result does not depend on the choice of excluded coordinate). The Fourier transforms (2.54) of connected or 1-irreducible Green functions contain only a single δ function for momentum conservation, and so there will not be any such δ functions in the corresponding F_{np}.

Equations (2.54)–(2.56) pertain to the case $n \geq 1$. For constant F_0 like connected vacuum loops (Sec. 2.3), in translationally invariant models it is natural to isolate the infinite system "volume" $\int dx = (2\pi)^d \delta(0)$ from each graph. The coefficient F_{0p} will be the exact analog of the other F_{np}, i.e., for $n = 0$ instead of (2.55) we set $F_0 = F_{0p} \int dx$. With these definitions, for any $n \geq 0$ the canonical dimensions of these functions are related as

$$d[\widetilde{F}_n(p_1...p_n)] = d[F_{np}(p_1...p_n)] - d = d[F_n(x_1...x_n)] - nd. \tag{2.57}$$

Let us explain the general rules for passing to momentum space in the graphs for the example of the φ^4 model. Let F_n be the coordinate representation of an individual graph of a connected or 1-irreducible Green function. Its Fourier transform $\widetilde{F}_n$ and F_{np} are defined by (2.54) and (2.55). The arguments p are referred to as external momenta, and $\sum p_s = 0$. The explicit expression for F_{np} is constructed according to the following rules: (1) each line Δ is assigned an independent d-dimensional integrated momentum k, with an arrow indicating its (arbitrary) direction; (2) the external momenta p_s are assumed to flow into the graph at the points where the corresponding free coordinates x_s are located [for the graphs of (2.37) these are the ends of the legs Δ, and for the graphs of (2.50) the momenta p_s flow directly to the interaction vertices]; (3) each vertex, including the ends of the legs in (2.37), is associated with a momentum-conserving δ function ensuring that the sum of the incoming momenta is equal to the sum of the outgoing momenta; the resulting factor $\delta(\sum p_s)$ is discarded, because the quantity analogous to F_{np} in (2.55) is calculated; (4) each line Δ with momentum k_i is associated with a factor $\Delta(k_i)$ with $\Delta(k)$ from (2.34) [in general, the Fourier transform (1.35b) of the coordinate line $\Delta(x, x')$], and each interaction vertex is associated with a factor of $-g$; (5) all d-dimensional vectors k_i are integrated over, with the coefficient $(2\pi)^{-dl}$ multiplying the integral; here l is the number of loops, i.e., independent integration momenta in the graph; (6) the result is completely symmetrized over all permutations of the external momenta p.

When the fields carry discrete indices, additional index structures appear in the lines and vertices of the graphs (see the example of the O_n φ^4 model in Sec. 2.10), and the symmetrization is then done with respect to simultaneous permutations of the external momenta and the corresponding indices.

Let us explain Rule 5. The momentum-conserving δ functions at the vertices eliminate some of the integrations over the k_i [in particular, momentum conservation at external one-leg vertices of the graphs in (2.37) leads to the replacement $k \to p$ in the legs Δ]. By definition, l is the number of remaining independent integration momenta. For a connected or 1-irreducible graph, the number of loops l is equal to the number of lines minus the number of vertices [including the ends of legs for connected graphs in (2.37)] plus one. The one is added because one function $\delta(\sum p_s)$ is isolated explicitly in (2.55). The coefficient given in Rule 5 is made up from all the factors of 2π in the Fourier transforms and from (2.55).

The direction of the external momenta can be changed arbitrarily, assuming that some flow into the graph, while others flow out of it. This corresponds to a change of signs in (2.54)–(2.56), which, of course, must be taken into account when symmetrizing over the external momenta. For objects like W_2, Γ_2, and so on, the function $F_{2p}(p, -p)$ is usually denoted simply as $F_2(p)$ and coincides with the Fourier transform from (1.35). When a uniform external field h is present in the graphs of W_n, one-leg vertices with coefficients $v_1(x) = h$ appear (Sec. 2.3). Since h is uniform, zero momenta flow in these, and so the momenta of the corresponding legs Δ will be zero. Zero external momenta also flow through the corresponding tails in the graphs of $\Gamma_n(\alpha)$ with uniform $\alpha(x) = $ const (Sec. 2.6).

As an illustration, let us give the contributions to the functions Γ_{np} from the graphs of (2.50), specifying the choice of the ordering and directions of all the momenta:

$$= (-g)(2\pi)^{-d}\int dk\ \Delta(k), \qquad (2.58)$$

$$= (-g)^2(2\pi)^{-2d}\int\int dkdq\ \Delta(k)\Delta(q)\Delta(p-k-q), \qquad (2.59)$$

$$= (-g)^2(2\pi)^{-2d}\int\int dkdq\ \Delta^2(k)\Delta(q), \qquad (2.60)$$

$$= (-g)^2(2\pi)^{-d}\mathrm{Sym}\int dk\ \Delta(k)\Delta(p_1+p_2+k). \qquad (2.61)$$

All the variables are d-dimensional, $\Delta(k) \equiv (k^2 + \tau)^{-1}$, and Sym in (2.61) denotes the usual symmetrization with respect to permutations of $p_1...p_4$. In the graphs of Γ_n the external lines indicate only the points where the momenta p_s enter, and additional factors of $\Delta(p_s)$ from each leg appear in going to the graphs of the connected functions (2.37).

2.8 The saddle-point method. Loop expansion of $W(A)$

Let us consider the procedure for calculating the integral (2.30) with the action (2.13) by the saddle-point method. In this method one seeks the saddle point ψ of the argument of the integrated exponential, in our case, of the functional $S(\varphi) + A\varphi$. After contraction with $\Delta = K^{-1}$, the corresponding saddle-point equation $-K\psi + \delta V(\psi)/\delta\psi + A = 0$ takes the form

$$\psi = \Delta[A + V_1], \quad \text{with } V_n = V_n(\psi) \equiv \delta^n V(\psi)/\delta\psi^n, \qquad (2.62)$$

where the arguments x in the $V_n(\psi)$ analogous to (2.21) are understood. The iteration solution of (2.62) represents $\psi = \psi(x; A)$ as an infinite series of loopless graphs, commonly called *tree graphs*. For example, for the φ^4 model

$$\psi = \quad\text{---}\bullet\ +\ \frac{1}{6}\ \text{---}\!\!<\!\!<\ +\ \frac{1}{12}\ \text{---}\!\!<\!\!<\!\!<\ +\ ... \qquad (2.63)$$

with lines Δ and vertex factors (2.35). The coordinate x of the free end of the left-hand line Δ is the argument of $\psi(x)$.

We shift the integration variable in (2.30) $\varphi \to \varphi + \psi$ and then expand the exponent $S(\varphi + \psi) + A(\varphi + \psi)$ in a functional Taylor series in the new variable φ [using the notation of (2.28) and (2.62)]:

$$S(\varphi + \psi) + A(\varphi + \psi) = \sum_{n \neq 1} \tfrac{1}{n!} \overline{v}_n \varphi^n,$$

$$\overline{v}_n \equiv \frac{\delta^n [S(\psi) + A\psi]}{\delta \psi^n} = \left\{ \begin{array}{ll} S(\psi) + A\psi & \text{for } n = 0 \\ 0 & \text{for } n = 1 \\ -\Delta^{-1} + V_2 & \text{for } n = 2 \\ V_n & \text{for } n > 2 \end{array} \right\}, \tag{2.64}$$

where the arguments x and the required integrations over them are understood. There is no $n = 1$ contribution in the sum (2.64) owing to the saddle-point condition $\overline{v}_1 = 0$.

The configurations φ in the original integral (2.30) fluctuate about ψ (Sec. 1.13), and so the new variable φ after the shift fluctuates about zero, i.e., always $\varphi(\infty) = 0$ for it.

We thus obtain the following representation for the integral (2.30):

$$G(A) = \exp W(A) = c \int D\varphi \exp[\sum_{n \neq 1} \tfrac{1}{n!} \overline{v}_n \varphi^n] \tag{2.65}$$

with integration over fields φ falling off at infinity. The integral (2.65) is similar to (2.28), but has "incorrect" normalization: according to (2.28), the correct normalization $\overline{c}$ for (2.65) is given by $1/\overline{c} = \int D\varphi \exp[\tfrac{1}{2} \overline{v}_2 \varphi^2]$. Multiplying and dividing the right-hand side of (2.65) by $\overline{c}$ and then taking the logarithm of the equation, we obtain

$$W = \overline{W} + \ln(c/\overline{c}), \qquad \overline{W} \equiv \ln \left\{ \overline{c} \int D\varphi \exp[\sum_{n \neq 1} \tfrac{1}{n!} \overline{v}_n \varphi^n] \right\}. \tag{2.66}$$

According to Rules 2 and 3 of Sec. 2.2, $\overline{W}$ is the sum of all connected graphs with the standard symmetry coefficients, the lines

$$\overline{\Delta} \equiv -1/\overline{v}_2 = [\Delta^{-1} - V_2]^{-1}, \tag{2.67}$$

and the vertex factors $\overline{v}_n$ with $n \neq 1, 2$ from (2.64), i.e.,

$$\overline{W} = \bullet + \frac{1}{8}\, \bigcirc\!\!\bigcirc + \frac{1}{12}\, \ominus + \frac{1}{8}\, \bigcirc\!\!-\!\!\bigcirc + \dots . \tag{2.68}$$

The double lines stand for the propagator (2.67), represented graphically as

$$\overline{\Delta} \equiv =\!\!\!= \; = \; -\!\!- \; + \; -\!\!\bullet\!\!- \; + \; -\!\!\bullet\;\bullet\!\!- \; + \dots , \tag{2.69}$$

the points correspond to the vertex factor V_2, and the single lines to Δ.

The ratio $c/\overline{c}$ of the known (see above) Gaussian integrals is calculated using (2.16) and (2.4), from which, with (2.67), we find

$$\ln(c/\overline{c}) = -\tfrac{1}{2} \text{tr} \ln[\Delta/\overline{\Delta}] = -\tfrac{1}{2} \text{tr} \ln[1 - \Delta V_2], \tag{2.70}$$

or, graphically,

$$\ln(c/\bar{c}) = \frac{1}{2} \left[\bigcirc + \frac{1}{2} \bigcirc + \frac{1}{3} \bigcirc + \ldots \right]. \qquad (2.71)$$

The lines and vertices of the graphs in (2.68) have a functional dependence on ψ, and the substitution of $\psi = \psi(A)$ in the form of graphs of the type (2.63) takes us back to the ordinary graphs of Sec. 2.3 for the functional $W \equiv W(A)$. We shall classify them according to the number of loops (Sec. 2.7): $W = W^{(0)} + W^{(1)} + \ldots$, where $W^{(l)}$ is the infinite sum of all the ordinary l-loop graphs. Equations (2.66) and (2.70) actually solve the problem of summing all the ordinary graphs with a given number of loops. In fact, substitution of tree graphs of the type (2.63) into the lines and vertices of the graphs in (2.68) does not change the number of loops, and so all the loopless graphs are contained in the simple point of (2.68), i.e., $W^{(0)} = \bar{v}_0 = S(\psi) + A\psi$. The one-loop contribution $W^{(1)}$ is $\ln(c/\bar{c})$ in (2.66), known from (2.70) and (2.71), and $W^{(l)}$ with $l \geq 2$ is the sum of the contributions of all the l-loop graphs in (2.68). There is a finite number of them owing to the absence of a one-leg vertex in (2.64). For example, there are only three two-loop graphs of $\overline{W}$ (for any action S); they are all given in (2.68).

In conclusion, we note that it is natural to use the saddle-point method for calculating the integral $\int D\varphi \exp[NF(\varphi)]$ with large parameter N. Then each vertex of the graphs acquires a factor of N, and each line a factor of $1/N$, so that a factor of $1/N$ raised to the power $l - 1$ (the number of lines minus the number of vertices), where l is the number of loops, is associated with a connected graph. In this case the loop expansion will be an expansion in $1/N$.[*]

2.9 Loop expansion of $\Gamma(\alpha)$

Let $S(\varphi)$ be the action of the form (2.13) for the problem with zero external field h, $W(A)$ the corresponding functional from (2.30), and $\Gamma(\alpha)$ its Legendre transform (Sec. 2.5). We recall that in the variational formulation the role of h is played by the argument A itself in the final expressions.

Using the known (Sec. 2.8) loop expansion of W, from (2.46) and (2.47) we can easily construct the analogous expansion of Γ. We rewrite (2.47) as

$$\Gamma(\alpha) = S_0(\alpha) + \text{1-ir part of } \left[W - \tfrac{1}{2} A \Delta A \right] \big|_{A = \Delta^{-1}\alpha} \qquad (2.72)$$

and substitute into it the loop expansion $W = W^{(0)} + W^{(1)} + \ldots$. Since the replacement $A \to \Delta^{-1}\alpha$ does not change the number of loops, from (2.72)

---

[*]In quantum field theory, S does actually have the dimension of an action and so $SN = S/\hbar$, where $\hbar$ is Planck's constant, is dimensionless. For such functional integrals the loop expansion corresponds to the semiclassical expansion.

we find that

$$\Gamma^{(0)}(\alpha) = S_0(\alpha) + \text{1-ir part of } [W^{(0)} - \tfrac{1}{2}A\Delta A]|_{A=\Delta^{-1}\alpha}, \qquad (2.73)$$

$$\Gamma^{(l)}(\alpha) = \text{1-ir part of } W^{(l)}|_{A=\Delta^{-1}\alpha}, \qquad \text{for all } l \geq 1. \qquad (2.74)$$

We know (Sec. 2.8) that all the $W^{(l)}$ with $l \geq 1$ depend on A only through the variable $\psi = \psi(A)$, represented graphically by tree graphs like those in (2.63). In them, lines with a point at the end correspond to the combination ΔA [in general $\Delta(h + A)$, but here $h = 0$]. It therefore follows from (2.62) that $\psi(A = \Delta^{-1}\alpha) = \alpha$ plus 1-reducible graphs. The latter can be discarded upon substitution into (2.74), since they generate graphs that are certainly reducible. From (2.74) we then find that

$$\Gamma^{(l)}(\alpha) = \text{1-ir part of } W^{(l)}|_{\psi=\alpha}, \qquad \text{for all } l \geq 1. \qquad (2.75)$$

In (2.73) the quantity $W^{(0)} = S(\psi) + A\psi = S_0(\psi) + V(\psi) + A\psi$ with $\psi = \psi(A)$ from (2.62) is known (Sec. 2.8). All the arguments justifying going from (2.74) to (2.75) remain valid for the term $V(\psi)$, and so its contribution to $\Gamma^{(0)}$ is $V(\psi = \alpha)$, which together with $S_0(\alpha)$ from (2.73) forms the total action $S(\alpha)$. After removing V, in the square brackets in (2.73) we are left with the expression $S_0(\psi) + A\psi - \tfrac{1}{2}A\Delta A$ with $S_0(\psi) = -\tfrac{1}{2}\psi\Delta^{-1}\psi$. Substituting $\psi = \Delta(A + V_1)$ from (2.62) into it, we easily verify that this expression is reduced to the form $-\tfrac{1}{2}V_1\Delta V_1$. The line Δ joining two blocks V_1 cannot be contracted after the replacement $\psi \to \alpha$+tree graphs (see above), and so this expression consists only of 1-reducible graphs and does not contribute to the 1-irreducible part. Finally, in the loopless approximation

$$\Gamma^{(0)}(\alpha) = S(\alpha). \qquad (2.76)$$

Equations (2.75) and (2.76) give the general solution of the problem.

Let us briefly discuss the graph representation of the $\Gamma^{(l)}$. We know (Sec. 2.8) that the one-loop contribution $W^{(1)} = \ln(c/\bar{c})$ in (2.66) is represented by the 1-irreducible graphs (2.71), while the other $W^{(l)}$ with $l \geq 2$ are finite sums of l-loop graphs (2.68) with lines (2.67) and vertex factors $V_n = V_n(\psi)$ from (2.62), which become $V_n(\alpha)$ after the replacement $\psi \to \alpha$ is made in (2.75). When the 1-irreducible part is chosen in (2.75) the graphs of (2.68) which are reducible with respect to the double lines are discarded. Therefore,

$$\Gamma^{(1)}(\alpha) = \ln(c/\bar{c}) = \tfrac{1}{2}\text{tr} \ln[\overline{\Delta}/\Delta] = -\tfrac{1}{2}\text{tr} \ln[1 - \Delta V_2], \qquad (2.77)$$

$$\Gamma^{(2)}(\alpha) = \frac{1}{8} \;\; \text{} \;\; + \;\; \frac{1}{12} \;\; \text{} \qquad (2.78)$$

and so on, where in the lines $\overline{\Delta} = [\Delta^{-1} - V_2]^{-1}$ and the vertex factors V_n, $n \geq 3$, of these graphs it is understood that $V_n \equiv V_n(\alpha) \equiv \delta^n V(\alpha)/\delta\alpha^n$.

2.10   Loop calculation of $\Gamma(\alpha)$ in the O_n φ^4 model

As an example, let us consider the O_n-symmetric generalization of the model (1.63) with the field $\varphi = \{\varphi_a(x),\ a = 1, ..., n\}$ and action ($h = 0$)

$$S(\varphi) = \int dx[-\tfrac{1}{2}\varphi_a K\varphi_a - \tfrac{1}{24}g(\varphi^2)^2], \qquad K \equiv -\partial^2 + \tau, \qquad (2.79)$$

where $\tau = T - T_{c0}$, T_{c0} is the critical temperature in the Landau approximation, $\varphi^2 = \varphi_a\varphi_a$, and repeated indices are always summed over.

The general rules of the graph technique given above are valid for any model if universal notation is used, where x includes all the arguments of φ (the convention of Sec. 1.9). In this section we shall assume that x includes only spatial coordinates and shall explicitly isolate any indices. It should be recalled that in the usual symmetry relations, permutations of all the arguments of φ, i.e., coordinates and indices simultaneously, are understood.

The graphs of the O_n φ^4 model differ from those of the simple φ^4 model only by the appearance of additional index structures in the lines and vertices:

$$a\underline{\quad}b = \delta_{ab}, \qquad \overset{a}{\underset{d}{\diagdown}}\!\!\overset{b}{\underset{c}{\diagup}} = \frac{1}{3}\,[\delta_{ab}\delta_{cd} + \delta_{ac}\delta_{bd} + \delta_{ad}\delta_{bc}\,] \equiv v_{abcd}, \qquad (2.80)$$

so that $v_{abcd}\varphi_a\varphi_b\varphi_c\varphi_d = (\varphi^2)^2$. It is necessary to symmetrize v with respect to the indices because the coordinate vertex factor (2.35) is symmetric separately.

The graphs of the loop expansion of Γ contain the vertex factors $V_n = \delta^n V(\alpha)/\delta\alpha^n$. For our interaction $V(\alpha) = -\tfrac{g}{24}\int dx(\alpha^2(x))^2$ they are all local, i.e., they contain δ functions making all the coordinates coincide. We shall drop these and give only the index structures of the vertex factors V_n:

$$V_1^a = -\tfrac{g}{6}\alpha^2\alpha_a, \qquad V_2^{ab} = -\tfrac{g}{6}(\alpha^2\delta_{ab} + 2\alpha_a\alpha_b),$$

$$V_3^{abc} = -\tfrac{g}{3}(\delta_{ab}\alpha_c + \delta_{ac}\alpha_b + \delta_{bc}\alpha_a), \qquad V_4^{abcd} = -g v_{abcd} \qquad (2.81)$$

with v_{abcd} from (2.80). The placement of the indices in these expressions (raised or lowered) does not have any meaning but is chosen simply for convenience of the notation.

The double line in graphs like (2.78) corresponds to the propagator $\overline{\Delta}_{ab}$. In our case, from (2.67) and (2.81) we have

$$(1/\overline{\Delta})_{ab} = K\delta_{ab} + \tfrac{g}{6}(\alpha^2\delta_{ab} + 2\alpha_a\alpha_b) \qquad (2.82)$$

with K from (2.79) [in this notation K in (2.13) is the matrix $\delta_{ab}K$].

Later on we will need to calculate the variational derivatives with respect to α of various contributions to Γ. Vertices in graphs like (2.78) are differentiated according to the rule $\delta V_n/\delta\alpha = V_{n+1}$ (following from the definition of V_n), lines according to the rule $\delta\overline{\Delta}/\delta\alpha = \overline{\Delta}V_3\overline{\Delta}$ [following from the preceding expression and (2.67)], and tr ln... in (2.77) according to the rule $\delta\,\mathrm{tr}\ln\overline{\Delta}/\delta\alpha = -\mathrm{tr}(V_3\overline{\Delta})$ [following from (2.6)]. Graphically, for (2.77) and (2.78) we have

$$\delta\Gamma^{(1)}/\delta\alpha_a(x) \;=\; \tfrac{1}{2}V_3^{abc}(x)\overline{\Delta}_{bc}(x,x) \;=\; \tfrac{1}{2}\;\text{—}\bigcirc \;,$$

$$\delta\Gamma^{(2)}/\delta\alpha_a(x) \;=\; \tfrac{1}{4}\;\text{—}\bigcirc\!\bigcirc \;+\; \tfrac{1}{4}\;\text{—}\oplus \;+\; \tfrac{1}{6}\;\text{—}\ominus \;. \tag{2.83}$$

Here the leg plays the same role as in the graphs of (2.50), i.e., it is included only for defining the type of vertex and indicates that its arguments (in this case a and x) are free.

A general method of calculating the connected Green functions $W_n(A)$ and the quantity $W(A)$ in terms of $\Gamma(\alpha)$ and A has been described in detail in Secs. 2.5 and 1.11. It is usually only various thermodynamical characteristics and the dressed propagator (correlator) $D \equiv W_2$ that are calculated. The magnetization α (for definiteness we shall use magnet terminology) is determined from the saddle-point equation (1.51) for a given uniform external field A, and the propagator D, susceptibility χ, free energy $\mathcal{F}$, and specific heat C are calculated from the expressions

$$D = -\Gamma_2^{-1}, \qquad \chi = \partial\alpha/\partial A = D(p)\,|_{p=0},$$

$$\mathcal{F} = -W(A) - \ln Z_0 = \alpha A - \Gamma(\alpha) - \ln Z_0, \tag{2.84}$$

$$\ln Z_0 = -\tfrac{1}{2}\mathrm{tr}\ln(\Delta^{-1}/2\pi), \qquad C = -\partial_T^2\mathcal{F},$$

where Γ_2 is the matrix of second derivatives of Γ with respect to α. All quantities in (2.84) are taken at the saddle point $\alpha = \alpha(A)$.

Let us explain the expression for the free energy $\mathcal{F}$. For a magnet it is given in terms of the partition function Z as (see Sec. 1.4) $\ln Z = -\beta\mathcal{F}$, $\beta = 1/kT$. In the theory of critical behavior β is replaced by $\beta_c = 1/kT_c$ and becomes an unimportant normalization factor [as in (1.58)]. It is therefore usual to simply take $\mathcal{F} = -\ln Z$, understanding $\mathcal{F}$ as the free energy in units of kT_c. This normalization is also consistent with the definition of χ in (2.84), which differs from (1.36) by a coefficient. Then from the definitions (1.60), (2.16), and (2.30) we have $W(A) = \ln Z(A) - \ln Z_0$, where $Z(A)$ is the partition function for a system with external field A, and $Z_0 = c^{-1}$ is the known partition function (2.16) for the free system without interaction and the field. The quantity $W(A)$ is given in terms of $\Gamma(\alpha)$ by the first expression in (2.44). The contribution $\ln Z_0$ in (2.84) is naturally grouped together with the one-loop contribution (2.77):

$$\Gamma^{(1)} + \ln Z_0 = \tfrac{1}{2}\mathrm{tr}\ln(2\pi\overline{\Delta}) \cong \tfrac{1}{2}\mathrm{tr}\ln(\overline{\Delta}). \tag{2.85}$$

The factors of 2π can be dropped, since they contribute only an additive constant which is parameter-independent and therefore unimportant. The term $\ln Z_0$ in (2.84) thus eliminates the normalization factor Δ^{-1} inside the tr ln in (2.77), which affects the specific heat, since Δ depends on T

through $\tau = T - T_{c0}$. The elimination of Δ is particularly important for $\tau < 0$, where Δ is unsuitable because its sign is indefinite.

Let us go directly to the description of the calculational procedure. We shall need the first derivative $\Gamma_1^a \equiv \delta\Gamma/\delta\alpha_a(x)$ for substitution into the stationarity equation (1.51) and the second derivative $\Gamma_2^{ab} \equiv \delta^2\Gamma/\delta\alpha_a(x)\delta\alpha_b(y)$ to determine the propagator $D = -\Gamma_2^{-1}$. Representing Γ for the action (2.79) as a loop expansion, we find that

$$\Gamma_1^a = -K\alpha_a - g\alpha^2\alpha_a/6 + \frac{1}{2}\;\underset{a}{\bigcirc}\; + \ldots = -A_a, \qquad (2.86)$$

$$\Gamma_2^{ab} = -(1/\overline{\Delta})_{ab} + \frac{1}{2}\;\underset{a}{\bigcirc}_b\; + \frac{1}{2}\;\underset{a}{\overset{\bigcirc}{\frown}}_b\; + \ldots, \qquad (2.87)$$

where we have omitted the coordinate arguments, and the ellipsis denotes the contributions of graphs with two and more loops. The one-loop graphs in (2.87) were obtained by differentiating the graph in (2.86) with respect to α using the rules given above.

In actual calculations the external field A and the corresponding magnetization α [from this point on we take α to be not an arbitrary argument of Γ, but the solution of (2.86)] are usually assumed to be uniform. Then, owing to the symmetry of Γ, all the contributions on the left-hand side of (2.86) have the form $\alpha_a \Phi(\alpha^2)$, and so for $A \neq 0$ the vectors A and α must have the same direction. We take

$$A_a = Ae_a, \quad \alpha_a = \alpha e_a, \quad P_{ab}^{\|} = e_a e_b, \quad P_{ab}^{\perp} = \delta_{ab} - e_a e_b, \qquad (2.88)$$

where e_a is the unit vector in the direction of the external field, the new A and α are the moduli of the corresponding vectors, and P are projectors in which objects with two indices like D, $\overline{\Delta}$, and Γ_2 are expanded. For uniform α it is convenient to pass to momentum space everywhere. In particular, for the lines $\overline{\Delta}$ of the graphs (2.86) and (2.87), from (2.82) we obtain

$$\overline{\Delta}_{ab}(p) = (p^2 + \tau + \tfrac{1}{6}g\alpha^2)^{-1}P_{ab}^{\perp} + (p^2 + \tau + \tfrac{1}{2}g\alpha^2)^{-1}P_{ab}^{\|}. \qquad (2.89)$$

Let us first consider the loopless approximation, which is equivalent to the simple Landau theory. It is obtained by discarding all the graphs in (2.86) and (2.87). In this approximation $D = \overline{\Delta}$ and $T_c = T_{c0}$, and (2.86) for the homogeneous problem in the notation of (2.88) takes the form $\tau\alpha + \tfrac{1}{6}g\alpha^3 = A$, and for $A \neq 0$ uniquely determines $\alpha = \alpha(A, T)$ (of the three roots of the cubic equation, the one corresponding to the absolute minimum of the energy is chosen; see Sec. 1.13). For $A = 0$ and $\tau > 0$ the energy minimum corresponds to the solution $\alpha = 0$, and for $A = 0$ and $\tau < 0$

$$\alpha_a(0, T) = e_a(-6\tau/g)^{1/2}, \quad \tau = T - T_c < 0 \qquad (2.90)$$

with arbitrary direction of the spontaneous magnetization.

The loopless approximation for the correlator D is obtained by substituting $\alpha = \alpha(A, T)$ into (2.89). In particular, at zero field

$$D_{ab}(p)|_{\tau > 0} = \delta_{ab}(p^2 + \tau)^{-1},$$
$$D_{ab}(p)|_{\tau < 0} = p^{-2}P_{ab}^{\perp} + (p^2 - 2\tau)^{-1}P_{ab}^{\parallel}. \tag{2.91}$$

From this for the susceptibility in the isotropic phase we find that $\chi^{\parallel} = \chi^{\perp} = 1/\tau$, while in the anisotropic phase $\chi^{\perp} = \infty$, $\chi^{\parallel} = -1/2\tau$. These are the usual expressions of the Landau theory with the canonical values of the critical exponents. The masslessness of the transverse modes in the correlator (2.91) is a consequence of the general Goldstone theorem (Sec. 1.11). We note that the ratio of the amplitudes in a law like (1.22) for the longitudinal susceptibility is equal to 2 in this approximation.

Let us now briefly discuss the technique of calculating the corrections to the loopless approximation. Of course, it is impossible to solve (2.86) exactly taking into account the graphs. However, the exact solution is not needed, because for practical purposes (for example, for calculating the coefficients of $4 - \varepsilon$ expansions by the RG method) it is sufficient to solve (2.86) iteratively in the number of loops. This procedure can be given a precise meaning by taking g to be a small parameter and α a large one, such that $g\alpha^2 \cong \tau$ and $\tau\alpha \cong A$ for $A \neq 0$. Then all the contributions in the loopless saddle-point equation are of the same order of magnitude. We note that the condition $g\alpha^2 \cong \tau$ is automatically satisfied for spontaneous magnetization (2.90). With these assumptions the loop contribution in (2.86) is a small correction of order g compared to the loopless contribution, two-loop graphs give a smaller contribution of order g^2, and so on. Here we are discussing only the general picture; actual calculations are performed using the renormalized theory, where g is the dimensionless renormalized charge (Ch. 1). When the results are used to obtain $4 - \varepsilon$ expansions $g \to g_*(\varepsilon)$, where g_* is the RG fixed point (Sec. 1.33). The formal smallness of $g_*(\varepsilon) \sim \varepsilon$ justifies the iteration procedure.

In iterations the graph contributions in (2.86) are written as $\alpha_a \Phi(\alpha^2)$. Approximations are made in Φ, and the overall factor α_a is assumed to be exact. To obtain the one-loop correction to α in the function $\Phi(\alpha^2)$, it is necessary to replace α^2 in the one-loop graph (2.86) by the known loopless value, which is equivalent to adding a known term to τ (among other things, this leads to a shift of T_c). To find the propagator in this approximation, it is necessary to include in the loopless contribution of (2.87) the one-loop correction to α found in (2.86), while in the one-loop graphs of (2.87) it is sufficient to use the loopless approximation for α. At the next stage it is necessary to add to (2.86) the two-loop graphs from (2.83) with α in the loopless approximation and take into account the one-loop corrections to α in the one-loop graph of (2.86), and so on. It is essential that in this procedure the lines in the graphs always remain positive-definite even for $T < T_c$, in contrast to the graphs of simple perturbation theory.

Of course, this technique is needed only for nontrivial problems, for example, for calculations in an arbitrary external field $A \neq 0$, or for $A = 0$ but in the anisotropic phase, where ordinary perturbation theory is inapplicable (in the latter case the technique of Sec. 2.4 can also be used). For the calculation in zero field at $T \geq T_c$ it is simpler to use the graph technique of ordinary perturbation theory.

2.11 The Schwinger equations

The term Schwinger equations is generally used for any relations expressing invariance of the measure $D\varphi$ under translations $\varphi(x) \rightarrow \varphi(x) + \omega(x)$ by arbitrary fixed functions $\omega \in E$ with $\omega(\infty) = 0$ (Sec. 1.13). Such translations do not change the integration region specified by the condition $\varphi(\infty) = \varphi_0$, and so the quantity $\int D\varphi F(\varphi + \omega)$ is independent of ω for any F. Equality of its first variation with respect to ω to zero is expressed as

$$\int D\varphi \, \delta F(\varphi)/\delta\varphi(x) = 0. \tag{2.92}$$

In particular, for the integral (2.30) with $F = \exp[S + A\varphi]$, from (2.92) we have

$$\int D\varphi [\delta S(\varphi)/\delta\varphi(x) + A(x)] \exp[S(\varphi) + A\varphi] = 0. \tag{2.93}$$

Multiplication by φ inside the integral (2.30) is equivalent to differentiating the integral with respect to A, and so (2.93) can be rewritten as

$$\left[\delta S(\varphi)/\delta\varphi(x)|_{\varphi=\delta/\delta A} + A(x) \right] G(A) = 0. \tag{2.94}$$

When there is spontaneous symmetry breakdown (Sec. 2.4), each functional $G_\kappa(A)$ satisfies (2.94), because the statement (2.92) is valid (see above) for any integral of the type (2.38).

By substituting $G = \exp W$ from (2.94), we can obtain the equivalent equation for $W(A)$, and from it using the expressions of Sec. 2.5 we find the equation for $\Gamma(\alpha)$. All these equations are of finite order (for polynomial action) in the variational derivatives, and are equivalent to an infinite chain of coupled equations for the exact Green functions—the expansion coefficients of the corresponding functionals.

As an example, let us consider the φ^4 model (1.63) with $h = 0$. For it $\delta S/\delta\varphi(x) = -K\varphi(x) - \frac{g}{6}\varphi^3(x)$ with $K = -\partial^2 + \tau$. After substitution into (2.94) and contraction (for convenience) with $\Delta = K^{-1}$, we obtain the equation $\delta G(A)/\delta A(x) = \int dx' \Delta(x, x')[A(x') - \frac{g}{6}\delta^3/\delta A^3(x')]G(A)$. Substituting $G = \exp W$ into this and cancelling the exponential after differentiating, we find

$$W_1 \equiv \delta W/\delta A = \Delta A - \frac{g}{6}\Delta[W_3 + 3W_2 W_1 + W_1^3], \tag{2.95}$$

where W_n are the connected functions (2.42) of the theory with external field A, and for brevity we have omitted the arguments x and the integrations over them. The replacement $W(A) \to \Gamma(\alpha)$ in (2.95) is easily done using the equations of Sec. 2.5: $W_1 = \alpha$ becomes an independent functional argument, $A = -\Gamma_1$, $W_2 = -\Gamma_2^{-1}$, and W_3 is expressed in terms of Γ by (2.45). Finally, we have

$$\Gamma_1 \equiv \delta\Gamma/\delta\alpha = -\Delta^{-1}\alpha - \tfrac{g}{6}[W_3 + 3\alpha W_2 + \alpha^3]. \tag{2.96}$$

We have multiplied (2.95) by Δ^{-1}, rearranged the contributions, and retained the notation for $W_{2,3}$. However, now we assume that these quantities are expressed in terms of Γ (see above). We write out (2.95) graphically:

$$-\!\!\underbrace{}_{} = \underbrace{}_{\Delta} + \frac{1}{6}\underbrace{}_{\Delta} + \frac{1}{2}\underbrace{}_{\Delta} + \frac{1}{6}\underbrace{}_{\Delta} \; . \tag{2.97}$$

A circle with a number n together with its external lines denotes a connected function W_n, the bare lines Δ are labeled, a local 4-vertex corresponds to the standard factor $-g$, the one-leg vertex in the first graph corresponds to the factor A, and the coordinate x of the free left-hand end of a line Δ is the argument of the function $W_1(x)$ on the left-hand side of (2.97).

Equation (2.97) is an exact expression relating the first three functions W_n of the theory with external field A. By successively differentiating with respect to A, from (2.97) we obtain the equations for the higher functions W_n. Differentiation transforms W_n into W_{n+1} and introduces a new independent argument x, and so after the first differentiation from (2.97) we find

$$-\!\!\underbrace{}_{} = \underbrace{}_{\Delta} + \frac{1}{6}\underbrace{}_{\Delta} + \frac{1}{2}\underbrace{}_{\Delta} + \frac{1}{2}\underbrace{}_{\Delta} + \frac{1}{2}\underbrace{}_{\Delta} \; . \tag{2.98}$$

The equations simplify in zero field ($A = 0$) above T_c, when, owing to the $\varphi \to -\varphi$ symmetry, all the odd Green functions are zero. In this special case (2.98) takes the following form in the more convenient graphical notation of Sec. 2.5:

$$-\!\!\!\!- = -\!\!\!\!- + \frac{1}{6}\underbrace{}_{} + \frac{1}{2}\underbrace{}_{} \; , \tag{2.99}$$

where the heavy line denotes the dressed propagator $W_2 \equiv D$, and the shaded circle is the 1-irreducible 4-leg function Γ_4. Multiplying both sides of (2.99) on the left by Δ^{-1} and on the right by D^{-1}, we arrive at the Dyson equation (2.51) with the free energy

$$\Sigma = \frac{1}{6}\underbrace{}_{} + \frac{1}{2}\underbrace{}_{} \; . \tag{2.100}$$

This is a closed expression for the self-energy Σ in the φ^4 model with zero external field and $\tau \geq 0$, when the $\varphi \to -\varphi$ symmetry is not spontaneously broken. Relations of this type, which are also often referred to as Schwinger equations, can be obtained in any model by direct summation of the graphs of perturbation theory. However, it is much simpler to derive them using the general rule (2.93), because then it is not necessary to analyze the symmetry coefficients. Moreover, equations like (2.95) are more general, because they contain an infinite set of relations like (2.99) obtained from (2.95) by subsequent differentiation with respect to A and then taking $A = 0$.

We conclude by giving the graphical form of (2.96):

$$\Gamma_1 = \delta\Gamma/\delta\alpha = -\Delta^{-1}\alpha + \frac{1}{6} \ \text{} \ + \ \frac{1}{2} \ \text{} \ + \ \frac{1}{6} \ \text{} \ , \qquad (2.101)$$

where the tail denotes α and the rest of the notation is the same as in (2.100).

2.12   Solutions of the equations of motion

By iterating the equations (2.97) and (2.101) in the form of series in g, it is possible to obtain all the graphs of the corresponding functional except for the vacuum loops, since these equations determine the solution only up to an arbitrary additive constant. To obtain all the graphs it is sufficient to add to the Schwinger equation a second equation involving the variational derivative with respect to the line $\Delta(x, x')$. We shall assume that the symmetric kernel $\Delta(x, x')$ in $S_0 = -\frac{1}{2}\varphi\Delta^{-1}\varphi$ is an independent functional argument whose specific value is assigned only after differentiation. Using the rule for differentiating an inverse operator (see Sec. 2.5), it is easy to show that $2\delta S_0/\delta\Delta(x, y) = (\Delta^{-1}\varphi)_x(\Delta^{-1}\varphi)_y$, where each expression in parentheses is a convolution like (2.2). Differentiating the integral in (2.30) with respect to Δ (the normalization factor c is assumed constant and is not differentiated) and taking the resulting preexponential factor outside the integral as a differential operator ($\varphi \to \delta/\delta A$), we obtain the desired equation:

$$2\delta G/\delta\Delta(x, y) = (\Delta^{-1}\delta/\delta A)_x(\Delta^{-1}\delta/\delta A)_y G, \qquad (2.102)$$

which by standard manipulations (Sec. 2.11) is easily rewritten in terms of W and Γ. The Schwinger equation together with (2.102) in the required form make up the complete set of *equations of motion*. The iterative solution of this system of equations in the form of a series in g reproduces all the graphs of the functional under study.

The equations of motion are useful for proving statements of a topological nature such as (2.27) and (2.47). By performing one or two steps of the iteration procedure, it is easy to see that iterations in (2.97) generate only connected graphs, while in (2.101) they generate only 1-irreducible graphs. All the graphs of the functionals $W(A)$ and $\Gamma(\alpha)$ except for the vacuum loops $W(0)$ and $\Gamma(0)$ can be found in this way. The latter are then found using the constraint equation (2.102) rewritten in suitable form, and retain the properties of connectedness (for W) or 1-irreducibility (for Γ). It is difficult to find the explicit form of the graphs and their coefficients by this method, but the necessary topological property is a consequence of the general structure of the equations, and it is easy to establish. The 1-irreducibility of all the graphs of Γ with reference to the equations of motion was essentially used in Sec. 2.6 in proving (2.47). Let us give a proof similar in spirit of the statement (2.27) for the functionals (2.30): W = the connected part of G. Let it be known (from the equations of motion) that all the graphs of W are connected, i.e., W = the connected part of W. Since $G = 1+$ graphs, $W = \ln G$ is only graphs, and so all powers W^n with $n > 1$ are only disconnected graphs, i.e., their connected parts vanish. It follows from this and from the equation $G = \exp W = 1 + W + \frac{1}{2}W^2 + ...$ that the connected part of G is equal to the connected part of W. Together with the known (see above) equation W = the connected part of W, this gives the desired relation W = the connected part of G.

Therefore, use of the information obtained from the equations of motion allows statements like (2.27) to be proved without analyzing the symmetry coefficients, as is done in the usual proof (Sec. 2.2).

The equations can be iterated not only in powers of g, but also in other parameters, for example, the number of loops. The loopless approximation for Γ corresponds to discarding all graphs with loops on the right-hand side of (2.101). The one-loop graph in (2.101) with the line in the known loopless approximation is included in the following order, and so on. By integrating the corresponding contributions over α [using (2.6) for the one-loop term], it is easy to obtain $\Gamma^{(0)}$ and $\Gamma^{(1)}$ from (2.101) (Sec. 2.9). However, it is already difficult to obtain the next, two-loop, approximation $\Gamma^{(2)}$ in this way. In fact, this is not necessary, because the loop expansions of W and Γ are already known (Secs. 2.8 and 2.9) from other considerations.

As emphasized in Sec. 2.11, each functional W_κ from (2.38) also satisfies the equations of motion, being an anomalous solution (i.e., a solution that cannot be represented as a series in integer powers of g). Such solutions arise only for $\tau < 0$, when the normal solution (a series in g) becomes meaningless owing to the loss of definite sign of the bare propagator $\Delta(k) = (k^2 + \tau)^{-1}$. We recall (see Sec. 1.9) that positive-definiteness is a general property of the dressed propagator (correlator) D, and as such must be preserved in the leading order of its expansion in any parameter, in particular, in the bare correlator Δ, i.e., in zeroth order in g. The loss of the required sign definiteness formally indicates instability: for $\tau < 0$ the normal solution corresponds to the energy maximum in the Landau theory, and the solutions W_κ correspond to local minima, with the parameter κ labeling (Sec. 2.4) the individual pure phases. These phases are equivalent only for zero external field h, while for $h \neq 0$ the correct solution will be the one with definite $\kappa(h)$ corresponding to the absolute minimum of the energy in the Landau theory. We note that the uniform field h is always distinguishable from the localized (see the definitions and rules in Sec. 1.13) artificial functional argument A: although formally $W(A, h) = W(A + h, 0)$, A and h are uniquely reconstructed from the given $\widetilde{A} = A + h$, namely, $h = \widetilde{A}(\infty)$ and $A = \widetilde{A} - h$.

The equation $W(A, h) = W_{\kappa(h)}(A, h)$ determines (see above) the correct solution $W(A, h)$ for all $h \neq 0$, $\tau < 0$, which is obtained by joining together the solutions $W_\kappa(A, h)$ corresponding to the individual pure phases. This functional $W(A, h)$ is defined for all $h \neq 0$ and, in contrast to the individual W_κ, preserves the symmetry of the problem, as must occur when W is defined by taking the rigorous limit from the theory in a finite volume (Sec. 1.11). However, $h = 0$ is a singular point: the limits of $W(A, h)$ as h tends to zero along different directions are different, and coincide with the individual solutions $W_\kappa(A, 0)$.

This discussion is useful only for understanding the general picture. In practice, it is more convenient to work with anomalous solutions within the variational formulation, i.e., in terms of the functional Γ, which always preserves the symmetry of the problem and is independent of the parameter κ. In this language, the different phases correspond to different saddle points of the functional Γ.

2.13 Green functions with insertion of composite operators

A composite operator $\widehat{F}$ is any local construction formed from the field $\widehat{\varphi}(x)$ and its derivatives, for example, $\widehat{\varphi}^n(x)$, $\partial_i \widehat{\varphi}^n(x)$, $\widehat{\varphi}(x)\partial^2\widehat{\varphi}(x)$, and so on. The term is borrowed from quantum field theory, where $\widehat{\varphi}$ and anything constructed from it are actually operators in Hilbert space. For us they are random quantities. The replacement $\widehat{\varphi}(x) \rightarrow \varphi(x)$ transforms the random quantity $\widehat{F} = F(\widehat{\varphi})$ into a classical functional $F = F(\varphi)$, which we shall also refer to as a composite operator, distinguishing these objects only by the hat. In detailed notation $F = F(x; \varphi)$. We shall often omit the numerical (x) and functional (φ) arguments, or indicate only one of them.

Let $F_i(x; \varphi)$ be a given set of composite operators, where i is a label. The integral analogous to (2.30)

$$G(A, a) = c\int D\varphi \exp[S(\varphi) + aF(\varphi) + A\varphi] \qquad (2.103)$$

with arbitrary "sources" $a_i(x)$ in the linear form

$$aF(\varphi) \equiv \sum_i \int dx\, a_i(x)F_i(x; \varphi) \qquad (2.104)$$

is the generating functional of unnormalized full Green functions involving any number of fields $\widehat{\varphi}$ and operators $\widehat{F}$, and these functions themselves are the coefficients of the expansion of the functional in the set of sources A and a. We transform to normalized functions $\langle ... \rangle$ by dividing the functional (2.103) by the constant $G_0 = G(0, 0)$ (Sec. 2.3). *Definition:* We shall assume that $\widehat{F} = 0$ if all the full Green functions $\langle \widehat{F}\widehat{\varphi}...\widehat{\varphi}\rangle$ with one $\widehat{F}$ and any number $n \geq 0$ of factors $\widehat{\varphi}$ are equal to zero. Then all the Green functions with the given $\widehat{F}$, fields $\widehat{\varphi}$, and any composite operators constructed from the $\widehat{\varphi}$ will automatically vanish.

The graph representation of the functional (2.103) is defined by the same rules as for (2.30), and the contribution $aF(\varphi)$ is assumed to be added to the interaction $V(\varphi)$. The additional vertices that it generates are called *insertions of composite operators*. For example, the operators $F_1 = \varphi^2(x)$, $F_2 = \varphi^3(x)$, and $F_3 = \varphi(x)\partial^2\varphi(x)$ correspond to the local (with a single x) vertices

$$F_1 \sim \begin{matrix} \diagdown \\ \bullet \!\!\!< \\ \diagup \end{matrix} \quad ; \quad F_2 \sim \begin{matrix} \diagdown \\ < \!\!\! < \\ \diagup \end{matrix} \quad ; \quad F_3 \sim \frac{1}{2}\left[\begin{matrix} \diagdown \\ \bullet\!\!\!< \\ \diagup \end{matrix} + \begin{matrix} \diagdown \\ \bullet\!\!\!< \\ \diagup \end{matrix}\right]. \tag{2.105}$$

The line segments depict the ends of the lines Δ connected to a given vertex, and in F_3 the double slash on the lines denotes the action of the operator ∂^2 on their argument x. The half-sum is the required symmetrization. In the Green functions the argument x of the vertices (2.105) is free, and in the graphs of the functional (2.103) it is contracted with the argument of the corresponding source $a(x)$, which then plays the role of the vertex factor (Sec. 2.2).

Definition: The generating functional of connected Green functions of fields and operators is $W(A,a) = \ln G(A,a)$, and that of 1-irreducible functions is its Legendre transform $\Gamma(\alpha, a)$ with respect to the variable A at fixed a. When the replacement $W \to \Gamma$ is made in the various expressions, equations like (1.53) involving derivatives with respect to a play an important role. The expressions for the normalized full functions in terms of the connected ones for fields and operators are the usual ones (Sec. 1.10), in particular, $\langle \widehat{F} \rangle = \langle \widehat{F} \rangle_{\text{conn}}$, $\langle \widehat{F}\widehat{\varphi} \rangle = \langle \widehat{F}\widehat{\varphi} \rangle_{\text{conn}} + \langle \widehat{F} \rangle_{\text{conn}}\langle \widehat{\varphi} \rangle_{\text{conn}}$, and so on. It then follows that $\widehat{F} = 0$ if all the connected Green functions with a single $\widehat{F}$ and any number of simple fields $\widehat{\varphi}$ are zero, i.e., $\widehat{F}$ is uniquely specified by the generating functional of such functions, which we shall denote as $W_F = W_F(x; A)$. By definition, for any $n \geq 0$,

$$\langle \widehat{F}(x)\widehat{\varphi}(x_1)...\widehat{\varphi}(x_n) \rangle_{\text{conn}} = \delta^n W_F(x; A)/\delta A(x_1)... \, \delta A(x_n)|_{A=0}. \tag{2.106}$$

This allows the concept of composite operator to be generalized: since W_F determines $\widehat{F}$ uniquely, it can be assumed that the specification of any particular functional W_F defines a composite operator $\widehat{F}$ via the relations (2.106). This $\widehat{F}$ is defined independently of its representation in terms of $\widehat{\varphi}$, which may not exist at all.

It follows from the definition of $W(A,a)$ (see above) that for ordinary operators $F = F(x;\varphi)$ the logarithmic derivative of the functional (2.103) with respect to the corresponding source $a(x)$ at $a = 0$ plays the role of W_F. For such functionals it is convenient to use the notation

$$\langle\!\langle F(\varphi) \rangle\!\rangle \equiv \frac{\int D\varphi F(\varphi) \exp[S(\varphi) + A\varphi]}{\int D\varphi \exp[S(\varphi) + A\varphi]}, \tag{2.107a}$$

$$\langle F(\widehat{\varphi}) \rangle = \langle F(\varphi) \rangle \equiv \frac{\int D\varphi F(\varphi) \exp S(\varphi)}{\int D\varphi \exp S(\varphi)}. \tag{2.107b}$$

The functional averages $\langle...\rangle$ and $\langle\langle...\rangle\rangle$ are thereby defined: the first denotes the functional average with weight $\exp S(\varphi)$, and the second the average with weight $\exp[S(\varphi) + A\varphi]$. The A dependence of $\langle\langle...\rangle\rangle$ is not indicated explicitly but is understood. We recall that the symbol $\langle...\rangle$ for random quantities denotes the statistical average (1.17). The first equation in (2.107b) is a consequence of the definitions (1.60).

From the above discussion of local operators $F(x; \varphi)$ it follows that for them $W_F = \langle\langle F(x; \varphi)\rangle\rangle$ in the notation of (2.107). In the same notation the Schwinger equation (2.93) can be written as

$$\langle\langle u(x; \varphi)\rangle\rangle = -A(x), \quad \text{where} \quad u(x; \varphi) \equiv \delta S(\varphi)/\delta\varphi(x). \tag{2.108}$$

Equality of the generating functionals W_F of two operators is equivalent (see above) to equality of the operators themselves. The left-hand side of (2.108) is the functional W_F for the operator $F = u(x; \varphi)$. The right-hand side is not interpreted in terms of φ, but it is possible to introduce the formal operator $\widehat{F} \equiv \widehat{A}(x)$, defining (see above) all its Green functions (2.106) by $W_F(x; A) = A(x)$. The introduction of $\widehat{A}$ allows us to write the Schwinger equation (2.108) as an operator equation:[*]

$$u(x; \widehat{\varphi}) = -\widehat{A}(x). \tag{2.109}$$

In conclusion, we note that the relation

$$\langle\langle F(x; \varphi)\rangle\rangle = \exp(-W(A))F(x; \delta/\delta A)\exp W(A) \tag{2.110}$$

following from the definitions (2.107) and (2.30) allows any equation involving the average $\langle\langle...\rangle\rangle$ to be rewritten as an equation in variational derivatives.

---

[*]In quantum field theory, $\widehat{\varphi}(x)$ is not a random quantity, but an operator in Hilbert space. This (Heisenberg) operator satisfies the classical equation of motion $u(x; \widehat{\varphi}) = 0$, as is easily proved using the operator equations of motion and the commutation relations for the field operators [35]. Nevertheless, there also an analog of (2.109) is valid. Let us explain. The correlation functions of random quantities in statistical physics correspond in field theory to Green functions — vacuum expectation values of T-products, where T stands for ordering in the time t, containing time Θ functions like $\Theta(t - t')$. For operators of the form $\partial_t \widehat{F}$ involving external derivatives with respect to the time t, it is possible to distinguish the Dyson and Wick T-products and the corresponding Green functions. The Dyson T-product involves any operator as a whole, while in the Wick T-product, by definition, the derivatives ∂_t from $\partial_t \widehat{F}$ are taken outside the T-product. They therefore begin to act on the time Θ functions, generating additional contributions which distinguish the Wick T-product from the Dyson one. For a composite operator with external derivatives ∂_t equal to zero, all the Dyson Green functions involving this operator are zero, but the Wick functions may be nonzero. In quantum theory, functional averages with weight $\exp S$ are Wick, and not Dyson, Green functions, and so the equation (2.109) obtained by the functional technique determines the Wick Green functions with the operator $u(x; \widehat{\varphi})$. This is the reason that a nonzero value of its right-hand side does not contradict the equation $u(x; \widehat{\varphi}) = 0$. In the theory of a random field no such "paradoxes" arise, because there is no analog of the Dyson Green functions: the correlation functions are analogs of the Wick functions.

2.14 Summary of definitions of various Green functions

It is convenient to collect all the fundamental definitions and formulas together in one place. We shall denote the set of all fields of a model by the symbol $\varphi(x)$ (the convention of Sec. 1.9). A model is specified by the action functional $S(\varphi) = S_0(\varphi) + V(\varphi)$, with $S_0(\varphi) = -\frac{1}{2}\varphi K\varphi$. The symmetric operator kernel $\Delta = K^{-1}$ plays the role of the lines in the graphs of perturbation theory, and the interaction V defines the vertices of the graphs. The reduction operator $P \equiv P_\varphi$ is introduced by (2.14), and any expression involving P can be rewritten in the language of functional integrals using the identity (2.18).

Infinite families of symmetric Green functions are specified by the corresponding generating functionals of the type (1.41). This notation for the expansion is general, but the functional argument can be denoted in different ways (A, α, φ, and so on). We have introduced the following families of Green functions: the full Green functions $G_n(x_1, ..., x_n)$ (the arguments x will henceforth be omitted), the connected Green functions W_n, the 1-irreducible Green functions Γ_n, and the S-matrix Green functions H_n, where the normalized and unnormalized full Green functions are distinguished (Sec. 2.3). The generating functional of the unnormalized full Green functions $G(A)$ is defined by (2.30). The normalized full functions (1.33) are obtained by dividing this by $G_0 = G(0)$, and the connected functions by taking the logarithm: $W(A) = \ln G(A)$. The generating functional of 1-irreducible functions $\Gamma(\alpha)$ is the Legendre transform of $W(A)$ with respect to A (Sec. 2.5), and the generating functional of the S-matrix functions $H(\varphi)$ is defined in (2.19). Let us give the basic relations (see Secs. 2.3 and 2.5), replacing the argument $\alpha(x)$ of Γ by $\varphi(x)$:

$$H(\varphi) = P \exp V(\varphi), \quad G(A) = H(\Delta A) \exp[\tfrac{1}{2}A\Delta A],$$

$$W(A) = \ln G(A) = \text{conn. part of } G(A), \tag{2.111}$$

$$G_{\text{norm}}(A) = G(A)/G(0), \quad \Gamma(\varphi) = S_0(\varphi) + \text{1-ir part of } H(\varphi).$$

The graphs of the functions G_n and their connected parts W_n end in external lines Δ (Sec. 2.3), while those of Γ_n and H_n end directly in vertices (Secs. 2.2 and 2.6). By amputating the legs Δ in the graphs of G_n, we obtain the graphs of H_n. By selecting only the 1-irreducible graphs of the latter, we obtain the graphs of Γ_n. This all follows from (2.111).

All the definitions given above and Eq. (2.111) remain valid also when composite operators are inserted by adding (2.104) to the interaction $V(\varphi)$. The sources a in (2.104) are then treated as additional functional arguments, and the Taylor series of the functionals are written down for the set of all arguments. When defining 1-irreducible functions with composite operators, the Legendre transform $W \to \Gamma$ is performed only with respect to the arguments A, and the sources a are treated as fixed parameters.

2.15 Symmetries, currents, and the energy–momentum tensor

In this section we will take x to be only d-dimensional spatial coordinates, whose components will be labeled by the indices i, k, and s. The field components will be labeled by the indices a, b, c, and d, and the components of small parameters ω of infinitesimal transformations independent of x will be labeled by the index α. Repeated indices will always be summed over. The position of the indices (i.e., raised or lowered) in the expressions is irrelevant, and is chosen only to make the notation convenient.

Let $\varphi \to \varphi'$ be a Lie group of transformations of the form[*]

$$\varphi(x) \to \varphi'(x) = v(x)\varphi(x'), \quad x' = x'(x), \tag{2.112}$$

where $v(x)$ is a given matrix in the field indices and $x'(x)$ is a given coordinate transformation. For infinitesimal transformations,

$$x'(x) \equiv x_\omega = x + \delta_\omega x, \quad \varphi'(x) \equiv \varphi_\omega(x) = \varphi(x) + \delta_\omega \varphi(x),$$

$$\delta_\omega x_i = \omega_\alpha t_i^\alpha, \quad \delta_\omega \varphi_a(x) = \omega_\alpha T_{ab}^\alpha \varphi_b(x), \tag{2.113}$$

where T^α are the generators corresponding to the parameters ω_α, linear operators on the fields φ. For a given group, the T^α and t_i^α in (2.113) are known quantities. Let us describe some specific groups of the type (2.112).

1. An O_n *rotation* of an n-component field $\varphi = \{\varphi_a\}$: $\varphi'(x) = v\varphi(x)$, where v are orthogonal rotation matrices in n-dimensional ("isospin") space. In (2.113) $\delta_\omega x_i = 0$, $\delta_\omega \varphi_a(x) = \omega_{ab} \varphi_b(x)$ with arbitrary antisymmetric matrix ω, $\alpha = (cd)$ with $c < d$, $t_i^{cd} = 0$, and $T_{ab}^{cd} = \partial \omega_{ab}/\partial \omega_{cd} = \delta_{ac}\delta_{bd} - \delta_{ad}\delta_{bc}$.

2. *Spatial rotations* of a set of scalar fields $\varphi = \{\varphi_a\}$: $\varphi'_a(x) = \varphi_a(x')$, $x' = vx$, where v are orthogonal matrices of d-dimensional coordinate rotations. In (2.113) $\delta_\omega x_i = \omega_{ik} x_k$, $\delta_\omega \varphi_a(x) = \omega_{ks} x_s \partial_k \varphi_a(x)$ with arbitrary antisymmetric matrix ω, $\alpha = (ks)$ with $k < s$, $t_i^{ks} = x_s \delta_{ik} - x_k \delta_{is}$, and $T_{ab}^{ks} = \delta_{ab}\partial(\omega_{ks} x_s \partial_k)/\partial \omega_{ks} = \delta_{ab}(x_s \partial_k - x_k \partial_s)$.

3. *Translations* $\varphi'_a(x) = \varphi_a(x')$, $x' = x + \omega$, parametrized by the vector ω_i. In (2.113) $\alpha = k$, $t_i^k = \delta_{ik}$, and $T_{ab}^k = \delta_{ab}\partial_k$.

4. *Scale transformations* $\varphi'_a(x) = \lambda^\Delta \varphi_a(x')$, $x' = \lambda x$, with given (not necessarily canonical) dimension of the field Δ and arbitrary $\lambda > 0$. For infinitesimal transformations $\lambda = 1 + \omega$, $\delta_\omega x_i = \omega x_i$, the index α is absent in (2.113), $t_i = x_i$, and $T_{ab} = \delta_{ab}(x_k \partial_k + \Delta)$.

5. *Special conformal transformations* of a set of scalar fields $\{\varphi_a\}$ with given dimension Δ, parametrized, like translations, by the vector ω_i: $\varphi'_a(x) = Q^\Delta \varphi_a(x')$, $x'_i = (x_i + x^2 \omega_i)Q$, where $Q = (1 + 2\omega x + \omega^2 x^2)^{-1}$ (here $\det[dx'/dx] = Q^d$). From this for infinitesimal transformations we

---

[*]We treat (2.112) as an independent group, and not as a representation of a group of coordinate transformations $x \to x'$. The expression more natural for representation theory, $\varphi'(x') = v(x)\varphi(x)$, can be reduced to (2.112) by a change of notation.

have $\delta_\omega x_i = (x^2 \delta_{is} - 2x_i x_s)\omega_s$. In (2.113) $\alpha = k$, $t_i^k = x^2 \delta_{ik} - 2x_i x_k$, and $T_{ab}^k = \delta_{ab}(x^2 \partial_k - 2x_k x_s \partial_s - 2\Delta x_k)$.

A conformal transformation can be written as a product $rt_\omega r$, where t_ω is an ordinary translation and r is the following discrete ($r^2 = 1$) transformation corresponding to *conformal inversion*:

$$\varphi_a(x) \rightarrow \varphi_a'(x) = x^{-2\Delta}\varphi_a(x'), \qquad x_i' = x_i/x^2. \qquad (2.114)$$

Therefore, to have conformal invariance when there is translational invariance, it is sufficient (and, for arbitrary dimension d, also necessary[*]) to have invariance under the inversion (2.114).

Above we assumed that $\{\varphi_a\}$ is a set of scalar fields. In the general case of fields with spin, the generators of spatial rotations T^{ks} defined above are supplemented by spin contributions Σ^{ks}, which are known (from the type of field) matrices in the indices a and b and have the same commutation relations as the T^{ks}. The generators T^k of conformal transformations then acquire the additional term $2\Sigma^{ks}x_s$. The form of the conformal inversion (2.114) is changed accordingly: in the case of vector or tensor fields, the right-hand side of (2.114) is multiplied by the matrix $g_{ik} = \delta_{ik} - 2x_i x_k/x^2$ for each vector index. In the case of d-dimensional Dirac spinors ψ, $\overline{\psi}$ the factor $\widehat{n} \equiv \gamma_i n_i$ is added on the left for ψ and the factor $-\widehat{n}$ on the right for $\overline{\psi}$, where $n_i \equiv x_i/|x|$ and γ_i are the Dirac matrices obeying the anticommutation relation $\gamma_i \gamma_k + \gamma_k \gamma_i = 2\delta_{ik}$ in the Euclidean version of the formalism under discussion (see Sec. 4.43 for more details).

Now that we have finished our description of specific groups, let us return to the general analysis. Henceforth, if not stated otherwise, we shall assume that the action S is a functional on a space E of fields which fall off quickly at infinity (Sec. 1.13), which guarantees the absence of surface terms in the integral when the action is varied. Then variations of the field in (2.113) give

$$\delta_\omega S(\varphi) = \int dx\, u_a(x;\varphi)\delta_\omega \varphi_a(x), \qquad u_a(x;\varphi) \equiv \delta S(\varphi)/\delta \varphi_a(x). \qquad (2.115)$$

For invariant action S the integral (2.115) vanishes for any $\varphi \in E$. For reasonable models, the integrand in (2.115) for $\varphi \in E$ will also be a function which falls off quickly at infinity. In this situation, the following statement is valid. *Statement 1*: The equation $\int dx F(x) = 0$ with some function $F \in E$ guarantees that the function can be written as $F(x) = \partial_i J_i(x)$ with certain "currents" $J_i \in E$. In fact, $F(x) \in E$ has Fourier transform $\widetilde{F}(k)$ which is analytic in the neighborhood of zero (Sec. 1.13). By condition, $\widetilde{F}(0) = 0$, and for all the other monomials in the Taylor expansion of $\widetilde{F}(k)$ in k it is always possible to isolate one of the factors k_i to obtain the desired representation $\widetilde{F}(k) \sim k_i J_i(k)$. It follows from Statement 1 as applied to the integral (2.115) that if the action S is invariant under the transformations (2.113), there exist

---

[*]The exceptions are certain models in odd-dimensional spaces, for which the inversion gives an additional factor of -1 which cancels in the combination $rt_\omega r$.

currents J_i^α for which

$$u_a(x; \varphi) T_{ab}^\alpha \varphi_b(x) = -\partial_i J_i^\alpha(x; \varphi), \tag{2.116}$$

where the minus sign corresponds to the traditional notation. The equation $\partial_i J_i = 0$ is called *current conservation*, and the statement (2.116) is *Noether's theorem*: every independent parameter w_α in the symmetry group of the action is associated with a current J_i^α that is conserved on classical solutions [i.e., on φ which satisfy the classical equations of motion $u_a(x; \varphi) = 0$].

This derivation of Noether's theorem (2.116) is valid for any invariant action, including a nonlocal one. In specific models, the integrand in (2.115) is a linear combination of a finite number of local composite operators (Sec. 2.13), i.e., local monomials constructed from the field and its derivatives. All the currents in (2.116) can be regarded as such, because the following statement is valid. *Statement 2*: The equation $\int dx F(x; \varphi) = 0$ for any linear combination $F(x; \varphi) = \sum c_a F_a(x; \varphi)$ of a finite number of local monomials F_a guarantees that the function can be written as $F(x; \varphi) = \partial_i J_i(x; \varphi)$ with several local composite operators $J_i(x; \varphi)$. To prove this, it is sufficient to note that the Fourier transforms $\widetilde{F}_a(k; \varphi)$ for different monomials are independent for any k, and so the equation $\widetilde{F}(0; \varphi) = 0$ can hold only when an overall factor of k is present in the combination $\widetilde{F}(k; \varphi)$.

For local polynomials, (2.116) is naturally extended to any smooth $\varphi(x)$, which makes the statement about "current conservation on solutions" meaningful [in the space E of functions that fall off quickly at infinity, the classical equation $u_a(x; \varphi) = 0$ usually has only the solution equal to zero].

The equation (2.116) defines the currents only up to arbitrary, purely transverse additions J_i' with $\partial_i J_i' = 0$ for all φ. In specific models with local action $S = \int dx L(x)$ it is usual to use definite *canonical currents*, constructed from the Lagrangian density $L(x) = \mathcal{L}(\varphi(x), \partial\varphi(x))$ specified explicitly as a function $\mathcal{L}$ of the field $\varphi(x)$ and all its first (but no higher) derivatives $\partial\varphi(x)$. Invariance of the action under the transformations (2.112) is usually ensured by the equation

$$\mathcal{L}(\varphi'(x), \partial\varphi'(x))dx \equiv L'(x)dx = L(x')dx'. \tag{2.117}$$

Strictly speaking, this is not a necessary but only a sufficient condition for invariance, because in general we can add to the right-hand side an arbitrary expression of the form $(\partial_i \phi_i)dx$ that does not contribute to the integral over x. However, (2.117) is usually satisfied; the expressions for the currents obtained below are valid only if it is satisfied, which is easy to check for specific $\mathcal{L}$ and a specific group.

Let us use (2.117) for infinitesimal transformations (2.113). We immediately note (and this will be needed later) that for such transformations, $dx_w/dx \simeq 1 + \partial_i(\delta_w x_i)$, according to the last equation in (2.4). The first variation of $\mathcal{L}(\varphi_w(x), \partial\varphi_w(x))$ with respect to w can be found in two ways:

directly from variations of its arguments $\varphi(x)$ and $\partial\varphi(x)$, or using the equation

$$\mathcal{L}(\varphi_\omega(x), \partial\varphi_\omega(x)) = L(x_\omega)(dx_\omega/dx) \simeq L(x) + \partial_i[L(x)\delta_\omega x_i]$$

with $L(x) = \mathcal{L}(\varphi(x), \partial\varphi(x))$, which follows from (2.117) and the expression for dx_ω/dx given above. Equating the expressions obtained by these two methods, we find Eq. (2.116) with the currents

$$J_i^\alpha = \phi_{ia}T_{ab}^\alpha\varphi_b - \mathcal{L}t_i^\alpha, \qquad \phi_{ia} \equiv \partial\mathcal{L}/\partial(\partial_i\varphi_a) \tag{2.118}$$

and

$$u_a(x;\varphi) = \delta S(\varphi)/\delta\varphi_a(x) = \partial\mathcal{L}/\partial\varphi_a - \partial_i\phi_{ia}$$

in the same notation.

The currents for specific groups are obtained by substituting the known quantities t_i^α and T_{ab}^α into (2.118). The set of four currents (2.118) for translations is called the *energy–momentum tensor* (in the terminology of field theory) and denoted as $J_i^k = \vartheta_i^k$:

$$\vartheta_i^k = \phi_{ia}\partial_k\varphi_a - \mathcal{L}\delta_{ik} \tag{2.119}$$

with ϕ_{ia} from (2.118).

Let us give the form of the currents (2.118) in the same notation for the other groups described above. For O_n rotations,

$$J_i^{ab} = \phi_{ic}T_{cd}^{ab}\varphi_d = \phi_{ia}\varphi_b - \phi_{ib}\varphi_a,$$

for spatial rotations,

$$J_i^{ks} = x_s\vartheta_i^k - x_k\vartheta_i^s + \phi_{ia}\Sigma_{ab}^{ks}\varphi_b,$$

for scale transformations,

$$J_i = x_k\vartheta_i^k + \Delta\phi_{ia}\varphi_a,$$

and for conformal transformations,

$$J_i^k = (x^2\delta_{ks} - 2x_kx_s)\vartheta_i^s + 2\phi_{ia}[x_s\Sigma_{ab}^{ks} - \Delta x_k\delta_{ab}]\varphi_b.$$

The currents are given for the general case of fields with spin. If all the fields are scalar, then $\Sigma^{ks} = 0$. In this case, symmetry of the tensor (2.119) with respect to ik is the necessary and sufficient condition for the Lagrangian density $\mathcal{L}$ to be invariant under spatial rotations. In fact, using the equations $\phi_{ia}\partial_k\varphi_a = \phi_{ka}\partial_i\varphi_a$ with ϕ_{ia} from (2.118), understood as a system of linear, homogeneous partial differential equations for $\mathcal{L}$, it is easily proved that the derivatives $\partial_i\varphi_a$ for each component φ_a enter into $\mathcal{L}$ only as the gradient squared, i.e., symmetry of ϑ_i^k implies isotropy of $\mathcal{L}$. The reverse is obvious.

It is impossible to give more specific expressions for the currents without additional information about $\mathcal{L}$. If, as is usual, $\mathcal{L} = -\frac{1}{2}\partial_i\varphi_a \cdot \partial_i\varphi_a +$ contributions not containing derivatives, then in the expressions given above $\phi_{ia} \equiv \partial\mathcal{L}/\partial(\partial_i\varphi_a) = -\partial_i\varphi_a$, and so, for example, for O_n rotations,

$$J_i^{ab} = \varphi_a\partial_i\varphi_b - \varphi_b\partial_i\varphi_a. \tag{2.120}$$

We note that the symmetry of ϑ_i^k in this case is obvious, and that the other currents depend on the contributions in $\mathcal{L}$ without derivatives only through the ϑ_i^k.

The definition of the currents (2.118) remains meaningful also in the absence of symmetry, but then the left-hand side of (2.116) acquires an additional term that spoils the current conservation on the classical solutions. For translations and spatial rotations the invariance is violated by the appearance of an additional explicit x dependence in $\mathcal{L}$. Canonical scale invariance (1.67) with field dimensions $\Delta = d_\varphi$ is always present, but there both the fields and all the dimensional parameters of the model are transformed simultaneously. Invariance under scale transformations of only the fields themselves and its consequence — conservation of the corresponding dilatation current — occur only when the action does not contain any dimensional parameters, for example, in the massless φ_4^4 model. However, any regularization (Sec. 1.18) unavoidably introduces dimensional parameters, and so the symmetry is broken. The situation with conformal invariance is exactly the same. The violation of a symmetry due only to regularization is commonly called an *anomaly* in field theory, in contrast to *explicit violation* by noninvariant contributions to S.

Let us make some concluding remarks. (1) A symmetry of the action implies the existence of a current conserved on the classical solutions, but the reverse is not true in general. For current conservation on solutions [i.e., for $u_a(x;\varphi) = 0$] it is sufficient to have $\partial_i J_i = c_a u_a$ with arbitrary coefficients c_a. However, symmetry of the action follows from this equation only for completely defined [as in (2.116)] coefficients $c_a \sim \delta_\omega\varphi_a$. (2) In models with nonscalar fields, the canonical tensor (2.119) is not always symmetric in ik,[*] and if not it is symmetrized by a suitable addition conserved on the classical solutions, taking advantage of the natural arbitrariness in the definition of the concepts of energy and momentum. It should be born in mind that (2.116) may not be satisfied for a symmetric tensor ϑ_i^k obtained in this manner.

Finally, in models with massless fields, for which canonical scale and conformal invariance are violated only by regularization, instead of the canonical expression (2.119) the *conformal energy–momentum tensor* $\overline{\vartheta}_i^k$ and the corresponding *dilatation current* $\overline{J}_i = \overline{\vartheta}_i^k x_k$ are often used. For the latter, $\partial_i\overline{J}_i = \overline{\vartheta}_i^i = 0$ on classical solutions. In scalar models like the φ^4 model

---

[*]For example, in electrodynamics.

the quantities $\overline{\vartheta}_i^k$ and $\overline{J}_i$ differ from the corresponding canonical expressions by terms with the structure $(\partial^2 \delta_{ik} - \partial_i \partial_k)\varphi^2$ for ϑ_i^k and $\partial_k(x_i \partial_k - x_k \partial_i)\varphi^2$ for J_i, which are purely transverse for any φ (and not only on solutions).

2.16   Ward identities

The Ward identities are various relations following from an exact or approximate symmetry of the action. The simplest is the statement that the Green functions are invariant, and it is conveniently stated in the language of the corresponding generating functionals (Sec. 2.14). For transformations linear in φ (2.112), the equation $A'\varphi' = A\varphi$ for a form of the type (1.48) defines the transformation of the source $A \to A'$ corresponding to (2.112). For infinitesimal transformations (2.113) with $\delta_\omega \varphi = \omega T \varphi$ we then obtain $\delta_\omega A = -(\omega T)^\top A = -A\omega T$, where $\top$ denotes the transpose of a linear operator (Sec. 2.1). Therefore, the invariance of the connected Green functions under transformations (2.112) can be expressed by the following equivalent relations:

$$W(A') = W(A) \Leftrightarrow \partial[\delta_\omega W(A)]/\partial \omega_\alpha = \int dx \, W_x[T^\alpha] = 0 \quad \text{for all } \alpha. \quad (2.121)$$

Here we have introduced the convenient notation

$$W_x[T] \equiv A_a(x)T_{ab}\delta W(A)/\delta A_b(x), \quad (2.122)$$

where the subscript x on $W[T]$ indicates that this quantity depends on x. The functional dependence on A is not indicated but understood.

The second equation in (2.121) is the infinitesimal form of the first. Relations of this type are called *global Ward identities*. They express the invariance of the functional $W(A)$ under global (i.e., involving parameters ω independent of x) transformations of the type (2.112).

The invariance (2.121) is guaranteed if the action S, measure $D\varphi$, and integration region E_{int} as a whole in the integral (2.30) are invariant under the transformations (2.112). The latter condition is not satisfied when this symmetry is spontaneously broken, and then instead of (2.121) we have

$$W_{\kappa'}(A') = W_\kappa(A), \quad (2.123)$$

i.e., the group transformations relate the functionals W_κ for different pure phases to each other, and for each individual phase with a given κ the symmetry is broken: $W_\kappa(A') \neq W_\kappa(A)$. By differentiating (2.123) with respect to the parameters ω_α we obtain the infinitesimal form.

This method [differentiation of $W(A')$ with respect to ω_α] of obtaining the global Ward identities is convenient for transformations like (2.112) linear in φ, for which from the equation $A'\varphi' = A\varphi$ it is easy to introduce the source transformation $A \to A'$ corresponding to $\varphi \to \varphi'$. In the general case of an arbitrary continuous group that does not change the integration region E_{int} as a whole, the Ward identities analogous to (2.121) are derived as follows. We consider an arbitrary infinitesimal transformation $\varphi(x) \to \varphi_\omega(x) = \varphi(x) + \delta_\omega \varphi(x)$ and then claim that, since the integral is independent of the notation used for the integration variable, the quantity $\int D\varphi_\omega \exp[S(\varphi_\omega) + A\varphi_\omega]$ is independent of the parameters ω, so that its first variation with respect to ω must be zero. In the notation of (2.107) this can be written as

$$\langle\!\langle \mathrm{tr}[\delta(\delta_\omega \varphi)/\delta\varphi] + \delta_\omega S(\varphi) + A\delta_\omega \varphi \rangle\!\rangle = 0, \qquad (2.124)$$

where the first term is the contribution of the Jacobian $D\varphi_\omega/D\varphi = \det[\delta\varphi_\omega/\delta\varphi] \simeq 1 + \mathrm{tr}[\delta(\delta_\omega \varphi)/\delta\varphi]$ with the required accuracy in ω [a consequence of the last expression in (2.4)]. In particular, for the groups (2.112) with variation linear in φ (2.113) we have (omitting the indices for brevity)

$$\mathrm{tr}[\delta(\delta_\omega \varphi)/\delta\varphi] = \mathrm{tr}(\omega T) = \mathrm{tr}(\omega \, \mathrm{Sym} \, T), \quad \mathrm{Sym} \, T \equiv \frac{1}{2}(T + T^\top), \qquad (2.125)$$

since the trace of the linear operator T is determined (Sec. 2.1) by only its symmetric part $\mathrm{Sym} \, T$.

If the action and the measure are invariant, then only the last term remains on the left-hand side of (2.124). The corresponding equation $\langle\!\langle A\delta_\omega \varphi \rangle\!\rangle = 0$ for the variations (2.113) taking into account (2.110) is equivalent to the Ward identity (2.121).

We see from (2.125) that invariance of the measure for groups of the type (2.112) is ensured by antisymmetry of the generators T. Therefore, the measure $D\varphi$ is invariant under translations and various rotations (Sec. 2.15) whose generators are antisymmetric (we recall that $\partial^\top = -\partial$). However, for scale and conformal transformations the generators T^α contain a nontrivial symmetric part [for example, for the dilatation generator $T = \Delta + x_k\partial_k$ taking into account the relations $\partial^\top = -\partial$ and $(\partial_k x_k) = d$, we find $T^\top = \Delta - \partial_k x_k = \Delta - d - x_k\partial_k$, from which we have $\mathrm{Sym} \, T = \Delta - d/2 \neq 0$ for $\Delta \neq d/2$], and so the measure $D\varphi$ is not invariant under these transformations. The interpretation of the quantity $\langle\!\langle \delta_\omega S \rangle\!\rangle$ in (2.124) for these groups is also nontrivial. We shall return to this below after discussing the local Ward identities.

Local Ward identities differ from global identities like (2.121) in that there is no integration over x, and they can be obtained in different ways for any specific model. One way is to use the trick described above in the derivation of (2.124) for the local analog of the transformation in question with x-dependent parameters $\omega_\alpha(x)$. Equation (2.124) remains valid, and by variational differentiation of it with respect to the parameters $\omega_\alpha(x)$ we obtain the desired local relations:

$$\frac{\delta}{\delta\omega_\alpha(x)} \langle\!\langle \mathrm{tr}[\delta(\delta_\omega \varphi)/\delta\varphi] + \delta S_\omega(\varphi) + A\delta_\omega \varphi \rangle\!\rangle = 0, \qquad (2.126)$$

which in models with (globally) invariant action will have the meaning of *local Ward identities*.

For the important special case of the transformations (2.112) we can obtain a more explicit form of equations like (2.126) by a different method. We use the following equation from the general rule (2.92):

$$0 = \int D\varphi \; \delta\{\varphi_b(x) \exp[S(\varphi) + A\varphi]\}/\delta\varphi_a(x'),$$

which in the notation of (2.107) and (2.115) takes the form

$$\langle\!\langle \delta_{ab}\delta(x - x') + [u_a(x';\varphi) + A_a(x')]\varphi_b(x)\rangle\!\rangle = 0.$$

Transforming the contribution involving A by the rule (2.110), we find

$$\langle\!\langle -\delta_{ab}\delta(x - x') - u_a(x';\varphi)\varphi_b(x)\rangle\!\rangle = A_a(x')\delta W(A)/\delta A_b(x). \qquad (2.127)$$

We contract this equation with the generators T^α_{ab} of the group in question (notation of Sec. 2.15) acting on the variable x and then set $x' = x$. The quantity $T^\alpha_{ab}\delta(x - x')$ is the kernel (Sec. 2.1) of the linear operator T^α. Only the symmetric part of the operator T, Sym $T \equiv \frac{1}{2}(T + T^\top)$, contributes to the diagonal element with $x' = x$ and $b = a$, and so from (2.127) and (2.122) we find

$$\langle\!\langle -\text{Sym } T^\alpha_{aa}\delta(x - x')|_{x'=x} - u_a(x;\varphi)T^\alpha_{ab}\varphi_b(x)\rangle\!\rangle = W_x[T^\alpha]. \qquad (2.128)$$

By analogy with the passage from (2.108) to (2.109), Eq. (2.128) can also be interpreted as an equality of composite operators. The functional on the right-hand side of (2.128) defines, via (2.106) with $W_F = W_x[T^\alpha]$, a composite operator not expressed explicitly in terms of φ, similar to $\widehat{A}$ on the right-hand side of (2.109).

Equation (2.128) is valid independently of the symmetry of the action. In the presence of this symmetry, from (2.128) and (2.116) we find

$$\langle\!\langle -\text{Sym } T^\alpha_{aa}\delta(x - x')|_{x'=x} + \partial_i J^\alpha_i(x;\varphi)\rangle\!\rangle = W_x[T^\alpha]. \qquad (2.129)$$

This is the desired form of the local Ward identity for groups like (2.112). Like any relation obtained from (2.92), it remains valid for each individual pure phase when there is spontaneous symmetry breakdown (Sec. 2.11), in contrast to the corresponding global Ward identity, which does not hold for the individual phases (see above).

Let us explain the relation between the local and global Ward identities in more detail for the example of O_n symmetry. For it Sym $T^\alpha = 0$, and so the first term on the left-hand side of (2.129) vanishes. For $T > T_c$ the functional integration in (2.30) and (2.107) runs over the space $E_{\text{int}} = E$ of fields that fall off rapidly at infinity (Sec. 1.13). The currents (2.118) constructed from φ will also possess this property, which guarantees the absence of surface terms in the integral over x from the divergence of the current. Therefore, the left-hand side of (2.129) vanishes after integration over x, which leads to the global Ward identity (2.121).

When there is spontaneous symmetry breakdown $(T < T_c)$, the functional integration in (2.107) runs over the region $E_{int} = E_{int}(\varphi_0)$, with $\varphi_0 \equiv \kappa u$ [see the definition (2.38) and also Rule 3 and the following text in Sec. 1.13]. The local equation (2.116) and its consequence (2.129) are satisfied also in this case, but the global relation (2.115) becomes meaningless, because for a functional $S(\varphi)$ defined on this region E_{int} global transformations taking us outside E_{int} are inadmissible [local transformations with parameters $\omega_\alpha(x)$ vanishing for $|x| \to \infty$ are allowed, which is essential for justifying (2.126)]. It can therefore no longer be stated that the integral over x of the divergence of the current reduces to variation of the action, and therefore vanishes owing to the invariance of the latter. This argument becomes meaningless, and the integral may be nonzero owing to nonvanishing surface terms. This does in fact occur, as can be verified directly by a straightforward evaluation of the surface terms for the current (2.120) with $\varphi \in E_{int}(\varphi_0)$. Here it is important not only that the configurations $\varphi(x) \in E_{int}$ reach a uniform asymptote φ_0, but also that the falloff of the difference $\varphi(x) - \varphi_0$ for $|x| \to \infty$ in general is not rapid owing to the presence of massless Goldstone fluctuations. In fact, within the perturbation theory for the integral (2.38) and its O_n analog, the structure of E_{int} is specified by Remark 4 in Sec. 2.1: in the integral (2.38), $E_{int} = \kappa u + \widetilde{K}^{-1} E$, where κ is a unit vector directed along the spontaneous magnetization $\varphi_0 = \kappa u$, u is its absolute value in the loopless approximation, $\widetilde{K}$ is the kernel of the free action $\widetilde{S}_0$ in (2.39), and $\widetilde{K}^{-1} E$ is the linear space of all functions of the form $\widetilde{K}^{-1} A$, $A \in E$. Owing to spontaneous symmetry breakdown, the operator $\widetilde{K}$ is massless $(\sim \partial^2)$ for directions transverse to $\varphi_0 = \kappa u$. Therefore, in general, the falloff of the functions $\varphi(x) - \varphi_0 \equiv \psi(x) \in \widetilde{K}^{-1} E$ will not be rapid, but only power-law falloff, just as is needed to ensure finiteness of the surface terms in the limit of unbounded integration volume in the integral of the current divergence (2.120). In momentum space this is equivalent to finiteness of the limit $k_i \widetilde{J}_i^\alpha(k)$ for $k \to 0$, which is ensured by the presence of Goldstone singularities of the type k_i / k^2 in the composite operator $\widetilde{J}_i^\alpha(k)$ (i.e., in the Green functions involving this operator).

The local Ward identities with the required modification are also useful in the absence of symmetry, particularly when the symmetry-violating terms in the action are fairly simple. Equation (2.128) is also valid in the absence of symmetry, and instead of (2.116) we then have

$$-u_a(x; \varphi) T_{ab}^\alpha \varphi_b(x) = \partial_i J_i^\alpha(x; \varphi) + N^\alpha(x; \varphi), \qquad (2.130)$$

where N^α are additional local composite operators generated by the symmetry-violating terms in the action. The N^α can always be found explicitly for a specific model and group. The required local Ward identity for an explicitly broken symmetry is obtained by substituting (2.130) into (2.128), i.e., by adding the symmetry-violating operators N^α to $\partial_i J_i^\alpha$ in (2.129).

Let us now discuss the role of the contributions involving $\delta(0)$, which for brevity we shall refer to as δ *terms*. They are constants independent of φ

that are nonzero only for groups with Sym $T^\alpha \neq 0$ (of the symmetries listed in Sec. 2.15, only the scale and conformal symmetries) and are generated, as seen from (2.125), by the noninvariance of the measure $D\varphi$ under these transformations. Since the δ terms are independent of φ, the brackets $\langle\!\langle ... \rangle\!\rangle$ can everywhere be omitted for them. The role of these terms is understood most simply by taking $A = 0$ in (2.128) (we recall that this functional argument is understood everywhere). Then the right-hand side of this equation vanishes owing to the definition (2.122), and the brackets $\langle\!\langle ... \rangle\!\rangle$ on the left-hand side become the average $\langle ... \rangle$ [see (2.107)]. We then obtain

$$\langle -u_a(x;\varphi)T^\alpha_{ab}\varphi_b(x)\rangle = \text{Sym } T^\alpha_{aa}\delta(x - x')|_{x'=x}. \tag{2.131}$$

From this we see that the inclusion of δ terms in relations like (2.128) and (2.129) is equivalent to subtracting from the nontrivial composite operators $F(\varphi)$ inside the brackets $\langle\!\langle ... \rangle\!\rangle$ the average values — the constants $\langle F(\varphi)\rangle$. This is in fact necessary, because the right-hand sides of these expressions vanish for $A = 0$ according to the definition (2.122).

Let us discuss (2.131) in more detail. Its right-hand side does not contain parameters like the charges g, and so it must be generated only by contributions of order zero in g on the left-hand side. These can be checked explicitly: the free action $S_0(\varphi) = -\frac{1}{2}\varphi K\varphi$ contributes to $u_a \equiv \delta S/\delta\varphi_a$ the term $-(K\varphi)_a = -K_{ac}\varphi_c$, the contribution of which to the left-hand side of (2.131) in zeroth-order perturbation theory is $(K_{ac})_{x'}(T^\alpha_{ab})_x\Delta_{cb}(x',x)|_{x'=x}$, where $\Delta_{cb}(x',x) \equiv \langle\widehat{\varphi}_c(x')\widehat{\varphi}_b(x)\rangle_0$ is the unperturbed propagator and the indices x, x' on the operators specify the variables on which they act. Since $\Delta = K^{-1}$ [see (2.13)], we have $(K_{ac})_{x'}\Delta_{cb}(x',x) = \delta_{ab}\delta(x - x')$ in the above expression, and so the latter reduces to $T^\alpha_{aa}\delta(x - x')|_{x'=x}$ and coincides with the right-hand side of (2.131) when the equivalence of T^α and Sym T^α for diagonal elements is taken into account.

We have thus directly verified the validity of (2.131). Nevertheless, it may appear paradoxical, because the presence of the symmetry reduces the composite operator on the left-hand side of (2.131) to a current divergence owing to (2.116), and its integral over x contracted with the parameters w_α leads to the variation of the action (2.115), i.e., in this case

$$\langle \partial_i J^\alpha_i(x;\varphi)\rangle = \text{Sym } T^\alpha_{aa}\delta(x - x')\,|_{x'=x}, \tag{2.132}$$

$$\langle -\delta_w S(\varphi)\rangle = \langle w_\alpha \int dx \, \partial_i J^\alpha_i(x;\varphi)\rangle = \text{tr}(w \text{ Sym } T). \tag{2.133}$$

The nonzero value of the right-hand side of (2.133) obviously contradicts the assumed symmetry, which requires $\delta_w S(\varphi) = 0$. At the same time, there is no doubt of the validity of the local equation (2.132), because it is equivalent to (2.131) checked explicitly above.

Here it might be said that the symmetries discussed above (for definiteness, we shall take the case of scale symmetry) are always violated by regularization. However, the problem also occurs in the massless free theory with action of the type $S(\varphi) = -\frac{1}{2}\varphi K\varphi$, $K = -\partial^2$, which is explicitly scale-invariant (for transformations with canonical dimension $\Delta = d_\varphi = d/2 - 1$) and, at first glance, does not require regularization. Equation (2.116) and its consequence (2.132) for the dilatation current in this theory are satisfied, and the equation $\delta_\omega S(\varphi) = -\omega \int dx \partial_i J_i(x; \varphi)$ is also true for any $\varphi \in E_{\text{int}} = K^{-1}E$ (see Remark 4 in Sec. 2.1). Here scale transformations do not take us outside the space E_{int}, and the falloff of the functions $\varphi(x) \in E_{\text{int}}$ for $|x| \to \infty$, while not rapid since K is massless, is still sufficiently fast that the surface terms in the integral of the current divergence vanish. Therefore, the arguments used above for spontaneous symmetry breakdown are not applicable here — there is no doubt that $\delta_\omega S(\varphi)$ exists and is equal to zero.

The only reasonable explanation of this paradox at the formal level is that here the operations of taking the average (2.107) and integrating over x do not commute. It was the interchange of these two operations in going from (2.132) to (2.133) which led to the seemingly paradoxical result that the average of zero is not zero. The right-hand side of (2.133) is actually not the average of the integral, but the integral of the average, and when this is understood there is no contradiction.

When dealing with noncommuting operators the question of their ordering always arises. The answer is simple, because we have no choice: using the graph technique, we can calculate the various Green functions, i.e., functional averages of field products, and only then can we perform other operations like differentiation and integration on the arguments x of these fields. In other words, the functional integration [or averaging (2.107)] is always done first. This is how quantities like $\langle\langle \delta_\omega S(\varphi) \rangle\rangle$ in (2.124) must be interpreted in doubtful cases: for such transformations it does not follow from the invariance, i.e., from the equation $\delta_\omega S(\varphi) = 0$, that $\langle\langle \delta_\omega S(\varphi) \rangle\rangle = 0$. The correct relation is the one following from (2.133):

$$\langle\langle \delta_\omega S(\varphi) \rangle\rangle = \langle \delta_\omega S(\varphi) \rangle = -\text{tr}(\omega \text{ Sym } T). \qquad (2.134)$$

This constant cancels the contribution of the Jacobian in (2.124) [see (2.125)], and only with this interpretation do we obtain from (2.124), for example, the required global Ward identity for the scale invariance of a massless free theory.

This also explains, for example, the following paradox. The scale transformation in the integral (2.16) for the massless free theory formally leads to the equation $c^{-1} = Jc^{-1}$, where the constant $J \neq 1$ is the Jacobian of the transformation. Simply stating that the quantity (2.16) is "nonexistent" is, of course, completely correct in the strict mathematical sense, but it is not at all satisfactory, because the experience gained in working with such "nonexistent" objects shows that, when understood correctly, the results obtained formally are correct and self-consistent. In the present case, the correct interpretation is given by (2.134), and then it is easily verified that the infinitesimal form of the expression obtained above reduces to the simple identity $c^{-1} = c^{-1}$, as it should.

The primary cause of these problems is very simple. Even in a free theory, the objects studied, in particular, the Green functions of composite operators like the divergence of a current, contain power-law UV divergences, and so their correct definition requires regularization with a cutoff Λ, which violates the symmetry. Of course, there must not be any contradictions in exact expressions involving Λ, and we could have stated this from the start. However, the introduction of Λ greatly complicates the expressions, and so we have tried to formulate an interpretation of these relations which would allow formal contradictions to be avoided even without including Λ. In this regard, we also note that in calculations using the formal scheme of dimensional regularization, all the δ terms are discarded, as they are regular contributions (Sec. 1.20). In general, they are constants which affect only the averages of operators and are needed to make these averages consistent in the resulting equations.

Returning to the Ward identities (2.129), we note that, since in their original form they are equations for A-dependent functionals [see the definitions (2.107) and (2.122)], they are equivalent to equations for the corresponding composite operators [like the passage from (2.108) to (2.109)] and can always be rewritten using (2.110) as equations involving variational derivatives for the functional $W(A)$. As an example, let us give the equivalent of (2.129) for the O_n-symmetry current (2.120) (Sym $T^\alpha = 0$):

$$\mathrm{As}_{ab}\left\{A_a(x)\frac{\delta W(A)}{\delta A_b(x)} - \partial_i\left[\frac{\delta}{\delta A_a(x)}\partial_i\frac{\delta}{\delta A_b(x)}W(A) + \frac{\delta W(A)}{\delta A_a(x)}\partial_i\frac{\delta W(A)}{\delta A_b(x)}\right]\right\}=0,$$

(2.135)

where As_{ab} is the operator that antisymmetrizes under permutations of the indices a and b. Equation (2.135) can be rewritten in terms of the functional $\Gamma(\alpha)$ using standard techniques (see Sec. 2.11).

2.17   The relation between scale and conformal invariance

In this section, as in the preceding ones, we shall understand scale invariance to be not the symmetry (1.67), but invariance under transformations of only the fields with the parameters fixed. Formally, this symmetry with the canonical dimensions of the fields is present in any action that does not contain dimensional constants. However, the UV divergences of such "logarithmic" (Sec. 1.16) theories require the introduction of regularization (Sec. 1.18), which always leads to violation of this symmetry. Therefore, in what follows we shall speak of critical rather than canonical scale invariance. This is a property of the renormalized Green functions of the theory taken right at the critical point (with all parameters like masses equal to zero), and, in addition, at $g = g_*$, i.e., at a fixed point of the renormalization group.

For $g \neq g_*$ only the IR asymptotes of the Green functions (Sec. 1.33) possess this property. For this symmetry it is not the canonical but the critical dimension of the field that plays the role of Δ in the generators of scale transformations (Sec. 2.15).

It turns out that in some models (including the O_n φ_d^4 model), critical scale invariance of the Green functions together with translational and rotational invariance automatically leads to conformal invariance. The version of the proof presented below is taken from [77]. All the objects considered (the action and everything constructed from it) pertain to the renormalized theory, since we are speaking of the properties of the renormalized Green functions. Rotations will always be understood to be spatial rotations. We shall assume that the model is local and invariant under translations and rotations, that it contains only scalar fields, and that the parameters in its (renormalized, as will always be the case here) action functional S satisfy the conditions for critical scale invariance, i.e., all parameters like masses are equal to zero and $g = g_*$ (for $g \neq g_*$ we would have to deal with not the Green functions themselves, but their IR asymptotes). Strictly speaking, the renormalized action S itself for $g = g_*$ is poorly defined owing to singularities of the renormalization constants Z at this point (Sec. 1.27). However, this is unimportant for the proof given below, because we shall deal only with well defined objects — the renormalized Green functions of the fields and composite operators like the energy–momentum tensor. We recall that invariance under rotations leads to symmetry of the energy–momentum tensor (2.119).

In the notation of (2.122), invariance of the Green functions under a group with generators T^α is expressed as

$$\int dx \, W_x[T^\alpha] = 0. \tag{2.136}$$

Let us recall the form of the generators T^α (Sec. 2.15) for the groups of interest to us:

$$\text{translations: } T^k = \partial_k; \tag{2.137a}$$

$$\text{spatial rotations: } T^{ks} = x_s \partial_k - x_k \partial_s; \tag{2.137b}$$

$$\text{scale: } T = x_k \partial_k + \Delta; \tag{2.137c}$$

$$\text{conformal: } T^k = x^2 \partial_k - 2x_k x_s \partial_s - 2\Delta x_k. \tag{2.137d}$$

We have omitted the spin contribution (Sec. 2.15), because we assume all the fields to be scalar. If there are many fields, all the generators are assumed to be multiples of the unit matrix in the corresponding label.

We want to show that sometimes the first three symmetries in (2.137) automatically lead to the fourth. We use the Ward identity (2.129) for the special case of the energy–momentum tensor $J_i^\alpha = \vartheta_i^k$ from (2.119) (for it, Sym $T^\alpha =$ Sym $\partial_k = 0$):

$$\langle\langle \partial_i \vartheta_i^k \rangle\rangle = W_x[\partial_k]. \tag{2.138}$$

From this and the symmetry of ϑ_i^k we obtain

$$\langle\!\langle\partial_i[x_k\vartheta_i^k]\rangle\!\rangle = \langle\!\langle x_k\partial_i\vartheta_i^k + \vartheta_i^i\rangle\!\rangle = W_x[x_k\partial_k] + \langle\!\langle\vartheta_i^i\rangle\!\rangle,$$

$$\langle\!\langle\partial_i[(x^2\delta_{ks} - 2x_kx_s)\vartheta_i^s]\rangle\!\rangle = \langle\!\langle(x^2\delta_{ks} - 2x_kx_s)\partial_i\vartheta_i^s - 2x_k\vartheta_i^i\rangle\!\rangle$$

$$= W_x[x^2\partial_k - 2x_kx_s\partial_s] - \langle\!\langle 2x_k\vartheta_i^i\rangle\!\rangle.$$

Integrating these equations over x, we find that the left-hand sides vanish, as they are integrals of total derivatives, except for possible δ terms (Sec. 2.16), and we obtain

$$0 = \int dx\{W_x[x_k\partial_k] + \langle\!\langle\vartheta_i^i\rangle\!\rangle|_*\}, \tag{2.139}$$

$$0 = \int dx\{W_x[x^2\partial_k - 2x_kx_s\partial_s] - 2x_k\langle\!\langle\vartheta_i^i\rangle\!\rangle|_*\}. \tag{2.140}$$

Here and below the symbol $|_*$ denotes the subtraction, generated by the δ terms, of the average, i.e., the constant $\langle F\rangle$, from the operator F inside the brackets $\langle\!\langle...\rangle\!\rangle$ (it should be noted that this does not affect the nontrivial connected Green functions involving a given composite operator).

Now let us use the assumed scale invariance of the Green functions, expressed by (2.136) with the generator (2.137c). From this equation and (2.139) we find that

$$0 = \int dx\{W_x[\Delta] - \langle\!\langle\vartheta_i^i\rangle\!\rangle|_*\}. \tag{2.141}$$

It follows from (2.127) that the quantity $W_x[T]$ involving an arbitrary local linear operator T can be represented as the average $\langle\!\langle F\rangle\!\rangle$ of a local composite operator F. Therefore, the entire integrand in (2.141) can be represented in this manner, and the corresponding local operator F is, according to Statement 2 in Sec. 2.15, the divergence of a local current. Equation (2.141) then gives

$$\langle\!\langle\vartheta_i^i + \partial_iJ_i\rangle\!\rangle\big|_* = W_x[\Delta] \tag{2.142}$$

involving an unknown local (this is important) composite operator J_i. When (2.142) is substituted into (2.140), the operators in the argument of $W_x[...]$ combine to form the generators T^k of conformal transformations (2.137d), and we find

$$0 = \int dx\{W_x[T^k] + 2x_k\langle\!\langle\partial_iJ_i\rangle\!\rangle|_*\}. \tag{2.143}$$

If the term involving J_i were absent, this would imply, according to (2.136), conformal invariance of the Green functions.

It can sometimes be shown that the term involving J_i in (2.143) certainly does not contribute. This happens when a vector local composite operator J_i with the required symmetry and canonical dimension [for ∂_iJ_i as for ϑ_i^i, according to (2.142)] can be constructed only as a gradient ∂_iF. Then in (2.143) we have $x_k\partial_iJ_i = x_k\partial_i\partial_iF$, and this contribution vanishes upon integration over x, since one factor of x cannot cancel two derivatives, and the possible δ terms are eliminated by subtraction. In this case the conformal invariance is a consequence of the other invariances. This will always occur in models with a single scalar field, because for a single field a local vector can be constructed only in the form of a gradient. However, for two scalar fields it is possible to have constructions like $J_i = \varphi_1 \partial_i \varphi_2$ that do not reduce to $\partial_i F$. In special cases such as the O_n φ^4 model, such constructions are forbidden by the stipulation that J_i have the "required symmetry": J_i, like ϑ_i^i in (2.142), must be an O_n scalar, i.e., an expression of the form $\varphi_a \partial_i \varphi_a$ reducing to $\partial_i(\varphi_a \varphi_a)$. Therefore, also for this model the conformal invariance is a consequence of the other invariances.

This does not occur in general, and study of the symmetry properties in the critical regime requires analysis of the divergences of the corresponding currents taking into account renormalizations (Ch. 3). As a rule, there is no conformal invariance in the critical regime for models involving nonscalar fields.

2.18 Conformal structures for dressed propagators and triple vertices

It turns out that the four symmetries (2.137) together define the first three connected Green functions up to normalizations. This is sometimes used (for example, in Ch. 4) to calculate the critical dimensions. As in Sec. 2.17, we shall consider only scalar fields, always understanding the fields $\widehat{\varphi}_a(x)$ and their Green functions to be renormalized objects.

The symmetry of the Green functions under transformations of the type (2.112) studied in Sec. 2.15 is expressed as

$$\langle \widehat{\varphi}'_{a_1}(x_1)...\widehat{\varphi}'_{a_n}(x_n)\rangle = \langle \widehat{\varphi}_{a_1}(x_1)...\widehat{\varphi}_{a_n}(x_n)\rangle. \tag{2.144}$$

It follows from the symmetry under translations and rotations that $\langle \widehat{\varphi}_a(x)\rangle$ is independent of x, and the higher functions depend only on the relative separations $r_{ab} \equiv |x_a - x_b|$. In what follows we shall assume that $\langle \widehat{\varphi} \rangle = 0$, because this can always be obtained by a shift $\widehat{\varphi} \to \widehat{\varphi}+$ const. We note that $\langle \widehat{\varphi} \rangle = 0$ is always assumed to be true at a critical point (Sec. 1.9), and this always follows from scale invariance if the critical dimension Δ_φ is nonzero.

For $W_1 = \langle \widehat{\varphi} \rangle = 0$ the connected functions $W_{2,3}$ coincide with the normalized full functions (1.33) for two and three fields (Sec. 1.10). It follows from their invariance under translations, rotations, and scale transformations (Sec. 2.15) that in general (see the end of this section for exceptions)

$$\langle \widehat{\varphi}_1(x_1)\widehat{\varphi}_2(x_2)\rangle = Ar_{12}^{-\Delta_1-\Delta_2}, \tag{2.145}$$

$$\langle \widehat{\varphi}_1(x_1)\widehat{\varphi}_2(x_2)\widehat{\varphi}_3(x_3)\rangle = r_{12}^{-\Delta_1-\Delta_2-\Delta_3} f(r_{13}/r_{12},\ r_{23}/r_{12}), \tag{2.146}$$

where Δ_a are the given dimensions of the fields φ_a, $r_{ab} \equiv |x_a - x_b|$, and the constant A and function f are arbitrary.

Now let us impose the additional requirement that (2.144) be invariant under conformal inversion (2.114) (we recall that this is the necessary and sufficient condition for conformal invariance when there is translational invariance; see Sec. 2.15). It is easily checked that under conformal inversion of the coordinates $x \to x' \equiv x/|x|^2$, the separations $r_{ab} \equiv |x_a - x_b|$ transform as $r'_{ab} \equiv |x'_a - x'_b| = r_{ab}/|x_a||x_b|$, and it is impossible to construct a conformally invariant combination of the three separations. Using this, it is easy to see that the representation (2.145) is compatible with conformal invariance only for $\Delta_1 = \Delta_2$, while in (2.146) the function f for any given dimensions of the fields Δ_a is determined uniquely up to normalization by the requirement of conformal invariance [78]:

$$
\langle \widehat{\varphi}_1(x_1)\widehat{\varphi}_2(x_2) \rangle = \begin{cases} A r_{12}^{-2\Delta} & \text{for } \Delta_1 = \Delta_2 \equiv \Delta, \\[2mm] 0 & \text{for } \Delta_1 \neq \Delta_2, \end{cases} \tag{2.147a}
$$

$$
\langle \widehat{\varphi}_1(x_1)\widehat{\varphi}_2(x_2)\widehat{\varphi}_3(x_3) \rangle = A' r_{12}^{\Delta_3 - \Delta_1 - \Delta_2} \, r_{13}^{\Delta_2 - \Delta_1 - \Delta_3} \, r_{23}^{\Delta_1 - \Delta_2 - \Delta_3}. \tag{2.147b}
$$

For four and more fields, in contrast to three, it is possible to construct conformally invariant combinations of the r_{ab}, namely, $r_{ab}r_{cd}/r_{a'b'}r_{c'd'}$, where $a'b'c'd'$ is any permutation of the original set of indices $abcd$. The requirement that these functions be conformally invariant therefore does not determine them uniquely, but only up to an arbitrary function of all the independent conformal invariants (of which there are two for four fields).

Conformal Ansätze analogous to (2.147) are also known for nonscalar fields (spinor, vector, and tensor fields). However, in practice they can be used only when the conformal invariance itself is guaranteed, which for models involving nonscalar fields is more often the exception than the rule.

Finally, (2.145)–(2.147) are valid for the general case of arbitrary dimensions Δ. In some special cases the symmetry allows the appearance of "exceptional" structures involving δ functions instead of powers. For the correlators (2.145) and (2.147a) a special case is that of the shadow relation $\Delta_1 + \Delta_2 = d$, which corresponds to the conformally invariant exceptional structure $\delta(x_1 - x_2)$. For the vertices (2.147b), exceptional structures of the type $\delta(x_1 - x_2)r_{13}^{-2\Delta_3}$ in the special case $\Delta_1 + \Delta_2 = d + \Delta_3$ and $\delta(x_1 - x_2)\delta(x_2 - x_3)$ for $\Delta_1 + \Delta_2 + \Delta_3 = 2d$ are also possible.

2.19   The large-n expansion in the O_n φ^4 model for $T \geq T_c$

In the O_n φ^4 model (2.79), each leg of a graph is assigned an independent free index, and the corresponding *index structure factors* for a given graph are calculated from the *index graph* of the same form with the lines and vertices (2.80). Any of the graphs of W_2 is a multiple of δ_{ab} in the two external indices ab (here we shall discuss only the symmetric phase with $h = 0$,

$T > T_c$), and for the higher functions there is only symmetry with respect to simultaneous permutations of the external indices and the corresponding coordinates or momenta.

Contractions $\delta_{aa} = n$ can arise in calculating the index factors, and so for each graph these factors are polynomials in n. The calculation of the index factors for complicated graphs is a purely technical but complicated task which can always be done by computer. The calculations for simple graphs are not complicated, and there it is convenient to use the following representation of the index vertex factor (2.80):

$$\times \quad = \quad \frac{1}{3}\left[\; \succ\!\!\cdots\!\!\prec \;+\; \overline{\underline{\mathrm{I}}} \;+\; \underline{\rlap{\diagdown}{\diagup}}\!\!\diagdown \;\right]. \tag{2.148}$$

The dashed "interaction" line does not carry indices, and the solid lines and triple vertices are associated with δ_{ab}. Therefore, each chain of solid lines actually corresponds to a single index, summation over which in closed cycles of solid lines gives the factor n. Let us present a simple example of calculating an index graph using the representation (2.148):

$$\overset{\displaystyle\bigcirc}{\underset{\bullet}{\rule{0pt}{0pt}}} \quad = \quad \frac{1}{3}\;\overset{\displaystyle\bigcirc}{\underset{\perp}{\rule{0pt}{0pt}}} \;+\; \frac{2}{3}\;\overset{\frown}{\rule{0pt}{0pt}} \;=\; \frac{n+2}{3}\;-\!-\!- . \tag{2.149}$$

Analysis of the procedure for calculating the index factors in this model shows that in the graphs of all the connected functions W_{2m}, the degree of the polynomial in n (the highest power of n) is no greater than the number of loops l and is necessarily equal to l in some graphs. It then follows that $W_{2m} = \sum_l g^{l+m-1} P_{ml}(n)$, where $P_{ml}(n)$ is a polynomial in n of degree l (the other arguments are understood), and the summation over the number of loops l begins at two for W_0 and zero for all the other functions. Therefore, the problem of the large-n expansion can be stated as follows: it is necessary to find the $n \to \infty$ asymptote for all the functions W_{2m} for $g = \lambda/n$, $\lambda = \text{const}$. The leading term of the asymptote of W_{2m} is then of order n^{1-m} and is made up of all the graphs of the ordinary perturbation theory from the contributions of only the highest powers n^l in the polynomials $P_{ml}(n)$. The inclusion of the next highest contributions gives a systematic series of corrections in powers of $1/n$.

It turns out that this seemingly difficult problem is quite easy to solve by the technique of functional integration. Let us present the solution. We write the functional (2.30) for our model (2.79) with $g = \lambda/n$ as

$$G(A) = c\!\int\! D\varphi \exp\left[-\tfrac{1}{2}\varphi_a K \varphi_a - \tfrac{\lambda}{24n}(\varphi^2)^2 + A_a\varphi_a\right], \tag{2.150}$$

where $K = -\partial^2 + \tau$ and $\tau = T - T_{c0}$. For brevity, in (2.150) and below we do not explicitly indicate the required integrations over coordinates and summation over repeated indices.

Equation (2.1) can be used to write the contribution of the interaction in (2.150) as a Gaussian integral over an auxiliary O_n-scalar field $\psi = \psi(x)$

(without an index):

$$\exp[-\tfrac{\lambda}{24n}(\varphi^2)^2] = c'\int D\psi \exp[\tfrac{3n}{2\lambda}\psi^2 - \tfrac{1}{2}\psi\varphi^2]. \qquad (2.151)$$

To avoid misunderstandings, the argument of the integrated exponential in detailed notation is

$$\int dx[\tfrac{3n}{2\lambda}\psi^2(x) - \tfrac{1}{2}\psi(x)\varphi_a(x)\varphi_a(x)].$$

The coefficient of $\psi\varphi^2$ in (2.151) can be changed by rescaling ψ; we have chosen the standard symmetry coefficient $\tfrac{1}{2}$, and the minus sign is introduced for convenience. The formal convergence of the integral (2.151) can be ensured by the replacement $\psi \to i\psi$. However, for deriving the large-n expansion by the graph technique this is not important, and we shall use the form (2.151), which is more convenient for technical reasons.

Substituting (2.151) into (2.150), we write $G(A)$ as

$$G(A) = cc'\int\int D\varphi D\psi \exp[-\tfrac{1}{2}\varphi_a(K+\psi)\varphi_a + \tfrac{3n}{2\lambda}\psi^2 + A_a\varphi_a]. \qquad (2.152)$$

Calculating the Gaussian integral over φ using the rule (2.1) and taking into account the definition of the normalization constant c in (2.150) (Sec. 2.3), we find

$$G(A) = c'\int D\psi \exp[nF(\psi) + \tfrac{1}{2}A_a\overline{\Delta}A_a]. \qquad (2.153)$$

Here and below

$$\overline{K} \equiv K+\psi, \quad \overline{\Delta} \equiv 1/\overline{K}, \quad F(\psi) \equiv \tfrac{3}{2\lambda}\psi^2 - \tfrac{1}{2}\mathrm{tr}\ln(\overline{K}/K). \qquad (2.154)$$

Let us explain the derivation of (2.153). The term quadratic in φ in (2.152) has the form $-\tfrac{1}{2}\varphi L\varphi$ with diagonal block matrix L having n identical blocks $\overline{K}$ on the diagonal. Therefore, $\det L = (\det\overline{K})^n = \exp[n\,\mathrm{tr}\ln\overline{K}]$ according to (2.4), and the inclusion of the normalization constant c in front of the integral (2.150) reduces to the substitution $\det\overline{K} \to \det\overline{K}/\det K = \det(\overline{K}/K)$. We recall that $K = -\partial^2 + \tau$ and that the contribution ψ in $\overline{K} = K+\psi$ is understood as multiplication by $\psi(x)$ and in the notation of Sec. 2.1 corresponds to the kernel $\psi(x)\delta(x-x')$. Moreover, there is understood to be integration over x in the term involving ψ^2 in (2.154), and $\mathrm{tr}\ln(\overline{K}/K) = \mathrm{tr}\ln\overline{K} - \mathrm{tr}\ln K$ according to the rules (2.4).

In the argument of (2.153) we have explicitly isolated the "large" parameter n, and so the desired large-n expansion can be obtained by calculating the integral (2.153) by the saddle-point method (Sec. 2.8). The term involving A in (2.153) does not have the coefficient n, and so we must seek the saddle point ψ_0 of only the functional $F(\psi)$ in the argument of (2.153):

$$\delta F(\psi)/\delta\psi(x)|_{\psi=\psi_0} = \tfrac{3}{\lambda}\psi_0(x) - \tfrac{1}{2}\overline{\Delta}(x,x)|_{\psi=\psi_0} = 0. \qquad (2.155)$$

Let us explain. According to (2.6), variation of ψ gives

$$\delta\mathrm{tr}\ln\overline{K} = \mathrm{tr}(\delta\overline{K}/\overline{K}) = \mathrm{tr}(\overline{\Delta}\delta\psi) = \int dx\,\overline{\Delta}(x,x)\delta\psi(x)$$

owing to the locality of the multiplication by ψ. Therefore,

$$\delta \operatorname{tr} \ln(K + \psi)/\delta\psi(x) = \overline{\Delta}(x, x), \quad \overline{\Delta} = 1/\overline{K} = (K + \psi)^{-1}. \quad (2.156)$$

We shall seek a solution of (2.155) in the class $\psi_0 = \text{const}$, because we do not expect spontaneous breakdown of translational invariance. Then in (2.155) we can go to momentum space (1.35) for $\overline{\Delta} = (-\partial^2 + \tau + \psi_0)^{-1}$, which gives

$$\psi_0 = \frac{\lambda}{6(2\pi)^d} \int\limits_{|k| \le \Lambda} \frac{dk}{(k^2 + \tau + \psi_0)}, \quad (2.157)$$

with integration over the d-dimensional momentum k. The divergence of the integral at large momenta for $d \ge 2$ requires the introduction of the cutoff Λ.

Assuming the existence and uniqueness of the solution of (2.157) (more on this later), we proceed according to the general scheme of Sec. 2.8. We make the shift $\psi \to \psi + \psi_0$ in the integral (2.153) and then expand the argument of the exponential in a Taylor series in the new variable ψ. After this shift,

$$\overline{\Delta} = (K + \psi_0 + \psi)^{-1} = \Delta - \Delta\psi\Delta + \Delta\psi\Delta\psi\Delta - ..., \quad \Delta \equiv (K + \psi_0)^{-1}. \quad (2.158)$$

The $\operatorname{tr} \ln ...$ is calculated using (2.4):

$$\operatorname{tr} \ln(K + \psi_0 + \psi) - \operatorname{tr} \ln(K + \psi_0) = \operatorname{tr} \ln[\Delta(\Delta^{-1} + \psi)] = \operatorname{tr} \ln(1 + \Delta\psi),$$

and the result is expanded in a series like (2.7). Then for the functional F from (2.154) we obtain

$$F(\psi_0 + \psi) = F(\psi_0) + \tfrac{3}{2\lambda}\psi^2 + \sum_{k=2}^{\infty} \tfrac{1}{2k} \operatorname{tr}(-\Delta\psi)^k, \quad (2.159)$$

where the contribution linear in ψ is absent owing to the stationarity condition (2.155). As a result, the integral (2.153) can be written as

$$G(A) = c' \int D\psi \exp[nF(\psi_0) - \tfrac{1}{2}\psi\Delta_\psi^{-1}\psi + U(\psi) + \tfrac{1}{2}A_a\overline{\Delta}A_a], \quad (2.160)$$

where $\overline{\Delta}$ is the series (2.158), the quantity $\Delta_\psi^{-1} = -n[\tfrac{3}{\lambda} + \tfrac{1}{2}\Delta^2]$ is determined by the contributions in (2.159) quadratic in ψ, and $U(\psi)$ is the series from (2.159) without the $k = 2$ contribution. In detailed notation,

$$\Delta_\psi^{-1}(x, x') = -n[\tfrac{3}{\lambda}\delta(x - x') + \tfrac{1}{2}\Delta^2(x, x')]. \quad (2.161)$$

The graph representation of the integral (2.160) is determined by Rule 2 in Sec. 2.2, but in order to use (2.28) it is necessary to first multiply and divide (2.160) by the normalization factor c'', given by (2.16) for the form quadratic in ψ from (2.160). After taking the logarithm we obtain

$$W(A) = \ln G(A) = \gamma + \ln[P_\psi \exp V]|_{\psi=0}, \quad (2.162)$$

where P_ψ is the reduction operator (2.14) with the replacement $\varphi \to \psi$, $\Delta \to \Delta_\psi$, and

$$\gamma = nF(\psi_0) + \ln(c'/c''), \qquad V = U + \tfrac{1}{2}A_a\overline{\Delta}A_a. \qquad (2.163)$$

Substituting the explicit expressions for the normalization factors of the integrals (2.151) and (2.160) known from the general rule (2.16), using (2.154) we find that

$$\gamma = \tfrac{n}{2}[\tfrac{3}{\lambda}\psi_0^2 - \operatorname{tr}\ln((K+\psi_0)/K)] - \tfrac{1}{2}\operatorname{tr}\ln(-\tfrac{\lambda}{3n}\Delta_\psi^{-1}). \qquad (2.164)$$

Here all the contributions are multiples of the infinite volume of the system $\int dx$, because, according to (2.5), this factor always appears when the $\operatorname{tr}\ln\ldots$ of translationally invariant operators is taken. Integration over x in the first term of (2.164) is understood (see above).

Let us now turn to the second term on the right-hand side of (2.162). This is a standard construction, easily interpreted in terms of graphs following the general rules of Sec. 2.2. We write out the functional V graphically:

$$V = \frac{1}{2}\left[\ \times\!\!-\!\!\times\ -\ \times\!\!-\!\!\bullet\!\!-\!\!\times\ +\ \times\!\!-\!\!\bullet\!\!-\!\!\bullet\!\!-\!\!\times\ -\ \ldots\right] + \sum_{k=3}^{\infty}\frac{(-1)^k}{2k}\ \overset{k}{\bigcirc}_{\ldots}. \qquad (2.165)$$

The tails correspond to factors of ψ, the crosses to A, the solid lines to the matrix propagator $\Delta_{ab} = \delta_{ab}\Delta$ with Δ from (2.158) for the fundamental field φ_a [it is no longer present in the integral (2.160), but its trace remains in the graph constructions], and the triple vertices to factors of unity.

In this notation, (2.161) takes the form

$$\Delta_\psi^{-1} = -\frac{3n}{\lambda} - \frac{1}{2}\ \bigcirc\ . \qquad (2.166)$$

Contractions of the index factors δ_{ab} in closed cycles of solid lines give a factor of n, and so the contributions of the loops in (2.165) and (2.166) are proportional to n. The quantity Δ_ψ inverse to (2.161) is the propagator of the auxiliary field ψ, and we shall represent it graphically by a dashed line. Upon expanding $\exp V$ in (2.162) in a series, we obtain a sum of disconnected graphs with connected components from (2.165) (chains and cycles). The action of the operator P_ψ on them corresponds to pairwise contraction of the tails ψ to give dashed lines in all possible ways. Owing to the condition $\psi = 0$ in (2.162), all the fields ψ must be contracted. Taking the logarithm in (2.162) corresponds to selecting only connected graphs (Sec. 2.2). In this way we obtain the graphs of the functional $W(A)$, and the coefficients of $A^{2m}/(2m)!$ in its expansion are the connected functions W_{2m}. The functions W_{2m} are determined graphically by graphs containing $2m$ crosses A. Such graphs have m chains (2.165) and any number of cycles. According to these rules we obtain

$$W_0 = \gamma + \frac{1}{4}\ \bigcirc\!\!\!\bigcirc\ + \frac{1}{8}\ \bigotimes\ + \frac{1}{8}\ \bigcirc\!\!-\!\!\bigcirc\ + \frac{1}{12}\ \bigcirc\!\!-\!\!\bigcirc\ + \ldots;$$

$$W_2 = -\!\!\!\bullet\!\!+ \cdots + \tfrac{1}{2}\cdots + \ldots; \quad W_4 = 3\cdots + \ldots \quad (2.167)$$

and so on. We recall that a solid line in these graphs (which is understood to carry indices) depicts the propagator of the fundamental field φ_a, a dashed line depicts the propagator of the auxiliary field ψ, and a triple vertex corresponds to a factor of unity and a Kronecker δ carrying the indices:

$$\underset{a\quad\;\; b}{\rule[2pt]{2em}{0.4pt}} \equiv \delta_{ab}\Delta, \;\; \Delta \equiv [-\partial^2 + \tau + \psi_0]^{-1}; \quad \text{----} \equiv \Delta_\psi; \quad \underset{b}{\overset{a}{\rule{0pt}{0pt}}}\!\!\!\text{---} \equiv \delta_{ab}. \quad (2.168)$$

The graphs in (2.167) have exactly the same form and coefficients as in the theory of two fields φ, ψ with the interaction $\frac{1}{2}\psi\varphi^2$ (this is rigorously justified in Sec. 2.20), but they form only a subset of the latter, because the sum (2.165) does not contain contributions from single and double cycles. Since there are no contracted or "multiplied" lines (several lines of the same type connecting any given pair of vertices) in the graphs of (2.167), all the factors of two and factorials in (2.23) are absent, and so the coefficient of any graph of the functional $W(A)$ is simply the inverse symmetry number of the graph. In transforming to the graphs of the functions W_{2m} this gets multiplied by $(2m)!$, which gives the coefficients in (2.167).

The order of a graph in $1/n$ is determined by the following simple considerations. A solid line is of order unity, a dashed line is of order $1/n$, and every closed cycle of solid lines gives a factor n from summation over the indices. The leading contributions to W_0 of order n and 1 are concentrated in the quantity γ known from (2.164), the graphs of W_0 are of order $1/n$ and higher, and (2.167) contains all the graphs of order $1/n$. The leading contribution to W_2 (solid line) is a quantity of order 1, and the first correction $\sim 1/n$ is given by the two graphs in (2.167). It should be noted that already in the first steps of the large-n expansion the number of graphs grows with order much more quickly than in ordinary perturbation theory: for example, for the propagator W_2 there are only the two graphs given in (2.167) in order $1/n$, while in the next order $1/n^2$ there are already 20 (finding these graphs and their coefficients is an excellent exercise). The rapid increase in the number of graphs is one of the main difficulties in calculating large-n expansions of critical exponents in higher orders (Ch. 4). However, it is important that in any given order in $1/n$ the number of graphs is finite. Therefore, the problem is solved in principle if the saddle point ψ_0 exists, as we have assumed. This solution corresponds to only the symmetric phase with $\langle\hat\varphi\rangle = 0$, since the external field does not enter into the stationarity equation (2.157), and the integral over φ in (2.152) is calculated without shifting φ. Therefore, the vanishing of the solution of the stationarity equation (2.157) at some value of T corresponds in this case to going to the nonsymmetric phase with nonzero spontaneous magnetization $\langle\hat\varphi\rangle$. For a unified description of the two phases within the large-n expansion it is usually convenient to use the language of the Legendre transform Γ (Sec. 2.21).

Let us discuss the saddle-point equation (2.157) in more detail. It is convenient to change from ψ_0 to the new variable $\xi \equiv \tau + \psi_0$. It must be nonnegative, because in momentum space $\delta_{ab}(k^2 + \xi)^{-1}$ is the correlator of the field φ_a in leading order of the large-n expansion (see above). We recall that the correlator of any observable must be positive-definite, and this property must be preserved in the leading order of any regular expansion (Sec. 1.9). The analogous argument does not apply to the artificially introduced field ψ.

Substituting $\xi \equiv \tau + \psi_0$, $\tau \equiv T - T_{c0}$ into (2.157), where T_{c0} is the critical temperature in the Landau approximation, we can rewrite this equation as

$$\xi - T + T_{c0} = \frac{\lambda}{6(2\pi)^d} \int\limits_{|k| \leq \Lambda} \frac{dk}{(k^2 + \xi)} \equiv I(\xi) \qquad (2.169)$$

with integration over the d-dimensional momentum k. The function $I(\xi)$ defined in (2.169) falls off monotonically on the semiaxis $\xi \geq 0$, vanishing as $1/\xi$ for $\xi \to \infty$. It is maximal at $\xi = 0$, and its maximum value $I(0)$ is infinite for $d \leq 2$ and finite for $d > 2$. Therefore, in the first case a solution $\xi = \xi(T)$ of (2.169) exists for any value of T [it is easily found graphically, and is unique because $I(\xi)$ is monotonic]. In the second case a solution exists only for $T \geq T_{c0} - I(0) \equiv T_c$. The right-hand side of this inequality is the value of the critical temperature T_c in leading order of the large-n expansion. The vanishing of the solution of (2.169) for $T < T_c$ corresponds to the transition to the nonsymmetric phase with the appearance of spontaneous magnetization (see above).

These conclusions are consistent with the well-known general Mermin–Wagner theorem [71] forbidding a phase transition with the appearance of an average $\langle \hat{\varphi} \rangle \neq 0$ at a finite value of T in dimension $d \leq 2$. The predictions of the large-n expansion are correct in this regard, and this favorably distinguishes it from the simple Landau theory (equivalent to the self-consistent field approximation, which is equivalent to the loopless approximation for Γ), which predicts a phase transition with finite T_c in any dimension d.

2.20 A simple method of constructing the large-n expansion

The method presented in the preceding section for obtaining the large-n expansion by the graph technique can be substantially simplified. We shall start from the representation (2.152) in terms of the theory of two fields. We make the shift $\psi \to \psi_0 + \psi$ in this integral, after which the action functional in the exponent takes the form

$$S = -\tfrac{1}{2}\varphi_a(K + \psi_0 + \psi)\varphi_a + \tfrac{3n}{2\lambda}(\psi^2 + 2\psi_0\psi + \psi_0^2). \qquad (2.170)$$

We define the shift parameter ψ_0 from the saddle-point condition (2.155), which we rewrite as

$$\frac{3n\psi_0}{\lambda} = \frac{1}{2} \bigcirc .\tag{2.171}$$

The summation over the index a in the contracted line $\delta_{aa}\Delta(x,x)$ gives a factor n, which was cancelled explicitly in (2.155). Defining Δ_ψ^{-1} by (2.166), we can rewrite the action functional (2.170) using (2.171) as

$$S = \frac{3n\psi_0^2}{2\lambda} - \frac{1}{2}[\varphi_a \Delta^{-1}\varphi_a + \psi\Delta_\psi^{-1}\psi + \psi\varphi^2] + \frac{1}{2}\,\text{$\sim\!\!\bigcirc$} - \frac{1}{4}\,\text{$\sim\!\!\bigcirc\!\!\sim$} \tag{2.172}$$

[using the notation of (2.165), where the tails in the graphs denote ψ]. Performing the φ integration in integral of the type (2.152) with the action (2.172), we obtain the representation (2.160) of the functional $G(A)$. The graphs involved in (2.172) cancel the cycles with $k = 1, 2$ in the tr ln ..., and so they do not appear on the right-hand side of (2.165).

Discarding the contributions of the two graphs in the action (2.172), we obtain the usual theory of two fields with given propagators and interaction $-\frac{1}{2}\psi\varphi^2$ (plus an unimportant constant), which will henceforth be referred to as the simple $\psi\varphi^2$ model. Our model (2.172) differs from the simple one only by the elimination of certain graphs, namely, those containing at least one single or double cycle of solid lines. This is actually the simplest formulation of the graph technique for the large-n expansion, as the graphs of the $\psi\varphi^2$ model with certain lines and without certain graphs. The unnecessary graphs are eliminated by those of (2.172), which should be viewed as an addition to the interaction, not to the free part, even though the first of these graphs is linear and the second is quadratic in ψ.

In the simple model the problem of the large-n expansion is nontrivial, because the following insertions (and only these insertions) into the lines do not change the order in $1/n$:

$$ \text{—} \rightarrow \text{⊷} ; \quad \text{- - - -} \rightarrow \text{- -⊶- -} \tag{2.173}$$

(we recall that each cycle gives a factor n, a line ψ is of order $1/n$, and a line φ is of order 1). The number of graphs of a given order in $1/n$ in the simple model is infinite, because any number of insertions (2.173) does not change the order of the graph in $1/n$. The elimination of all graphs with single and double cycles by the graphs in (2.172) excludes the dangerous insertions (2.173), thereby automatically solving the problem of the large-n expansion.

The final prescription can be stated as follows. We go from the original integral (2.150) to the representation (2.152) in terms of two fields and perform the shift $\psi \rightarrow \psi_0 + \psi$ in it, choosing ψ_0 so as to cancel the contribution of the single cycle in the graphs, which is equivalent to using the saddle-point equation (2.155). We then write $S = S - \psi L\psi + \psi L\psi$, associating $-\psi L\psi$

with the free part and $\psi L \psi$ with the interaction, and choose the kernel L so as to cancel the double cycle in the graphs. The usual graph technique of the theory of two fields with nontrivial propagators and without dangerous cycles obtained in this way will then be the desired graph technique of the large-n expansion. This prescription is also applicable to various generalizations of the $O_n \varphi^4$ model studied below (Secs. 2.27 and 2.28).

2.21 The large-n expansion of the functionals W and Γ for $A \sim \alpha \sim n^{1/2}$

Up to now we have studied the $n \to \infty$ asymptote for $g = \lambda/n$, $\lambda = \text{const}$, of the connected Green functions W_{2m} for the model with interaction $\sim g\varphi^4$ in zero external field. In leading order $W_{2m} \sim n^{1-m}$ (Sec. 2.19), and so for A of order unity the individual contributions to the expansion of $W(A)$ in A are of different orders in $1/n$. However, for $A \sim n^{1/2}$ they become of the same order, and the problem arises of obtaining the large-n expansions of the functional $W(A)$ for $A \sim n^{1/2}$ and its Legendre transform $\Gamma(\alpha)$ for $\alpha \sim n^{1/2}$. This statement of the problem is needed for constructing solutions in an arbitrary external field $h \sim n^{1/2}$ (as usual, h is assumed to be included in A), and is necessary for $T < T_c$, because in the large-n expansion the spontaneous magnetization α_0 is automatically of order $n^{1/2}$ [see, for example, (2.90) with $g = \lambda/n$]. Therefore, in the large-n expansion it is natural to take $h \sim n^{1/2}$, so that the corrections to α_0 from h are of the same order of magnitude.

This problem is easily solved by a modification of the expressions in Sec. 2.19. For $A \sim n^{1/2}$ (henceforth always understood), both terms in the exponent in (2.153) are of order n. It is therefore now necessary to seek the saddle point of the entire exponent, i.e., the solution of the equation [notation of (2.165)]

$$\frac{3n\psi_0}{\lambda} = \frac{1}{2} \bigcirc + \frac{1}{2} \prec . \tag{2.174}$$

All the contributions are of the same order in n, and for $A = 0$ Eq. (2.174) reduces to (2.171), equivalent to (2.155). Equation (2.174) is the self-consistency equation for ψ_0, which also enters [see (2.168)] into the lines of the graphs in (2.174). In the expansion of the exponent in (2.153) at the saddle point we now must now include the contribution of the term involving A in Δ_ψ^{-1}. Therefore, instead of (2.166) we now have

$$\Delta_\psi^{-1} = \frac{-3n}{\lambda} - \frac{1}{2} \bigcirc - \overset{\times \quad \times}{\overset{\frown}{}} . \tag{2.175}$$

The vertices (points) in these graphs correspond to the free arguments x, x', and the first term is a multiple of the unit operator with kernel $\delta(x - x')$. The solution ψ_0 of (2.174) and the quantity (2.175) now become functionals of A, and are inhomogeneous for $A(x) \neq \text{const}$. Equations (2.162) and (2.164) with the changed ψ_0 and Δ_ψ^{-1} remain valid, but now in (2.165) the contributions of chains with one and two ψ vanish: the first goes into the saddle-point equation (2.174), and the second into the kernel (2.175). The first term in the progression (2.165) not containing ψ gets added to $W(A)$. Therefore,

$$W(A) = \gamma + \tfrac{1}{2} A_a \Delta A_a + O(1/n), \tag{2.176}$$

where γ is given by (2.164), $\Delta \equiv (-\partial^2 + \tau + \psi_0)^{-1}$, and $O(1/n)$ are the contributions of nontrivial graphs generated by the cycles and chains in (2.165) with three or more tails ψ in each. The first of these graphs are of order $1/n$: they are the 4 vacuum loops from (2.167) and the other 11 graphs with the crosses A, which the reader can find independently as an exercise.

Equations (2.176) and (2.164) explicitly determine the contributions of order n and 1 to the functional $W(A)$, which allows its Legendre transform $\Gamma(\alpha)$ to be found using the equations of Sec. 2.5 with the same accuracy for $\alpha = \delta W / \delta A \sim n^{1/2}$. The result, which will be derived below, is

$$\Gamma(\alpha) = \gamma - \tfrac{1}{2} \alpha_a \Delta^{-1} \alpha_a + O(1/n). \tag{2.177}$$

The quantities entering here are given by (2.174) and (2.175) with the replacement $A_a \to \Delta^{-1} \alpha_a$, which transforms a line with a cross at the end into the independent argument α of Γ:

$$\frac{3n\psi_0}{\lambda} = \frac{1}{2} \bigcirc + \frac{1}{2} \, \reflectbox{$\curlywedge$} \, , \tag{2.178}$$

$$\Delta_\psi^{-1} = -\frac{3n}{\lambda} - \frac{1}{2} \bigcirc - \, \}{\cdot}\{ \, , \tag{2.179}$$

where the tail denotes α, not ψ as in (2.165). The analytic expressions are

$$\tfrac{3n}{\lambda} \psi_0(x) = \tfrac{1}{2} n \Delta(x, x) + \tfrac{1}{2} \alpha^2(x),$$

$$\Delta_\psi^{-1}(x, x') = -\tfrac{3n}{\lambda} \delta(x - x') - \tfrac{1}{2} n \Delta^2(x, x') - \alpha_a(x) \Delta(x, x') \alpha_a(x'),$$

with the propagator Δ from (2.168). The solution ψ_0 of (2.178) and the kernel (2.179) are now functionals of α and are substituted into (2.164) and (2.177).

To prove (2.177) we write $W = W^{(0)} + W^{(1)}$, where $W^{(0)}$ is the leading contribution of order n [i.e., the quadratic form from (2.176) plus the first term in (2.164)], and $W^{(1)}$ is a correction of order unity [the second term in (2.164)] plus higher-order contributions unimportant for what follows. According to the definition of Sec. 2.5, we have $\alpha_a = \delta W / \delta A_a$. The two contributions to W depend on A explicitly and also implicitly through ψ_0. $W^{(0)}$ is essentially the value of the exponent in (2.160) at the saddle point, and so its partial derivative with respect to ψ_0 is zero owing to the saddle-point condition (2.174). It then follows that when differentiating $W^{(0)}$ with respect to A it is necessary to take into account only the explicit A dependence of the quadratic form in (2.176), and so $\alpha_a = \Delta A_a + \delta W^{(1)}/\delta A_a$. Solving this equation iteratively for A, we find $A_a = \Delta^{-1}\alpha_a + \delta A_a$. The explicit form of the correction δA_a is not needed, because it cancels in the final result (to the required accuracy in $1/n$).

We have thus found $A = A(\alpha, \psi_0)$, where the ψ_0 dependence enters through the propagator $\Delta = (-\partial^2 + \tau + \psi_0)^{-1} \equiv \Delta(\psi_0)$. Substitution of this expression for A into (2.174) gives an equation implicitly determining $\psi_0 = \psi_0(\alpha)$ and, thereby, $A = A(\alpha, \psi_0) = A(\alpha)$, which is needed for constructing the Legendre transform.

Let $\overline{\psi}_0 = \overline{\psi}_0(\alpha)$ be the solution of (2.178) obtained from the exact equation (2.174) by substitution of the zeroth-order approximation $A_a = \Delta^{-1}\alpha_a$ neglecting the correction δA, and let $\overline{A}_a = \Delta^{-1}(\overline{\psi}_0)\alpha_a \equiv \overline{A}_a(\alpha)$ be the corresponding zeroth-order approximation for A. The exact values $A(\alpha)$ and $\psi_0(\alpha)$ differ from $\overline{A}$ and $\overline{\psi}_0$ introduced above by corrections δA and $\delta\psi_0$ of order $1/n$. By definition, we have $\Gamma(\alpha) = W(A) - \alpha A = W^{(0)}(A, \psi_0) + W^{(1)}(A, \psi_0) - \alpha A$. The arguments $A(\alpha)$ and $\psi_0(\alpha)$ are needed in the term $W^{(1)}$ only in the zeroth-order approximation (including corrections in the arguments exceeds the accuracy), and so $W^{(1)}(\overline{A}, \overline{\psi}_0)$ is a known function of α [the second term in (2.164) with the quantities defined in (2.178) and (2.179)]. For the leading contribution $W^{(0)}(A, \psi_0) = W^{(0)}(\overline{A} + \delta A, \overline{\psi}_0 + \delta\psi_0)$ we need to include the correction terms in the arguments in first order. The correction $\delta\psi_0$ does not contribute owing to the saddle-point condition (see above), and only the contribution δA remains in the explicit A dependence. Only the quadratic form (2.176) depends explicitly on A, and with the required accuracy the corresponding correction from δA has the form $\overline{A}_a \Delta \delta A_a = \alpha_a \delta A_a$ using the equation $\overline{A}_a = \Delta^{-1}\alpha_a$. This correction cancels exactly with the analogous correction to the term $-\alpha A$ in the Legendre transform (see above): $-\alpha A = -\alpha_a(\overline{A}_a + \delta A_a) = -\alpha_a \Delta^{-1}\alpha_a - \alpha_a \delta A_a$. Therefore, in this approximation

$$\Gamma(\alpha) = W^{(0)}(\overline{A}, \overline{\psi}_0) + W^{(1)}(\overline{A}, \overline{\psi}_0) - \alpha_a \Delta^{-1}\alpha_a,$$

which, taking into account (2.176) and the definitions of $\overline{A}$ and $\overline{\psi}_0$, is equivalent to the desired relations (2.177)–(2.179).

2.22 The solution for arbitrary A, T in leading order in $1/n$

Like the ordinary loop expansion, the graph technique of the large-n expansion for Γ is applicable for calculating thermodynamical and correlation functions at arbitrary values of the external field A and temperature T. We recall that the results of Sec. 2.19 pertain only to the symmetric phase with $A = 0$ and $T \geq T_c$. They can be generalized to the case $A \neq 0$ for any value of T using the technique for $W(A)$ described in Sec. 2.21, and then the solution for $A = 0$, $T < T_c$ can be obtained by taking the limit $A \to 0$. However, it is simpler to do this in the language of $\Gamma(\alpha)$, as explained in detail in Sec. 2.10 for the example of the loop expansion.

The explicit expressions (2.177) and (2.164) can be used to find the solution not only in leading order in $1/n$, but also with the first correction. Here we shall restrict ourselves to the leading order, in which

$$\Gamma(\alpha) = -\tfrac{1}{2}\alpha_a(K + \psi_0)\alpha_a + \tfrac{n}{2}[\tfrac{3}{\lambda}\psi_0^2 - \mathrm{tr}\,\ln((K + \psi_0)/K)],$$

$$K \equiv -\partial^2 + \tau, \quad \tau \equiv T - T_{c0}, \tag{2.180}$$

$$\Delta \equiv (K + \psi_0)^{-1} = (-\partial^2 + \xi)^{-1}, \quad \xi \equiv \tau + \psi_0.$$

For convenience, we have reminded the reader of the basic notation. We also recall that $\psi_0 = \psi_0(x; \alpha)$ is the solution of the saddle-point equation (2.178) depending functionally on α, the solid lines in all the graphs denote the matrix propagators $\delta_{ab}\Delta$ of the field φ_a, and the required coordinate integrations are always understood.

Equation (2.180) for Γ contains all the contributions of order n from graphs with any number of loops. The full loopless approximation (2.76) for the action (2.79) enters into (2.180), because for $\alpha^2 \sim n$ and $g \sim 1/n$ all the contributions in (2.76) are of order n. Calculations in the large-n expansion are usually performed for arbitrary dimension $2 < d < 4$. The inequality $2 < d$ is the necessary condition for the existence of a critical point with $T_c > 0$, and $d < 4$ is the condition for the critical behavior to be nontrivial. Formulas for d-dimensional integrals can be found below in Sec. 3.15.

The general calculational scheme in terms of Γ was described in detail in Sec. 2.10. We need the first and second variational derivatives of Γ with respect to α, $\Gamma_1^a \equiv \delta\Gamma/\delta\alpha_a$ and $\Gamma_2^{ab} \equiv \delta^2\Gamma/\delta\alpha_a\delta\alpha_b$. The functional (2.180) depends on α both explicitly and implicitly through ψ_0. In the first differentiation it is necessary to take into account only the explicit dependence, since the partial derivative of the functional (2.180) with respect to ψ_0 is equal to zero owing to the saddle-point equation (2.178). Therefore, instead of (2.86) and (2.87) we now have

$$\Gamma_1^a = (\partial^2 - \xi)\alpha_a = -A_a, \quad \xi \equiv \tau + \psi_0, \tag{2.181}$$

$$\Gamma_2^{ab} = (\partial^2 - \xi)\delta_{ab} - \alpha_a\delta\xi/\delta\alpha_b, \tag{2.182}$$

where the arguments x are understood. Differentiating (2.178) with respect to α_b, we find that

$$\left[\frac{3n}{\lambda} + \frac{1}{2}\,\bigcirc\!\!\!-\!\bullet\,\right]\frac{\delta\xi}{\delta\alpha_b} = \alpha_b. \tag{2.183}$$

Solving for $\delta\xi/\delta\alpha_b$ and substituting the result into (2.182) gives

$$\Gamma_2^{ab} = (\partial^2 - \xi)\delta_{ab} - \alpha_a\left[\frac{3n}{\lambda} + \frac{1}{2}\,\bigcirc\!\!\!-\!\bullet\,\right]^{-1}\alpha_b. \tag{2.184}$$

From now on we shall take α to be not an arbitrary argument of the functional Γ, but the solution of the saddle-point equation (2.181). For uniform A, α, and $\xi = \tau + \psi_0$, (2.181) takes the form $\xi\alpha_a = A_a$ and together with (2.178) forms a system of two equations implicitly determining the dependence of ξ and α on A and T. In the notation of (2.88) and (2.169) this system can be rewritten as

$$\xi = A/\alpha, \quad \xi - T = I(\xi) - T_{c0} + \lambda A^2/6n\xi^2. \tag{2.185}$$

From the second equation we find $\xi = \xi(A, T)$, and then from the first $\alpha = A/\xi = \alpha(A, T)$. The graphical solution of the second equation in (2.185) is shown in Fig. 2.1.

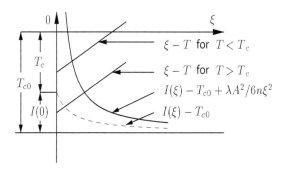

Figure 2.1 Graphical solution of the second equation in (2.185).

From the graphs in this figure we see that for $A \neq 0$ and any T, a solution $\xi(A, T)$ exists, is unique, and depends smoothly on A and T. A difference between the phases appears, as usual, only for $A \to 0$: if $T > T_c = T_{c0} - I(0)$, then $\xi(A, T)$ has a finite limit for $A \to 0$, the solution of (2.169). If $T < T_c$, then $\xi(A, T) \to 0$, which is also obvious from the first equation in (2.185). In this section we shall restrict ourselves to the case $A \neq 0$.

Equations (2.185), which implicitly give the equation of state (Sec. 1.4), can be used to find $\alpha = \alpha(A, T)$. The other needed quantities are then calculated from (2.84). The inclusion of $\ln Z_0$ in the expression for the free energy in this case reduces to removing the normalization factor K^{-1} from the $\text{tr}\ln\ldots$ in (2.180). For the inverse correlator $D^{-1} = -\Gamma_2$ [see (2.84)] in momentum space, (2.88), (2.184), and (2.185) give

$$(D^{-1})_{ab} = \delta_{ab}(p^2 + \xi) + 2\lambda\alpha^2 e_a e_b/n(6 + \lambda\Pi), \tag{2.186}$$

where ξ and α are the solutions of the system (2.185), and Π is the scalar loop:

$$\Pi \equiv \Pi(p, \xi) = (2\pi)^{-d}\int dk[k^2 + \xi]^{-1}[(p - k)^2 + \xi]^{-1}. \tag{2.187}$$

From (2.186) we find the coefficients of the projectors (2.88) in D:

$$D^{\perp} = (p^2 + \xi)^{-1}, \quad D^{\parallel} = [p^2 + \xi + 2\lambda\alpha^2/n(6 + \lambda\Pi)]^{-1}. \tag{2.188}$$

Setting $p = 0$, we obtain the corresponding susceptibilities:

$$\chi^\perp = \xi^{-1}, \quad \chi^\| = [\xi + 2\lambda\alpha^2/n(6 + \lambda\Pi)]^{-1}\big|_{p=0}. \tag{2.189}$$

We recall that the equation $\chi^\perp = \xi^{-1} = \alpha/A$ is a simple consequence of the symmetry (Sec. 1.6). For $A \neq 0$ these equations determine a unique solution depending smoothly on A and T.

Strictly speaking, the cutoff Λ in the theory must be general, and so it must also be taken into account in the UV-convergent (for $2 < d < 4$) integrals for the loop (2.187) and the difference $I(\xi) - I(0)$ (an integral with subtraction) in (2.169). However, for small $p \ll \Lambda$ and $\xi \ll \Lambda^2$, the Λ dependence in these objects can be neglected, and then from the formulas in Sec. 3.15 we find that

$$\Pi\big|_{\xi=0} = c_1 p^{d-4}, \quad \Pi\big|_{p=0} = c_2 \xi^{d/2-2},$$
$$\tag{2.190}$$
$$I(\xi) = I(0) - \lambda c_3 \xi^{d/2-1},$$

where $c_2 = (4\pi)^{-d/2}\Gamma(2-d/2)$, $c_1 = c_2\Gamma^2(d/2-1)/\Gamma(d-2)$, $c_3 = c_2/3(d-2)$, and $\Gamma(z)$ is the gamma function. For arbitrary p and ξ the loop (2.187) even without the cutoff cannot be expressed in terms of elementary functions. Equations (2.190) for the integrals without the cutoff are exact, and when Λ is included they determine the leading contributions in the region $p \ll \Lambda$, $\xi \ll \Lambda^2$.

2.23 The $A \to 0$ asymptote. Singularity of the longitudinal susceptibility for $T < T_c$

Let us first consider the $A \to 0$ asymptote in the anisotropic phase [$T < T_c = T_{c0} - I(0)$]. Then $\xi = A/\alpha \to 0$, and in (2.185) we can use the asymptotic representation (2.190) for $I(\xi)$:

$$\xi = A/\alpha, \quad \xi + T_c - T + \lambda c_3 \xi^{d/2-1} = \lambda\alpha^2/6n. \tag{2.191}$$

From this it follows that for small A,

$$\alpha(A, T) = \alpha(0, T) + cA + c'A^{d/2-1} + \ldots,$$
$$\tag{2.192}$$
$$\alpha(0, T) = [6n(T_c - T)/\lambda]^{1/2}$$

with coefficients c, c' known from (2.190). Equation (2.192) for the spontaneous magnetization $\alpha(0, T)$ in leading order in $1/n$ is the same as in the self-consistent field approximation (2.90), the only difference being the shift of T_c. The $A \to 0$ asymptote of the longitudinal susceptibility $\chi^\| = \partial\alpha/\partial A$ can be found by differentiating the first equation in (2.192) or by substituting (2.190) into (2.189). The part of (2.192) linear in A contributes a constant to $\chi^\|$ corresponding to the self-consistent field approximation. However, the leading contribution for $A \to 0$ in dimension $d < 4$ is the term $\sim A^{d/2-1}$ generating the divergence of $\chi^{\|}$.

Therefore, for $A \to 0$ and $T < T_c$, in leading order in $1/n$ we have

$$\chi^{\perp}(A) \sim A^{-1}, \quad \chi^{\|}(A) \sim A^{d/2-2}, \tag{2.193}$$

so that both susceptibilities are infinite on the phase-coexistence line $A = 0$. We recall (Sec. 2.10) that in the self-consistent field approximation the longitudinal susceptibility remained finite.

The correlators (2.188) on the line $A = 0$, $T < T_c$ take the form

$$D^{\perp} = p^{-2}, \quad D^{\|} = [p^2 + 2\lambda\alpha^2/n(6 + \lambda\Pi)]^{-1}|_{\xi=0}, \tag{2.194}$$

where $\alpha = \alpha(0, T)$ is the spontaneous magnetization from (2.192). At small $p \ll \Lambda$ for $\Pi|_{\xi=0}$ in (2.194) we can use the corresponding asymptote (2.190). For $2 < d < 4$ the loop contribution $\Pi \sim p^{d-4}$ then dominates in (2.194) for $D^{\|}$, corresponding to the asymptote

$$D^{\perp}\Big|_{\substack{A=0 \\ p\to 0}} \sim p^{-2}, \quad D^{\|}\Big|_{\substack{A=0 \\ p\to 0}} \cong n\Pi/2\alpha^2\Big|_{\xi=0} \sim p^{d-4}. \tag{2.195}$$

Let us now consider the line $A = 0$, $T \geq T_c$. On it $\alpha = 0$, and so from (2.186) and (2.189) we have

$$D^{\perp} = D^{\|} = (p^2 + \xi)^{-1}, \quad \chi^{\perp} = \chi^{\|} = \xi^{-1}, \tag{2.196}$$

where $\xi = \xi(T)$ is the solution of (2.169). The correlator (2.196) corresponds to the first term $\delta_{ab}\Delta$ of the graph expansion (2.167) for $D \equiv W_2$.

Equations (2.196) are valid for any $T \geq T_c$, $A = 0$. By definition, $\xi(T_c) = 0$ (Sec. 2.19), and so ξ is small for small $\Delta T \equiv T - T_c \ll \Lambda^2$. Then in (2.169) we can use the asymptote (2.190) for $I(\xi)$:

$$\Delta T = \xi + \lambda c_3 \xi^{d/2-1} \Rightarrow \xi(T) \sim (\Delta T)^{2/(d-2)} \text{ for } \Delta T \to 0. \tag{2.197}$$

The susceptibility (2.196) therefore diverges for $\Delta T \to 0$ according to a standard power law (Sec. 1.5) $\chi \sim (\Delta T)^{-\gamma}$ with exponent $\gamma = 2/(d - 2)$, in contrast to the result $\gamma = 1$ in the Landau theory (Sec. 2.10).

2.24   Critical behavior in leading order in $1/n$

The expressions obtained in the two preceding sections suggest the existence of critical scaling with definite critical dimensions Δ_F of the various quantities F (Sec. 1.3). The critical region is, by definition, the region of small p, A, and $\Delta T \equiv T - T_c$, and also, as a consequence, small α and ξ (we assume that F is small if $|F| \ll \Lambda^{d[F]}$, where Λ is the cutoff and $d[F] \equiv d_F$ is the canonical dimension of F). In the critical region the contribution of ξ

in (2.197) is small compared to $\xi^{d/2-1}$ (we always assume that $2 < d < 4$), and in (2.179) and (2.188) the contribution of the constant is small compared to the loop contribution Π divergent at $\xi \sim p^2 \to 0$. When studying the leading critical singularities these small corrections must be discarded, and then (2.179), (2.185), and (2.188) take the form

$$\xi = A/\alpha, \quad -\Delta T + \lambda c_3 \xi^{d/2-1} = \lambda \alpha^2/6n ,$$

$$D^\perp = (p^2 + \xi)^{-1}, \quad D^\| = [p^2 + \xi + 2\alpha^2/n\Pi]^{-1}, \qquad (2.198)$$

$$\Delta_\psi^{-1} = -n\Pi/2 - \alpha^2/(p^2 + \xi).$$

For fixed parameters λ and Λ, and, consequently, $T_c = T_c(\lambda, \Lambda)$, (2.198) describes scaling with definite dimensions Δ, which are easily found using the equivalence relations following from (2.198): $\xi \sim A/\alpha$, $\Delta T \sim \xi^{d/2-1} \sim \alpha^2$, $D^{-1} \sim p^2 \sim \xi$, $\Delta_\psi^{-1} \sim \Pi \sim \alpha^2 p^{-2}$ (here $a \sim b$ denotes equality of the critical dimensions of a and b), and the standard normalization condition (1.11). The critical dimensions Δ of all the quantities involved are determined uniquely from these expressions. From the known dimensions of the correlators D and Δ_ψ we can then find the dimensions of the corresponding fields φ and ψ using the general expression $2\Delta_\phi = d - \Delta[D]$ following from (1.35), where Δ_ϕ is the dimension of the arbitrary field $\phi(x)$ and $\Delta[D]$ is the dimension of its correlator in momentum space.

The information on the dimensions obtained in this manner from (2.198) is summarized in Table 2.1. In the first line we list the various quantities F, in the second we give their critical dimensions Δ_F in the leading order of the $1/n$ expansion under discussion, and for comparison in the third line we give the usual canonical dimensions of the same quantities for the action functional in the integral (2.152).

Table 2.1 Values of the canonical (d_F) and critical (Δ_F) dimensions in leading order in $1/n$ in the $O_n \, \varphi^4$ model.

F	p	x	Λ	λ	T_c	$T - T_c$	$\alpha, \, \varphi(x)$	$\xi, \, \psi(x)$	Π
Δ_F ($n = \infty$)	1	-1	0	0	0	$d - 2$	$d/2 - 1$	2	$d - 4$
d_F	1	-1	1	$4 - d$	2	2	$d/2 - 1$	2	$d - 4$

The difference between d_F and Δ_F should particularly be noted. Canonical scaling (scale invariance) with dimensions d_F always exists, but this has nothing to do with the properties of the solutions of a particular model with fixed λ and Λ, because these parameters are also transformed in (1.67). In other words, canonical scale invariance (1.67) relates the solutions of different models to each other. By assigning zero critical dimensions to the parameters λ and Λ, we thereby assume them to be fixed under transformations like (1.4), and so critical scaling with dimensions Δ_F is a property of the solutions of a particular model.

Using the dimensions Δ_F given in Table 2.1 and the general rules (1.19) and (1.40) (where now $h = A$, $\tau = \Delta T = T - T_c$), we can find all the traditional critical exponents (Sec. 1.5) in leading order in $1/n$. The results are summarized in Table 2.2.

Table 2.2 Critical exponents of the O_n φ^4 model in leading order in $1/n$.

Exponent	α	β	γ	δ	η	ν	ω
Value at $n = \infty$	$\frac{d-4}{d-2}$	$\frac{1}{2}$	$\frac{2}{d-2}$	$\frac{d+2}{d-2}$	0	$\frac{1}{d-2}$	$4-d$

In this table we have included the exponent ω, which determines the leading correction to critical scaling (Sec. 1.3). In our case the corrections are the discarded contributions ξ in the first expression in (2.197) and the constants (compared to $\lambda\Pi$) in (2.179) and (2.188). It is easily checked using the results in Table 2.1 that the critical dimension Δ_F of any of these discarded correction terms exceeds the dimension of the leading contribution kept in (2.198) by $4 - d$, and this extra amount is the exponent ω (Sec. 1.3).

In Ch. 1 the RG technique was used to show that there is critical scaling in this model for any $d < 4$. The corresponding critical dimensions depend on d and n. In practice, they can be calculated only as asymptotic series: a $4 - \varepsilon$ expansion in the parameter $4 - d$, as discussed in detail in Ch. 1, then a $2 + \varepsilon$ expansion in the parameter $d - 2$, which will be discussed in Ch. 4, and, finally, a large-n expansion, under discussion now. The values of Δ_F given in Table 2.1 are the first terms of the corresponding large-n expansions. The corrections will be given in Ch. 4.

2.25 A simplified field model for calculating the large-n expansions of critical exponents

In Secs. 2.22–2.24 we studied the solution only in leading order in $1/n$. In principle, we know not only the leading term but also the first correction term in (2.177). However, direct generalization of the technique described above to higher orders in $1/n$ leads to very complicated calculations. This is unavoidable if we want complete information about the critical behavior, including the scaling functions taking into account all the variables: the momenta, temperature, and external field. For this reason there are very few results in this area (see Ch. 4). However, if we are interested only in the critical dimensions, the calculation can be greatly simplified. First, it is not necessary to prove the existence of critical scaling, since this is already done by the RG method in Ch. 1. The required relations (Sec. 1.5) between the various critical exponents are thus guaranteed, and it is sufficient to calculate only two independent critical dimensions, for example, Δ_φ and Δ_τ. The others are expressed in terms of them using the standard expressions of the scaling hypothesis. Second, the desired dimensions Δ_φ and Δ_τ can be found from the Green functions in zero field for $T \geq T_c$, which is much simpler than the calculation with a field or for $T < T_c$. It is then convenient to take (2.170) as the initial action functional, rewriting it as

$$S = -\tfrac{1}{2}\varphi(-\partial^2 + \xi + \psi)\varphi + \tfrac{3n}{\lambda}(\xi - T + T_{c0})\psi + \tfrac{3n}{2\lambda}\psi^2, \qquad (2.199)$$

where $\xi = \xi(T)$ is the solution of (2.169), equivalent to (2.155). In going from (2.170) to (2.199) we have discarded the unimportant constant $\sim \psi_0^2$ and the (understood) index a, and have used $K = -\partial^2 + \tau$ and $\psi_0 = \xi - \tau$. In these expressions $\tau = T - T_{c0}$, and in what follows we shall use τ to denote $\Delta T \equiv T - T_c$.

As explained in Sec. 2.20, to obtain the large-n expansion by the graph technique, the action (2.199) must be rewritten as

$$S = S - \frac{1}{2}\psi L\psi + \frac{1}{2}\psi L\psi, \quad L = -\frac{1}{2}\,\bigcirc\; = -\frac{n}{2}\Pi, \qquad (2.200)$$

where Π is the scalar loop (2.187), depending on T through ξ. The contribution $-\frac{1}{2}\psi L\psi$ in (2.200) plus the last term in (2.199) form the free action of the field ψ, and the term $\frac{1}{2}\psi L\psi$ in (2.200) is an addition to the interaction which cancels insertions of double cycles like (2.173). The single cycles are canceled by the term in (2.199) linear in ψ.

If we are not interested in the corrections to scaling, the last term in (2.199) can be dropped, because we know (Sec. 2.24) that its contribution is negligible in the critical region compared to that of the loop Π. However, even after this simplification the calculation of the graphs in the model (2.199) is rather complicated. It is greatly simplified right at the critical point $T = T_c$, because the fundamental field then becomes massless $[\xi(T_c) = 0]$. This theory without corrections corresponds to the action

$$S|_{\tau=0} = -\tfrac{1}{2}(\partial\varphi)^2 - \tfrac{1}{2}\psi\varphi^2 + h_c\psi - \tfrac{1}{2}\psi L\psi + \tfrac{1}{2}\psi L\psi \qquad (2.201)$$

with constant $h_c = \frac{3n}{\lambda}(T_{c0} - T_c)$ and kernel L from (2.200) for the massless lines in the loop. The dimension Δ_φ can be found from the IR asymptote of the φ correlator in the model (2.201).

Let us show that within the model (2.201) it is also possible to calculate the second exponent Δ_τ, the critical dimension of the parameter $\tau \equiv T - T_c$. Clearly, it can be found from $\partial_\tau \ln Z|_{\tau=0}$, where $Z \equiv \int D\phi \exp S(\phi)$ is the partition function (1.60) for the model (2.199), and $\phi \equiv \varphi, \psi$ is the set of all the fields of the model. The T dependence enters into (2.199) both explicitly and implicitly through ξ. For what follows it is important that at small τ, ξ behaves, according to (2.197), as a power τ^γ with exponent $\gamma = 2/(d-2) > 1$, and so $\partial_\tau \xi = 0$ for $\tau = 0$. It follows that the implicit τ dependence from ξ in (2.199) can be neglected when calculating the derivative $\partial_\tau \ln Z|_{\tau=0}$, and we find

$$\partial_\tau \ln Z|_{\tau=0} = -\tfrac{3n}{\lambda} \int dx \langle \widehat{\psi}(x) \rangle|_{\tau=0}. \tag{2.202}$$

We have used the notation (1.60) and explicitly indicated the x integration understood in (2.199). Equation (2.202) together with the zero critical dimension of the partition function Z and the coefficient $3n/\lambda$ allow the desired value $\Delta_\tau = 1/\nu$ to be related to the critical dimension of the auxiliary field ψ in the model (2.201):

$$\Delta_\tau \equiv 1/\nu = d - \Delta_\psi. \tag{2.203}$$

It is also possible to calculate the correction critical exponent ω in this model. It is related to the last term in (2.199) discarded in the action (2.201), and is determined by the critical dimension of the corresponding IR-irrelevant composite operator ψ^2 (this is shown rigorously in Ch. 4):

$$\omega = \Delta[\psi^2(x)] - d. \tag{2.204}$$

Therefore, all the needed critical exponents can be calculated directly in the massless model (2.201).

There is yet another simplification important for practical applications. As explained in Sec. 2.20, the terms linear in ψ in (2.199) and (2.201) exactly cancel the insertions of single cycles of the type (2.173). Therefore, the answers are not changed if both are discarded simultaneously. In the massless model (2.201) a single cycle corresponds to the integral $I(0)$ in (2.169), a multiple of Λ^{d-2}. These contributions are discarded in calculations using the formal scheme of dimensional regularization (Sec. 1.20). Then it is necessary to also discard the contribution $h_c\psi$ in (2.201), setting

$$S|_{\tau=0} = -\tfrac{1}{2}(\partial\varphi)^2 - \tfrac{1}{2}\psi L\psi - \tfrac{1}{2}\psi\varphi^2 + \tfrac{1}{2}\psi L\psi. \tag{2.205}$$

The first two terms form S_0, and the last two form the interaction V. There are no numerical parameters in the action (2.205), and the canonical dimensions of the fields calculated for it according to the standard rules in Sec. 1.15

$$d_\varphi = d/2 - 1, \qquad d_\psi = 2 \tag{2.206}$$

coincide with the corresponding Δ_F in Table 2.1. The equation $2d_\varphi + d_\psi = d$ following from (2.206) ensures that the vertex $\psi\varphi^2$ and the quadratic form $\psi L\psi$ simultaneously have zero dimension. The first is obvious, and the second follows from the definition of L in (2.200).

Specific calculations in the model (2.205) and their results are discussed in Ch. 4.

2.26 The classical Heisenberg magnet and the nonlinear σ model

In the model of the classical Heisenberg magnet, each node i of a given spatial lattice is associated with a random classical n-component spin vector of unit length (Sec. 1.6). It will play the role of the field and is denoted as $\widehat{\varphi} \equiv \{\widehat{\varphi}_{ia}\}$, where i is the lattice index, $a = 1, ..., n$ is the spin index, and $\widehat{\varphi}_i^2 \equiv \sum_a \widehat{\varphi}_{ia}^2 = 1$. The Hamiltonian of the model is (with summation over repeated indices)

$$\widehat{H} = -\tfrac{1}{2}\widehat{\varphi}_{ia} v_{ia,sb} \widehat{\varphi}_{sb} - h_{ia}\widehat{\varphi}_{ia} \equiv -\tfrac{1}{2}\widehat{\varphi}v\widehat{\varphi} - h\widehat{\varphi}, \tag{2.207}$$

where h is the external field and v is the symmetric (in pairs of indices) exchange-interaction matrix. For a spin-isotropic model $v_{ia,sb} = v_{is}\delta_{ab}$. In the special case of a nearest-neighbor interaction we take $v_{is} = J$ if the pair is are nearest neighbors on the lattice, and $v_{is} = 0$ otherwise. The constant J is the exchange integral; for a ferromagnet $J > 0$ and for an antiferromagnet $J < 0$.

Using G to denote the partition function in an arbitrary external field, we have

$$G = \operatorname{tr}\exp[-\beta\widehat{H}] = \operatorname{tr}\exp[\tfrac{1}{2}\widehat{\varphi}\Delta\widehat{\varphi} + A\widehat{\varphi}],$$

where $\beta = 1/kT$, $\Delta = \beta v$, $A = \beta h$, and tr denotes integration over the directions of all the φ_i. Multiplication by φ inside the tr is equivalent to differentiation with respect to A, and so

$$G = \exp\left[\frac{1}{2}\cdot\frac{\delta}{\delta A}\Delta\frac{\delta}{\delta A}\right]\exp V, \quad V = \ln\operatorname{tr}\exp A\widehat{\varphi}. \tag{2.208}$$

The functional V is interpreted as the logarithm of the partition function without the exchange interaction. It reduces to a sum over nodes and can therefore be considered known. In the special case $n = 1$ (the Ising model) we have $V = \sum_i \ln(2\cosh A_i)$. Equation (2.208) is a special case of (2.19), and so the partition function admits the standard graph representation of Sec. 2.2 and can be written as a functional integral using (2.18).

Another representation is the definition of the partition function itself:

$$G \sim \int D\varphi\,\delta(\varphi^2 - 1)\exp[-\beta H(\varphi)], \tag{2.209}$$

where $\varphi = \{\varphi_{ia}\}$ is the field on the lattice, $D\varphi \equiv \Pi_{ia}d\varphi_{ia}$, and $\delta(\varphi^2 - 1) \equiv \Pi_i\delta(\varphi_i^2 - 1)$.

Up to now we have been dealing with the exact microscopic model. When studying the critical behavior of a spin-isotropic ferromagnet it is usual to go to the continuous analog of (2.209), replacing $\varphi_{ia} \equiv \varphi_a(x_i)$ by $\varphi_a(x)$, and the quadratic form in (2.207) by $\int dx(\partial\widehat{\varphi})^2$. Let us explain. For an unbounded periodic lattice with translationally invariant (i.e., depending only on the difference of node coordinates) matrix v_{is} we can go to momentum space on the lattice, $\varphi v\varphi = \int dk\varphi_a(-k)v(k)\varphi_a(k)$, with integration over a restricted region — the first Brillouin zone (for a simple cubic lattice with spacing L this is the cube $|k_\alpha| \leq \pi/L$ for all components). In the case of a ferromagnet the maximum of the function $v(k)$ is reached at $k = 0$, which corresponds to spatial uniformity of the spin configuration in the ground state. The symmetry of the matrix v_{is} implies that the function $v(k)$ is even, and so near the maximum $v(k) = v(0) - c_{\alpha\beta}k_\alpha k_\beta + ...$, where $c_{\alpha\beta}$ is a positive-definite matrix in the spatial indices α, β. The first term $v(0)$ corresponds to a multiple of δ_{is} in coordinate space, i.e., a constant in (2.207) owing to the condition $\varphi_i^2 = 1$. Therefore, the contribution $v(0)$ can be eliminated by adding a suitable constant to the Hamiltonian (2.207), which does not essentially change the model. Then the quadratic form in k becomes the leading term in the expansion of $v(k)$, and the contributions of higher powers of k are discarded because they are IR-irrelevant (Sec. 1.16). Any positive-definite quadratic form can be brought to the isotropic form $k_\alpha k_\alpha = k^2$ by a suitable rotation and rescaling of the spatial coordinates, which explains the choice of the model Hamiltonian. The spatial isotropy is often a consequence of the symmetry, and then the coordinate transformation is not required. For a nearest-neighbor interaction with a single J (see above), the form will be isotropic for any lattice with cubic symmetry (Sec. 1.14). Invariance under rotations by $60°$ for the planar triangular lattice also leads to isotropy, and so on. However, isotropy is not the general case. A simple counterexample is the nearest-neighbor interaction on a simple cubic lattice with different values of the exchange integral J for different axes. Here a suitable rescaling is required in order to restore the spatial isotropy.

This explains the origin of the universality of critical behavior and simultaneously makes the meaning of this term precise. Universality is manifested only for the correct choice of variables, in particular, the scales on different axes in the example given above.

Therefore, in the fully isotropic continuum model

$$H(\varphi) = L^{-d} \int dx [\alpha L^2 (\partial\varphi)^2 - h\varphi], \qquad (2.210)$$

the lattice constant L makes dx and ∂ dimensionless, the constant α in (2.210) is a dimensionless positive coefficient of order unity (in principle, calculable from v_{is}), and, as usual, $(\partial\varphi)^2 = \partial_i\varphi_a \cdot \partial_i\varphi_a$, where now i is a spatial index. Then (2.209) is understood as a functional integral, and the δ function as a continuous product for all x. By construction, the UV cutoff $\Lambda \sim 1/L$ is assumed in the theory. Substitution of (2.210) and rescaling of φ brings (2.209) to the form

$$G \sim \int D\varphi \delta(\varphi^2 - cT^{-1}) \exp[S(\varphi) + A\varphi], \quad S(\varphi) = -\tfrac{1}{2}(\partial\varphi)^2, \qquad (2.211)$$

where $A \sim h$, c is a positive constant, and integration over x in the exponent is understood. In field theory, (2.211) is called the nonlinear σ model. The constant in the δ function can be arbitrarily changed without changing the exponent in (2.211) by a suitable simultaneous rescaling of φ, A, and x. However, in any case to ensure canonical scale invariance (Sec. 1.15) the right-hand side of the constraint condition $\varphi^2(x) = \text{const}$ must contain a parameter (here, T^{-1}) which can be assigned the canonical dimension of φ^2.

By rescaling $\varphi \to \varphi T^{-1/2}$ we can go to the new canonically dimensionless field φ. Then the factor T^{-1} appears in front of the action $\sim (\partial\varphi)^2$ in the exponent. The calculation of the resulting integral by the saddle-point method (Sec. 2.8) naturally leads to low-temperature perturbation theory in powers of T. The $2 + \varepsilon$ expansion of the critical exponents of the σ model can be obtained within this perturbation theory (Ch. 4).

2.27 The large-n expansion in the nonlinear σ model

Representing the δ function in (2.211) as a Fourier functional integral

$$\delta(F) \equiv \prod_x \delta(F(x)) \sim \int D\psi \exp\left[i\int dx\,\psi(x)F(x)\right] \equiv \int D\psi \exp i\psi F, \quad (2.212)$$

we pass to the theory of two fields φ and ψ.

In view of the practical importance of (2.212), let us make a few explanatory remarks. First, this expression has an explicit meaning in the lattice model. Second, the equation $\delta(F)\exp(tF) = \delta(F)$ can be used to add to $i\psi(x)$ in (2.212) an arbitrarily fixed real part $t(x)$, which is sometimes done to ensure convergence of the resulting integrals. Finally and most importantly, if we deal with statements that can be verified only perturbatively (by rearrangement of the graph series and so on), then when manipulating functional integrals the question of their convergence can be entirely neglected. Every result obtained in this manner will certainly be correct at the graph level, and it is difficult to expect more in field theory. It is only important that the answer itself be meaningful. In Sec. 2.4 we changed the functional integral not because it is nonexistent in ordinary perturbation theory in g owing to divergences at $\tau < 0$, but because in perturbation theory the answer itself is poor: the required positive-definiteness of the correlator is violated. As an example, consider the integral (2.151). It diverges, but the results obtained using it are correct in the sense that they are exactly the same as those obtained by direct summation of the contributions of the required order in $1/n$ in the graphs of ordinary perturbation theory.

At this level of rigor, we can make any variable substitutions in the functional integrals, in particular, in (2.212) $i\psi \to \psi \to c\psi$ with arbitrary constant c. Therefore, in what follows we shall take the equation

$$\delta(F) \equiv \prod_x \delta(F(x)) \sim \int D\psi \exp[c(\psi F)] \quad (2.213)$$

to be valid, where ψF is a scalar product like (1.48) and c is an arbitrary number. When (2.213) is used in expressions like (2.211), the contribution $c\psi F$ is interpreted as an addition to the action, and $c\psi(x)$ as the Lagrange multiplier of the constraint $F(x) = 0$.

Returning to the σ model (2.211), we replace the constant cT^{-1} in the δ function by nT^{-1} (see the remark at the end of Sec. 2.26), where n is the number of field components, because the large-n expansion in this model is meaningful only for $\varphi^2 \sim n$ (see below). The condition $\varphi^2 \sim n$ is the analog of $g \sim 1/n$ in the model (2.79). Then, writing the constraint as an integral like (2.213) with coefficient $c = -1/2$, we obtain

$$
\begin{aligned}
G(A) &= \int D\varphi \, \delta(\varphi^2 - nT^{-1}) \exp[-\tfrac{1}{2}(\partial\varphi)^2 + A\varphi] = \\
&= \iint D\varphi \, D\psi \exp[-\tfrac{1}{2}(\partial\varphi)^2 - \tfrac{1}{2}\psi(\varphi^2 - nT^{-1}) + A\varphi].
\end{aligned}
\tag{2.214}
$$

In this case all the normalization factors in front of the integrals are independent of essential parameters like T, and so they are not of interest and can just be included in the measure $D\varphi D\psi$. The integral on the right-hand side of (2.214) is analogous to the integral (2.152); we have only omitted the index a for brevity. The role of K in (2.214) is played by the operator $K = -\partial^2$ without the τ contribution (which anyhow would lead to an unimportant constant owing to the constraint). It should also be noted that the action (2.214) does not contain the contribution quadratic in ψ, but in return it does contain the contribution linear in ψ, which was not the case in (2.152).

The desired large-n expansion for the model (2.214) in the symmetric phase ($T \geq T_\mathrm{c}$, A of order unity) is constructed exactly as for the integral (2.152). Integrating over φ using (2.1) (which is valid only for $T \geq T_\mathrm{c}$, i.e., in the absence of spontaneous magnetization), we obtain an integral of the type (2.153) with $nF(\psi) = \tfrac{n}{2}[\psi T^{-1} - \mathrm{tr}\ln(-\partial^2 + \psi)]$ and $\overline{\Delta} = (-\partial^2 + \psi)^{-1}$. Therefore, instead of (2.171) and (2.166) we now have

$$
nT^{-1} = \bigcirc, \quad \Delta_\psi^{-1} = -\frac{1}{2}\bigcirc, \quad \text{where} \quad —— \equiv \delta_{ab}(-\partial^2 + \psi_0)^{-1}. \tag{2.215}
$$

The first equation is the saddle-point equation, and the second determines the propagator of the auxiliary field ψ. We now see the reason for choosing the constraint $\varphi^2 \sim n$ instead of $\varphi^2 \sim 1$: the constant in the δ function must be of order n, otherwise it would not contribute to the saddle-point equation (2.215). The left-hand side would vanish, and the equation would not have any solutions.

With the usual assumption that the solution $\psi_0 = \xi$ is homogeneous, the saddle-point equation (2.215) can be rewritten as $T^{-1} = (2\pi)^{-d} \int dk(k^2+\xi)^{-1} \equiv I(\xi)$ with the same momentum cutoff as in (2.169). A solution $\xi \geq 0$ exists and is unique for all $T \geq I_{\max}^{-1} = I(0)^{-1}$; the right-hand side of this inequality is T_c in leading order in $1/n$.

At the critical point $T = T_\mathrm{c}$ the propagator $\delta_{ab}(k^2 + \xi)^{-1}$ of the field φ_a becomes massless, and the loops in (2.166) and (2.215) diverge as p^{d-4} as the external momentum p tends to zero (we assume that $d < 4$). Therefore, the contribution of the constant in (2.166) is IR-irrelevant compared to the loop contribution, as noted in Sec. 2.24. After the shift $\psi \to \psi_0 + \psi$ the action functionals in the integrals (2.152) and (2.214) will differ only by an

IR-irrelevant term $\sim \psi^2$. This leads to the following important statement: *The O_n φ^4 and nonlinear σ models belong to the same universality class, i.e., they have the same critical behavior.*

Strictly speaking, this statement is justified above only within the large-n expansion for arbitrary spatial dimension, and so differences in contributions with identically zero large-n expansions are formally allowed. However, the general opinion is that there are no such contributions and the equivalence between these two models is complete, of course, only as far as the critical behavior is concerned. This belief is supported by the mutual consistency of the three types of expansion of the critical exponents: $4-\varepsilon$, $2+\varepsilon$, and large-n. The coefficients of the first can be calculated only for the φ^4 model (Ch. 1), those of the second only for the σ model (Ch. 4), and those of the third are the same for both models. The results of these calculations indicate (see Ch. 4 for more details) that these three types of expansion are mutually consistent, i.e., the coefficients of the double expansions in ε and $1/n$ obtained by different methods coincide. Therefore, they really can be viewed as different types of expansion of the same critical exponents, which is consistent with the above statement. We again note that the equivalence statement concerns only the critical behavior; in all other respects these two models differ significantly.

Equations (2.203) and (2.204) remain valid for the exponents of the σ model, and (2.203) can be proved in the σ model even more easily than in Sec. 2.25. In fact, it is obvious that the addition of the term $h\psi$ to the exponent in (2.214) is equivalent to changing T. It then follows that for $T = T_c$ the coefficient h in the addition $h\psi$ plays the role of $\tau = T - T_c$ up to an insignificant coefficient. The critical dimensions of h and ψ are related by the standard shadow relation (Sec. 1.10), which also leads to (2.203).

2.28 Generalizations: the CP^{n-1} and matrix σ models

One of the simplest generalizations of the σ model is the CP^{n-1} model [79]. It is described by the complex field $\varphi_a(x)$, $a = 1, ..., n$, with the constraint

$$\varphi_a^+(x)\varphi_a(x) = nT^{-1} \equiv \overline{n} \tag{2.216}$$

and action functional (summation over repeated indices and the required x integrations in the action are always understood)

$$S = -(\nabla_i \varphi_a)^+ \cdot (\nabla_i \varphi_a) = -(\partial_i \varphi_a)^+ \cdot (\partial_i \varphi_a) - \tfrac{1}{4\overline{n}}Q_i Q_i, \tag{2.217a}$$

where ∇_i is the "covariant derivative":

$$\nabla_i \equiv \partial_i - \tfrac{1}{2\overline{n}}Q_i, \quad Q_i \equiv \partial_i \varphi_a \cdot \varphi_a^+ - \varphi_a \cdot \partial_i \varphi_a^+. \tag{2.217b}$$

A more complicated generalization of (2.211) is the $n \times p$ matrix nonlinear σ model [80]. It is described by the complex field φ_a^b with $a = 1, ..., n$ for the lower indices and $b = 1, ..., p$ for the upper ones, with the constraint

$$(\varphi_a^b(x))^+ \cdot \varphi_a^c(x) = \overline{n}\delta^{bc}, \qquad \overline{n} \equiv nT^{-1}, \tag{2.218}$$

and action functional

$$S = -(\nabla_i^{bc}\varphi_a^c)^+(\nabla_i^{bd}\varphi_a^d) = -(\partial_i\varphi_a^b)^+(\partial_i\varphi_a^b) - \tfrac{1}{4\overline{n}}Q_i^{bc}Q_i^{cb}, \tag{2.219a}$$

in which

$$\nabla_i^{bc} \equiv \delta^{bc}\partial_i - \tfrac{1}{2\overline{n}}Q_i^{bc}, \quad Q_i^{bc} \equiv (\partial_i\varphi_a^b)(\varphi_a^c)^+ - (\varphi_a^b)(\partial_i\varphi_a^c)^+. \tag{2.219b}$$

For $p = 1$ the model (2.219) becomes (2.217).

The model (2.217) is invariant under the Abelian $U(1)$ group of local gauge transformations $[\varphi_a(x) \to u(x)\varphi_a(x)$ with arbitrary phase factor $u(x)$, $|u(x)| = 1]$, and the model (2.219) has a global $U(n)$ symmetry in the lower index of the field φ and a local $U(p)$ gauge symmetry in the upper index [we recall that $U(m)$ is the standard notation for the group of all unitary $m \times m$ matrices].

Large-n expansions can be constructed for the two models using the technique described above. Two auxiliary fields must be introduced: the Hermitian scalar field ψ or ψ^{ab} as the Lagrange multiplier of the corresponding constraint (2.216) or (2.218), and the anti-Hermitian vector field B_i or B_i^{bc} in the representation

$$\exp\left[-\tfrac{1}{4\overline{n}}\mathrm{tr}(Q_iQ_i)\right] - \mathrm{const} \int DB \exp \mathrm{tr}[\overline{n}B_iB_i + B_iQ_i] \tag{2.220}$$

[with a trace in the upper indices for the model (2.219)], which must be used to transform the contributions involving Q^2 in the original functional integral with the action (2.217) or (2.219).[*]

2.29   The large-n expansion for $(\varphi^2)^3$-type interactions

In this section we shall briefly explain how the technique described in Sec. 2.19 can be generalized to the case of an interaction which is an arbitrary polynomial in $\varphi^2(x) \equiv \varphi_a(x)\varphi_a(x)$, for example, $\varphi^6 \equiv (\varphi^2)^3$ and so on. We have seen (Sec. 2.19) that to ensure the correctness of the large-n expansion, the coupling constant g_4 in the φ^4 interaction must be assumed a multiple of $1/n$. By the same arguments, the constant g_6 in the φ^6 interaction must be

---

[*]From the viewpoint of the critical behavior, the model (2.217) with auxiliary fields is equivalent to scalar electrodynamics, and (2.219) is equivalent to scalar chromodynamics exactly in the same sense in which the σ model (2.211) is equivalent to the φ^4 model. In the terminology of chromodynamics, in the model (2.219) p is the number of "colors" and n is the number of "flavors." For large (compared to p) n, chromodynamics loses the property of ultraviolet asymptotic freedom, and so it is not the large-n but the large-p expansion which is relevant. However, unfortunately, it is not possible to construct the latter.

a multiple of $1/n^2$, the constant g_8 a multiple of $1/n^3$, and so on. In general, this statement of the problem corresponds to the action functional

$$S(\varphi) = \int dx \left[-\tfrac{1}{2} \partial \varphi_a(x) \cdot \partial \varphi_a(x) - nP(\varphi^2(x)/n) \right] \equiv$$
$$\equiv \tfrac{1}{2} \varphi_a \partial^2 \varphi_a - nP(\varphi^2/n), \tag{2.221}$$

in which $P(z)$ is a given monomial or polynomial in z with coefficients independent of n (we have also included the contribution $\tau \varphi^2$ in it).

To obtain the large-n expansion of the logarithm of the integral analogous to (2.150) (omitting all inessential normalization factors)

$$G(A) = \int D\varphi \exp \left[\tfrac{1}{2} \varphi_a \partial^2 \varphi_a - nP(\varphi^2/n) + A_a \varphi_a \right], \tag{2.222}$$

it is necessary to introduce the following expansion of unity:

$$1 = \int D\overline{\psi} \delta(\overline{\psi} - \varphi^2/n) = \int\int D\overline{\psi} D\psi \exp \left[\tfrac{1}{2} \psi (n\overline{\psi} - \varphi^2) \right] \tag{2.223}$$

involving two auxiliary scalar fields $\overline{\psi}(x)$ and $\psi(x)$. The second of these is the analog of ψ in (2.152) and is introduced as a Lagrange multiplier in the representation of the δ functional from (2.223) using an expression like (2.213).

Inserting the expansion of unity (2.223) inside the integral (2.222) and interchanging the integrations, we can then make the replacement $\varphi^2/n \to \overline{\psi}$ in the argument of P owing to the presence of the δ function, which gives

$$G(A) = \int D\varphi D\overline{\psi} D\psi \exp \left[\tfrac{1}{2} \varphi_a \partial^2 \varphi_a - nP(\overline{\psi}) + \tfrac{1}{2} \psi (n\overline{\psi} - \varphi^2) + A_a \varphi_a \right]. \tag{2.224}$$

Performing the Gaussian integration over φ, we obtain

$$G(A) = \int D\overline{\psi} D\psi \exp \left[-nP(\overline{\psi}) + \tfrac{n}{2} \psi \overline{\psi} - \tfrac{n}{2} \operatorname{tr} \ln(-\partial^2 + \psi) + \tfrac{1}{2} A_a \frac{1}{(-\partial^2 + \psi)} A_a \right]. \tag{2.225}$$

The calculation of this integral by the saddle-point method leads to the desired large-n expansion. In the special case of the φ^4 interaction, the integral over $\overline{\psi}$ in (2.225) is Gaussian. Doing it, we obtain the representation (2.153) up to an insignificant shift of ψ.

2.30   Systems with random admixtures

In real problems it is often necessary to include the effect of randomly distributed admixtures. Let φ be the fundamental field (system of fields) and ψ the field of the admixture. "Melted" and "frozen" admixtures are distinguished. The former are essentially in thermodynamical equilibrium with the other degrees of freedom, and then ψ is simply an additional field that is included in the general scheme on an equal footing with the other fields. Frozen admixtures are ones for which the distribution of ψ does not reach thermodynamical equilibrium with the other degrees of freedom φ, because its relaxation time is large compared to that for φ. When calculating the statistical averages with respect to φ for such admixtures the configuration ψ is considered to be fixed, and then the answer is averaged over ψ with a given normalized weight $\rho(\psi)$:

$$\langle F(\phi) \rangle = \overline{\langle F(\phi) \rangle}_\psi = \int D\psi \rho(\psi) \frac{\int D\varphi F(\phi) \exp S(\phi)}{\int D\varphi \exp S(\phi)}. \tag{2.226}$$

Here $\phi \equiv \varphi, \psi$ is the set of all fields; $S(\phi) = S(\varphi, \psi)$ is the action functional for φ, depending parametrically on ψ; the brackets $\langle ... \rangle_\psi$ denote the statistical (i.e., with weight $\exp S$) average over φ for fixed ψ; the overline denotes the average over ψ with given weight $\rho(\psi)$ normalized to unity; and the brackets $\langle ... \rangle$ denote the full average. There is no fundamental difference between the two cases at the level of the definition (2.226): it is easily checked that the right-hand side of (2.226) is the usual statistical average over ϕ with weight $\exp S'(\phi)$, where $S'(\phi) = S(\phi) + \ln \rho(\psi) - \ln \int D\varphi \exp S(\phi)$ and $\int D\phi \exp S'(\phi) = 1$ owing to the normalization condition for ρ, $\int D\psi \rho(\psi) = 1$.

Equation (2.226) uniquely determines the full Green functions $\langle \varphi ... \varphi \rangle = \overline{\langle \varphi ... \varphi \rangle}_\psi$ (the arguments x are omitted) and their generating functional $G(A) = \overline{G(A, \psi)}$, where $G(A, \psi)$ is the functional (1.62) with fixed ψ. However, the connected functions are not determined uniquely. For example, the correlator can be defined as $\overline{\langle \varphi\varphi \rangle}_\psi - \overline{\langle \varphi \rangle}_\psi \cdot \overline{\langle \varphi \rangle}_\psi$ or as $\overline{\langle \varphi\varphi \rangle_\psi - \langle \varphi \rangle_\psi \langle \varphi \rangle_\psi}$. The first definition corresponds to the generating functional $\ln \overline{G(A, \psi)}$, and the second to $\overline{\ln G(A, \psi)}$. The two definitions are meaningful, but correspond to different experiments: if the measured quantity is expressed directly in terms of the full function $\langle \varphi\varphi \rangle_\psi$ the first definition is natural, while if it is expressed in terms of the connected function $\langle \varphi\varphi \rangle_\psi - \langle \varphi \rangle_\psi \langle \varphi \rangle_\psi$ the second is. In practice, Green functions are measured as the response of a system to external perturbations. They are expressed in terms of connected rather than full functions. A typical example is critical opalescence, where the connected function — the correlator of density fluctuations — is measured from the intensity of the scattered light. Therefore, for a frozen admixture we take the second definition:

$$W(A) = \overline{\ln G(A, \psi)} = \overline{W(A, \psi)} \equiv \int D\psi \rho(\psi) W(A, \psi). \tag{2.227}$$

Regarding experiments, we also note that in a real situation the average over the ensemble of all possible configurations of the field ψ is determined by a single measurement for large (compared to the characteristic scale r_c of the fluctuations of the fundamental field φ) sample size. For example, in the scattering of light, each individual event occurs in a region of size r_c inside which the configuration of the admixture $\psi(x)$ can be assumed fixed.

The full intensity is the sum of the contributions from scattering on different "elementary regions," which leads to the average over random configurations of ψ.

Up to now we have dealt with the normalized functional (1.62), which determines only the A-dependent factor in the partition function with external field A (Sec. 1.10). Its logarithm determines the A-dependent part of the free energy $\mathcal{F} = -kT \ln Z$ (in magnet terminology). Equation (2.227) is the definition of this part of the free energy for a system with frozen admixtures. Taking this definition for the part makes it natural to use it for the whole, i.e., for the total free energy $\mathcal{F}$. The latter is defined as $\mathcal{F} - \mathcal{F}_0 = -kT \ln G$, where G is the functional (2.30), and $\mathcal{F}_0$ is the ψ-independent free energy of the unperturbed system with action S_0. This implies that (2.227) should be viewed as valid also for the unnormalized functionals G and W defined by (2.30).

Let us briefly discuss the graph representation of the functional (2.227). By definition, the overline in (2.227) denotes the average over configurations of $\psi(x)$ with given normalized weight $\rho(\psi)$, i.e.,

$$\overline{F(\psi, ...)} \equiv \int D\psi \rho(\psi) F(\psi, ...), \tag{2.228}$$

where F is an arbitrary functional and the dots are other arguments. The weight $\rho(\psi)$ in (2.228) can always be written as

$$\rho(\psi) = \alpha_\psi c_\psi \exp S'(\psi), \qquad S'(\psi) \equiv -\tfrac{1}{2}\psi \Delta_\psi^{-1} \psi + V'(\psi). \tag{2.229}$$

The first term in the exponent is the part of $\ln \rho(\psi)$ quadratic in ψ, the contribution $V'(\psi)$ is everything else (it is absent for Gaussian weight), the factor c_ψ is defined from the quadratic part by (2.16), and α_ψ is an additional factor ensuring that the integral of $\rho(\psi)$ is normalized to unity [c_ψ^{-1} is the free and $(\alpha_\psi c_\psi)^{-1}$ is the exact partition function for the "action" $S'(\psi)$ in (2.229)]. From (2.227)–(2.229) and (2.18) we have

$$W(A) = \alpha_\psi P_\psi [W(A, \psi) \exp V'(\psi)]|_{\psi=0}. \tag{2.230}$$

Since $W(A, \psi)$ is the usual functional from (2.30) with parametric dependence on ψ, it has the standard graph representation of Sec. 2.3. We shall represent the free propagator of the field φ in the graphs by a solid line and that of ψ by a dashed line, associating the latter with Δ_ψ from (2.229). The quantity $W(A, \psi)$ is the sum of connected graphs of solid lines containing external tails corresponding to the admixure field ψ. When operated on with P_ψ in (2.230), these external tails are pairwise contracted in all possible ways to form dashed lines Δ_ψ, which can be connected by interaction vertices $V'(\psi)$ if the weight $\rho(\psi)$ is non-Gaussian. In the end we obtain graphs consisting of solid and dashed lines, where the subgraph of solid lines in all the graphs of the functional (2.230) is necessarily connected. The general rules of Secs. 2.2 and 2.3 completely determine the graphs of the functional (2.230) and their coefficients.

2.31   The replica method for a system with frozen admixtures

There is another method of obtaining the graph representation of the functional (2.227), called the replica method [81]. It is based on the simple formula $\ln a = \partial_n a^n|_{n=0}$, allowing (2.227) to be written as

$$W(A) = \partial_n \overline{G^n(A, \psi)}|_{n=0}. \tag{2.231}$$

According to the definition (2.30), $G(A, \psi) = c \int D\varphi \exp[S(\varphi, \psi) + A\varphi]$ and G^n is a power of this "single" integral, which can be written as an "n-fold" integral with factors of the same type. For this we introduce the n-component field $\overline{\varphi} \equiv \{\overline{\varphi}_a\}$ and the source $\overline{A} \equiv \{\overline{A}_a\}$ with $a = 1, ..., n$, and from the given $S(\varphi, \psi)$ we define the functional $\overline{S}(\overline{\varphi}, \psi) = \sum_{a=1}^{n} S(\overline{\varphi}_a, \psi)$. Clearly,

$$G^n(A, \psi) = \overline{c} \int D\overline{\varphi} \exp[\overline{S}(\overline{\varphi}, \psi) + \overline{A}\overline{\varphi}], \tag{2.232}$$

where $\overline{c} = c^n$, $D\overline{\varphi} = \Pi_a D\overline{\varphi}_a$, and $\overline{A}\overline{\varphi} = A \sum_a \overline{\varphi}_a$ is a linear form with source of the particular form $\overline{A}_a = A$ for all $a = 1, ..., n$. From (2.228), (2.229), and (2.232) we find

$$\overline{G^n(A, \psi)} = \alpha_\psi c_\psi \overline{c} \int D\phi \exp[S(\phi) + \overline{A}\overline{\varphi}] \equiv \alpha_\psi \widetilde{G}, \tag{2.233}$$

where $\phi \equiv \overline{\varphi}, \psi$; $D\phi \equiv D\overline{\varphi} D\psi$; and

$$S(\phi) = S'(\psi) + \sum_{a=1}^{n} S(\overline{\varphi}_a, \psi). \tag{2.234}$$

The functional $\widetilde{G}$ defined in (2.233) and its logarithm $\widetilde{W}$ are the usual objects of the type (2.30) in the theory of the field ϕ with action (2.234) and therefore have the standard graph representations of Sec. 2.3. Substituting (2.233) into (2.231), we obtain

$$W(A) = \alpha_\psi(\partial_n \widetilde{W}) \exp \widetilde{W}|_{n=0}, \quad \widetilde{W} \equiv \ln \widetilde{G}. \tag{2.235}$$

Let us analyze the n dependence of the graphs of $\widetilde{W}$, representing, as before, propagators of the field φ by solid lines and of ψ by dashed lines. The field $\overline{\varphi}$ carries the index a, and the field ψ has no index. It follows from (2.234) that the line $\overline{\varphi}$ corresponds to a factor δ_{ab}, and at all vertices involving the fields $\overline{\varphi}$ there is a Kronecker δ making the indices of all the fields $\overline{\varphi}$ coincide. Owing to these δ, each connected subgraph of solid lines actually corresponds to a single index a, and summation over a gives a factor n, because the one-leg vertex factors $\overline{A}_a = A$ (see above) are independent of a. Therefore, summation over indices gives a factor n from each connected subgraph of solid lines in the full graph. The factor α_ψ in (2.235) ensures that the weight (2.229) is normalized to unity, i.e., it exactly cancels all the vacuum loops consisting of only dashed lines (there are none at all for Gaussian weight). This means that there are no such graphs in the product $\alpha_\psi \widetilde{G}$. All the other graphs contain at least one subgraph of solid lines and, therefore, a factor n, and so they vanish for $n = 0$. Equation (2.235) then simplifies to

$$W(A) = \partial_n \widetilde{W}|_{n=0}. \tag{2.236}$$

This is the final result. We recall that $\widetilde{G}$ and $\widetilde{W}$ are the usual objects (2.30) for the action (2.234) with linear form of the special type $A \sum \overline{\varphi}_a$ (i.e., without a source of the field ψ and with identical $\overline{A}_a = A$ for all components $\overline{\varphi}_a$). They have the standard graph representations, and $\widetilde{W}$ is the sum of all connected graphs with the standard coefficients. It then follows that only graphs of $\widetilde{W}$ with a single connected subgraph of solid lines contribute to the final result (2.236).

Equation (2.236) is also useful in that it allows the various general relations obtained for functionals like (2.30), for example, the renormalization-group equations in the case of multiplicative renormalizability of the model (2.234), to be carried over to the problem of frozen admixtures without additional effort.

Chapter 3

Ultraviolet Renormalization

In this chapter we shall describe the technique of ultraviolet (UV) renormalization used in field theory to isolate and then eliminate UV divergences of Green functions. This technique was developed in the late 1940s and early 1950s by Feynman, Dyson, Schwinger, and others. The complete solution of the problem is given by the Bogolyubov–Parasyuk R operation (1958), and important improvements were later made by Zav'yalov et al. (in and after 1965), Hepp (in and after 1966), Zimmermann (in and after 1970), and Epstein and Glaser (1973). In 1968 Speer introduced an analytic regularization useful for calculations [82]. The even more valuable dimensional regularization was developed by 't Hooft and Veltman [83] in 1972, and the minimal subtraction scheme by 't Hooft [61] in 1973. A detailed bibiography can be found in [35], [60], [84], [85].

We have already discussed the ideas behind UV renormalization and its relation to the theory of critical phenomena in Ch. 1. In the present chapter we shall concentrate on the actual technique, trying to explain and illustrate it as clearly as possible by examples of actual calculations.

The fundamental statements of renormalization theory can be divided into two groups: analytical and combinatorial. We shall not present the complicated proofs of the statements of the first group, but just present the formulation and explanations, as this subject is dealt with fairly completely in the books referred to above, and knowledge of the proofs is not required for practical applications. The statements of the second group, the combinatorial ones, will be fully proved with a more than usual extensive use of the functional technique of Ch. 2, which simplifies the formulation and the proofs. Our discussion will also be original in that we shall include vacuum loops on an equal footing within the general scheme of renormalization theory.

3.1 Preliminary remarks

Let us begin by defining the important concept of a *subgraph* of a given graph. When discussing graphs, we shall distinguish *free* (unlabeled), *labeled* (with labeled vertices), and *completely labeled* (with labeled vertices and lines) graphs. We shall define a subgraph as a part of a completely labeled graph, i.e., a set of vertices and the lines connecting them, where set means a set of numbers. Two subgraphs are equal if their sets of vertices and lines are equal (i.e., completely coincide), and different if they differ (in any way). They do not intersect if the sets do not intersect, i.e., if there are no common vertices or lines. *By convention, the empty set and the graph as a whole are not considered as subgraphs.*

Together with a line, a subgraph necessarily contains the vertices to which the line is connected, but, in general, it does not have to contain along with a pair of vertices all the lines of the graph connecting these vertices. The latter property is possessed by subgraphs of a special type, which we shall term *complete*. A complete subgraph is uniquely specified by the set of its vertices, since all the lines of the graph connecting these vertices are included by definition.

Our terminology coincides with that of [60], but differs from that of [35], [84], [85], where a different axiomatics of the R operation is used that involves only complete subgraphs, referred to in [35] simply as subgraphs (incomplete subgraphs are not considered at all there). Let us explain the difference for the example of the graph in (2.59). In our language, its subgraphs are the two simple vertices (points), and also the three different, nontrivial, vertex subgraphs (two vertices plus any two of the three internal lines), whereas in the language of [35] this graph has no nontrivial subgraphs differing from simple vertices.

Let us discuss the terminology a bit more. In this chapter, the dimension of a quantity F will always be understood as its canonical (Sec. 1.15) dimension d_F, denoted by $d[F]$ for complicated F. By convergence (divergence) of a graph we will mean the convergence (divergence) of the integral corresponding to the graph over the momenta of the internal lines. The term "divergent" will, unless stated otherwise, always mean UV-divergent, i.e., divergence of the integral at large momenta.

The presence of divergences requires some regularization to make the graphs meaningful. From the technical point of view it is very important that the regularization be unique for all the graphs, i.e., that the theory as a whole be regularized. This is achieved by the simple cutoff Λ in (1.64) and, of course, by other methods, for example, a suitable change of a line in the graphs (i.e., the free action), ensuring more rapid falloff at large momenta. Dimensional regularization will play an especially important role in what follows.

We shall illustrate the general theory of renormalization by the example of the φ^4 model (1.63) with zero external field and $T \geq T_c$. Its action functional will be written compactly as

$$S(\varphi) = -\tfrac{1}{2}(\partial\varphi)^2 - \tfrac{1}{2}\tau\varphi^2 - \tfrac{1}{24}g\varphi^4, \qquad (3.1)$$

omitting the (implicit) integration over d-dimensional x. Using quantum field terminology, parameters like τ will be called mass parameters, a model with $\tau > 0$ a massive model, and a model with $\tau = 0$ a massless model (or critical model in statistical physics). First, we shall study the renormalization of the model (3.1) in its logarithmic dimension $d = 4$ with cutoff Λ (the φ_4^4 model), then we shall consider dimensional regularization with $d = 4 - 2\varepsilon$, which allows the use of the results of renormalization theory for real (in our problems) dimension $d < 4$. The renormalization for the case $\tau < 0$, i.e., for a system with spontaneous symmetry breakdown, is discussed in Sec. 3.36.

3.2 Superficially divergent graphs. Classification of theories according to their renormalizability

Let $\Gamma_n(x_1...x_n)$ be the 1-irreducible Green functions in zero field of a model [any model, not necessarily (3.1)] studied in dimension d, and

$$\widetilde{\Gamma}_n(p_1...p_n) = (2\pi)^d \delta(\textstyle\sum p_i)\Gamma_{np}(p_1...p_n) \qquad (3.2)$$

be their Fourier transforms of the type (2.55). Unless stated otherwise, in the present chapter Γ_n will always be the function $\Gamma_{np}(p_1...p_n)$ from (3.2) for $n \geq 1$, and the quantity Γ_{0p} defined in Sec. 2.7 for $n = 0$: the coefficient of the volume $V = \int dx$ in vacuum loops. The dimensions of the functions Γ_n are easily found from (1.70) and (2.57):

$$d[\Gamma_n] \equiv d[\Gamma_{np}(p_1...p_n)] = d - nd_\varphi \qquad \text{for all } n \geq 0. \qquad (3.3)$$

For simplicity, we shall take a single field φ and a single charge g. Generalizations are trivial, and even the outward form of the expressions remains unchanged if the multiple-index notation is used.

Let $\gamma_{nv} = (-g)^v I_{nv}$ be the contribution to Γ_n of an individual graph with v vertices and I_{nv} be its "momentum integral," i.e., the value of the graph without explicitly isolated vertex factors $-g$. Its canonical dimension

$$\omega \equiv d[I_{nv}] = d[\Gamma_n] - vd_g = d - nd_\varphi - vd_g \qquad (3.4)$$

is called the exponent or, more precisely, the *formal UV exponent* (superficial degree of divergence) of the graph. For $\omega \geq 0$, the integrand of I_{nv} has as many momenta in the numerator, including the differentials dk, as in the denominator, and if they are all integration momenta k, the integral certainly diverges at large k (however, for $\omega < 0$ convergence is not guaranteed; see below). The adjective "formal" is important for models with derivatives at vertices: sometimes (we shall give examples below) some of these derivatives must be taken out of the momentum integral as a factor proportional to the external momenta. Then the convergence is decided by the dimension of the remaining integral, which will be the *real UV exponent of divergence*. In models of the type (3.1) without derivatives at vertices, the concepts of the real and formal UV exponents coincide.

Definitions. A graph is termed *superficially divergent* (SD) when, first of all, it has loops, second, it is 1-irreducible, and, third, it has nonnegative real UV exponent. A model is *superrenormalizable* if it has only a finite number of such graphs, *renormalizable* if it has an infinite number of them, but they are present only in a finite set of functions Γ_n, i.e., the number of types of SD graphs is finite, and *nonrenormalizable* if the number of types of SD graphs is infinite.

We recall that 1-irreducible graphs are necessarily connected (Sec. 2.6).

If the real exponent coincides with the formal one and $d_\varphi > 0$, the class of the theory is determined simply by the sign of d_g: since the number of vertices v in (3.4) for a given n can grow without bound, the theory is nonrenormalizable for $d_g < 0$, superrenormalizable for $d_g > 0$, and renormalizable for $d_g = 0$, i.e., in this case the concept of renormalizability actually coincides with the concept of logarithmicity introduced in Sec. 1.16. For a specific model, the dimensions $d_{\varphi,g}$ are determined from the form of the action functional (Sec. 1.15) and are expressed in terms of d, for example, $d_\varphi = d/2 - 1$, $d_g = 4 - d$ for the model (3.1). We see from these dimensions that the φ^4 model is nonrenormalizable for $d > 4$, renormalizable in its logarithmic dimension $d^* = 4$, and superrenormalizable for $2 \leq d < 4$, while for $d < 2$ it has no UV divergences at all. We note that the UV convergence improves as d is decreased, and so a model that is superrenormalizable at some $d = d_0$ remains such for any $d < d_0$ independently of the sign of d_φ.

3.3   Primitive and superficial divergences

The number of integrals over internal (d-dimensional) momenta k is equal to the number of loops in the graph. In any model the convergence of the one-loop integral I is completely determined by its dimension $d[I] \equiv \omega$: obviously, for $\omega < 0$ the integral converges, while for $\omega \geq 0$ it diverges. Differentiation of the integral with respect to the external momenta p and mass parameters τ lowers the dimension, and so derivatives of sufficiently high order will be convergent. It then follows that the divergent part of such an integral is a polynomial (since it vanishes upon differentiation) in p and τ with divergent coefficients. Divergences defined in this manner in terms of the dimensions will be termed *primitive*. Their structure is found from the rule

$$\text{Prim. div. } I = \text{polynomial } P(p, \tau) \text{ with } d[\text{monomials}] \leq d[I] \qquad (3.5)$$

(for $d[I] < 0$ there are no divergences). This is a special case of a broader class of *superficial divergences* whose structure is given by the rule

$$\text{Superf. div. } I = \text{polynomial } P(p) \text{ with } d[\text{monomial}] \leq d[I]. \qquad (3.6)$$

The τ dependence of the coefficients of the polynomial in (3.6) is not specified, whereas in (3.5) it is necessarily a polynomial dependence. Divergences of the type (3.6) are wholly contained in the initial segment of length $d[I]$ of the Taylor expansion of I with respect to p, while divergences of the type (3.5) are contained in that with respect to p and τ, because the polynomial does not contribute to the higher coefficients of the series. This is true for any choice of expansion point with the condition that its coefficients are meaningful. There is an important caveat: it is usually possible to make the expansion in p about the origin for $\tau > 0$, but the coefficients of the double expansion in p and τ at the origin do not exist owing to IR divergences, and then the point about which the expansion in p or τ is done must be shifted. Therefore, the polynomial in (3.5) cannot be interpreted as the initial segment of the expansion in p, and τ about the origin.

It was shown above that the divergences of one-loop graphs are primitive, i.e., they are determined by the rule (3.5). The more general *Statement 1* is valid: any graph without SD subgraphs has only primitive divergences (3.5).

Looking ahead, we note that for graphs γ with SD subgraphs, the structure of the divergences becomes simple only after subtraction of the "subdivergences," i.e., the divergences of subgraphs, by means of a certain "R' operation." Here $\gamma \to R'\gamma$ and only superficial divergences (3.6) remain in $I = R'\gamma$. There exist different subtraction schemes, and the rule (3.6) for the divergences of $R'\gamma$ is valid for any of them. The only difference is in the τ structure: some schemes preserve the primitiveness of the divergences, and then for any graph γ the divergences of $R'\gamma$ have the structure (3.5), while in other schemes the polynomial behavior in τ is lost.

In order to explain the problems that arise, let us focus for a while on the φ_4^4 model (3.1) with cutoff Λ. For $\tau \geq 0$ only even functions Γ_{2n} are nonzero, and we shall refer to functions with $n \geq 3$ as "higher-order." All the dimensions are known:

$$d = 4, \quad d_\varphi = 1, \quad d_\tau = 2, \quad d_g = 0, \quad d[\Gamma_{2n}] = 4 - 2n.$$

From these and (3.5) we find the structure of the primitive divergences:

$$\begin{aligned}
&\text{Prim. div. } \Gamma_0 = c\Lambda^4 + c\Lambda^2\tau + c\tau^2, \\
&\text{Prim. div. } \Gamma_2 = c\Lambda^2 + c\tau + cp^2, \\
&\text{Prim. div. } \Gamma_4 = c, \\
&\text{Prim. div. } \Gamma_{2n} = 0 \text{ for all } n \geq 3,
\end{aligned} \tag{3.7}$$

where all the c are dimensionless coefficients independent of p and τ—functions of $\ln\Lambda$ somehow made dimensionless. As an example, let us calculate explicitly for $d = 4$ the momentum integral I of the graph in (2.58), i.e., the integral in (2.58) without the vertex factor $-g$ (the angular integration for $d = 4$ gives $2\pi^2$):

$$I\big[\,\bigcirc\,\big] = [\Lambda^2 - \tau\ln(1 + \Lambda^2\tau^{-1})]/16\pi^2. \tag{3.8}$$

Operating with $-\partial_\tau$ in (3.8) gives the analogous integral I for the vertex graph (2.61) at zero external momenta:

$$I\big[\,\bigtimes\,\big]_{p=0} = [\ln(1 + \Lambda^2\tau^{-1}) - \Lambda^2/(\Lambda^2 + \tau)]/16\pi^2, \tag{3.9}$$

where $p = 0$ for the graphs of Γ_4 will always mean $p_1 = p_2 = p_3 = p_4 = 0$.

We see that the results (3.8) and (3.9) agree with the rule (3.5), but are not expanded in τ. Therefore, to isolate the divergent part as a polynomial in τ we need an additional dimensional parameter like the renormalization mass μ (Sec. 1.20) that allows $\ln(\Lambda^2/\tau)$ to be written as $\ln(\Lambda^2/\mu^2)+\ln(\mu^2/\tau)$.

The integral I for the graph (2.60) is the product of (3.8) and (3.9). Clearly, its divergent part will no longer be a polynomial in τ, and so Statement 1 above for graphs with SD subgraphs is clearly incorrect.

3.4 Renormalization of the parameters τ and g in the one-loop approximation

The model (3.1) belongs to the class of multiplicatively renormalizable models, for which all the divergences of the Green functions Γ_n with $n \neq 0$ can be eliminated by renormalization of the fields and parameters. Divergences of the vacuum loops Γ_0 cannot be eliminated in this manner, and we shall delay their study until Sec. 3.8. The general idea of multiplicative renormalization has already been described in Ch. 1, and so we shall only recall it briefly: the original action of the model is assumed to be unrenormalized, and its parameters are bare and will be denoted as $\{\tau_0, g_0\} \equiv e_0$. Instead of (3.1) we shall write

$$S(\varphi, e_0) = -\tfrac{1}{2}(\partial\varphi)^2 - \tfrac{1}{2}\tau_0\varphi^2 - \tfrac{1}{24}g_0\varphi^4. \qquad (3.10)$$

The renormalized action is defined from the unrenormalized one (3.10) by the general expression (1.84), and its (i.e., renormalized) Green functions are expressed in terms of the new renormalized parameters $e \equiv \{\tau, g\}$. For the renormalized theory the parameters e are assumed to be independent variables, and the bare parameters e_0 and renormalization constant Z_φ are certain functions of e and the cutoff, subject to definition from the requirement that divergences be eliminated in the renormalized Green functions. All objects are constructed as series in the renormalized charge g, in particular,

$$\tau_0 = \tau + \sum_{n=1}^{\infty} g^n \alpha_n, \quad g_0 = g[1 + \sum_{n=1}^{\infty} g^n \beta_n], \quad Z_\varphi = 1 + \sum_{n=1}^{\infty} g^n \lambda_n. \quad (3.11)$$

It is stated that by suitable choice of the coefficients of these series, it is possible to eliminate in each order in g all the divergences from the renormalized functions $\Gamma_{nR} = Z_\varphi^n \Gamma_n$ with $n \neq 0$ defined in (1.81) (we recall that we write Z_φ instead of the usual $Z_\varphi^{1/2}$).

Let us check this statement for the model (3.10) in lowest orders in g. From (1.81) and the graph representations of Sec. 2.6 we have

$$\Gamma_{2\mathrm{R}} = Z_\varphi^2 \left\{ -p^2 - \tau_0 + \tfrac{1}{2} \; \bigcirc \!\!\!\!\! \smallsmile \; + \tfrac{1}{4} \; \substack{\bigcirc \\ \bigcirc} \; + \tfrac{1}{6} \; \ominus \!\!\!- \; + \; \ldots \right\},$$

$$(3.12)$$

$$\Gamma_{4\mathrm{R}} = Z_\varphi^4 \left\{ -g_0 + \tfrac{3}{2} \; \times\!\!\bigcirc\!\!\times \; + \tfrac{3}{4} \times\!\!\bigcirc\!\!\bigcirc\!\!\times \; + 3 \times\!\!\bigcirc\!\!\!\!\square \; + \tfrac{3}{2} \times\!\!\bigotimes\!\!\times \; + \; \ldots \right\}.$$

Here the graphs represent the unrenormalized functions $\Gamma_{2,4}$, and all the parameters in them are bare, i.e., a vertex corresponds to a factor $-g_0$, and a line to the bare propagator $\Delta_0 = (k^2 + \tau_0)^{-1}$. In (3.12) we give all the 1- and 2-loop graphs with their symmetry coefficients; the external momenta p and the symmetrization over them for Γ_4 are understood.

All quantities are constructed as series in g, with the orders conveniently classified according to the number of loops, shown by a superscript on Γ. In the loopless approximation we have $e_0 = e$, $Z_\varphi = 1$, $\Gamma_{2\mathrm{R}}^{(0)} = -p^2 - \tau$, and $\Gamma_{4\mathrm{R}}^{(0)} = -g$, and all the higher-order functions are zero, since their expansions begin with one-loop graphs.

In next order there are contributions from one-loop graphs with $e = e_0$ [including the corrections (3.11) in them exceeds the accuracy] and the first-order corrections (3.11) in the loopless contributions (3.12):

$$\Gamma_{2\mathrm{R}}^{(1)} = g\left\{ -2\lambda_1(p^2 + \tau) - \alpha_1 - \tfrac{1}{2}I[\,\bigcirc\,]\right\},$$

$$(3.13)$$

$$\Gamma_{4\mathrm{R}}^{(1)} = g^2\left\{ -4\lambda_1 - \beta_1 + \tfrac{3}{2}I[\,\times\!\!\bigcirc\!\!\times\,]\right\}.$$

Here and below, $I[\gamma]$ will denote the momentum integral of the graph γ, the value of the graph with unit factors at the vertices and renormalized τ in the lines.

The one-loop graphs $\Gamma_{2,4}$ in (3.13) have only primitive divergences (3.7), which can obviously be cancelled in the combinations (3.13) by suitable choice of the coefficients α_1, β_1, and λ_1. The special feature of the one-loop approximation in this model is the absence of a primitive divergence $\sim p^2$ in Γ_2 (the one-loop graph of Γ_2 is completely independent of p). We can therefore take $\lambda_1 = 0$, i.e., in this approximation renormalization of the field is not required.

3.5   Various subtraction schemes. The physical meaning of the parameter τ

The requirement that the combinations (3.13) be UV-finite does not determine the first coefficients of the series (3.11) uniquely, but only up to arbitrary finite additions. To determine them uniquely, additional conditions are required; this is referred to as the choice of subtraction scheme (for subtracting the divergent parts). There are many such schemes, and all are related by UV-finite renormalization and are equivalent from the viewpoint of the RG (Ch. 1). Therefore, the choice of scheme is a matter of convenience. In a massive theory ($\tau > 0$), the scheme of *subtractions at zero-momenta* is often used. It is specified by the normalization conditions:

$$\Gamma_{2R}\big|_{p=0} = -\tau, \quad \partial\Gamma_{2R}/\partial p^2\big|_{p=0} = -1, \quad \Gamma_{4R}\big|_{p=0} = -g , \quad (3.14)$$

i.e., in Γ_{2R} the first two terms of the expansion in p^2 at zero are fixed, and in Γ_{4R} the value at zero (of all momenta) is fixed. This gives three conditions for determining the coefficients of the three series in (3.11), in particular, from (3.13) and (3.14) we find that

$$\alpha_1 = -\frac{1}{2}I\big[\,\bigcirc\,\big], \quad \beta_1 = \frac{3}{2}I\big[\,\bowtie\,\big]_{p=0} , \quad \lambda_1 = 0. \quad (3.15)$$

The inclusion of these quantities in (3.13) is equivalent to subtracting the terms of the Taylor expansions corresponding to the conditions (3.14) from the graphs. We note that this is a general rule for graphs with loops, since the loopless approximation for Γ_{nR} already satisfies the conditions (3.14).

The theory with $\tau = 0$ (and not with $\tau_0 = 0$) is massless. In it $\Gamma_{2R}(p = 0) = 0$, but the other two conditions (3.14) are not generalized directly, since the quantities involved them do not exist owing to the IR singularities. When the point around which the momentum expansion is done is shifted away from zero, it is necessary to introduce an additional dimensional parameter: the renormalization mass μ (also called the normalization point). At $\tau = 0$ instead of (3.14) it is then usual to require that

$$\Gamma_{2R}\big|_{p=0} = 0, \quad \partial\Gamma_{2R}/\partial p^2\big|_{p\sim\mu} = -1, \quad \Gamma_{4R}\big|_{p\sim\mu} = -g , \quad (3.16)$$

where $p \sim \mu$ for Γ_2 denotes $p^2 = \mu^2$, while for Γ_4 it denotes the symmetric point in the space of four external momenta whose sum is zero: $(p_i p_k) = \frac{1}{3}\mu^2(4\delta_{ik} - 1)$.

Returning to the massive theory, we shall term coefficients of series like (3.11) *primitive* if they are polynomials in τ of dimension consistent with the rule (3.5), i.e., in our case

$$\alpha = c\Lambda^2 + c\tau, \quad \beta = c, \quad \lambda = c, \quad (3.17)$$

where all the c are different dimensionless coefficients independent of τ. Since the divergences of the one-loop graphs (3.12) are primitive, they can always be canceled by the primitive coefficients α_1, β_1, and λ_1. This then ensures that the two-loop superficial divergences are primitive (see Sec. 3.6 for more detail), and thus that they can be cancelled by the primitive second coefficients of the series (3.11), and so on, i.e., the primitiveness of the divergences and coefficients of the series (3.11) will be guaranteed in all orders. The scheme (3.14) does not preserve the primitiveness, because even the first coefficients (3.15) are complicated functions of τ, known explicitly from (3.8) and (3.9). To ensure the structure (3.17) in the normalization conditions it is necessary to fix the coefficients of the expansion not in p for arbitrary τ, as in (3.14), but of the expansion in both p and τ. Expansion in p and τ at zero is impossible owing to the IR singularities, but it can be done, for example, at the point $\{p = 0, \tau = \mu^2\} \equiv *$, fixing the conditions

$$\Gamma_{2R}\big|_* = -\mu^2, \quad \partial\Gamma_{2R}/\partial p^2\big|_* = -1,$$

$$\partial\Gamma_{2R}/\partial\tau\big|_* = -1, \quad \Gamma_{4R}\big|_* = -g. \tag{3.18}$$

There are as many of them as there are independent terms in the primitive divergences (3.7) of the functions $\Gamma_{2,4}$. From (3.13) and (3.18) we find $\lambda_1 = 0$, and α_1 and β_1 will now be the needed segments of the Taylor expansions in τ at the point μ^2 of the coefficients (3.15).

Up to now we have been discussing the φ_4^4 model with cutoff Λ. In practical calculations for the φ_d^4 model with $d = 4 - 2\varepsilon$ it is usual to use the MS scheme (Sec. 1.22). It satisfies conditions like (3.17), i.e., it ensures that the divergences are primitive in all orders.

Let us now discuss the physical meaning of the parameter τ in various schemes. It is related to a finite, experimentally measurable quantity, the temperature T, by the first equation (3.11), where T enters through $\tau_0 = a(T - T_{c0})$, where a is a coefficient independent of T [in (1.63) it was included in T], and T_{c0} is the critical temperature in the Landau approximation. We recall that the unrenormalized action (3.10) has the meaning of the Landau functional, and the form of the function $\tau_0(T)$ given above is a consequence of the general principles by which it was constructed (Sec. 1.12).

Denoting the entire right-hand side of the first equation in (3.11) by $\alpha(\tau)$ (the other arguments are understood), we rewrite it as

$$\tau_0(T) = a(T - T_{c0}) = \alpha(\tau). \tag{3.19}$$

For a given subtraction scheme, $\alpha(\tau)$ is a known (as a series in g) function, and (3.19) implicitly determines $\tau = \tau(T)$.

The parameter τ is always chosen such that $\tau = 0$ corresponds to the massless theory, i.e., $\Gamma_{2R}(p = 0, \tau = 0) = 0$; on the other hand, the exact (within some model) value of T_c is determined from the unrenormalized function as $\Gamma_2(p = 0, T = T_c) = 0$. Since Γ_2 and $\Gamma_{2R} = Z_\varphi^2\Gamma_2$ are proportional to each other, they vanish simultaneously, i.e., $\tau(T_c) = 0$. Taking this into account, from (3.19) we obtain

$$a(T_c - T_{c0}) = \alpha(0), \quad a(T - T_c) = \alpha(\tau) - \alpha(0). \tag{3.20}$$

The first equation determines the shift of T_c in going beyond the Landau theory (i.e., the loopless approximation), and the second determines the τ

dependence of $T - T_c$. It is simple for schemes like (3.18) or the MS scheme, which preserve the primitiveness of the divergences in higher orders. Then $\alpha(\tau)$ is a linear function owing to (3.17), and so from (3.20) it follows that $\tau \sim T - T_c$. Making the appropriate choice of the still arbitrary coefficient a in (3.19), it can be assumed that in such schemes $\tau = T - T_c$.

In schemes that do not preserve the primitiveness of the divergences, the relation between τ and $T - T_c$ is more complicated. For example, for the scheme (3.14) in the one-loop approximation, from (3.8), (3.15), and (3.20) we obtain $a(T - T_c) = \tau[1 + cg \ln(1 + \Lambda^2/\tau)] \cong \tau(\Lambda^2/\tau)^{cg}$, where $c = 1/32\pi^2$. Calculations in higher orders and the RG analysis show that in this scheme the leading term of the asymptote $\tau \to 0$ in $T - T_c$ is actually some fractional power τ^λ with exponent $\lambda = 1+$series in g. We found the first term of this series above. In this case the parameter τ is simply related not to the temperature, but to the susceptibility χ, according to (1.36).

3.6 The two-loop approximation

Let us return to the analysis of Sec. 3.4 and study the next highest two-loop approximation. Then in (3.12) we must include the next highest terms of the same order in g: (1) the two-loop graphs with $e_0 \to e$, (2) the first corrections (3.11) in the bare parameters e_0 of the one-loop graphs, which we already know, and (3) the still undetermined second-order correction terms of the series (3.11) in the loopless terms (3.12). The highest-order Green functions contain only contributions (1) and (2). The statement is that in the highest-order functions the divergences cancel automatically, while in (3.12) they can be made to cancel by a suitable choice of the still undetermined second coefficients of the series (3.11). This follows from the basic theorem about the R-operation, which will be discussed in detail later on. For now we only specify the structure of the individual contributions in order to explain the meaning of the statement.

Contribution (1) comes from two-loop graphs with renormalized parmeters. Contribution (2) is generated by the inclusion (for now, no more than single) of the first-order corrections $\sim \alpha_1$, β_1 (we assume that $\lambda_1 = 0$) in the vertex factors $-g_0 = -g(1 + g\beta_1 + ...)$ and the propagators $\Delta_0(k) = (k^2 + \tau_0)^{-1} = (k^2 + \tau + g\alpha_1 + ...)^{-1}$ of unrenormalized one-loop graphs. In first order we have $\Delta_0(k) = \Delta(k) - g\alpha_1\Delta^2(k) + ...$, i.e., each line of the unrenormalized graphs is replaced by

$$\text{———} \; - \; g\alpha_1 \; \text{—•—} \; + \; ... \; ,$$

where a new line corresponds to $\Delta(k) = (k^2 + \tau)^{-1}$, and the point on it to the insertion of unity. These arguments determine contribution (2) in (3.12). Contribution (3) is determined by the terms $-Z_\varphi^2(p^2 + \tau_0)$ in Γ_{2R} and $-Z_\varphi^4 g_0$

in Γ_{4R}. The series (3.11) should be substituted into them and the terms of the needed order in g selected, which for $\lambda_1 = 0$ gives $-g^2[\alpha_2 + 2\lambda_2(p^2 + \tau)]$ in Γ_{2R} and $-g^3(\beta_2 + 4\lambda_2)$ in Γ_{4R}. Finally, for the sum of contributions (1)–(3) in (3.12) in the needed order in g we obtain

$$\Gamma_{2R}^{(2)} = g^2 \left\{ -\alpha_2 - 2\lambda_2(p^2 + \tau) + \right.$$

$$\left. + I\left[\tfrac{1}{2}\alpha_1 \; \bigcirc\!\!\!\!\!\frown \; - \tfrac{1}{2}\beta_1 \; \bigcirc\!\!\!\!\!\frown \; + \tfrac{1}{4} \; \bigcirc\!\!\!\!\!\bigcirc \; + \tfrac{1}{6} \; \bigcirc\!\!\!\!\!\!\!\!\frown \; \right] \right\},$$

$$\Gamma_{4R}^{(2)} = g^3 \left\{ -\beta_2 - 4\lambda_2 + \right.$$

$$\left. + I\left[-3\alpha_1 \; \rightthreetimes\!\!\bigcirc\!\!\leftthreetimes \; + 3\beta_1 \; \rightthreetimes\!\!\bigcirc\!\!\leftthreetimes \; - \tfrac{3}{4} \; \rightthreetimes\!\!\bigcirc\!\!\bigcirc\!\!\leftthreetimes \; - 3 \; \rightthreetimes\!\!\bigcirc\!\!\!- \; - \tfrac{3}{2} \; \rightthreetimes\!\!\bigcirc\!\!\leftthreetimes \; \right] \right\},$$

(3.21)

where I denotes the "momentum integral": by convention, in the graphs inside of I, the vertices are associated with factors of unity, and the lines with the propagator $\Delta(k) = (k^2 + \tau)^{-1}$ with the renormalized parameter τ.

If all the graphs of (3.21) had only superficial divergences of the type (3.6), i.e., if their "divergent parts" were second-order polynomials in p for Γ_2 and zero-order ones for Γ_4, they would obviously be eliminated in the combinations (3.21) by suitable choice of the coefficients α_2, β_2, and λ_2. However, the two-loop graphs contain SD subgraphs, and so their divergences have a more complicated structure. The basic statement is essentially that in the sum of the contributions (1) and (2), i.e., in the sum of 2- and 1-loop graphs, the divergences with complex structure cancel out and only superficial divergences of the type (3.6) remain. This means that there will be no divergences at all in the higher-order Green functions, while the divergences eliminated by choice of α_2, β_2, and λ_2 remain in (3.12). And so on: the sum of the next-highest l-loop graphs and all the correction graphs with a smaller number of loops will contain only the superficial divergences (3.6), eliminated by appropriate choice of the following coefficients of the series (3.11).

The UV-finite arbitrariness in the latter is fixed by the choice of subtraction scheme, on which the structure of the divergences in τ depends. The following nontrivial *statement* is true: in the induction process, the primitiveness (Sec. 3.5) of all the coefficients (3.11) constructed earlier guarantees the primitiveness of the following superficial divergences and, consequently, the possibility of eliminating them by the following primitive coefficients in (3.11). The first step of the induction is ensured by the primitiveness of the one-loop divergences.

Taking this on faith, it is easy to understand why nonprimitiveness of the coefficients in (3.11) leads to violation of primitiveness of the divergences in higher orders. Let us compare (3.21) for two schemes: one with primitive coefficients α_1, β_1, and the other with new coefficients $\alpha_1' = \alpha_1 + \Delta\alpha_1$, $\beta_1' =$

$\beta_1 + \Delta\beta_1$, containing UV-finite nonprimitive additions $\Delta\alpha_1$ and $\Delta\beta_1$. Let A be the sum of all 1- and 2-loop graphs (3.21) in the first scheme, and A' be that in the second. According to the above statement, the divergences of A are primitive. We shall show that this is not true for A'. Actually, the difference $A - A'$ is a combination of one-loop graphs with coefficients that are multiples of $\Delta\alpha_1$ and $\Delta\beta_1$. The divergences of the one-loop graphs are primitive, but they are multiplied by the nonprimitive UV-finite coefficients $\Delta\alpha_1$, $\Delta\beta_1$, and so the divergences of $A' - A$, and, accordingly, of A', are certainly nonprimitive.

3.7 The basic action and counterterms

In this and the following sections we shall describe the general theory. We shall turn to the φ^4 model only as an example.

Let $S(\varphi, e_0)$ be the original unrenormalized action, e_0 be the set of its bare parameters, and e their renormalized analogs. The dependence on the cutoff Λ or the renormalization mass μ and the parameter ε from (1.86) in dimensional regularization will not be indicated explicitly, but is always understood. We shall assume that the model is multiplicatively renormalizable (Sec. 1.18), and then its renormalized action $S_R(\varphi, e)$ is associated with the unrenormalized action by (1.84): $S_R(\varphi, e) = S(Z_\varphi \varphi, e_0)$, $Z_\varphi = Z_\varphi(e)$, $e_0 = e_0(e)$ (for now we do not consider the renormalization of the vacuum loops and the corresponding vacuum counterterms; see the remarks in Sec. 1.18).

Let us introduce some new concepts. The quantities $e_0(e)$ in lowest nontrivial order in the renormalized charge [for example, zero order for τ_0 and first order for g_0 in (3.11)] will be called the *basic parameters* $e_B = e_B(e)$, the action obtained by the simple replacement $e_0 \to e_B$ from the unrenormalized action $S_B(\varphi, e) = S(\varphi, e_B(e))$ will be called the *basic action*, and the difference $S_R(\varphi, e) - S_B(\varphi, e) = \Delta S(\varphi, e)$ will be called *counterterms*.

Let us explain these general definitions for the example of the φ^4 model. For it, $e_0 = \{\tau_0, g_0\}$, $e = \{\tau, g\}$, and from (3.11) and the definition of e_B it follows that for the φ_4^4 model with cutoff Λ, the basic parameters coincide with the renormalized ones. However, in dimensional regularization with $d = 4 - 2\varepsilon$, instead of (3.11) we have (Sec. 1.21) $\tau_0 = \tau + O(g)$ and $g_0 = g\mu^{2\varepsilon} + O(g^2)$, i.e., the equality $\tau_B = \tau$ is preserved, but now $g_B = g\mu^{2\varepsilon}$, and so the basic charge g_B will in general differ from the renormalized charge g. The basic action is obtained from the unrenormalized action (3.10) by the replacement $\tau_0 \to \tau_B = \tau$, $g_0 \to g_B = g\mu^{2\varepsilon}$, i.e.,

$$S_B(\varphi, e) = -\tfrac{1}{2}(\partial\varphi)^2 - \tfrac{1}{2}\tau\varphi^2 - \tfrac{1}{24}g_B\varphi^4, \tag{3.22}$$

and the renormalized action is obtained according to the rule (1.84):

$$S_R(\varphi, e) = -\tfrac{1}{2}Z_\varphi^2(\partial\varphi)^2 - \tfrac{1}{2}\tau_0 Z_\varphi^2\varphi^2 - \tfrac{1}{24}g_0 Z_\varphi^4\varphi^4. \tag{3.23}$$

The counterterms $\Delta S(\varphi, e)$ are determined by the expression

$$S_{\mathrm{R}}(\varphi, e) = S_{\mathrm{B}}(\varphi, e) + \Delta S(\varphi, e), \qquad (3.24)$$

and in this case from (3.22) and (3.23) we obtain

$$\Delta S(\varphi, e) = \tfrac{1}{2} c_1 (\partial \varphi)^2 + \tfrac{1}{2} c_2 \varphi^2 + \tfrac{1}{24} g_{\mathrm{B}} c_3 \varphi^4$$

with the coefficients $c_1 = 1 - Z_\varphi^2$, $c_2 = \tau - \tau_0 Z_\varphi^2$, and $c_3 = 1 - g_{\mathrm{B}}^{-1} g_0 Z_\varphi^4$. The bare parameters τ_0 and g_0 are everywhere assumed to be expressed in terms of τ and g [by (3.11) for $d = 4$] as series in g, the series for all the introduced c_i begin with terms of first order in g, and the contributions of order g^l in these series correspond to l-loop counterterms in ΔS.

Earlier we calculated the functions $\Gamma_{n\mathrm{R}}$ for a given model from the graphs of the unrenormalized theory (3.10) followed by substitution of the expressions (3.11). It is clear that the same functions $\Gamma_{n\mathrm{R}}$ can also be calculated directly from the graphs of the basic theory, by treating the counterterms ΔS in (3.24) as additions to the basic interaction. The results must be the same, since we are dealing with the Green functions of the same model (specified by the action S_{R}), represented in the same form as series in g. In lowest order only the loopless contributions of S_{B} are taken into account, while in the next order we include the one-loop graphs of S_{B} and the contributions of the one-loop counterterms (see the definitions above), then the two-loop graphs of S_{B} [contribution (1) in the terminology of Sec. 3.6], the corrections of the same order in g from one-loop counterterms [contribution (2)] and from two-loop counterterms [contribution (3)], and so on.

Above we defined the basic action and the counterterms assuming that the unrenormalized model is the primary object and that it is multiplicatively renormalizable. In the general theory of renormalization, first developed in relativistic quantum field theory, the statement of the problem is different. The primary object specifying the model is assumed to be the basic action, and the counterterms are constructed from it using the R operation to eliminate UV divergences (it will be discussed in the following sections). The renormalized action is defined as the sum of the basic action and the counterterms. This scheme is more general, because it can happen that the number of independent types of counterterms exceeds the number of fields and parameters in the basic action, and then they cannot be reproduced by the standard procedure of multiplicative renormaliation of the unrenormalized action, which has the same form as the basic action. This is natural in relativistic field theory, because there quantities that are directly observable are always related to the renormalized parameters and Green functions, and the unrenormalized action and bare parameters, if they can be introduced, are assumed to be secondary objects that are unobservable. The situation is different in the theory of critical phenomena. Here the unrenormalized action and its (i.e., bare) parameters are the primary objects, and all others are secondary, and so the standard RG technique described in Ch. 1 can be used only when the model is multiplicatively renormalizable. However, it is convenient to explain the technique for calculating the counterterms and renormalization constants in higher orders in g using the general theory of renormalization, in which the basic action is taken as the primary object. Let us now turn to the description of this general theory.

3.8   The operators L, R, and R'

In field theory, it is proved that the addition of counterterms to a given basic action S_B can be replaced by a certain R operation acting on the graphs of the basic theory. Here we shall introduce the definitions needed for this. The contributions of the individual 1-irreducible graphs for the given action S_B of the basic theory will be denoted as γ_n, or simply γ if no further specification is required, where γ_n simultaneously denotes both the graph and the analytic expression corresponding to it: the contribution to $\Gamma_n \equiv \Gamma_{np}(p_1...p_n)$ (Sec. 3.2) without the symmetry coefficient. The vacuum loops γ_0 are treated on the same footing as the other γ_n.

The operators L, R, and R' are defined to act on graphs (any graph, not only 1-irreducible ones) and are extended by linearity to act on sums of graphs with symmetry coefficients, i.e., on Green functions, and via them on the corresponding generating functionals. We immediately note that the property $L(a\gamma) = aL\gamma$ is guaranteed only for a simple scalar, parameter-independent factor a like a symmetry coefficient; otherwise, this relation may not be satisfied (this will be discussed further later on).

The primary operator is the *counterterm operator L*. By definition, L operating on a graph gives a counterterm of the graph, some function of the same variables and of the same dimension as the graph itself. Graphs associated with nonzero counterterms will be termed *relevant for L*, and the others will be termed *irrelevant*. Everywhere below we shall assume that

$$L[\text{1-reducible graph}] = 0, \tag{3.25}$$

i.e., the only relevant graphs can be 1-irreducible graphs γ (and not necessarily all of them). To specify L it is necessary to indicate the set of relevant graphs γ and the quantity $L\gamma$, or the rule for constructing it for each one. In what follows the R operation, constructed using L, will be used to eliminate UV divergences, which presupposes a definite (although nonunique) choice of L. However, the equations of this section are purely combinatorial and are true for any operator L satisfying the condition (3.25).

The operators R and R' on an arbitrary (not necessarily 1-irreducible) graph are determined from a given operator L by the equations [86]

$$R' = 1 - \sum_{\gamma} L_{\gamma} + \sum_{\gamma,\gamma'} L_{\gamma} L_{\gamma'} - ... \; ; \quad R = R' - L. \tag{3.26}$$

The graph is taken to be fully labeled (see Sec. 3.1), the letters γ, γ'..., denote its relevant subgraphs (the graph itself is not a subgraph!), the action of L_{γ}

on a graph reduces to the replacement of a subgraph γ by its counterterm $L\gamma$, where here γ and $L\gamma$ are understood as functions of the integration momenta in the graph that are external with respect to it. The single summation in (3.26) runs over all the different (for a completely labeled graph) relevant subgraphs, and the double sum runs over all the different pairs of nonintersecting relevant subgraphs, then over triplets of them, and so on, and for any particular graph the series terminates. For the R operation the contribution of the graph itself as a whole is added to the single sum in (3.26) if the graph is relevant; this is the only difference between R and R'. The summation in (3.26) can also run over all connected subgraphs, and the operator L itself selects the relevant ones.

Let us illustrate the rule (3.26) for the R' operation by specific examples for the φ^4 model, assuming that only the SD graphs $\gamma_{0,2,4}$ (γ_0 do not occur as subgraphs) are relevant, and denoting the action of the operator L_γ in (3.26) by a dashed line encircling the subgraph γ. One-loop graphs do not have SD-subgraphs, and so for them $R' = 1$. For the two-loop graphs in (3.12) we have

$$(3.27)$$

The second graph in (3.27) contains three different vertex subgraphs (see Sec. 3.1). Their contributions are equivalent, and so a single graph with coefficient 3 remains. Let us give some more examples; the last two are for vacuum loops:

$$R' \;\; \ldots \; = \text{itself} - \ldots - \ldots - $$

$$ - \ldots + \ldots \; ; $$

$$R' \;\; \ldots \; = \text{itself} - \ldots - $$

$$ - \ldots - \ldots \; ; $$

$$R' \;\; \ldots \; = \text{itself} - \ldots - $$

$$ - \ldots - \ldots - \ldots - $$

$$ - \ldots + \ldots + \ldots + $$

$$ + \ldots + \ldots + \ldots - $$

$$ - \ldots \; ; \tag{3.28}$$

$$R' \;\; \ldots \; = \text{itself} - 3 \, \ldots - \ldots \; ; $$

$$R' \;\; \ldots \; = \text{itself} - 2 \, \ldots \; ; $$

$$R' \;\; \ldots \; = \text{itself} - 6 \, \ldots - 4 \, \ldots \; . $$

The examples are sufficient for explaining the rules for the arrangement of the dashed-line circles in the R' operation. The R operation differs from the R' one only by an addition on the right-hand side of the graph itself with a minus sign, encircled by a dashed line if the graph itself is relevant (if it is not, then $R = R'$).

The action of the R operation on some object will be referred to as the renormalization of the object. The following important *property of renormalization factorization* is valid. Let the original graph (or sum of graphs) be represented as a product AB, with the graph itself being irrelevant, and all its relevant subgraphs being located either entirely in A or in B, and the relevant subgraphs of different factors never intersect each other. It then follows from

(3.26) that $R(AB) = RA \cdot RB$, i.e., in this case the renormalization of the product reduces to renormalization of the cofactors.

Corollary: The action of the R operator on the dressed propagator (2.52) of the basic theory is equivalent to the renormalization of the self energy Σ [symbolically, $RD(\Sigma) = D(R\Sigma)$]; the action of the R operator on connected functions is equivalent to the renormalization of all the dressed lines and vertices in (2.45), and for the full functions it is equivalent to the renormalization of all the connected cofactors.

Two statements follow from this.

Statement 1. The R operator commutes with the operators that select the connected and 1-irreducible parts of the generating functionals represented by graphs, and also with the operator that amputates external lines Δ, transforming the graphs of the full Green functions G_n into the graphs of the S-matrix functions H_n (see Sec. 2.14).

Statement 2. If the R operator eliminates the UV divergences from some particular set of Green functions (full, or connected, or S-matrix, or 1-irreducible), then this automatically ensures the UV finiteness of all the other Green functions.

3.9 The Bogolyubov–Parasyuk R operation

The books [35], [84], [85] and many other studies use the Bogolyubov–Parasyuk operator, whose form differs from that of $R(L)$. We shall denote it as $\overline{R}(L)$. Formally, $\overline{R}'(L)$ and $\overline{R}(L) = \overline{R}'(L) - L$ are defined by the same rule (3.26), if the term "subgraph" is understood in the sense of [35], i.e., as a subset of vertices. In our language this implies that the summation in (3.26) for $\overline{R}'(L)$ runs not over all subgraphs, as for $R'(L)$, but only over complete ones (see the definition in Sec. 3.1). These are different but equivalent forms of the R operation: for any counterterm operator L it is possible to construct a new operator $\overline{L}$ such that $R(L) = \overline{R}(\overline{L})$, namely, in the notation of (3.26),

$$\overline{L}\gamma = \left[L + \sum_{\gamma'} L_{\gamma'} \right] \gamma, \qquad R(L) = \overline{R}(\overline{L}), \qquad (3.29)$$

with summation over all subgraphs $\gamma' \subset \gamma$, relevant for L, containing all vertices γ. The term L is isolated in (3.29) only because in Sec. 3.1 we have stipulated that the graph itself not be viewed as a subgraph. If the graph γ has no subgraphs relevant for L (for example, any one-loop graph), then for it $L = \overline{L}$.

Let us explain (3.29) for $\overline{L}$ by means of examples, denoting the action of L by a dashed line, as in Sec. 3.8, and $\overline{L}$ by a dotted line:

$$\overline{L} \;\bigcirc \;\equiv\; \bigcirc \;=\; \bigcirc \;+\; 6 \;\bigcirc \;+\; 4 \;\bigcirc\; ;$$

$$\overline{L} \;\bigcirc\bigcirc\bigcirc \;\equiv\; \bigcirc\bigcirc\bigcirc \;=\; \bigcirc\bigcirc\bigcirc \;+$$

$$+\; 2 \;\bigcirc\bigcirc\bigcirc \;+\; \bigcirc\bigcirc\bigcirc \;. \tag{3.30}$$

Let us also explain the equation $R\gamma \equiv R(L)\gamma = \overline{R}(\overline{L})\gamma$ by examples:

$$R \;\bigcirc\; =\; \bigcirc \;-\; 3 \;\bigcirc \;-\; \bigcirc \;=$$

$$=\; \bigcirc \;-\; \bigcirc \;;$$

$$R \;\bigcirc\; =\; \bigcirc \;-\; 6 \;\bigcirc \;-\; 4 \;\bigcirc \;-\; \bigcirc \;=$$

$$=\; \bigcirc \;-\; \bigcirc \;;$$

$$R \;\bigcirc\bigcirc\bigcirc \;=\; \bigcirc\bigcirc\bigcirc \;-\; 2 \;\bigcirc\bigcirc\bigcirc \;-$$

$$-\; \bigcirc\bigcirc\bigcirc \;-\; 2 \;\bigcirc\bigcirc\bigcirc \;+\; \bigcirc\bigcirc\bigcirc \;-$$

$$-\; \bigcirc\bigcirc\bigcirc \;=\; \bigcirc\bigcirc\bigcirc \;-\; 2 \;\bigcirc\bigcirc\bigcirc \;+$$

$$+\; \bigcirc\bigcirc\bigcirc \;-\; \bigcirc\bigcirc\bigcirc \;.$$

These equations become obvious when (3.30) is taken into account. To prove the statement (3.29) in the general case, it is sufficient to note that each summation over connected, fully labeled subgraphs in (3.26) [we write out the sum over 1-irreducible subgraphs γ only because we shall immediately use the condition (3.25)] can be written as a double sum, with summation first over connected, completely labeled subgraphs with a given set of vertices, and then over all variants of choosing this set. The inner summation transforms L into $\overline{L}$, and the outer one is equivalent to summation over complete subgraphs.

We have described the Bogolyubov–Parasyuk operator $\overline{R}(L)$ only for completeness; we shall not use it. Everywhere in what follows, the R operation will be understood to be the operation $R(L)$ defined in Sec. 3.8. It is more convenient for practical calculations, in particular, because in schemes like the MS scheme the counterterms $L\gamma$ are polynomials in the mass parameters τ, while this property is lost for $\overline{L}\gamma$.

3.10 Recursive construction of L in terms of the subtraction operator K

For a given set of relevant graphs γ, the relation

$$L\gamma = \mu^\delta K \mu^{-\delta} R'\gamma, \quad \delta \equiv d[\gamma] - \text{nearest integer}, \quad (3.31)$$

where μ is the renormalization mass, expresses the counterterm operator L in terms of the simpler *subtraction operator* K. We have given the general statement, applicable also for dimensional-type regularizations. For them, in contrast to (3.7), the canonical dimensions $d[\gamma]$ of the graphs γ differ from integers by a small amount δ. The factor $\mu^{-\delta}$ compensates for this difference, so that the operator K in (3.31) always acts on a function with integer dimensions [the operators (3.26) do not change the dimensions].

After specifying an arbitrary linear (with the same stipulation as for L in Sec. 3.8) operator K on a set of functions of the needed type [we emphasize the fact that these are functions, not graphs, since the quantities $R'\gamma$ in (3.31) are not simply graphs], we define L in terms of K recursively from (3.31) and (3.26). This is done by ordinary induction in the number of loops: for loopless graphs γ, i.e., simple point vertices, $R' = 1$ owing to the absence of subgraphs, and so $L\gamma$ for them is determined in terms of K from (3.31). In one-loop graphs γ, the subgraphs are loopless, and L is already defined on them, so that $R'\gamma$ is defined from (3.26), then it is used to define $L\gamma$ from (3.31), and so on.

In what follows, as a rule, we shall assume that there are no loopless counterterms (this does not restrict the generality, since such counterterms only redefine the charges), and so $R' = 1$ for one-loop graphs.

Assuming enumeration of the relevant graphs by part of the definition of the operator K, it can be stated that it completely specifies the operators L, R, and R' by (3.31) and (3.26). In formal constructions, K is arbitrary, but if we want to eliminate divergences by the R operation, K must be "correctly" constructed. *Definition*: the operator K will be termed *correctly constructed* if for it all SD graphs are relevant, and for any one-loop[*] graph γ, the counterterm $L\gamma = \mu^\delta K \mu^{-\delta} \gamma$ is a polynomial in the external momenta containing the entire "divergent part" of γ, defined by the rule (3.6) for

---

[*]More precisely, simplest: in the φ^6 model, for example, the number of loops in SD graphs γ is always even, and two-loop graphs will be the simplest (the φ^6 model is studied in Ch. 4).

$I = \mu^{-\delta}\gamma$ (i.e., there are no divergences in $R\gamma = \gamma - L\gamma$). If here $L\gamma$ has the structure (3.5), the operator K and the corresponding operators R, L, and R' will be termed *primitive*.

The divergences of one-loop graphs are primitive (Sec. 3.3), and so they can always be eliminated by the primitive operator K. This is the most economical method, because only what needs to be subtracted is subtracted. However, it is possible to subtract more than necessary (less is impossible), and so there are operators K that are correctly constructed, but not primitive. The choice of a particular operator K amounts to the choice of subtraction scheme (Sec. 3.5). In some schemes K commutes with the factors μ^{δ}, and then they can be dropped in (3.31), while in other schemes this is not so.

It is usually only SD graphs that are considered relevant, but this is not mandatory — the class of relevant graphs can be larger. Vacuum loops γ_0 are always SD graphs, but they cannot occur as subgraphs, and so they have no effect on the renormalization of the other γ_n and can, if desired, be included in or omitted from the general scheme. In field theory they are usually simply discarded, which can be done formally by the R operation, taking $K = 1$ for the graphs γ_0 in (3.31), and then $R\gamma_0 = 0$ according to (3.31) and (3.26). However, in the theory of critical behavior vacuum loops have a physical meaning (Sec. 2.3), and so it is desirable to renormalize the γ_0 along with the other γ_n.

Let us give several examples of particular operators K. We recall that p always denotes the set of all external momenta, and τ denotes the mass parameters ($d_\tau = 2$) of the internal lines. We use $K_p^{(n)}(p_*)$ to denote the operator that selects all the first terms up to monomials of dimension n inclusive of the Taylor expansion in p at the point p_*, and $K_{p,\tau}^{(n)}(p_*, \tau_*)$ to denote the analogous operator for the expansion in p and τ. The following two operators K are expressed in terms of these; we shall give them below for the φ^4 model. It is assumed that only SD graphs are relevant for them, i.e., the graphs $\gamma_{0,2,4}$, and both of these operators K commute with the factor μ^{δ} in (3.31), and so they are specified directly on the functions $R'\gamma_n \equiv f_n$.

1. *Subtractions at zero momenta*:

$$K f_0 = K_p^{(0)} f_0 = f_0, \quad K f_2 = K_p^{(2)} f_2, \quad K f_4 = K_p^{(0)} f_4, \tag{3.32}$$

where $K_p^{(n)} \equiv K_p^{(n)}(0)$. This operator corresponds to the conditions (3.14) for $\gamma_{2,4}$ and to complete elimination of the vacuum loops γ_0.

2. *Primitive subtractions* corresponding to the conditions (3.18) at the point $p_* = 0$, $\tau_* = \mu^2$ [everywhere, $K_{p,\tau}^{(n)} \equiv K_{p,\tau}^{(n)}(0, \mu^2)$, $f_n \equiv R'\gamma_n$]:

$$K f_0 = K_{p,\tau}^{(4)} f_0, \quad K f_2 = K_{p,\tau}^{(2)} f_2, \quad K f_4 = K_{p,\tau}^{(0)} f_4. \tag{3.33}$$

3. *Minimal subtractions* (MS) in any models with dimensional-type regularizations (Sec. 1.19). This operator K is specified on arbitrary functions $f(\varepsilon, ...)$ expandable in a Laurent series in the regularizer ε (the dots denote

other arguments) as

$$K \sum_{n=-\infty}^{+\infty} \varepsilon^n f_n(...) = \sum_{n=-\infty}^{-1} \varepsilon^n f_n(...), \qquad (3.34)$$

i.e., K selects only poles in ε, which in this regularization play the role of UV divergences. The quantities $\mu^{-\delta} R' \gamma \equiv f$ in (3.31) are meromorphic functions of ε with integer dimensions. The exponent δ is usually a multiple of ε, and so the factors μ^δ do not commute with the operator K in (3.34), and in this case cannot be canceled in (3.31). The class of relevant graphs is automatically determined from (3.34) (those for which f contains poles in ε are relevant) and coincides with the class of graphs that are SD graphs at $\varepsilon = 0$.

The operator (3.32) is not primitive, while (3.33) and (3.34) are. For (3.33) this is obvious from the definition, and for (3.34) it is a nontrivial property of the MS scheme (see Sec. 3.16 for more details).

3.11   The commutativity of L, R', and R with ∂_τ-type operators

Let $\mathcal{D}$ be a linear operator on graphs γ, acting on their elements (lines and vertices), for example, ∂_τ or ∂_p, the derivatives with respect to τ or the external momenta p, or a generator of a global symmetry group, and so on. They all have the common property of being representable as $\mathcal{D} = \sum \mathcal{D}_i$, where i is an index labeling the elements (lines and vertices) of the graph, and $\mathcal{D}_i$ is the action of $\mathcal{D}$ on an individual element i. If $\mathcal{D}$ acts on certain variables z in the elements i (like τ, p, vector indices, and so on), then to formally isolate $\mathcal{D}_i$ it is possible, by analogy with (2.12), to first introduce different z_i, replacing them by the single z after acting with the operator $\mathcal{D}$ on the argument z_i. The quantity $\mathcal{D}_i \gamma$ will be represented graphically by a fully labeled graph γ with the element i marked in some manner (to indicate the action of $\mathcal{D}$ on that element). We shall define the operators R and R' on $\mathcal{D}_i \gamma$ by the usual topological rules of Sec. 3.8, without regard to a mark on an element in the arrangement of dashed-line circles, for example,

$$(3.35)$$

where we have marked a single line. The stated rule together with (3.31) completely defines all the operations of Sec. 3.8 on the graphs $\mathcal{D}_i \gamma$ via the subtraction operator K, if it is defined not only on functions $\mu^{-\delta} R' \gamma \equiv f$, but also on functions $\mathcal{D}_i f$. The definitions generalize in an obvious manner to objects of the type $\mathcal{D}_i \mathcal{D}_k ... \gamma$ with several elements marked. The following lemma will prove very useful in applications.

Lemma. If $\mathcal{D}$ commutes with the factors μ^δ and operators K in (3.31), i.e., if K is defined on functions of the form $\mathcal{D}f$ with $f \equiv \mu^{-\delta}R'\gamma$ and

$$K\mathcal{D}f = \mathcal{D}Kf, \qquad (3.36)$$

then $\mathcal{D}$ commutes with all the operators L, R', R.

As usual, this statement is proved by induction in the number of loops l. Let T be any of L, R, and R', and T_l be operators on l-loop graphs. We assume (only for simplicity) that $L_0 = 0$, i.e., there are no counterterms for simple vertices. Then $R'_1 = 1$ and from (3.31) and the conditions of the lemma we have $L_1\mathcal{D}_i\gamma = \mathcal{D}_iL_1\gamma$, i.e., all the $\mathcal{D}_i$ and their sum $\mathcal{D}$ commute with the operators L_1 and $R_1 = 1 - L_1$. Graphically, commutativity with L_1 means that in the graphs $\mathcal{D}_iR'_2\gamma$ the operators $\mathcal{D}_i$ can be inserted inside the dashed lines encircling the one-loop subgraphs (Sec. 3.8), which gives $R'_2\mathcal{D}_i\gamma$ by definition (see above). It is thereby proved that $\mathcal{D}_i$ commutes with the two-loop operator R'_2 and, therefore [from (3.31) and the conditions of the lemma] with L_2 and $R_2 = R'_2 - L_2$. It is thereby proved that $\mathcal{D}_i$ can also be inserted into the dashed lines encircling the two-loop subgraphs, which leads to the commutativity of $\mathcal{D}_i$ with R'_3 and so on. The lemma is proved. It has an obvious generalization to products of the type $\mathcal{D}^2$, $\mathcal{D}\mathcal{D}'$, and so on.

Corollary. The operator K of minimal subtractions (3.34) and the corresponding L, R, and R' commute with ∂_τ and ∂_p, with generators of global symmetries of the type O_n, and so on; in general, with any operator $\mathcal{D}$ of the above type which does not act on the variable ε in (3.34).

The operator (3.32) commutes with ∂_τ, but not with ∂_p [a counterexample is $0 = \partial_p K_p^{(2)} p^3 \neq K_p^{(2)} \partial_p p^3 - \partial_p p^3]$, and the operator (3.33) does not commute with either ∂_p or ∂_τ.

3.12 The basic statements of renormalization theory

For convenience, we shall list in one place the complete set of basic statements of renormalization theory, including those formulated above. As applied to regularizations of the type $d = 4-2\varepsilon$, the class of superficially divergent graphs is defined by their dimensions at $\varepsilon = 0$, and when referring to the rules (3.5) and (3.6) it is kept in mind that the role of I in these is played by the object studied with an additional factor of $\mu^{-\delta}$, $\delta \sim \varepsilon$, which restores the integer dimension (see Sec. 3.16 for more details on dimensional regularization). The statements given below are valid for any field model with a local interaction, and nonlocality in the free action of the type of a dipole force contribution (Sec. 1.14) is allowed. Therefore:

(1) The graphs γ without superficially divergent subgraphs have only primitive divergences with the structure (3.5).

(2) Let K be any correctly (Sec. 3.10) constructed subtraction operator, and L, R, and R' be constructed in terms of K. Then for any graph γ,

including those containing superficially divergent subgraphs, the quantities $R'\gamma$ have only superficial divergences (3.6), and $R\gamma = R'\gamma - L\gamma$ has no divergences. If here the operator K is primitive (Sec. 3.10), then the divergences in $R'\gamma$ for all the graphs, and also their counterterms $L\gamma$, are also primitive.

(3) The application of the R operation to the graphs of the basic theory is completely equivalent to adding to its action $S_B(\varphi) = S_0(\varphi) + V(\varphi)$ the counterterms

$$\Delta S(\varphi) = -LP \exp V(\varphi) = -L\Gamma_B(\varphi), \tag{3.37}$$

where $P \exp V(\varphi)$ is the S-matrix functional (2.19) of the basic theory, and $\Gamma_B(\varphi)$ is the generating functional of 1-irreducible Green functions, related to it by the last expression in (2.111). When substituting (2.111) into (3.37) it is understood that $LS_0(\varphi) = 0$. The first equation in (3.37) is more general (Sec. 3.14), and the second is a consequence of the first and (3.25).

We shall not give the general proofs of the analytic statements (1) and (2); they can be found in [35], [60], [84], [85]. The combinatorial statement (3) will be proved in Sec. 3.14, and here we shall limit ourselves to discussion. Let us begin with statement (3). It implies that the transition from the original basic theory with action $S_B(\varphi)$ to the renormalized theory (3.24) with the counterterms (3.37) is equivalent to applying the R operation corresponding to L in (3.37) to the graphs of the basic theory, i.e., $\Gamma_R(\varphi) = R\Gamma_B(\varphi)$, as for any other Green functions (full, connected, etc.).

Taking $\Gamma_B(\varphi)$ as an example, let us explain in more detail how to understand the action of the operators L, R', and R on the functional. We have

$$\Gamma_B(\varphi) = \Gamma_0 \int dx$$
$$+ \sum_{n=1}^{\infty} \frac{1}{n!} \int ... \int dx_1 ... dx_n \Gamma_n(x_1...x_n) \varphi(x_1)...\varphi(x_n). \tag{3.38}$$

We have changed the notation of the vacuum contribution compared to the expressions of Ch. 2: in this chapter we are taking (Sec. 3.2) $\Gamma_0 \equiv \Gamma_{0p}$, while in Ch. 2 we used Γ_0 to denote the entire vacuum contribution $\Gamma_{0p} \int dx$. We must go to momentum space in the integrals (3.38), after which the coordinate functions there are replaced by their Fourier transforms (3.2). The operators L, R' and R act on the functions $\Gamma_n \equiv \Gamma_{np}(p_1...p_n)$, which determines their action on the functional (3.38). The polynomial form of $L\Gamma_n$ in the momenta corresponds to locality of the counterterms, for which $p \to i\partial$. In our normalization (Sec. 2.7), the constant a in $L\Gamma_n$ corresponds to the local contribution $a \int dx \varphi^n(x)/n!$ in $L\Gamma_B(\varphi)$, and the term ap^2 in $L\Gamma_2$ corresponds to $\frac{a}{2} \int dx \varphi (i\partial)^2 \varphi = \frac{a}{2} \int dx (\partial \varphi)^2$. All this will be needed later on. The counterterm functional (3.37) is the sum of the contributions of all the graphs relevant for L (Sec. 3.8). This class includes all superficially divergent graphs, but can be larger (depending on the choice of K). For the φ^4 model all three of the operators K introduced in Sec. 3.10 generate counterterms

ΔS only from superficially divergent graphs, namely, a vacuum counterterm of the form const $\int dx$ from $L\Gamma_0$, local counterterms $\sim \varphi^2$ and $(\partial\varphi)^2$ from the independent structures 1 and p^2 in $L\Gamma_2$, and a counterterm $\sim \varphi^4$ from $L\Gamma_4$.

Statement (3) is nontrivial, and the following exercise is useful for understanding it. We act with the operator R' according to the rules (3.27) on the two-loop graphs in (3.21). The one-loop subgraphs in (3.27) encircled by the dashed lines are, by definition, replaced by the corresponding counterterms, which in the scheme (3.32) are simply the values of these subgraphs at zero external momenta and are proportional to (3.15). Expressing the counterterms in terms of these constants α_1 and β_1, it can be verified that after the reduction of similar terms taking into account all the symmetry coefficients, we exactly obtain the one-loop contributions (3.21) with the correct coefficients.

Now let us consider statements (1) and (2). The first implies that nontrivial divergences with complicated structure can be generated in the graph γ only by its superficially divergent subgraphs. They are eliminated by the R' operation, the action of which on γ is equivalent to including all the counterterms for the subgraphs and thereby eliminating their divergences. If γ itself is not a superficially divergent graph, then $R'\gamma$ will no longer contain any UV divergences. If γ is a superficially divergent graph, then $R'\gamma$ will contain only superficial divergences of the graph as a whole with the structure (3.6), which are eliminated by the subsequent subtraction of the R operation: $R\gamma = R'\gamma - L\gamma$, where the counterterm $L\gamma$ generates the next contribution in (3.37). If here the counterterms of all the subgraphs are primitive, i.e., polynomials not only in the momenta, but also in the masses τ, then the divergences of $R'\gamma$ will also be primitive. Therefore, since the divergences of the one-loop graphs (Sec. 3.3) are primitive and if the chosen operator K (Sec. 3.10) is also primitive, it is guaranteed that all the superficial divergences and counterterms are primitive.

The general proofs of these statements are rather complicated, but for practical applications it is only essential to know that they are correct. In actual calculations (examples of which will be given below), the quantities $R'\gamma$ and $L\gamma$ must be calculated explicitly, so that the validity of the above statements is verified directly each time. In conclusion, we recall that the use of the R operation to eliminate UV divergences from 1-irreducible Green functions results in their elimination from the theory as a whole (Statement 2 in Sec. 3.8).

3.13 Remarks about the basic statements

1. *The structure of the counterterms.* The form of the counterterms depends on the choice of the operator K, i.e., on the subtraction scheme. In principle, the divergences can always be eliminated by counterterms with the structure (3.5), i.e., by polynomials in the momenta and τ. Polynomial behavior in the momenta corresponds to locality of the counterterms, which is the general property of any subtraction scheme for a local interaction.

Polynomial behavior in τ is a specific property of primitive (Sec. 3.10) subtraction operations of the type (3.33) or minimal subtraction, and is spoiled in schemes like (3.32). In models having a complicated index structure (vector and tensor fields, derivatives at interaction vertices, and so on), to determine the structure of the primitive divergences and the corresponding counterterms in general it is necessary to include in (3.5) and (3.6) all the monomials allowed by the dimensions and symmetry.

2. *Nonlocal contributions.* Owing to the locality of the counterterms, nonlocal contributions to the free action, if there are any (for example, from dipole forces), *are not renormalized*. Here and below the term "not renormalized" will mean that the contribution in question is the same in the basic and the renormalized action, i.e., the corresponding renormalization constant $Z = 1$.

3. *An external field.* Even if an external field h is present in a model, it may not enter into the counterterms owing to (3.25) and the 1-reducibility of all the graphs of γ with insertions h (Sec. 2.3). Therefore, the contribution $h\varphi$ in the basic action is not renormalized, which corresponds to the rule (1.85).

4. *Symmetry.* If the regularized basic theory possesses some symmetry (an invariance), then the primitive divergences (3.5) of the one-loop graphs will also be invariant. In this case it is always possible to choose an operator K that preserves the symmetry (it is important that the "finite part" of this operator not violate the invariance). Using statements like the Lemma in Sec. 3.11, it is possible to prove the invariance of all the multiloop counterterms and, therefore, of the renormalized action. A typical example is the O_n symmetry of the model (2.79), which can always be preserved by the counterterms.

It should be emphasized that we are speaking of a symmetry of the regularized basic theory. Without regularization, all these constructions become meaningless. None of the above discussion is applicable to the formal symmetry S_B violated by the regularization. There are symmetries for which no invariant regularization exists at all,[*] for example, the scale or conformal invariance of the massless model (3.1) in dimension $d = 4$. This invariance is a consequence of the absence of dimensional parameters in S_B, and any regularization necessarily introduces such parameters.

5. *Nonrenormalizable theories.* Introducing a counterterm for each superficially divergent graph according to the rule (3.37), all the divergences can be eliminated from the theory independently of whether it is renormalizable or not (Sec. 3.2). The only difference is that the number of types of counterterm (independent monomials in φ and derivatives) is finite for a renormalizable theory, while it is infinite for a nonrenormalizable one, and so the number of normalization conditions needed to fix the subtraction scheme is also infinite [one condition is needed for each independent counterterm, for example, three

---

[*]The gauge invariance of quantum electrodynamics is preserved by dimensional regularization, and some quantum field models which are chirally symmetric do not have any invariant regularization, hence "chiral anomalies": the violation of the given symmetry in the renormalized theory.

conditions (3.14) for the three counterterms $(\partial\varphi)^2$, φ^2, and φ^4 of the standard model]. Although formally in the MS scheme normalization conditions are in general not necessary, the presence of such parameters is understood implicitly and used to justify the possibility of going to the formal scheme (Sec. 1.19) in which divergences which are powers of Λ are discarded.

6. *The multiplicativity of the renormalization.* The scheme described in Secs. 3.8 to 3.12 forms the basis of the general renormalization formalism: specification of a basic theory S_B with a regularization $\rightarrow$ analysis of its primitive divergences $\rightarrow$ choice of a suitable operator K $\rightarrow$ construction of the corresponding counterterms ΔS and the renormalized action $S_R = S_B + \Delta S$. If the functional S_R obtained in this way can be reproduced up to the vacuum counterterm by (1.84) with unrenormalized action S differing from S_B by only the values of the parameters, then the model is termed *multiplicatively renormalizable*; otherwise, it is said that the *renormalization is nonmultiplicative.* When the usual notation is used for the action, the vacuum counterterm is never reproduced by the multiplicative renormalization procedure,[*] and so above it is said "up to ...". For the renormalization to be multiplicative it is necessary that all the independent structures appearing in the counterterms already be contained in the original S_B, necessarily with independent relative coefficients. Let us explain this by examples. The model (3.22) is multiplicatively renormalizable, since its counterterms of the type $(\partial\varphi)^2$, φ^2, and φ^4 are already contained in (3.22) with independent coefficients. Another example is the model $S_B = -\frac{1}{2}\partial\varphi_a\cdot\partial\varphi_a - \frac{\tau}{2}\varphi_a\varphi_a - g\varphi_1^2\varphi_2^2$ with two-component field $\varphi = \{\varphi_a, a = 1, 2\}$, with summation over the repeated index. The free action in S_B is isotropic, and the interaction $V = -g\varphi_1^2\varphi_2^2$ has only a smaller cubic symmetry (Sec. 1.14). The model is similar to (3.22) in dimensions, and so the counterterms are also similar: φ^2 and $(\partial\varphi)^2$ from $L\Gamma_2$ and φ^4 from $L\Gamma_4$. However, now their index structure is important. For quadratic forms, cubic symmetry implies complete isotropy, and so the counterterms quadratic in φ coincide with the terms of S_B. However, for the counterterm of the type φ^4, the cubic symmetry admits two independent structures: the purely cubic one $V_1 \equiv \varphi_1^4 + \varphi_2^4$ and the isotropic one $V_2 \equiv (\varphi_1^2 + \varphi_2^2)^2$, and the original interaction V is a multiple of $V_1 - V_2$. It is easily seen from the first graphs that the counterterm $L\Gamma_4$ contains the two structures $V_{1,2}$ in a combination which is not a multiple of the original V, and so the vertex counterterms are not reproduced by multiplicative renormalization.

Multiplicativity would have been guaranteed if the original interaction V had contained the two structures $V_{1,2}$ with independent coefficients. However, there are also other possibilities. If $V \sim V_2$, the model is isotropic, and this symmetry can be preserved by the counterterms, thus ensuring that the renormalization is multiplicative. It will also be multiplicative for $V \sim V_1$, although here the absence of the other counterterm V_2 is due not to symmetry

---

[*]This would become possible if the constant $c_0 \int dx$ were added to S, where the vacuum counterterm would correspond to the renormalization of c_0.

(the cubic symmetry of the basic V_1 does not prohibit an isotropic form of V_2), but to specific features of the graphs of Γ_4 (see the discussion of the index structure in Sec. 2.31). This last example shows that the statement that "in general, all counterterms allowed by dimension and symmetry are present" implies the possibility, but not the necessity, that in a specific situation the counterterms may also be fewer.

3.14 Proof of the basic combinatorial formula for the R operation

Let $S_{\mathrm{B}}(\varphi) = S_0(\varphi) + V(\varphi)$ be a given basic action, $H(\varphi) = P \exp V(\varphi)$ the corresponding S-matrix functional (2.19), and $R = R(L)$ be defined by (3.26) from a given arbitrary counterterm operator L satisfying a condition weaker than (3.25):

$$L[\text{disconnected graph}] = L[1] = 0 \qquad (3.39)$$

(unity is the first term in the expansion of H). We shall prove that for $R = R(L)$

$$RP \exp V = P \exp[V - LP \exp V], \qquad (3.40)$$

i.e., the action of the R operator on the S-matrix Green functions is completely equivalent to adding the counterterms (3.37) to the basic interaction V [the second equation in (3.37) is a consequence of (3.40) and (2.111) when the stronger condition (3.25) for L is satisfied]. In this section, the reduction operator $P = P_\varphi$ (2.14) will be written as

$$P = \exp \Delta \mathcal{D}, \qquad 2\mathcal{D} \equiv \delta^2/\delta\varphi\delta\varphi, \qquad (3.41)$$

including the factor of $\frac{1}{2}$ in $\mathcal{D}$ and omitting the (understood) arguments x and the integration over them. The expressions in the proof given below are similified if in addition to (3.39) we require that there be no counterterms at simple vertices (points) of the graphs of H:

$$LV = 0, \qquad LV_k = 0 \quad \text{for all } k, \qquad (3.42)$$

where V_k are the vertex factors (2.21). The second equation in (3.42) follows from the first, because the operators L and $\delta/\delta\varphi$ commute, as they act on different variables: the first acts on the coefficient functions in the expansion of the functional, in this case $V(\varphi)$, and the second acts on its argument φ. The counterterms at simple vertices are never needed to eliminate the divergences, and so the requirement (3.42) is actually always satisfied. It is not required for the statement (3.40) to be valid, and will be used only in the first step of the proof to simplify the expressions, and then the restriction (3.42) will be lifted.

Let us turn directly to the proof of the statement (3.40) [87]. If it is true, then since V is arbitrary it must be satisfied in each order in V. In first order (3.40) takes the form

$$RPV = P[V - LPV]. \qquad (3.43)$$

Let us prove this. Expanding $P = \exp \Delta \mathcal{D}$ in a series, we have

$$PV = \left[\sum_{m=0}^{\infty} (\Delta \mathcal{D})^m / m!\right] V = \sum_{m=0}^{\infty} \frac{1}{m!} \ \text{✿} \ . \tag{3.44}$$

A vertex is associated with the factor $\mathcal{D}^m V$, i.e., $2^{-m} V_{2m}$ in the notation of (2.21). The connected subgraphs of the m-petal graph in (3.44) are the central vertex (point) and all its s-petal subgraphs with $1 \le s \le m-1$. There are no nonintersecting pairs of subgraphs, and so the action of the R operator (3.26) on the m-petal graph (3.44) can be depicted as

$$R \ \text{✿} \ = \ \text{✿} \ - \sum_{s=1}^{m} C_m^s \ \text{✿} \ - \ \text{✿} \ . \tag{3.45}$$

In the sum over s a dashed line encircles the s-petal subgraph and the central point in the last term. The coefficient $C_m^s = m!/s!(m-s)!$ is the number of equivalent but different (for a fully labeled graph) s-petal subgraphs in an m-petal graph.

Let us first assume that the condition (3.42) is satisfied. Then the contribution of the last term on the right-hand side of (3.45) is equal to zero, and this equation can be written symbolically as

$$R\Delta^m = \Delta^m - \sum_{s=1}^{m} C_m^s [L\Delta^s] \Delta^{m-s}, \tag{3.46}$$

where $\lfloor L\Delta^s \rfloor$ corresponds to the counterterm of the s-petal subgraph. For simple interactions like $g \int dx \varphi^n(x)$, when the condition (3.42) is satisfied the term "corresponds" can be interpreted as "is equal to." However, in more complicated cases, for example, when there are derivatives at the vertex, there is only correspondence, because the derivatives in the appropriate proportions must be included also in the counterterm $[L\Delta^s]$. Let us explain. In the functional $V(\varphi)$ the derivatives are distributed symmetrically between all the factors of φ, and in the graphs of the functional $H(\varphi)$ between the ends of all the lines Δ connected to the vertex and arguments φ remaining in the vertex factors (2.21). In this case it must be understood that in (3.46) there are derivatives uniformly distributed between the ends of all the lines Δ. Obviously, this is of no importance for the combinatorics, and (3.46) can be used symbolically.

Now, we note that it can be reproduced by replacing the R operator (3.46) by the differential operator $\widetilde{R}$:

$$R\Delta^m = \widetilde{R}(\partial_\Delta)\Delta^m, \quad \widetilde{R}(\partial_\Delta) \equiv 1 - \sum_{s=1}^{\infty} \frac{1}{s!}[L\Delta^s]\partial_\Delta^s, \tag{3.47}$$

where $\partial_\Delta \equiv \partial/\partial\Delta$ represents formal differentiation with respect to the line Δ. The equation (3.47) is valid for any $m \ge 0$, and so the replacement

$R \to \tilde{R}(\partial_\Delta)$ can be made directly on the left-hand side of (3.43), and then the operator $P = \exp \Delta \mathcal{D}$ can be moved to the left through $\tilde{R}$ following the rule (2.9). The replacement $\partial_\Delta \to \mathcal{D} + \partial_\Delta \to \mathcal{D}$ then occurs in the argument of $\tilde{R}$, the later because the symbols Δ no longer remain to the right of $\tilde{R}$. As a result, we obtain

$$RPV = \tilde{R}(\partial_\Delta)PV = P\tilde{R}(\mathcal{D})V = P[V - U], \qquad (3.48)$$

where U is the contribution from the sum in (3.47):

$$U = \left\{ \sum_{s=1}^{\infty} \frac{1}{s!} [L\Delta^s]\mathcal{D}^s \right\} V = L[P - 1]V = LPV. \qquad (3.49)$$

In calculating U we have taken L outside the summation, after which it is grouped together to form $P - 1$. The last equation in (3.49) follows from (3.42). Substitution of (3.49) into (3.48) gives the desired equation (3.43).

Let us show that it remains valid also when the restriction (3.42) is lifted, i.e., when it is assumed that there are counterterms at the simple vertices (points). Then the last term in (3.45) is also nonzero, and for accurate notation the vertex factors must now be retained inside L, i.e., instead of (3.46) we now must write

$$R[\Delta^m \mathcal{D}^m V] = \Delta^m \mathcal{D}^m V - \sum_{s=0}^{m} C_m^s [L\Delta^s \mathcal{D}^m V]\Delta^{m-s}. \qquad (3.50)$$

The contribution with $s = 0$ corresponds to the last term in (3.45). The powers of the operator $\mathcal{D}$ defined in (3.41) can be taken outside of L [see the discussion following (3.42)]. Here we can put $\mathcal{D}^m$ either on the left or on the right of V, understanding that in any case $\mathcal{D}^m$ acts on V. This obviously does not affect the combinatorics. With this convention, (3.50) can be rewritten as

$$R[\Delta^m \mathcal{D}^m V] = \Delta^m \mathcal{D}^m V - \left\{ \sum_{s=0}^{\infty} \frac{1}{s!} [L\Delta^s V]\partial_\Delta^s \right\} \mathcal{D}^m \Delta^m. \qquad (3.51)$$

The desired quantity RPV is obtained by dividing (3.51) by $m!$ and summing over all $m \geq 0$. Then the operator $P = \exp \Delta \mathcal{D}$ appears on the right, and the $\mathcal{D}$ in it acts on the V on the left. As before, it is moved using (2.9) to the left through ∂_Δ^s, which leads (see above) to the replacement $\partial_\Delta \to \mathcal{D}$, and in the end from (3.51) we obtain

$$RPV = PV - P \sum_{s=0}^{\infty} \frac{1}{s!} [L\Delta^s V]\mathcal{D}^s. \qquad (3.52)$$

Inserting the factor $\mathcal{D}^s/s!$ and summation inside $[L...V]$ and then summing to obtain $P = \exp \Delta \mathcal{D}$, we arrive at the desired equation (3.43), thereby proving that it is valid even without the assumption (3.42). This will happen below: in proving (3.40) in higher orders in V, to simplify the notation we shall first use the condition (3.42) allowing us to take V outside L, and then we shall show that it is actually insignificant.

We have proved (3.40) in first order in V. Turning to the general term of the series for $\exp V$, we write PV^n on the left-hand side (3.40) in the form of (2.24):

$$PV^n = P[V_1...V_n]\,|_*, \qquad P = \exp \Delta\mathcal{D} = \exp \sum_a \Delta_a \mathcal{D}_a. \tag{3.53}$$

The left-hand side is the sum of the contributions of all free graphs with n vertices. The expression on the right-hand side transforms the result into the language of labeled graphs (Sec. 2.2): the vertices are assigned numbers $i = 1, 2, ..., n$, and instead of a single φ, n different independent arguments φ_i are introduced. On the right-hand side of (3.53), $V_i \equiv V(\varphi_i)$, and the symbol $|_*$ indicates that all the φ_i are replaced by φ after the differentiations. The form of the operator P on the right-hand side of (3.53) is exactly the same as in (2.24), but now we can easily change the notation: the index $a \equiv (i \leq k)$ numbers the independent pairs of vertices, Δ_a is the line joining the pair a, and $\mathcal{D}_a$ is the corresponding differential operator, namely, $\mathcal{D}_{ik} = \delta^2/\delta\varphi_i\delta\varphi_k$ for pairs with $i < k$ and $2\mathcal{D}_{ii} = \delta^2/\delta\varphi_i\delta\varphi_i$. As in (3.41), the coefficient of $\frac{1}{2}$ for diagonal pairs is included in the definition of $\mathcal{D}_{ii}$.

Writing the operator P on the right-hand side of (3.53) as the product $\prod \exp \Delta_a \mathcal{D}_a$ and expanding each factor in a series (see Sec. 2.2), we find

$$PV^n - P[V_1...V_n]\,|_* = \left\{ \sum_m \frac{1}{m!} \Delta^m \mathcal{D}^m \right\} V_1...V_n\,|_* . \tag{3.54}$$

Here and below, $m \equiv \{m_a\}$ denotes a multiple index: the set of all m_a, and the sum over m denotes multiple summation over all m_a, each taking all values from 0 to ∞, and factorials and powers are understood as

$$m! \equiv \prod_a m_a!, \qquad \Delta^m \equiv \prod_a \Delta_a^{m_a}, \qquad \mathcal{D}^m \equiv \prod_a \mathcal{D}_a^{m_a}. \tag{3.55}$$

As explained in Sec. 2.2, each term with a given m in the sum (3.54) identically corresponds to some labeled graph with n vertices, m_a is the number of equivalent lines Δ_a connecting a pair of vertices a, and the diagonal m_{ii} correspond to contracted lines ($m_{ik} = \pi_{ik}$ in the notation of Sec. 2.2).

Let us consider the action of the R operator (3.26) on an arbitrary graph in (3.54), assuming that the condition (3.42) is satisfied. It allows us not to consider simple vertices (points) as relevant subgraphs. Then all the relevant subgraphs γ, γ', ... in (3.26) have lines and can be specified by multiple indices s, s', ... analogous to m, and the action of the R operator on an arbitrary graph (3.54) is obviously equivalent to the action of the differential operator

$$\widetilde{R}(\partial_\Delta) = 1 - \sum_s \frac{1}{s!}[L\Delta^s]\partial_\Delta^s + \sum_{s,s'} \frac{1}{s!s'!}[L\Delta^s][L\Delta^{s'}]\partial_\Delta^{s+s'} - ... . \tag{3.56}$$

All the factorials and powers are understood as (3.55). According to the definition (3.26), the single sum in (3.56) runs over all multiple indices s that correspond to connected (inside themselves) blocks of lines Δ^s inside L (Δ^s is the block that is replaced by the corresponding counterterm $[L\Delta^s]$), the double sum runs over pairs s, s' corresponding to connected (inside themselves) counterterm blocks that do not intersect each other, then similarly for the triple, quadruple, etc. sums. It should also be clear that (3.26) is to be understood in the language of fully labeled graphs (Sec. 3.1), and (3.54) and (3.56) are written in the language of labeled graphs. A given subgraph labeled by the multiple index s corresponds to many different but equivalent, fully labeled subgraphs, namely, $\prod_a C^{s_a}_{m_a}$, where each factor is the number of ways of choosing s_a lines of a subgraph from m_a equivalent lines Δ_a in the graph. The needed numbers of combinations are correctly reproduced by the action of the operators $\partial^{s_a}_{\Delta_a}/s_a!$ in (3.56) on the factors $\Delta^{m_a}_a$ of the graphs (3.54).

Let us mentally perform the operations (3.56) on the expression $P[V_1...V_n]$ from (3.54). As before, $P = \exp \Delta \mathcal{D}$ can be moved to the left through the operator (3.56) using (2.9), which brings it to the replacement $\partial_\Delta \to \mathcal{D}$:

$$RPV^n = \widetilde{R}(\partial_\Delta)P[V_1...V_n]\,|_* = P\widetilde{R}(\mathcal{D})[V_1...V_n]\,|_*, \qquad (3.57)$$

where

$$\widetilde{R}(\mathcal{D}) = 1 - \sum_s \frac{1}{s!}[L\Delta^s]\mathcal{D}^s + \sum_{s,s'} \frac{1}{s!s'!}[L\Delta^s][L\Delta^{s'}]\mathcal{D}^{s+s'} - ...\,. \qquad (3.58)$$

Let us consider the contribution of the single sum (3.58) to (3.57):

$$\left\{ \sum_s \frac{1}{s!}[L\Delta^s]\mathcal{D}^s \right\}[V_1...V_n]\,|_*\,. \qquad (3.59)$$

If we calculate it by analogy with (3.49), i.e., if we take L outside the summation and sum over all $s \neq 0$, then the sum, like in (3.49), forms $P - 1$, the contribution of unity can be discarded owing to (3.42), and in the end we obtain the result $LP[V_1...V_n]|_*$ for (3.59). However, it is clearly incorrect, since owing to (3.39) inside L in $P[V_1...V_n]|_*$ there actually remain only the contributions of connected graphs, i.e., those for which all the vertices $V_1...V_n$ are connected by lines. In the original sum (3.59) (see above) this is not so: there only the connectedness of a block of lines Δ^s, inside L, inside itself is important. This block connected inside itself can connect only some of the vertices of a graph, and the others remain as isolated points in (3.59). By moving L from (3.59) by a general factor, we have illegally replaced the requirement that the block Δ^s be connected inside itself by the more rigorous requirement that a graph specified by the multiple index s be connected as a whole, which leads to an incorrect result [this problem did not occur for the single vertex in (3.49)].

The summation over all blocks Δ^s connected inside themselves in (3.59) can be done correctly as follows. We take an arbitrary subset v of the set of all vertices and call the graph a v subgraph in which all the vertices not entering into v remain isolated points. We perform the summation over all s in (3.59) corresponding to the v subgraphs, taking L to be an operator acting only on elements of the v subgraph. It is then possible to sum over all v subgraphs without exception, including those disconnected inside v, because the operator L itself selects the needed [relevant for L, and so necessarily connected inside v owing to (3.39)] graphs. Finally, we obtain the expression

$$\left[LP\prod_{i\in v}V_i \right] \cdot \left[\prod_{i\notin v}V_i \right], \tag{3.60}$$

that must then be summed over all ways of choosing subsets of vertices v in order to obtain the complete answer (3.59). The contributions to (3.57) of the multiple sums (3.58) can be calculated similarly, summing over all pairs v, v', then over triplets, and so on, of nonintersecting subsets of the set of n labeled vertices. In the final results it is necessary to replace all the φ_i by φ, and $V_i \equiv V(\varphi_i)$ by $V \equiv V(\varphi)$ (see above), which corresponds to erasing the numbers of the vertices and changing over to the language of free graphs (we recall that this is indicated by the symbol $|_*$ in the expressions). Then (3.60) becomes $Q_{|v|}V^{n-|v|}$, where $|v| \geq 1$ is the number of vertices in the subset v, and $Q_k \equiv LPV^k$. The entire expression $\widetilde{R}(\mathcal{D})[V_1...V_n]|_*$ in (3.57) is then written as

$$V^n - \sum_v Q_{|v|}V^{n-|v|} \;\mid\; \sum_{v,v'} Q_{|v|}Q_{|v'|}V^{n-|v|-|v'|} - \dots . \tag{3.61}$$

Collecting similar terms and writing $|v| \equiv k \geq 1$, we obtain

$$V^n - \sum_k C_n^k Q_k V^{n-k} + \sum_{k,k'} C_n^{k,k'} Q_k Q_{k'} V^{n-k-k'} - \dots, \tag{3.62}$$

where $C_n^{k,k',\dots}$ is the number of ways of choosing nonintersecting subsets of k, k', ... elements from n elements (vertices). For a single subset C_n^k is the number of combinations, and we can write a general expression. However, for clarity it is better to give the explicit expressions only for the first few n:

$$\widetilde{R}(\mathcal{D})V_1|_* = V - Q_1,$$

$$\widetilde{R}(\mathcal{D})[V_1 V_2]|_* = V^2 - Q_2 - 2VQ_1 + Q_1^2,$$

$$\widetilde{R}(\mathcal{D})[V_1 V_2 V_3]|_* = V^3 - Q_3 - 3VQ_2 - 3V^2Q_1 + 3Q_1Q_2 + 3VQ_1^2 - Q_1^3,$$

$$\widetilde{R}(\mathcal{D})[V_1 V_2 V_3 V_4]|_* = V^4 - Q_4 - 4VQ_3 - 6V^2Q_2 - 4V^3Q_1 + 6V^2Q_1^2$$
$$+ 12VQ_1Q_2 + 4Q_1Q_3 + 3Q_2^2 - 4VQ_1^3 - 6Q_1^2Q_2 + Q_1^4,$$

and so on. The coefficients are interpreted as the numbers of ways of arranging the dashed-line circles according to the rules of Sec. 3.8 on a set of n isolated points. A circle enclosing k points corresponds to a factor $-Q_k$, and the points remaining outside circles correspond to original factors of V. It is easy to see [and to prove, by writing down the general expression for the coefficients in (3.62)] that the expressions given above correctly reproduce the following general expression:

$$\tilde{R}(\mathcal{D})[V_1...V_n]\,|_* = \left\{ \exp\left[-\sum_{k=1}^{\infty}\frac{1}{k!}Q_k\partial_V^k\right]\right\} V^n, \qquad (3.63)$$

in which $\partial_V \equiv \partial/\partial V$ denotes symbolic differentiation with respect to the vertex. From (3.57) and (3.63), by summing over n with the coefficient $1/n!$ we obtain

$$RP\exp V = P\exp\left[-\sum_{k=1}^{\infty}\frac{1}{k!}Q_k\partial_V^k\right]\exp V. \qquad (3.64)$$

The factor $\exp V$ can again be moved through the differential operator in (3.64) using the rule (2.9), which leads to the replacement $\partial_V \to 1 + \partial_V \to 1$ in it, the last one because there are no longer any V symbols to the right of the differential operator. In the end we obtain

$$RP\exp V = P\exp\left[V - \sum_{k=1}^{\infty}\frac{1}{k!}Q_k\right] \equiv P\exp V_{\mathrm{R}}, \qquad (3.65)$$

where [using the condition (3.39) and the definition $Q_k \equiv LPV^k$]

$$V_{\mathrm{R}} \equiv V - \sum_{k=1}^{\infty}\frac{1}{k!}[LPV^k] = V - LP\exp V. \qquad (3.66)$$

This is the desired statement (3.40).

We have obtained this expression assuming that the condition (3.42) holds. If it does not, the only consequence is that the simple point is added to the relevant one-vertex v-subgraphs of the type (3.44). This obviously has no effect on the validity of the proof given above, although it considerably complicates the equations [first, owing to the impossibility of taking V outside of L and, second, to the impossibility of specifying one-vertex point subgraphs by a multiple index s in the R-operation (3.56)].

Let us conclude by discussing the equivalent of relation (3.40) for the Bogolyubov–Parasyuk R operator $\overline{R}(L)$ (Sec. 3.9). In axiomatics [35] it is considered natural to study interactions in quantum field theory with normal ordering, which leads to the exclusion of all graphs with contracted lines. In functional language this corresponds to the following change of notation for the functional (2.19) (see Remark 1 in Sec. 2.2):

$$H = P\exp[P^{-1}\overline{V}], \quad \text{where} \quad \overline{V} \equiv PV. \qquad (3.67)$$

In [35], the primary object is considered to be not V, but the reduced interaction $\overline{V}$ (in Sec. 2.2 it was denoted V_*), and the R operator is written as $\overline{R}(\overline{L})$ with a given arbitrary $\overline{L}$. The analog of (3.40) is the following statement, proved in [84] [Eq. (65) in [84]]: $\overline{V}_R = \overline{V} - \overline{L}H$. If $\overline{L}$ is defined in terms of a given L by (3.29), then $R(L) = \overline{R}(\overline{L})$ are different forms of the same R operator. Then from (3.40) and the definition of $\overline{V}$, we have $\overline{V}_R \equiv P V_R = P[V - LH] = \overline{V} - PLH$, so that to obtain the statement (3.40) from the expression $\overline{V}_R = \overline{V} - \overline{L}H$ proved in [84], it is necessary to also prove the equation $\overline{L}H = PLH$ (we note that it does not imply that $\overline{L} = PL$, since the counterterm $\overline{L}\gamma$ for an individual graph $\gamma \in H$ when written in the form PLH is generated by contributions from both γ itself, and other graphs of H). We shall not need this, since we have proved the statement (3.40) independently.

3.15   Graph calculations in arbitrary dimension

Let us describe the tricks that can be used to calculate graphs in arbitrary dimension d, including noninteger and complex dimensions. This is essentially always an analytic continuation: for example, it is well known that an integral over angles in d-dimensional space for any integer $d \geq 1$ can be represented as $2\pi^{d/2}/\Gamma(d/2)$, where $\Gamma(z)$ is the gamma function. This can be continued to noninteger d. The continuation procedure has been discussed in detail in Sec. 1.19. All the reference formulas given below for d-dimensional integrals are formal results in the terminology of Sec. 1.19. These formulas are used in practical calculations in the formal dimensional regulariztion scheme (Sec. 1.20).

For convenience, let us collect in a single place the notation which will be used in what follows in order to simplify the awkward expressions involving gamma functions:

$$\Gamma(z) \equiv \|z\|, \quad \|\alpha\|\|\beta\|\|\gamma\|... \equiv \Gamma(\alpha)\Gamma(\beta)\Gamma(\gamma)...,$$

$$H(z) \equiv \|d/2 - z\|/\|z\|, \quad H(\alpha, \beta, \gamma, ...) \equiv H(\alpha)H(\beta)H(\gamma)... \ . \tag{3.68}$$

The gamma function $\|z\| \equiv \Gamma(z)$ and its logarithmic derivative $\psi(z) \equiv \Gamma'(z)/\Gamma(z)$ are meromorphic and have first-order poles at the points $z = -n$, $n = 0, 1, 2, ...$ The residues at these poles are given by $\| - n + \Delta\| = (-1)^n/n!\Delta + ...$, $\psi(-n + \Delta) = -1/\Delta + ...$, where Δ is a small parameter. The gamma function has no zeros. A very useful representation is

$$\|z + \Delta\| = \|z\| \exp\{\Delta\psi(z) + \tfrac{1}{2}\Delta^2\psi'(z) + \tfrac{1}{3!}\Delta^3\psi''(z) + ...\}, \tag{3.69}$$

where the primes denote derivatives of ψ. The relations

$$z\|z\| = \|z + 1\|, \quad \|z\|\|1 - z\| = \pi/\sin(\pi z), \quad \psi(z + 1) = \psi(z) + 1/z \tag{3.70}$$

are valid, as are their consequences for derivatives of ψ. Special values are

$$\|n+1\| = n!, \quad \|1/2\| = \pi^{1/2}, \quad \psi^{(n)}(1) = (-1)^{n+1}n!\zeta(n+1),$$

$$\psi^{(n)}(1/2) = [2^{n+1} - 1]\psi^{(n)}(1), \quad \psi(1) = -\mathcal{C}, \quad \psi'(1) = \pi^2/6,$$

$$\psi(1/2) = \psi(1) - 2\ln 2, \quad \psi'(1/2) = \pi^2/2, \quad (3.71)$$

where $\mathcal{C} = 0.5772...$ is the Euler constant, $\zeta(z)$ is the Riemann zeta function, and in the expressions involving $\psi^{(n)}$ we have assumed that $n \geq 1$.

Let us give several useful expressions and integrals in arbitrary dimension d. We always use the notation of (3.68).

The Fourier transform of a power function follows:

$$(2\pi)^{-d} \int dk k^{-2\lambda} \exp(ikx) = (4\pi)^{-d/2} H(\lambda)(2/x)^{d-2\lambda},$$

$$\int dx x^{-2\lambda} \exp(-ikx) = \pi^{d/2} H(\lambda)(2/k)^{d-2\lambda}. \quad (3.72)$$

The mutual consistency of the direct and inverse Fourier transforms is ensured by the relation following from the definitions (3.68):

$$H(d/2 - z) = 1/H(z) \Leftrightarrow H(z, d/2 - z) = 1. \quad (3.73)$$

Here are some useful one-loop integrals:

$$(2\pi)^{-d} \int dk (k^2 + \tau)^{-\alpha} = (4\pi)^{-d/2}\tau^{d/2-\alpha}\|\alpha - d/2\|/\|\alpha\|, \quad (3.74)$$

$$(2\pi)^{-d} \int dk (k^2 + \tau)^{-\alpha}k^{-2\beta} = \tau^{d/2-\alpha-\beta}\frac{\|\alpha + \beta - d/2\|d/2 - \beta\|}{(4\pi)^{d/2}\|\alpha\|d/2\|}, \quad (3.75)$$

$$(2\pi)^{-d} \int dk (k - p)^{-2\alpha}k^{-2\beta} = (4\pi)^{-d/2}p^{d-2\alpha-2\beta}H(\alpha, \beta, d - \alpha - \beta). \quad (3.76)$$

A two-loop integral is

$$(2\pi)^{-2d} \int\int dk_1 dk_2 (k_1^2 + \tau)^{-\alpha}(k_2^2 + \tau)^{-\beta}(k_1 + k_2)^{-2\gamma} =$$

$$= \tau^{d-\alpha-\beta-\gamma}\frac{\|\alpha + \gamma - d/2\|\beta + \gamma - d/2\|\alpha + \beta + \gamma - d\|d/2 - \gamma\|}{(4\pi)^d\|d/2\|\alpha\|\beta\|\alpha + \beta + 2\gamma - d\|}, \quad (3.77)$$

and an l-loop integral with arbitrary power of the quadratic form is

$$(2\pi)^{-dl} \int ... \int \frac{dk_1...dk_l}{[v_{is}k_i k_s + 2a_i k_i + c]^{\alpha}} =$$

$$= \frac{(4\pi)^{-dl/2}\|\alpha - dl/2\|(\det v)^{-d/2}}{\|\alpha\| [c - (v^{-1})_{is}a_i a_s]^{\alpha-dl/2}}. \quad (3.78)$$

All the variables k, p, x, k_i, and a_i are d-dimensional vectors, $k_1 + k_2$ in (3.77) and $k - p$ in (3.76) are sums of vectors, everywhere $dk \equiv d^d k$, the quantities kx in (3.72) and $k_i k_s$, $a_i k_i$, and $a_i a_s$ in (3.78) are scalar products of d-dimensional vectors, v in (3.78) is an $l \times l$ matrix, v^{-1} is the inverse matrix, c is a constant, and repeated indices in (3.78) are summed from 1 to l. The arguments of power functions with fractional powers are assumed to be positive (this restriction can be lifted, but then it is necessary to make a consistent choice of branches), and expressions like k^λ for vectors are always understood as $(k^2)^{\lambda/2}$.

Let us also give the Fourier transform of a massive line:

$$(2\pi)^{-d} \int dk \frac{\exp(ikx)}{(k^2 + m^2)^\alpha} = \left| \frac{2m}{x} \right|^\nu \frac{2K_\nu(|mx|)}{(4\pi)^{d/2}\|\alpha\|}, \qquad \nu = d/2 - \alpha, \qquad (3.79)$$

where $K_\nu(z) = K_{-\nu}(z)$ is the Macdonald function. For noninteger ν,

$$K_\nu(z) = \frac{\|\nu\|1-\nu\|}{2}[I_{-\nu}(z) - I_\nu(z)],$$

$$(3.80a)$$

$$I_\nu(2z) = z^\nu \sum_{n=0}^{\infty} \frac{z^{2n}}{n!\|\nu + n + 1\|},$$

and for integer $n \geq 0$,

$$K_n(2z) = \frac{1}{2}\sum_{s=0}^{n-1}(-1)^s \frac{(n - s - 1)!}{s!z^{n-2s}} +$$

$$(3.80b)$$

$$+(-1)^{n+1}\sum_{s=0}^{\infty} \frac{z^{n+2s}}{s!(n+s)!}\left[\ln(z) - \frac{1}{2}\psi(s + 1) - \frac{1}{2}\psi(n + s + 1)\right].$$

The following representation is also useful:

$$K_{n-\varepsilon}(2z) = \frac{\|n-\varepsilon\|1-n+\varepsilon\|}{2}\left\{\sum_{s=0}^{n-1}\frac{z^{2s-n+\varepsilon}}{s!\|1-n+s+\varepsilon\|} + \right.$$

$$(3.81)$$

$$\left. +\sum_{s=0}^{\infty}\left[\frac{z^{2s+n+\varepsilon}}{(n+s)!\|s+1+\varepsilon\|} - \frac{z^{2s+n-\varepsilon}}{s!\|s+n+1-\varepsilon\|}\right]\right\},$$

which goes into (3.80b) for $\varepsilon \to 0$. For $n = 0$ the first sum in (3.80b) and (3.81) is absent.

All these integrals are well defined (convergent) in some region of parameter values — the exponents and d, and outside this region they are understood as the analytic continuation of the right-hand side from the region of convergence. These expresssions are used practically everywhere, including outside the region of convergence, which corresponds to the formal dimensional regularization scheme (Sec. 1.20).

The Feynman formula is also very useful:

$$A_1^{-\lambda_1}...A_n^{-\lambda_n} = \frac{\|\Sigma\lambda_i\|}{\Pi\|\lambda_i\|} \int_0^1 ...\int_0^1 du_1...du_n \frac{\delta(\Sigma u_i - 1)\Pi u_i^{\lambda_i - 1}}{[\Sigma A_i u_i]^{\Sigma\lambda_i}} \qquad (3.82)$$

(the sums and products are over all $i = 1, ..., n$), as is the expression

$$A^{-\lambda} = \frac{1}{\|\lambda\|} \int_0^\infty du u^{\lambda-1} \exp(-Au), \qquad (3.83)$$

which is often called the α representation, since the integration variable u in (3.83) is often denoted as α. The expressions given above allow the calculation of graphs in arbitrary spatial dimension. For models like the φ^4 model, the integrand in a graph is a product of propagators $\Delta_i = A_i^{-1}$, where A_i are quadratic forms (in general, inhomogeneous) in the integration momenta. Representing products of factors A_i^{-1} by (3.82) with $\lambda_i = 1$, we can then perform the momentum integral using (3.78) (the expression $\sum A_i u_i$ is a quadratic form in the set of all integration momenta). Then only the integral over the variables u with parametric dependence on d remains, i.e., an expression that is meaningful for any value of d. Another method is the following. By using the α representation (3.83) for all the propagators, we arrive at a Gaussian integral over the set of momenta, which can also be performed for any d using (2.1). These tricks are easily generalized to any model with more complicated propagators and vertices. The additional formulas needed in that case can be obtained by differentiating the integrals given above with respect to various parameters.

In calculating objects with free vector indices, the results are expressed in terms of the symbol δ_{is} and the external momenta with indices. Therefore, the question of what is a d-dimensional vector when d is not an integer does not arise in practice (see Ch. 4 of [60] for more details). The trace of δ_{is} in d-dimensional space is given by $\delta_{ii} = \delta_{is}\delta_{si} = d$. Here it is necessary to keep in mind the fact that *the minimal subtraction operation (3.34) does not commute with the operation of taking the trace of a tensor*: for example, for $d = 4-2\varepsilon$ we have $\delta_{is}K[\varepsilon^{-1}\delta_{si}] = \varepsilon^{-1}\delta_{is}\delta_{si} = d/\varepsilon = (4-2\varepsilon)/\varepsilon \neq K[\delta_{is}\varepsilon^{-1}\delta_{si}] = K[d/\varepsilon] = 4/\varepsilon$. This example is sufficient to clearly illustrate the meaning of the above statement.

The following expressions can be used to calculate integrals with factors of k_i:

$$\int dk f(k^2)k_i k_s = \delta_{is}\frac{1}{d} \int dk f(k^2)k^2,$$

$$\int dk f(k^2)k_i k_s k_l k_m = [\delta_{is}\delta_{lm} + \delta_{il}\delta_{sm} + \delta_{im}\delta_{sl}]\frac{1}{d(d+2)} \int dk f(k^2)k^4$$

$$(3.84)$$

and so on. In general, integrals like (3.84) with an odd number of k_i are zero; for an even number $2n$ of these factors the symmetrized product of δ symbols will contain $(2n-1)!!$ terms, and the denominator of the coefficient will be the product $d(d+2)...(d+2n-2)$.

3.16 Dimensional regularization and minimal subtractions

Dimensional regularization (1.86) for the φ^4 model corresponds to $d = 4 - 2\varepsilon$. Then in the unrenormalized action (3.10) $d[\tau_0] = 2$, $d[g_0] = 2\varepsilon$, $d_\varphi = 1 - \varepsilon$ and from (3.3) we have

$$d[\Gamma_{2n}] = 4 - 2n + 2\varepsilon(n - 1). \tag{3.85}$$

By definition (Sec. 3.7), the dimensions of the basic parameters in (3.22) and the bare parameters in (3.10) coincide. In order to have integer dimensions $d_\tau = 2$, $d_g = 0$ for the renormalized parameters τ and g, it is usual to take $\tau_B = \tau$ [this was taken into account in (3.22)] and

$$g_B = g\mu^{2\varepsilon} \qquad (d = 4 - 2\varepsilon), \tag{3.86}$$

where μ is the renormalization mass, an additional dimensional parameter of the renormalized theory with $d_\mu = 1$. As explained in detail in Sec. 1.20, dimensional regularization does not eliminate the divergences in the functions $\Gamma_{0,2}$ which are powers of the cutoff Λ, but they give only insignificant regular contributions (see below). If desired, these power-law divergences can be eliminated by a suitable auxiliary R operator. In practice, they are usually simply ignored, and the following *convention* is adopted: all the calculations are performed within the formal scheme (Sec. 1.20) without the cutoff Λ, i.e., equations like those given in Sec. 3.15 are used even for integrals that are divergent in the strict sense of the word. The poles in ε are then assumed to be UV divergences.

In dimension $d = 4 - 2\varepsilon$ the exponents of Λ in equations analogous to (3.7) are slightly reduced (by $2\varepsilon l$ for graphs with l loops), remaining positive. Ignoring these contributions in the formal scheme is equivalent to loss of the constant in Γ_2 determining only the shift of T_c and the contributions to the vacuum loops of Γ_0 regular in τ. From the viewpoint of physics all these losses are unimportant, because fluctuation models themselves of the theory of critical phenomena usually cannot exactly determine these quantities (Sec. 1.12).

As an example, let us give the integrals analogous to (3.8) and (3.9), which are easily calculated in the formal scheme using the rule (3.74):

$$I\left[\,\bigcirc\,\right] = \frac{1}{(2\pi)^d} \int \frac{dk}{k^2 + \tau} = \frac{\|1 - d/2\|\tau^{d/2-1}}{(4\pi)^{d/2}}, \tag{3.87a}$$

$$I\left[\,\bowtie\,\right]_{p=0} = \frac{1}{(2\pi)^d} \int \frac{dk}{(k^2 + \tau)^2} = \frac{\|2 - d/2\|\tau^{d/2-2}}{(4\pi)^{d/2}}. \tag{3.87b}$$

For $d = 4 - 2\varepsilon$ the two integrals have only a simple pole in ε.

In Sec. 3.2 we defined superficially divergent graphs as graphs of the functions Γ with nonnegative dimensions of the momentum integrals, having in mind an exact theory with cutoff Λ. In the formal scheme for integrals that are divergent in this sense it is possible to have finite results, divergences appearing only at poles in ε. The momentum integrals I of the graphs of the basic theory depend on the external momenta p and the renormalized parameters e having positive integer dimensions. Such integrals have poles for $d_I = d_M$, where M are all possible monomials constructed from p and e that can enter into the expansion of the function Γ (Sec. 1.19). For $\varepsilon = 0$ for the graphs of any function Γ we have $d_I = d[\Gamma]$ (Sec. 3.2). Therefore the criterion in Sec. 3.2 for an arbitrary model in the formal scheme should be replaced by the following: *in general, only graphs of those functions Γ for which*

$$d[\Gamma]|_{\varepsilon=0} = \left[d - \sum_\varphi n_\varphi d_\varphi \right] |_{\varepsilon=0} = \text{ one of } d_M = $$
$$= \text{ a nonnegative integer} \tag{3.88}$$

can be superficially divergent graphs. Here the summation runs over all fields φ entering into the function Γ, n_φ is the number of fields of a given type, d_φ are the canonical dimensions of these fields, ε is a parameter characterizing the deviation from logarithmic behavior, and d_M are the canonical dimensions of all possible monomials M constructed from the external momenta contained in Γ and the renormalized parameters e with positive integer dimensions which can enter into the expansion of Γ. Henceforth, as before, we shall take $e = \tau$ to be specific.

A necessary but not sufficient condition for superficial divergence is that the right-hand side of (3.88) be an integer. In some cases (including all scalar models of the φ^n type), the dimensions d_M of the monomials M in the expansion of Γ can only be even, and then graphs with odd positive dimensions do not diverge (for example, if Γ is a scalar and is expanded in τ and p^2). There is also another reason that the statements "may be" instead of "are". For a one-loop integral I with $d_I = d_M$, there is necessarily a divergence, but multiloop integrals can be reduced to products of more simple ones. If the cofactors do not contain any divergences according to the criterion (3.88), there also will not be any in the product, although the necessary conditions (3.88) may be satisfied (see the example of the φ^6 model in Ch. 4).

In the absence of the parameter Λ, the structure of the primitive divergences (3.5) of the one-loop graphs of γ simplifies:

Prim. div. I = polynomial $P(p, \tau)$ with $d[\text{monomial}] = d[I]$, $\qquad$ (3.89)

where $I = \mu^{-\delta}\gamma$, the factor $\mu^{-\delta}$ with $\delta \sim \varepsilon$ restores the integer dimension. In contrast to (3.5), in (3.89) there are no monomials of lower dimension, since without a dimensional parameter like Λ their coefficients are not constructed [integer powers of μ cannot be formed from a finite number of factors (3.86) in a graph].

Dimensional regularization still does not determine the subtraction scheme; any one can be used, for example, the subtractions (3.32) or (3.33). The most convenient for practical calculations is the MS scheme with the subtraction operator (3.34). It preserves the primitive nature of the divergences in all orders (Secs. 3.10 and 3.12), so that in the MS scheme for any graph γ,

$$\text{Div. part } R'\gamma = \qquad\qquad\qquad\qquad\qquad\qquad (3.90)$$
$$= \mu^\delta g^v \{\text{polynomial } P(p,\tau) \text{ with } d[\text{monomial}] = d[\gamma] - \delta\},$$

where v is the number of vertices in γ, and δ is the deviation of the dimension $d[\gamma]$ from an integer (for example, $\delta = 2\varepsilon(n-1)$ for the graphs in (3.85)]. The dimensionless coefficients of the polynomial $P(p,\tau)$ depend only on ε and in the MS scheme are simple polynomials in $1/\varepsilon$. Experience shows that their order cannot exceed the number of loops in γ.

The rules we have written are common to various models and various dimensional-type regularizations. They are obvious only for one-loop graphs, and in general are nontrivial and express the basic statement (2) in Sec. 3.12 as applied to regularizations of the dimensional type. Additional explanation of the rule (3.90) can be found in Sec. 3.22.

3.17   Normalized functions

From this point on we shall deal particularly with the φ^4 model with renormalization in the MS scheme. It is convenient to move the operators L, R, and R' to the new *normalized functions* $\overline{\Gamma}$ and their graphs $\overline{\gamma}$ with integer [see (3.85) and (3.86)] dimensions, defining them in this model (differently in other models) by the relations

$$\overline{\Gamma}_{2n} = (-g_B)^{1-n}\Gamma_{2n}, \qquad d[\overline{\Gamma}_{2n}] = 4 - 2n. \qquad (3.91)$$

The graphs $\overline{\gamma}_{2n}$ of the functions $\overline{\Gamma}_{2n}$ have exactly one factor $-g_B$ per loop for any n. This is a consequence of the topological relations $l = N - v + 1$ and $4v = 2N + 2n$, where l is the number of loops, v is the number of vertices, N is the number of internal lines, and $2n$ is the number of external lines. Therefore, in the graphs $\overline{\gamma}$ it is convenient to assign the factors $-g_B$ to loops rather than vertices, as we shall do. We define the operator L on the graphs $\overline{\gamma}$ by (3.31) with $\delta = 0$ and K from (3.34)

$$L\overline{\gamma} = KR'\overline{\gamma}, \qquad\qquad\qquad\qquad (3.92)$$

and R and R' by the general rule (3.26) with L from (3.92). The latter implies that the factors L_γ in (3.26) now act on the subgraphs $\overline{\gamma}$, containing, by convention, one factor $-g_B$ per loop. This refinement is important, because the factors (3.86) do not commute with the minimal subtraction operator (3.34), i.e., they cannot be removed from inside the dashed-line circles in expressions like (3.27), and so it is necessary to exactly indicate the number of factors $-g_B$ inside each circle. Earlier, in the graphs γ this was fixed by the rule of one per vertex, and now for the graphs $\overline{\gamma}$ it is fixed as one per loop. We can return to the usual operators L, R, and R' on Γ_{2n} using the relations

$$T\Gamma_{2n} = (-g_B)^{n-1}T\overline{\Gamma}_{2n}, \quad T \equiv \text{any of } L, R, R'. \tag{3.93}$$

These are nontrivial statements and not simple consequences of linearity, since the MS operations do not commute with multiplication by g_B. The expressions (3.93) are valid because in (3.31) the operator KR' actually acts on the $\overline{\gamma}$ since they differ from $\mu^{-\delta}\gamma$ in (3.31) only by factors $-g$ which commute with all the operators.

The rule (3.90) for operators on the graphs $\overline{\gamma}$ takes the form

$$\text{Div. part } R'\overline{\gamma} = g^l P(p,\tau) \text{ with } d[\text{monomial}] = d[\overline{\gamma}], \tag{3.94}$$

where l is the number of loops in γ, and the dimensionless coefficients of the polynomials $P(p,\tau)$ are polynomials in $1/\varepsilon$ of order l (in special cases it can be smaller, but not larger). It follows from (3.91) and (3.94) that

$$\text{Div. part } R'\overline{\gamma}_0 = c_0\tau^2, \qquad \text{Div. part } R'\overline{\gamma}_2 = c_1\tau + c_2 p^2,$$
$$\text{Div. part } R'\overline{\gamma}_4 = c_3, \qquad \text{Div. part } R'\overline{\gamma}_{2n} = 0 \text{ for all } n \geq 3, \tag{3.95}$$

where the c_i are dimensionless constants, polynomials in $1/\varepsilon$. It is convenient to introduce the four auxiliary operators K_a and L_a with $a = 0, 1, 2, 3$, defining them as

$$L_a = K_a R', \quad L_0\overline{\gamma}_0 = c_0, \quad L_1\overline{\gamma}_2 = c_1, \quad L_2\overline{\gamma}_2 = c_2, \quad L_3\overline{\gamma}_4 = c_3 \tag{3.96}$$

with the constants c_a from (3.95). Since the definitions (3.96) are so important, let us restate them verbally. The four operators K_a are defined on functions $R'\overline{\gamma}_n \equiv f_n$ with $n = 0, 2, 4$ having primitive divergences with the structure (3.95). By definition, the operator K_0 selects the coefficient of τ^2 in the divergent part of the function $R'\overline{\gamma}_0 \equiv f_0(\tau)$ (we indicate only the needed arguments); the operator K_1 selects the coefficient of τ in the divergent part of the function $R'\overline{\gamma}_2 \equiv f_2(\tau, p^2)$; and the operator K_2 selects the coefficient of p^2 in the same. Finally, the operator K_3 selects the entire divergent part, a constant independent of the arguments τ, $p_1...p_4$, in the function $R'\overline{\gamma}_4 \equiv f_4(\tau, p_1...p_4)$.

The counterterm operation (3.92) on the normalized functions (3.91) (and similarly on their graphs $\overline{\gamma}$) is expressed in terms of the elementary operators (3.96) as

$$L\overline{\Gamma}_0 = \tau^2[L_0\overline{\Gamma}_0], \quad L\overline{\Gamma}_2 = \tau[L_1\overline{\Gamma}_2] + p^2[L_2\overline{\Gamma}_2],$$
$$L\overline{\Gamma}_4 = [L_3\overline{\Gamma}_4], \quad L\overline{\Gamma}_{2n} = 0 \text{ for all } n \geq 3, \tag{3.97}$$

where p is the external momentum in the graphs of $\overline{\Gamma}_2$. The quantities $[L_a\overline{\Gamma}]$ are the sums of contributions $[L_a\overline{\gamma}]$ of individual graphs with symmetric coefficients. They have the structure

$$[L_a\overline{\gamma}] = g^l \times \text{polynomial in } 1/\varepsilon \text{ of order } l, \tag{3.98}$$

where l is the number of loops in a graph γ. The coefficients of the polynomials (3.98) are numbers independent of any parameters.

The counterterm functional (3.37) can be expressed in terms of $[L_a\overline{\Gamma}_{2n}]$ using (3.91), (3.93), and (3.97) and the rules for going from the functions Γ_n to the functional [see the discussion following (3.38)]:

$$\Delta S(\varphi) = -L\Gamma_{\mathrm{B}}(\varphi) = -L\left\{\Gamma_0 + \tfrac{1}{2}\Gamma_2\varphi^2 + \tfrac{1}{4!}\Gamma_4\varphi^4\right\} =$$

$$= g_{\mathrm{B}}^{-1}L\overline{\Gamma}_0 - \tfrac{1}{2}L\overline{\Gamma}_2\varphi^2 + \tfrac{1}{4!}g_{\mathrm{B}}L\overline{\Gamma}_4\varphi^4 =$$

$$= \int dx\left\{g_{\mathrm{B}}^{-1}\tau^2\left[L_0\overline{\Gamma}_0\right] - \tfrac{1}{2}\left[L_2\overline{\Gamma}_2\right](\partial\varphi)^2 - \tfrac{1}{2}\tau\left[L_1\overline{\Gamma}_2\right]\varphi^2 + \tfrac{1}{4!}\left[L_3\overline{\Gamma}_4\right]g_{\mathrm{B}}\varphi^4\right\}.$$
$$(3.99)$$

All the Γ_n are Green functions of the basic theory, the intermediate expressions in (3.99) are written symbolically, with the required integrations in them understood, and the integration is indicated explicitly in the final local expression. The contribution involving $[L_0\overline{\Gamma}_0]$ is the vacuum counterterm, which was ignored in the expressions of Sec. 1.18, since there we did not study vacuum loops and their renormalization.

To calculate the RG functions we need the renormalization constants. They will be calculated in terms of the quantities [...] in (3.99), and the latter directly from the graphs. Let us briefly describe the general calculational scheme; a detailed calculation will be given later. All the figures in this section will denote graphs $\overline{\gamma}$; their vertices correspond to unit factors, and a factor of $-g_{\mathrm{B}}$ is assigned to loops. The quantities $R'\overline{\gamma}$ in $[L_a\overline{\gamma}] = [K_aR'\overline{\gamma}]$ are constructed according to the rules of Sec. 3.8. The superficially divergent subgraphs of the type $\overline{\gamma}_{2,4}$ encircled by the dashed lines are replaced by the corresponding counterterms, in particular, the vertex subgraph $\overline{\gamma}_4$ is replaced by a simple vertex (point) with coefficient $[L_3\overline{\gamma}_4]$, and the self-energy subgraph $\overline{\gamma}_2(k)$ by the counterterm $[L_2\overline{\gamma}_2]k^2 + \tau[L_1\overline{\gamma}_2]$, where k is the integration momentum of the original graph flowing through $\overline{\gamma}_2$. Graphically,

$$\begin{array}{c}\includegraphics{}\end{array} = [L_3\overline{\gamma}_4]\ \times\ ;\quad {}_k\ \includegraphics{}\ {}_k = \tau[L_1\overline{\gamma}_2]\ \ \bullet\!\!-\!\!-\ \ +[L_2\overline{\gamma}_2]\ \ -\!\!\!\times\!\!-\ ,\qquad (3.100)$$

where $\overline{\gamma}$ is a subgraph enclosed by a circle, and for $\overline{\gamma}_2(k)$ the point denotes the insertion of unity and the cross denotes the factor k^2 (instead of the subgraph $\overline{\gamma}_2$). The simple vertex in the counterterm (3.100) corresponds to a factor of unity [but this is only because we have changed over to the graphs of the normalized functions $\overline{\gamma}$. For the original graphs γ it would be natural to assign the standard vertex factor $-g_{\mathrm{B}}$ to the vertex counterterm in the first expression in (3.100)].

The quantities [...] are calculated recursively. For one-loop superficially divergent graphs they are found directly from the definitions (3.96) with $R' = 1$ and the explicit expressions (3.87). According to the general rule of one per loop, they must be multiplied by $-g_{\mathrm{B}} = -g\mu^{2\varepsilon}$, and then the pole in ε in the graph $\overline{\gamma}_4$ and the pole part of the coefficient of τ in the graph $\overline{\gamma}_2$ should be selected. This gives

$$\left[L_1\, \bigcirc\, \right] = -\left[L_3\, \bigotimes\, \right] = g/16\pi^2\varepsilon, \quad \left[L_2\, \bigcirc\, \right] = 0. \tag{3.101}$$

Going to the two-loop term $R'\bar{\gamma}$, let us specify the individual terms on the right-hand sides of (3.27) using the rule (3.100):

$$
\begin{aligned}
&= \tau\left[L_1\, \bigcirc\, \right]\, \bigcirc\, ; \\
&= \left[L_3\, \bigotimes\, \right]\, \bigcirc\, ; \\
&= \left[L_3\, \bigotimes\, \right]\, \bigcirc\, ; \\
&= \ = \left[L_3\, \bigotimes\, \right]\, \bigotimes\, ; \\
&= \left[L_3\, \bigotimes\, \right]\, \bigotimes\, ; \\
&= \tau\left[L_1\, \bigcirc\, \right]\, \bigotimes\, .
\end{aligned}
\tag{3.102}
$$

The one-loop coefficients [...] entering here are already known from (3.101), and so the right-hand sides of (3.102) are known functions of the external momenta and τ (it is assumed that we can calculate the contributions of the ordinary graphs). The operators K_a acting on them are determined by (3.96), and so it is possible to calculate the two-loop quantities $[L_a\bar{\gamma}] = [K_a R'\bar{\gamma}]$ and so on.

To make this clearer, let us give several more examples of interpreting the individual contributions on the right-hand sides of (3.28) for the three-loop terms of $R'\bar{\gamma}$:

$$
\begin{aligned}
&= \left[L_3\, \bigotimes\bigotimes\, \right]\, \bigotimes\, ; \\
&= \left[L_3\, \bigotimes\, \right]\, \bigotimes\, ;
\end{aligned}
$$

$$= [L_3 \, \text{}]^2 \, \text{};$$

(3.103)

$$\text{} = \tau [L_1 \, \text{}] \, \text{} + [L_2 \, \text{}] \, \text{}.$$

We recall that the insertion of unity is depicted by a point on the line $\Delta(k) = (k^2 + \tau)^{-1}$ and a factor of k^2 by a cross, i.e., lines with a point are associated with the quantity $\Delta^2(k)$ and with a cross $k^2 \Delta^2(k)$. The coefficients on the right-hand sides of (3.103) are one- and two-loop quantities [...], which must be calculated before going to the three-loop graphs. For now we limit ourselves to these general remarks; specific calculations will be done in Secs. 3.20 and 3.21.

In conclusion, we note the following useful *statement*: if a graph consists of subgraphs A and B connected by a single vertex [for example, the first, third, and fifth graphs in (3.27), and the first, second and next-to-last in (3.28)], its momentum integral is the product AB, and for our R operation of the MS scheme

$$R(AB) = RA \cdot RB, \qquad L(AB) = -LA \cdot LB. \tag{3.104}$$

The validity of this statement can be checked by simple examples; the general proof is left to the reader.

3.18   The renormalization constants in terms of counterterms in the MS scheme

According to the general rule (3.24), the sum of the basic action (3.22) with $g_B = g\mu^{2\varepsilon}$ and the counterterms (3.99) is the renormalized action

$$S_R(\varphi) = \Delta S_{\text{vac}} - \tfrac{1}{2} Z_1 (\partial \varphi)^2 - \tfrac{1}{2} Z_2 \tau \varphi^2 - \tfrac{1}{24} Z_3 g_B \varphi^4 \tag{3.105}$$

[as in (3.22), the integral is understood] with the vacuum counterterm

$$\Delta S_{\text{vac}} = V \tau^2 Z_0 / 2 g_B, \qquad V \equiv \int dx \tag{3.106}$$

and the renormalization constants

$$Z_1 = 1 + [L_2 \overline{\Gamma}_2], \qquad Z_3 = 1 - [L_3 \overline{\Gamma}_4], \tag{3.107}$$

$$Z_2 = 1 + [L_1 \overline{\Gamma}_2], \qquad Z_0 = 2 [L_0 \overline{\Gamma}_0]. \tag{3.108}$$

The model is multiplicatively renormalizable: up to the vacuum counterterm, the action (3.105) is obtained from (3.10) by the transformation (1.84) using

$$\tau_0 = \tau Z_\tau, \qquad g_0 = g_B Z_g = g\mu^{2\varepsilon} Z_g, \tag{3.109}$$

$$Z_1 = Z_\varphi^2, \quad Z_2 = Z_\varphi^2 Z_\tau, \quad Z_3 = Z_\varphi^4 Z_g. \tag{3.110}$$

The renormalization constants of the field and parameters are needed to calculate the RG functions using the expressions in Sec. 1.25. They are calculated using (3.110) in terms of the action constants (3.105), and those in turn are calculated in terms of the quantities [...] directly from the graphs of $\bar\gamma$. The constants Z in the MS scheme are independent of τ [they are sums of contributions like (3.98)], and so in principle they can be calculated for $\tau = 0$. Experience shows that this greatly simplifies the calculation. However, the operators L_1 and L_0 in (3.108) become meaningless for $\tau = 0$ [see the definition (3.96)]. This difficulty can be avoided by using the commutativity of the operators ∂_τ and L in the MS scheme, proved in Sec. 3.11 [changing over to operations on normalized functions (3.91) has no effect on the validity of that proof]. It leads, in particular, to the following equations:

$$\partial_\tau L \bar\Gamma_2 = L \partial_\tau \bar\Gamma_2, \quad \partial_\tau^2 L \bar\Gamma_0 = L \partial_\tau^2 \bar\Gamma_0. \tag{3.111}$$

The operator ∂_τ acts on lines of the graphs $\bar\gamma$ according to the rule

$$\partial_\tau \quad \underline{\quad\quad} = - \quad \underline{\quad}\bullet\underline{\quad} \tag{3.112}$$

and reduces the canonical dimension by 2. Therefore, the superficial divergences of the graphs $\partial_\tau \bar\Gamma_2$ and $\partial_\tau^2 \bar\Gamma_0$ are the same as in the vertex graphs of $\bar\Gamma_4$ (logarithmic). They are selected by the operator $L_3 = K_3 R'$ defined in (3.96), i.e., L on the right-hand side of (3.111) can be replaced by L_3. On the other hand, the left-hand sides of (3.111) involve the quantities (3.97) with known τ dependence (we recall that the factors [...] in (3.97) are independent of τ). Therefore, from (3.97) and (3.111) it follows that

$$[L_1 \bar\Gamma_2] = [L_3 \partial_\tau \bar\Gamma_2], \quad 2[L_0 \bar\Gamma_0] = [L_3 \partial_\tau^2 \bar\Gamma_0]. \tag{3.113}$$

These expressions are very useful, because it is much simpler to calculate their right-hand sides than the desired objects on the left-hand sides. From (3.108) and (3.113) we have

$$Z_2 = 1 + [L_3 \partial_\tau \bar\Gamma_2], \quad Z_0 = [L_3 \partial_\tau^2 \bar\Gamma_0]. \tag{3.114}$$

These will be the working expressions for calculating the renormalization constants $Z_{0,2}$, instead of the original expressions (3.108).

3.19   The passage to massless graphs

All the quantities [...] in (3.107) and (3.114) are independent of τ, and so they can be calculated at $\tau = 0$. We stress the fact that we are speaking of the calculation of objects in the massive theory with $\tau \geq 0$; the passage to $\tau = 0$ in the graphs is only a technical simplification [the objects (3.108) become meaningless in the massless theory]. Equations (3.107) and (3.114) involve only the operators $L_{2,3}$ defined, in contrast to $L_{0,1}$ in (3.108), also for $\tau = 0$. We recall [see (3.96)] that L_2 selects the coefficient of p^2 in the superficial divergence of the graphs $\overline{\gamma}_2$ (the superficial divergence $\equiv$ the intrinsic divergence $\equiv$ divergence of $R'\overline{\gamma}_2$), and L_3 selects the entire superficial divergence of the logarithmic graphs $\overline{\gamma}_4$, $\partial_\tau\overline{\gamma}_2$, and $\partial_\tau^2\overline{\gamma}_0$. The latter one is a constant independent of the external momenta and the mass $\tau \geq 0$, or the various masses $\tau_i \geq 0$ of the internal lines in the general case. From this we find an important rule.

Rule 1. In calculating $[L_3\overline{\gamma}] = [K_3R'\overline{\gamma}]$ for a given logarithmically divergent graph $\overline{\gamma}$, the configuration of the external momenta and the mass of the internal lines in $\overline{\gamma}$ can be chosen arbitrarily, taking into account only the following constraints: (1) the choice must be the same for all the terms of $R'\overline{\gamma}$, i.e., it must be uniquely determined by fixing the external momenta and masses in the original graph $\overline{\gamma}$; (2) IR divergences must not appear in any of the terms $R'\overline{\gamma}$ for this choice. Here it is even possible to artificially introduce additional external momenta flowing into the internal vertices of the graph.

Let us explain the last remark. The vertices of the original graph are split into internal (integration vertices) and external ones, the latter being the ones the external momenta flow into. The value of the function $\overline{\gamma}$ corresponding to the graph does not change if external outgoing lines are added to the internal vertices, assuming that zero momenta flow along them. The graph of a different Green function of the given (and maybe also another) model is then obtained in the external form. If the graph is logarithmic, its superficial divergence does not depend on the configuration of the external momenta. Therefore, in calculating this divergence, the zero external momenta in the artificially added outgoing lines can, if desired, be replaced by nonzero momenta. This trick is sometimes very useful. In particular, it allows the graphs of $\partial_\tau\overline{\gamma}_2$ and $\partial_\tau^2\overline{\gamma}_0$ to be identified with the vertex graphs $\overline{\gamma}_4$ if we imagine that a pair of external outgoing lines is added to the point in (3.112):

$$\text{[graph]} \rightarrow \text{[graph]} \; ; \; \text{[graph]} \rightarrow \text{[graph]} \; . \tag{3.115}$$

From this we find another rule that is very useful in practice.

Rule 2. The quantities $[L_3\partial_\tau\overline{\Gamma}_2]$ and $[L_3\partial_\tau^2\overline{\Gamma}_0]$ in (3.114) are calculated from the same vertex graphs $\overline{\gamma}_4$ as $L_3\overline{\Gamma}_4$ in (3.107); the only difference is in the symmetry coefficients.

The calculations can be simplified by freely choosing the configuration of the external momenta (which henceforth will be referred to as the momentum flows) and the masses of the internal lines for each individual logarithmically divergent graph. It is usually easiest to calculate massless graphs like self-energy graphs with the simple flow of a single momentum, flowing into one of the vertices and out of another. In addition, care should be taken that the masslessness of the lines does not lead to the appearance of IR divergences. This is usually done successfully. In Sec. 3.21 we present a complete calculation of the quantities [...] of all the graphs of the φ^4 model through three loops (13 altogether), and only one needs to be calculated with masses. In connection with this, we note that the graphs of $\partial_\tau \overline{\gamma}_2$ and $\partial_\tau^2 \overline{\gamma}_0$ with the original configuration of the external momenta (in $\overline{\gamma}_0$ they are all zero, and in $\overline{\gamma}_2$ there is a simple flow) do not admit direct passage to massless lines, since then they acquire [see (3.112)] the square of the massless propagator $\Delta^2(k) = k^{-4}$, which generates an IR divergence in $d = 4 - 2\varepsilon$ dimensions. It vanishes if we artificially introduce a nonzero external momentum flowing into the "point" (3.112), which is possible when the graphs are identified with the vertex graphs according to the rules (3.115).

We note that in the formal scheme of dimensional regularization (see Sec. 1.20 and the convention in Sec. 3.16) for IR-divergent integrals at $\varepsilon \neq 0$ finite results are usually obtained in which the IR divergence is also manifested as poles in ε. These additional IR poles mix with the UV poles we are interested in, and this is what makes IR divergences dangerous in the formal scheme (and in the rigorous scheme IR-divergent integrals simply do not exist).

Another important simplification in the calculation of the quantities $[L_{2,3}\overline{\gamma}]$ is related to the fact that in the formal scheme we adopt (with the convention of Sec. 3.16), we can use the relations

$$\bigcirc \sim \tau \,, \qquad \bigcirc \bigg|_{\tau=0} = 0 \,. \tag{3.116}$$

The first is valid up to factors $(\mu^2/\tau)^\varepsilon$, which are expanded in ε in the calculation of [...] and cannot compete with the factor τ in (3.116) for $\tau \to 0$. In the rigorous scheme with cutoff Λ (Sec. 1.20), (3.116) are satisfied after a power-law primitive divergence $\sim \Lambda^2$ is subtracted from the graphs, which corresponds to going to the exact value of T_c in $\tau = T - T_c$ (Sec. 3.5).

Subgraphs like (3.116) generate factors of τ that cannot be completely cancelled owing to the IR singularity of the coefficient. In fact, the insertion of n such subgraphs into the line $\Delta(k) = (k^2 + \tau)^{-1}$ gives, on the one hand, a factor τ^n from the subgraphs (all with accuracy τ^ε), and, on the other, generates a power of the line $\Delta^{n+1}(k)$ dangerous at $\tau \to 0$, which upon integration over k gives only a weaker IR singularity $\sim \tau^{1-n}$. From this it follows that the insertion of any number $n \geq 1$ of subgraphs like (3.116) into a single line effectively gives a single factor of τ. The presence of such a factor implies that the graph is irrelevant when operated on by $L_{2,3}$, which select [see (3.96)] only divergences independent of τ. The same arguments are also valid for the individual terms $R'\overline{\gamma}$ in $[L_{2,3}\overline{\gamma}] = [K_{2,3}R'\overline{\gamma}]$. Contractions in them can lead to the appearance of the subgraphs (3.116) in some terms $R'\overline{\gamma}$, even if there were none in the original graph $\overline{\gamma}$. Such terms $R'\overline{\gamma}$ are unimportant when operated on by $K_{2,3}$ owing to the presence of the factor τ in them. This leads to the following rule.

Rule 3. In calculating the quantities $[L_2\overline{\Gamma}_2]$ and $[L_3\overline{\Gamma}_4]$ the graphs $\overline{\gamma}_{2,4}$ with subgraphs like (3.116) and individual terms $R'\overline{\gamma}_{2,4}$ with such subgraphs need not be taken into account, as they do not contribute to the objects calculated.

In $[L_3\partial_\tau\overline{\Gamma}_2]$ and $[L_3\partial_\tau^2\overline{\Gamma}_0]$ up to differentiations with respect to τ it is necessary to include all graphs $\overline{\gamma}_{0,2}$, including ones containing the subgraphs (3.116), since the latter can vanish upon differentiation. Rule 3 can be used for these objects only after differentiating with respect to τ according to the rule (3.112) and subsequent identification of (3.115) of the resulting graphs with vertex graphs. In addition, Rule 3 can also be justified using the second equation in (3.104): one of the factors LA or LB in this case will necessarily be zero.

Usually when calculating massless integrals in dimensional regularization one simply postulates the rule

$$I \equiv \int dk \cdot k^{-2\alpha} = 0 \qquad (dk \equiv d^d k). \tag{3.117}$$

Strictly speaking, objects like (3.117) only appear because of carelessness, for example, when divergences that are powers of Λ are ignored in dimensional regularization with small parameter ε, i.e., when the formal scheme (Sec. 1.20) is made absolute. When done rigorously, integrals like (3.117) are always somehow regularized (by a cutoff, masses, and so on), and their exact meaning is then clear. A typical example is the subgraphs like (3.116) discussed above. We understand them as objects with an auxiliary R operator removing divergences which are powers of Λ, and subsequently take the limit $\Lambda \to \infty$. In the absence of Λ, positive dimension of an integral like (3.116) can be realized only as a positive power of τ, and so the integral vanishes for $\tau \to 0$. This is the essence of the statement (3.117) for the integral with essentially positive dimension $d_I = d - 2\alpha > 0$. However, this justification of the rule (3.117) is inapplicable to the case $d_I < 0$. The appearance of such integrals indicates unsatisfactory treatment of the IR divergences, which must then be dealt with specially. It is also impossible to blindly use the postulate (3.117) for integrals with nearly zero (of order ε) dimension. For example, this could lead to the incorrect conclusion that in the massless theory the vertex loop (3.87b) has no primitive divergence, since at zero external momenta it corresponds to an integral like (3.117).

We conclude by noting that the group that performed the very difficult 5-loop calculations in the φ^4 model developed a special R^*-operator technique [88] which significantly extends the possibilities of this computational method. Their technique makes it possible to avoid the restriction, stated in Rule 1, on the choice of momentum flow forbidding IR-dangerous flows (leading to IR singularities). As mentioned above, IR divergences are important because they also generate poles in ε which mix with the calculated UV poles. In the R^* technique, first a certain infrared R operator is applied to a graph in order to eliminate all the IR poles in ε without affecting the UV poles. When this IR operator is used the flow can be chosen completely arbitrarily, without any danger from the (eliminated) IR divergences. Complete freedom of choice of the momentum flow is very useful in calculating the UV poles, since some graphs are integrated explicitly only with such an IR-dangerous flow (which is thus forbidden by Rule 1).

The technique of the R^* operation is quite complicated, and its effectiveness is actually manifested only in multiloop (4 and 5 loops in the φ^4 model) calculations. We shall therefore not use it in the three-loop calculation of Sec. 3.21, and shall not discuss it in more detail, referring the interested reader to [88] and references therein.

3.20 The constants Z in three-loop order in the MS scheme for the $O_n \varphi^4$ model

Let us begin by presenting tables of the results.

In Table 3.1 we give all the graphs $\gamma_{2,4}$ through three-loop order needed to calculate (3.107) and (3.114) which do not have the subgraphs (3.116) (Rule 3 in Sec. 3.19). Altogether there are 13 such graphs; for convenience, they are numbered in Table 3.1 and the same numbering is used in Table 3.2. The constants $Z_{1,2,3}$ are calculated from the expression

$$Z = 1 + \sum_\gamma c_\gamma r_\gamma [L_a \overline{\gamma}], \quad [L_a \overline{\gamma}] = (-u)^l \xi_\gamma, \tag{3.118}$$

and similarly for Z_0 without the unit term. For Z_1 the summation in (3.118) runs over self-energy graphs $\gamma = \gamma_2$ (there are only two, Nos. 2 and 5), and then in (3.118) $L_a = L_2$. For $Z_{0,2,3}$ it runs over the vertex graphs $\gamma = \gamma_4$, and then $L_a = L_3$. The quantity c_γ in (3.118) is the symmetry coefficient of the graph γ in the calculated object, namely [see (3.107) and (3.114)], c_γ in Z_1 is the coefficient for a given $\gamma = \gamma_2$ in Γ_2 (positive); in Z_2 it is the coefficient for a given $\gamma = \gamma_4$ in $\partial_\tau \Gamma_2$ [negative owing to the minus sign in (3.112)]; in Z_3 it is the coefficient for a given $\gamma = \gamma_4$ in Γ_4 with the opposite sign (negative); in Z_0 it is the coefficient for a given $\gamma = \gamma_4$ in $\partial_\tau^2 \Gamma_0$ (positive), and here the graphs in $\partial_\tau \Gamma_2$ and $\partial_\tau^2 \Gamma_0$ are identified with vertex graphs according to the rule (3.115). All these c_γ are easily determined from the known (Ch. 2) symmetry coefficients of the graphs in the functions Γ_n and the differentiation rule (3.112), and Rule 3 of Sec. 3.19 must be taken into account in selecting the graphs.

The values of r_γ in (3.118), given in Table 3.2 for all thirteen graphs in Table 3.1, are additional factors from the index structures of the O_n-symmetric model (Sec. 2.10). They are all normalized by the condition $r_\gamma(n = 1) = 1$ and are expressed in terms of the simple factors $r_i(n)$ given in Table 3.3. Let us briefly explain how the r_γ are calculated. Each graph γ in $\Gamma_{0,2,4}$ is associated with an index graph with lines and vertices (2.80) according to the rules of Sec. 2.10. After symmetrizing over external indices (there are none in the graphs γ_0, two in γ_2, four in γ_4, and symmetrization is actually needed only in γ_4), the index graph becomes proportional to the corresponding bare structure (2.80), with the desired r_γ as the proportionality factor. The

Table 3.1 Contributions of graphs to the renormalization constants Z in the MS scheme for the simple φ^4 model with basic action $S_{\mathrm{B}} = -\frac{1}{2}(\partial\varphi)^2 - \frac{1}{2}\tau\varphi^2 - \frac{g}{24}\mu^{2\varepsilon}\varphi^4$ in dimension $d = 4 - 2\varepsilon$, $u \equiv g/16\pi^2$.

No.	Graph γ	$[L_a\bar{\gamma}] = (-u)^l \times \xi_\gamma$		Symmetry coefficient c_γ of graph γ in constants			
		$(-u)^l$	ξ_γ	Z_0	Z_1	Z_2	Z_3
1		$-u$	$1/\varepsilon$	0	0	$-1/2$	$-3/2$
2		u^2	$-1/4\varepsilon$	0	$1/6$	0	0
3		u^2	$-1/\varepsilon^2$	$1/4$	0	$-1/4$	$-3/4$
4		u^2	$-(1-\varepsilon)/2\varepsilon^2$	0	0	$-1/2$	-3
5		$-u^3$	$(1-\varepsilon/2)/6\varepsilon^2$	0	$1/4$	0	0
6		$-u^3$	$1/\varepsilon^3$	$1/8$	0	$-1/8$	$-3/8$
7		$-u^3$	$(1-\varepsilon)/2\varepsilon^3$	0	0	$-1/4$	$-3/2$
8		$-u^3$	$(1-9\varepsilon/4)/6\varepsilon^2$	$1/6$	0	$-1/6$	$-1/2$
9		$-u^3$	$(1-\varepsilon-\varepsilon^2)/3\varepsilon^3$	0	0	$-1/4$	$-3/2$
10		$-u^3$	$(1-\varepsilon-\varepsilon^2)/3\varepsilon^3$	0	0	0	$-3/2$
11		$-u^3$	$(1-3\varepsilon+4\varepsilon^2)/6\varepsilon^3$	0	0	-1	-6
12		$-u^3$	$2\zeta(3)/\varepsilon$	0	0	0	-1
13		$-u^3$	$(1-2\varepsilon+\varepsilon^2)/3\varepsilon^3$	$1/4$	0	$-1/4$	$-3/4$

Table 3.2 Additional structure factors r_γ of the graphs γ from Table 3.1 in the $O_n \, \varphi^4$ model.

Graph No.	Factor r_γ in constants			
	Z_0	Z_1	Z_2	Z_3
1	$-$	$-$	r_1	r_2
2	$-$	r_1	$-$	$-$
3	nr_1	$-$	r_1^2	r_4
4	$-$	$-$	r_1	r_3
5	$-$	$r_1 r_2$	$-$	$-$
6	nr_1^2	$-$	r_1^3	r_7
7	$-$	$-$	r_1^2	r_5
8	nr_1	$-$	r_1^2	$r_1 r_2$
9	$-$	$-$	$r_1 r_2$	r_5
10	$-$	$-$	$-$	r_6
11	$-$	$-$	$r_1 r_2$	r_6
12	$-$	$-$	$-$	r_3
13	nr_1	$-$	r_1^2	r_5

operator ∂_τ does not affect the index structure, and so the factors r_γ for the graphs $\partial_\tau \Gamma_2$ are calculated using the graphs Γ_2, and those for $\partial_\tau^2 \Gamma_0$ for the graphs Γ_0.

The objects $[L_a \bar{\gamma}]$ involve graphs $\bar{\gamma}_{2,4}$ of normalized functions, having one factor $-g_B = -g\mu^{2\varepsilon}$ per loop (Sec. 3.17). It is therefore possible to isolate the factor $(-g)^l$ from $[L_a \bar{\gamma}]$, where l is the number of loops in γ. Experience with calculations shows that in this model each factor of g is naturally accompanied by the coefficient $1/16\pi^2$. Therefore, in (3.118) and in Table 3.1 a factor $(-u)^l$ with $u \equiv g/16\pi^2$ is isolated from $[L_a \bar{\gamma}]$. The remaining expression is denoted by ξ_γ and is given in a separate column of Table 3.1. By definition, ξ_γ is the usual quantity $[L_a \bar{\gamma}]$ for graphs γ normalized by the condition of a factor of $16\pi^2 \mu^{2\varepsilon}$ per loop. The calculation of these quantities is the most difficult part of the problem and is explained in the next section.

Table 3.3 Values of the factors $r_i(n)$ in Table 3.2.

i	$r_i(n)$	i	$r_i(n)$
1	$(n+2)/3$	5	$(3n^2 + 22n + 56)/81$
2	$(n+8)/9$	6	$(n^2 + 20n + 60)/81$
3	$(5n+22)/27$	7	$(n^3 + 8n^2 + 24n + 48)/81$
4	$(n^2 + 6n + 20)/27$		

3.21 Technique for calculating the ξ_γ

In Table 3.1 there are only 13 graphs. In this section we shall denote them as γ_i, and the corresponding quantities ξ_γ in the counterterms $[L_a\overline{\gamma}] = (-u)^l \xi_\gamma$ will be denoted as ξ_i, where $i = 1, ..., 13$ is the number of the graph. The line over γ will be dropped, the rule of a factor of $16\pi^2\mu^{2\varepsilon}$ per loop being understood. Depending on the context, γ is understood either as a graph, or as the corresponding analytic expression. Earlier we used a subscript on γ to denote the type of graph (γ_{2n} were the graphs of Γ_{2n}), and now, to avoid confusion, we shall speak of the graphs of Γ_2 (or Γ_4) and write $\gamma \in \Gamma_2$ (or $\in \Gamma_4$). In this notation $\gamma_{2,5} \in \Gamma_2$, and all the others are graphs of Γ_4. For the former ξ is the divergent part (the poles in ε) of the coefficient of p^2 in $R'\gamma$, and for the latter it is simply the pole part of $R'\gamma$. All the ξ_i are independent of the external momenta and the masses of internal lines, and the choice of these quantities in the calculations is a matter of convenience. We shall calculate all the ξ_i except for ξ_{13} in the massless theory. The quantities $R'\gamma$ for all γ are easily found from the rules in Sec. 3.8 taking into account (3.100) and Rule 3 in Sec. 3.19:

$$
\begin{aligned}
R'\gamma_1 &= \gamma_1, \\
R'\gamma_2 &= \gamma_2, \\
R'\gamma_3 &= \gamma_3 - 2\xi_1\gamma_1, \\
R'\gamma_4 &= \gamma_4 - \xi_1\gamma_1, \\
R'\gamma_5 &= \gamma_5 - 2\xi_1\gamma_2, \\
R'\gamma_6 &= \gamma_6 - 3\xi_1\gamma_3 - 2\xi_3\gamma_1 + \xi_1^2\gamma_1, \\
R'\gamma_7 &= \gamma_7 - \xi_1(\gamma_3 + \gamma_4) - \xi_4\gamma_1 + \xi_1^2\gamma_1, \\
R'\gamma_8 &= \gamma_8 - \xi_2\gamma_1, \\
R'\gamma_9 &= \gamma_9 - 2\xi_1\gamma_4 - \xi_3\gamma_1, \\
R'\gamma_{10} &= \gamma_{10} - 2\xi_1\gamma_4 + \xi_1^2\gamma_1, \\
R'\gamma_{11} &= \gamma_{11} - \xi_1\gamma_4 - \xi_4\gamma_1, \\
R'\gamma_{12} &= \gamma_{12}, \\
R'\gamma_{13} &= \gamma_{13} - \xi_1\gamma_3 - 2\xi_4\gamma_1.
\end{aligned}
\tag{3.119}
$$

Let us explain $R'\gamma_8$. For $\tau = 0$ the self-energy counterterm (3.100) contains only the second term. The cross on it corresponds to a factor of k^2, which cancels the single massless propagator $1/k^2$, transforming the graph γ_8 with insertion of the counterterm into the graph γ_1.

The calculation of the γ_i in (3.119) as functions of the masses and external momenta (henceforth referred to briefly as calculation of the graphs) is the most difficult part of the entire procedure. In general, there is no guarantee that a given graph γ can be completely calculated. The expressions in Sec. 3.15 guarantee only that γ can be represented as a finite-fold integral over the auxiliary parameters. This problem is not academic. Even the relatively simple graph γ_2 with massive lines and $p \neq 0$ can only be reduced to a single integral that cannot be computed analytically. Our problem is simplified by the fact that we can arbitrarily choose the masses of the lines and the configuration of the external momenta so as to maximally simplify the calculation of γ. As a rule, massless graphs are easier to calculate than massive ones, and so everywhere possible we shall work with the former. They necessarily contain external momenta, and it is desirable that their configuration in the vertex graphs be maximally simple, otherwise γ cannot be calculated. It will be shown below that all the graphs in Table 3.1 except for γ_{13} can be completely calculated for zero masses and simple flow of a single momentum p into one of the external vertices of the graph and out of another. Rule 1 of Sec. 3.19 must be observed in choosing the momentum flow. It forbids, for example, "vertical" momentum flow (from the lower right-hand external vertex to the upper right-hand one) in the massless graphs $\gamma_{4,7,9}$, but allows "horizontal" flow from the left-hand external vertex to any of the right-hand ones. A suitable choice of flow exists for all our graphs and will be indicated during the calculation.

It is convenient to perform the calculations in coordinate space and to transform to momentum space using (2.56) only in the final step. For simple flow, (2.56) involves the factor $\exp[ip(x'' - x')]$, where x' is the coordinate of the vertex to which the momentum p flows in and x'' is the coordinate of the vertex from which the momentum p flows out. We shall assume that x'' is the eliminated coordinate in (2.56) ("without one" will mean "without x''"); the answer is independent of it, and so we can simply take $x'' = 0$. Then the desired γ_s is written as

$$\gamma_s \equiv \gamma_s(p) = \int dx \tilde{\gamma}_s(x) \exp(-ipx), \qquad (3.120)$$

where $\tilde{\gamma}_s(x)$ is the coordinate graph with two fixed external vertices with coordinates $x'' = 0$, $x' = x$, and the coordinates of all the other vertices of $\tilde{\gamma}_s(x)$ are integrated over. We stress the fact that in $\tilde{\gamma}_s(x)$ the external vertices of the graphs of Γ_4 in which there is no incoming or outgoing momentum p do not differ at all from the internal integration vertices, i.e., all graphs with simple flow are similar to the self-energy graphs Γ_2. We shall first calculate the quantities $\tilde{\gamma}_s(x)$, and then use them to find the desired γ_s from (3.120).

According to (3.72), the massless propagator $\Delta(k) = k^{-2}$ in dimension d corresponds in coordinate space to the expression

$$\Delta(x', x'') = a|x' - x''|^{-2\alpha} \equiv a \overset{x'}{\underset{\alpha}{\bullet\!\!-\!\!-\!\!-\!\!\bullet}}\overset{x''}{} ,$$

$$a = H(1)/4\pi^{d/2}, \qquad \alpha = d/2 - 1. \tag{3.121}$$

We use the notation of (3.68) everywhere. For $d = 4 - 2\varepsilon$ we have

$$\alpha = 1 - \varepsilon, \qquad a = H(1)/4\pi^{2-\varepsilon} = \|1 - \varepsilon\|/4\pi^{2-\varepsilon}. \tag{3.122}$$

In this section the letters α and a will denote only the quantities in (3.122). In calculating massless graphs in coordinate space, it is convenient to represent the factor $|x' - x''|^{-2\lambda}$ by a line with index λ joining the points with coordinates x' and x'' (the latter will usually not be shown explicitly). In this notation

$$\tag{3.123}$$

For power-law lines it is also easy to calculate chains with integration over the coordinates of the internal vertices for fixed external ones. In coordinate space the chains correspond to contractions of power-law lines, while in momentum space they become simple products. All these operations do not take us outside the class of power-law functions and are easily performed using the expressions for the Fourier transform (3.72). Passing to momentum space in each link of the chain, multiplying the resulting Fourier transforms, and returning to coordinate space, we easily obtain the expressions

$$\underset{\lambda_1 \quad \lambda_2}{\bullet\!\bullet\!\bullet} = \underset{\lambda_1 + \lambda_2 \text{ - } d/2}{\bullet\!-\!\bullet} \cdot \pi^{d/2} H(\lambda_1, \lambda_2, d - \lambda_1 - \lambda_2),$$

$$\tag{3.124}$$

$$\underset{\lambda_1 \quad \lambda_2 \quad \lambda_3}{\bullet\!\bullet\!\bullet\!\bullet} = \underset{\lambda_1 + \lambda_2 + \lambda_3 \text{ - } d}{\bullet\!-\!\bullet} \cdot \pi^{d} H(\lambda_1, \lambda_2, \lambda_3, 3d/2 - \lambda_1 - \lambda_2 - \lambda_3),$$

which are valid for arbitrary dimension d and arbitrary indices λ_1, λ_2, ... of the links of the chain.

Most of our graphs with simple flow will reduce to combinations of chains and loops. The quantities $\tilde{\gamma}_s(x)$ for them are easily calculated using (3.123) and (3.124), and up to a known factor reduce to a simple line with index λ. The final answer (3.120) is calculated as the Fourier transform (3.72) of the resulting power-law line. This operation is also conveniently represented graphically as

$$\underset{\lambda}{\bullet\!-\!\bullet} \underset{\text{Fourier}}{\longrightarrow} \pi^{d/2} H(\lambda)(2/p)^{d-2\lambda}, \tag{3.125}$$

where p is the external momentum flow. We immediately note that the index λ of the resulting line in (3.125) is easily found by power counting. It is known from the definitions that the quantity $\overline{\gamma}(p)$ for a graph of $\overline{\Gamma}_{2n}$ has dimension $4 - 2n$ and one factor of $16\pi^2\mu^{2\varepsilon}$ per loop, i.e., $\overline{\gamma} \sim p^{4-2n}(\mu/p)^{2\varepsilon l}$, where l is the number of loops in γ. This determines the power of the momentum on the right-hand side of (3.125):

$$2\lambda - d = 4 - 2n - 2\varepsilon l, \quad \lambda \equiv \lambda_{nl} = 4 - n - \varepsilon(l + 1). \tag{3.126}$$

Turning now to the actual calculation of the γ_s, we first collect all the simple factors—the powers of 2, π, p, and μ. A graph $\gamma \in \Gamma_{2n}$ with l loops contains $N = 2l + n - 2$ internal lines and $v = l + n - 1$ vertices. The factor we are interested in is made up of the following: (1) $16\pi^2\mu^{2\varepsilon}$ for each loop; (2) the amplitude factor a from (3.122) for each internal line; (3) a factor of $\pi^{d/2}$ for each integration vertex in expressions like (3.124), used to reduce the original graph to a simple line (the number of such vertices is $v - 2$); (4) the factors from (3.125) with λ from (3.126). Altogether this gives $[16\pi^2\mu^{2\varepsilon}]^l a^N \pi^{(v-1)d/2}(2/p)^{d-2\lambda}$. Expressing N, v, and λ in terms of n and l (see above), we obtain the expression $[16\pi^2\mu^{2\varepsilon}a^2\pi^{d/2}(2/p)^{2\varepsilon}]^l \cdot [4a\pi^{d/2}p^{-2}]^{n-2}$. Substituting $d = 4 - 2\varepsilon$ and a from (3.122) and using the equation $H(\alpha)H(1) = 1$ following from (3.73), we easily reduce the resulting expression to the form

$$t^{\varepsilon l}[p^2 H(\alpha)]^{2-n}, \quad \text{where } t \equiv \|1 - \varepsilon\|^{2/\varepsilon}(4\pi\mu^2/p^2). \tag{3.127}$$

It is important that $\ln t$ does not have any poles in ε. This follows from the representation (3.69) for the gamma function $\|1-\varepsilon\|$. The factor t^ε containing a fractional power of the momentum p drops out of the final expressions for ξ (see below for details), along with the coefficient $\|1 - \varepsilon\|^2$ introduced into t^ε, and so there is no point in isolating it.

Isolating from the result expected for the graphs $\gamma_s \in \Gamma_{2n}$ with l loops the coefficient (3.127) and the factor of H from (3.125) with λ from (3.126), we obtain the representation

$$\gamma_s(p) = t^{\varepsilon l}[p^2 H(\alpha)]^{2-n} H(\lambda_{nl})q_s. \tag{3.128}$$

The remaining unknown q_s is, by definition, the product of all the factors of H appearing in the reduction of the graph γ_s to a simple line in coordinate space.

Let us now turn to the calculation of the q_s. For most of our graphs they are easily calculated using (3.123) and (3.124). The first graphs $\gamma_{1,2}$ are reduced using (3.123) to a simple line without integrations, and so for them $q_1 = q_2 = 1$. We shall not discuss the next graph γ_3 or $\gamma_{6,7}$, since they reduce to products of the former: $\gamma_3 = \gamma_1^2$, $\gamma_6 = \gamma_1^3$, and $\gamma_7 = \gamma_1\gamma_4$. For them $\xi_3 = -\xi_1^2$, $\xi_6 = \xi_1^3$, and $\xi_7 = -\xi_1\xi_4$ owing to the second relation in (3.104).

For γ_4 we have

$$\text{[diagram]} \longrightarrow \text{[diagram]}, \qquad q_4 = H(\alpha, 2\alpha, d - 3\alpha). \tag{3.129}$$

The black dots show the points into and out of which the external momentum flows (the selected flow always satisfies the requirements of Rule 1 in Sec. 3.19). All the lines in the original graph are ordinary propagators (3.121), and we shall not keep track of the amplitude factors as they are all already included in the coefficient of q_s in (3.128). We are interested only in the factors of H appearing in the reduction to a simple line and making up q_s. The loop in the original graph reduces to a line with index 2α according to the rule (3.123), and factors of H arise only from integration of the upper chain using the rule (3.124), which gives the coefficient q_4 in (3.129) [the notation of (3.68) is used everywhere].

Similarly, using (3.123) and (3.124) we find

$$\gamma_5 = \text{[diagram]} \longrightarrow \text{[diagram]} \;, \quad q_5 = H(2\alpha, 2\alpha, d - 4\alpha);$$

$$\gamma_8 = \text{[diagram]} \longrightarrow \text{[diagram]} \;, \quad q_8 = H(\alpha, \alpha, 3\alpha, 3d/2 - 5\alpha);$$

$$\gamma_9 = \text{[diagram]} \longrightarrow \text{[diagram]} \;, \quad q_9 = H(\alpha, 2\alpha, 2\alpha, 3d/2 - 5\alpha);$$

$$\gamma_{10} = \text{[diagram]} \longrightarrow \text{[diagram]} \;, \quad q_{10} = q_9.$$

(3.130)

The reduction of the graph γ_{11} to a simple line requires double integration of chains. The transformation scheme is

$$\gamma_{11} = \text{[diagram]} \longrightarrow \text{[diagram]} \longrightarrow \text{[diagram]} \longrightarrow \text{[diagram]}, \tag{3.131}$$

and it corresponds to the factor

$$q_{11} = H(\alpha, 2\alpha, d - 3\alpha, \alpha, 4\alpha - d/2, 3d/2 - 5\alpha). \tag{3.132}$$

The first three arguments of H originate from integration of the first chain with link indices α, 2α, and the last three from integration of the upper chain in the last graph in (3.131).

For the remaining graphs we have

$$\gamma_{12} = \text{[diagram]} \longrightarrow \text{[diagram]} \;; \quad \gamma_{13} = \text{[diagram]} \longrightarrow \text{[diagram]} . \tag{3.133}$$

For the general graph of this type, power counting gives

$$\text{[diagram]} \quad = \quad \frac{\bullet \!-\!\!-\!\!-\!\!-\! \bullet}{\sum \alpha_i - d} \; \Pi(d; \alpha_1, ..., \alpha_5), \tag{3.134}$$

where Π is a numerical factor depending on the spatial dimension d and the indices of all five edges α_i. For arbitrary α_i the coefficient Π cannot be expressed explicitly in terms of a finite number of gamma functions and their derivatives. However, in some special cases this is possible, and the class of such "completely integrable" graphs is quite large [89]. In particular, the graph of (3.134) is integrable if all the edges of one of the triangles (left or right) have index $\alpha = d/2 - 1$, and the two other indices β and γ are arbitrary:

$$\text{[diagram]} \quad \sim \quad \Pi(d; \alpha, \beta, \gamma, \alpha, \alpha) \tag{3.135}$$

$$= \frac{\pi^d H(d-2)}{\|\alpha\|} \left\{ \frac{H(\beta, 2-\beta)}{(\gamma-1)(2-\beta-\gamma)} + \frac{H(\gamma, 2-\gamma)}{(\beta-1)(2-\beta-\gamma)} + \frac{H(\beta+\gamma-1, 3-\beta-\gamma)}{(\beta-1)(\gamma-1)} \right\}.$$

This expression was first obtained in [90] by summing infinite series of Gegenbauer polynomials; a simple derivation is given in [89] (see Sec. 4.36 for more details). The graph γ_{12} in (3.133) is a special case of (3.135) with $\beta = \gamma = \alpha \equiv d/2 - 1$. Substitution of these values of β, γ, and $d = 4 - 2\varepsilon$ into (3.135) gives [π^d is included in the coefficient of q in (3.128)]

$$q_{12} = \frac{H(2-2\varepsilon)}{\varepsilon^2 \|1-\varepsilon\|} \left[H(1-2\varepsilon, 1+2\varepsilon) - H(1-\varepsilon, 1+\varepsilon) \right]. \tag{3.136}$$

Expressing all the H here in terms of the gamma function (see Table 3.4 below) and using the expansion (3.69) for them, it is easily checked that for $\varepsilon \to 0$, (3.136) is finite owing to mutual cancellations inside the square brackets in (3.136) of the contributions of order 1, ε, and ε^2:

$$q_{12} = -3\psi''(1) + O(\varepsilon) = 6\xi(3) + O(\varepsilon). \tag{3.137}$$

The corrections of order ε and higher in (3.137) are not needed to calculate ξ_{12}.

Therefore, we have calculated the quantities q_s in (3.128) for all the graphs γ_s except γ_{13} in the massless theory with simple flow of a single momentum. The graph like (3.134) in γ_{13} has index 2α on the diagonal [see (3.133)] and does not belong to the class of fully integrable graphs, i.e., γ_{13} cannot be calculated by the tricks described above. The massless graph γ_{13} would be elementary to calculate if in (3.133) the horizontal momentum flow were replaced by vertical flow, which, in principle, is allowed in calculating ξ (see Rule 1 in Sec. 3.19). However, this does not help in the present case, because vertical flow is forbidden by the other conditions

Table 3.4 Expressions for $H(z)$ in terms of the gamma function $\|z\| \equiv \Gamma(z)$ for $d = 4 - 2\varepsilon$ $[\alpha = 1 - \varepsilon, \lambda_{nl} = 4 - n - \varepsilon(l+1)]$.

i	z_i	$H(z_i) \equiv \|d/2 - z_i\|/\|z_i\|$
1	$2\epsilon = d - 4\alpha$	$2\epsilon\|2 - 3\epsilon\|/\|1 + 2\epsilon\|$
2	$1 - 2\epsilon$	$\|1 + \epsilon\|/\|1 - 2\epsilon\|$
3	$1 - \epsilon = \alpha$	$1/\|1 - \epsilon\|$
4	$1 + \epsilon = d - 3\alpha$	$\|1 - 2\epsilon\|/\|1 + \epsilon\|$
5	$1 + 2\epsilon = 3d/2 - 5\alpha$	$\|1 - 3\epsilon\|/\|1 + 2\epsilon\|$
6	$2 - 4\epsilon = \lambda_{23} = 6\alpha - d$	$\|1 + 3\epsilon\|/3\epsilon\|2 - 4\epsilon\|$
7	$2 - 3\epsilon = \lambda_{22} = 4\alpha - d/2$	$\|1 + 2\epsilon\|/2\epsilon\|2 - 3\epsilon\|$
8	$2 - 2\epsilon = \lambda_{21} = 2\alpha$	$\|1 + \epsilon\|/\epsilon\|2 - 2\epsilon\|$
9	$3 - 4\epsilon = \lambda_{13} = 5\alpha - d/2$	$-\|1 + 3\epsilon\|/3\epsilon(1 - 3\epsilon)\|3 - 4\epsilon\|$
10	$3 - 3\epsilon = \lambda_{12} = 3\alpha$	$-\|1 + 2\epsilon\|/2\epsilon(1 - 2\epsilon)\|3 - 3\epsilon\|$

of the same rule: for this choice of flow, IR divergences arise in the massless graph γ_{13}.

The terms of the Laurent expansion in ε for the massless graph γ_{13} needed to calculate ξ_{13} can be found using special tricks which work only for the ε expansion. However, it is simpler to leave the diagonal lines of γ_{13} in (3.133) massless and to make the replacement $k^2 \to k^2 + \tau$ with the mass τ in the four external edges (this is sufficient to eliminate the IR divergences in all the terms of $R'\gamma_{13}$), using τ instead of the external momentum as the regularizer. Then the graph γ_{13} at zero external momenta is easily calculated from (3.77). In the end we obtain

$$\gamma_{13}^\tau = t_1^{3\epsilon} \frac{\|1 - \varepsilon\|^2 \|2\varepsilon\|^2 \|\varepsilon\| \|3\varepsilon\|}{\|4\varepsilon\| \|2 - \varepsilon\|}, \qquad t_1 \equiv \frac{4\pi\mu^2}{\tau}, \qquad (3.138)$$

where the index τ recalls that the regularizer is now τ and not the external momentum. According to Rule 1 in Sec. 3.19, all the other graphs γ in $R'\gamma_{13}$ must then be calculated with the same regularizer (ξ does not depend on this choice). In this case, from (3.119) using $\gamma_3 = \gamma_1^2$ we have

$$R'\gamma_{13}^\tau = \gamma_{13}^\tau - 2\xi_4 \gamma_1^\tau - \xi_1(\gamma_1^\tau)^2, \qquad \gamma_1^\tau = t_1^\epsilon \|\varepsilon\|, \qquad (3.139)$$

where γ_1^τ is the graph γ_1 with massive lines and zero external momenta. The expression for it in (3.139) is obtained simply by multiplying the known momentum integral (3.87b) by the factor $16\pi^2\mu^{2\varepsilon}$ corresponding to a single loop (with our normalization).

The final step is to use the known γ_s to calculate the quantities

$$\xi_s = K_2 R' \gamma_s \text{ for } s = 2, 5; \qquad \xi_s = K_3 R' \gamma_s \text{ for other } s. \qquad (3.140)$$

We recall that $\gamma_{2,5}$ are the (self-energy) graphs of Γ_2, and all the other γ are the (vertex) graphs of Γ_4; for the former K_2 is the operator that selects poles in ε in the coefficient of p^2, and for the latter K_3 is the operator that selects poles in ε in the expression itself, i.e., K from (3.34).

The calculation of the quantities in (3.140) is elementary, but can be messy in higher orders. All the factors of H in γ_s are expressed explicitly in terms of the gamma function [see (3.68)], and the expansion (3.69) is used for them. The needed poles in ε are selected in the expression obtained for $R'\gamma_s$. The lower ξ_s enter into the higher $R'\gamma_s$ as coefficients, and so the ξ_s must be calculated successively, from lower to higher. For convenience, the explicit expressions for all the $H(z)$ encountered in (3.128) in terms of gamma functions are given in Table 3.4.

In the functions H we have explicitly isolated all the poles and zeros in ε using relations like $\|\Delta\| = \|1 + \Delta\|/\Delta$, $\| - 1 + \Delta\| = \|\Delta\|/(-1 + \Delta) = \|1 + \Delta\|/\Delta(-1 + \Delta)$.

Let us illustrate the procedure for calculating ξ_s by several examples. For the first graph γ_1 in (3.128) we have $n = 2$, $l = 1$, and $q_1 = 1$, from which $\gamma_1 = t^\varepsilon H(\lambda_{21})$, i.e. (see Table 3.4),

$$\gamma_1 = t^\varepsilon \frac{\|1 + \varepsilon\|}{\varepsilon\|2 - 2\varepsilon\|}, \qquad \xi_1 = K_3\gamma_1 = \frac{1}{\varepsilon}, \qquad (3.141)$$

because the expansions (3.69) of the two gamma functions and t^ε begin with unity, and for the given graph $R' = 1$. The same answer for ξ_1 can be obtained by selecting the poles in ε in (3.139) for γ_1^τ.

For the graph γ_2 in (3.128) we have $n = 1$, $l = 2$, and $q_2 = 1$, and so $\gamma_2 = p^2 t^{2\varepsilon} H(\alpha, \lambda_{12})$. For it also $R' = 1$ (owing to the masslessness of the lines using Rule 3 in Sec. 3.19), and so

$$\xi_2 = K_2\gamma_2 = K_2\left[-p^2 t^{2\varepsilon}\frac{\|1 + 2\varepsilon\|}{2\varepsilon(1 - 2\varepsilon)\|3 - 3\varepsilon\|1 - \varepsilon\|}\right] = -\frac{1}{4\varepsilon}. \qquad (3.142)$$

For the next graph $\gamma_3 = \gamma_1^2$ from (3.119) we have $R'\gamma_3 = \gamma_3 - 2\xi_1\gamma_1$, and substitution of the quantities known from (3.141) gives

$$\xi_3 = K_3 R' \gamma_3 = K_3\left[t^{2\varepsilon}\frac{\|1 + \varepsilon\|^2}{\varepsilon^2\|2 - 2\varepsilon\|^2} - \frac{2}{\varepsilon}t^\varepsilon\frac{\|1 + \varepsilon\|}{\varepsilon\|2 - 2\varepsilon\|}\right]. \qquad (3.143)$$

The operator K_3 selects only poles in ε, and so when the expansions (3.69) are substituted into (3.143), only terms of order 1 and ε in the exponents and the expansions of the exponentials need to be included. The desired ξ_3 is the pole part of the expression [the explicit form of the expansions (3.69) is used for the gamma functions in (3.143)]:

$$\frac{1}{\varepsilon^2} \left\{ \exp\left[\varepsilon \left(2\ln t + 2\psi(1) + 4\psi(2)\right)\right] - 2\exp\left[\varepsilon \left(\ln t + \psi(1) + 2\psi(2)\right)\right] \right\}.$$

The terms linear in ε in the difference of the exponentials cancel (along with the $\ln t$), from which we find $\xi_3 = -1/\varepsilon^2$. This answer could have been obtained directly using the rule (3.104).

For the next graph from (3.128) we have $\gamma_4 = t^{2\varepsilon} H(\lambda_{22}) q_4$, where q_4 is known from (3.129), all the needed H are given in Table 3.4, $R'\gamma_4$ is known from (3.119), and γ_1 and ξ_1 are known from (3.141). In the end we have

$$R'\gamma_4 = t^{2\varepsilon} \frac{\|1 - 2\varepsilon\|\|1 + 2\varepsilon\|}{2\varepsilon^2\|1 - \varepsilon\|\|2 - 2\varepsilon\|\|2 - 3\varepsilon\|} - t^\varepsilon \frac{\|1 + \varepsilon\|}{\varepsilon^2\|2 - 2\varepsilon\|} \cong$$

$$\cong \frac{1}{2\varepsilon^2} \left\{ \exp\left[\varepsilon \left(2\ln t + \psi(1) + 5\psi(2)\right)\right] - 2\exp\left[\varepsilon \left(\ln t + \psi(1) + 2\psi(2)\right)\right] \right\} \cong$$

$$\cong \frac{1}{2\varepsilon^2} \left\{ -1 + \varepsilon[\psi(2) - \psi(1)] \right\} = \frac{1}{2\varepsilon^2}(-1 + \varepsilon).$$

The symbol $\cong$ denotes equalities that are valid only through the pole terms of interest to us. The final expression is the desired ξ_4, and (3.70) for ψ was used to obtain it.

We have therefore found ξ_s for all the 1- and 2-loop graphs. The calculation of the three-loop ξ_s is similar but more complicated, since now third-order poles in ε can occur, and so terms of order ε^2 must also be kept in expansions like (3.69). To illustrate the degree of complexity, let us explicitly calculate ξ_9. From (3.128) and (3.130) we find $\gamma_9 = t^{3\varepsilon} H(\lambda_{23}) q_9$; the expressions for H in terms of gamma functions are given in Table 3.4, and $R'\gamma_9$ is known from (3.119). It also involves γ_1 and γ_4, and the explicit expression for γ_4 is the first term in the above expression for $R'\gamma_4$, while γ_1 is known from (3.141). Substitution of the known quantities gives

$$R'\gamma_9 = \frac{1}{3\varepsilon^3} \left[t^{3\varepsilon} \frac{\|1 + \varepsilon\|^2\|1 - 3\varepsilon\|\|1 + 3\varepsilon\|}{\|2 - 2\varepsilon\|^2\|1 - \varepsilon\|\|1 + 2\varepsilon\|\|2 - 4\varepsilon\|} - \right.$$

$$\left. - 3t^{2\varepsilon} \frac{\|1 - 2\varepsilon\|\|1 + 2\varepsilon\|}{\|1 - \varepsilon\|\|2 - 2\varepsilon\|\|2 - 3\varepsilon\|} + 3t^\varepsilon \frac{\|1 + \varepsilon\|}{\|2 - 2\varepsilon\|} \right]. \tag{3.144}$$

The gamma functions appearing here must be expanded through order ε^2. Using F to denote the expression in square brackets in (3.144), with the required accuracy we have

$$F \cong \exp\{\varepsilon[3\ln t + \psi(1)(2 - 3 + 3 + 1 - 2) + \psi(2)(4 + 4)] +$$

$$+ (\varepsilon^2/2)[\psi'(1)(2 + 9 + 9 - 1 - 4) + \psi'(2)(-8 - 16)]\} -$$

$$-3\exp\{\varepsilon[2\ln t + \psi(1)(-2+2+1) + \psi(2)(2+3)] +$$
$$+(\varepsilon^2/2)[\psi'(1)(4+4-1) + \psi'(2)(-4-9)]\} +$$
$$+3\exp\{\varepsilon[\ln t + \psi(1) + 2\psi(2)] + (\varepsilon^2/2)[\psi'(1) - 4\psi'(2)]\}.$$

Here we have shown explicitly how the coefficients of $\psi(z)$ and $\psi'(z)$ are formed: reading from left to right, first we give the numerator of the fraction in (3.144), then the denominator. Each factor $\|z+n\varepsilon\|$ of the numerator gives a term n in the coefficient of $\psi(z)$ and a term n^2 in the coefficient of $\psi'(z)$. The contributions from the factors in the denominator are taken with the opposite sign, and those from the squared gamma functions are doubled. Collecting like terms and reducing all the arguments of ψ and ψ' to unity using (3.70) for ψ and its analog for ψ', we find

$$F \cong \exp\{\varepsilon(3\ln t + 9\psi(1) + 8) + (\varepsilon^2/2)(-9\psi'(1) + 24)\} -$$
$$-3\exp\{\varepsilon(2\ln t + 6\psi(1) + 5) + (\varepsilon^2/2)(-6\psi'(1) + 13)\} +$$
$$+3\exp\{\varepsilon(\ln t + 3\psi(1) + 2) + (\varepsilon^2/2)(-3\psi'(1) + 4)\} \cong$$
$$\cong 1 + \varepsilon(3\ln t + 9\psi(1) + 8) +$$
$$+(\varepsilon^2/2)[(3\ln t + 9\psi(1) + 8)^2 - 9\psi'(1) + 24] -$$
$$-3\{1 + \varepsilon(2\ln t + 6\psi(1) + 5) +$$
$$+(\varepsilon^2/2)[(2\ln t + 6\psi(1) + 5)^2 - 6\psi'(1) + 13]\} +$$
$$+3\{1 + \varepsilon(\ln t + 3\psi(1) + 2) +$$
$$+(\varepsilon^2/2)[(\ln t + 3\psi(1) + 2)^2 - 3\psi'(1) + 4]\} =$$
$$= 1 - \varepsilon - \varepsilon^2$$

This expression with the coefficient $1/3\varepsilon^3$ from (3.144) is ξ_9. The other three-loop ξ_s are calculated from the known γ_s in exactly the same way. Let us make a few concluding remarks.

1. It follows from our discussion that only the calculation of the actual graph γ is nontrivial. In our approximation (three loops), all graphs in dimension $d = 4 - 2\varepsilon$ can be calculated exactly for a suitable choice of external momenta or internal masses, but this is not true in general. By now, the calculations in the φ^4 model have been carried out through five loops [67], [68]. Many higher graphs cannot be calculated for arbitrary ε, but it is possible to find all the pole terms (in ε) in them needed to calculate the ξ. Even this is not guaranteed above five loops.

2. Verification of the cancellation of quantities like $\ln t$ nonanalytic in the external momentum in divergent contributions amounts to direct verification of the fundamental theorem of the R operator (Statement 2 in Sec. 3.12) for the graphs in question. On the basis of this statement, such contributions can in principle be discarded immediately, but it is safer to keep them. The verification of the cancellations will then be a method of checking the calculations.

3. In the φ^4 model, all contributions containing $\psi(1)$ and $\psi'(1)$ also cancel along with $\ln t$. The first nontrivial quantity of this type remaining in the

results is the second derivative $\psi''(1) \sim \zeta(3)$. The quantities $\zeta(4)$ and $\zeta(5)$ appear in the four-loop approximation, and $\zeta(6)$ and $\zeta(7)$ in the five-loop one.

4. Table 3.1 can be used to calculate the renormalization constants in any model of the φ^4 type with a different index structure (examples will be given in Ch. 4). The graphs γ_s and quantities ξ_s for them remain unchanged; only structure factors like r_γ in Table 3.2 change. They can easily be calculated for any index structure, and, in any case, their calculation always reduces to a programmable algorithm, in contrast to the calculation of the ξ_γ. Therefore, in any model it is desirable to present the results of calculating the renormalization constants Z in the form of a table like Table 3.1, containing data on the contributions of the individual graphs. The results presented in this form can then be used in any model of the same type with arbitrary additional index structure.

3.22   Nonmultiplicativity of the renormalization in analytic regularization

The massless model (3.22) can be regularized by a shift of the exponent in the massless propagator, $k^2 \rightarrow k^{2+2\varepsilon}$, instead of the usual shift of the spatial dimension. This is referred to as analytic regularization. The model is then studied directly in $d = 4$, and the basic action is taken to be $S_B(\varphi) = -\frac{1}{2}\varphi k^{2+2\varepsilon}\varphi - \frac{1}{24}g_B\varphi^4$, where the quadratic term is written symbolically and its exact meaning is clear. The canonical dimensions are read off from the action (Sec. 1.15): $d_\varphi = 1 - \varepsilon$, $d[g_B] = 4\varepsilon$, and so we take $g_B = g\mu^{4\varepsilon}$ with the dimensionless renormalized charge g. For the simple scalar φ^4 model $(n = 1)$ we then have

$$\Gamma_2 = -p^{2+2\varepsilon} + \gamma_2/6 + \gamma_5/4 + ..., \qquad \Gamma_4 = -g\mu^{4\varepsilon} + 3\gamma_1/2 + ..., \qquad (3.145)$$

where γ_i are the graphs of Table 3.1 of Sec. 3.20 with their symmetry coefficients $(n = 1$, and so all the $r_\gamma = 1)$. We shall not change over to the normalized functions (3.91); all the γ_i are ordinary graphs with lines $\Delta(k) = k^{-2-2\varepsilon}$ and factors of $-g_B = -g\mu^{4\varepsilon}$ at the vertices. In coordinate space a line corresponds to the former expression (3.121), but now in it $d = 4$, $\alpha = 1 - \varepsilon$, and $a = H(1+\varepsilon)/\pi^2 4^{1+\varepsilon}$. All the graphs in (3.145) with such lines and simple flow of momentum p are easily calculated using (3.123)–(3.125), and the result in the notation of (3.68) with $u \equiv g/16\pi^2$ is

$$\gamma_1 = gu\mu^{8\varepsilon}p^{-4\varepsilon}H^2(1+\varepsilon)H(2-2\varepsilon),$$

$$\gamma_2 = p^2u^2\mu^{8\varepsilon}p^{-6\varepsilon}H^3(1+\varepsilon)H(3-3\varepsilon),$$

$$\gamma_5 = -p^2u^3\mu^{12\varepsilon}p^{-10\varepsilon}H^5(1+\varepsilon)H(2-2\varepsilon, 2-2\varepsilon, 4\varepsilon, 3-5\varepsilon).$$

Substituting $H(z) = \|2 - z\|/\|z\|$ and selecting the poles of the gamma functions, we find

$$\gamma_1 \cong \frac{gu\mu^{8\varepsilon}}{2\varepsilon p^{4\varepsilon}}, \qquad \gamma_2 \cong -\frac{p^2u^2\mu^{8\varepsilon}}{6\varepsilon p^{6\varepsilon}}, \qquad \gamma_5 \cong \frac{p^2u^3\mu^{12\varepsilon}}{10\varepsilon^2 p^{10\varepsilon}}. \qquad (3.146)$$

These expressions are approximate, because they include only the contributions of the leading poles in ε, which are all we shall need below.

To eliminate the poles in ε in the functions (3.145) we need to construct a suitable R operator, which is equivalent to including the counterterms $L\gamma$ from all superficially divergent graphs and subgraphs γ. A principal of renormalization theory is the requirement of locality (polynomial behavior in p) of all the counterterms, and when this condition is satisfied in MS-type schemes the property (3.90) is guaranteed. When the dimensions of the functions (3.145) are taken into account, locality implies that the counterterms must have the following structure in the variables p and μ: $L\gamma \sim p^2\mu^{2\varepsilon}$ for $\gamma \in \Gamma_2$ and $L\gamma \sim \mu^{4\varepsilon}$ for $\gamma \in \Gamma_4$. Here the renormalization is certainly nonmultiplicative, since the counterterms of Γ_2 differ in structure from the bare term $\sim p^{2+2\varepsilon}$.

If we want to preserve the multiplicativity, we can try to cancel the divergences of $\gamma \in \Gamma_2$ by nonlocal counterterms $\sim p^{2+2\varepsilon}$, while preserving the locality of the vertex counterterms, since also in S_B the vertex is local. We shall show below that this problem cannot be solved, and at the same time explain the meaning of (3.90).

We see from (3.146) that the first graph $\gamma_2 \in \Gamma_2$ has a simple pole in ε, which is equally well eliminated by a local counterterm $L\gamma_2 = -p^2u^2\mu^{2\varepsilon}/6\varepsilon$ or a nonlocal one $L\gamma_2 = -u^2p^{2+2\varepsilon}/6\varepsilon$. Let us now consider the next three-loop graph γ_5. To isolate its superficial divergence it is necessary to first use the R' operator to eliminate the divergences of the two equivalent vertex subgraphs: $R'\gamma_5 = \gamma_5 - 2\xi_1\gamma_2$ as in (3.119). However, now the γ_i are the usual (unnormalized) graphs with factors $-\gamma_\mathrm{B} = -g\mu^{4\varepsilon}$ at the vertices, including at the counterterm vertex (3.100), and so $\xi_1 = K[\gamma_1/(-g_\mathrm{B})]$ with the operator K from (3.34). From (3.146) for γ_1 we then find $\xi_1 = -u/2\varepsilon$ and

$$R'\gamma_5 = p^2u^3F, \quad \text{where } F \cong \frac{\mu^{12\varepsilon}}{10\varepsilon^2p^{10\varepsilon}} - \frac{\mu^{8\varepsilon}}{6\varepsilon^2p^{6\varepsilon}} \tag{3.147}$$

where only contributions of order $1/\varepsilon^2$ are retained.

The quantity F has dimension 2ε. In a scheme with a local counterterm we shall try to cancel its divergence by a quantity like $\mu^{2\varepsilon}$·const, and in a nonlocal scheme by a quantity like $p^{2\varepsilon}$·const. We shall show that the first is possible, and the second is not. In fact, from (3.147) we have $F = \mu^{2\varepsilon}A = p^{2\varepsilon}B$, where $A = t^{10\varepsilon}/10\varepsilon^2 - t^{6\varepsilon}/6\varepsilon^2$ and $B = t^{12\varepsilon}/10\varepsilon^2 - t^{8\varepsilon}/6\varepsilon^2$ with $t \equiv \mu/p$. The difference between A and B is obvious: the powers of t are such that in A the contribution $\varepsilon^{-1}\ln t$ cancels, while in B it does not, i.e., the "divergent part" of A is, as required, a constant independent of the momentum p, while this is not so for B. It then follows that the superficial divergence of γ_5 can be eliminated by a local counterterm $\sim p^2\mu^{2\varepsilon}$ and not by a nonlocal one $\sim p^{2+2\varepsilon}$ owing to the appearance in the divergent part of a contribution with $\ln t = \ln(\mu/p)$ in the second case. We note that the difference between the two renormalization variants is quite subtle, and the fact that one is inapplicable is seen only in the three-loop graph γ_5. In the two-loop approximation, not to speak of the one-loop approximation, we would not notice anything wrong with the procedure of multiplicative renormalization using a nonlocal counterterm. We also note that among all the three-loop graphs of Table 3.1, the defectiveness of a nonlocal counterterm is seen only in the single graph γ_5. In the superficial divergence of the graph γ_8, the contributions $\varepsilon^{-1} \ln t$ are cancelled by both the local and the nonlocal counterterm of the subgraph γ_2.

The general conclusion is that *when constructing counterterms, the excess (above the integer) dimension should always be isolated in the form of the factor μ^δ and not p^δ*, as only in this case does the structure of the superficial divergences in higher orders remain simple and given by the rule (3.90).

3.23   The inclusion of composite operators

Composite operators $F(x; \varphi)$ (the argument x will usually be omitted) and the various Green functions involving them were defined in Secs. 2.13 and 2.14. We recall that for all the generating functionals the definitions are the usual ones but with the form (2.104) added to the interaction $V(\varphi)$. The model with such an addition $aF(\varphi)$ will be termed *extended*. Therefore, the generating functional of the S-matrix Green functions of the fields and operators is (Sec. 2.14)

$$H(\varphi, a) = P \exp[V(\varphi) + aF(\varphi)], \tag{3.148}$$

and for 1-irreducible functions it is the functional

$$\Gamma(\varphi, a) = S_0(\varphi) + \text{1-ir part of } P \exp[V(\varphi) + aF(\varphi)]. \tag{3.149}$$

We have agreed (Sec. 2.13) to call both the random quantity $F(\widehat{\varphi})$ and the corresponding classical functional $F(\varphi)$ composite operators, distinguishing them by the hat. The same functional $F(\varphi)$ can correspond to different random quantities, since upon the replacement $\varphi(x) \to \widehat{\varphi}(x)$ the field $\widehat{\varphi}$ can have a different meaning (unrenormalized, renormalized, or basic). It is determined by the meaning of the action functional $S = S_0 + V$ in the construction [a functional integral or expressions like (3.148) and (3.149)] into which $F(\varphi)$ enters: whatever S is, so is $\widehat{\varphi}$. When studying renormalization according to the scheme of Secs. 3.8–3.12, the original theory is assumed to be the basic theory. For it, $V(\varphi)$ in (3.148) and (3.149) is the basic interaction and $aF(\varphi)$ is the addition (2.104) to it with given operators $F \equiv \{F_i(\varphi)\}$. The corresponding *basic sources* $a \equiv \{a_i\}$ *will always be identified as renormalized sources*, assuming that they are independent finite functional arguments of the renormalized and basic theories. Their bare analogs will be introduced later.

Since all the main statements of renormalization theory are formulated for an arbitrary local interaction, they remain valid also for the extended model with interaction $V + aF$. This observation contains the solution in principle of all the problems which arise in the renormalization of local composite operators, and all of the following amounts to only refinements and explanation.

The first step is to analyze the dimensions. To each local monomial $F(x)$ constructed from the field $\varphi(x)$ and its derivatives we can assign a definite canonical dimension d_F, the sum of the dimensions of the fields and derivatives entering into F (for example, for the φ_4^4 model, $d_\varphi = d_\partial = 1$, and then d_F is the total number of fields and derivatives in F). The dimension of the source $a(x)$ conjugate to a given $F(x)$ in the form (2.104) is determined from the requirement that it be dimensionless, which leads to the standard shadow relation:

$$d_a + d_F = d, \quad d_a \equiv d[a(x)], \quad d_F \equiv d[F(x)]. \tag{3.150}$$

According to the definitions of Sec. 1.16, operators with $d_F \leq d$ are called IR-relevant, while those with $d_F > d$ are irrelevant (accordingly, UV-irrelevant and relevant). Here the question of competition between the contribution of aF and the original interaction V does not arise, i.e., aF is always treated as a perturbation. The dimensions of 1-irreducible coordinate functions $\Gamma_x \equiv \Gamma(x_1...x_n)$ [the coefficients of the expansion of the dimensionless functional (3.149) in the set of sources φ, a] are found from the rule (1.69):

$$d[\Gamma_x] = \sum_\varphi (d - d_\varphi) + \sum_F (d - d_a) = \sum_\varphi (d - d_\varphi) + \sum_F d_F \tag{3.151}$$

with summation over all the fields and operators entering into the given function. All the operators of Sec. 3.8 are defined on functions $\Gamma \equiv \Gamma_p(p_1...p_n)$ [the analogs of Γ_{np} from (3.2)], and their dimensions are found from (3.151) and (2.57):

$$d_\Gamma \equiv d[\Gamma_p] = d - \sum_\varphi d_\varphi - \sum_F (d - d_F). \tag{3.152}$$

The superficially divergent graphs of the extended model are defined in the usual way (Secs. 3.2 and 3.16) according to the values of the dimensions (3.152) in the logarithmic theory, i.e., for $\varepsilon = 0$ in dimensional regularization. It is convenient to use the example of composite operators to explain the difference between the formal and real UV exponents ω (Sec. 3.2): in the logarithmic theory $\omega_{\text{form}} = d_\Gamma$, but the real exponent can be smaller. Let $F = \partial F'$. Then in the graphs of Γ with operator F the exterior derivative ∂ becomes a factor of p (the external momentum), and so the convergence of the momentum integral of the graph must be judged from the dimension of F' and not of F: for $\varepsilon = 0$,

$$\omega_{\text{real}} = d_\Gamma - \text{the number of exterior derivatives } \partial. \tag{3.153}$$

The set of superficially divergent graphs of the extended model consists of all the SD graphs of the original model and the new ones involving the operators F. Let us consider, for example, the φ_4^4 model with a single operator F in the form aF. Then from (3.152) taking into account $d_\varphi = 1$ we have $d_\Gamma = 4 - n_\varphi - (4 - d_F)n_F$, where $n_{\varphi,F}$ are the numbers of factors of φ and F in Γ. It follows from the expression for d_Γ that for the operator

$F = \varphi^2(x)$ with $d_F = 2$, the new SD graphs will be all the 1-irreducible graphs of $\langle \widehat{F} \rangle$, $\langle \widehat{F}\widehat{\varphi}\widehat{\varphi} \rangle$, and $\langle \widehat{F}\widehat{F} \rangle$ (the $\varphi \to -\varphi$ symmetry has been taken into account); for $F = \varphi^3(x)$ with $d_F = 3$, the new SD graphs are the 1-irreducible graphs of $\langle \widehat{F}\widehat{\varphi} \rangle$, $\langle \widehat{F}\widehat{\varphi}\widehat{\varphi}\widehat{\varphi} \rangle$, $\langle \widehat{F}\widehat{F} \rangle$, $\langle \widehat{F}\widehat{F}\widehat{\varphi}\widehat{\varphi} \rangle$, $\langle \widehat{F}\widehat{F}\widehat{F}\widehat{\varphi} \rangle$, and $\langle \widehat{F}\widehat{F}\widehat{F}\widehat{F} \rangle$; for $F = \varphi^4(x)$ or $\varphi(x)\partial^2\varphi(x)$ with $d_F = 4$, the new SD graphs will be all the 1-irreducible graphs with $n_\varphi \le 4$ and any number of $\widehat{F}$, while for $F = \partial^2\varphi^2(x)$ they will be the same as for $F = \varphi^2(x)$ owing to the rule (3.153).

We see from (3.152) that the insertion of an additional operator with $d_F < d$ decreases d_Γ, while that of one with $d_F > d$ increases it, and so the presence of at least one operator with $d_F > d$ in aF leads to nonrenormalizability of the extended model according to the classification of Sec. 3.2. However, this does not hinder eliminating all of its divergences (see Sec. 3.13) by introducing the needed counterterms $L\gamma$ for each SD graph γ of the extended model. Here the renormalization of Green functions with a given finite number of composite operators requires only a finite number of additional counterterms, of course, as long as the original model is renormalizable, which will always be assumed. In addition, for simplicity we shall assume that the extended model is renormalized using dimensional regularization and the MS scheme. Then there will not be any problem with fixing the renormalization arbitrariness even when there are infinitely many counterterms. We note that the condition (3.25) is satisfied automatically in the MS scheme.

3.24 The renormalized composite operator

Let (3.148) and (3.149) be constructions in the basic theory and $V_R(\varphi)$ be the corresponding renormalized interaction (3.66) of the original model. *Definition:* the composite operator $F(\varphi)$ is termed UV-finite if the functional $P[F(\varphi)\exp V_R(\varphi)]$ is finite.

Let us explain this definition.

1. Comparison with (3.148) taking into account the rule for the correspondence between functionals and operators [see the discussion following (3.149)] shows clearly that UV finiteness of the functional indicated in the definition is equivalent to UV finiteness of all the S-matrix Green functions with a single operator $\widehat{F} = F(\widehat{\varphi}_R) \equiv F(\varphi)|_{\varphi = \widehat{\varphi}_R}$ and any number $n \ge 0$ of factors $\widehat{\varphi}_R$, where $\widehat{\varphi}_R = Z_\varphi^{-1}\widehat{\varphi}$ is the renormalized field operator from (1.74).

2. From the definitions of Sec. 2.14 and Statement 2 in Sec. 3.8, it follows that UV finiteness of the system of all the S-matrix Green functions with a single $\widehat{F}$ is equivalent to UV finiteness of any other analogous [i.e., with a single $\widehat{F} = F(\widehat{\varphi}_R)$ and any number of fields $\widehat{\varphi}_R$] system of Green functions — full, connected, or 1-irreducible. Therefore, UV finiteness of $F(\varphi)$ is equivalent to UV finiteness of the functional $\langle\langle F(\varphi) \rangle\rangle$ defined by (2.107a) with $S = S_R$.

3. Since the operators P and ∂ commute (Sec. 2.1), UV finiteness of $F(\varphi)$ implies that of $\partial F(\varphi)$.

4. The operators $F(\varphi) = 1$ and $F(\varphi) = \varphi$ are UV-finite.

5. UV finiteness of Green functions with a single $\widehat{F}$ does not guarantee UV finiteness of functions with two or more $\widehat{F}$.

Let $F(\varphi)$ be the local monomial of interest, $aF(\varphi)$ be the addition with a single given $F(\varphi)$ to the basic interaction $V(\varphi)$ in (3.148), and L be the counterterm operator of the MS scheme associating each SD graph γ of the extended model with a counterterm $L\gamma$. By construction, the corresponding R operator eliminates all the divergences of the extended model, i.e., it ensures the UV finiteness of all Green functions with any number of fields and operators. For the S-matrix functions this implies the UV finiteness of the functional $RP\exp[V + aF]$. Using the general statement (3.40) for the extended model, we obtain

$$R(L)P\exp[V + aF] = P\exp[V + aF - LP\exp(V + aF)]. \tag{3.154}$$

Equating the coefficients of the first power of a, we find

$$R(L)P[F\exp V] = P\{[F - LP(F\exp V)]\exp(V - LP\exp V)\}.$$

The left-hand side is UV-finite by construction, and so the local functional (we recall that the argument x of F is understood)

$$[F(\varphi)]_{\mathrm{R}} \equiv F(\varphi) - LP[F(\varphi)\exp V(\varphi)] \tag{3.155}$$

is a UV-finite composite operator in the sense of the above definition. It is called the *renormalized operator* $F(\varphi)$, and the additions $\Delta F = LP[F\exp V]$ are the *operator counterterms* of $F(\varphi)$.

It follows from (3.155) that $[\partial F(\varphi)]_{\mathrm{R}} = \partial[F(\varphi)]_{\mathrm{R}}$, since the operator ∂ commutes with P by definition and with L owing to the lemma of Sec. 3.11. It is also clear that $[\varphi]_{\mathrm{R}} = \varphi$, since for the operator $F(\varphi) = \varphi(x)$ all the graphs acted on by L in (3.155) are 1-reducible, and $L = 0$ on them according to the condition (3.25). *Warning: the operator $[\widehat{\varphi}]_{\mathrm{R}} = \widehat{\varphi}$ should not be confused with the renormalized field $\widehat{\varphi}_{\mathrm{R}} = Z_\varphi^{-1}\widehat{\varphi}$ from (1.74).*

For $F = 1$ all the graphs of $P[F\exp V]$ are disconnected and $L = 0$ on them, i.e., $[1]_{\mathrm{R}} = 1$. Here it is important that L is defined as an operator on graphs (Sec. 3.8) and not on functionals: the quantities $P[\exp V]$ and $P[1\exp V]$ as functionals are equal, but graphically they are different, since the operator $F = 1$ associates an isolated vertex with the unit factor.

3.25 Renormalization of the action and Green functions of the extended model

Equation (3.154) is obviously valid also for an arbitrary form (2.104) with any set of local monomials $F_i(x; \varphi)$, and the exponent on the right-hand side is then the renormalized interaction (3.66) of the corresponding extended (by the addition of composite operators) model:

$$(V + aF)_{\mathrm{R}} = V + aF - LP\exp(V + aF). \tag{3.156}$$

Its basic action is the functional $S_B = S_0 + V + aF$, and the renormalized action is the functional $S_R = S_0 + (V + aF)_R$. Comparing this with the definition (3.155), we see that the contributions in (3.156) linear in a are grouped to form the renormalized operators (3.155). Therefore, from (3.156) we obtain

$$S_R(\varphi, a) = S_R(\varphi) + \sum_i a_i [F_i(\varphi)]_R -$$
$$-\tfrac{1}{2} \sum_{ik} a_i a_k LP[F_i(\varphi) F_k(\varphi) \exp V(\varphi)] - \dots . \tag{3.157}$$

Here $S_R(\varphi) = S_0(\varphi) + V_R(\varphi)$ is the renormalized action of the original model. The terms linear in a include the addition $aF(\varphi)$ to the basic action and the counterterms from SD graphs with a single F [which transform the original monomials $F_i(\varphi)$ into the corresponding renormalized operators (3.155)], and the higher-order terms of the expansion in a in (3.157) correspond to the counterterms from SD graphs with two and more composite operators F. All the F_i and a_i have one argument x each and integration over it is understood in (3.157). Owing to the general property of locality of the counterterms, all the contributions to (3.157) actually contain only a single integral over x. To renormalize the F themselves, or, equivalently, the Green functions with a single F, only the terms linear in a in (3.157) are needed. To renormalize the Green functions with two F the counterterms quadratic in a are also needed, and so on. Therefore, for a single $F(\varphi)$,

$$R\langle F(\widehat{\varphi})\widehat{\varphi}...\widehat{\varphi}\rangle_B = \langle [F(\widehat{\varphi}_R)]_R \widehat{\varphi}_R...\widehat{\varphi}_R\rangle. \tag{3.158}$$

For the two arbitrary operators (local monomials) $F_1(\varphi)$ and $F_2(\varphi)$ from (3.157) we obtain

$$R\langle F_1(\widehat{\varphi})F_2(\widehat{\varphi})\widehat{\varphi}...\widehat{\varphi}\rangle_B = \langle [F_1(\widehat{\varphi}_R)]_R [F_2(\widehat{\varphi}_R)]_R \widehat{\varphi}_R...\widehat{\varphi}_R\rangle -$$
$$-\langle LP[F_1(\varphi)F_2(\varphi) \exp V(\varphi)]\big|_{\varphi=\widehat{\varphi}_R} \cdot \widehat{\varphi}_R...\widehat{\varphi}_R\rangle \tag{3.159}$$

and so on; all the operators and individual factors of $\widehat{\varphi}$ are understood to have independent arguments x. The subscript B on the left-hand side of (3.158) and (3.159) indicates that the R operator acts on the graphs of the Green functions of the basic theory (not to be confused with the unrenormalized theory).

3.26 Structure of the operator counterterms

In (3.155), L operates on the quantity

$$P[F(\varphi) \exp V(\varphi)] = \delta H(\varphi, a)/\delta a|_{a=0}, \tag{3.160}$$

where H is the functional (3.148) and a and F are understood to have the argument x. Equation (3.160) is interpreted as the generating functional of S-matrix Green functions with a single $F(\widehat{\varphi})$ and any number of factors $\widehat{\varphi}$ in the basic theory (i.e., in this case $\widehat{\varphi} \equiv \widehat{\varphi}_B$), and these functions themselves are the coefficients of the expansion of the functional (3.160) in φ. Only the 1-irreducible graphs of the functional (3.160) contribute inside the counterterm operator L in (3.155) [according to (2.111), the 1-irreducible part of (3.160) is the generating functional of 1-irreducible Green functions with a single F and any number of fields φ]. Let us explain the form of these graphs using two examples for the φ^4 model:

1-ir part of $P\left[\varphi^2(x)\exp V(\varphi)\right] = \varphi^2(x) +$ $+ \frac{1}{2}$ $+$

$$+ \frac{1}{4} \text{(graph)} + \frac{1}{2} \text{(graph)} + \frac{1}{2} \text{(graph)} + \frac{1}{4} \text{(graph)} + \tag{3.161}$$

$$+ \frac{1}{2} \text{(graph)} + \frac{1}{6} \text{(graph)} + \frac{1}{4} \text{(graph)} + \frac{1}{4} \text{(graph)} + \dots ,$$

1-ir part of $P\left[\varphi^3(x)\exp V(\varphi)\right] = \varphi^3(x) + 3\varphi(x)[\text{graphs of }\varphi^2(x)]$

$$\tag{3.162}$$

$$+ \text{(graph)} + \frac{3}{2} \text{(graph)} + \frac{3}{2} \text{(graph)} + \frac{3}{2} \text{(graph)} + \dots .$$

The large black dot denotes the composite-operator vertex. It corresponds to the free argument x and a special vertex factor, which for all the graphs in (3.161) and (3.162) is equal to unity. The other vertices and all the lines correspond to the standard elements of the basic theory, and the tails (wavy lines) to factors of φ connected directly to the vertices. All the 1-irreducible graphs of order 1, g, and g^2 are given with their symmetry coefficients. Some of the graphs in (3.162) are obtained simply by joining an additional tail φ to a composite-operator vertex in the graphs of (3.161) and tripling the coefficient. In (3.162) we explicitly give only the new graphs.

The symmetry coefficients of the graphs of the functional (3.160) can be found, for example, from the recursion formula (2.26), using a graph technique in the intermediate stage like that in Sec. 2.2 with the vertex factors (2.21) and their analogs for the vertex of the composite operator F. When (2.26) is used, the original graph can be taken to be the disconnected graph consisting of isolated points, one of which is associated with the factor $F(\varphi)$, and all the others with $V(\varphi)$. Its symmetry coefficient (2.23) is equal to $1/n!$, where n is the number of V vertices.

The contribution of the 1-irreducible graph γ with n tails φ to the functional (3.160) has the form

$$\int \dots \int dx_1 \dots dx_n \gamma(x; x_1 \dots x_n)\varphi(x_1) \dots \varphi(x_n). \tag{3.163}$$

In transforming to momentum space, we shall assume that independent external momenta p_s (which are symmetrized over) flow out of the graph at the points x_s, while $\sum p_s$ flows into a composite-operator vertex x. The analog of (2.56) will be

$$\gamma_p(p_1...p_n) = \int ... \int dx_1...dx_n \gamma(x; x_1...x_n) \exp\left[i\sum_{s=1}^{n} p_s(x_s - x)\right]. \quad (3.164)$$

The quantity γ_p is constructed directly from the graph using the usual rules for 1-irreducible graphs γ (Sec. 2.7). The only change is the nonstandard vertex factor at the composite-operator vertex, which is determined from the form of F.

Only the SD graphs of the functional (3.160) contribute to the counterterms (3.155). They are found by power counting (Sec. 3.23), and the number of tails φ in the graphs corresponds to the number of factors of $\widehat{\varphi}$ in the Green functions. Therefore, the SD graphs are all the graphs in (3.161) with 0 and 2 tails φ (taking into account the $\varphi \to -\varphi$ symmetry) and all the graphs in (3.162) with 1 and 3 tails (Sec. 3.23). The counterterm operator of the MS scheme is defined in the usual manner (Sec. 3.10) on the functions (3.164): we eliminate the divergences of the SD subgraphs by operating with R', multiply the result by $\mu^{-\delta}$ [see (3.31)] to obtain a quantity with integer dimension, apply the operator (3.34) to it to select the part with poles in ε, and multiply the result by μ^{δ} to restore the correct dimension. The counterterm $L\gamma_p$ obtained in this manner will be a polynomial $P(p_1...p_n)$ in all the independent external momenta and τ (the τ dependence is not shown). In going from γ_p to the coordinate function γ by means of the inverse transformation (3.164), this polynomial is transformed into $P(i\partial_1...i\partial_n)\prod_s \delta(x_s - x)$ with $\partial_s \equiv \partial/\partial x_s$ and then, upon substitution into (3.163), gives the local monomial $P(-i\partial_1...-i\partial_n)\varphi(x_1)...\varphi(x_n)|_{...}$. The ellipsis denotes $x_1 = ... = x_n = x$ after the differentiations are performed. The contribution to the counterterms (3.155) from an individual graph of the functional (3.160) is thereby determined, and then the contributions from all the SD graphs are added up with the required coefficients.

Let us use examples to illustrate the action of the R' operator on several graphs in (3.161) and (3.162) (with the notation of Sec. 3.8):

$$R'\ \text{(graph)} = \text{itself} \ - \ \text{(graph)} \ - \ \text{(graph)}\ ,$$

$$R'\ \text{(graph)} = \text{itself} \ - \ \text{(graph)} \ - \ \text{(graph)}\ , \qquad (3.165)$$

$$R'\ \text{(graph)} = \text{itself} \ - \ \text{(graph)} \ - \ \text{(graph)} \ - \ \text{(graph)}\ .$$

All the operators in Sec. 3.8 act on the coefficients of the tails φ, which play the same role as the external lines in the 1-irreducible graphs γ, i.e., indicating only the points out of which the external momenta flow.

The procedure for constructing the operator counterterms ΔF in (3.155) uniquely determines their general structure in the MS scheme. All the counterterms are local operator monomials, and their coefficients can contain only nonnegative integer powers of τ and factors of μ^δ with $\delta \sim \varepsilon$. By dimension $\Delta F \sim F$, and so ΔF can contain only those monomials which in the logarithmic theory ($\varepsilon = 0$) have dimension not exceeding that of the original operator F. In general, ΔF contains all the monomials allowed by the dimension and symmetry. Let us explain using examples from the φ^4 model ($d = 4 - 2\varepsilon$, $d_\varphi = 1 - \varepsilon$, and ∂^2 is the Laplace operator):

$$\Delta[\varphi^2(x)] = c_1\varphi^2(x) + c_2\tau\mu^{-2\varepsilon},$$

$$\Delta[\varphi^3(x)] = c_1\varphi^3(x) + c_2\tau\mu^{-2\varepsilon}\varphi(x) + c_3\mu^{-2\varepsilon}\partial^2\varphi(x),$$

and

$$\Delta[\varphi^4(x)] = c_1\varphi^4(x) + c_2\tau\mu^{-2\varepsilon}\varphi^2(x) + c_3\mu^{-2\varepsilon}\partial^2\varphi^2(x)$$
$$+ c_4\mu^{-2\varepsilon}\varphi(x)\partial^2\varphi(x) + c_5\tau^2\mu^{-4\varepsilon},$$

where all the c_i are different dimensionless coefficients — sums of expressions like (3.98) from all the SD graphs contributing to a given coefficient. The contribution $\partial\varphi \cdot \partial\varphi$ is absent in the counterterms $\Delta[\varphi^4(x)]$ because it is expressed algebraically in terms of the operators $\partial^2\varphi^2(x)$ and $\varphi(x)\partial^2\varphi(x)$ already included: $\partial^2\varphi^2(x) = 2[\partial\varphi(x) \cdot \partial\varphi(x) + \varphi(x)\partial^2\varphi(x)]$.

The generalization of these rules to other models and subtraction schemes is obvious. Any parameter like the mass and also Λ in cutoff schemes can play the role of τ in the coefficients. The polynomial behavior of the counterterms in τ is a specific feature of primitive subtraction schemes like the MS scheme (Sec. 3.10), and for nonprimitive subtractions like (3.32) the τ dependence is complicated.

3.27 An example of calculating operator counterterms

To illustrate the general technique, let us give the detailed calculation of the counterterms of the scalar $F = \varphi^2(x)$ and tensor $F_{is} = \varphi\partial_i\partial_s\varphi$ local operators in the O_n φ^4 model (both are scalars under the O_n group, i.e., $\varphi^2 \equiv \varphi_a\varphi_a$, and similarly in F_{is}) in the one-loop approximation for F_{is} and the two-loop approximation for φ^2. All the counterterms for the operator φ^2 and almost all those for $\varphi\partial_i\partial_s\varphi$ can actually be found by another method without resorting to graphs (see Secs. 3.30 and 3.32). However, this is not possible for all operators. In general, it is necessary to analyze the graphs, and to show this we shall consider these two operators as the simplest examples. First we give the table of results, and then the explanations.

The renormalized operator (3.155) is calculated using

$$[F]_{\mathrm{R}} = F - \sum_\gamma c_\gamma r_\gamma L\gamma, \tag{3.166}$$

Table 3.5 Contributions of graphs to the counterterms of the O_n-scalar operators $\varphi^2(x)$ and $\varphi(x)\partial_i\partial_s\varphi(x)$ in the O_n φ^4 model ($g_B \equiv g\mu^{2\varepsilon}$, $u \equiv g/16\pi^2$).

No.	Graph γ	c_γ	r_γ	$L\gamma$ for $F = \varphi^2$	$L\gamma$ for $F_{is} = \varphi\partial_i\partial_s\varphi$
1		1	n	$-\dfrac{\tau u}{\varepsilon g_B}$	$-\delta_{is}\dfrac{\tau^2 u}{4\varepsilon g_B}$
2		$\dfrac{1}{2}$	r_1	$-\dfrac{u}{\varepsilon}\varphi^2(x)$	$\dfrac{u}{12\epsilon}[(\partial^2 - 6\tau)\delta_{is} - 4\partial_i\partial_s]\,\varphi^2(x)$
3		$\dfrac{1}{4}$	r_2	0	$-\delta_{is}\dfrac{ug_B}{4\varepsilon}\varphi^4(x)$
4		$\dfrac{1}{2}$	nr_1	$-\dfrac{\tau u^2}{\varepsilon^2 g_B}$	—
5		$\dfrac{1}{4}$	r_1^2	$-\dfrac{u^2}{\varepsilon^2}\varphi^2(x)$	—
6		$\dfrac{1}{2}$	r_1^2	0	—
7		$\dfrac{1}{2}$	r_1	$-\dfrac{u^2(1-\varepsilon)}{2\varepsilon^2}\,\varphi^2(x)$	—

where $L\gamma$ is the counterterm of an individual graph γ, c_γ is its symmetry coefficient in the functional (3.160), and r_γ is an additional structure factor of the O_n group, expressed in terms of the r_i given in Table 3.3 of Sec. 3.20. The graphs γ themselves and the coefficients c_γ for our operators are the same as in (3.161), and the only difference between the two operators is in the factors at the F vertex: for φ^2 this factor is unity, and for $\varphi\partial\partial\varphi$ it is determined by the two derivatives at the vertex and will be given below.

The counterterm $L\gamma$ for the operator φ^2 is nonzero only for graphs γ with 0 and 2 tails φ, and for the operator $\varphi\partial\partial\varphi$ it is nonzero only for graphs with 0, 2, and 4 tails. In Table 3.5 we give all the one-loop graphs with 0, 2, and 4 tails and all the two-loop graphs with 0 and 2 tails. The first column contains the number of the graph, and in what follows it is used as the subscript on γ. In this table we give all the quantities needed in (3.166) to find $[\varphi^2]_R$ through two loops and $[\varphi\partial\partial\varphi]_R$ through one loop. The two-loop calculation of $[\varphi\partial\partial\varphi]_R$ would require 6 more two-loop graphs with four tails φ, which are not given in Table 3.5.

Let us explain the procedure for calculating the counterterms $L\gamma$. It is not complicated for φ^2: $L\gamma_{3,6} = 0$ from power counting [for γ_6 owing to the presence of the factor of τ from a subgraph like (3.116)], and the coefficient of $\varphi^2(x)$ in $L\gamma_{2,5,7}$ can simply be taken from Table 3.1 in Sec. 3.20. The counterterms of the graphs $\gamma_{1,4}$ are proportional to τ owing to subgraphs of the type (3.116). They must be eliminated by differentiation with respect to τ (Sec. 3.19), which again leads to the graphs of Table 3.1 but missing a factor of $-g_B$, since there was one per loop. The quantity $\partial_\tau L\gamma = L\partial_\tau\gamma$ (see the Lemma of Sec. 3.11) for these graphs is therefore equal to the corresponding result in Table 3.1 divided by $-g_B$.

Let us turn to the tensor operator $\varphi\partial_i\partial_s\varphi$, restricting ourselves to the one-loop graphs $\gamma_{1,2,3}$. Now we shall use γ to mean the graphs of the functional (3.160), and denote the corresponding function (3.164) by $\tilde\gamma$. In our case the graphs of $\tilde\gamma$ have the following general structure:

$$\tilde\gamma = \quad\text{(3.167)}$$

The external momenta $p_1...p_n$ correspond to tails φ, $p = \sum p_i$ is the momentum flowing into the composite-operator vertex, and k is the circulating integrated momentum. In practice, it is necessary to calculate the counterterms $L\tilde\gamma$, which are polynomials in the external momenta and τ. The degree of the polynomial is equal to the dimension $d[\tilde\gamma]$ for $\varepsilon = 0$. For our operator this is $4 - n_\varphi$, where n_φ is the number of tails φ, i.e., of external momenta $p_1...p_n$ in (3.167). The three graphs $\tilde\gamma_{1,2,3}$ have dimensions 4, 2, and 0, respectively, at $\varepsilon = 0$, which determines the required accuracy of the expansion in the external momenta in calculating $L\tilde\gamma$. The graph $\tilde\gamma_1$ is independent of the external momenta, and so $L\tilde\gamma_1 \sim \tau^2$, $L\tilde\gamma_2$ contains contributions like τ and p^2, and $L\tilde\gamma_3$ contains only a constant. After calculating the counterterms $L\tilde\gamma$, the desired operator counterterms $L\gamma$ taking into account the form and dimensions of our graphs can then be found from the expressions

$$L\gamma_1 = L\tilde\gamma_1, \qquad L\gamma_2 = L\tilde\gamma_2|_{p=i\partial} \cdot \varphi^2(x), \qquad L\gamma_3 = L\tilde\gamma_3 \cdot \varphi^4(x). \qquad \text{(3.168)}$$

The calculation of the $L\tilde\gamma$ is not complicated. Our operator $\varphi\partial_i\partial_s\varphi$ corresponds to the vertex factor $v_{is} = -(q_iq_s + q_i'q_s')/2$ in the graphs of (3.167), where $q \equiv p/2 + k$, $q' \equiv p/2 - k$, the overall minus sign is from the two factors of i in the replacement $\partial \to i\times$momentum, and the half sum is the required symmetrization of $\partial\partial$ over two lines (Sec. 2.13). After substituting q and q' into v_{is}, we find that

$$v_{is} = -\tfrac{1}{4}p_ip_s - k_ik_s. \qquad \text{(3.169)}$$

Therefore, for the graphs $\tilde\gamma_{1,3}$ at zero external momenta (see above) and $\tilde\gamma_2$ at arbitrary p we have

$$\tilde\gamma_n = (-g_B)^{n-1}\int Dk\Delta^n(k)(-k_ik_s), \qquad n = 1, 3, \qquad \text{(3.170)}$$

$$\tilde{\gamma}_2 = -g_{\mathrm{B}} \int Dk \Delta(p/2+k)\Delta(p/2-k)v_{is}, \qquad (3.171)$$

where v_{is} is the factor (3.169) and everywhere

$$Dk \equiv dk/(2\pi)^d, \qquad \Delta(q) \equiv (q^2+\tau)^{-1}. \qquad (3.172)$$

In calculating the counterterm, the integrand in (3.171) can be expanded in p through terms of order p^2. With this accuracy the product of two Δ in the integral (3.171) can be replaced by the expression

$$\Delta^2[1 - \tfrac{1}{2}p^2\Delta + (pk)^2\Delta^2], \qquad \Delta \equiv \Delta(k). \qquad (3.173)$$

Multiplying by the vertex factor (3.169) and discarding the unnecessary terms $\sim p^4$, for $\tilde{\gamma}_2$ we obtain the following expression with the required accuracy:

$$g_{\mathrm{B}} \int Dk\{\tfrac{1}{4}p_i p_s + k_i k_s[1 - \tfrac{1}{2}p^2\Delta + (pk)^2\Delta^2]\}\Delta^2. \qquad (3.174)$$

The tensor integrals in (3.170) and (3.174) are reduced to scalar integrals using the expressions (3.84). As a result, we obtain scalar integrals with the integrands Δ^2, $k^4\Delta^4$, and $k^2\Delta^n$ with $n = 1, 2, 3$. The substitution $k^2 = k^2 + \tau - \tau$ reduces them to linear combinations of integrals of Δ^n with $n = 0, 1, 2, 3, 4$. We are only interested in the poles in ε, which occur only in the integrals with $n = 1, 2$ known from (3.87). In the exact scheme it would be necessary to also take into account the power-law Λ divergences in the integrals with $n = 0, 1$, but calculations are usually done using the formal scheme (Sec. 3.16), where all the Λ divergences are simply discarded. The justification of this procedure for ordinary Green functions was discussed earlier, and for composite operators it reduces to the following two remarks: (1) For the results to be self-consistent, the computational procedure must be unique; (2) Power-law Λ divergences do not affect the critical dimensions of the composite operators, the calculation of which is the ultimate goal of the renormalization theory under discussion as applied to problems of the type we are interested in (see Ch. 4 for more details).

After selecting the poles in ε in the expressions obtained for $\tilde{\gamma}$ by the operator L of the MS scheme ($R' = 1$ for all our graphs), we find the quantities $L\tilde{\gamma}_i$, and then from them using (3.168) the desired $L\gamma_i$. The results are given in Table 3.5.

The two-loop calculation of the counterterms for $\varphi\partial\partial\varphi$ is rather complicated technically, and we shall not present it. We only note that, owing to the difference of the dimensions of φ^2 and $\varphi\partial\partial\varphi$, the R' operator takes different forms for graphs of the same type of these two operators. For example, for $\varphi\partial\partial\varphi$

$$R' \; \vcenter{\hbox{\includegraphics}} \; = \text{ itself } - \; \vcenter{\hbox{\includegraphics}} \; - \; 2 \; \vcenter{\hbox{\includegraphics}} \;, \qquad (3.175)$$

while for φ^2 the last term on the right-hand side is absent.

3.28 Matrix multiplicative renormalization of families of operators

We shall use d_F^* to denote the canonical dimension d_F for a given operator F in the logarithmic theory, i.e., for $\varepsilon = 0$. All local monomials are naturally grouped according to symmetry and the value of d_F^* into *elementary sets*, for example, 1; φ; φ^2; φ^3, $\partial^2\varphi$; φ^4, $\varphi\partial^2\varphi$, $\partial^2\varphi^2$, and so on for the scalar operators of the φ^4 model (the sets are separated by the semicolon, and the operators within a set by a comma). The counterterms ΔF of a given monomial F consist of monomials with the same or smaller (by an even number for the φ^4 model) value of d_F^*. A finite family of monomials $\{F_i\}$ that mix only among themselves under renormalization will be termed *closed*. By definition, such a family together with any F_i contains all the monomials entering into its counterterms ΔF_i. In practice, this means that it wholly contains an elementary set with the maximum d_F^* and all the preceding ones with smaller d_F^* (decreased by 2, 4,... for the φ^4 model). For a closed family, (3.155) takes the matrix form

$$[F_i(\varphi)]_{\mathrm{R}} = \sum_k Q_{ik} F_k(\varphi), \quad \text{where}$$

$$Q_{ik} = \delta_{ik} - \{\text{the coefficient of } F_k(\varphi) \text{ in } LP[F_i(\varphi)\exp V(\varphi)]\}. \tag{3.176}$$

This *mixing matrix* Q is calculated directly from the graphs involving composite operators. Its elements depend on μ and the renormalized parameters e. For example, for the φ^4 model, $e = \{\tau, g\}$ and $Q_{ik} = \delta_{ik} + \mu^*\tau^* c_{ik}(g)$, where c_{ik} is a dimensionless coefficient containing the poles in ε, τ^* is an integer, nonnegative power of τ, and μ^* is a power of μ with exponent of order ε. All these exponents are uniquely determined from the dimensions of the operators F: $d[\mu^*\tau^*] = d[F_i] - d[F_k]$.

Let aF in (3.156) be the form (2.104) for our system $\{F_i\}$. We rewrite (3.157) for the renormalized action in more detail, indicating the dependence on the parameters (integration over x in linear forms is always understood):

$$S_{\mathrm{R}}(\varphi, a, e, \mu) = S_{\mathrm{R}}(\varphi, e, \mu) + \sum_i a_i[F_i(\varphi)]_{\mathrm{R}} + ..., \tag{3.177}$$

where the ellipsis stands for the contributions of counterterms of order a^2 and higher.

We shall assume that the original model without aF is multiplicatively renormalizable, i.e., up to the vacuum counterterm (Sec. 3.13) $S_{\mathrm{R}}(\varphi, e, \mu) = S(Z_\varphi\varphi, e_0)$ for it, where $S(\varphi, e_0)$ is the unrenormalized action, e_0 is the set of its bare parameters, and Z_φ is the field renormalization constant. The expression for multiplicative renormalization through terms of first order in a can be generalized to the extended model:

$$S_{\mathrm{R}}(\varphi, a, e, \mu) = S(Z_\varphi\varphi, a_0, e_0) + ... \tag{3.178}$$

For this we must introduce its unrenormalized action,

$$S(\varphi, a_0, e_0) = S(\varphi, e_0) + \sum_i a_{0i} F_i(\varphi), \qquad (3.179)$$

defining the set of bare sources $a_0 \equiv \{a_{0i}(x)\}$ by

$$a_{0k}(x) = \sum_i a_i(x) Z_a^{ik}, \qquad (a_0 = aZ_a) \qquad (3.180)$$

(the compact notation is given in parentheses), with the matrix of renormalization constants

$$Z_a^{ik} = Q_{ik}(Z_\varphi)^{-n_k}, \qquad Z_a^{-1} \equiv Z_F, \qquad (3.181)$$

where Q is the mixing matrix from (3.176), n_k is the number of factors of φ in the monomial $F_k(\varphi)$, and we have introduced the notation Z_F for the matrix inverse of Z_a (which will be needed below). The matrix indices ik on the matrix Z_a (and Z_F) will be written as superscripts only for convenience.

Equations (3.176), (3.180), and (3.181) lead to the equation $\sum_i a_{0i} F_i(Z_\varphi \varphi)$ $= \sum_i a_i [F_i(\varphi)]_R$, which shows that the functionals (3.177) and (3.179) are actually related by (3.178).

It follows from the definitions (3.176) and (3.181) that

$$F_i(Z_\varphi \varphi) = \sum_k Z_F^{ik} [F_k(\varphi)]_R. \qquad (3.182)$$

By the replacement $\varphi \rightarrow \varphi_R = Z_\varphi^{-1} \varphi$ we obtain the following operator equivalent of the functional formula (3.182):

$$\widehat{F}_i = \sum_k Z_F^{ik} \widehat{F}_{kR}. \qquad (3.183)$$

Here and below by definition,

$$\widehat{F} \equiv F(\varphi)\big|_{\varphi = \widehat{\varphi}}, \qquad \widehat{F}_R \equiv [F(\varphi)]_R\big|_{\varphi = \widehat{\varphi}_R}, \qquad (3.184)$$

where $\widehat{\varphi}$ is the unrenormalized field from (1.74), $\widehat{\varphi}_R = Z_\varphi^{-1} \widehat{\varphi}$ is the renormalized field, and the arguments x of all the F are understood.

The matrices Z_F and Z_a, which are the inverses of each other, are called the matrices of the renormalization constants of the operators F and the corresponding sources a. The field itself is a special case of a system of a single operator $F(\varphi) = \varphi$, and for it in (3.176) $Q = 1$ (Sec. 3.24), and so from (3.181) we have $Z_F = Z_\varphi$, as required.

For the full and connected Green functions, it follows from (3.183) and (1.74) that

$$\langle \widehat{F}_i \widehat{\varphi}...\widehat{\varphi} \rangle = Z_\varphi^n \sum_k Z_F^{ik} \langle \widehat{F}_{kR} \widehat{\varphi}_R ... \widehat{\varphi}_R \rangle, \qquad (3.185)$$

and from this for the 1-irreducible Green functions (see Secs. 1.18 and 2.13) we have

$$\langle \widehat{F}_i \widehat{\varphi}...\widehat{\varphi} \rangle_{1-\text{ir}} = Z_\varphi^{-n} \sum_k Z_F^{ik} \langle \widehat{F}_{kR} \widehat{\varphi}_R...\widehat{\varphi}_R \rangle_{1-\text{ir}}, \tag{3.186}$$

where n is the number of simple factors of $\widehat{\varphi}$, and all the fields and operators are understood to have independent arguments x.

Equations (3.178) and (3.180) allow the functional arguments a to be included in the general scheme on an equal footing with the parameters e, so that RG equations analogous to (1.110) can be written down for the extended model. Like (3.178), this will all be valid only through first order in a. However, this is sufficient for defining the concepts of critical dimensions of the operators F and sources a and calculating them using the standard RG technique (see Ch. 4).

3.29 UV finiteness of operators associated with the renormalized action and conserved currents

The generating functionals of the unnormalized full $[G(A)]$ and connected $[W(A)]$ Green functions in any theory (unrenormalized, renormalized, basic) are determined by the general expression (2.30) with the corresponding action, and the Schwinger equations in the form (2.94) or (2.108) follow from it. In universal notation (the convention of Sec. 1.9), the expression (2.108) is applicable to all systems, but in the present section to be more specific we shall indicate the discrete field indices explicitly, taking $\varphi \equiv \{\varphi_a(x)\}$ and understanding x to be only the spatial coordinates. In this notation the Schwinger equation (2.108) for the renormalized theory takes the form

$$\langle\!\langle u_{aR}(x; \varphi) \rangle\!\rangle = -A_a(x), \qquad u_{aR}(x; \varphi) \equiv \delta S_R(\varphi)/\delta\varphi_a(x). \tag{3.187}$$

The index a should be omitted for a one-component field.

We shall also use (2.127) for $x' = x$ and its consequences (2.128) and (2.129). These expressions involve the function $\delta(x' - x)$ or its derivatives at $x' = x$. Such objects correspond to divergences which are purely powers of Λ, which we have agreed to discard (Sec. 3.16). *Rule: In calculations using the formal scheme of dimensional regularization where divergences which are powers of Λ are discarded, all contributions containing coordinate delta functions $\delta(0)$ are discarded.* Using this convention, (2.127) at $x' = x$ for the renormalized theory takes the form

$$\langle\!\langle u_{aR}(x; \varphi)\varphi_b(x) \rangle\!\rangle = -A_a(x)\delta W_R(A)/\delta A_b(x), \tag{3.188}$$

and from (2.128) taking into account the definition (2.122) we have

$$\langle\!\langle u_{aR}(x; \varphi)T_{ab}\varphi_b(x) \rangle\!\rangle = -A_a(x)T_{ab}\delta W_R(A)/\delta A_b(x), \tag{3.189}$$

where T_{ab} are arbitrary (independent of φ) linear operators on the fields. If they are the generators T_{ab}^α of some exact symmetry group of the renormalized action (Sec. 2.15), then (3.189) becomes the local Ward identities (2.129):

$$\langle\!\langle \partial_i J_{i\mathrm{R}}^\alpha(x;\varphi)\rangle\!\rangle = A_a(x)T_{ab}^\alpha \delta W_\mathrm{R}(A)/\delta A_b(x), \tag{3.190}$$

where $J_{i\mathrm{R}}^\alpha$ are the canonical Noether currents for the renormalized action (Sec. 2.15).

Repeating word for word the derivation of (2.127) for the renormalized theory with replacement of the preexponential $\varphi_b(x)$ by an arbitrary renormalized operator $[F(x;\varphi)]_\mathrm{R}$, we obtain the following generalization of (3.188):

$$\langle\!\langle u_{a\mathrm{R}}(x;\varphi)[F(x;\varphi)]_\mathrm{R}\rangle\!\rangle = -A_a(x)\langle\!\langle [F(x;\varphi)]_\mathrm{R}\rangle\!\rangle. \tag{3.191}$$

Finally, yet another series of useful relations can be obtained by differentiating (2.30) for the renormalized theory with respect to any numerical parameter λ, which gives [in the notation of (2.107)]

$$\langle\!\langle \partial_\lambda[S_\mathrm{R}(\varphi) + \ln c_\mathrm{R}]\rangle\!\rangle = \partial_\lambda W_\mathrm{R}(A), \tag{3.192}$$

where c_R is the normalization constant in (2.30) for the renormalized theory. For example, for the O_n φ^4 model, from (2.5) and (2.16) we have

$$\ln c_\mathrm{R} = \frac{Vn}{2(2\pi)^d} \int dk \ln\left[\frac{k^2 + \tau}{2\pi}\right], \tag{3.193}$$

where $V = \int dx$ is the infinite volume of the system, and the factor of n appears in taking the trace over the field indices in (2.5).

All these relations can be written down for any theory — unrenormalized, renormalized, or basic. We have written them for the renormalized theory, where $W_\mathrm{R}(A)$ is a UV-finite renormalized functional, and the symbol $\langle\!\langle ...\rangle\!\rangle$ denotes the average (2.107a) with $S = S_\mathrm{R}$ in the exponent. As explained in Sec. 3.24, UV finiteness of this functional $\langle\!\langle F(\varphi)\rangle\!\rangle$ is equivalent to UV finiteness of the composite operator $F(\varphi)$. In our case this is guaranteed by the UV-finiteness of the right-hand sides of all the equations (3.187)–(3.192), following from the UV-finiteness of the functional $W_\mathrm{R}(A)$ [the operators acting on it on the right-hand sides of (3.188)–(3.190) and (3.192) do not spoil the UV finiteness if λ in (3.192) is any UV-finite parameter of the renormalized theory, and T_{ab} in (3.189) is any φ-independent UV-finite linear operator]. It then follows that all the composite operators inside the $\langle\!\langle ...\rangle\!\rangle$ on the left-hand sides of these equations are UV-finite.*

In what follows we shall deal specifically with the MS scheme, in which the separation into UV-finite and infinite parts is unique (the infinite part contains

---

*In quantum field theory, the replacement $\varphi \to \widehat{\varphi}_\mathrm{R}$ in $u_{a\mathrm{R}}(x;\varphi)$ leads to a composite operator which is not simply UV-finite, but equal to zero. The right-hand sides of (3.187)–(3.190) are nonzero only because they involve Wick, not Dyson, Green functions (see the remark in Sec. 2.13).

only poles in ε, and the finite part contains all the nonpole contributions), and all the counterterms, including the operator counterterms in (3.155), consist only of UV-divergent contributions. It can therefore be stated that *in the MS scheme, every UV-finite operator coincides with its UV-finite part obtained by discarding all the UV-divergent contributions.* The functional $u_{aR} = \delta S_R / \delta \varphi_a$ is a linear combination of unrenormalized local monomials. Clearly, its UV-finite part is obtained, first, by discarding all the counterterm contributions of the action S_R, and, second, by replacing the unrenormalized local monomials by renormalized ones. The first operation leads to the replacement $u_{aR} \rightarrow u_{aB} \equiv \delta S_B / \delta \varphi_a$, where S_B is the basic action. The second leads to the replacement $u_{aB} \rightarrow [u_{aB}]_R$. Therefore, the UV finiteness of all the quantities in (3.187) implies that

$$u_{aR}(x; \varphi) \equiv \delta S_R(\varphi) / \delta \varphi_a(x) = [u_{aB}(x; \varphi)]_R. \qquad (3.194)$$

Similarly, the UV finiteness of (3.188)–(3.191) implies that

$$u_{aR}(x; \varphi) \varphi_b(x) = [u_{aB}(x; \varphi) \varphi_b(x)]_R, \qquad (3.195)$$

$$u_{aR}(x; \varphi) T_{ab} \varphi_b(x) = [u_{aB}(x; \varphi) T_{ab} \varphi_b(x)]_R, \qquad (3.196)$$

$$\partial_i J_{iR}^\alpha(x; \varphi) = [\partial_i J_{iB}^\alpha(x; \varphi)]_R, \qquad (3.197)$$

$$u_{aR}(x; \varphi) [F(x; \varphi)]_R = [u_{aB}(x; \varphi) F(x; \varphi)]_R, \qquad (3.198)$$

where J_{iB}^α is the conserved current (Sec. 2.15) for the basic theory, T_{ab} in (3.196) is any φ-independent UV-finite linear operator, and $F(x; \varphi)$ in (3.198) is any local monomial.

The role of λ in (3.192) can be played by μ or any of the renormalized parameters e. The statement that an object in the brackets $\langle\!\langle ... \rangle\!\rangle$ is UV-finite is valid for any of these parameters, but simple consequences like (3.194) can be obtained only for $\lambda \neq \mu$. The point is that μ always enters as μ^ε, and so a factor of ε appears upon differentiation with respect to μ, which causes the counterterms with first-order poles in ε to also contribute to the UV-finite parts. Factors of ε do not arise upon differentiation for the other parameters $\lambda = e$. Therefore, as above, the UV-finite part of $[\partial_e S_R(\varphi)] = [\partial_e S_B(\varphi)]_R$, and from (3.192) with $\lambda = e$ it follows that $\partial_e S_R(\varphi) + \partial_e \ln c_R = $ the UV-finite part of the same $= [\partial_e S_B(\varphi)]_R + $ the UV-finite part of $\partial_e \ln c_R$. Cancelling the contribution from the UV-finite part of $\partial_e \ln c_R$ in this equation, for any $\lambda = e$ (but not μ) we obtain

$$\partial_e S_R(\varphi) + \text{UV-div. part of } \partial_e \ln c_R = [\partial_e S_B(\varphi)]_R. \qquad (3.199)$$

The above discussion regarding ∂_μ shows that the equation (3.198) that is valid for any monomial $F(x; \varphi)$ can be generalized to polynomials only when they do not have any factors of ε in their coefficients.

3.30   The O_n φ^4 model: renormalization of scalar operators with $d_F^* = 2, 3, 4$

The general expressions in Sec. 3.29 can sometimes be used to find the operator counterterms without calculating graphs involving composite operators. The results are then expressed in terms of the renormalization constants of the action, which are assumed to be known. As an example, let us consider the renormalization in the O_n φ^4 model of a system of scalar operators containing the monomials $\varphi^2(x)$, $\varphi^3(x)$, and $\varphi^4(x)$. We shall drop the argument x, and assume that all even forms (i.e., $\varphi^2 \equiv \varphi_a \varphi_a$, $\varphi^4 = (\varphi^2)^2$, and so on) are scalars and odd forms ($\varphi \equiv \varphi_a$, $\varphi^3 \equiv \varphi_a \varphi^2$, and so on) are vectors under the O_n group. The basic action of the model has the form (3.22), and the renormalized action the form (3.105).

In studying the renormalization, each of the monomials in question must first be augmented to form a closed (Sec. 3.28) system $F = \{F_i\}$. In our case these will be the systems $\{1, \varphi^2\}$, $\{\varphi, \partial^2\varphi, \varphi^3\}$, and $\{1, \varphi^2, \partial^2\varphi^2, \varphi\partial^2\varphi, \varphi^4\}$. In each of them we shall number the operators sequentially, i.e., $F_1 = 1$, $F_2 = \varphi^2$ in the first system and analogously in the others. Complete information about the renormalization of the system F is contained in the matrix Q from (3.176), and it is used to find the matrices Z_a and Z_F from (3.181). They are all usually triangular or block-triangular matrices, because in the renormalization, monomials of lower dimension can mix with those of higher, but not vice versa.

It follows from (3.176) that knowledge of the specific form of the operator $[F_i(\varphi)]_R$ is equivalent to knowing the ith line of the matrix Q. If $[F_i(\varphi)]_R = F_i(\varphi)$, the corresponding line of Q is trivial: unity on the diagonal and zeros elsewhere. In our case this will be true for the operators 1, φ, and $\partial^2\varphi$ (we recall that $[\partial F(\varphi)]_R = \partial[F(\varphi)]_R$).

Let us consider the system $F = \{1, \varphi^2\}$ with 2×2 matrix Q. Its first line is trivial (see above), and we use (3.199) with $e = \tau$ to determine the second. From (3.105) we find $2\partial_\tau S_R(\varphi) = \int dx[-Z_2\varphi^2(x) + 2g_B^{-1}\tau Z_0]$, where the second term is the contribution of the vacuum counterterm (3.106). Furthermore, from (3.193) we have $2\partial_\tau \ln c_R = n \int dx \Delta(x, x)$, where $\Delta(x, x)$ is the contracted basic propagator, i.e., (3.87a). Equation (3.199) involves its divergent part, which is selected by the counterterm operator (3.31) of the MS scheme and is equal to $-\tau\mu^{-2\varepsilon}/16\pi^2\varepsilon = -\tau u/\varepsilon g_B$ with the usual $u \equiv g/16\pi^2$ and $g_B \equiv g\mu^{2\varepsilon}$. Substituting the resulting expressions into (3.199) leads to the equation $\int dx[Z_2\varphi^2(x) - 2g_B^{-1}\tau Z_0 + n\tau u/\varepsilon g_B] = \int dx[\varphi^2(x)]_R$. This system contains no operators of the form ∂F, the contributions of which vanish upon integration over x. Therefore, equality of the integrals in this case implies equality of the integrands, namely,

$$[\varphi^2(x)]_R = Z_2\varphi^2(x) - 2g_B^{-1}\tau \overline{Z}_0, \quad \overline{Z}_0 \equiv Z_0 - nu/2\varepsilon. \qquad (3.200)$$

The second line of the matrix Q is thereby (see above) determined: $Q_{21} = -2g_B^{-1}\tau\overline{Z}_0$, $Q_{22} = Z_2$. The known Q and (3.181) are used to find $Z_a = Z_F^{-1}$, namely, $Z_a^{11} = 1$, $Z_a^{12} = 0$, $Z_a^{21} = Q_{21}$, and $Z_a^{22} = Q_{22}Z_\varphi^{-2}$. Taking into account (3.110), we obtain $Z_a^{22} = Z_2 Z_\varphi^{-2} = Z_\tau$, which is natural, because the source $a(x)$ of the composite operator $\varphi^2(x)$ enters into the basic action of the extended (Sec. 3.23) model in the combination $a(x) - \tau/2$. In the massless theory the operator $F = \varphi^2(x)$ is renormalized multiplicatively with $Z_F = Z_a^{-1} = Z_\tau^{-1}$.

Let us turn to the system $F = \{\varphi, \partial^2\varphi, \varphi^3\}$. The first and second lines of the corresponding 3×3 matrix Q are trivial (see above), and we use (3.194) to find the third. Substituting into it the equations obtained from (3.22) and (3.105) for the variational derivatives $u = \delta S/\delta\varphi$, we have $Z_1\partial^2\varphi - \tau Z_2\varphi - g_B Z_3\varphi^3/6 = [\partial^2\varphi]_R - \tau[\varphi]_R - g_B[\varphi^3]_R/6$, from which using $[\varphi]_R = \varphi$ and $[\partial^2\varphi]_R = \partial^2\varphi$ we find $[\varphi^3]_R = Z_3\varphi^3 + 6\tau g_B^{-1}(Z_2 - 1)\varphi + 6g_B^{-1}(1 - Z_1)\partial^2\varphi$. We have thereby found the third row of the matrix Q: $Q_{31} = 6\tau g_B^{-1}(Z_2 - 1)$, $Q_{32} = 6g_B^{-1}(1 - Z_1)$, and $Q_{33} = Z_3$. Using the known matrix Q and (3.181) we find all the nonzero matrix elements Z_a: $Z_a^{11} = Z_a^{22} = Z_\varphi^{-1}$, $Z_a^{31} = Q_{31}Z_\varphi^{-1}$, $Z_a^{32} = Q_{32}Z_\varphi^{-1}$, and $Z_a^{33} = Q_{33}Z_\varphi^{-3}$ [we recall that $Z_1 = Z_\varphi^2$ according to (3.110)].

Now we consider the system $F = \{1, \varphi^2, \partial^2\varphi^2, \varphi\partial^2\varphi, \varphi^4\}$. The first line of the corresponding 5×5 matrix Q is trivial, and the second and third are known from (3.200) and the equation $[\partial^2\varphi^2]_R = \partial^2[\varphi^2]_R$. We determine the fourth and fifth lines using (3.195) and (3.199) with $e = g$. In our case, (3.195) takes the form $Z_1\varphi\partial^2\varphi - \tau Z_2\varphi^2 - g_B Z_3\varphi^4/6 = [\varphi\partial^2\varphi]_R - \tau[\varphi^2]_R - g_B[\varphi^4]_R/6$. Using (3.176) to express the renormalized operators on the right-hand side in terms of the unrenormalized ones and equating the coefficients of each of them (they are all independent) on the two sides of the equation, we obtain the following system of equations:

$$Q_{4i} - \tau Q_{2i} - g_B Q_{5i}/6 = c_i, \quad i = 1, ..., 5, \tag{3.201}$$

in which c_i are the known coefficients of the $F_i(\varphi)$ on the left-hand side: $c_1 = 0$, $c_2 = -\tau Z_2$, $c_3 = 0$, $c_4 = Z_1$, and $c_5 = -g_B Z_3/6$. The Q_{2i} in (3.201) are also known, and so (3.201) for each i is a single equation for the two unknowns Q_{4i} and Q_{5i}. To obtain the second equation we use (3.199) with $e = g$. The constant (3.193) is independent of g, and so (3.199) for the action (3.105) takes the form

$$\partial_g \int dx\{Z_0\tau^2/2g_B + \tfrac{1}{2}Z_1\varphi\partial^2\varphi - \tfrac{1}{2}\tau Z_2\varphi^2 - \tfrac{1}{24}g_B Z_3\varphi^4\} =$$
$$= [\partial_g \int dx(-\tfrac{1}{24}g_B\varphi^4)]_R = -\tfrac{1}{24}\mu^{2\varepsilon}\int dx[\varphi^4]_R,$$

where we have used the fact that inside the integral $(\partial\varphi)^2 \cong -\varphi\partial^2\varphi$. It is important that the operator $[...]_R$ on the right-hand side of (3.199) acts after ∂_g (if the opposite order were taken, ∂_g would operate also on the g-dependent counterterms of the operator, which does not occur for us). In this equation it is possible, as above, to use (3.176) to express the renormalized operators in terms of unrenormalized ones and then to equate the coefficients of the latter on both sides of the equation. The only difference is that the contribution of the operator $F_3 = \partial^2 \varphi^2$ vanishes in the integral over x, and so it is impossible to obtain any information about its coefficients. For all the other $i \neq 3$ the above equation gives

$$-\tfrac{1}{24}\mu^{2\varepsilon}Q_{5i} = c'_i, \quad i \neq 3, \tag{3.202}$$

where c'_i are the known coefficients of the $F_i(\varphi)$ in the integrand on the left-hand side: $c'_1 = \mu^{-2\varepsilon}\tau^2\partial_g(Z_0/2g)$, $c'_2 = -\tau\partial_g Z_2/2$, $c'_4 = \partial_g Z_1/2$, and $c'_5 = -\mu^{2\varepsilon}\partial_g(gZ_3)/24$. Therefore, from (3.202) we immediately find all the elements Q_{5i} with $i \neq 3$, and then from (3.201) the elements Q_{4i} with $i \neq 3$. For Q_{43} and Q_{53} there is only one relation from (3.201), and no second equation. Therefore, to completely determine the matrix Q, one of these two elements must be calculated explicitly from the massless graphs of the composite operator F_4 or F_5, where only the admixture of F_3 in the counterterms is of interest. We see from the graphs that F_3 cannot mix with the operator F_5 in first order in g, i.e., the expansion of Q_{53} begins at least with g^2, and $Q_{43} = g_\mathrm{B}Q_{53}/6$ begins at least with g^3.

The expressions of Sec. 3.29 are also useful for analyzing the renormalization of higher-dimensional operators. For example, for a system containing φ^6 we can obtain information from (3.198) with $F(\varphi) = \varphi^3$ and from (3.196) with the operator $T_{ab} = \delta_{ab}\partial^2$.

3.31   Renormalization of conserved currents

If the regularized basic theory possesses an exact symmetry and the latter is not broken by the counterterms (Sec. 3.13), then the corresponding conserved canonical Noether currents (Sec. 2.15) satisfy (3.197), which is sometimes formulated as the statement that divergences of conserved currents are not renormalized. For the actual currents, (3.197) gives

$$J^\alpha_{i\mathrm{R}}(x;\varphi) = [J^\alpha_{i\mathrm{B}}(x;\varphi)]_\mathrm{R} + \overline{J}^\alpha_i(x;\varphi), \quad \partial_i\overline{J}^\alpha_i = 0, \tag{3.203}$$

where $\overline{J}^\alpha_i(x;\varphi)$ are composite operators, transverse on any φ (and not only on classical solutions), with the same symmetry and dimension as the currents. Purely transverse terms do not contribute to the Ward identities (2.129), and so in the renormalized theory it is always possible in principle to use the UV-finite quantities $[J^\alpha_{i\mathrm{B}}]_\mathrm{R}$ instead of the canonical $J^\alpha_{i\mathrm{R}}$ as the conserved currents. However, now we shall discuss the properties of the latter.

The explicit form of the addition $\overline{J}^\alpha_i$ for any particular current can be found by analyzing the renormalization of all the local monomials entering into it. It can sometimes be stated immediately that there are no additions with the required properties, and then $\overline{J}^\alpha_i = 0$ in (3.203). This will occur, for example, for the current (2.120) corresponding to O_n rotations in the basic $O_n\ \varphi^4$ model. The current $J^{ab}_{i\mathrm{R}}$ for the renormalized action (3.105) differs from the basic current (2.120) by only an additional factor of Z_1. It is impossible to construct any other operator with the dimension of a current that is transverse in the index i and antisymmetric in ab. The given current is therefore renormalized multiplicatively (it does not mix with anything), and the UV finiteness of its divergence implies the UV finiteness of the current itself, i.e.,

$$J_{i\mathrm{R}}^{ab} = Z_1(\varphi_a \partial_i \varphi_b - \varphi_b \partial_i \varphi_a) = [\varphi_a \partial_i \varphi_b - \varphi_b \partial_i \varphi_a]_\mathrm{R} = [J_{i\mathrm{B}}^{ab}]_\mathrm{R}. \qquad (3.204)$$

Denoting the basic current (2.120) as $F(\varphi)$, from (3.204) we conclude that the given operator is multiplicatively renormalized with $Q = Z_1$ in (3.176). Therefore, from the definitions (3.181) taking into account (3.110) we have $Z_a = Z_F^{-1} = QZ_\varphi^{-2} = 1$, i.e., $\widehat{F} = Z_F \widehat{F}_\mathrm{R}$ with $Z_F = 1$ in the notation of (3.183) and (3.184). It is the equation $Z_F = 1$ that we have in mind when we say that a given conserved current is not renormalized.

As a second example, we consider the canonical energy-momentum tensor in the same model. For it from (3.197) we find that

$$\vartheta_{i\mathrm{R}}^k(\varphi) = [\vartheta_{i\mathrm{B}}^k(\varphi)]_\mathrm{R} + c(\partial^2 \delta_{ik} - \partial_i \partial_k)\varphi^2 + c' \delta_{ik}, \qquad (3.205)$$

where the addition is the general form of a composite operator of the required dimension and symmetry that is transverse in the index i, and c and c' are unknown coefficients. From the definition (2.119) for the basic action (3.22) we have

$$\vartheta_{i\mathrm{B}}^k(\varphi) = -\partial_i \varphi \cdot \partial_k \varphi + \delta_{ik}[\tfrac{1}{2}(\partial \varphi)^2 + \tfrac{1}{2}\tau \varphi^2 + \tfrac{1}{24}g_\mathrm{B} \varphi^4] \qquad (3.206)$$

(all contributions are scalars under the O_n group), and for the renormalized action (3.105) we have

$$\vartheta_{i\mathrm{R}}^k(\varphi) = -Z_1 \partial_i \varphi \cdot \partial_k \varphi + \delta_{ik}[\tfrac{1}{2}Z_1(\partial \varphi)^2 + \tfrac{\tau}{2}Z_2 \varphi^2 + \tfrac{g_\mathrm{B}}{24}Z_3 \varphi^4 - \tfrac{\tau^2}{2g_\mathrm{B}}Z_0], \qquad (3.207)$$

where the last term is the contribution of the vacuum counterterm (3.106) per unit volume. Equation (3.205) gives information about the renormalization of the tensor local monomials entering into (3.206).

3.32   Renormalization of tensor operators with $d_F^* = 4$ in the O_n φ^4 model

The system of interest now consists of seven operators: the five scalars studied in Sec. 3.30, 1, φ^2, $\partial^2 \varphi^2$, $\varphi \partial^2 \varphi$, and φ^4 with coefficient δ_{ik}, and the two new operators $\partial_i \partial_k \varphi^2$ and $\varphi \partial_i \partial_k \varphi$. As before, we denote the old operators multiplied by δ_{ik} by $F_{1,2,3,4,5}$, and as the two independent new ones we take the transverse operator $F_6 = (\delta_{ik}\partial^2 - \partial_i \partial_k)\varphi^2$ and $F_7 = \varphi \partial_i \partial_k \varphi$. Using the relations $\partial_i \varphi \cdot \partial_k \varphi = \partial_i \partial_k \varphi^2/2 - \varphi \partial_i \partial_k \varphi$ and $(\partial \varphi)^2 = \partial^2 \varphi^2/2 - \varphi \partial^2 \varphi$, all the monomials in (3.206) can be expressed in terms of the operators introduced above:

$$\vartheta_{i\mathrm{B}}^k = F_7 + \tfrac{1}{2}F_6 - \tfrac{1}{2}F_4 - \tfrac{1}{4}F_3 + \tfrac{1}{2}\tau F_2 + \tfrac{1}{24}g_\mathrm{B} F_5.$$

Writing the operator (3.207) in the same form and substituting everything into (3.205), we obtain

$$Z_1\left(F_7 + \tfrac{1}{2}F_6 - \tfrac{1}{2}F_4 - \tfrac{1}{4}F_3\right) + \tfrac{1}{2}\tau Z_2 F_2 + \tfrac{1}{24}g_B Z_3 F_5 - \tfrac{\tau^2}{2g_B}Z_0 F_1 =$$
$$= [F_7 + \tfrac{1}{2}F_6 - \tfrac{1}{2}F_4 - \tfrac{1}{4}F_3 + \tfrac{1}{2}\tau F_2 + \tfrac{1}{24}g_B F_5]_R + \text{const} \cdot F_6 + \text{const} \cdot F_1,$$

with unknown coefficients of $F_{1,6}$ on the right-hand side. Expressing, as usual, the renormalized operators $[F_i]_R$ on the right-hand side in terms of the unrenormalized ones using (3.176) and equating the coefficients of the unrenormalized F_i with $i \neq 1, 6$ [the coefficients of $F_{1,6}$ in (3.205) remain undetermined] on both sides of the equation, we find that

$$c_i'' = Q_{7i} + \tfrac{1}{2}Q_{6i} - \tfrac{1}{2}Q_{4i} - \tfrac{1}{4}Q_{3i} + \tfrac{1}{2}\tau Q_{2i} + \tfrac{1}{24}g_B Q_{5i}, \qquad i \neq 1, 6, \qquad (3.208)$$

where c_i'' are the known coefficients of the F_i on the left-hand side: $c_2'' = \tau Z_2/2$, $c_3'' = -Z_1/4$, $c_4'' = -Z_1/2$, $c_5'' = g_B Z_3/24$, and $c_7'' = Z_1$.

The matrix Q in (3.176) for our system has the following form (the crosses denote the known nonzero elements, and the question marks denote the unknown nonzero elements):

$$Q = \begin{array}{c|c|c|c|c|c|c|c|}
 & 1 & 2 & 3 & 4 & 5 & 6 & 7 \\
\hline
1 & \times & 0 & 0 & 0 & 0 & 0 & 0 \\
\hline
2 & \times & \times & 0 & 0 & 0 & 0 & 0 \\
\hline
3 & 0 & 0 & \times & 0 & 0 & 0 & 0 \\
\hline
4 & \times & \times & ? & \times & \times & 0 & 0 \\
\hline
5 & \times & \times & ? & \times & \times & 0 & 0 \\
\hline
6 & 0 & 0 & 0 & 0 & 0 & \times & 0 \\
\hline
7 & ? & \times & ? & \times & \times & ? & \times \\
\hline
\end{array} \qquad (3.209)$$

The first 5×5 block coincides with the matrix for the system containing φ^4 discussed in Sec. 3.30, the sixth line is uniquely determined by the renormalization of φ^2 in (3.200) (i.e., $Q_{66} = Z_2$ and the other elements equal to zero), and the elements Q_{7i} with $i = 2, 4, 5, 7$ are found from (3.208), in particular, $Q_{77} = Z_1$. We do not obtain any information on Q_{71} and Q_{76} from (3.208), and Q_{73} is expressed in terms of Q_{43} and Q_{53}. From (3.208) and (3.201) we have two relations for these three elements, and so to completely determine the matrix Q from the graphs, it is necessary to have one of these three elements, and also Q_{71} and Q_{76}.

The counterterms F_7 were calculated in the one-loop approximation as an example in Sec. 3.27. From (3.166) and (3.176) and the results in Table 3.5 of Sec. 3.27 in the one-loop approximation we find $Q_{71} = n\tau^2 u/4\varepsilon g_B$, $Q_{73} =$

$(n + 2)u/24\varepsilon$, and $Q_{76} = -(n + 2)u/18\varepsilon$. We note that from (3.201) and (3.208) with $Q_{33} = Z_2$ we have $Q_{73} = (Z_2 - Z_1)/4 + g_B Q_{53}/24$, which, using $Q_{53} = O(g^2)$ (Sec. 3.30), allows us to find the contributions of order g and g^2 to Q_{73} without calculating graphs.

Since the matrix (3.209) is block-triangular, the critical dimensions associated with the operators $F_{6,7}$ will be determined (Ch. 4) by the diagonal elements $Q_{66} = Z_2$ and $Q_{77} = Z_1$ (see above), for which from (3.181) and (3.110) we obtain $Z_a^{66} = Z_2 Z_\varphi^{-2} = Z_\tau$ and $Z_a^{77} = Q_{77} Z_\varphi^{-2} = 1$.

We shall limit ourselves to these examples, and here conclude our discussion of the technique for renormalizing composite operators.

3.33 The Wilson operator expansion for short distances

The Wilson operator or short-distance expansion (SDE) is a method of studying the $\xi \to 0$ asymptote for operators of the form

$$\widehat{F}_\xi = \widehat{\varphi}_R(x + \xi)\widehat{\varphi}_R(x - \xi), \tag{3.210}$$

or, in general, for the product of any number of renormalized composite operators with different arguments $x_i = x + \xi_i$ for $\xi \equiv \{\xi_i\} \to 0$. In the theory of critical phenomena, knowledge of this asymptote gives, in particular, important additional information about the behavior of the scaling functions that is inaccessible in ordinary ε expansions (Ch. 4).

In formulating the problem, we first decide upon the desired accuracy of determining the asymptote, i.e., the maximum power ξ^N of the included terms. All contributions of order ξ^n with $n \leq N$ will be termed ξ-*relevant*, and ones with $n > N$ will be termed ξ-*irrelevant*. We stress the fact that we are speaking here only of integer powers of ξ. Additional factors of the type $(\mu\xi)^\varepsilon$ in dimensional regularization (which are logarithmic at $\varepsilon = 0$) are assumed to be of order unity and must be exactly included everywhere. The condition $(\mu\xi)^\varepsilon \cong 1$ corresponds formally to $\mu\xi \gg 1$, $\varepsilon \ll 1$, i.e., this is the region of critical scaling (Sec. 1.35) in the ε expansion, and the smallness of ξ is only relative: $\tau\xi^2 \ll 1$, $p\xi \ll 1$, where p are any momenta conjugate to coordinates different from ξ in Green functions with operator of the type (3.210). For the operator itself in coordinate space we must formally take $\partial\xi \ll 1$ for $\partial \equiv \partial/\partial x$.

We shall illustrate the general rules for the example of the operator (3.210) in the simple φ^4 model ($n = 1$) with $d = 4 - 2\varepsilon$ in the MS scheme, studying its asymptote through order ξ^2, in order not to limit ourselves to the lowest order $\xi^0 = 1$. We pass from operator notation to functional notation (Sec. 2.13), introducing

$$F_\xi \equiv \varphi(x + \xi)\varphi(x - \xi). \tag{3.211}$$

When we return to operator notation $\varphi \to \widehat{\varphi}_R$, which corresponds to placing the functional equations inside brackets $\langle\!\langle ... \rangle\!\rangle$ with $S = S_R$ in the definition (2.107a).

Let us now turn to the solution of this problem. In general, the original operator F_ξ can be expanded in a Taylor series in ξ, writing

$$F_\xi = H_\xi + O_\xi, \tag{3.212}$$

where H_ξ is the initial segment of the series containing all the ξ-relevant contributions (the length of this segment is determined by the arbitrarily specified accuracy ξ^N) — a linear combination of unrenormalized local monomials, and O_ξ is the remainder of the series. In our example [the operator (3.211) and accuracy through ξ^2], after multiplying the series $\varphi(x \pm \xi) = [1 \pm (\xi\partial) + (\xi\partial)^2/2 + ...]\varphi(x)$ and using the relation $\partial\varphi\partial\varphi = \partial\partial\varphi^2/2 - \varphi\partial\partial\varphi$, we obtain

$$H_\xi = [1 - \tfrac{1}{2}(\xi\partial)^2]\varphi^2(x) + 2\xi\xi\varphi(x)\partial\partial\varphi(x). \tag{3.213}$$

The symbols ξ and $\partial \equiv \partial/\partial x$ carry pairwise-contracted vector indices. In detailed notation, $H_\xi = \varphi^2 + \xi_i\xi_s[2\varphi\partial_i\partial_s\varphi - \partial_i\partial_s\varphi^2/2]$ is a linear combination of scalar φ^2 and tensor $\partial\partial\varphi^2$, $\varphi\partial\partial\varphi$ local monomials.

The graphical representations of the functional (3.160) for all three operators in (3.213) are identical and coincide with the representation of F_ξ with $\xi = 0$. The only difference is in the vertex factors v of the composite operator. In our example (3.211), this will be the graphs of (3.161), and the vertex factor $v[F_\xi]$ for the choice (3.167) of momentum configuration is given by $\exp[i(p/2 + k)(x + \xi) + i(p/2 - k)(x - \xi)] = \exp(ipx)\exp(2ik\xi)$. The first exponential is standard and vanishes upon transformation to momentum space. The second is the desired vertex factor $v[F_\xi] = \exp(2ik\xi)$. The first terms of its series in ξ make up $v[H_\xi]$, and the rest $v[O_\xi]$. With our accuracy

$$v[O_\xi] = \exp(2ik\xi) - 1 - 2ik\xi - \tfrac{1}{2}(2ik\xi)^2. \tag{3.214}$$

As usual, all the calculations are performed in momentum space, and in the final answers the external momenta become local operators $i\partial$ (Sec. 3.26). To avoid misunderstandings, we again note that the quantities $v[...]$ given above are the vertex factors of the operators from (3.212) for graphs in momentum space with the configuration (3.167). The coordinate variable ξ enters into these vertex factors as a parameter.

At first glance it appears that the desired asymptote is given simply by the operator H_ξ in (3.212). However, this actually is not true owing to UV divergences: for $\xi \neq 0$ the operator F_ξ on the left-hand side of (3.212) is UV-finite, but both terms on the right-hand side have divergences (as does H_ξ, since it is a linear combination of local monomials; they must therefore also occur in O_ξ for cancellation). Calculations show (Sec. 3.34) that, owing to UV divergences in the Green functions, the operator O_ξ contains the ξ-relevant contributions, although formally O_ξ consists only of the ξ-irrelevant terms of the Taylor series. We note the important fact that everywhere in what follows, O_ξ is treated not as an infinite sum, but as a single expression

— the remainder of a series, and this is consistent with the way its vertex factor (3.214) is written.

According to the Wilson hypothesis [91], which was later rigorously justified [92], the desired asymptote gives not (3.212), but the relation

$$F_\xi = H_\xi^* + O_\xi^*, \qquad H_\xi^* = \sum_i B_i(\xi, e, \mu)[F_i(x; \varphi)]_R, \qquad (3.215)$$

where O_ξ^* is the UV-finite and ξ-irrelevant residual term, and H_ξ^* is the needed segment of the operator expansion — a linear combination of renormalized local monomials with x-independent UV-finite Wilson coefficients $B_i(\xi, e, \mu) = B_i(\xi, g, \tau, \mu)$ containing only ξ-relevant contributions [then the representation (3.215) is unique]. The set of operators F_i in (3.215) is determined using a simple rule: it is the minimal finite closed system (Sec. 3.28) containing all the local monomials from H_ξ. The number of operators in the set $\{F_i\}$ increases as the specified accuracy is raised owing to the successive inclusion of operators with growing canonical dimensions. For example, for the operator (3.213)

$$\{F_i\} = \{1, \varphi^2, \partial^2 \varphi^2, \varphi \partial^2 \varphi, \varphi^4, \partial\partial\varphi^2, \varphi\partial\partial\varphi\}, \qquad (3.216)$$

where $\partial\partial$ are tensor objects (the indices are understood) and ∂^2 is the Laplacian. Operators with $d_F^* = 4$ enter into (3.216) only because we have decided to also consider correction terms of order ξ^2. If we were interested only in the leading contributions of order unity (and $\sim \xi^{-2}$ in the constant), then we would have to restrict ourselves to the contribution $\varphi^2(x)$ in H_ξ and to the first two monomials in (3.216).

For clarity, let us also give the operator form of (3.215) for our example (3.210):

$$\widehat{\varphi}_R(x + \xi)\widehat{\varphi}_R(x - \xi) = \sum_i B_i(\xi, e, \mu)\widehat{F}_{iR}(x) + \widehat{O}_{\xi R}, \qquad (3.217)$$

where $e = g, \tau$ and the exact meaning of symbols like $\widehat{F}_R$ is given in (3.184).

The representation (3.215) is justified by the methods of renormalization theory. The renormalized operators $[H_\xi]_R$ and $[O_\xi]_R$ are defined by the general expression (3.155), and the role of L for H_ξ is played by the principal counterterm operator used in renormalizing the local monomials (for us, the MS operator), while for O_ξ it is played by a new operator $L \equiv L_\xi$. It is defined (more details follow) so as to ensure the ξ-irrelevance and simultaneously the UV finiteness of the renormalized operator $[O_\xi]_R$, which will play the role of O_ξ^* in (3.215).

The formal derivation of the representation (3.215) is simple. From the definition (3.155) we have $H_\xi = [H_\xi]_R + LP[H_\xi \exp V]$, $O_\xi = [O_\xi]_R + L_\xi P[O_\xi \exp V]$, and, substituting these expressions into (3.212), we obtain a representation like (3.215) with $O_\xi^* = [O_\xi]_R$ and

$$H_\xi^* = [H_\xi]_R + LP[H_\xi \exp V] + L_\xi P[O_\xi \exp V] = \sum_i B_i[F_i(x; \varphi)]_R. \quad (3.218)$$

For the appropriate choice of L_ξ, the operator (3.218) will be the desired segment of the Wilson expansion. This is reflected by the second equation in (3.218), in which the arguments of the coefficients B have been dropped.

The most important point is the construction of the needed operator L_ξ. It is defined as an operator on the 1-irreducible graphs and subgraphs γ of the operator O_ξ [i.e., the functional (3.160) with $F = O_\xi$] as follows. If γ is a subgraph not containing the vertex O_ξ, then for it $L_\xi = L$ is the main counterterm operator (for us, the MS operator). If γ is a graph or subgraph with vertex O_ξ, then for it $L_\xi = K_\xi R'$, where K_ξ is the operator that subtracts all the ξ-relevant contributions, i.e.,

$$K_\xi f(\xi, ...) = \xi\text{-relevant part of } f(\xi, ...) \qquad (3.219)$$

(the ellipsis indicates other arguments). This definition is a special case of (3.31), since the introduced operator commutes with the factors of μ^δ.

By construction, the corresponding (to a given L_ξ) R operator effects the subtraction of all the ξ-relevant contributions from graphs containing O_ξ, thereby ensuring the ξ-irrelevance of the renormalized operator $[O_\xi]_R$. Its counterterms, like those of H_ξ, and therefore the entire operator (3.218) consist only of ξ-relevant contributions. These properties determine K_ξ uniquely.

The following statements are true for the introduced operator L_ξ.

(1) The subtraction of UV divergences occurs simultaneously with the subtraction of all the ξ-relevant contributions, i.e., $[O_\xi]_R$ is UV-finite.

(2) $L_\xi \gamma \neq 0$ for those and only those graphs and subgraphs γ which are SD graphs (Sec. 3.26) of at least one operator monomial in H_ξ (in practice, they are determined by the monomial of highest dimension).

(3) If a primitive subtraction operator like the MS operator (Sec. 3.10) is used as the basic one, then for any graph or subgraph with vertex O_ξ the counterterm $L_\xi \gamma = K_\xi R' \gamma$ is a polynomial in p and τ (the external momenta and internal masses) corresponding to the initial segment of needed length of the Taylor expansion of the function $f \equiv R' \gamma$ at the origin in the set of p, τ. The polynomial behavior of the counterterms in τ (but not in p) is spoiled if a nonprimitive subtraction operator like (3.32) is used as the basic operator. *Corollary:* If the basic operator is primitive (in particular, the MS operator), then K_ξ from (3.219) can also be defined as the operator that selects the initial segment of needed length of the Taylor expansion at the origin in the set of p, τ.

Let us briefly explain the above statements. The first is almost obvious: if $A = B + C$ is the sum of expressions that are somehow uniquely distinguishable, then the UV finiteness of A implies the UV finiteness of B and C separately. In our case, $A \equiv F_\xi$ is UV-finite, $B = H_\xi^*$ contains only ξ-relevant contributions, and $C = O_\xi^*$ contains only ξ-irrelevant ones. Clearly, B and C cannot have mutually cancelling UV divergences, since cancellation is impossible owing to the different behavior in ξ.

The second statement is also clear. The ξ-relevant contributions to O_ξ appear only owing to UV divergences, and those, in turn, are generated only by the subtracted terms at the vertex (3.214), i.e., they are the same as in H_ξ. Therefore, ξ-relevant contributions are generated only by those γ that are the SD graphs of at least one operator monomial in H_ξ.

Now let us discuss Statement 3. It is easily proved for one-loop graphs γ with vertex O_ξ: their ξ-relevant parts $K_\xi\gamma = L_\xi\gamma$ are nonzero only when UV divergences are present, and so they vanish along with the latter in derivatives with respect to p and τ of sufficiently high order. They are therefore polynomials $P(p,\tau)$ with the structure (3.5). These primitive divergences along with all the ξ-relevant contributions are wholly contained (Sec. 3.3) in the initial segment of needed length of the Taylor expansion in the set p, τ for any choice of expansion point p_*, τ_* if the needed first coefficients of the expansion exist. The latter condition for ordinary graphs made it impossible to choose $p_* = \tau_* = 0$ owing to IR divergences (Sec. 3.3). This now becomes possible, since vertex factors like (3.214) of the operator O_ξ have a zero of sufficiently high order in the integration momentum k: in general, $v[O_\xi] \sim k^{N+1}$, where N is the specified accuracy [$N = 2$ for us in (3.214)]. This weakens the IR singularities in the integrals over k by just the amount required for the existence of the needed first coefficients of the expansion about the origin in the set p, τ, and the following coefficients no longer exist.

The choice $p_* = \tau_* = 0$ is distinguished by the absence of additional dimensional parameters. The coefficients of the polynomial $P(p,\tau)$ in this case are powers of ξ and they are all ξ-relevant by dimension, i.e., in this polynomial there are no ξ-irrelevant contributions, and so it really is the ξ-relevant part of γ. For any other choice of expansion point, the corresponding polynomial will contain the entire ξ-relevant part, but will not coincide with it owing to the presence of additional ξ-irrelevant contributions.

We have demonstrated that L_ξ is primitive on one-loop graphs γ and that $L_\xi\gamma$ can be interpreted as the initial segment of needed length of the Taylor expansion about the origin in p, τ, the essential point being that the needed first few coefficients of this expansion exist. When we go over to multiloop graphs, the main counterterm operator L on subgraphs without the vertex O_ξ will be the composite part L_ξ, and so for L_ξ to be primitive, it is in general necessary that L be. The meaning of Statement 3 is that this is sufficient, i.e., in this case the counterterms $L_\xi\gamma$ of any multiloop graph possess the same properties as the one-loop counterterms. This is the analog of the basic Statement 2 of Sec. 3.12. As usual, we shall omit the general proof and instead present the calculation of the Wilson coefficients for the operator (3.211) with the chosen accuracy (ξ^2) in the one-loop approximation (Sec. 3.34) along with additional explanations (Sec. 3.35).

In conclusion, we note that the operator expansion (SDE) is often written as an infinite series, omitting the residual term. This notation is symbolic; the rigorous notation is that of (3.215) with residual term that can be obtained for arbitrarily high specified accuracy ξ^N. As N grows the system $\{F_i\}$ is augmented by operators of higher and higher canonical dimension, and the Wilson coefficients B_i are simultaneously made more accurate for the operators existing earlier owing to the increased accuracy with which they are calculated (the inclusion of higher powers of $\tau\xi^2$ in the φ^4 model). If the contributions of such expansions are classified not according to the operators, but according to the powers of the variable ξ [assuming $(\mu\xi^\varepsilon \cong 1)$, the answer can be formally represented as an infinite series, since the contributions of a given order ξ^n are objective: since they were obtained from the representation (3.215) with a certain $N \geq n$, they obviously are not changed if we go to another representation (3.215) with higher accuracy $(N' > N)$. In other words, an increase of the specified accuracy in the representation (3.215) allowing new corrections to be obtained does not affect the contributions obtained earlier with lower accuracy, and only in this sense can the result be treated as an infinite series.

Clearly, the leading terms of the $\xi \to 0$ asymptote are generated in (3.215) by the contributions of operators with the lowest dimensions and lowest (zero) powers of τ in the coefficients. The inclusion of operators of higher dimensions and contributions with higher powers of τ corresponds to the successive inclusion of corrections with extra powers of ξ.

3.34 Calculation of the Wilson coefficients in the one-loop approximation

The results are given in Table 3.6.

Table 3.6 Contributions of one-loop graphs to the operator expansion $F_\xi = \varphi(x + \xi)\varphi(x-\xi)$ through order ξ^2 $[g_B \equiv g\mu^{2\varepsilon}, u \equiv g/16\pi^2, c \equiv \|1-\varepsilon\|/(1+\varepsilon), t \equiv 4\pi\mu^2\xi^2]$.

No.	$LP[H_\xi \exp V]_\gamma$	$L_\xi P[O_\xi \exp V]_\gamma$
1	$\dfrac{\tau u}{2\varepsilon g_B}[-2 - \tau\xi^2]$	$\dfrac{cut^\varepsilon \xi^{-2}}{2\varepsilon g_B} \cdot$ $\cdot\{2(1+\varepsilon)(\varepsilon + \tau\xi^2) + (\tau\xi^2)^2\}$
2	$\dfrac{u}{6\varepsilon} \cdot$ $\cdot[-6-6\tau\xi^2-(\xi\partial)^2+\xi^2\partial^2]\varphi^2$	$\dfrac{cut^\varepsilon}{6\varepsilon} \cdot$ $\cdot\{(1+\varepsilon)[6+(\xi\partial)^2]+6\tau\xi^2-\xi^2\partial^2\}\varphi^2$
3	$-\dfrac{ug_B}{2\varepsilon}\xi^2\varphi^4(x)$	$\dfrac{cut^\varepsilon g_B}{2\varepsilon}\xi^2\varphi^4(x)$

The desired segment of the operator expansion (3.215) is

$$H_\xi^* = [H_\xi]_\mathrm{R} + \sum_\gamma c_\gamma \left\{ LP[H_\xi \exp V]_\gamma + L_\xi P[O_\xi \exp V]_\gamma \right\}, \qquad (3.220)$$

where the index γ denotes the contribution of an individual graph γ to the counterterm functionals and c_γ are the symmetry coefficients. The graphs γ and coefficients c_γ for the operator (3.211) are the same as in Table 3.5 of Sec. 3.27, and so in Table 3.6 we indicate only the number of the graph and its contribution to the counterterms of H_ξ (second column) and O_ξ (third column). The operator $[H_\xi]_\mathrm{R}$ in (3.220) does not need to be calculated; it is obtained from (3.213) by the simple replacement $\varphi^2 \to [\varphi^2]_\mathrm{R}$, $\varphi\partial\partial\varphi \to [\varphi\partial\partial\varphi]_\mathrm{R}$.

Let us explain the procedure for calculating the counterterm contributions. For H_ξ everything is simple, because the counterterms of the local operators φ^2 and $\varphi\partial\partial\varphi$ entering into it were calculated earlier in Sec. 3.27. It follows from (3.213) that $L\gamma$ for H_ξ is the sum of the counterterms $L\gamma$ for $F = \varphi^2$ and $F_{is} = \varphi\partial_i\partial_s\varphi$ given in Table 3.5 of Sec. 3.27 with the coefficients $[1 - (\xi\partial)^2/2]$ and $2\xi_i\xi_s$, respectively. In this manner, for the first three graphs of Table 3.5 we obtain the expressions given in the second column of Table 3.6.

The calculation of the counterterms $L_\xi\gamma$ for the residual term O_ξ is a new feature. According to the general rules of Sec. 3.33 for O_ξ, it is necessary to consider the same graphs γ and the same accuracy of the expansion in the external momenta as for the operator of maximum dimension in H_ξ, in the present case the operator $F_{is} = \varphi\partial_i\partial_s\varphi$ in Sec. 3.27. Therefore, the quantities $\widetilde{\gamma}$ needed in the calculation (the values of γ in momentum space neglecting the factors of φ) for the first three graphs of Table 3.5 are obtained from the corresponding expressions (3.170) and (3.171) by simple replacement of the vertex factor (3.169) [in (3.170) we use it at $p = 0$] by the factor (3.214):

$$\widetilde{\gamma}_n = (-g_\mathrm{B})^{n-1} \int Dk\Delta^n(k)[\exp(2ik\xi) - \ldots], \qquad n = 1, 3,$$

$$\widetilde{\gamma}_2 = -g_\mathrm{B} \int Dk\Delta(p/2 + k)\Delta(p/2 - k)[\exp(2ik\xi) - \ldots] \qquad (3.221)$$

with the earlier notation (3.172). For $\widetilde{\gamma}_2$ it is sufficient to limit ourselves to the approximation (3.173) for the product of propagators.

The ellipsis in (3.221) denotes the subtractions indicated explicitly in (3.214). Their role is simple: they eliminate all the ξ-relevant analytic contributions, i.e., the ξ-relevant nonnegative integer powers of ξ. This is a useful observation that justifies the following simple recipe for calculating integrals like (3.221): first the integral is calculated without subtractions, and then all the ξ-relevant analytic (in ξ) terms are simply discarded in the result. Henceforth, we shall denote the last operation by the symbol $|_\mathrm{sub}$.

We introduce the auxiliary functions

$$Q_n \equiv \int Dk\Delta^n(k)\exp(2ik\xi), \qquad Q_n' \equiv Q_n|_\mathrm{sub}, \qquad (3.222)$$

in terms of which the integrals (3.221) are expressed:

$$\widetilde{\gamma}_n = (-g_{\mathrm{B}})^{n-1} Q_n|_{\mathrm{sub}} = (-g_{\mathrm{B}})^{n-1} Q_n', \qquad n = 1, 3;$$

$$\widetilde{\gamma}_2 = -g_{\mathrm{B}} [Q_2 - \tfrac{1}{2} p^2 Q_3 - \tfrac{1}{4} (p\partial/\partial\xi)^2 Q_4]|_{\mathrm{sub}}.$$
(3.223)

In $\widetilde{\gamma}_2$ we have replaced the product of propagators by the approximate expression (3.173). The factors of k in the term involving $(pk)^2$ for the integral without subtractions are reproduced by the operator $\partial/\partial(2i\xi)$, and the needed subtractions are made after calculating the integral, i.e., after differentiating with respect to ξ in (3.223).

The explicit expression for Q_n is easily obtained from (3.79)–(3.81):

$$Q_n(\xi, \tau) = \sum_{m=0}^{\infty} (\tau\xi^2)^m \left[\alpha_{nm} \xi^{2n-4+2\varepsilon} + \beta_{nm} \tau^{2-n-\varepsilon} \right],$$
(3.224)

where α and β are numerical factors [notation of (3.68)]:

$$\alpha_{nm} = \frac{\|2 - n - \varepsilon\| n - 1 + \varepsilon\|}{(4\pi)^{2-\varepsilon} \|n\| m + 1\| n + m - 1 + \varepsilon\|},$$

$$\beta_{nm} = -\frac{\|2 - n - \varepsilon\| n - 1 + \varepsilon\|}{(4\pi)^{2-\varepsilon} \|n\| m + 1\| 3 + m - n - \varepsilon\|}.$$
(3.225)

We shall call the terms in (3.224) with the coefficients α and β the α and β terms. The former are analytic in τ but not in ξ, and the latter in ξ but not in τ, and so the expandability in τ about the origin is violated by the β terms. In addition, the functions (3.224) are UV-finite: the overall coefficient in the numerator of (3.225) always contains a pole in ε, but it is cancelled by the mutual cancellation at $\varepsilon = 0$ of the α and β terms with identical powers of ξ, and those (and only those) terms in (3.224) without such pairs have the zero $\sim \varepsilon$ cancelling pole in the coefficient.

The operator $|_{\mathrm{sub}}$ in (3.222) and (3.223) eliminates all the ξ-relevant β terms and simultaneously ensures the existence of the needed initial segment of the expansion in τ at the origin:

$$Q_n' = \xi^{2n-4+2\varepsilon} \sum_{m=0}^{\infty} \alpha_{nm} (\tau\xi^2)^m + \xi\text{-irrelevant } \beta\text{-terms}.$$
(3.226)

At small n (for us, $n \leq 3$) the sum (3.226) contains ξ-relevant α terms and, consequently, UV divergences owing to the violation of the balance between the α and β terms which earlier (see above) ensured cancellation of the UV divergences. The subsequent subtractions of ξ-relevant α terms effected by the $R(L_\xi)$ operator restore the balance violated in (3.226) and along with it the UV finiteness. These remarks clarify the general discussion of Sec. 3.33.

Let us return to the calculations. The counterterms of interest $L_\xi\widetilde{\gamma} = K_\xi\widetilde{\gamma}$ of the one-loop graphs in (3.223) ($R' = 1$) are obtained, according to the definition (3.219), by selecting the ξ-relevant part of (3.226). All the ξ-relevant β terms have already been eliminated by the subtractions, and so only the ξ-relevant α terms, namely, the first three terms of the series (3.226) for Q_1, the first two terms for Q_2, and the first term for $Q_{3,4}$, contribute to the counterterms $L_\xi\tilde\gamma$. The first term of Q_4 becomes ξ-relevant after double differentiation with respect to ξ in (3.223), which is performed using the formula $(p\partial/\partial\xi)^2\xi^{2\lambda} = 2\lambda\xi^{2\lambda-4}[2(\lambda-1)(p\xi)^2 + p^2\xi^2]$ with $\lambda = 2+\varepsilon$ for the first α term of Q_4. As a result, from (3.223) and (3.226) we obtain

$$L_\xi\tilde\gamma_1 = \xi^{-2+2\varepsilon}[\alpha_{10} + \alpha_{11}\tau\xi^2 + \alpha_{12}(\tau\xi^2)^2],$$

$$L_\xi\tilde\gamma_2 = -g_B\xi^{2\varepsilon}\{\alpha_{20} + \alpha_{21}\tau\xi^2 - (1+\varepsilon)(2+\varepsilon)\alpha_{40}(p\xi)^2 - $$
$$-\tfrac{1}{2}p^2\xi^2[\alpha_{30} + (2+\varepsilon)\alpha_{40}]\}, \tag{3.227}$$

$$L_\xi\tilde\gamma_3 = g_B^2\xi^{2+2\varepsilon}\alpha_{30}.$$

The counterterms $L_\xi\gamma$ are then found from (3.168) with the replacement $L \to L_\xi$. After substituting the α coefficients known from (3.225) into (3.227), performing some simple algebra, and changing the notation as indicated in the caption of Table 3.6, we obtain the expressions for the counterterms $L_\xi\gamma$ for the operator O_ξ listed in the last column of that table.

To obtain the final result, the Wilson coefficients B_i for the renormalized operators (3.216) in (3.220), it is necessary to substitute into (3.220) the known values of c_γ (Table 3.5 in Sec. 3.27) and the counterterm contributions of the graphs (Table 3.6). The latter contain the unrenormalized operators (3.216), which in our one-loop approximation should simply be replaced by the renormalized operators, since in this case the inclusion of corrections would exceed the accuracy.

The UV finiteness of the resulting one-loop coefficients B_i is ensured by the mutual cancellation of the poles in ε of the counterterm contributions of H_ξ and O_ξ for each of the three graphs (see Table 3.6). This is a feature specific to the one-loop approximation. There will be no such cancellation in higher orders, and the remaining contributions with poles in ε will build up the unrenormalized operators (3.216) in the combination of counterterms (3.220) to form renormalized operators.

3.35 Expandability of multiloop counterterms $L_\xi\gamma$ in p and τ

Let us use examples to explain the meaning of Statement 3 in Sec. 3.33 for multiloop graphs. According to this statement, for an arbitrary graph γ with vertex O_ξ and primitive basic operator L (for example, the MS operator), the ξ-relevant part of the expression $R'\tilde\gamma$ ($\tilde\gamma$ is the graph γ in momentum space with the tails φ replaced by simple external lines) has no IR divergences in the coefficients of the initial segment of needed length of the Taylor expansion in p and τ at the origin. For brevity, we shall refer to this property simply as *expandability*. Since expandability in p for $\tau > 0$ is guaranteed, only expandability in τ at zero external momenta is nontrivial. The coefficient of the highest power of τ, determined by the specified accuracy (Sec. 3.33), is the most dangerous from the viewpoint of IR divergences.

The expandability of one-loop graphs is ensured by the suppression of the IR singularities by the vertex factor O_ξ (Sec. 3.33). For multiloop graphs the mechanism of eliminating IR divergences is nontrivial, and we shall explain it by several examples for the operator O_ξ discussed above (Secs. 3.33 and 3.34) with the vertex factor (3.214).

The first example is

$$R' \,\bullet\!\!\infty = \,\bullet\!\!\infty \; - \; \bullet\!\!\infty \; - \; \bullet\!\!\infty$$

$$= \Big[\bullet\!\!\rightthreetimes \; - \; \rightthreetimes \Big]\, \bigcirc \; - \; \bullet\!\!\rightthreetimes \cdot \bigcirc \; . \tag{3.228}$$

Only the factors with vertex O_ξ, shown by the large black dot, depend on ξ. According to the definition (3.219) of the operator L_ξ, the expression in square brackets in (3.228) is ξ-irrelevant, and so it does not contribute to the counterterm:

$$L_\xi \,\bullet\!\!\infty = K_\xi R' \,\bullet\!\!\infty = - \Big[K_\xi \; \bullet\!\!\rightthreetimes \Big] \cdot \bigcirc \; . \tag{3.229}$$

The first factor on the right-hand side is expandable (we assume that it is known for one-loop graphs), and the expandability of the second is ensured by the primitive nature of the basic operator L (in the MS scheme the second factor is a multiple of τ).

We note that for nonprimitive subtractions at zero momenta (3.32), this second factor is a multiple of $\tau^{1-\varepsilon}$, and the expandability of the counterterm (3.229) is lost.

Another example is the following. With our accuracy (through order ξ^2)

$$R' \,\bullet\!\!\Subset = \,\bullet\!\!\Subset \; - \; 2 \,\bullet\!\!\Subset \; - \; \bullet\!\!\Subset \; , \tag{3.230}$$

where the external lines replace the tails φ. The last graph on the right-hand side has structure similar to that of the last graph on the right-hand side of (3.228), and so the expandability of its ξ-relevant part for primitive operator L is guaranteed (see above).

Let us therefore consider separately the sum of the first two graphs on the right-hand side of (3.230). We wish to show that its ξ-relevant part at zero external momenta is expandable in τ at the origin through order τ (with our accuracy). In other words, this expression itself and its first derivative with respect to τ have a finite limit for $\tau \to 0$. The first is obvious, and so let us consider the derivative with respect to τ. According to the rule for differentiating a line (3.112), we obtain

$$\partial_\tau \left[\; \text{⟨graph⟩} \; - \; 2 \; \text{⟨graph⟩} \; \right] = - \; \text{⟨graph⟩} \; -$$

$$- \; \text{⟨graph⟩} \; - \; 2 \; \text{⟨graph⟩} \; + \; 2 \; \text{⟨graph⟩} \; . \tag{3.231}$$

We have taken into account the fact that the lines of the subgraph enclosed by the dashed line do not need to be differentiated, since the corresponding counterterm does not depend on τ with our accuracy (it is known explicitly from Table 3.6 of Sec. 3.34). The lines with insertion of a point in the graphs of (3.231) correspond to factors of $(k^2 + \tau)^{-2}$ that generate singularities $\sim \tau^{-\varepsilon}$ from small integration momenta k which are dangerous for $\tau \to 0$. The k dependence of the other lines can be neglected in determining the coefficient of $\tau^{-\varepsilon}$, and so the part of the sum of the last two graphs on the right-hand side of (3.231) singular for $\tau \to 0$ coincides with the singular part of the expression

$$-2 \left[\; \text{⟨graph⟩} \; - \; \text{⟨graph⟩} \; \right]_{p=0} \cdot \; \text{⟨loop⟩} \; . \tag{3.232}$$

This expression has a dangerous singularity $\sim \tau^{-\varepsilon}$ in the second cofactor (the vertex loop) at zero external momenta. However, it does not contribute to the counterterm $L_\xi \gamma$, since it does not have ξ-relevant contributions owing to their complete cancellation inside the square brackets in (3.232) because of the definition of the operator L_ξ.

We still need to consider the first two graphs on the right-hand side of (3.231). The standard (as in one-loop graphs) mechanism of suppression of the IR singularity by the vertex factor O_ξ (Sec. 3.33) operates for each of them, and so they cannot have dangerous singularities $\sim \tau^{-\varepsilon}$.

We have therefore shown that the ξ-relevant part of the sum of graphs on the right-hand side of (3.231) has no contributions like $\tau^{-\varepsilon}$, which proves the property of expandability of the counterterm $L_\xi \gamma$ of the two-loop graph in (3.230).

These examples illustrate the meaning of Statement 3 in Sec. 3.33 and at the same time show that this statement is nontrivial.

3.36 Renormalization in the case of spontaneous symmetry breaking

So far we have restricted outselves to the case $\tau \geq 0$, when the ordinary graphs with the lines $(k^2 + \tau)^{-1}$ are meaningful. For $\tau < 0$ they become meaningless and the perturbation series must be rearranged. This can be done in two ways: either by shifting to the correct energy minimum in the functional integrals (Sec. 2.4), or by changing over to the language of the Legendre transform Γ using the loop expansion or another suitable one (for example, the $1/n$ expansion) instead of simple perturbation theory (Sec. 2.10). These constructions adapted to the case $\tau < 0$ and the corresponding series will be termed rearranged. The concept of correct rearrangement needs to be made precise for the free energy (vacuum loops). We shall discuss this in detail below, after we formulate the general idea of renormalization for $\tau < 0$.

Since the entire renormalization theory described above essentially uses simple graph expansions, at first glance it appears that the passage to the case $\tau < 0$, when these expansions become meaningless, must lead to serious difficulties. This will be the case in poorly chosen subtraction schemes of the type (3.32), in which the counterterms can contain, for example, $\ln \tau$. However, no problems actually arise in primitive subtraction schemes of the MS type with counterterms that are polynomials in τ.

Statement. For any primitive subtraction scheme, including the MS scheme, all the UV divergences of constructions correctly rearranged for $\tau < 0$ are completely eliminated by the same counterterms that are polynomials in τ as for the case $\tau > 0$. In other words, the counterterms constructed for $\tau > 0$ are suitable for either sign of τ.

To prove this statement, we replace the mass term $\tau \int dx \varphi^2(x)/2$ in the basic action by the insertion of the composite operator $aF \equiv \int dx a(x)\varphi^2(x)/2$ (Sec. 2.13) and consider all objects as series in a. In momentum space, inhomogeneity of the arbitrary coefficient $a(x)$ corresponds to the presence of external momenta flowing into the vertices of the composite operator, which is needed for eliminating the IR divergences in the terms of the series in a (it is therefore impossible to simply expand in the constant τ). To eliminate all the UV divergences of the massless theory with the addition aF in the action, we must add all the counterterms determined by (3.156). They will be polynomials in a, since, owing to the IR-relevance ($d_F < d$) of the operator $F = \varphi^2(x)/2$, graphs with a large number of insertions aF do not have superficial divergences (Sec. 3.23). For example, for the φ^4 model only counterterms of order a and a^2 are needed. After adding the counterterms, we obtain a theory in which all objects are represented as series in a with coefficients which do not contain UV divergences.

Now let us make the change $a(x) \to \tau$. This can lead to new problems, but the crucial point is that they now have an infrared, not ultraviolet, origin. Since the action functional is a polynomial in a, no problems arise: polynomials in a become the usual action counterterms which are polynomials in τ, and the sign of τ makes no difference. However, the change $a(x) \to \tau$ in the renormalized Green functions is nontrivial even for $\tau > 0$, since it leads to the usual expansions in τ containing growing IR divergences. Therefore, the series in a must first be rearranged such that later there are no IR divergences for $a \to \tau$. For $\tau > 0$ it is sufficient to use a rearrangement corresponding to the passage to the usual massive lines. For $\tau < 0$ a different resummation is needed [as in Sec. 2.4, or in the lines of the loop expansion for the functional $\Gamma(\alpha)$]. However, it is important that in any case the problem of the correct rearrangement of the series no longer has any relation to the problem of UV divergences: they are absent in the coefficients of the series in a and therefore do not appear in the rearranged objects (the correct rearrangement is an IR and not a UV problem).

Therefore, in schemes of the MS type it is possible to construct counterterms in the usual manner for $\tau > 0$ and then use these expressions for either sign of τ. The above arguments show that these counterterms will automatically eliminate all the UV divergences in any structures correctly constructed for the case $\tau < 0$.

Let us now define the term "correctly constructed." For Green functions given by expressions like (1.62) and (1.82) without vacuum loops, the correct rearrangement for $\tau < 0$ reduces only to change of the functional integration space E_{int} according to the general rules of Sec. 1.13. Since E_{int} is uniquely determined by the form of the action S (Rule 3 in Sec. 1.13), for an unrenormalized theory E_{int} will be renormalization-invariant in the terminology of Sec. 1.24, and this will ensure the preservation of the form of RG equations of the type (1.107) for $\tau < 0$. The contributions of the vacuum counterterm (Sec. 3.16) cancel in the ratio (1.82), and so in schemes of the MS type all the UV divergences of Green functions without vacuum loops for $\tau < 0$ are eliminated by multiplicative renormalization (1.84) with the same constants Z as for $\tau > 0$.

The vacuum loops $W_0 \equiv W(A = 0)$ for functionals defined by expressions like (2.30) represent a special case. For $\tau > 0$ the UV-finite quantity $W_{0R} \equiv W_R(A = 0)$ is determined by (2.30) with $S = S_R$ and the normalization constant $C = C_R$ from (2.16) with renormalized parameter τ in the operator $K = -\partial^2 + \tau$. For example, in the O_n φ^4 model, for the quantity $\ln C_R = \frac{1}{2}\operatorname{tr}\ln(-\partial^2 + \tau)$ in dimension $d = 4 - 2\varepsilon$ the second expression in (2.5) gives the following in the formal scheme of dimensional regularization (Sec. 3.16):

$$\ln C_R = \frac{Vn}{2} \int \frac{dk}{(2\pi)^d}\ln(k^2 + \tau) = \frac{Vn\|-1+\varepsilon\|\tau^{2-\varepsilon}}{2(2-\varepsilon)(4\pi)^{2-\varepsilon}} \tag{3.233}$$

up to an additive constant that is independent of τ and therefore irrelevant [in (3.233) $V \equiv \int dx$ is the volume and n is the number of components of the field φ]. Differentiation with respect to τ reduces the integral in (3.233) to the known integral (3.87a), and only the integral over τ remains to be done.

The quantity W_{0R} does not admit direct continuation to the case $\tau < 0$, since its definition involves not only the action S_R which is a polynomial in τ, but also the constant C_R, which becomes meaningless for $\tau < 0$ [the result (3.233) would contain a fractional power of a negative number]. The correct object that can be continued to $\tau < 0$ is not W_{0R}, but the free energy $\mathcal{F} \equiv -\ln Z$, defined in terms of the functional integral (see Secs. 1.13 and 2.10) by

$$\exp(-\mathcal{F}) = \int D\varphi \exp S(\varphi), \tag{3.234}$$

and similarly for the renormalized theory with S replaced by S_R. Comparing the definitions (2.30) and (3.234), we easily find the relation between $\mathcal{F}_R$ and the vacuum loops W_{0R} for $\tau > 0$:

$$\mathcal{F}_R = -\ln \int D\varphi \exp S_R(\varphi) = -W_{0R} + \ln C_R. \tag{3.235}$$

For the action S_R with the vacuum counterterm (Sec. 3.18), the quantity W_{0R} in (3.235) is UV-finite, but $\ln C_R$ contains a UV-divergent part that is a polynomial in τ. For example, for (3.233) in the MS scheme (Sec. 3.16) we have

$$\text{UV-div. part of } \ln C_R = -\frac{Vn\tau^2\mu^{-2\varepsilon}}{4\varepsilon(4\pi)^2}, \tag{3.236}$$

where μ is the renormalization mass.

If (3.236) is subtracted from (3.235), we obtain a quantity that is UV-finite for $\tau > 0$:

$$\left.\begin{aligned}
\mathcal{F}_R' &= \mathcal{F}_R - \text{UV-div. part of } \ln C_R \\
&= -[W_{0R} - \text{UV-finite part of } \ln C_R] \\
&= -\ln \int D\varphi \exp[S_R(\varphi) + \text{UV-div. part of } \ln C_R],
\end{aligned}\right\} \tag{3.237}$$

which can now be continued to the region $\tau < 0$, since it contains only S_R and the expression (3.236) which are polynomials in τ. The statement is that in MS-type schemes, the UV finiteness of the quantity (3.237) for $\tau > 0$ automatically ensures its UV finiteness for $\tau < 0$ with the same counterterms as for $\tau > 0$. The contribution (3.236) to the argument of the exponential in (3.237) should also be understood as a one-loop vacuum counterterm that preserves the form of (3.236) for either sign of τ, like all the other counterterms entering into S_R. In (3.237) the contribution (3.236) is naturally grouped with the vacuum counterterm in S_R of the type (3.106). For the O_n φ^4 model this leads to the replacement $Z_0 \to \overline{Z}_0$ with $\overline{Z}_0$ from (3.200).

The quantities $\mathcal{F}_R$ and $\mathcal{F}_R'$ differ only by the contribution (3.236) regular in τ, which is unimportant for analyzing the critical singularities (it is regular just because it is the same for either sign of τ). Therefore, the UV-finite quantity (3.237) is a correctly constructed object that can be analyzed using the RG technique when studying the critical singularities of the free energy $\mathcal{F}_R$ and the specific heat $C = -\partial_\tau^2 \mathcal{F}_R$. The critical dimensions of these quantities are known (Sec. 1.3), but the RG technique can also be used to calculate other quantities of interest, for example, the universal ratio of amplitudes in a law like (1.22) for the specific heat. Owing to the presence of vacuum counterterms, the RG equations for the free energy (3.237) differ from the usual RG equations like (1.107) by the presence of inhomogeneities; they will be discussed in Ch. 4.

We conclude by noting that all local relations like the Schwinger equation (Sec. 3.29), including the local Ward identities (3.190) (in contrast to the global ones) with invariant renormalized action functional, are valid for any sign of τ (Sec. 2.16).

Chapter 4

Critical Statics

Statics deals with problems in equilibrium statistical physics and thermo-dynamics, i.e., problems without time dependence. This was the type of problem discussed in Ch. 1 in our exposition of the RG technique. In the present chapter we expand upon the material of Ch. 1 and describe the techniques for practical calculations using specific examples. The justification of the quantum field models that we use and comparison of our results to experiment lie outside the scope of this book; instead, we refer the reader to the corresponding specialized literature.

4.1 General scheme for the RG analysis of an arbitrary model

A particular quantum field model chosen on the basis of physical considerations is specified by its action functional, and the goal of the RG analysis of the model is the demonstration of critical scaling and, if it exists, the calculation of the universal characteristics of the critical behavior, i.e., the critical exponents and the normalized scaling functions (Sec. 1.33). Only results independent of the arbitrariness of the finite renormalization are of objective value; in particular, these are the coefficients of the ε expansions in the deviation $d^* - d \equiv 2\varepsilon$ of the spatial dimension d from the logarithmic dimension d^* (a particular one for each model) or the corresponding logarithmic corrections for $\varepsilon = 0$ (Sec. 1.43). They are calculated using the RG technique following the standard scheme consisting of the following steps:

1. Determination of the canonical dimensions of all the quantities in the action functional (Sec. 1.15) and of the corresponding logarithmic dimension d^*, and maximal simplification of the model by discarding all interactions that are IR-irrelevant compared to the leading one, if such exist (Sec. 1.16). A simplified model constructed in this manner is considered below.

2. The use of power counting and symmetry arguments to determine the structure of the primitive divergences of the graphs of all 1-irreducible functions Γ_n for $d = d^*$ and, thereby, the structure of all the counterterms needed to eliminate them in dimensional regularization $d = d^* - 2\varepsilon$. Divergences that are powers of the UV-cutoff parameter Λ can be ignored if they lead only to the redefinition of quantities like T_c, which are unimportant in the theory

of critical behavior (Sec. 1.20). This means that all the calculations can be carried out within the formal scheme of dimensional regularization.

At this stage the multiplicative renormalizability of the model is checked. All the obtained counterterms, except for the vacuum one, must be reproduced by the multiplicative renormalization of the fields and redefinition of the parameters (see Secs. 3.13 and 1.18, especially the footnotes in the latter).

The standard RG analysis can be completed only if multiplicative renormalizability is present. Its absence implies that the model is constructed incorrectly from the viewpoint of the general theory of critical behavior, i.e., it does not include all the interactions allowed by power counting and symmetry that can affect the critical behavior of the system. We note that in the original technique of Wilson recursion relations (Sec. 1.1), all interactions allowed by power counting and symmetry arise in the iteration process, and this is analogous to the appearance of counterterms in the renormalization.

Therefore, a "good" model must be multiplicatively renormalizable. If the original model is not, it must be extended to include all the interactions needed to ensure multiplicative renormalizability. When this is achieved, the further analysis proceeds as follows:

3. The general rules of Sec. 1.24 are used to derive the RG equations for the renormalized objects, and the equations expressing the RG functions in terms of the renormalization constants Z are obtained.

4. The graphs of Γ_n are used within a selected subtraction scheme to calculate all the needed renormalization constants Z in the form of initial segments of series in the charge (or charges) g to the desired accuracy.

Technically, this is the most difficult part of the work, and it can be further complicated by an unfortunate choice of subtraction scheme. The objective results do not depend on subtraction scheme (Sec. 1.40), and so it is always desirable to use the simplest one. In regularizations of the dimensional type this is usually the MS scheme, which also leads to a simpler form of the RG equations.

5. The RG functions β and γ (the β functions of all the charges and the anomalous dimensions γ of all the needed quantities) are calculated in the form of initial segments of series in g using equations that express them in terms of the renormalization constants Z calculated earlier.

6. The β functions are used to calculate the coordinates of the fixed points g_* and the corresponding exponents ω in the form of initial segments of the ε expansion, and the points g_* are classified according to their IR stability (Sec. 1.42).

In single-charge models there can be only one fixed point $g_* \sim \varepsilon$, while in multi-charge ones there may be several. If there are no IR-stable points among the g_*, it must be concluded that the system does not possess critical scaling. If there are such points, the following is done:

7. For each IR-stable point $g_* \sim \varepsilon$ all the needed critical anomalous dimensions $\gamma(g_*)$ and the corresponding critical exponents are calculated in the form of initial segments of the ε expansions. The results are then refined,

if possible, by means of Borel summation (Sec. 1.38).

All the possible critical regimes and the critical exponents for each are thereby enumerated. If there are several of them (which is possible only in multi-charge models), the problem arises of finding the basin of attraction of each one, i.e., of determining the set of initial data in the space of charges g such that starting from these values, a given IR-stable point g_* is reached at the IR asymptote $s \rightarrow 0$ by the phase trajectory given by (1.188).

In complicated models it is in practice possible to calculate only one or two orders of the ε expansion of the exponents and to determine the general qualitative behavior of the phase trajectories. Only in exceptional cases can the calculations be pushed farther. The best is the φ^4 model, for which the efforts of many authors over a long period of time have raised the accuracy of the calculations through order ε^5 (Sec. 1.37).

In addition to calculating exponents, the RG technique can be used to formulate and solve some additional problems, in particular:

8. The calculation of the initial segments of the ε expansions of various scaling functions (Sec. 1.36);

9. The analysis of their singularities outside the framework of the ε expansion by the combined use of the RG technique and the Wilson operator expansion (Sec. 3.33).

10. The analysis of the renormalization and calculation of the critical dimensions of various systems of composite operators. In particular, it can be used to calculate the next-to-leading (calculable from the β functions) correction critical exponents ω (Sec. 1.3) corresponding to the simplest IR-irrelevant corrections discarded in constructing the model.

All of this will be explained below by specific examples. For some models different versions of perturbation theory and the corresponding exponent expansions are possible. For example, for the O_n φ^4 model, three variants of these expansions are known: $4 - \varepsilon$, $2 + \varepsilon$, and $1/n$, and all will be studied in this chapter.

4.2   The O_n φ^4 model: the constants Z, RG functions, and $4 - \varepsilon$ expansion of the exponents

The technique for calculating the constants Z in the renormalized action (3.105) of the O_n φ^4 model was described in detail in Ch. 3, and the contributions of individual graphs to the constants Z and the corresponding structure factors of the O_n group are given in the tables of Sec. 3.20. According to these tables and the general rule (3.118), we find ($d = 4 - 2\varepsilon$)

$$Z_1 = 1 - \frac{u^2}{24\varepsilon}r_1 - \frac{u^3}{48\varepsilon^2}(2 - \varepsilon)r_1 r_2 + ...,$$

$$Z_2 = 1 + \frac{u}{2\varepsilon}r_1 + \frac{u^2}{4\varepsilon^2}r_1^2 + \frac{u^2}{4\varepsilon^2}(1-\varepsilon)r_1 + ...,$$

$$Z_3 = 1 + \frac{3u}{2\varepsilon}r_2 + \frac{3u^2}{4\varepsilon^2}r_4 + \frac{3u^2}{2\varepsilon^2}(1-\varepsilon)r_3 + ..., \tag{4.1}$$

$$Z_0 = nr_1\left\{-\frac{u^2}{4\varepsilon^2} - u^3\left[\frac{r_1}{8\varepsilon^3} + \frac{1-9\varepsilon/4}{36\varepsilon^2} + \frac{1-2\varepsilon+\varepsilon^2}{12\varepsilon^3}\right]\right\} + ...,$$

where $u \equiv g/16\pi^2$ and the r_i are the known factors from Table 3.3 in Sec. 3.20. All the Z can be found with three-loop accuracy (through u^3) using the tables of Sec. 3.20, but we want only to explain the calculations, and so we have limited ourselves to the two-loop approximation in $Z_{2,3}$.

Let us now turn to the calculation of the RG functions (1.113), using this example to explain the calculations in detail so that it will not be necessary to return to these questions later. We first note that it is usually convenient to change the normalization of the renormalized charge, using not the original g, but the parameter $u = cg$, in terms of which the constants Z take on simpler forms. In this case $u = g/16\pi^2$. The quantities $\mathcal{D}_g \equiv g\partial_g$, β_g/g, and $\beta_g\partial_g$ are invariant under the replacement $g \to cg \equiv u$. In particular, for our model the first equation in (1.114) gives

$$\beta_u \equiv \widetilde{\mathcal{D}}_\mu u = -2\varepsilon u - u\gamma_g, \quad (\beta_g\partial_g = \beta_u\partial_u). \tag{4.2}$$

The RG functions γ_a in (1.113) are expressed in terms of the constants Z_a related to the action constants (3.105) by (3.110). From them and (4.1) we can calculate all the Z_a, but it is technically simpler to proceed in a different manner, introducing the quantity $\gamma \equiv \widetilde{\mathcal{D}}_\mu \ln Z$ for any constant Z and using the relations following from (3.110):

$$\gamma_1 = 2\gamma_\varphi, \quad \gamma_2 = \gamma_\tau + 2\gamma_\varphi, \quad \gamma_3 = \gamma_g + 4\gamma_\varphi. \tag{4.3}$$

If we are not concerned with controlling the accuracy of the calculations by checking the cancellation of the poles in ε in the RG functions γ, they can be found simply using (1.120), expressing the γ in terms of the residues $A(u)$ of the first-order poles in ε of the corresponding constants $Z = 1+A(u)/\varepsilon+...$, namely, $\gamma = \widetilde{\mathcal{D}}_\mu \ln Z = -2\mathcal{D}_u A$. For the first three constants in (4.1) we find the corresponding residues $A_1 = -u^2 r_1/24 + u^3 r_1 r_2/48 + ...$, $A_2 = ur_1/2 - u^2 r_1/4 + ...$, and $A_3 = 3ur_2/2 - 3u^2 r_3/2 + ...$, from which for $\gamma_i = -2\mathcal{D}_u A_i$ we obtain

$$\gamma_1 = u^2 r_1/6 - u^3 r_1 r_2/8+...; \quad \gamma_2 = -ur_1+u^2 r_1+...; \quad \gamma_3 = -3ur_2+6u^2 r_3+.... \tag{4.4}$$

We note that in this calculational method we do not need the contributions of graphs which do not contain first-order poles in ε, in particular, graphs 3, 6, and 7 from Table 3.1 in Sec. 3.20. Moreover, in obtaining (1.120) we used the explicit form of the charge renormalization formula (1.103). In other models the factor μ can enter into the analogous formula with a different exponent, which leads to the following change of the form of (1.120):

$$\{g_0 = g\mu^{k\varepsilon}Z_g, \ Z = 1 + A(g)/\varepsilon + ...\} \Rightarrow \gamma \equiv \widetilde{\mathcal{D}}_\mu \ln Z = -k\mathcal{D}_g A, \qquad (4.5)$$

where Z is an arbitrary renormalization constant like (1.104) and the ellipsis denotes contributions of higher order in $1/\varepsilon$. The rule (4.5) will often be used below for various models.

Returning to the φ^4 model, from (4.2), (4.3), and (4.4) we find all the needed RG functions:

$$\gamma_\varphi = \frac{u^2}{12}r_1 - \frac{u^3}{16}r_1r_2 + ..., \ \gamma_\tau = -ur_1 + \frac{5}{6}u^2r_1 + ...,$$

$$\beta \equiv \beta_u = -2\varepsilon u + 3u^2r_2 - 6u^3r_3 + u^3r_1/3 + ... \tag{4.6}$$

We do not give γ_g since this RG function is easily reconstructed from (4.2) using $\beta \equiv \beta_u$.

Substituting into (4.6) the known factors from Table 3.3 in Sec. 3.20, we obtain the final expressions for the RG functions of the O_n φ^4 model. Then from the equation $\beta(u_*) = 0$ we can find the initial segment of the ε expansion of the coordinate of the fixed point $u_* = g_*/16\pi^2$. The accuracy in (4.6) is not very high, but we shall quote the RG functions to five-loop order, the most accurate obtained so far, from the corrected data of Ref. 68 ($2\varepsilon = 4 - d$ everywhere, and $\zeta(z)$ is the Riemann zeta function):

$$\gamma_\varphi = \frac{u^2(n+2)}{36} - \frac{u^3(n+2)(n+8)}{432} + \frac{5u^4(n+2)}{5184}\left[-n^2 + 18n + 100\right] -$$

$$-\frac{u^5(n+2)}{186624}\left[39n^3 + 296n^2 + 22752n + 77056-\right.$$

$$\left.-48\zeta(3)(n^3 - 6n^2 + 64n + 184) + 1152\zeta(4)(5n + 22)\right] + ..., \tag{4.7a}$$

$$\gamma_\tau = -\frac{u(n+2)}{3} + \frac{5u^2(n+2)}{18} - \frac{u^3(n+2)(5n+37)}{36} +$$

$$+\frac{u^4(n+2)}{7776}\left[-n^2 + 7578n + 31060+\right.$$

$$\left.+48\zeta(3)(3n^2 + 10n + 68) + 288\zeta(4)(5n + 22)\right] -$$

$$-\frac{u^5(n+2)}{186624}\Big[21n^3 + 45254n^2 + 1077120n + 3166528+$$

$$+48\zeta(3)(17n^3 + 940n^2 + 8208n + 31848)-$$

$$-768\zeta^2(3)(2n^2 + 145n + 582) + 288\zeta(4)(-3n^3+ \qquad (4.7b)$$

$$+29n^2 + 816n + 2668) + 768\zeta(5)(-5n^2 + 14n + 72)+$$

$$+9600\zeta(6)(2n^2 + 55n + 186)\Big] + ...,$$

$$\beta \equiv \beta_u \quad = -2\varepsilon u + \frac{u^2(n+8)}{3} - \frac{u^3(3n+14)}{3}+$$

$$+\frac{u^4}{216}\Big[33n^2 + 922n + 2960 + 96\zeta(3)(5n+22)\Big]-$$

$$-\frac{u^5}{3888}\Big[-5n^3 + 6320n^2 + 80456n + 196648+$$

$$+96\zeta(3)(63n^2 + 764n + 2332) - 288\zeta(4)(5n^2 + 62n+$$

$$+176) + 1920\zeta(5)(2n^2 + 55n + 186)\Big]+$$

$$+\frac{u^6}{62208}\Big[13n^4 + 12578n^3 + 808496n^2 + 6646336n+ \qquad (4.7c)$$

$$+13177344 + 16\zeta(3)(-9n^4 + 1248n^3 + 67640n^2+$$

$$+552280n + 1314336) + 768\zeta^2(3)(-6n^3 - 59n^2 + 446n+$$

$$+3264) - 288\zeta(4)(63n^3 + 1388n^2 + 9532n + 21120)+$$

$$+256\zeta(5)(305n^3 + 7466n^2 + 66986n + 165084)-$$

$$-9600\zeta(6)(n+8)(2n^2 + 55n + 186)+$$

$$+112896\zeta(7)(14n^2 + 189n + 526)\Big] + ...$$

From now on we restrict ourselves to four loops:

$$u_* \quad \equiv g_*/16\pi^2 = 2\varepsilon\frac{3}{(n+8)} + (2\varepsilon)^2\frac{9(3n+14)}{(n+8)^3}+$$

$$+\frac{3(2\varepsilon)^3}{8(n+8)^5}\Big[-33n^3 + 110n^2 + 1760n + 4544-$$

$$-96\zeta(3)(n+8)(5n+22)\Big]+$$

$$+\frac{(2\varepsilon)^4}{16(n+8)^7}\Big[-5n^5-2670n^4-5584n^3+52784n^2+$$

$$+309312n+529792+96\zeta(3)(63n^4+422n^3-4452n^2- \tag{4.7d}$$

$$-39432n-72512)-288\zeta(4)(n+8)^3(5n+22)+$$

$$+1920\zeta(5)(n+8)^2(2n^2+55n+186)\Big]+....$$

For convenience we also give the RG functions of the simple φ^4 model ($n=1$):

$$\gamma_\varphi = \frac{u^2}{12}-\frac{u^3}{16}+\frac{65u^4}{192}+...,$$

$$\gamma_\tau = -u+\frac{5u^2}{6}-\frac{7u^3}{2}+\frac{3u^4}{2}\left[\frac{159}{16}+\zeta(3)+2\zeta(4)\right]+...,$$

$$\beta = -2u\varepsilon+3u^2-\frac{17u^3}{3}+u^4\left[\frac{145}{8}+12\zeta(3)\right]+ \tag{4.8}$$

$$+u^5\left[-\frac{3499}{48}-78\zeta(3)+18\zeta(4)-120\zeta(5)\right]+...,$$

$$u_* \equiv g_*/16\pi^2=\frac{2\varepsilon}{3}+(2\varepsilon)^2\frac{17}{81}+\frac{(2\varepsilon)^3}{27}\left[\frac{709}{648}-4\zeta(3)\right]+$$

$$+\frac{(2\varepsilon)^4}{81}\left[\frac{10909}{11664}-\frac{106}{9}\zeta(3)-6\zeta(4)+40\zeta(5)\right]+....$$

These expressions can be used to find the $4-\varepsilon$ expansions given in Sec. 1.37 of the critical exponents $\eta=2\gamma_\varphi^*$, $1/\nu=2+\gamma_\tau^*$, and $\omega=\beta'(u_*)$, and then all the other traditional exponents are calculated from (1.165) using η and ν.

4.3 Renormalization and the RG equations for the renormalized functional $W_R(A)$ including vacuum loops

In Ch. 1 we did not study vacuum loops, and took $G=\exp W$ to mean the generating functional of normalized full Green functions (1.33) defined in Sec. 1.10, and so all the RG equations of Sec. 1.24 pertained only to Green functions with $n\geq1$. Beginning in Ch. 2, we changed over to the new definition (2.30), including in the general formalism vacuum loops, connected graphs with no external lines $W_0 = \ln G_0$ (Sec. 2.3). The usual multiplicative renormalization (1.84) is insufficient for eliminating their UV divergences; it is necessary to also add the constant ΔS_{vac}, the vacuum counterterm (Sec. 3.13), to the right-hand side of (1.84). Then from the expression $S_{\text{R}}(\varphi) = S(Z_\varphi \varphi) + \Delta S_{\text{vac}}$ and the definitions, in the usual manner (Sec. 1.18) instead of (1.79) for the normal phase with $\tau > 0$ we obtain the following expressions (the passage to $\tau < 0$ will be discussed at the end of this section):

$$G_{\text{R}}(A) = (C_{\text{R}}/C)\det Z_\varphi^{-1} G(Z_\varphi^{-1}A)\exp(\Delta S_{\text{vac}}), \tag{4.9a}$$

$$W_{\text{R}}(A) = W(Z_\varphi^{-1}A) + \Delta W_0, \quad \Gamma_{\text{R}}(\alpha) = \Gamma(Z_\varphi \alpha) + \Delta W_0, \tag{4.9b}$$

$$\Delta W_0 \equiv \Delta S_{\text{vac}} + \ln(C_{\text{R}}/C) - \ln\det Z_\varphi, \tag{4.9c}$$

where C and C_{R} are the normalization factors for the corresponding integrals in (2.30), and $\det(Z_\varphi^{-1})$ is the Jacobian of the variable substitution $\varphi \to Z_\varphi^{-1}\varphi$ which must be made to obtain (4.9a) (see the analogous derivation in Sec. 1.18). Equations (4.9) replace (1.79) for the new definition (2.30) of the functionals G and W. All the additions in (4.9) are independent of the source A, and so they do not spoil the validity of (1.80) and (1.81) with $n \geq 1$ and are important only for the renormalization of vacuum loops $W_0 = W(A = 0)$, which were not considered in Ch. 1.

The constants C in (4.9) are defined by the general rule (2.16) and are calculated using (2.4) and (2.5). The explicit expression for $\ln C_{\text{R}}$ in the $O_n \varphi^4$ model is given in (3.193) (we assume that $\tau > 0$), and the expression for $\ln C$ is obtained from it by the replacement $\tau \to \tau_0$. Similarly, for the constant $\ln\det Z_\varphi$ in (4.9), from (2.5) for this model we have

$$\ln\det Z_\varphi = \ln Z_\varphi \cdot Vn(2\pi)^{-d}\int dk, \quad V \equiv \int dx. \tag{4.10}$$

Dimensional regularization $d = 4 - 2\varepsilon$ does not eliminate the UV divergences of such integrals, and so the cutoff $|k| \leq \Lambda$ must be understood.

If the model were renormalized taking into account all the UV divergences, including powers of Λ, all the contributions on the right-hand sides of (4.9) would be relevant. However, if, as usual, the counterterm ΔS_{vac} is calculated in the formal scheme discarding all divergences that are powers of Λ (Secs. 1.20 and 3.16), which will always be assumed below, then all the other contributions to (4.9) must also be calculated in the same way. This means that a quantity (4.10) that is purely a power of Λ should simply be discarded, and expressions like (3.193) for $d = 4 - 2\varepsilon$ should be understood as an integral without a cutoff and with the subtraction of the first two terms of the τ expansion containing divergences that are powers of Λ [for the $O_n \varphi^4$ model this leads to the result (3.233)]. Therefore, below we shall assume that in (4.9)

$$\Delta W_0 \equiv \Delta S_{\text{vac}} + \ln(C_{\text{R}}/C), \tag{4.11}$$

understanding quantities like (3.193) as indicated above. Then only poles in ε are assumed to be UV divergences. In (3.233) a pole occurs only in the term quadratic in τ and is selected by the the counterterm operator L of the MS scheme given by (3.31), which leads to (3.236) for the "UV-divergent" part of $\ln C_R$.

Making the replacement $A \to Z_\varphi A$, $\alpha \to Z_\varphi^{-1}\alpha$ in (4.9b) and then acting on both sides of the equation with the operator $\widetilde{\mathcal{D}}_\mu = \mathcal{D}_{RG}$ from (1.106), taking into account (4.11) and the renormalization invariance of all unrenormalized objects (Sec. 1.24), we obtain the RG equations generalizing (1.110) and (1.111) to the case of the functionals $W_R \equiv W_R(A; e, \mu)$ and $\Gamma_R \equiv \Gamma_R(\alpha; e, \mu)$ with vacuum loops:

$$[\mathcal{D}_{RG} + \gamma_\varphi \mathcal{D}_A]W_R = V\xi, \quad [\mathcal{D}_{RG} - \gamma_\varphi \mathcal{D}_\alpha]\Gamma_R = V\xi, \tag{4.12a}$$

$$V\xi \equiv \widetilde{\mathcal{D}}_\mu[\Delta S_{\text{vac}} + \ln C_R], \tag{4.12b}$$

where $V \equiv \int dx$ is the volume and ξ is a new RG function, the UV-finiteness of which is guaranteed by the UV-finiteness of the left-hand side of (4.12a). Setting $A = 0$ in the first expression in (4.12a), we obtain the RG equation $\mathcal{D}_{RG}W_{0R} = V\xi$ for vacuum loops $W_{0R} \equiv W_R(A = 0)$; by subtracting it from (4.12a) we obtain the usual homogeneous equations (1.110) and (1.111) for the normalized functionals (denoted in Ch. 1 by the same letters).

It is technically more convenient to rewrite (4.12) as

$$[\mathcal{D}_{RG} + \gamma_\varphi \mathcal{D}_A]W_R' = V\xi', \quad [\mathcal{D}_{RG} - \gamma_\varphi \mathcal{D}_\alpha]\Gamma_R' = V\xi', \tag{4.13a}$$

$$W_R' \equiv W_R\text{–UV-finite part of } \ln C_R, \Gamma_R' \equiv \Gamma_R\text{–UV-finite part of } \ln C_R, \tag{4.13b}$$

$$V\xi' \equiv \widetilde{\mathcal{D}}_\mu[\Delta S_{\text{vac}} + \text{UV-divergent part of } \ln C_R], \tag{4.13c}$$

grouping the UV-finite part of the constant $\ln C_R$ with the functionals, and the divergent part with the vacuum counterterm. This addition reproduces the one-loop contribution absent in ΔS_{vac}. The RG function ξ' in (4.13) (see Sec. 4.4 for the explicit expression) always has the same structure in the variables τ and μ as the counterterm ΔS_{vac}, in particular, it is a polynomial in τ in the MS scheme, in contrast to the more complicated RG function ξ in (4.12).

The main advantage of (4.13a) in MS-type schemes is the fact that the quantities W_R' and Γ_R' entering into them, in contrast to W_R and Γ_R, can be directly continued to the region $\tau < 0$ with preservation of their UV finiteness and the form of all the counterterms, including the contribution of the UV-divergent part of $\ln C_R$, understood as a one-loop addition to ΔS_{vac}. This was shown in Sec. 3.36 in discussing the renormalization of the free energy. Let us recall the main argument. The definition of W_R' in (4.13b) leads to a representation analogous to (3.237):

$$\exp W_R'(A) = \int D\varphi \exp [S_R(\varphi) + \text{UV-divergent part of } \ln C_R + A\varphi], \tag{4.14}$$

which remains meaningful also for $\tau < 0$ owing to the elimination of the contribution of the UV-finite part of $\ln C_R$. The functional integration space E_{int} in (4.14) is determined by the general rules of Sec. 1.13 from the form of the action $S_R(\varphi) = S(Z_\varphi \varphi) + \Delta S_{vac}$.

Therefore, (4.14) with a suitable choice of integration space defines the quantity $W'_R(A)$ for either sign of τ. Using the representation (4.14), it is easy to verify that the functional $W'_R(A)$ [like its Legendre transform $\Gamma'_R(\alpha)$] satisfies the same RG equation (4.13a) for both $\tau < 0$ and $\tau > 0$. The proof depends only on the relation between S and S_R (see above) and the fact that the unrenormalized quantity $\int D\varphi \exp[S(\varphi) + A(\varphi)]$ is renormalization-invariant (Sec. 1.24) for either sign of $\tau_0 \sim \tau$.

4.4 The O_n φ^4 model: renormalization and the RG equation for the free energy

As explained in Sec. 3.36, the quantity $\mathcal{F}'_R$ defined by (3.237) and differing from the renormalized free energy $\mathcal{F}_R = -\ln Z_R$ by only the contribution of the UV-divergent part of $\ln C_R$ (3.236) regular in τ (and therefore irrelevant for the critical singularities) is a UV-finite object which can be directly continued to the region $\tau < 0$. In addition, the unrenormalized ($\mathcal{F}$) and renormalized ($\mathcal{F}_R$) quantities defined by (3.234) are related as $\mathcal{F}_R = \mathcal{F} - \Delta S_{vac} + \ln \det Z_\varphi$, and so they differ only by regular contributions and either can be used to analyze the critical singularities.

Comparing (3.237) and (4.14), we see that

$$\mathcal{F}'_R = -W'_R(A = 0) \equiv -W'_{0R}. \tag{4.15}$$

From (4.14) and (4.15) we have

$$\mathcal{F}'_R = -\ln \int D\varphi \exp[S_R(\varphi) + \text{UV-divergent part of } \ln C_R]. \tag{4.16}$$

Setting $A = 0$ in (4.13a), we obtain the RG equation for the free energy valid for either sign of τ:

$$\mathcal{D}_{RG}\mathcal{F}'_R = -V\xi'. \tag{4.17}$$

As an example, let us consider (4.17) for the O_n φ^4 model without an external field. We note that the generalization to the case of a system in an external field $h \neq 0$ is effected by the simple replacement $W'_R(A = 0) \rightarrow W'_R(A = h)$, and then in (4.13a) it is necessary to make the replacement $\mathcal{D}_A \rightarrow \mathcal{D}_h$.

The explicit expression for the UV-divergent part of $\ln C_R$ is given in (3.236). It is easily checked that the addition of (3.236) in (4.13c) to the vacuum counterterm (3.106) is equivalent to making the replacement $Z_0 \rightarrow \overline{Z}_0$ with $\overline{Z}_0$ from (3.200). In this model ΔS_{vac} and the addition (3.236) are multiples of $\tau^2 \mu^{-2\varepsilon}$. Since the operator $\widetilde{\mathcal{D}}_\mu$ ($= \mathcal{D}_{RG}$ in renormalized variables) preserves this dependence, it is natural to introduce a new UV-finite RG function $\gamma_0 = \gamma_0(g)$ analogous to the other γ, defining it as

$$\widetilde{\mathcal{D}}_\mu[\Delta S_{\text{vac}} + \text{UV-divergent part of } \ln C_{\text{R}}] \equiv V\xi' = V\tau^2\mu^{-2\varepsilon}\gamma_0(g)/2 \tag{4.18}$$

(the factor of $1/2$ on the right-hand side is convenient). Substituting (3.106) with the replacement $Z_0 \to \overline{Z}_0$ (see above) and taking into account the explicit form (1.122) of the operator $\widetilde{\mathcal{D}}_\mu = \mathcal{D}_{\text{RG}}$ and the first expression in (1.114), we obtain

$$\gamma_0 = \tau^{-2}\mu^{2\varepsilon}\widetilde{\mathcal{D}}_\mu(\tau^2\mu^{-2\varepsilon}g^{-1}\overline{Z}_0) = g^{-1}[\gamma_g - 2\gamma_\tau - (2\varepsilon + \gamma_g)\mathcal{D}_g]\overline{Z}_0(g). \tag{4.19}$$

We have used the following commutation rule for the operator (1.122):

$$\widetilde{\mathcal{D}}_\mu[\mu^a\tau^bg^c\ldots] = \mu^a\tau^bg^c[a - b\gamma_\tau - c(2\varepsilon + \gamma_g) + \widetilde{\mathcal{D}}_\mu]\ldots. \tag{4.20}$$

Here a, b, and c are arbitrary exponents, and the ellipsis denotes any additional term.

Substituting into (4.19) the quantities known from (4.1), (4.7), and (3.200) through three loops we find

$$\gamma_0(g) = \frac{n}{16\pi^2}\left[1 + \frac{(n+2)}{24}u^2 + \ldots\right]. \tag{4.21}$$

All the poles in ε cancel, as required.

If these cancellations are not checked, the calculation of γ_0 can be considerably simplified by keeping only the contribution $-2\varepsilon\mathcal{D}_g$ in the square brackets in (4.19), and only the contributions of first-order poles in ε in $\overline{Z}_0$. This is equivalent to a relation analogous to (1.120)

$$\gamma_0 = -\text{UV-finite part of } [2\varepsilon\partial_g\overline{Z}_0], \tag{4.22}$$

which follows from (4.19) when the (known) ε dependence of all quantities is taken into account.

4.5 General solution of the inhomogeneous RG equation for the free energy of the φ^4 model and the amplitude ratio A_+/A_- in the specific heat

Using (4.17), (4.18), and (1.122), we obtain the following inhomogeneous RG equation for the specific (per unit volume) free energy (4.16) $\overline{\mathcal{F}}_{\text{R}} \equiv \mathcal{F}'_{\text{R}}V^{-1}$:

$$[\mathcal{D}_\mu + \beta\partial_g - \gamma_\tau\mathcal{D}_\tau]\overline{\mathcal{F}}_{\text{R}}(\tau, g, \mu) = -\tau^2\mu^{-2\varepsilon}\gamma_0(g)/2, \tag{4.23}$$

valid for any sign of τ. To take into account the canonical scale invariance, it is convenient, as usual (Ch. 1), to use dimensionless quantities in (4.23). The free energy is dimensionless, and its specific value $\overline{\mathcal{F}}$ has dimension $d = 4 - 2\varepsilon$ (Sec. 1.3). Therefore, (4.23) can be rewritten as

$$\overline{\mathcal{F}}'_{\mathrm{R}}(\tau, g, \mu) = |\tau|^{2-\varepsilon} \Phi(s, g), \quad s \equiv |\tau|/\mu^2, \tag{4.24a}$$

$$[-(2 + \gamma_\tau)\mathcal{D}_s + \beta \partial_g - (2 - \varepsilon)\gamma_\tau]\Phi(s, g) = -s^\varepsilon \gamma_0(g)/2. \tag{4.24b}$$

Equation (4.24b) was obtained by substituting (4.24a) into (4.23). It differs from the standard equation of the type (1.142) by only the inhomogeneous term on the right-hand side. The particular solution of the inhomogeneous equation (4.24b) can be sought in the form $s^\varepsilon b(g)$. Substituting this into (4.24b), we obtain the following equation for the unknown function b:

$$[-2\varepsilon - 2\gamma_\tau + \beta \partial_g]b(g) = -\gamma_0(g)/2. \tag{4.25}$$

It can be solved explicitly in quadratures, or using perturbation theory when b is sought in the form of a series in g in terms of the known series for the RG functions. We note that the coefficients in the b series are found uniquely from (4.25) and contain poles in ε, even though the solution of (4.24b) as a whole must be UV-finite in each order in g.

To the particular solution $s^\varepsilon b(g)$ we must add the general solution of the homogeneous equation (4.24b) with zero right-hand side. Division by $2 + \gamma_\tau$ reduces it to the standard form (1.142). The correpsonding invariant charge $\overline{g}(s, g)$ is determined by (1.179), and the general solution is given by (1.148), with Φ being only the solution of the homogeneous equation. Taking into account the inhomogeneous term in (4.24b), the statement (1.148) of the general solution must be replaced by

$$\Phi(s, g) = s^\varepsilon b(g) + [\Phi(1, \overline{g}) - b(\overline{g})] \exp\left\{-(2 - \varepsilon)\int\limits_{g}^{\overline{g}} dx \gamma_\tau(x)/\beta(x)\right\}, \tag{4.26}$$

where $b(g)$ is any solution of (4.25).

We have studied the RG equation (4.23) in detail only in order to display the role of an inhomogeneous addition to the right-hand side of equations like (1.142) for a specific example. If the critical dimension $\Delta_\tau = 2 + \gamma_\tau^*$ of the variable τ is assumed to be already known (Sec. 4.2), we do not obtain any new information about the critical exponents from (4.23). In fact, the only new feature compared to the analysis of Ch. 1 is the appearance of additional constants in (4.9) in the renormalization of the vacuum loops. From the viewpoint of the RG their role reduces to the generation of an inhomogeneous term in (4.24b) and the corresponding contribution $\sim s^\varepsilon$ to the general solution. When substituted into (4.24a), this contribution becomes τ^2, i.e., it falls into the class of irrelevant (and not controlled by the theory of critical behavior) regular terms in the free energy. In this sense the effect of the inhomogeneous term can be neglected, and the general analysis of Ch. 1 remains in force, which proves the existence of critical scaling as a consequence of the

RG equations (1.110). The critical dimensions of all the needed quantities are known *a priori* from general considerations in this case: $\Delta_\tau = 2 + \gamma_\tau^*$ and $\Delta[\overline{\mathcal{F}}_R'] = d = 4 - 2\varepsilon$ (Sec. 1.3). Therefore, the asymptote of the singular part of the function $\overline{\mathcal{F}}_R(\tau, g, \mu)$ for $s = |\tau|\mu^{-2} \to 0$ must have the form $|\tau|^{d/\Delta_\tau}$, from which we obtain $C = -\partial_\tau^2 \overline{\mathcal{F}}_R \sim |\tau|^{(d-2\Delta_\tau)/\Delta_\tau}$ for the specific heat.

Of course, all this can also be obtained from the representation (4.26) using an expression like (1.157) for the asymptote of the integral in the exponential. However, in (1.157) the canonical form (1.142) of the RG equation is assumed, which is obtained in (4.26) by dividing by $2 + \gamma_\tau$. Therefore, the role of γ^* in (1.157) for the integral in the exponential of (4.26) will now be played by $\gamma_\tau^*/(2 + \gamma_\tau^*)$, and so (4.26) and (1.157) will give $\Phi \sim s^{-(2-\varepsilon)\gamma_\tau^*/(2+\gamma_\tau^*)}$. Upon substitution into (4.24a) this gives the asymptote $\overline{\mathcal{F}}_R' \sim |\tau|^{d/\Delta_\tau}$, in agreement with the general arguments given above.

A new universal characteristic of the critical behavior that can be calculated using the RG representation (4.26) is the amplitude ratio A_+/A_- in the law (1.22) for the free energy or (equivalently) the specific heat $C = -\partial_\tau^2 \overline{\mathcal{F}}_R'$. From (4.25) and (4.26) it follows that

$$\frac{A_+}{A_-} = \frac{b(g_*) - \Phi_+(1, g_*)}{b(g_*) - \Phi_-(1, g_*)}, \qquad b(g_*) = \frac{\gamma_0^*}{4(\varepsilon + \gamma_\tau^*)}, \qquad (4.27)$$

where Φ_+ and Φ_- are the values of the function Φ defined in (4.24a) for the two signs of τ. Equation (4.27) for $b(g_*)$ was obtained by making the substitution $g = g_*$ in (4.25), and then the contribution involving the β function vanishes.

The functions Φ are calculated using the representation (4.16). The g series for $\Phi_+(s, g)$ begins with the one-loop contribution of order unity from the UV-finite part of (3.233), and for $\Phi_-(s, g)$ it begins with the contribution of order $1/g$ from the term $S_R(\varphi_0)$ appearing in (4.16) for $\tau < 0$ from the shift $\varphi \to \varphi + \varphi_0$ to the stationarity point $\varphi_0 \sim g^{-1/2}$ (Sec. 2.4). Then comes the one-loop correction of the type (3.233) with the replacement $\tau \to -2\tau$ and $n \to 1$ (the massless Goldstone modes do not contribute), and so on. Therefore, the ε expansion of $\Phi_+(1, g_*)$ begins with a contribution of order unity, and for $\Phi_-(1, g_*)$ and $b(g_*)$ it begins with a pole $\sim 1/\varepsilon$. The calculation through one loop gives $(d = 4 - 2\varepsilon)$

$$A_+/A_- = n \cdot 2^{\alpha-2} [1 + 2\varepsilon + ...], \qquad (4.28)$$

where $\alpha = \varepsilon(4-n)/(n+8) + ...$ is the exponent of the specific heat (Sec. 1.37), and the ellipsis denotes contributions of higher order in ε.

In conclusion, we note that when calculating the specific heat C in terms of the correlator of two composite operators $[\varphi^2]_R$, an inhomogeneity will be generated in the RG equation by the conterterm for the operator pair (Sec. 3.25). It can be expressed in terms of the vacuum counterterm using (4.23) and the relation $C = -\partial_\tau^2 \overline{\mathcal{F}}_R'$.

4.6   RG equations for composite operators and coefficients of the Wilson operator expansion

In this section we shall use the notation and material of Sec. 3.28. Let $F \equiv \{F_i(x; \varphi), i = 1, 2, ...\}$ be a finite closed family of local monomials that mix only with each other in renormalization, $a_0 \equiv \{a_{0i}(x)\}$ be the set of corresponding bare sources in the unrenormalized action (3.179), and $a \equiv \{a_i(x)\}$ be the set of renormalized sources related to a_0 by (3.180). As explained in Sec. 3.28, the generalization of the RG equations of Sec. 1.24 to the case of Green functions with a single (and no more) composite operator and any number of simple fields is obtained simply by including a_0 in the general scheme on an equal footing with the other bare parameters e_0. This means that in the differentiation in (1.106) it is necessary to also assume that the sources a_0 are fixed. Then in (1.110) W_R acquires an additional functional argument a, and the operator (1.108) acquires the additional term

$$(\widetilde{\mathcal{D}}_\mu a) \frac{\delta}{\delta a} \equiv \int dx [\widetilde{\mathcal{D}}_\mu a_i(x)] \frac{\delta}{\delta a_i(x)}. \tag{4.29}$$

Here and below the repeated indices of sources and operators are always understood to be summed over (in contrast to the notation of Sec. 3.28). For brevity, in what follows we shall use matrix notation, understanding F, a, and a_0 to be vectors (in the index i) and writing (3.180) and (3.183) as

$$a_0 = aZ_a, \qquad \widehat{F} = Z_F \widehat{F}_R, \qquad Z_F = Z_a^{-1}, \tag{4.30}$$

where Z_a and Z_F are mutually inverse matrices of the renormalization constants of the parameters a and operators F. The exact meaning of the unrenormalized $(\widehat{F})$ and renormalized $(\widehat{F}_R)$ operators is given in (3.184). We also recall that $Mx = xM^\top$ for arbitrary matrix M and vector x.

The UV-finite quantities

$$\gamma_a = -\gamma_F, \qquad \gamma_a \equiv \widetilde{\mathcal{D}}_\mu Z_a \cdot Z_a^{-1}, \qquad \gamma_F \equiv Z_F^{-1} \cdot \widetilde{\mathcal{D}}_\mu Z_F \tag{4.31}$$

analogous to the RG functions (1.113) are called the matrices (of the RG functions) of the anomalous dimensions of the sources a and operators F, and the relation $\gamma_a = -\gamma_F$ follows from their definitions (4.31) and the last equation in (4.30).

The operator (4.29) involves the quantity $\widetilde{\mathcal{D}}_\mu a$, which is easily found by acting on both sides of the first equation in (4.30) with the operator $\widetilde{\mathcal{D}}_\mu$. On the left-hand side we obtain zero owing to the definition of $\widetilde{\mathcal{D}}_\mu$ (see above), and so $0 = \widetilde{\mathcal{D}}_\mu a Z_a + a \widetilde{\mathcal{D}}_\mu Z_a$, from which $\widetilde{\mathcal{D}}_\mu a = -a \widetilde{\mathcal{D}}_\mu Z_a \cdot Z_a^{-1} = -a\gamma_a$ according to (4.31). Therefore, the desired generalization of (1.110) to the case of the generating functional of the renormalized connected Green functions of the fields and operators F can be written as

$$\left[\mathcal{D}_{RG} - a\gamma_a \frac{\delta}{\delta a} + \gamma_\varphi \mathcal{D}_A \right] W_R(A, a, e, \mu) = 0. \tag{4.32}$$

Here $\mathcal{D}_{\mathrm{RG}}$ is understood to be the former operator (1.108), the extra term (4.29) has been isolated, and in the latter the required integration over x and summation over source indices are understood (i.e., $a\gamma_a\delta/\delta a$ is a quadratic form with matrix γ_a and "vectors" a and $\delta/\delta a$). The analog of (1.111) will be the equation

$$\left[\mathcal{D}_{\mathrm{RG}} - a\gamma_a\frac{\delta}{\delta a} - \gamma_\varphi\mathcal{D}_\alpha\right]\Gamma_{\mathrm{R}}(\alpha, a, e, \mu) = 0, \qquad (4.33)$$

obtained from (4.32) using (1.50), (1.51), and (1.53).

Equations (4.32) and (4.33) are not rigorous; they are valid only to terms of first order in a, which correspond to Green functions with a single composite operator and any number of simple fields, since the original multiplicative renormalization formula (3.178) is only valid with this accuracy. The stipulation "accurate to terms of first order in a" will always be understood when writing equations like (4.32) and (4.33), as well as the second stipulation "except for vacuum loops," so as not to complicate simple equations like (1.110) by the introduction of inhomogeneities on the right-hand side (Sec. 4.3).

Equations (4.32) and (4.33) can be rewritten as equations for the corresponding Green functions with a single $\widehat{F}_{\mathrm{R}}$ and any number of simple factors $\widehat{\varphi}_{\mathrm{R}}$, similar to how (1.110) and (1.111) can be rewritten in the form (1.107) and (1.109). This is done most simply using the fact that all the RG equations in Sec. 1.24 for connected or full Green functions of the field $\widehat{\varphi}_{\mathrm{R}}$ are equivalent to a single equation

$$\mathcal{D}_{\mathrm{RG}}\widehat{\varphi}_{\mathrm{R}} = -\gamma_\varphi\widehat{\varphi}_{\mathrm{R}} \qquad (4.34)$$

for the field $\widehat{\varphi}_{\mathrm{R}}$ itself (as it is a fluctuating quantity, it depends implicitly on all the parameters of the action functional specifying the distribution function). Equation (4.34) expresses the renormalization invariance $\widetilde{\mathcal{D}}_\mu\widehat{\varphi} = 0$ of the unrenormalied field $\widehat{\varphi} = Z_\varphi\widehat{\varphi}_{\mathrm{R}}$ (Sec. 1.24) and is easily obtained using (1.108) and the definition of γ_φ in (1.107):

$$\mathcal{D}_{\mathrm{RG}}\widehat{\varphi}_{\mathrm{R}} \equiv \widetilde{\mathcal{D}}_\mu\widehat{\varphi}_{\mathrm{R}} = \widetilde{\mathcal{D}}_\mu(Z_\varphi^{-1}\widehat{\varphi}) = (\widetilde{\mathcal{D}}_\mu Z_\varphi^{-1})\widehat{\varphi} = -\gamma_\varphi Z_\varphi^{-1}\widehat{\varphi} = -\gamma_\varphi\widehat{\varphi}_{\mathrm{R}}.$$

In the same way, from the second equation in (4.30) and the definition of γ_F in (4.31) we find

$$\mathcal{D}_{\mathrm{RG}}\widehat{F}_{\mathrm{R}} = -\gamma_F\widehat{F}_{\mathrm{R}} \quad (\mathcal{D}_{\mathrm{RG}}\widehat{F}_{i\mathrm{R}} = -\gamma_F^{ik}\widehat{F}_{k\mathrm{R}}), \qquad (4.35)$$

with the more detailed notation given in parentheses. As in the expressions of Sec. 3.28, for convenience the matrix indices ik are written as superscripts, but now we drop the symbol for summation over the repeated index.

When the linear differential operator $\mathcal{D}_{\mathrm{RG}}$ acts on the full Green function, the expectation value of a product of fields and operators, the result is the sum of contributions from each cofactor; for example, $\mathcal{D}_{\mathrm{RG}}\langle\widehat{\varphi}_{\mathrm{R}}...\widehat{\varphi}_{\mathrm{R}}\rangle = -n\gamma_\varphi\langle\widehat{\varphi}_{\mathrm{R}}...\widehat{\varphi}_{\mathrm{R}}\rangle$ and, similarly,

$$\mathcal{D}_{\mathrm{RG}}\langle\widehat{F}_{i\mathrm{R}}\widehat{\varphi}_{\mathrm{R}}...\widehat{\varphi}_{\mathrm{R}}\rangle = -[n\gamma_\varphi\delta^{ik} + \gamma_F^{ik}]\langle\widehat{F}_{k\mathrm{R}}\widehat{\varphi}_{\mathrm{R}}...\widehat{\varphi}_{\mathrm{R}}\rangle, \qquad (4.36)$$

where n is the number of factors $\widehat{\varphi}_{\mathrm{R}}$, and, as usual, arguments x are understood everywhere. In going from full to connected Green functions the form of RG equations like (4.36) does not change, and in going to 1-irreducible functions the replacement $Z_\varphi \to Z_\varphi^{-1}$ in the renormalization formulas [see (1.81) and (3.186)] corresponds to the replacement $\gamma_\varphi \to -\gamma_\varphi$ in the RG equations while preserving the sign of γ_F. We also note that (4.36) generalizes directly to the case of any number of factors $\widehat{F}_i$. However, we should bear in mind that for two or more factors (in contrast to a single one) quantities like $\langle\widehat{F}_{\mathrm{R}}\widehat{F}'_{\mathrm{R}}\widehat{\varphi}_{\mathrm{R}}...\widehat{\varphi}_{\mathrm{R}}\rangle$ are not the same as the corresponding renormalized Green functions owing to the absence of counterterm contributions for pairs, triplets, etc. of composite operators [see (3.159)]. Since the counterterms are local, their contributions vanish if the arguments x of all the operator factors are different. Therefore, the critical asymptotic behavior at large (relative) separations for renormalized Green functions with several $\widehat{F}_{\mathrm{R}}$ will be exactly the same as that for simple expectation values $\langle\widehat{F}_{\mathrm{R}}\widehat{F}'_{\mathrm{R}}\widehat{\varphi}_{\mathrm{R}}...\widehat{\varphi}_{\mathrm{R}}\rangle$, i.e., it will be determined by the critical dimensions of the cofactors.

Let us conclude by giving the RG equations following from (4.35) for the coefficients of the Wilson operator expansion (Sec. 3.33) for the example of (3.217). We write it compactly in the form $\widehat{\varphi}_{\mathrm{R}}\widehat{\varphi}_{\mathrm{R}} = B\widehat{F}_{\mathrm{R}} + \widehat{O}_{\xi\mathrm{R}}$ and operate on both sides with the linear operator $\mathcal{D}_{\mathrm{RG}}$. Taking into account (4.34) and (4.35), we then obtain

$$\mathcal{D}_{\mathrm{RG}}B\cdot\widehat{F}_{\mathrm{R}} - B\gamma_F\widehat{F}_{\mathrm{R}} + \mathcal{D}_{\mathrm{RG}}\widehat{O}_{\xi\mathrm{R}} = -2\gamma_\varphi\widehat{\varphi}_{\mathrm{R}}\widehat{\varphi}_{\mathrm{R}} = -2\gamma_\varphi(B\widehat{F}_{\mathrm{R}} + \widehat{O}_{\xi\mathrm{R}}).$$

Choosing the ξ-relevant contributions (the operator $\mathcal{D}_{\mathrm{RG}}$ does not act on ξ and therefore does not affect the ξ relevancy), we obtain

$$\mathcal{D}_{\mathrm{RG}}B\cdot\widehat{F}_{\mathrm{R}} = B\gamma_{\mathrm{F}}\widehat{F}_{\mathrm{R}} - 2\gamma_\varphi B\widehat{F}_{\mathrm{R}}.$$

Since all the $\widehat{F}_{i\mathrm{R}}$ are independent, we can equate their coefficients on the two sides of the equation to find

$$(\mathcal{D}_{\mathrm{RG}} + 2\gamma_\varphi)B = \gamma_F^\top B \quad ((\mathcal{D}_{\mathrm{RG}} + 2\gamma_\varphi)B_i = \gamma_F^{ki}B_k). \qquad (4.37)$$

This is the desired equation for the Wilson coefficients $B_i \equiv B_i(\xi, e, \mu)$ in the expansion (3.217). As in (4.35), in the parentheses we give the more detailed notation.

4.7 Critical dimensions of composite operators

Let us assume that the critical asymptotic behavior of the original model is determined by an IR-stable fixed point g_* and that we are interested in the critical dimensions Δ_F of some given closed family of renormalized composite operators $\widehat{F}_{\mathrm{R}}$. The existence and type of point g_* are determined at the level of the β functions of the original model. Extension of the model by adding (2.104) to the action does not affect these since the addition is treated as a perturbation. The only new objects are the new Green functions involving composite operators. Their critical asymptotic behavior (at large relative separations or small momenta) is characterized by new critical dimensions that must be determined.

As explained in Sec. 1.33, information about critical scaling can be obtained simply by eliminating $\mathcal{D}_\mu$ from the system of two equations of the type (1.6), the first of which is the RG equation for $g = g_*$, while the second expresses the canonical scale invariance. In general, the RG operator (1.108) in MS-type schemes has the form

$$\mathcal{D}_{\mathrm{RG}} = \mathcal{D}_\mu + \sum_g \beta(g)\partial_g - \sum_{e \neq g} \gamma_e(g)\mathcal{D}_e, \qquad (4.38)$$

where g is one or several dimensionless charges and $e \neq g$ are other parameters of the model like τ in (1.122). For $g = g_*$ the contribution of the β functions vanishes, and $\gamma_e(g)$ becomes $\gamma_e(g_*) \equiv \gamma_e^*$, i.e.,

$$\mathcal{D}_{\mathrm{RG}}|_{g=g_*} = \mathcal{D}_\mu - \sum_{e \neq g} \gamma_e^* \mathcal{D}_e. \qquad (4.39)$$

In our case, from (4.35) and (4.39) for $g = g_*$ we find

$$\left[\mathcal{D}_\mu - \sum_{e \neq g} \gamma_e^* \mathcal{D}_e \right] \widehat{F}_{\mathrm{R}} = -\gamma_F^* \widehat{F}_{\mathrm{R}}. \qquad (4.40)$$

The canonical scaling equation for the operators $\widehat{F}_{\mathrm{R}}(x)$ in coordinate space is similar in form:

$$\left[\mathcal{D}_\mu - \mathcal{D}_x + \sum_{e \neq g} d_e \mathcal{D}_e \right] \widehat{F}_{\mathrm{R}} = d_F \widehat{F}_{\mathrm{R}}, \qquad (4.41)$$

where $d_{e,F}$ are the canonical dimensions of the corresponding quantities, the contribution $\mathcal{D}_g$ is absent because the charges are dimensionless, and $\mathcal{D}_x \equiv x\partial_x$ enters with the coefficient $d_x = -1$ (Sec. 1.15). Equations (4.40) and (4.41) and others of the same type are valid for any $\widehat{F}_{\mathrm{R}}$ inside the brackets $\langle...\rangle$, and this is the precise meaning of such equations.

Subtracting (4.40) from (4.41), we obtain the desired equation of critical scaling:

$$\left[-\mathcal{D}_x + \sum_{e \neq g} \Delta_e \mathcal{D}_e \right] \widehat{F}_{\mathrm{R}} = \Delta_F \widehat{F}_{\mathrm{R}}, \qquad (4.42)$$

in which $\Delta_e = d_e + \gamma_e^*$ are the critical dimensions of the parameters e, and

$$\Delta_F = d_F + \gamma_F^*, \qquad \gamma_F^* \equiv \gamma_F(g_*), \qquad (4.43)$$

is the matrix of critical dimensions of the operators $\widehat{F}_{\mathrm{R}}$.

The diagonal matrix d_F of the canonical dimensions $d_i \equiv d[\widehat{F}_{i\mathrm{R}}] = d[\widehat{F}_i]$ (which do not change in renormalization) in (4.43) is known, and γ_F^* is calculated in terms of Z_F or $Z_a = Z_F^{-1}$ from the definitions (4.31) and (4.43). If the latter were also diagonal, (4.42) would have the same structure as (1.6), and the diagonal elements of the matrix Δ_F would then be interpreted as the critical dimensions of the corresponding operators $\widehat{F}_{i\mathrm{R}}$. However, in general the matrix γ_F^* is nondiagonal (and asymmetric), and its off-diagonal elements have dimensions $(d[\gamma_F^{ik}] = d_i - d_k \neq 0)$, and so they must depend on the dimensional variables μ and $e \neq g$. Then the operators $\widehat{F}_{i\mathrm{R}}$ themselves do not possess definite critical dimensions; instead, certain linear combinations of them ("basis operators") do:

$$\widehat{F}'_{i\mathrm{R}} = U_{ik}\widehat{F}_{k\mathrm{R}}, \tag{4.44}$$

where U is some matrix which in general depends on the parameters e and μ [since $\Delta_F(e, \mu)$ cannot be diagonalized by a transformation (4.44) independent of e and μ]. The substitution $\widehat{F}_{\mathrm{R}} = U^{-1}\widehat{F}'_{\mathrm{R}}$ in (4.42) leads to a similar equation for $\widehat{F}'_{\mathrm{R}}$ with the new matrix

$$\Delta'_F = U\Delta_F U^{-1} + \sum_{e \neq g} \Delta_e \mathcal{D}_e U \cdot U^{-1}. \tag{4.45}$$

If the U making the matrix (4.45) diagonal is found, the problem is solved: the diagonal elements Δ'_F are the desired critical dimensions of the operators (4.44), and the original $\widehat{F}_{i\mathrm{R}}$ are known linear combinations $\widehat{F}_{i\mathrm{R}} = (U^{-1})_{ik}\widehat{F}'_{k\mathrm{R}}$ of basis operators with different critical dimensions. The IR asymptote of various Green functions involving one or more composite operators (see Sec. 4.6) is determined from the known critical dimensions by the general rules of similarity theory (Ch. 1). In the following sections we shall explain this using specific examples, while here we limit ourselves to a few general remarks.

1. Without additional constraints on the matrix U in (4.44) the critical dimensions cannot be uniquely determined, since, for example, new operators $\widehat{F}''_{i\mathrm{R}}$ with definite but different critical dimensions can be constructed by multiplying the $\widehat{F}'_{i\mathrm{R}}$ by any power of the dimensional variables $e \neq g$. This is unimportant, since by explicitly finding some system of basis operators $\widehat{F}'_{i\mathrm{R}}$ the common factors can be eliminated and the dimensions rescaled accordingly. We note that the factors μ are critically dimensionless, since $\mathcal{D}_\mu$ does not enter into (4.42).

2. Knowledge of any operator $\widehat{F}'_{i\mathrm{R}}$ is equivalent to the determination of the corresponding row of the matrix U in (4.44).

3. Owing to the presence of the second term on the right-hand side of (4.45), the critical dimensions in general are not eigenvalues of the matrix Δ_F.

4. The system of operators F usually breaks up into groups with different $d_F^* \equiv d[F]|_{d=d^*}$ (Sec. 3.28). The corresponding matrix Δ_F is block-triangular,

because the senior monomials (with maximum d_F^*) cannot admix under renormalization into the junior monomials (Sec. 3.28). Assuming that the critical dimensions of the junior operators are known, for the new senior operators they can then be determined simply as the eigenvalues of the corresponding diagonal block in Δ_F. The other blocks of Δ_F are needed only to determine the corresponding basis operators (4.44).

5. The system F can be redefined by multiplying the unrenormalized monomials by suitable powers of the dimensional bare parameters $e_0 \neq g_0$ (which does not spoil the renormalization invariance), and the renormalized ones by powers of the corresponding renormalized parameters e in order to make all the quantities d_F^* coincide (in the φ^4 model only τ_0 and τ can serve as the parameters e_0 and e). The off-diagonal elements Δ_F of the new system will contain only factors of μ^ε; there will no longer be other parameters $e \neq g$, and so the second contribution to the right-hand side of (4.45) will be absent. In this formulation the problem reduces to the simple diagonalization of the matrix Δ_F by a similarity transformation, and the critical dimensions are then the eigenvalues of the matrix Δ_F. The latter is also true, of course, in massless models because there the junior operators do not mix with the senior ones.

6. In general, in the diagonalization of the matrix Δ_F by a similarity transformation "$U\Delta_F U^{-1} =$ diagonal" the eigenvectors $\mathbf{x}^a$ of the matrix Δ_F, which are nonorthogonal for asymmetric Δ_F, form the columns of the matrix U^{-1}, and the system $\mathbf{y}^b$ biorthogonal to $\mathbf{x}^a$ ($\mathbf{x}^a\mathbf{y}^b = \delta^{ab}$) form the desired rows of the matrix U.

7. It is theoretically possible (though we know of no real examples) that the transformation (4.44) can take an asymmetric matrix Δ_F to only Jordan form rather than diagonal form. Then in the equations of critical scaling for $\widehat{F}'_{iR}$ the usual power-law solutions would be distorted by the appearance of additional logarithms.

8. In the MS scheme, when ε enters into the constants Z only as poles, the calculation of the UV-finite matrix $\gamma_F = -\gamma_a$ in (4.31) can be simplified by analogy with the derivation of (1.120). Computing the matrices $Z_{a,F}$ in the form of series in g and then grouping together the contributions of the same order in $1/\varepsilon$, we obtain $Z_a = 1 + A_{1a}/\varepsilon + A_{2a}/\varepsilon^2 + \ldots$ and similarly for $Z_F = Z_a^{-1}$. The matrix residues A_{na} independent of ε are series in g and may, in contrast to their analogs in (1.119), also depend on other parameters e, μ, for example, on τ and μ in the φ^4 model. In the MS scheme the operator (1.108) depends explicitly on ε only through the contributions to the β functions linear in ε (Sec. 1.25), and so by acting with the operator $\widetilde{\mathcal{D}}_\mu = \mathcal{D}_{\mathrm{RG}}$ on residues A_{na} independent of ε it is possible to obtain no more than a single factor of ε. It then follows that when the certainly UV-finite (like every RG function) matrix $\gamma_F = -\widetilde{\mathcal{D}}_\mu Z_a \cdot Z_F$ is calculated from (4.31), the higher-order poles A_{na}/ε^n with $n \geq 2$ cannot give UV-finite contributions to $\widetilde{\mathcal{D}}_\mu Z_a$. All such contributions are contained in $\widetilde{\mathcal{D}}_\mu A_{1a}/\varepsilon$ and obviously do not

change upon subsequent multiplication by Z_F. Therefore, in the MS scheme for any matrix constant $Z_a = Z_F^{-1} = 1 + A_{1a}/\varepsilon + \dots$ we have

$$\gamma_F = -\text{UV-finite part of } \left[\widetilde{\mathcal{D}}_\mu Z_a\right] = -\text{UV-finite part of } [\mathcal{D}_{\text{RG}} A_{1a}/\varepsilon],$$
(4.46)

where "UV-finite part of" stands for the operator that selects the UV-finite part.

In the φ^4 model the operator $\widetilde{\mathcal{D}}_\mu = \mathcal{D}_{\text{RG}}$ on functions of g, τ, and μ has the form (1.122), and when substituting into (4.46) it is necessary to keep only contributions generating a factor of ε, i.e., $\mathcal{D}_\mu$ (the parameter μ can enter only as μ^ε) and $-2\varepsilon\mathcal{D}_g$ from the first term in the β function (1.114). Therefore, from (4.46) for this model we have

$$\gamma_F = -\text{UV-finite part of } \left[\widetilde{\mathcal{D}}_\mu Z_a\right] = (2\varepsilon\mathcal{D}_g - \mathcal{D}_\mu)\cdot[\text{contribs} \sim 1/\varepsilon \text{ to } Z_a].$$
(4.47)

We note that the operator $\mathcal{D}_\mu - 2\varepsilon\mathcal{D}_g$ commutes with the factors τ and $g_{\text{B}} = g\mu^{2\varepsilon}$, and also with the operators $\mathcal{D}_g \equiv g\partial_g$ and $\mu^{-2\varepsilon}\partial_g$, and these simple properties are useful in calculations.

Relations like (4.46) and (4.47) demonstrate the ε independence of all the matrices γ_F in the MS scheme and considerably simplify the calculation of these matrices, especially for complicated systems.

9. In the language of the functional W_{R} from (4.32), the canonical scaling equation equivalent to (4.41) for the fields and operators can be written as

$$\left[\mathcal{D}_\mu - \mathcal{D}_x^{A,a} + \sum_{e\neq g} d_e\mathcal{D}_e - d_\varphi\mathcal{D}_A - d_F\mathcal{D}_a\right]W_{\text{R}} = 0,$$
(4.48)

where $\mathcal{D}_A$ is the operator from (1.112), $\mathcal{D}_a$ is its analog with $A \to a$, and

$$\mathcal{D}_x^{A,a} \equiv \int dx \left[A(x)\mathcal{D}_x\frac{\delta}{\delta A(x)} + a(x)\mathcal{D}_x\frac{\delta}{\delta a(x)}\right].$$
(4.49)

When there are several fields and operators, the sources A and a carry the appropriate indices and all operations include summing over them. The analog of (4.40) will be (4.32) for $g = g_*$; subtracting it from (4.48) we obtain the equation of critical scaling,

$$\left[-\mathcal{D}_x^{A,a} + \sum_{e\neq g} \Delta_e\mathcal{D}_e - \Delta_\varphi\mathcal{D}_A - a\Delta_F\frac{\delta}{\delta a}\right]W_{\text{R}}\big|_{g=g_*} = 0,$$
(4.50)

in which the equation $\gamma_a^* = -\gamma_F^*$ following from (4.31) has been used, and the contribution of sources to the operator is written as in (4.32). If the matrix of critical dimensions of the sources $\Delta_a = d_a + \gamma_a^*$ analogous to Δ_F in (4.43) is introduced, then using (3.150) and the equation $\gamma_a^* = -\gamma_F^*$ it can be concluded that the canonical and critical dimensions of the operators and the corresponding sources always satisfy a shadow relation analogous to (1.46):

$$d_a + d_F = \Delta_a + \Delta_F = d, \tag{4.51}$$

where $d_{a,F}$ are diagonal matrices of the corresponding canonical dimensions, and d is a multiple of the unit matrix.

4.8 Correction exponents ω associated with IR-irrelevant composite operators

As explained in Sec. 1.16, in the construction of any fluctuation model of critical behavior, one first discards all the IR-irrelevant contributions to the action, i.e., all the IR-irrelevant composite operators, in the terminology we are now using. If we wish to evaluate the associated corrections to scaling, these operators should be added to the action in the form (2.104) with uniform sources $a_i(x) = \text{const}$, taking steps, of course, to make sure they form a closed system. The matrix Δ_F calculated for these operators can then be used to find from (4.51) the matrix Δ_a of critical dimensions of the parameters a. Only linear combinations a_i' of parameters a corresponding to the basis operators (4.44) will have definite dimensions $\Delta[a_i'] = d - \Delta[\widehat{F}_{iR}']$. The inclusion of the perturbation aF in first order in a leads to the appearance in the results of critically (and canonically) dimensionless relative correction factors of the type $1 + \sum c_i a_i' \cdot p^{-\Delta[a_i']} + ...$ with critically dimensionless (but, in general, canonically dimensional) coefficients c_i independent of a. The exponents of the momentum p (or another IR-relevant parameter taking into account its critical dimension) that make the contributions of the a_i' dimensionless are, by definition (Sec. 1.3), the corresponding correction exponents:

$$\omega_i = -\Delta[a_i'] = \Delta[\widehat{F}_{iR}'] - d. \tag{4.52}$$

The fundamentally new ones are those corresponding to the subset of senior operators with maximum d_F^* (Sec. 4.7). For the junior operators the corresponding parameters a' will be expressions like $a_i' = \tau$ times the preceding a_i' (i.e., the a_i' of the system F preceding in the dimension), and so the corresponding ω_i are not new. There are no "unnecessary" ω_i in massless models, because the junior operators do not mix with senior ones.

4.9 Example: the system $F = \{1, \varphi^2\}$ in the O_n φ^4 model

Let us consider the very simple example of two operators $F_1 = 1$ and $F_2 = \varphi^2(x) = \varphi_a(x)\varphi_a(x)$ in the O_n φ^4 model. The corresponding 2×2 matrix of renormalization constants Z_a was calculated in Sec. 3.30 [see the discussion following (3.200)]:

$$Z_a^{11} = 1, \quad Z_a^{12} = 0, \quad Z_a^{21} = -2\tau\mu^{-2\varepsilon}g^{-1}\overline{Z}_0, \quad Z_a^{22} = Z_\tau \tag{4.53}$$

with $\overline{Z}_0$ from (3.200). Inverting the 2×2 matrix using the rule

$$\begin{vmatrix} a & b \\ c & d \end{vmatrix}^{-1} = \frac{1}{(ad - bc)} \begin{vmatrix} d & -b \\ -c & a \end{vmatrix}, \tag{4.54}$$

we find $Z_F = Z_a^{-1}$ and then $\gamma_F = -\widetilde{\mathcal{D}}_\mu Z_a \cdot Z_F$, namely, $\gamma_F^{11} = \gamma_F^{12} = 0$, $\gamma_F^{21} = (\gamma_\tau - \widetilde{\mathcal{D}}_\mu) Z_a^{21}$, and $\gamma_F^{22} = -\gamma_\tau$ using the definition of γ_τ in (1.113). The nontrivial element γ_F^{21} is expressed in terms of the RG function (4.19) because it follows from the rule (4.20) that $(\gamma_\tau - \widetilde{\mathcal{D}}_\mu) Z_F^{21} = -\tau^{-1} \widetilde{\mathcal{D}}_\mu (\tau Z_a^{21})$, and the quantity τZ_a^{21} coincides up to a coefficient with the expression operated on by $\widetilde{\mathcal{D}}_\mu$ in (4.19). Finally, we obtain

$$\gamma_F^{11} = \gamma_F^{12} = 0, \quad \gamma_F^{21} = 2\tau\mu^{-2\varepsilon}\gamma_0(g), \quad \gamma_F^{22} = -\gamma_\tau(g). \tag{4.55}$$

Setting $g = g_*$ and adding the diagonal matrix of canonical dimensions $d_1 = d[1] = 0$, $d_2 = d[\varphi^2] = 2 - 2\varepsilon$, we find the matrix Δ_F:

$$\Delta_F^{11} = \Delta_F^{12} = 0, \quad \Delta_F^{21} = 2\tau\mu^{-2\varepsilon}\gamma_0^*, \quad \Delta_F^{22} = 2 - 2\varepsilon - \gamma_\tau^*, \tag{4.56}$$

where γ^* are the values of the corresponding RG functions at the fixed point, γ_τ^* are the values known (Sec. 1.37) to five loops, and γ_0^* are the values from (4.21) and (4.7d) to three loops.

In this case the system (4.42) reduces to just a single nontrivial equation:

$$[-\mathcal{D}_x + \Delta_\tau \mathcal{D}_\tau] \widehat{F}_{2\mathrm{R}} = \Delta_F^{21} + \Delta_F^{22} \widehat{F}_{2\mathrm{R}}. \tag{4.57}$$

Clearly, the basis operators (4.44) must now be sought in the form $\widehat{F}'_{1\mathrm{R}} = \widehat{F}_{1\mathrm{R}} = 1$, $\widehat{F}'_{2\mathrm{R}} = \widehat{F}_{2\mathrm{R}} + c\tau\mu^{-2\varepsilon}$, with the constant c chosen from the condition

$$\Delta_\tau \mathcal{D}_\tau (c\tau\mu^{-2\varepsilon}) = \Delta_\tau c\tau\mu^{-2\varepsilon} = \Delta_F^{22} c\tau\mu^{-2\varepsilon} - \Delta_F^{21},$$

which when satisfied for $\widehat{F}'_{2\mathrm{R}}$ gives the homogeneous equation (4.57) without the contribution Δ_F^{21}. Taking into account the explicit form of the matrix elements (4.56) and the equation $\Delta_\tau = 2 + \gamma_\tau^*$, from the above equation we find the constant c and the corresponding operator

$$\widehat{F}'_{2\mathrm{R}} = \widehat{F}_{2\mathrm{R}} - \tau\mu^{-2\varepsilon}\gamma_0^*/(\varepsilon + \gamma_\tau^*). \tag{4.58}$$

It is this and not the original $\widehat{F}_{2\mathrm{R}} = [\widehat{\varphi}_\mathrm{R}^2]_\mathrm{R}$ that has definite critical dimension (obeying the shadow relation with Δ_τ):

$$\Delta_2 = \Delta_F^{22} = d - \Delta_\tau = 2 - 2\varepsilon - \gamma_\tau^*. \tag{4.59}$$

The second operator $\widehat{F}'_{1\mathrm{R}} = 1$ of the system is trivial, and $\Delta_1 = 0$.

Substituting the known values of γ^*, we obtain the ε expansion of the constant c in (4.58). It will begin at $1/\varepsilon$, because the ε expansion of γ_0^*

obtained from (4.21) begins with a constant, and that of γ_τ^* begins at terms of order ε. In leading order,

$$\widehat{F}'_{2\mathrm{R}} = [\widehat{\varphi}_\mathrm{R}^2]_\mathrm{R} - \tau\mu^{-2\varepsilon}\frac{n(n+8)}{16\pi^2\varepsilon(4-n)}. \tag{4.60}$$

For $n = 4$ the terms of order ε in the denominator of (4.58) cancel out and ε^2 becomes the leading term, and so the expansion of the constant in (4.58) begins at $1/\varepsilon^2$. We note that the poles in ε arise as a result of the diagonalization and do not occur in the original matrix (4.56). The impossibility of taking the limit $\varepsilon = 0$ in this case does not imply nonrenormalizability, but is simply an indication of a change of regime: for $\varepsilon = 0$ the power-law asymptotes are replaced by logarithmic ones.

4.10 Second example: scalar operators with $d_F^* = 4$

Let us now consider, in the same model, the following closed system of operators:

$$F = \{1, \varphi^2, \partial^2\varphi^2, \varphi\partial^2\varphi, \varphi^4\} \tag{4.61}$$

with canonical dimensions $d_F = \{0, 2 - 2\varepsilon, 4 - 2\varepsilon, 4 - 2\varepsilon, 4 - 4\varepsilon\}$. Its renormalization has been discussed in Sec. 3.30. The corresponding 5×5 matrix Z_a (and therefore all others) has the following block structure:

$$
Z_a =
\begin{array}{c|ccccc}
 & 1 & 2 & 3 & 4 & 5 \\
\hline
1 & \times & 0 & 0 & 0 & 0 \\
2 & \times & \times & 0 & 0 & 0 \\
3 & 0 & 0 & \times & 0 & 0 \\
4 & \times & \times & ? & \times & \times \\
5 & \times & \times & ? & \times & \times \\
\end{array}
. \tag{4.62}
$$

We have explicitly indicated all the zero elements, the known nonzero elements are indicated by a cross, and the unknown ones by a question mark. We also show the breakup into blocks of junior (1,2) and senior (3,4,5) operators. The diagonal 1,2 block coincides with the matrix (4.53), and the renormalization of the senior operator F_3 is determined by the renormalization of the junior one F_2, so that elements 33 and 22 coincide. For the fourth and fifth rows the elements of the matrix Q in (3.181) for this system are determined by (3.201) and (3.202), and then Q is used with the general rule (3.181) to find the desired elements of the matrix Z_a. As a result, we obtain

$$Z_a^{11} = 1; \quad Z_a^{21} = -2\tau\mu^{-2\varepsilon}g^{-1}\overline{Z}_0; \quad Z_a^{22} = Z_a^{33} = Z_\tau;$$

$$Z_a^{41} = -2\tau^2\mu^{-2\varepsilon}\partial_g\overline{Z}_0; \quad Z_a^{51} = -12\tau^2\mu^{-4\varepsilon}\partial_g(g^{-1}\overline{Z}_0);$$

$$Z_a^{42} = 2\tau g Z_\varphi^{-2}\partial_g Z_2; \quad Z_a^{52} = 12\tau\mu^{-2\varepsilon}Z_\varphi^{-2}\partial_g Z_2; \tag{4.63}$$

$$Z_a^{44} = Z_\varphi^{-2}(1 - 2g\partial_g)Z_1; \quad Z_a^{54} = -12\mu^{-2\varepsilon}Z_\varphi^{-2}\partial_g Z_1;$$

$$Z_a^{45} = \mu^{2\varepsilon}g^2 Z_\varphi^{-4}\partial_g Z_3/6; \quad Z_a^{55} = Z_\varphi^{-4}\partial_g(g Z_3)$$

with the standard constants Z (Sec. 3.18) and $\overline{Z}_0$ from (3.200). Regarding elements 43 and 53, it is known that Z_a^{43} depends only on g (from power counting) and that $Z_a^{53} = 6Z_a^{43}/g\mu^{2\varepsilon}$ owing to (3.201). The value of Z_a^{43} itself is not determined from the general arguments of Sec. 3.30 (it is only known that its expansion begins at g^3 or higher), and if needed it can be calculated directly from the graphs. It is not needed for determining the critical dimensions, but enters into the matrix U from (4.44).

We wish to use the known Z_a to calculate the matrices $\gamma_F = -\widetilde{\mathcal{D}}_\mu Z_a \cdot Z_F$ and $\Delta_F = d_F + \gamma_F^*$. The calculations are awkward but straightforward if we begin directly from the definition (4.31) and check the cancellation of ε poles in γ_F during the calculations. It is much simpler to find γ_F using (4.47). We shall explain the calculations only briefly and present the results. For all the constants Z_i with $i = 1, 2, 3, g, \tau, \varphi$ in (4.63) it is possible to introduce the corresponding RG functions $\gamma_i = \widetilde{\mathcal{D}}_\mu \ln Z_i(g) = \beta\partial_g \ln Z_i(y)$. They are all independent of ε and are related by expressions (4.3) equivalent to (3.110), and $\beta = -g(2\varepsilon + \gamma_g)$ according to (1.114). The definition of γ_i leads to the equations UV-finite part of $\widetilde{\mathcal{D}}_\mu Z_i = \gamma_i$ and $\partial_g Z_i = \gamma_i Z_i/\beta = -\gamma_i/2\varepsilon g +$ higher-order poles in $1/\varepsilon$. Using these expressions, from (4.47) we easily find the desired elements γ_F^{ik}. For several contributions Z_a^{ik} it is more convenient to use the first equation in (4.47) and for others the second, which is allowed owing to the linearity of (4.47). For example, for $Z_a^{55} = Z_g + Z_\varphi^{-4}g\partial_g Z_3$ we have $\gamma_F^{55} = -$UV-finite part of $\widetilde{\mathcal{D}}_\mu Z_g + (2\varepsilon\mathcal{D}_g - \mathcal{D}_\mu)$ times a contribution $\sim 1/\varepsilon$ to $g\partial_g Z_3 = -\gamma_g + 2\varepsilon\mathcal{D}_g[g(-\gamma_3/2\varepsilon g)] = -\gamma_g - \mathcal{D}_g\gamma_3 = -\gamma_g - \mathcal{D}_g(\gamma_g + 4\gamma_\varphi)$. The result for the elements in (4.63) containing $\overline{Z}_0$ is expressed in terms of the RG function (4.19). The calculations are simplified when we take into account the commutativity of the operator $2\varepsilon\mathcal{D}_g - \mathcal{D}_\mu$ in (4.47) with the factors τ and $g\mu^{2\varepsilon}$ and the operators $\mathcal{D}_g \equiv g\partial_g$ and $\mu^{-2\varepsilon}\partial_g$. In the end we obtain

$$\gamma_F^{11} = 0, \quad \gamma_F^{21} = 2\tau\mu^{-2\varepsilon}\gamma_0, \quad \gamma_F^{22} = \gamma_F^{33} = -\gamma_\tau,$$

$$\gamma_F^{41} = 2\tau\mu^{-2\varepsilon}(\gamma_0 + g\gamma_0'), \quad \gamma_F^{51} = 12\tau^2\mu^{-4\varepsilon}\gamma_0',$$

$$\gamma_F^{42} = -2\tau g(\gamma_\tau' + 2\gamma_\varphi'), \quad \gamma_F^{52} = -12\tau\mu^{-2\varepsilon}(\gamma_\tau' + 2\gamma_\varphi'), \tag{4.64}$$

$$\gamma_F^{43} = \text{UV-finite part of } [2\varepsilon \mathcal{D}_g Z_a^{43}], \quad \gamma_F^{53} = 6\gamma_F^{43}/g\mu^{2\varepsilon},$$

$$\gamma_F^{44} = 4g\gamma_\varphi', \quad \gamma_F^{54} = 24\mu^{-2\varepsilon}\gamma_\varphi', \quad \gamma_F^{45} = -\mu^{2\varepsilon}g^2(\gamma_g' + 4\gamma_\varphi')/6,$$

$$\gamma_F^{55} = -\gamma_g - g(\gamma_g' + 4\gamma_\varphi'),$$

where the primes denote derivatives with respect to g and all the elements missing in (4.64) are zero. Setting $g = g_*$ everywhere [at this point $\gamma_g = -2\varepsilon$ and $g\gamma_g' = -\beta' = -\omega$ according to (1.114) and the definition of ω from (1.156)] and adding to γ_F^* the diagonal matrix of canonical dimensions given at the beginning of this section, we obtain the desired matrix Δ_F. It can be used to find the critical dimensions Δ_i and the corresponding basis operators (4.44) which have these dimensions. The answers for the first three operators are actually known from the results of Sec. 4.9: $\widehat{F}_{1R}' = \widehat{F}_{1R} = 1$, $\widehat{F}_{2R}'$ is the operator (4.58), and $\widehat{F}_{3R}' = \partial^2\widehat{F}_{2R}' = \widehat{F}_{3R}$ with the critical dimensions $\Delta_1 = 0$, $\Delta_2 = d - \Delta_\tau$, and $\Delta_3 = 2 + d - \Delta_\tau$, respectively (everywhere $d = 4 - 2\varepsilon$ and $\Delta_\tau = 2 + \gamma_\tau^*$ is the critical dimension of τ). For the senior operators 4,5 the corresponding $\Delta_{4,5}$ can be found as the eigenvalues of the two-dimensional 4,5 block in the matrix Δ_F and turn out to be d and $d + \omega$. The operator

$$\widehat{F}_{4R}' = \widehat{F}_{4R} - \tau\widehat{F}_{2R} - g_*\mu^{2\varepsilon}\widehat{F}_{5R}/6 \tag{4.65}$$

has dimension $\Delta_4 = d$, and the operator

$$\widehat{F}_{5R}' = \widehat{F}_{5R} + \frac{6\mu^{-2\varepsilon}}{(\omega - 4g\gamma_\varphi')}\left[4\gamma_\varphi'\widehat{F}_{4R} + \frac{\gamma_F^{43}\omega}{g(\Delta_\tau + \omega - 2)}\widehat{F}_{3R} - 2(\gamma_\tau' + 2\gamma_\varphi')\tau\widehat{F}_{2R} + \frac{2\tau^2\mu^{-2\varepsilon}(\omega\gamma_0' - 2\gamma_0\gamma_\tau')}{(d + \omega - 2\Delta_\tau)}\widehat{F}_{1R}\right]\Bigg|_{g=g_*} \tag{4.66}$$

has dimension $\Delta_5 = d + \omega$. To avoid confusion, we recall (Sec. 3.28) the definition of the renormalized operators corresponding to (4.61): $\widehat{F}_{1R} = 1$, $\widehat{F}_{2R} = [\widehat{\varphi}_R^2]_R$, $\widehat{F}_{3R} = \partial^2[\widehat{\varphi}_R^2]_R$, $\widehat{F}_{4R} = [\widehat{\varphi}_R\partial^2\widehat{\varphi}_R]_R$, $\widehat{F}_{5R} = [\widehat{\varphi}_R^4]_R$, where the argument x and contraction in the index of the field φ_a in the O_n φ^4 model are understood everywhere. It can be verified directly from the matrix Δ_F known from (4.64) that the operators (4.65) and (4.66) do actually satisfy the diagonal equations of critical scaling following from (4.42): $[-\mathcal{D}_x + \Delta_\tau\mathcal{D}_\tau]\widehat{F}_{iR}' = \Delta_i\widehat{F}_{iR}'$ with $\Delta_4 = d$ and $\Delta_5 = d + \omega$.

Finally, we note that the expression for $\tau\widehat{F}_{2R}'$ has the same critical dimension d as the operator (4.65), and so they can mix with arbitrary coefficient.

4.11   Determination of the critical dimensions of composite operators following Sec. 3.29

In Sec. 4.10 we followed the standard scheme, but the form of the operators (4.65) and (4.66) and their critical dimensions could have been found more simply from the general expressions of Sec. 3.29.

In any model, the operator (3.195) is related to a functional W_R by (3.188). In the expressions of Ch. 3 W_R is taken to be a functional with vacuum loops, and the RG equation for it has the form (4.12). For $g = g_*$ the equation of critical scaling analogous to (4.50) with an additional inhomogeneity is obtained from it by the standard method (Sec. 4.7). Acting on it with the operator $A(x)\delta/\delta A(x)$, we obtain the analogous homogeneous equation for the functional $A(x)\delta W_R(A)/\delta A(x)$ entering into the right-hand side of (3.188). The fact that this equation is homogeneous implies that the functional in question has definite critical dimension. It is easily found: since the contribution of the inhomogeneity vanishes, the functional $W_R(A)$ itself in the construction $A(x)\delta W_R(A)/\delta A(x)$ can simply be assumed to be critically dimensionless. Its variational derivative $\delta W_R(A)/\delta A(x)$ then has dimension $d - \Delta_A$ (Sec. 1.15), and upon multiplication by $A(x)$ a factor of Δ_A is added to it. Therefore, for $g = g_*$ the expression on the right-hand side of (3.188) always has definite critical dimension d, and so the operator (3.195) must have the same dimension for $g = g_*$. It is easily checked that this operator coincides with (4.65) in our model.

The second nontrivial operator (4.66) is associated with the expression $\partial_g S_R(\varphi)$, which is the integral over x of a local operator. Equation (3.192) with $\lambda = g$ relates the corresponding composite operator to the quantity $\partial_g W_R = \partial_g W'_R$, where W'_R is the functional from (4.13) ($\ln C_R$ and its finite part are independent of g). Differentiating (4.13a) for W'_R with the operator (1.122) with respect to g and then setting $g = g_*$ and using the definition of γ_0 in (4.18), we find

$$L\partial_g W'_R + [\omega\partial_g - \gamma'_\tau \mathcal{D}_\tau + \gamma'_\varphi \mathcal{D}_A]W'_R = V\tau^2\mu^{-2\varepsilon}\gamma'_0/2. \qquad (4.67)$$

Here $\gamma' \equiv \partial_g\gamma$ for all γ, substitution of $g = g_*$ after differentiation with respect to g is understood everywhere, the definition $\omega = \beta'(g_*)$ has been used, and we have introduced the notation

$$L \equiv [\mathcal{D}_{RG} + \gamma_\varphi \mathcal{D}_A]\,|_{g=g_*} = \mathcal{D}_\mu - \gamma^*_\tau \mathcal{D}_\tau + \gamma^*_\varphi \mathcal{D}_A \qquad (4.68)$$

for the operator acting on W'_R at the fixed point in (4.13). This operator commutes with $\mathcal{D}_\tau$ and $\mathcal{D}_A$, and so for the corresponding derivatives from (4.13) and (4.18) we find the equations

$$L\mathcal{D}_\tau W'_R = V\tau^2\mu^{-2\varepsilon}\gamma_0, \qquad L\mathcal{D}_A W'_R = 0. \qquad (4.69)$$

From (4.67) and (4.69) we see that the functional

$$\Phi' = (\omega\partial_g - \gamma'_\tau \mathcal{D}_\tau + \gamma'_\varphi \mathcal{D}_A)W'_R|_{g=g_*}$$

satisfies the inhomogeneous equation $(L + \omega)\Phi' = cV\tau^2\mu^{-2\varepsilon}$ with known coefficient $c = \omega\gamma_0'/2 - \gamma_0\gamma_\tau'$ on the right-hand side. By adding a suitable constant to the functional Φ' it is possible to remove the inhomogeneity and thereby construct a functional Φ with definite critical dimension $\Delta[\Phi] = \omega$ (we recall that the equation of critical scaling for Φ is obtained by subtracting the RG equation $(L + \omega)\Phi = 0$ from the canonical scaling equation with $d[\Phi] = 0$ by analogy with the derivation of (4.50), from which it follows that $\Delta[\Phi] = \omega$). Omitting the straightforward calculations, the result of this procedure is

$$\Phi = (\omega\partial_g - \gamma_\tau'\mathcal{D}_\tau + \gamma_\varphi'\mathcal{D}_A)W_R' + V\tau^2\mu^{-2\varepsilon}\frac{(2\gamma_0\gamma_\tau' - \omega\gamma_0')}{2(d + \omega - 2\Delta_\tau)}. \qquad (4.70)$$

The equations in Sec. 3.29 allow us to associate this functional with the integral over x of some local composite operator, namely,

$$\Phi = \langle\!\langle[(\omega\partial_g - \gamma_\tau'\mathcal{D}_\tau - \gamma_\varphi'\varphi\frac{\delta}{\delta\varphi})S_B(\varphi)]_R + \text{const}\rangle\!\rangle \qquad (4.71)$$

with the same constant as in (4.70). We note that when operated on by ∂_g and $\mathcal{D}_A$ the difference between W_R and W_R' from (4.13) vanishes, and for $\mathcal{D}_\tau$ from (4.13), (3.192), and (3.199) with $\lambda = \tau$ we find the equation $\mathcal{D}_\tau W_R' = \langle\!\langle[\mathcal{D}_\tau S_B]_R\rangle\!\rangle$, which we used above in determining the corresponding functional (4.70) of the composite operator in (4.71).

We have shown that the operator inside the double brackets in (4.71) has definite critical dimension ω, and so the integrand corresponding to it when $S_B = \int dx\dots$ is substituted into (4.71) has dimension $d + \omega$. In our model it will be, up to a coefficient, the operator (4.66), but without the contribution $\widehat{F}_{3R} = \partial^2\widehat{F}_{2R}$, which vanishes upon integration over x. This contribution actually remains undetermined also in (4.66) without the explicit calculation of γ_F^{43} from graphs.

4.12 The O_n φ^4 model: calculation of the 1- and 2-loop graphs of the renormalized correlator in the symmetric phase

In the following sections we shall make a detailed study of the example of the scaling function of the correlator in the symmetric phase with two-loop accuracy (contributions of order 1, ε, and ε^2). Its ε expansion is constructed directly from the usual perturbation series for the renormalized object (Sec. 1.36). It is more convenient technically to calculate the graphs not of the correlator itself, but of the 1-irreducible function $\Gamma_{2R} = -D_R^{-1}$ related to it by (1.98). All its 1- and 2-loop graphs are given in (2.50), and in our model

$$\Gamma_{2R} = -p^2 - \tau + \frac{r}{2}R\gamma_1 + \frac{r^2}{4}R\gamma_2 + \frac{r}{6}R\gamma_3, \quad r = \frac{n+2}{3}, \qquad (4.72)$$

where γ_i are the graphs, numbered in the order in which they occur in (2.50), for the simple φ^4 model in the basic theory [a line corresponds to $(k^2 + \tau)^{-1}$ and the vertex factor is $-g_{\mathrm{B}} = -g\mu^{2\varepsilon}$], R stands for the R operator, and r are the additional structure factors of the O_n group (Sec. 3.20). The one-loop graph γ_1 must be calculated taking into account terms of zero and first order in ε, and the two-loop graphs $\gamma_{2,3}$ are calculated including only zeroth order. For the calculations in the present section we shall use the following notation:

$$u \equiv \frac{g}{16\pi^2}, \quad t \equiv \frac{4\pi\mu^2}{\tau}, \quad w \equiv \frac{p}{\tau^{1/2}}, \quad \alpha \equiv -\mathcal{C} + \ln t \qquad (4.73)$$

$[\mathcal{C} = -\psi(1)$ is the Euler constant], along with the notation and expressions from Sec. 3.15.

The calculation of the graphs $\gamma_{1,2}$ is elementary. The momentum integral of the graph γ_1 given in (3.87a); multiplying it by $-g\mu^{2\varepsilon}$ we obtain $\gamma_1 = -u\tau\| - 1 + \varepsilon\|t^\varepsilon$. Selecting the pole in ε gives the counterterm $L\gamma_1 = u\tau/\varepsilon$, and subtracting it from γ_1 we obtain

$$R\gamma_1 = -u\tau\left[\, \| - 1 + \varepsilon\| \, t^\varepsilon + 1/\varepsilon\right]. \qquad (4.74)$$

Expanding this expression in ε to the needed accuracy, we obtain $[\psi'(1) = \pi^2/6]$

$$R\gamma_1 = u\tau\left\{1 + \alpha + \frac{\varepsilon}{2}\left[\frac{\pi^2}{6} + \alpha^2 + 2\alpha + 2\right]\right\}. \qquad (4.75)$$

For the second graph the value of $R\gamma_2$, according to the general rule (3.104), is the product of (4.74) and the renormalized vertex loop [(3.87b) multiplied by $-g\mu^{2\varepsilon}$ minus the pole in ε], i.e.,

$$R\gamma_2 = u^2\tau[\, \| - 1 + \varepsilon\| \, t^\varepsilon + 1/\varepsilon][\, \|\varepsilon\| \, t^\varepsilon - 1/\varepsilon]. \qquad (4.76)$$

In zeroth order in ε

$$R\gamma_2 = -u^2\tau\alpha(1 + \alpha). \qquad (4.77)$$

Only the calculation of the graph γ_3 is nontrivial, and all the complications are associated with the fact that the lines are massive (the same massless graph in Sec. 3.21 was elementary to calculate). It is not possible to explicitly perform all the integrations in γ_3 for $\varepsilon \neq 0$ or even in $R\gamma_3$ for $\varepsilon = 0$, and so the problem is to reduce the answer to the simplest form possible, preferably a one-dimensional integral convenient for studying various asymptotes. Using (3.82) and (3.87), γ_3 can be written as a double integral over parameters, but it is difficult to reduce this to a single integral. We shall use the rule (1.35b) to calculate directly the Fourier transform of the coordinate-space graph, which is the simple cube of a line with coefficient $g^2\mu^{4\varepsilon}$. The expression for the line in coordinate space is obtained from (3.79) with $d = 4 - 2\varepsilon$, $\alpha = 1$, and $m = \tau^{1/2}$. This quantity depends only on the distance modulus, and so in calculating the Fourier transform it is possible to perform the integration over angles using

(1.87c). Finally, for $\gamma_3(p, \tau)$ we obtain the following representation by a single integral:

$$\gamma_3 = \tau u^2 t^{2\varepsilon} Q(\omega), \tag{4.78}$$

$$Q(\omega) \equiv 16\omega^{-1+\varepsilon} \int_0^\infty dx\, x^{-1+2\varepsilon} J_\nu(2\omega x) K_\nu^3(2x), \quad \nu = 1 - \varepsilon, \tag{4.79}$$

where J_ν is the Bessel function and K_ν is the Macdonald function. According to the general ideas (Sec. 1.19), the integral (4.79) should be understood as an analytic continuation in ε from the region $\varepsilon > 1/2$ in which it is well defined.

Poles in ε are generated by the divergences of (4.79) for $x = 0$. The expansion of the function K_ν at small x is known from (3.81):

$$K_{1-\varepsilon}(2x) = \frac{\|1 - \varepsilon\|}{2} \left\{ x^{-1+\varepsilon} + \sum_{s=0}^\infty \left[\frac{\|\varepsilon\|\, x^{2s+1+\varepsilon}}{\|s+2\|\|s+1+\varepsilon\|} - \frac{\|\varepsilon\|\, x^{2s+1-\varepsilon}}{\|s+1\|\|s+2-\varepsilon\|} \right] \right\}. \tag{4.80}$$

The integral (4.79) involves the cube of this expression. The cube of the first term in (4.80) plus three times the product of its square and the first term of the series in (4.80) with $s = 0$ will be called the "singular" part $K^3|_{\rm s}$, and the other contributions the "regular" part $K^3|_{\rm r}$, with the same terminology for the corresponding contributions to (4.78) and (4.79). All the UV divergences are concentrated in the singular contribution. In the regular one we can simply pass to the limit $\varepsilon = 0$, and then instead of (4.80) we have [see (3.80b) and (3.70)]

$$K_1(2x) = \frac{1}{2x} + \sum_{s=0}^\infty \frac{x^{2s+1}}{s!(s+1)!} \left[\ln x - \psi(s+1) - \frac{1}{2(s+1)} \right]. \tag{4.81}$$

From this for the regular part of (4.79) at $\varepsilon = 0$ we find that

$$Q_{\rm r}(\omega) = 16\omega^{-1} \int_0^\infty \frac{dx}{x} J_1(2\omega x) K_1^3(2x)|_{\rm r}, \tag{4.82}$$

where

$$K_1^3(2x)|_{\rm r} = K_1^3(2x) - \frac{1}{8x^3} - \frac{3}{4x}(\ln x + C - \frac{1}{2}). \tag{4.83}$$

Let us now consider the singular contribution, combining it with the counterterms generated by the R operator. Substituting the series (4.80) into (4.79) and keeping only the singular contributions (see above), term-by-term integration can be performed using the expression

$$\int_0^\infty dx\, x^\alpha J_\nu(2\omega x) = \frac{\omega^{-1-\alpha}}{2} \cdot \frac{\|(1+\nu+\alpha)/2\|}{\|(1+\nu-\alpha)/2\|}, \tag{4.84}$$

which ultimately leads to the following expression for the singular part of (4.78):

$$\gamma_3 \mid_s = \tau u^2 t^{2\varepsilon} \|1 - \varepsilon\|^3 \left\{ \frac{\| - 1 + 2\varepsilon\| \cdot \omega^{2-4\varepsilon}}{\|3 - 3\varepsilon\|} + \right.$$

$$\left. + 3\frac{\|\varepsilon\|2\varepsilon\| \cdot \omega^{-4\varepsilon}}{\|1 + \varepsilon\|2 - 3\varepsilon\|} - 3\frac{\|\varepsilon\|\varepsilon\| \cdot \omega^{-2\varepsilon}}{\|2 - \varepsilon\|2 - 2\varepsilon\|} \right\}. \tag{4.85}$$

Counterterms from subgraphs are generated by the R' operator in (3.27). In the notation of the present section we have $R'\gamma_3 = \gamma_3 - 3\gamma_1 \cdot L\gamma_1'$, where $L\gamma_1' = -u/\varepsilon$ is the counterterm of the vertex loop. Substituting the known (see the beginning of this section) expression for γ_1, we have the combination

$$R'\gamma_3 \mid_s = \gamma_3 \mid_s - 3u^2\tau\| - 1 + \varepsilon\|t^\varepsilon/\varepsilon \tag{4.86}$$

with $\gamma_3 \mid_s$ from (4.85). Expanding all the gamma functions in ε with the needed accuracy (as in the calculation of the three-loop counterterms in Sec. 3.21), it is easy to find the UV-divergent part (poles in ε) of (4.86), i.e., the counterterm

$$L\gamma_3 = u^2[-p^2/4\varepsilon + 3\tau(1 - \varepsilon)/2\varepsilon^2], \tag{4.87}$$

and also the contributions of zero order in ε in (4.86), which when added to the regular contribution [substitution of (4.82) into (4.78)] give the desired quantity:

$$R\gamma_3 \mid_{\varepsilon=0} = u^2 \left\{ p^2[\ln\omega - \alpha/2 - 13/8] + \right.$$

$$\left. + \tau[6\ln^2\omega - 3\alpha^2/2 - 6\alpha - 9/2] + \tau Q_r(\omega) \right\}. \tag{4.88}$$

The next terms of the expansion of $R\gamma_3$ in ε can also be found by this method, but we do not need them. We also note that the above expression (4.87) for the counterterm of the massive graph agrees with the results in Table 3.1 of Sec. 3.20.

Later on we will need the asymptote of the function $Q_r(\omega)$ in (4.88) at small and large ω. The $\omega \to \infty$ asymptote corresponds to the expansion of the function (4.83) in powers of x obtained by substituting the known series (4.81) into (4.83). The contributions of order x correspond in (4.83) to leading terms of order ω^{-2} [up to logarithms, which when (4.84) is used are reproduced by differentiation with respect to α]. Selecting them, we obtain

$$Q_r(\omega) \underset{\omega \to \infty}{=} 3\omega^{-2}[4\ln^2\omega + 3\ln\omega - 1/4] + \dots. \tag{4.89}$$

The contributions of order x^3, x^5, and so on in (4.83) correspond to corrections of order ω^{-4}, ω^{-6}, and so on denoted by the ellipsis in (4.89). If desired, they can be easily found by substituting the function (4.83) with expansion known from (4.81) into the integral (4.82) and selecting the needed powers of x.

It is more complicated to find the $\omega \to 0$ asymptote. The representation (4.82) itself is not convenient for finding it. It is simpler to return to the original expression (4.79), substituting into it the expansion of the Bessel function:

$$\omega^{-1+\varepsilon} J_{1-\varepsilon}(2\omega x) = \frac{x^{1-\varepsilon}}{\|2-\varepsilon\|} - \frac{\omega^2 x^{3-\varepsilon}}{\|3-\varepsilon\|} + \dots \qquad (4.90)$$

(we are interested in only these two terms). The integration in (4.79) can then be performed using the formula (Ref. 93)

$$16 \int_0^\infty dx \cdot x^{\alpha-1} K_\nu^3(2x) = [\|\nu\|\alpha_1\|\alpha_2\|\alpha_2\|\alpha_3\|/\|\alpha-\nu\|] \times$$

$$\times {}_3F_2(\alpha_1, \alpha_2, \alpha_3; 1-\nu, \alpha_4; 1/4) + [\| -\nu\|\alpha_1\|\alpha_1\|\alpha_2\|\alpha_5\|/\|\alpha+\nu\|] \times$$

$$\times {}_3F_2(\alpha_1, \alpha_2, \alpha_5; 1+\nu, \alpha_6; 1/4),$$

where $\alpha_1 = (\alpha+\nu)/2, \quad \alpha_2 = (\alpha-\nu)/2, \quad \alpha_3 = (\alpha-3\nu)/2,$

$$\alpha_4 = (1+\alpha-\nu)/2, \quad \alpha_5 = (\alpha+3\nu)/2, \quad \alpha_6 = (1+\alpha+\nu)/2,$$

and ${}_3F_2$ is the hypergeometric function, represented by a well convergent single series. Expanding all the quantities entering into the result in ε with the required accuracy, from the contributions in (4.90) to the integral (4.79) we obtain

$$Q(\omega) \underset{\omega\to0}{=} -\frac{3}{2\varepsilon^2} + \frac{3}{2\varepsilon}(2\mathcal{C}-3) + a_1 + \omega^2 \left[-\frac{1}{4\varepsilon} + a_0 \right] + \dots \qquad (4.91)$$

with the constants $a_0 = 0.694910244\dots$ and $a_1 = -5.256132\dots$, which are found by the numerical summation of two relatively uncomplicated single series. Substituting (4.91) into (4.78) and grouping the resulting expression with the counterterms from the subgraphs [given in (4.86)] and from the graph as a whole (4.87), we eliminate all the poles in ε and find the desired $\omega \to 0$ asymptote of the function $Q_r(\omega)$ in (4.88):

$$Q_r(\omega) \underset{\omega\to0}{=} -6\ln^2\omega + c_1 + \omega^2[c_0 - \ln\omega] + \dots,$$
$$c_0 = 2.031302412\dots, \quad c_1 = 0.5158604\dots. \qquad (4.92)$$

The constants $c_{0,1}$ are related simply to $a_{0,1}$ from (4.91) ($c_0 = a_0 - \mathcal{C}/2 + 13/8$, $c_1 = a_1 + \pi^2/4 + 3\mathcal{C}^2 - 9\mathcal{C} + 15/2$) and are four times the values of the constants Q_{00} and Q_{01} introduced in Ref. 94, where the scaling function of the correlator was first calculated with two-loop accuracy (see the historical review in Sec. 4.14 for more details).

4.13 ε expansion of the normalized scaling function

The reduced scaling function $f(z)$ in (1.161) for the critical asymptote of the correlator is determined by (1.149) and (1.163), which can be rewritten as

$$f(z) = p^2 D_R \big|_{u=u_*, p=\mu, \tau=z\mu^2}, \qquad (4.93)$$

where u_* is the coordinate of the fixed point (4.7d). The function inverse to (4.93) can be calculated directly from the graphs (4.72):

$$h(z) = 1/f(z) = -p^{-2}\Gamma_{2\mathrm{R}}\,|_{u=u_*,p=\mu,\tau=z\mu^2}. \tag{4.94}$$

Substituting into this the values of the graphs known from (4.75), (4.77), and (4.88) with $u \to u_*$ from (4.7d) and the variable substitutions given in (4.94) [for (4.73), $\omega \to z^{-1/2}$, $\alpha \to \ln(4\pi) - C - \ln z$], we obtain the following explicit expression for the first terms of the ε expansion:

$$h(z) = 1 + z - 2\varepsilon \frac{(n+2)}{2(n+8)} z(1 + \beta - \ln z) - (2\varepsilon)^2 \frac{(n+2)}{2(n+8)^2} \times$$

$$\times \left\{ I(z) - \frac{13 + 4\beta}{8} + z\left[\frac{n+8}{4}\left(\frac{\pi^2}{6} + 2\right) + \frac{3(3n+4)}{2(n+8)} + \frac{6(n+3)}{(n+8)}\beta - \right. \right.$$

$$\left. \left. - \frac{n+2}{4}\beta^2 + \left(\frac{n+2}{2}\beta - \frac{6(n+3)}{(n+8)}\right)\ln z + \frac{4-n}{4}\ln^2 z \right] \right\} + ...,$$

(4.95)

where $\beta \equiv \ln(4\pi) - C$, $I(z) \equiv zQ_\mathrm{r}(\omega)|_{\omega=z^{-1/2}}$. The asymptotes of the function $I(z)$ are known from (4.89) and (4.92):

$$I(z) \underset{z \to 0}{=} 3z^2\left[\ln^2 z - \frac{3}{2}\ln z - \frac{1}{4}\right] + O(z^3), \tag{4.96}$$

$$I(z) \underset{z \to \infty}{=} z\left[c_1 - \frac{3}{2}\ln^2 z\right] + c_0 + \frac{1}{2}\ln z + O(1/z). \tag{4.97}$$

For $p \to 0$, $\tau = \mathrm{const}$ the renormalized correlator and (1.161) for its critical asymptote, like their inverses, must be expanded in series in integer powers of p^2. Using the representation (1.161) for D_R, analyticity in p^2 of the function $\Gamma_{2\mathrm{R}} = -D_\mathrm{R}^{-1}$ in (4.94) corresponds to the asymptotic expansion

$$h(z) \underset{z \to \infty}{=} z^\gamma[h_0 + h_1 z^{-2\nu} + h_2 z^{-4\nu} + ...] \tag{4.98}$$

with the standard critical indices $\gamma = 1 + O(\varepsilon)$, $\nu = 1/2 + O(\varepsilon)$ (Sec. 1.37) and arbitrary coefficients h_n. Substituting into (4.98) the exponents and coefficients in the form of ε expansions (known for the exponents but not for the coefficients), the result can be compared to the asymptote of the function (4.95) known from (4.97). The accuracy of (4.97) is sufficient to control the first two terms of the expansion (4.98). Let us explain in more detail. Substituting the asymptote (4.97) into (4.95) and discarding all contributions involving logarithms, we must, according to (4.98), obtain the quantity $h_0 z + h_1$, which allows us to find the ε expansion of the coefficients $h_{0,1}$ through order ε^2 (both expansions begin with unity). From these results and the known ε expansions of the exponents γ and ν from the representation (4.98) we can then uniquely predict the coefficients of all contributions of the same type (z and 1) involving logarithms and then compare them to the coefficients obtained from substituting (4.97) into (4.95). All the required coefficients in (4.95) turn out to be correct, which indicates that (4.95) is consistent with the asymptote (4.98).

The coefficients $h_{0,1}$ in (4.98) change under transformations

$$h(z) \rightarrow h'(z) = c_1 h(c_2 z) \tag{4.99}$$

with arbitrary coefficients $c_{1,2}$, which correspond to arbitrary changes of the units in which the field and the various variables in its correlator are measured. Therefore, as already stated (Sec. 1.33), only normalized functions, in which arbitrariness like (4.99) is eliminated by some additional normalization conditions, are of objective value. Such conditions for the function under consideration here are usually taken to be

$$h_{\text{norm}}(z) \underset{z \to \infty}{=} z^\gamma [1 + z^{-2\nu} + h_2 z^{-4\nu} + ...], \tag{4.100}$$

i.e., h_{norm} is obtained from (4.98) by the transformation (4.99) ensuring that the first two coefficients $h_{0,1}$ are trivial. Clearly, the ε expansion of the normalized function is obtained simply by discarding in (4.95) all corrections to the zeroth-order approximation $h_0(z) = z + 1$ of order z and 1 without logarithms that are analytic at the asymptote $z \to \infty$, with the simultaneous rearrangement according to the representation (4.100) of the coefficient of $\varepsilon^2 z \ln z$, which gives

$$h_{\text{norm}}(z) = 1 + z + \frac{2\varepsilon(n+2)}{2(n+8)} z \ln z - \frac{(2\varepsilon)^2(n+2)}{2(n+8)^2} \times$$

$$\tag{4.101}$$

$$\times \left\{ I(z) - c_0 - c_1 z + \frac{4-n}{4} z \ln^2 z - \frac{n^2 + 22n + 52}{2(n+8)} z \ln z \right\}.$$

From this with the same accuracy we find the ε expansion of the normalized scaling function of the correlator itself:

$$f_{\text{norm}}(z) = 1/h_{\text{norm}}(z) = \frac{1}{1+z} - \frac{2\varepsilon\gamma_1 z \ln z}{(1+z)^2} + \frac{(2\varepsilon)^2}{(1+z)^2} \times$$

$$\tag{4.102}$$

$$\times \left\{ \eta_2 \left[I(z) - c_0 - c_1 z + \frac{4-n}{4} z \ln^2 z \right] - \gamma_2 z \ln z + \frac{\gamma_1^2 z^2 \ln^2 z}{1+z} \right\}.$$

Here and below, for economy of notation we use γ_l, η_l, and so on to denote the coefficients of $(2\varepsilon)^l$ in the known (Sec. 1.37) ε expansions of the corresponding exponents, in particular,

$$\gamma_1 = 2\nu_1 = \frac{n+2}{2(n+8)}; \quad \eta_2 = \frac{n+2}{2(n+8)^2};$$

$$\tag{4.103}$$

$$\gamma_2 = \frac{(n+2)(n^2 + 22n + 52)}{4(n+8)^3}.$$

4.14   Analysis of the $\tau \to 0$ asymptote using the Wilson operator expansion

Let us now consider the case $\tau \to 0$, $p = \text{const}$, which in the critical regime (1.161) corresponds to the asymptote $\tau p^{-\Delta_\tau} \to 0$ for simultaneously small (compared to μ) arguments p and τ (Sec. 1.34). In scaling functions like (4.93) this corresponds to the asymptote $z \to 0$, and since z and τ are proportional in the definition (4.93) the powers τ^λ in the correlator D_R correspond to z^λ in $f(z)$ and $f_{\text{norm}}(z)$.

In contrast to the $p \to 0$ case discussed above, the $\tau \to 0$ asymptote of the correlator D_R will no longer be a simple series in integer powers of the small variable τ: in addition to such contributions, it will contain fractional powers of τ, as first noted in Ref. 95. The set of corresponding exponents is easily found using the Wilson operator expansion, discussed in Ch. 3, for the product of two operators. The renormalized correlator is the expectation value of the operator (3.210). Upon averaging, the contributions of all operators with external derivatives ∂ in the expressions of Sec. 3.33 vanish (this corresponds to zero external momentum flowing into the vertex of the composite operator in the terminology of Sec. 3.27). Then in the notation of Sec. 3.33 the only small parameter of the operator expansion is $\tau \xi^2$, i.e., τp^{-2} for the Fourier transform of the correlator. In the MS scheme the coefficients B_i in the Wilson expansion (3.217) are polynomials in τ, so that all fractional powers of τ in the critical regime are generated by the expectation values of renormalized composite operators $\widehat{F}_R$. The expectation value of a local composite operator in our model depends only on a single IR-relevant (Sec. 1.15) parameter τ, and if $\widehat{F}_R$ has definite critical dimension $\Delta_F \equiv \Delta[\widehat{F}_R]$, then in the critical regime

$$\langle \widehat{F}_R \rangle \underset{\tau \to 0}{=} \text{const} \cdot \tau^{\Delta_F/\Delta_\tau}, \tag{4.104}$$

where $\Delta_\tau = 1/\nu$ is the critical dimension of the temperature.

The expansion (3.217) involves renormalized operators, which are in general certain linear combinations of operators with definite critical dimensions, where the coefficients can contain any integer power of τ. It is therefore clear that the $\tau \to 0$ asymptote of the renormalized correlator in the critical region has the general form

$$D_R \underset{\tau \to 0}{=} \sum_{n=0}^{\infty} \sum_F c_{nF} \cdot \tau^{n + \Delta_F/\Delta_\tau}, \tag{4.105}$$

where c_{nF} are coefficients independent of τ and $\sum_F \ldots$ denotes summation over all renormalized composite operators with definite critical dimensions Δ_F entering into the operator expansion. Some of these may have zero expectation value (4.104), and then the corresponding contributions are absent in (4.105).

In actual calculations the operators F are classified according to the value of $d_F^* \equiv d_F|_{\varepsilon=0}$ (Sec. 3.28) and fall into groups with $d_F^* = 0, 2, 4, \ldots$

[the expectation values (4.104) are nonzero only for operators with even values of d_F^*]. The values of Δ_F differ from d_F^* only by corrections of order ε, ε^2, and so on, and so for small ε the leading terms of the asymptote (4.105) are determined by the contributions with small n and d_F^*. Of course, if the ε expansions of Δ_F do not terminate and are not exactly known, we cannot say anything about the value of Δ_F at finite ε, and so we are forced to restrict ourselves to the standard ideology of small ε (Sec. 1.16). In this regard, we should add that the very fact of the existence of a finite correlator right at the critical point ($\tau = 0$) can be taken as an argument that all the Δ_F in (4.105) are nonnegative for the real value $2\varepsilon = 1$ [however, this does not constitute a proof, because when there are an infinite number of negative Δ_F, all the contributions diverging for $\tau \to 0$ must somehow be summed, and they may give a finite expression; only a finite number of negative Δ_F in (4.105) is strictly forbidden]. Actually, all the calculations in this model are based on the ε expansion, i.e., we always use the standard small-ε ideology.

Let us consider in more detail the first few terms of the asymptotic expansion (4.105). The set of values of $n + \Delta_F/\Delta_\tau$ with $n = 0, 1, 2, ...$ will be called the "series" generated by Δ_F, and the operators F will be classified according to the value of $d_F^* = 0, 2, 4, ...$ The only operators with $d_F^* = 0, 2$ are $F = 1$ and φ^2, respectively.

The correlator in question of the O_n φ^4 model in the symmetric phase is a multiple of δ_{ab} in the field indices. They can be contracted, so that we can limit ourselves to O_n-scalar operators. For $F_1 = 1$ we have $\Delta_F = 0$, and the operator $F_2 = \varphi^2$ is associated with the critical dimension $\Delta_F = d - \Delta_\tau$ (Sec. 4.9). These two operators are sufficient for determining the exact exponents of the contributions of order 1 and τ in (4.105) (by order here and below we mean the integer power of τ neglecting corrections $\sim \varepsilon, \varepsilon^2, ...$ in the exponent). Determination of the exponents of the higher contributions of order τ^2 requires all operators with $d_F^* = 4$. Here operators with external derivatives ∂ are irrelevant (see above), and so we need only the operators $F_4 = \varphi_a \partial^2 \varphi_a$ and $F_5 = (\varphi_a \varphi_a)^2$ from the system (4.61) studied in Sec. 4.10 and the operator $F_7 = \varphi_a \partial_i \partial_k \varphi_a$ from the tensor system in Sec. 3.32. The operators $F_{4,5}$ are associated with the critical dimensions d and $d + \omega$, respectively (Sec. 4.10). The first of these can be discarded, since it generates the series $n + d/\Delta_\tau$ wholly contained in the series $n + (d - \Delta_\tau)/\Delta_\tau$ of the operator φ^2 studied earlier [we also note that (3.188) leads to zero expectation value of an operator of this type in any model]. The critical dimensions of the operators of the system studied in Sec. 3.32 have not been discussed, but they are easily found. The first five are the same as for the system (4.61), and the dimensions of the operators $F_{6,7}$ are determined simply by the diagonal elements of the 6, 7 block of the matrix (3.209) because it is triangular. The corresponding diagonal elements of the matrix Z_a will be $Z_a^{66} = Z_\tau$ and $Z_a^{77} = 1$ (Sec. 3.32), from which we find (Sec. 4.7) $\gamma_F^{66} = -\gamma_\tau$ and $\gamma_F^{77} = 0$. Accordingly, for the operator of interest F_7 the critical dimension coincides with the canonical dimension ($\Delta_F = d_F = d$) and is irrelevant for the same reason as for F_4.

Therefore, in this approximation $(1, \tau, \tau^2)$, of all the operators with $d_F^* = 4$ a new series is generated only by F_5 with $\Delta_F = d + \omega$, and so we have three series, namely, n from $F = 1$, $n + (d - \Delta_\tau)/\Delta_\tau$ from φ^2, and $n + (d + \omega)/\Delta_\tau$ from $F_5 = \varphi^4$. To determine the asymptote (4.105) through order τ^2 we need the contributions with $n = 0, 1, 2$ in the first series, $n = 0, 1$ in the second, and only $n = 0$ in the third, i.e.,

$$D_R \underset{\tau \to 0}{=} A_1 + A_2\tau + A_3\tau^{1-\alpha} + A_4\tau^2 + A_5\tau^{2-\alpha} + A_6\tau^{2-\alpha+\omega\nu} + ..., \quad (4.106)$$

where $\nu = 1/\Delta_\tau$, $\alpha = 2 - d\nu$ is the exponent of the specific heat from (1.19), and all the A_i are amplitude factors independent of τ and containing known (from the known critical dimension D_R) negative powers of the momentum in the critical region.

Since z and τ are proportional to each other, the asymptote (4.106) corresponds to

$$f_{\text{norm}}(z) \underset{z \to 0}{=} C_1 + C_2 z + C_3 z^{1-\alpha} + C_4 z^2 + C_5 z^{2-\alpha} + C_6 z^{2-\alpha+\omega\nu} + ..., \quad (4.107)$$

where C_n are constants. Substituting into (4.107) all the exponents and the coefficients C_n in the form of ε expansions and selecting contributions of order 1, ε, and ε^2, we can compare the result with the $z \to 0$ asymptote of the function (4.102), in which it is necessary to limit ourselves to contributions of order 1, z, z^2 up to logarithms. In this manner it is easy to check that (4.102) and (4.107) are consistent and obtain information about the first few coefficients of the ε expansions of the constants C_n. It should be stressed that this is possible only owing to knowledge of the exact exponents in (4.107), i.e., the information obtained not from the RG equation for the object in question, but from other considerations, in the present case, from the Wilson operator expansion.

The results of this calculation then reduce to the following:

$$C_n = \sum_{l=0}^{\infty} (2\varepsilon)^l C_{nl}; \quad C_{10} = 1; \quad C_{20} = (n+2)/(4-n);$$

$$C_{30} = 6/(n-4); \quad C_{40} = C_{50} = 0; \quad C_{60} = 1;$$

$$C_{11} = 0; \quad C_{21} = -C_{31} = (n+2)(7n+20)/(n+8)(4-n)^2;$$

$$C_{41} + C_{51} + C_{61} = 0;$$

$$(n-4)C_{41} + 2(n+2)C_{61} = (-n^3 - 17n^2 + 22n + 392)/2(n+8)^2;$$

$$C_{12} = -\eta_2 c_0; \quad C_{22} + C_{32} = \eta_2(2c_0 - c_1);$$

$$C_{42} + C_{52} + C_{62} = \eta_2(2c_1 - 3c_0 - 3/4)$$

$$(4.108)$$

with η_2 from (4.103) and the constants $c_{0,1}$ from (4.92). Therefore, for the amplitude C_1 all the first three coefficients of the ε expansion are determined uniquely, for $C_{2,3}$ the first two plus one relation for the third coefficients are determined, and for $C_{4,5,6}$ only the first plus two relations for the second and one for the third are obtained. It is impossible to get anything more from the two-loop approximation (4.102). The three-loop calculation for the coefficients $C_{1,2,3}$ performed in Ref. 96 and for $C_{4,5,6}$ in Ref. 97 is much more complicated. Here we quote without derivation the results obtained in those studies, which supplement those in (4.108):

$$C_{22} = \eta_2[-\pi^2 R_1 - R_2 - (n+2)R_3 - \zeta(3)R_4]/2(n+8)(n-4)^3,$$

$$C_{32} = \eta_2[\pi^2 R_1 + R_2 + 6R_3 + \zeta(3)R_4]/2(n+8)(n-4)^3,$$

$$C_{13} = \eta_2[-4c_2 + c_0 R_5/4(n+8)^2], \quad C_{41} = \eta_2,$$

$$C_{51} = -2/(n+2), \quad C_{61} = 3(n+6)(n+14)/2(n+2)(n+8)^2,$$

$$C_{42} + 2(n+2)C_{52}/(n+8) = [R_6 + \zeta(3)R_7]/8(n+2)(n+8)^4,$$

$$C_{43} + C_{53} + C_{63} = \eta_2[-3c_1 + 8c_3 - 32/5 + R_5(12c_0 - 8c_1 + 3)/16(n+8)^2]$$
(4.109)

with two new constants $c_2 = 0.26825496...$, $c_3 = -0.51401...$, and the polynomials $R_1 \equiv (n+8)^2(n-4)^2$, $R_2 \equiv 2(n^4 + 50n^3 + 1332n^2 + 5696n + 7744)$, $R_3 \equiv 2(c_1 - 2c_0)(n+8)(n-4)^2$, $R_4 \equiv 48(n+8)(n-4)(5n+22)$, $R_5 \equiv n^2 - 56n - 272$, $R_6 \equiv -17n^4 + 440n^3 + 6756n^2 + 18896n + 736$, and $R_7 \equiv 192(n+2)(n+8)(5n+22)$.

Quantities related by transformations like (4.99) belong to the same equivalence class, which is characterized by values of the exponents and normalized scaling functions common to all the elements of the class. The renormalization transformations also have the form (4.99), and so the renormalized correlator calculated theoretically and the unrenormalized one measured experimentally must have identical universal characteristics, which allows theory to be compared with experiment. When analyzing the experimental data the correlator in the critical region is usually written as

$$D = At^{-\gamma}g(x), \quad x = p\xi, \quad \xi = \xi_0 t^{-\nu},$$
(4.110)

where γ and ν are standard exponents, p is the wave vector, A is the normalization constant, $t \equiv (T - T_c)/T_c$ (in contrast to $\tau \equiv T - T_c$), ξ is the correlation length, and $g(x)$ is the scaling function normalized by the condition $g(0) = 1$. The "true" correlation length ξ_{true} is determined by the location of the pole in the correlator in the complex p^2 plane. The pole is located on the negative semiaxis at the point $p^2 = -m^2$, and the equation $m = \xi_{\text{true}}^{-1}$ defines ξ_{true}. However, this definition is inconvenient in practice, and so experimentalists usually deal with the "effective" correlation length ξ

(differing only in the normalization ξ_0), defining ξ^2 as the coefficient of p^2 at the $p \to 0$ asymptote of the function $1/g(p\xi)$ from (4.110), represented by a series in integer powers of $x^2 = (p\xi)^2$ (Sec. 4.13). It follows from this definition of ξ that

$$1/g(x) \underset{x \to 0}{=} 1 + x^2 + \dots \qquad (4.111)$$

This corresponds to our normalization condition (4.100), and the function $g(x)$ from (4.110) is related to $f_{\text{norm}}(z)$ from (4.102) as

$$g(x) = x^{-2+\eta} f_{\text{norm}}(z) \Big|_{z=x^{-1/\nu}}. \qquad (4.112)$$

Therefore, the asymptote of the function $g(x)$ for $x \to \infty$ is determined from (4.107), and the only changes are the argument $(z \to x)$ and the exponents; the coefficients C_n remain unchanged.

Let us conclude this section with a brief historical note. As mentioned earlier, the fact that nonanalytic contributions of the type $\tau^{1-\alpha}$ must be present at the $\tau \to 0$ asymptote was first pointed out in [95]. The two-loop results for the first three coefficients $C_{1,2,3}$ given in (4.108) were obtained in [94], the three-loop calculation of $C_{1,2,3}$ was done in [96], and all the results for $C_{4,5,6}$ were obtained in [97]. The new version of the two-loop calculation presented in Sec. 4.12 is convenient for finding the correction terms at the $\tau \to 0$ asymptote, but it is difficult to generalize it to the three-loop graphs of Γ_2. In the three-loop calculation [96] the result is given as a double integral, for which it is relatively simple to find the contributions of order 1 and τ at the $\tau \to 0$ asymptote, which is sufficient to determine the coefficients $C_{1,2,3}$. However, already for the next terms of order τ^2 the calculation becomes very awkward, but, due to the lack of anything better, this was the method used to obtain the results [97] for $C_{4,5,6}$.

Scaling functions like $g(x)$ in (4.110) are measured experimentally using critical light scattering. Most experiments are performed for binary mixtures whose critical point belongs to the equivalence class of the Ising model and is described by the simple φ^4 model ($n = 1$). It is only such experiments that allow the direct measurement of η from the $x \to \infty$ asymptote of the function $g(x)$ [see (4.112)], i.e., $t \to 0$ in (4.110) (the wave vector p is fixed in the experiments). Since the region of large $x = p\xi$ is difficult to access experimentally, when analyzing the data it is important to know the correction terms (4.107), and also to appropriately choose the interpolation formula relating the asymptotes of the function $g(x)$ at large and small $x = p\xi$. These points are discussed in detail in [98] for the data analysis of a specific experiment. A general review of the data on this topic and additional references can be found, for example, in [52].

Finally, we note that the next two terms (x^4 and x^6) of the asymptote (4.111) in the three-loop approximation have also been calculated in [96]. The coefficients of these correction terms turn out to be very small, which explains why the very simple approximation $1/g(x) = 1 + x^2$ works at not too large $x \equiv (p\xi)$, a fact well known from analyzing the experimental data.

4.15 Goldstone singularities for $T < T_c$

In the preceding sections we studied the scaling function of the correlator of the $O_n \varphi^4$ model in the symmetric phase characterizing the critical asymptote $p \to 0$, $\tau \sim p^{\Delta_\tau} \to 0$.

In the present section we shall consider an example of the same model for a system with uniform external field $h \neq 0$ for $\tau < 0$ in arbitrary dimension $2 < d < 4$ and discuss the Goldstone asymptote

$$p \to 0, \quad h \sim p^2 \to 0, \quad \tau = \text{const}. \tag{4.113}$$

The problem is nontrivial only for $\tau < 0$ and only for models with continuous symmetry group $(n \geq 2)$ owing to the presence of Goldstone singularities (Sec. 2.4) generated by the vanishing for $h \to 0$ of the mass $\tau_\perp \sim h$ of the transverse modes in the decomposition:

$$\varphi_a(x) = \varphi_a^\perp(x) + \varphi_a^\parallel(x), \quad \varphi_a^\parallel(x) = (e\varphi(x))e_a. \tag{4.114}$$

Here e_a is the unit vector along the external field $h_a = he_a$ (we shall use the notation in Secs. 2.4 and 2.10, where in the latter $h \to A$).

Let us immediately state the main result [99], which reduces to the following two statements:

1. The Goldstone asymptote (4.113) of all the correlation functions of the fields (4.114) of the $O_n \varphi^4$ model can be obtained by the substitution

$$\varphi_a^\perp(x) = c\pi_a(x), \quad \varphi_a^\parallel(x) = c\sigma(x)e_a, \quad \sigma(x) = [1 - \pi^2(x)]^{1/2}, \tag{4.115}$$

where c is a normalization constant and π and σ are respectively the transverse (to e_a) and longitudinal fields of the nonlinear $O_n \sigma$ model (Sec. 2.26): $\pi^2(x) + \sigma^2(x) = 1$.

2. In determining the leading singular contribution of the asymptote (4.113) of the full correlation functions of the fields (4.114), π can be assumed a free field with correlator

$$\langle \pi_a \pi_b \rangle = P_{ab}^\perp [cp^2 + ch]^{-1} \tag{4.116}$$

in momentum space, while for the longitudinal field σ it is then possible to use the following approximation:

$$\sigma(x) = [1 - \pi^2(x)]^{1/2} \cong c + c\pi^2(x). \tag{4.117}$$

The letter c in (4.115)–(4.117) and other expressions in this section denotes various constants independent of h and p. It is understood that they depend on other parameters like τ, and so in general these constants can carry dimensions.

Statement 1 implies that the Goldstone asymptotes in the φ^4 and σ models coincide (up to normalization), and Statement 2, valid only for the leading singular contribution, is a further simplification reducing the problem to the theory of a free field π in which it is necessary to calculate the correlation functions of the field itself and the composite operator π^2. The first statement is valid up to correction factors like R_1 in (4.118), and the second up to correction factors like R_2:

$$R_1 = 1 + O(h, h^{d/2}), \qquad R_2 = 1 + O(h, h^{d/2-1}). \tag{4.118}$$

In (4.118) we have written only the leading (relative) regular and singular corrections taking into account $h \sim p^2$ in (4.113), i.e., for example, the presence of corrections $\sim h^{d/2-1}$ implies that there may also be corrections $\sim p^{d-2}$ plus higher powers.

These statements are valid for any dimension $2 < d < 4$. In this range of d corrections of order $h^{d/2}$ in R_1 are always "small" (smaller than the regular corrections of order h), while corrections $\sim h^{d/2-1}$ in R_2 depend on d: near $d = 4$ they are small ($\sim h^{1-\varepsilon}$ for $d = 4 - 2\varepsilon$), while near $d = 2$ they become nearly comparable to the leading singular contribution (an additional small factor $\sim h^\varepsilon$ for $d = 2 + 2\varepsilon$). In this sense Statement 1 is more general than Statement 2: reduction to a σ model is always possible, and further simplification of the σ model with "nearly the same" accuracy is possible only for $d = 4 - 2\varepsilon$, whereas for $d = 2 + 2\varepsilon$ if one wants to sum all the "not so small" corrections of order $h^{n\varepsilon}$ it is necessary to work with the exact σ model (Sec. 4.28). However, Statement 2 is always valid for the leading singular contribution to the full correlation functions neglecting corrections.

Turning now to the proof of the above statements, let us consider a standard integral like (2.38) for the O_n φ^4 model (1.63) in an external field $h_a = he_a$ for $\tau < 0$ and make the following variable substitution in it:

$$\varphi_a(x) = \rho(x)m_a(x), \ m^2(x) = 1, \ m_a^\perp \equiv \pi_a, \ m_a^\| \equiv \sigma e_a, \ \sigma^2 = 1 - \pi^2. \tag{4.119}$$

Here $\rho(x)$ is interpreted as the length of the n-dimensional vector $\varphi(x)$, $m(x)$ is the unit vector in its direction, ρ and π are treated as the new independent variables, and σ is a functional of π.

Substituting (4.119) into the integral (2.38), we obtain

$$G(A) \sim \int D\rho J(\rho) \int Dm \delta(m^2 - 1) \exp[S(\rho, m) + \rho(mA)], \tag{4.120}$$

where $J(\rho) \equiv \prod_x \rho^{n-1}(x)$ is the Jacobian, $\delta(m^2 - 1) \equiv \prod_x \delta[\pi^2(x) + \sigma^2(x) - 1]$ is the functional δ function, $Dm = D\pi D\sigma$, and

$$S(\rho, m) = -[(\partial\rho)^2 + \rho^2(\partial m)^2 + \tau\rho^2]/2 - g\rho^4/24 + h\rho\sigma,$$
$$(\partial m)^2 = (\partial\pi)^2 + (\partial\sigma)^2, \qquad (\partial\sigma)^2 = (\pi\partial\pi)^2/(1 - \pi^2). \tag{4.121}$$

The representation (4.120) is the analog of an ordinary d-dimensional integral, for which $dx = |x|^{d-1}d|x| \cdot dn$, and the angular integral over the directions $n = x/|x|$ of the d-dimensional vector x can be written as an ordinary d-dimensional integral with a δ function, i.e., as $\int dn\delta(|n| - 1)... = \int dn2\delta(n^2 - 1)....$ The unimportant contribution from the product of the factors of two in (4.120) has been omitted.

The meaning of functional δ functions was discussed in detail in Sec. 2.27, and a construction of the type $J(\rho)$ is formally interpreted using the relation (all integrals and δ functions are d-dimensional)

$$\prod_x f(x) = \exp[\delta(0)\int dx \ln f(x)], \quad \delta(0) \equiv \delta(x)\,|_{x=0} = (2\pi)^{-d}\int dk, \quad (4.122)$$

obtained by taking the limit from the lattice approximation. They give additions to the action which are multiples of Λ^d and serve only to cancel the analogous powers of Λ from the divergences in the graphs of the new perturbation theory obtained by the nonlinear variable substitution in the functional integral. In calculations using formal dimensional regularization contributions like (4.122) are discarded along with other divergences that are powers of Λ (Sec. 1.20).

For $\tau < 0$ the functional (4.121) has stationarity point corresponding to the energy minimum: $\pi_a(x) = 0$, $\sigma(x) = 1$, $\rho(x) = u$, where u is the solution of the first equation in (2.41). After shifting to this point $\rho(x) = u + \overline{\rho}(x)$, the mass parameters $\tau_{\rho,\pi}$ for the fields ρ and π will respectively become $\tau_{||}$ and $\tau_\perp$ from (2.41). At the asymptote (4.113) $\tau_\pi = \tau_\perp \sim h \to 0$, while $\tau_\rho = \tau_{||}$ remains a finite quantity of order τ, i.e., the radial field ρ in the loopless approximation remains massive for $h = 0$. This property is not spoiled even when loop corrections from the interaction of the fields $\overline{\rho}$ and π in (4.121) are taken into account, because π enters into this interaction only in the combinations $(\partial m)^2$ and $h\sigma$. The presence of two derivatives or a factor of h weakens the IR singularities of the self-energy graphs of the field ρ to the point that these graphs become finite at zero external momentum and $h = 0$, i.e., they lead only to a finite shift of the parameter τ_ρ. For this reason the leading singular contributions of the graphs $\langle \rho(x) \rangle = u + \langle \overline{\rho}(x) \rangle$ are multiples of $h^{d/2}$, so that including their h dependence corresponds to corrections like factors of R_1 from (4.118).

Performing the functional integration over ρ in (4.120), we obtain the representation

$$G(A) \sim \int Dm\; \delta(m^2 - 1) \exp \Phi(m, A) \quad (4.123)$$

with some functional Φ as the exponent. It follows from (4.120) and (4.121) that Φ depends on m and A only through the three local quantities $(\partial m(x))^2 \equiv F_1(x)$, $h\sigma(x) \equiv F_2(x)$, and $m(x)A(x) \equiv F_3(x)$. Since the field ρ is massive the dependence of Φ on these F_i is "analytic," i.e., Φ can be represented as a functional Taylor expansion in powers of the F_i themselves and all possible derivatives of them $\partial F_i, \partial\partial F_i, ...$

According to the general principles by which fluctuation models are constructed (Sec. 1.16), when evaluating the leading IR singularities we should neglect all contributions in the functional Φ that are certainly IR-irrelevant.

Here these are all higher powers and all derivatives of F_i compared to the simple contributions linear in F_i. In this approximation,

$$\Phi(m, A) = cu^2(\partial m)^2 + cuh\sigma + cu(mA). \qquad (4.124)$$

The factors of u have been isolated to ensure the canonically dimensionless nature of all the coefficients c, and, as usual, single integration over the d-dimensional argument x in (4.124) is understood. The representation (4.123) with exponent (4.124) is equivalent to the σ model.

The magnitude of the corrections from the IR-irrelevant contributions discarded in (4.124) can be estimated using the standard arguments of Sec. 1.16, but only after changing over (at least mentally) to the new field $\overline{m}(x) = um(x)$ needed to eliminate from the free part of the action in (4.124) the dimensional parameter u, which remains finite at the IR asymptote (4.113) under study. After this change all the local monomials inside the integral understood in (4.124) will have the standard dimension d, so that neglecting squares of them corresponds to corrections with additional dimension $p^d \sim h^{d/2}$, while neglecting derivatives like $\partial^2 F$ (since the expressions are scalars there are at least two derivatives) corresponds to corrections with additional dimension $p^2 \sim h$. These arguments determine the form of the correction factors R_1 in (4.118) characterizing the errors from changing over to the σ model (4.124).

We have justified Statement 1 of the beginning of this section: up to corrections like R_1 in (4.118), the Goldstone asymptote is described by the σ model. The arguments are not based on the specific form of the φ^4 interaction, and so they remain valid for any local O_n-invariant interaction, for example, for $(\varphi^2)^3$. This has an obvious generalization also to models with more complicated (for example, matrix) fields and symmetry groups. The following general statement can be made. For any local model of the type $S(\varphi) = -(\partial\varphi)^2/2 + V(\varphi)$ with action functional invariant under some continuous global group g of transformations of the field φ (a vector or a matrix), the Goldstone asymptote like (4.113) in the phase with spontaneous symmetry breaking is described by the corresponding generalized σ model, with the role of the Goldstone modes π being played by the basis in the space of states obtained by acting with local group transformations $g(x)$ on a specified "spontaneous magnetization vector" $\langle\varphi\rangle_0$. This space is locally isomorphic to the factor group g/g_0, where g_0 is the little group in the phase with specified $\langle\varphi\rangle_0$, corresponding to one of the minima of the potential energy (for a vector field, g_0 is the one-dimensional subgroup of rotations about the direction of the spontaneous magnetization). For more complicated models with tensor (matrix) fields φ, the structure of the manifold g/g_0 may be much more complicated, and this presents a natural method of introducing nontrivial matrix generalizations of the vector σ model [45]. Here the local invariant interaction $V(\varphi)$ of the original model never contributes to the interaction of the Goldstone modes, and so the action functional of any σ model is always generated only by the kinetic term $\sim (\partial\varphi)^2$. Incidentally, one of the matrix generalizations of the σ model is used in the quantum-field formulation of the theory of Anderson localization ([100]; see also the review [101] and [102], which significantly expands upon [100]).

Let us now turn to Statement 2. It can also be justified by a simple dimensional analysis. If the expansions of $(\partial m)^2$ and $h\sigma$ in π obtained from (4.119) and (4.121) are substituted into the σ-model action (4.124), we obtain a free contribution of the type $c(\partial\overline{\pi})^2 + c\overline{h}\overline{\pi}^2$ with $\overline{h} = h/u$, $\overline{\pi} = u\pi$, and dimensionless coefficients c plus interaction terms, the simplest of which have the form $(\overline{\pi}\partial\overline{\pi})^2$ and $h\overline{\pi}^4$ up to unimportant powers of u, while the others contain additional factors of $\overline{\pi}^2$. Clearly, even the simplest interactions are IR-irrelevant compared to the free contributions, because they differ from them by only additional "small" factors of $\overline{\pi}^2$ with positive dimension $d[\overline{\pi}^2] = d - 2$ (it is important that there are no interactions of the type $\overline{\pi}^{2n}$ without factors of ∂^2 or h that could compete with the free action).

Therefore, all interactions of the fields π in the functional (4.124) are IR-irrelevant contributions. According to the general theory (Sec. 1.16), such contributions only renormalize the coefficients of the IR-relevant (in our case, free) terms in the action and generate regular ($\sim h$) and singular ($\sim h^{d/2-1}$) corrections to the leading singularities of the studied IR asymptote. The latter are determined by only the IR-relevant part of the action, in our case, the free action $cu^2(\partial\pi)^2 + cuh\pi^2$ of the field π.

The correlation functions of the original fields (4.114) are determined by the variational derivatives of the functional (4.123) with exponent (4.124) involving the corresponding sources in the form $(mA) = \pi A^\perp + \sigma A^\parallel$, in particular, the correlators $D^\perp$ and $D^\parallel$ of the fields (4.114) are multiples of the connected parts of the expectation values $\langle\pi\pi\rangle$ and $\langle\sigma\sigma\rangle$, respectively. In leading order the field π can be assumed free, and so the correlator $D^\perp$ is a multiple of (4.116). For the magnetization $\langle\varphi^\parallel\rangle \sim \langle\sigma\rangle$ and the longitudinal correlator $D^\parallel$, a multiple of the connected part of $\langle\sigma\sigma\rangle$, in the expansion of σ in π^2 we need to keep only the constant and π^2 terms. The IR-irrelevant higher powers of π^2 can be taken into account by changing the coefficients in the terms that are kept and including the corrections.

Let us explain this in more detail. In calculating the magnetization $\langle\varphi^\parallel\rangle = c\langle\sigma\rangle$, the result is represented as a series in the expectation values $\langle\pi^{2m}(x)\rangle$ with free field π. For the simplest we have

$$\langle\pi^2(x)\rangle \sim \int dk(ck^2 + ch)^{-1} = c\Lambda^{d-2} + ch^{d/2-1} + ch\Lambda^{d-4} + \dots \quad (4.125)$$

with dimensionless coefficients c on the right-hand side. In the terminology of Sec. 1.18, the first term is the regular contribution, the second is the leading singular one, and the third is the first regular correction. For the free field π the expectation value $\langle\pi^{2m}(x)\rangle$ is a multiple of $\langle\pi^2(x)\rangle^m$. Any integer power of (4.125) contains, along with other terms, the leading singular contribution of the same form ($\sim h^{d/2-1}$) like expression (4.125) itself. This means that the leading singularity is built up from all the powers of π^2 in the expansion of $\sigma = [1 - \pi^2]^{1/2}$, but the inclusion of the contributions of all higher powers of π^2 is equivalent to a simple change of the coefficient of just π^2 itself.

The situation regarding the correlator $D^\parallel$, a multiple of the connected part of $\langle\sigma\sigma\rangle$, is exactly the same. Its leading singular contribution is determined by the correlator $\langle\pi^2\pi^2\rangle$ (the constant in the expansion of σ in π^2 does not contribute to the correlator), but singularities of the same type are present also in all the correlators $\langle\pi^{2m}\pi^{2n}\rangle$, leading to renormalization of the coefficient of π^2 in (4.117).

This explains Statement 2 pertaining to the field σ, and at the same time shows that the constants c in (4.117) are not simple coefficients of the expansion of σ in π^2. If desired, their values can be refined. For example, the coefficient c of π^2 in (4.117) is determined by the equation

$$\tfrac{1}{2}\langle\partial^2\sigma(x)/\partial\pi_a(x)\partial\pi_b(x)\rangle\,|_{h=0} = cP_{ab},\qquad(4.126)$$

which is obvious from analysis of the corresponding graphs using the technique of Sec. 2.2 [in the differentiation it must be assumed that $\partial\pi_a(x)/\partial\pi_b(x) = P_{ab} \equiv \delta_{ab} - e_a e_b$]. An analogous representation can be written down for the second (additive) constant c on the right-hand side of (4.117). Of course, in formal dimensional regularization such renormalizations of the coefficients are absent.

Summarizing this discussion of expectation values involving the field σ, for the Goldstone asymptote (4.113) of the magnetization $\alpha = \langle\varphi^\parallel\rangle$ we have

$$\alpha(h) \cong c\langle\sigma\rangle = c + ch^{d/2-1} + ch + ...,\qquad(4.127)$$

for the longitudinal susceptibility we have

$$\chi^\parallel(h) - \partial\alpha/\partial h = ch^{d/2-2} + c + ...,\qquad(4.128)$$

and for the longitudinal correlator in momentum space we have

$$D^\parallel(p,h) \cong c\langle\pi^2\pi^2\rangle_{\mathrm{conn}} = c\Pi(p,h) + c + ...,\qquad(4.129)$$

where $\Pi(p,h)$ is a simple loop of the type (2.187) with the lines (4.116), and the additive constant c in (4.129) is the leading regular correction. These asymptotes are valid for any dimension $2 < d < 4$ up to corrections like R_2 from (4.118). Comparison with the expressions of Sec. 2.24 shows that the asymptotes (4.127)–(4.129) are correctly reproduced by the leading order of the $1/n$ expansion.

All these expressions for the leading Goldstone singularities can be interpreted as *Goldstone scaling* (by analogy with critical scaling), in which the only IR-relevant parameters (Sec. 1.15) are h and $p \sim 1/x$ (but not τ), and the variables p, h, and the fields $\pi(x)$ and $\sigma(x) - \langle\sigma(x)\rangle|_{h=0} \equiv \bar\sigma(x)$ have definite "Goldstone dimensions" $\Delta^\Gamma[...]$:

$$\Delta^\Gamma[p] = 1,\quad \Delta^\Gamma[h] = 2,\quad \Delta^\Gamma[\pi(x)] = d/2 - 1,\quad \Delta^\Gamma[\bar\sigma(x)] = d - 2.\quad(4.130)$$

These dimensions, in contrast to the critical ones, are known exactly, as are the normalized scaling functions of the simplest correlators (4.116) and (4.129).

From (4.116) and (4.117) it is easy to find the explicit expressions for the leading singularities of all higher connected functions $\langle \sigma...\sigma \rangle$. The higher connected functions of the field π are zero in the leading approximation, and only the (IR-irrelevant) interaction of the fields π generates nontrivial contributions to them. In the terminology of Sec. 1.3, they are all corrections to scaling with the additional small factor $\sim h^{d/2-1}$ and can be calculated within simple perturbation theory for the interaction in the σ model (4.124).

We have studied the asymptote (4.113) for any fixed $\tau < 0$, and so the results are also valid in the critical region of small (on the scale of Λ) values of τ. In this region the Goldstone asymptote should be consistent with the formulas of critical scaling, which gives information about the corresponding scaling functions. Let us explain this for the simple example of the magnetization $\alpha(h, \tau)$ in the renormalized φ^4 model. According to the scaling arguments, in the critical region of small h and τ we have

$$\alpha(h, \tau) = c_1 |\tau|^{\Delta_\alpha/\Delta_\tau} f(z), \quad z = c_2 h |\tau|^{-\Delta_h/\Delta_\tau}, \quad (4.131)$$

where Δ_F are the critical dimensions of the corresponding quantities $(\Delta_\alpha = \Delta_\varphi = d - \Delta_h)$, $f(z)$ is the scaling function, and $c_{1,2}$ are nonuniversal normalization factors (Sec. 1.33) containing the powers of the renormalization mass μ needed to ensure the required canonical dimensions. The ambiguity associated with the arbitrariness of $c_{1,2}$ is, as usual, eliminated by changing over to the normalized scaling function.

Knowledge of the asymptote (4.127) for $h \to 0$, $\tau =$const gives information about the $z \to 0$ asymptote of the scaling function $f(z)$ in (4.131): $f(z) = c + cz^{d/2-1} + cz + ...$ The dimensionless constants c entering here can be calculated as $4 - \varepsilon$ expansions after the normalizations are fixed.

We can state this in another way: in the critical region the coefficients c in (4.127) are multiples of powers of $|\tau|$ with exponents known from the critical dimensions.

Let us conclude with a historical note. The key feature in justifying the Goldstone asymptote is the fact that in the $h \to 0$ limit the radial field ρ remains massive even when its interaction with Goldstone modes that are massless in this limit is taken into account. Therefore, in the variables (4.119) the problem is actually trivial and is solved by the standard arguments of Landau theory, allowing the problem to be reduced to a σ model. This was first clearly stated in [99]. The authors of [103] reached the same conclusion practically simultaneously on the basis of their "constant modulus principle," which states that it is possible to neglect fluctuations of the field $\rho \equiv |\varphi|$. This corresponds to the simple stationary phase approximation for the integral over ρ in (4.120), which also leads directly to a σ model, but without the additional renormalizations of the constants. Strictly speaking, the arguments of [103] are not completely correct, because, for example, for the rather strong φ^4 interaction the contributions of loop corrections to $\langle \rho \rangle$ are comparable in magnitude to the loopless approximation, i.e., the fluctuations of ρ cannot be assumed small. They are small only compared to fluctuations of the Goldstone modes π, because the latter become "infinite" in the $h \to 0$ limit. This corresponds to the inequality $\tau_\rho \gg \tau_\pi \sim h \to 0$, and the term "massive ρ" [99] is more precise than "constant modulus principle" [103].

The problem of Goldstone singularities, which is simple in the language of the fields (4.119), appears very complicated if the original φ^4 model is studied directly in its natural variables (4.114). For $h = 0$ the field $\varphi^\parallel$ in the loopless approximation (Landau theory) has finite mass $\tau_\parallel$ from (2.41), but when loop corrections are included it becomes massless [for $h = 0$ and $p \to 0$ the loop Π in (4.129) has IR divergence $\sim p^{d-4}$]. Therefore, the field $\varphi^\parallel$, in contrast to ρ, cannot simply be eliminated from the set of relevant variables, which makes the problem of studying the Goldstone singularities of the φ^4 model directly in terms of its fields (4.114) highly nontrivial [the Goldstone singularities appear more strongly in individual graphs of $D^\parallel$ than in the loop (4.129), and their cancellation must be demonstrated]. There are quite a few studies (see, for example, [104] and references therein) devoted to attempts at solving this problem by summing the Goldstone singularities of the scaling functions in the $4 - \varepsilon$ expansion. As far as we know, in none of these studies has it been possible to obtain a complete solution of the problem, i.e., to justify asymptotes like (4.129) in all orders in ε. However, this actually is not necessary, because the results [99] presented above completely solve the problem of the Goldstone singularities and certainly cover everything that can be obtained regarding this topic within the $4 - \varepsilon$ expansion.

4.16 The two-charge φ^4 model with cubic symmetry

So far we have discussed the RG technique mainly for the example of the O_n φ^4 model. Let us now consider one of its generalizations — the two-charge φ^4 model for the n-component field $\varphi = \{\varphi_a\}$ with unrenormalized action

$$S(\varphi) = -\frac{1}{2}(\partial\varphi)^2 - \frac{1}{2}\tau_0\varphi^2 - \frac{1}{24}[g_{10}V_1 + g_{20}V_2] \qquad (4.132)$$

with the ordinary local forms $(\partial\varphi)^2 \equiv \partial\varphi_a \cdot \partial\varphi_a$, $\varphi^2 \equiv \varphi_a\varphi_a$, $V_1 \equiv (\varphi^2)^2$ and a second "cubic" interaction $V_2 \equiv \sum_a \varphi_a^4$. This model is multiplicatively renormalizable (Sec. 3.13), and the analog of (3.105) is the renormalized action

$$S_{\mathrm{R}}(\varphi) = \Delta S_{\mathrm{vac}} - Z_1\frac{1}{2}(\partial\varphi)^2 - Z_2\frac{1}{2}\tau\varphi^2 - \frac{1}{24}\mu^{2\varepsilon}[Z_3^{(1)}g_1V_1 + Z_3^{(2)}g_2V_2]$$

$$(4.133)$$

with the usual renormalization formulas (Sec. 3.18) for φ and τ, and

$$g_{i0} = g_i\mu^{2\varepsilon}Z_g^{(i)}, \qquad Z_g^{(i)} = Z_3^{(i)}Z_\varphi^{-4}, \qquad i = 1, 2 \qquad (4.134)$$

for the two charges. We shall assume that the vaccum counterterm ΔS_{vac} in (4.133) is defined by (3.106) with $g_{\mathrm{B}} = g_{1\mathrm{B}} \equiv g_1\mu^{2\varepsilon}$ in the denominator.

As noted above (Sec. 3.21), the results given in Table 3.1 of Sec. 3.20 for the individual graphs of the simple φ^4 model can be used to calculate the renormalization constants and RG functions of any model of the φ^4 type. The specifics of the model are manifested only in the structure factors r_γ (Sec. 3.20), the calculation of which is an elementary problem compared to the calculation of the ε poles of the φ^4 model already performed. It can be awkward in higher orders, where it is useful to use a computer. It can be done by hand for the model (4.132) with three-loop accuracy. We shall write

$$u_i \equiv g_i/16\pi^2, \quad u \equiv u_1, \quad u_2/u_1 \equiv \alpha, \tag{4.135}$$

and assume that the factors of $(-u)^l$ (where l is the number of loops) are isolated from the graphs, as in Table 3.1. The additional structure factors r_γ taking into account in (3.118) the difference between this theory and the simple φ^4 model with a single charge $g = g_1$ will then be determined (Sec. 3.20) from the index graphs with lines δ_{ab} and vertices

$$v_{abcd} = v_{abcd}^{(1)} + \alpha \cdot v_{abcd}^{(2)}, \tag{4.136}$$

where $v^{(1)}$ is the usual O_n-symmetric vertex (2.80) and $v_{abcd}^{(2)} = \delta_{ab}\delta_{bc}\delta_{cd}$ corresponds to V_2. The corresponding index graph with lines δ_{ab} and vertices (4.136) is easily calculated for each of the graphs in Table 3.1. The constants Z_0 and Z_2 in Table 3.1 are calculated from the vertex graphs γ_4, identified with the graphs $\partial_\tau^2\gamma_0$ and $\partial_\tau\gamma_2$ (Sec. 3.20), and the index graph is then associated with the differentiated graph. For vacuum loops γ_0 without external indices the value of the index graph is some constant c, for the self-energy graphs we obtain $c \cdot \delta_{ab}$, for the vertex graphs after symmetrization in the external indices we obtain a linear combination of the vertex factors $c_1 v^{(1)} + c_2 v^{(2)}$, and all the scalar coefficients c are polynomials in n and α from (4.135). They play the role of the r_γ for the constants $Z_{0,1,2}$ and $Z_3^{(1)}$ in (4.133), and $r_\gamma = c_2/\alpha$ for $Z_3^{(2)}$.

Let us explain the last statement. The rule stated for the vertices follows from the general expression (3.37), from which in the present notation we find that $V_R = V_B - L\Gamma_4$, where V_B is the basic interaction, V_R is the renormalized one, and $L\Gamma_4$ are the contributions of the vertex counterterms. If all these quantities are sums of independent structures $v^{(i)}$ with coefficients $-g_{iB}$ in V_B and $-g_{iB}Z_3^{(i)}$ in V_R, then from the above expressions we find that

$$g_{iB} + [L\Gamma_4]_i = g_{iB}Z_3^{(i)}, \tag{4.137}$$

where $[L\Gamma_4]_i$ is the coefficient of $v^{(i)}$ in the vertex counterterms. In the one-charge model, (4.137) leads to (3.107) for Z_3, obtained by changing over to the normalized function according to the rule (3.91): $\Gamma_4 = -g_B\overline{\Gamma}_4$. If there are many structures $v^{(i)}$ and the corresponding charges and we use the relation $\Gamma_4 = -g_B\overline{\Gamma}_4$ to distinguish one of them, namely, $g_{1B} \equiv g_B$, from (4.137) for any $i = 1, 2, \ldots$ we obtain

$$Z_3^{(i)} = 1 - [L\overline{\Gamma}_4]_i \cdot g_{1B}/g_{iB}. \tag{4.138}$$

After isolating the factors $(-u)^l$ (see above), the vertices in the graphs of $\overline{\Gamma}_4$ will be the quantities (4.136). It follows from (4.138) that in defining the r_γ for the vertex graphs γ, the coefficients c_i of the structures $v^{(i)}$ obtained from the corresponding index graph with vertices (4.136) are accompanied by an additional factor of g_{1B}/g_{iB}. In our case it follows from the definitions (4.135) that this factor is unity for $Z_3^{(1)}$ and $1/\alpha$ for $Z_3^{(2)}$, and so $r_\gamma = c_1$ in the first case and $r_\gamma = c_2/\alpha$ in the second.

All the constants Z of the model (4.133) are calculated from the results in Table 3.1 of Sec. 3.20 and the general rule (3.118), but with different values of the r_γ.

These new values of r_γ in the model (4.133) for all 13 graphs of Table 3.1 are given in Tables 4.1 (for $Z_{0,1,2}$) and 4.2 (for $Z_3^{(i)}$). The numbering of the graphs and the factors $r_i = r_i(n)$ are everywhere the same as in the tables of Sec. 3.20. A graph number like 8,13 means that the two graphs have identical r_γ for all the Z of the given table, and the absence of a graph in a table means that it does not contribute to the given constants Z.

Table 4.1 Structure factors r_γ of the graphs γ from Table 3.1 in the renormalization constants $Z_{0,1,2}$ of the model (4.133); $\alpha \equiv g_2/g_1$.

Graph No.	Factor r_γ in the constants		
	Z_0	Z_1	Z_2
1	—	—	$r_1 + \alpha$
2	—	$r_1 + 2\alpha + \alpha^2$	—
3	$n(r_1 + \alpha)$	—	$(r_1 + \alpha)^2$
4	—	—	$r_1 + 2\alpha + \alpha^2$
5	—	$r_1 r_2 + \alpha(n+8)/3 + \\ +3\alpha^2 + \alpha^3$	—
6	$n(r_1 + \alpha)^2$	—	$(r_1 + \alpha)^3$
7	—	—	$r_1^2 + \alpha(n+2) + \\ +\alpha^2(n+8)/3 + \alpha^3$
8, 13	$n(r_1 + 2\alpha + \alpha^2)$	—	$r_1^2 + \alpha(n+2) + \\ +\alpha^2(n+8)/3 + \alpha^3$
9, 11	—	—	$r_1 r_2 + \alpha(n+8)/3 + \\ +3\alpha^2 + \alpha^3$

Table 4.2 Structure factors r_γ of the graphs γ from Table 3.1 in the renormalization constants $Z_3^{(i)}$ of the model (4.133), $\alpha \equiv g_2/g_1$.

Graph No.	Factor r_γ in the constants	
	$Z_3^{(1)}$	$Z_3^{(2)}$
1	$r_2 + \dfrac{2\alpha}{3}$	$\dfrac{4}{3} + \alpha$
3	$r_4 + \dfrac{\alpha(n+4)}{3} + \alpha^2$	$\dfrac{4}{3} + 2\alpha + \alpha^2$
4	$r_3 + \dfrac{4\alpha}{3} + \dfrac{\alpha^2}{3}$	$\dfrac{n+14}{9} + \dfrac{8\alpha}{3} + \alpha^2$
6	$r_7 + \dfrac{4\alpha(n^2+6n+12)}{27} +$ $+\dfrac{2\alpha^2(n+4)}{3} + \dfrac{4\alpha^3}{3}$	$\dfrac{32}{27} + \dfrac{8\alpha}{3} + \dfrac{8\alpha^2}{3} + \alpha^3$
7,13	$r_5 + \dfrac{4\alpha(3n+14)}{27} +$ $+\dfrac{\alpha^2(n+18)}{9} + \dfrac{2\alpha^3}{3}$	$\dfrac{4(n+10)}{27} + \dfrac{\alpha(n+34)}{9} +$ $+\dfrac{10\alpha^2}{3} + \alpha^3$
8	$r_1 r_2 + \dfrac{4\alpha(n+5)}{9} +$ $+\dfrac{\alpha^2(n+20)}{9} + \dfrac{2\alpha^3}{3}$	$\dfrac{4(n+2)}{9} + \dfrac{\alpha(n+10)}{3} +$ $+\dfrac{10\alpha^2}{3} + \alpha^3$
9	$r_5 + \dfrac{\alpha(9n+52)}{27} +$ $+\dfrac{5\alpha^2}{3} + \dfrac{\alpha^3}{3}$	$\dfrac{n^2+6n+40}{27} + \dfrac{\alpha(n+12)}{3} +$ $+\dfrac{11\alpha^2}{3} + \alpha^3$
10	$r_6 + \dfrac{4\alpha(n+12)}{27} + \dfrac{8\alpha^2}{9}$	$\dfrac{8(n+6)}{27} + \dfrac{2\alpha(n+22)}{9} +$ $+4\alpha^2 + \alpha^3$
11	$r_6 + \dfrac{2\alpha(2n+27)}{27} +$ $+\dfrac{13\alpha^2}{9} + \dfrac{\alpha^3}{3}$	$\dfrac{2(3n+22)}{27} + \dfrac{\alpha(n+40)}{9} +$ $+\dfrac{11\alpha^2}{3} + \alpha^3$
12	$r_3 + 16\alpha/9 + 2\alpha^2/3$	$\dfrac{4(n+14)}{27} + \dfrac{16\alpha}{3} +$ $+4\alpha^2 + \alpha^3$

4.17   RG functions and critical regimes

The general rule (3.118) can be used with the data of Tables 4.1 and 4.2 and those in Sec. 3.20 to calculate all the constants Z of the model (4.133) with three-loop accuracy, for example,

$$Z_1 = 1 - \frac{1}{24\varepsilon}\left[\frac{n+2}{3}u_1^2 + 2u_1u_2 + u_2^2\right] - $$

$$-\frac{(1-\varepsilon/2)}{24\varepsilon^2}\left[\frac{n^2+10n+16}{27}u_1^3 + \frac{n+8}{3}u_1^2u_2 + 3u_1u_2^2 + u_2^3\right] + \dots \tag{4.139}$$

and similarly for the other Z. It is convenient to use (Sec. 4.2) the u_i from (4.135) as the renormalized charges. If we want to check the calculations by seeing if the ε poles in the RG functions $\gamma_{u_i} = \widetilde{\mathcal{D}}_\mu \ln Z_g^{(i)}$ cancel, they must be calculated by solving the system of two linear equations obtained directly from the definitions of γ_{u_i} and $\beta_i \equiv \beta_{u_i} = \widetilde{\mathcal{D}}_\mu u_i$ taking into account the relation $\widetilde{\mathcal{D}}_\mu F(u) = [\beta_1\partial_{u_1} + \beta_2\partial_{u_2}]\cdot F(u)$ analogous to (1.115), which leads to analogs of (1.116). If we are not interested in checking the cancellation of the ε poles in the RG functions, they can be calculated much more simply, as in Sec. 4.2, using expressions like (4.5) with $\mathcal{D}_g$ in them replaced by $\mathcal{D}_u \equiv \mathcal{D}_{u_1} + \mathcal{D}_{u_2}$. Finally, from the known constants Z we find that

$$\gamma_\varphi = \frac{1}{12}\left[\frac{n+2}{3}u_1^2 + 2u_1u_2 + u_2^2\right] - \frac{1}{16}\left[\frac{n^2+10n+16}{27}u_1^3 + \right.$$

$$\left. +\frac{n+8}{3}u_1^2u_2 + 3u_1u_2^2 + u_2^3\right] + \dots; \tag{4.140a}$$

$$\gamma_\tau = -\frac{n+2}{3}u_1 - u_2 + \frac{5}{6}\left[\frac{n+2}{3}u_1^2 + 2u_1u_2 + u_2^2\right] + \dots;$$

$$\beta_1 = u_1\left[-2\varepsilon + \frac{n+8}{3}u_1 + 2u_2 - \frac{1}{3}[(3n+14)u_1^2 + 22u_1u_2 + 5u_2^2] + \dots\right];$$

$$\beta_2 = u_2\left[-2\varepsilon + 4u_1 + 3u_2 - \frac{1}{9}[(5n+82)u_1^2 + 138u_1u_2 + 51u_2^2] + \dots\right]. \tag{4.140b}$$

As in Sec. 4.2, we have given γ_φ in the three-loop approximation and all the other RG functions only in the two-loop one. We are not considering γ_0.

Let us briefly discuss possible critical regimes for this model within the ε expansion. They are determined (Sec. 1.42) by the IR-attractive fixed points $u_* = \{u_{1*}, u_{2*}\}$ located in the stability region of the basic interaction (the analog of $g > 0$ in the simple φ^4 model):

$$u_1 + u_2 \geq 0, \quad nu_1 + u_2 \geq 0. \tag{4.141}$$

The first of these inequalities ensures the required positivity of the form $u_1 V_1 + u_2 V_2$ for configurations of the particular form $\{\varphi_a\} = \{\varphi, 0, 0, ...\}$, and the second for $\{\varphi_a\} = \{\varphi, \varphi, \varphi, ...\}$. It is easily shown that the two conditions (4.141) together guarantee that the form $u_1 V_1 + u_2 V_2$ is positive definite for any configuration φ.

In leading order of perturbation theory in the renormalized charges g, the stability conditions of the type (4.141) for the unrenormalized and basic interactions are equivalent, because in this approximation $Z = 1$. Moreover, the exact criterion for stability is the correct convexity of the generating functional of the 1-irreducible Green functions $\Gamma(\varphi)$ or $\Gamma_R(\varphi)$; these are equivalent, since these two functionals are related by a simple dilatation of the argument (1.79). Like any exact property, it must be satisfied in leading order of the regular expansion of the functionals in any parameter, in particular, the expansion in the number of loops. In leading order of the loop expansion the renormalized functional $\Gamma_R(\varphi)$ coincides with the basic action (the counterterm contributions are grouped together with the loop corrections), which justifies the stability criterion (4.141).

In general, within the ε expansion the β functions (4.140b) have the following four fixed points (zeros of the β functions):

No. 1 (Gaussian): $u_{1*} = u_{2*} = 0$;
No. 2 (Heisenberg): $u_{1*} = 6\varepsilon/(n+8) + ..., \quad u_{2*} = 0$;
No. 3 (Ising): $u_{1*} = 0, \quad u_{2*} = 2\varepsilon/3 + ...;$ (4.142)
No. 4 (cubic): $u_{1*} = 2\varepsilon/n + ...,$
         $u_{2*} = 2\varepsilon(n-4)/3n + ...$

For $n \geq 1$ they all lie within the stability region (4.141).

To determine the nature of these points it is necessary to calculate the eigenvalues of the 2×2 matrix $\omega_{ik} = \partial\beta_i/\partial u_k$ at each (Sec. 1.42). This calculation shows that in the lowest order in ε these eigenvalues are -2ε, -2ε for point No. 1, 2ε, $2\varepsilon(4-n)/(n+8)$ for point No. 2, $-2\varepsilon/3$, 2ε for point No. 3, and 2ε, $2\varepsilon(n-4)/3n$ for point No. 4. Those points for which both eigenvalues are positive are IR-attractive (Sec. 1.42); here, these are only point No. 2 for $n < 4$ and point No. 4 for $n > 4$. In the critical regime an O_n symmetry arises for the first of these, while for the second there is only cubic symmetry. At $n = 4$ points No. 2 and 4 coincide in the one-loop approximation (4.142), so that the two-loop corrections must be included in the β functions (4.140b) in order to distinguish between them.

The location of the fixed points and the qualitative behavior of the phase trajectories [the solutions of (1.188) for the invariant charges] are shown in Fig. 4.1 for the two typical cases $n = 2 < 4$ and $n = 6 > 4$. The dashed lines show the boundaries of the stability region (4.141). We see from these pictures of the phase trajectories that not all of them lie in the basin of attraction of the IR-stable fixed point: some trajectories, bypassing u_*, leave the stability region (4.141), which is interpreted as a first-order phase transition (Sec. 1.42).

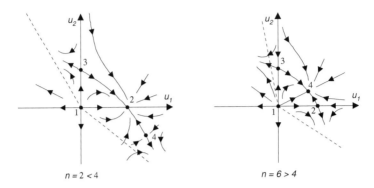

$n = 2 < 4$ $n = 6 > 4$

Figure 4.1 Behavior of the phase trajectories of the invariant charges $\bar{u}(s, u)$ for $s = p/\mu \to 0$ for the model (4.133). The dashed lines show the boundaries of the stability region (4.141).

The ε expansions of the critical exponents for a given critical regime are obtained, as usual, by substituting of the coordinates of the corresponding IR-stable fixed point u_* in the expansion of the RG functions (4.140a).

4.18 The Ising model (uniaxial magnet) with random impurities. $\varepsilon^{1/2}$ expansions of the exponents

The results given in the preceding section can be used to analyze the critical behavior of a uniaxial magnet with frozen random impurities [105], [106]. One of the commonly used methods of introducing random impurities into the Hamiltonian of the Ising model (1.16) is to make the replacement $\hat{s}_i \to \hat{s}_i \hat{p}_i$, where $\hat{p}_i$ is an additional random variable taking the value 0 (a defect or empty site) or 1 (a normal or occupied site) with given probabilities C and $1 - C$, respectively. The parameter $0 < C < 1$ is interpreted as the relative concentration of defects—nonmagnetic sites. Another method is to introduce an analogous random variable $\hat{p}_{ik}$ for each pair (ik) of adjacent lattice sites over which the summation in (1.16) runs, which corresponds to the "Ising model with random bonds."

The problem is to study the effect of the impurities on the critical behavior of the system. In the absence of impurities it is described by the simple φ^4 model (Ch. 1). It has been shown in [105] that in this language the inclusion of impurities of any of the two types described above for relatively low concentrations (below the "percolation threshold") is equivalent to adding to the usual action of the φ^4 model an interaction with the impurities $V_{\text{im}} = \psi\varphi^2 \equiv \int dx\psi(x)\varphi^2(x)$, where $\psi(x)$ is the Gaussian random field of the impurity with $\langle\psi\rangle = 0$ and correlator $\Delta_\psi(x, x') \equiv \langle\psi(x)\psi(x')\rangle = \lambda_0\delta(x - x')$ with bare constant $\lambda_0 > 0$ proportional to the impurity concentration C. The formal description of impurities by the replica method (Sec. 2.31) then leads to a model of an n-component field $\varphi = \{\varphi_a\}$ with unrenormalized action (2.234), which in the present case takes the form

$$S = -\frac{1}{2\lambda_0}\psi^2 - \frac{1}{2}\partial\varphi_a \cdot \partial\varphi_a - \frac{\tau_0}{2}\varphi_a\varphi_a - \frac{g_0}{24}\sum_a \varphi_a^4 + \psi\varphi_a\varphi_a \qquad (4.143)$$

(the repeated index a is summed over). If we are only interested in the Green functions of the basic field φ, then Gaussian functional integration over ψ can be performed in all the functional integrals, which leads to the model (4.132) with bare charges $g_{10} \sim -\lambda_0 \sim -C < 0$, $g_{20} = g_0 > 0$. These inequalities can be carried over to the basic parameters of the corresponding renormalized action (4.133) [see the discussion following (4.141)]:

$$u_1 < 0, \qquad u_2 > 0. \qquad (4.144)$$

Therefore, the critical behavior of the uniaxial magnet with frozen impurities is described by the model (4.133) with the conditions (4.144) for the charges and $n = 0$ in the final expressions according to the general rule of the replica method (Sec. 2.31). Since the charge $u_1 \sim g_1$ is proportional to the impurity concentration (see above), conditions (4.144) and (4.141) with $n = 0$ are compatible for low concentrations.

The case $n = 0$ for the β functions (4.140b) is special, because the coordinates of fixed point No. 4 in (4.142) contain n in the denominator, so that for $n \to 0$ the point goes off to infinity. This is a consequence of the accidental degeneracy of the one-loop contributions in (4.140b) for $n = 0$. Let us explain in more detail. The coordinates of fixed point No. 4 with $u_{1*} \neq 0$ and $u_{2*} \neq 0$ are determined from the system of equations $\beta_i/u_i = 0$, $i = 1, 2$. If u_* is taken to be a series in ε these equations in the one-loop approximation for $n = 0$ take the form $2\varepsilon = 8u_1/3 + 2u_2$, $2\varepsilon = 4u_1 + 3u_2$. Since the right-hand sides are proportional to each other (degeneracy), the desired solution with $u_* \sim \varepsilon$ does not exist. However, owing to this degeneracy it becomes possible [106] to seek a solution of the system $\beta_i/u_i = 0$ for u_* in the form of a series in powers of $\varepsilon^{1/2}$ rather than the usual ε. Such an "anomalous" solution actually exists. The two-loop approximation (4.140b) with $n = 0$ can be used to find explicitly only the first coefficients of the series:

$$u_{1*} = -3(3\varepsilon/53)^{1/2} + ..., \qquad u_{2*} = 4(3\varepsilon/53)^{1/2} + ... \qquad (4.145)$$

There is also a second solution differing by a sign, but we do not need it (see below). We note that the point (4.145) is located in the quadrant (4.144) and formally satisfies the conditions (4.141) with $n = 0$.

The eigenvalues of the matrix $\partial\beta_i/\partial u_k$ at the point (4.145) can be found from the β functions (4.140b) only in first order in $(\varepsilon)^{1/2}$. In this approximation one of them turns out to be positive and the other zero, so that its sign is not determined. Therefore, determination of the nature of the point (4.145) requires the next contributions of order ε, i.e., the three-loop corrections to the β functions (4.140b). The three-loop calculation was performed in [107] (the results presented in Tables 3.1, 4.1, and 4.2 are sufficient for it), where it was shown that the "impurity fixed point" (4.145) is actually IR-stable and "attracts" all phase trajectories with initial data like (4.144). It is therefore the point (4.145) that determines the critical behavior of the Ising model with impurities. Although the isotropic point No. 2 in (4.142) for $n = 0$ is also IR-stable, it is inaccessible for initial data of the type (4.144), as is the second "anomalous" fixed point differing from (4.145) by a sign. This is evident from the picture Fig. 4.1 gives of the phase trajectories, the behavior of which in the quadrant (4.144) remains qualitatively the same for $n = 0$ as for $n = 2$.

The three-loop results for the first terms of the $(\varepsilon)^{1/2}$ expansions of the critical exponents obtained in [107] have the form

$$\eta = -\frac{2\varepsilon}{106} + a \cdot \frac{216 + 63\zeta(3)}{(53)^2}(2\varepsilon)^{3/2} + ...,$$

$$\nu = \frac{1}{2} + \frac{a}{4}(2\varepsilon)^{1/2} + \frac{535 - 756\zeta(3)}{8 \cdot (53)^2}(2\varepsilon) + ...,$$

$$\gamma = 1 + \frac{a}{2}(2\varepsilon)^{1/2} + \frac{147 - 189\zeta(3)}{(53)^2}(2\varepsilon) + ...,$$ (4.146)

$$\alpha = -a(2\varepsilon)^{1/2} + \frac{1137 + 378\zeta(3)}{(53)^2}(2\varepsilon) + ...,$$

where $a \equiv (6/53)^{1/2}$, $2\varepsilon \equiv 4 - d$, and $\zeta(z)$ is the Riemann zeta function.

For $d = 3$, i.e., $2\varepsilon = 1$, the series (4.146) behave much worse than ordinary ε expansions: the corrections are larger than the first terms and sometimes even change the sign of the latter. This has stimulated attempts to calculate the exponents of this model numerically by the method of the real-space renormalization group (Sec. 1.41). The results can be found in [108].

4.19   Two-loop calculation of the ε expansions of the exponents for a uniaxial ferroelectric

Let us consider the model of a uniaxial ferroelectric (Sec. 1.17) in dimension $d = 3 - 2\varepsilon$, assuming one dimension is longitudinal and the other $2 - 2\varepsilon$ are transverse. The unrenormalized action (1.73) without an external field can be written symbolically as

$$S(\varphi) = -\frac{1}{2}\varphi\left[k^2 + \nu_0^2\frac{\omega^2}{k^2} + \tau_0\right]\varphi - \frac{\nu_0\lambda_0}{24}\varphi^4,$$ (4.147)

denoting $k_\perp \equiv k$ and $k_\parallel \equiv \omega$ and substituting $g_0 = \nu_0\lambda_0$, which is convenient from dimensional considerations (see below). The canonical dimensions of all quantities in (4.147) for $d = 3 - 2\varepsilon$ can be determined from Table 1.4 in Sec. 1.17. They are given in Table 4.3, where we also include the renormalization mass μ and all the renormalized parameters, whose dimensions are found from the renormalization formulas. The fully dimensionless parameter λ will henceforth play the role of the renormalized charge.

Table 4.3 Canonical dimensions of the fields and parameters of the model (4.147) for $d = 3 - 2\varepsilon$.

F	k, μ	ω	$\varphi(x)$	τ, τ_0	ν, ν_0	λ_0	λ
$d_F^{\perp}$	1	0	$-\epsilon$	2	2	2ϵ	0
$d_F^{\parallel}$	0	1	$1/2$	0	-1	0	0
d_F	1	2	$1 - \epsilon$	2	0	2ϵ	0

Analysis of the divergences of the graphs of this model shows that all the main conclusions of the general theory of renormalization (Ch. 3) remain valid as long as the term "dimension of F" is now taken to mean the total canonical dimension $d[F] = d^{\perp}[F] + 2d^{\parallel}[F]$. In particular, the condition of logarithmicity is that the total dimension of the bare charge be zero: $d[g_0] = 2\varepsilon = 0$ (isolation of a factor of ν or ν_0 with $d[\nu] = 0$ does not affect the value of $d[g_0]$). We see from Table 4.3 that φ, τ_0, and λ_0 for $d = 3 - 2\varepsilon$ have exactly the same total dimensions as the analogous quantities in the simple φ^4 model with $d = 4 - 2\varepsilon$. Therefore, as before, the superficially divergent graphs generating counterterms (Ch. 3) are the self-energy (γ_2) and vertex (γ_4) graphs (we shall not consider vacuum loops γ_0 and the corresponding counterterms). The structure of the counterterms is determined by their dimensions, which must both be exactly the same as those of the corresponding contributions to the action (4.147). Here it should be noted that in γ_2 the dimensions also allow a counterterm of the type $\nu\omega$ in addition to the usual k^2 and τ. However, this counterterm (which would spoil the multiplicative renormalizability) is forbidden by the obvious parity of all the graphs of γ_2 in the variable $\omega \equiv k_{\parallel}$.

The essentially new feature compared to the φ^4 model is the presence of a nonlocal contribution $\sim \omega^2/k^2$ in the action (4.147). It is known from the general theory of renormalization that this term should not be renormalized (i.e., $Z = 1$ for it) because nonlocal counterterms do not occur (Sec. 3.13).

It then follows that, neglecting the vacuum counterterm, the renormalized action for the model (4.147) takes the form

$$S_R(\varphi) = -\frac{1}{2}\varphi[Z_1 k^2 + \nu^2\omega^2/k^2 + Z_2\tau]\varphi - \frac{1}{24}Z_3\nu\lambda\mu^{2\varepsilon}\varphi^4. \qquad (4.148)$$

This corresponds to the multiplicative renormalization of the action (4.147):

$$\hat{\varphi} = \hat{\varphi}_R Z_\varphi, \quad \nu_0 = \nu Z_\nu, \quad \tau_0 = \tau Z_\tau, \quad \lambda_0 = \lambda \mu^{2\varepsilon} Z_\lambda,$$

$$Z_1 = Z_\varphi^2, \quad 1 = Z_\varphi Z_\nu, \quad Z_2 = Z_\tau Z_\varphi^2, \quad Z_3 = Z_\lambda Z_\nu Z_\varphi^4 = Z_\lambda Z_\varphi^3. \tag{4.149}$$

In the MS scheme the completely dimensionless constants Z depend only on the renormalized charge λ (since it is the only completely dimensionless renormalized parameter) and are calculated from the graphs of the basic theory, obtained from (4.148) by replacing all the Z by unity. The constants Z can be calculated from (3.107) and (3.114) after introducing the normalized functions $\bar{\Gamma}$ with integer dimensions analogous to (3.91) using the relations $\Gamma_2 = \bar{\Gamma}_2$ and $\Gamma_4 = -\nu\lambda\mu^{2\varepsilon}\bar{\Gamma}_4$. We recall that in the expressions of Sec. 3.18 we used L_3 to denote the operator selecting the superficial divergences (poles in ε and $R'\gamma$) of logarithmic graphs of the type γ_4, while L_2 selects the analogous divergence of the coefficient of p^2 in the quadratically divergent graphs γ_2.

As usual, in calculating the constants Z it is convenient to change over to massless graphs after differentiating with respect to τ in (3.114) for Z_2 (Sec. 3.19). It follows from (4.148) that a line in the massless graphs of the basic theory will correspond to the propagator $\Delta(k,\omega) = [k^2 + \nu^2\omega^2 k^{-2}]^{-1}$, and the vertex factor will be $-g_B = -\nu\lambda\mu^{2\varepsilon}$ (one per loop in the graphs $\bar{\gamma}$ of the normalized functions). It is technically more convenient to perform the calculations in k, t space after doing a one-dimensional Fourier transform (1.35a) with respect to the variable ω of the above propagator, which gives

$$\Delta(k,t) = (2\nu)^{-1}\exp[-k^2|t|/\nu]. \tag{4.150}$$

The form of the graphs γ themselves and their symmetry coefficients c_γ in the various constants Z are the same as in the simple φ^4 model and are given in Table 3.1 of Sec. 3.20.

Let us explain the calculation of $L\bar{\gamma} = KR'\bar{\gamma}$ for the example of the first graph γ_1 in Table 3.1. It diverges logarithmically, and so the calculation can be performed for any configuration of the transverse (p) and longitudinal (ω) external momenta (Sec. 3.19). It is convenient to take $\omega = 0$, $p \neq 0$, and then in k, t space for this graph we have

$$\bar{\gamma}(p) = -\nu\lambda\mu^{2\varepsilon}(2\pi)^{-2+2\varepsilon}\int dk\int dt\Delta(k,t)\Delta(p-k,t). \tag{4.151}$$

Substituting into this the expressions (4.150) for the propagators Δ and performing the integration over t, we arrive at an integral over the transverse momentum k with denominator $k^2 + (p-k)^2 = 2k^2 - 2pk + p^2$. When k is shifted this leads to the known integral (3.74), and in the end we find that

$$\bar{\gamma}(p) = -u(4\pi\mu^2/p^2)^\varepsilon\|\varepsilon\|, \quad u \equiv \lambda/16\pi. \tag{4.152}$$

Isolating the pole in ε, we have $KR'\bar{\gamma} = L_3\bar{\gamma} = -u/\varepsilon$.

It is relatively straightforward to calculate $L_a\bar\gamma$ also for the two-loop graphs Nos. 2, 3, 4 in Table 3.1. For the lines (4.150) we first perform all the integrals over t, and then the remaining two-loop integrals over the transverse momenta k are calculated using the Feynman parameterization (3.82) with the formulas of Sec. 3.15. For these four graphs we obtain

$$L_3\bar\gamma_1 = -u/\varepsilon, \qquad L_2\bar\gamma_2 = -2u^2/9\varepsilon,$$

$$L_3\bar\gamma_3 = -u^2/\varepsilon^2, \qquad L_3\bar\gamma_4 = -u^2(1-a\varepsilon)/2\varepsilon^2, \tag{4.153}$$

where $a \equiv 2/3 + \ln(4/3)$, $2\varepsilon \equiv 3 - d$, and $u \equiv \lambda/16\pi$. Below we shall see that it is this parameter u that is most convenient for use as the renormalized charge. Substituting (4.153), the coefficients c_γ from Table 3.1, and $r_\gamma = 1$ into (3.118), we find that

$$Z_1 = 1 - \frac{u^2}{27\varepsilon} + ...; \quad Z_2 = 1 + \frac{u}{2\varepsilon} + \frac{u^2}{4\varepsilon^2} + \frac{u^2}{4\varepsilon^2}(1-a\varepsilon) + ...;$$

$$Z_3 = 1 + \frac{3u}{2\varepsilon} + \frac{3u^2}{4\varepsilon^2} + \frac{3u^2}{2\varepsilon^2}(1-a\varepsilon) + ... \tag{4.154}$$

From this in the usual manner (Sec. 4.2) we find the RG functions:

$$\gamma_\varphi = 2u^2/27 + ..., \quad \gamma_\tau = -u + au^2 - 4u^2/27 + ...,$$

$$\beta \equiv \beta_u = u[-2\varepsilon + 3u - 6au^2 + 2u^2/9 + ...]. \tag{4.155}$$

The fixed point $u_* \sim \varepsilon$ is IR-stable, and from the RG functions (4.155) using the explicit expression $a \equiv 2/3 + \ln(4/3)$ we find that

$$\eta = 2\gamma_\varphi^* = 4(2\varepsilon)^2/243 + ...,$$

$$1/\nu = 2 + \gamma_\tau^* = 2 - \frac{2\varepsilon}{3} - \frac{(2\varepsilon)^2}{9}\left[\frac{20}{27} + \ln\left(\frac{4}{3}\right)\right] + ... \tag{4.156}$$

For $2\varepsilon = 1$, (4.156) are used for two-dimensional systems with dipole forces, and for $\varepsilon = 0$ the results of [38] can be reproduced using the known RG functions (4.155) and the general rules of Secs. 1.43 and 1.44.

The one-loop calculation of the ε expansions for this model was done in [58], and the two-loop one in [59] (with a different subtraction scheme). The results for the O_n generalization of the model (4.147) obtained by the replacement $\varphi \to \varphi_a$, $\varphi^4 \to (\varphi_a\varphi_a)^2$ are given in [59]. This leads only to the addition of known (from the tables of Sec. 3.20) structure factors r_γ for the graph contributions, and instead of (4.156) we obtain

$$\eta = \frac{4(n+2)}{9(n+8)^2}(2\varepsilon)^2 + ...,$$

$$1/\nu = 2 - 2\varepsilon\frac{(n+2)}{(n+8)} - (2\varepsilon)^2\frac{2(n+2)}{(n+8)^3}\left[\frac{20n+70}{9} + \frac{7n+20}{2}\ln\left(\frac{4}{3}\right)\right] + ... \tag{4.157}$$

The result of [59] for $1/\nu$ contains a misprint $[(n+8)^2$ instead of $(n+8)^3]$. The first few coefficients of the series (4.156) numerically differ little from the analogous ones of the usual ε expansions in the φ^4 model.

4.20   The $\tau^\alpha \varphi^4$ interaction (modified critical behavior)

Let us now consider a model of the φ^4 type with unrenormalized action [109]

$$S(\varphi) = -(\partial\varphi)^2/2 - \tau_0\varphi^2/2 - \lambda_0\tau_0^\alpha\varphi^4/24. \tag{4.158}$$

A possible origin for such an interaction in the real world was discussed in Sec. 1.16. There it was shown that for this model the upper critical dimension (above which the interaction becomes IR-irrelevant) is $4-2\alpha \equiv d^{**}$ and differs from the logarithmic dimension $d^* = 4$, which is interesting from the technical point of view. For $d = 4-2\varepsilon$ the condition for the interaction to be IR-relevant $d < d^{**}$ is equivalent to $\alpha < \varepsilon$, and so in the spirit of the ε expansion ε and α should be assumed to be small quantities of the same order of magnitude.

The model (4.158) is equivalent to the φ^4 model (1.96) with $g_0 = \lambda_0\tau_0^\alpha$; only the statement of the problem differs. Previously we were interested in the $\tau_0 \to 0$, $g_0 =$const asymptote, while for (4.158) we want $\tau_0 \to 0$, $\lambda_0 =$const, i.e., $g_0 \sim \tau_0^\alpha \to 0$. In order for the problem to be meaningful, τ_0 in (4.158) must, of course, be understood as the deviation from the actual critical temperature, denoted τ_0' in Sec. 1.21. The equation $\tau_0 = \tau_0'$ is ensured automatically by using formal dimensional regularization for the calculations (Sec. 1.20).

The model (4.158) is simply reduced to (1.96) by the substitution

$$\varphi(x) = \tau_0^a\widetilde{\varphi}(\widetilde{x}), \quad \widetilde{x} = \tau_0^b x, \quad a = \alpha(1-\varepsilon)/2\varepsilon, \quad b = \alpha/2\varepsilon, \tag{4.159}$$

transforming the action (4.158) to the form

$$\widetilde{S}(\widetilde{\varphi}) = -(\partial\widetilde{\varphi})^2/2 - \widetilde{\tau}_0\widetilde{\varphi}^2/2 - \lambda_0\widetilde{\varphi}^4/24, \quad \widetilde{\tau}_0 = \tau_0^c \tag{4.160}$$

with exponent $c = (\varepsilon - \alpha)/\varepsilon$. The condition that the interaction be IR-relevant $\alpha < \varepsilon$ is equivalent to $c > 0$, which ensures that $\widetilde{\tau}_0 = \tau_0^c \to 0$ for $\tau_0 \to 0$, i.e., the usual critical regime for the action (4.160). Therefore, all the usual predictions of the φ^4 model (Ch. 1) must hold for it, in particular, the formulas of critical scaling for the correlator (both the renormalized and the unrenormalized ones, which differ only by an insignificant normalization):

$$\langle\widetilde{\varphi}(\widetilde{x}')\widetilde{\varphi}(\widetilde{x}'')\rangle = \widetilde{r}^{-2\Delta_\varphi}f(\widetilde{\tau}_0 \cdot \widetilde{r}^{\Delta_\tau}), \tag{4.161}$$

where $\widetilde{r} \equiv |\widetilde{x}' - \widetilde{x}''|$, f is a scaling function, and $\Delta_{\tau,\varphi}$ are the usual critical dimensions of the φ^4 model. Rewriting (4.161) in the original variables φ, τ_0, and $r \equiv |x - x''|$ using (4.159) and (4.160), we obtain

$$\langle\varphi(x')\varphi(x'')\rangle = r^{-2\Delta_\varphi} \cdot \tau_0^{2a-2b\Delta_\varphi}f(\tau_0^{c+b\Delta_\tau} \cdot r^{\Delta_\tau}), \tag{4.162}$$

which corresponds to critical scaling with the new dimensions:

$$\Delta'_\varphi = (c\Delta_\varphi + a\Delta_\tau)/(c + b\Delta_\tau), \qquad \Delta'_\tau = \Delta_\tau/(c + b\Delta_\tau). \qquad (4.163)$$

This is our answer. The parameters a, b, and c contained in it are known explicitly (see above), and $\Delta_{\tau,\varphi}$ are the usual critical dimensions of the φ^4 model, known in the form of ε expansions.

It is instructive to analyze the same problem from a different angle, treating (4.158) as the φ^4 model with $g_0 \sim \tau_0^\alpha$. It follows from the renormalization expressions (1.103) with $\tau'_0 = \tau_0$ (see above) taking into account the finiteness of all Z for $\varepsilon \neq 0$ that in the renormalized variables we now have $g \sim \tau^\alpha$ instead of the usual $g =$const. We want to see how the above results can be obtained in the language of the renormalized φ^4 theory with $g \sim \tau^\alpha$.

This is fairly simple. In the RG equations for the renormalized Green functions it is convenient to take $s = \tau\mu^{-2}$ as the scaling variable. The corresponding invariant charge $\bar{g}(s,g)$ will then be determined by (1.179), which in the one-loop approximation leads to (1.154) with $s \to s^{1/2}$:

$$\bar{g}(s,g)\Big|_{\text{1-loop}} = gg_*/[g_*s^\varepsilon + g(1 - s^\varepsilon)]. \qquad (4.164)$$

Previously we were interested in the asymptote $s = \tau\mu^{-2} \to 0$, $g =$const, at which we had $\bar{g}(s,g) \to g_*$. Now we are interested in the case $s \to 0$, $g \sim s^\alpha \to 0$, and the behavior of $\bar{g}(s,g)$, as seen explicitly from (4.164) (and can also be justified without the one-loop approximation), depends on the relation between the parameters α and ε, namely, $\bar{g}(s,g) \to 0$ for $\alpha > \varepsilon$ and $\bar{g}(s,g) \to g_*$ for $\alpha < \varepsilon$. The first case corresponds to the asymptotically free theory, i.e., IR-irrelevance of the interaction, and the second to a nontrivial critical regime. This is consistent with the general conclusions in Sec. 1.16.

We still need to see how the rearrangement $\Delta \to \Delta'$ of the critical dimensions occurs for $\alpha < \varepsilon$. This is easily done by analyzing the derivation of the critical scaling formulas from the RG equations in Sec. 1.33. For $g =$const the quantities $\tilde{c}_i(g)$ defined in (1.158) are constants entering only into the unimportant normalization factors (1.160). However, if we set $g \sim s^\alpha \to 0$, then an additional logarithmic singularity $\sim \ln s$ from the lower limit appears in the integrals (1.158) for $\tilde{c}_i$. The corresponding additions $\sim \ln s$ in (1.157) transform the usual critical scaling formulas into new formulas with the exponents (4.163). This is the mechanism by which the exponents are rearranged.

4.21 The φ^6 model in dimension $d = 3 - 2\varepsilon$

Let us consider the standard MS renormalization of the simple φ^6 model with one-component scalar field $\varphi(x)$. Its unrenormalized and basic actions, and also the canonical dimensions of all the needed quantities are the following:

$$S(\varphi) = -(\partial\varphi)^2/2 - \tau_0\varphi^2/2 - g_0\varphi^6/6!, \qquad (4.165)$$

$$S_{\mathrm{B}}(\varphi) = -(\partial\varphi)^2/2 - \tau\varphi^2/2 - g_{\mathrm{B}}\varphi^6/6!, \qquad g_{\mathrm{B}} = g\mu^{4\varepsilon}. \qquad (4.166)$$

Table 4.4   Canonical dimensions d_F of various quantities F for the φ^6 model in dimension $d = 3 - 2\varepsilon$.

F	$\varphi(x)$	τ, τ_0	g_{B}, g_0	g	μ	Γ_{2n}
d_F	$1/2 - \varepsilon$	2	4ε	0	1	$3 - n + 2\varepsilon(n-1)$

We see from Table 4.4 that the functions $\Gamma_{0,2,4,6}$ have nonnegative dimensions for $\varepsilon = 0$, and so all their graphs would be superficially divergent for renormalization with cutoff Λ (Sec. 3.2). However, we are using formal MS renormalization, where SD graphs are defined by the criterion (3.88). The monomials M appearing here are scalars constructed from the momenta and τ, and so they have even integer dimensions $0,2,\dots$. It then follows that the graphs of Γ_0 and Γ_4 with odd (for $\varepsilon = 0$) dimensions are not SD graphs, i.e., they do not generate counterterms. They have only divergences that are powers of Λ, and which by convention we ignore [in the Λ scheme this implies exact cancellation of such divergences by the corresponding bare contributions, which should then be added to (4.165)], and they do not contain poles in ε. Therefore, counterterms are generated only by the SD graphs of the functions Γ_2 and Γ_6, i.e.,

$$S_{\mathrm{R}}(\varphi) = -Z_1(\partial\varphi)^2/2 - Z_2\tau\varphi^2/2 - Z_3 g_{\mathrm{B}}\varphi^6/6!. \qquad (4.167)$$

This corresponds to multiplicative (even without the vacuum counterterm) renormalization of the action (4.165):

$$Z_1 = Z_\varphi^2, \quad Z_2 = Z_\tau Z_\varphi^2, \quad Z_3 = Z_g Z_\varphi^6,$$

$$\widehat{\varphi} = \widehat{\varphi}_{\mathrm{R}} Z_\varphi, \quad \tau_0 = \tau Z_\tau, \quad g_0 = g\mu^{4\varepsilon} Z_g. \qquad (4.168)$$

The constants Z are determined by relations like (3.107) and (3.114):

$$Z_1 = 1 + [L_2\overline{\Gamma}_2], \quad Z_2 = 1 + [L_3\partial_\tau\overline{\Gamma}_2],$$

$$Z_3 = 1 - [L_3\overline{\Gamma}_6], \quad \Gamma_2 = \overline{\Gamma}_2, \quad \Gamma_6 = -g_{\mathrm{B}}\overline{\Gamma}_6, \qquad (4.169)$$

where $\overline{\Gamma}_{2n} = (-g_{\mathrm{B}})^{(1-n)/2}\Gamma_{2n}$ are the normalized functions with integer dimensions $d[\overline{\Gamma}_{2n}] = 3 - n$. The following graphs are sufficient for calculating

$Z_{1,2}$ through order g^3 and Z_3 through order g^2:

$$\Gamma_2 = -p^2 - \tau + \frac{1}{8} \text{（diagram）} + \frac{1}{32} \text{（diagram）} + \frac{1}{48} \text{（diagram）} +$$

$$+ \frac{1}{24} \text{（diagram）} + \frac{1}{120} \text{（diagram）} + \frac{1}{72} \text{（diagram）} + \frac{1}{192} \text{（diagram）} + \dots \,;$$

$$\Gamma_6 = \text{（diagram）} + \frac{5}{3} \text{（diagram）} + \frac{15}{4} \text{（diagram）} + \frac{5}{18} \text{（diagram）} + \tag{4.170}$$

$$+ \frac{5}{16} \text{（diagram）} + \frac{5}{8} \text{（diagram）} + 5 \, \text{（diagram）} + \frac{15}{8} \text{（diagram）} + \dots$$

$$+ ? \, \text{（diagram）} + ? \, \text{（diagram）} + ? \, \text{（diagram）} + ? \, \text{（diagram）} + \dots \,.$$

For the two functions $\Gamma_{2,6}$ we have given all the graphs in orders g and g^2, all the graphs without contracted lines in order g^3, and for Γ_6 only some graphs in order g^4 without the symmetry coefficients, which we do not need (see below). All the graphs in (4.170) have an even number of loops and one factor of $-g_B$ for two loops in the normalized functions $\overline{\Gamma}_{2,6}$. The results of calculating the contributions of the individual graphs are summarized in Table 4.5.

Let us first explain how the graphs are selected. A special feature of this model is the absence of one-loop SD graphs — all the graphs in (4.170) have an even number of loops. Accordingly, not all of them are SD graphs, even though their dimensions are the same: if any UV-finite (in the formal scheme) subgraph is isolated from a graph as a factor, then divergences can occur only in the other subgraphs, while the graph as a whole then certainly has no (superficial) divergence itself. In this case it is not an SD graph, i.e., it does not contribute to the counterterms.

In the graphs of (4.170), the UV-finite factors of this type are, in particular, the contracted lines and simple loops like (3.87b), if they can be factorized [in the integrals of (3.87) for $d = 3 - 2\varepsilon$ there are no poles in ε]. Therefore, any graph of (4.170) with such subgraphs (for example, the first four graphs in Γ_2 and the graphs of Γ_6 with the coefficients $15/4$ and $5/16$) is not an SD graph which generates counterterms. It is also clear that the derivatives with respect to τ of such "unnecessary" graphs of Γ_2 will also be "unnecessary" vertex graphs when they are identified as vertex graphs (Sec. 3.19) by the addition of four external outflows at the point (3.112). Therefore, such graphs of Γ_2 can be discarded before differentiation with respect to τ, in contrast to the graphs of Γ_2 with contracted lines in the φ^4 model, which may be SD graphs.

Table 4.5 Contributions of individual graphs to the renormalization constants Z of the model (4.167), $d = 3 - 2\varepsilon$, $u \equiv g/64\pi^2$.

No.	Graph γ	$[L_a\bar{\gamma}] = (-u)^{l/2}\xi_\gamma$		Symmetry coefficient c_γ in constants		
		$(-u)^{l/2}$	ξ_γ	Z_1	Z_2	Z_3
1		$-u$	$\dfrac{1}{\varepsilon}$	0	0	$-5/3$
2		u^2	$-\dfrac{1}{3\varepsilon}$	$1/120$	0	0
3		u^2	$-\dfrac{1}{\varepsilon}$	0	$-1/24$	$-5/8$
4		u^2	$\dfrac{-1+4\varepsilon}{2\varepsilon^2}$	0	0	-5
5		u^2	$\dfrac{\pi^2}{2\varepsilon}$	0	0	$-15/8$
6		u^2	$-\dfrac{1}{\varepsilon^2}$	0	0	$-5/18$
7		$-u^3$	$\dfrac{6-8\varepsilon}{27\varepsilon^2}$	$1/72$	0	0
8		$-u^3$	$\dfrac{2-8\varepsilon}{3\varepsilon^2}$	0	$-1/36$	?
9		$-u^3$	$\dfrac{1-12\varepsilon}{3\varepsilon^2}$	0	$-1/12$	?
10		$-u^3$	$-\dfrac{2\pi^2}{3\varepsilon}$	0	$-1/48$	?
11		$-u^3$	0	0	$-1/48$	?

After the unnecessary graphs in (4.170) are discarded, only the graphs listed in Table 4.5 remain, and the last four of these correspond to the derivative with respect to τ of the two SD graphs of order g^3 in Γ_2. They are used below only to calculate Z_2 in order g^3 (in this order we do not calculate the constant Z_3), and so we shall not need their symmetry coeficients in Γ_6 and do not give them in (4.170).

Let us now explain the calculation of the quantities given in Table 4.5. All the symmetry coefficients c_γ are easily found from the coefficients of (4.170) using (4.169) and (3.112). The quantities ξ_γ are calculated most simply, as usual, from the massless graphs with the simple flow of a single momentum. When choosing the latter it is important to avoid the IR divergences generated by a pair of massless lines with identical integration momentum (Sec. 3.19). Therefore, graphs 1, 2, 4, 5, 6, and 7 in Table 4.5 can be calculated with a horizontal flow, while all the others must have the external momentum flowing through the vertex with four external outflows. The point where this momentum enters in graphs 3, 8, and 9 will be assumed to be the leftmost vertex, while for graph 11 it will be the lowest one. Then all the graphs reduce to simple chains. For graph 10 the natural choice of lower vertex as the point where the momentum enters leads to a difficult graph like (3.134). Therefore, here it is convenient to use the trick described in Sec. 3.19, choosing the leftmost integration vertex as the point where the momentum enters (it is important that the graph be logarithmic), and then this graph also immediately reduces to simple chains.

The following expressions will then be the analogs of (3.122) and (3.128) for an arbitrary graph $\overline{\gamma} \in \overline{\Gamma}_{2n}$ with l loops (l even, $n = 1, 3$):

$$\alpha = 1/2 - \varepsilon, \quad a = \|1/2 - \varepsilon\|/4\pi^{3/2-\varepsilon}, \quad d = 3 - 2\varepsilon,$$

$$\overline{\gamma} = (-ut^\varepsilon \|1/2\|)^{l/2} [p^2 H(\alpha)]^{(3-n)/2} H(\lambda_{nl}) \cdot q,$$

$$u \equiv g/64\pi^2, \quad \lambda_{nl} = 3 - n/2 - \varepsilon(l+1),$$

$$t^\varepsilon \equiv (4\pi\mu^2/p^2)^{2\varepsilon} \left[\|1/2 - \varepsilon\| / \|1/2\| \right]^3,$$

(4.171)

where q, as usual, is the product of only the nontrivial factors H arising when a coordinate graph is reduced to a simple line with index λ_{nl} (Sec. 3.21). The quantities q for all the graphs in Table 4.5 with the above choice of momentum flow are easily calculated using (3.123) and (3.124), giving

$$q_1 = q_2 = 1, \quad q_3 = H(\alpha, 4\alpha, d - 5\alpha), \quad q_4 = H(2\alpha, 3\alpha, d - 5\alpha),$$

$$q_5 = H(2\alpha, 2\alpha, d - 4\alpha), \quad q_6 = q_7 = H(3\alpha, 3\alpha, d - 6\alpha),$$

(4.172)

$$q_8 = Bq_6, \quad q_9 = Bq_4, \quad q_{10} = Bq_5, \quad q_{11} = BH(2\alpha, 4\alpha, d - 6\alpha),$$

where $B \equiv H(\alpha, 7\alpha - d/2, 3d/2 - 8\alpha)$.

Subdivergences occur only in graphs 4, 6, 7, 8, and 9 and come only from vertex subgraphs like graph 1; as usual, they are eliminated by the R' operator (Sec. 3.8). The counterterm for graph 6 can be obtained without calculation using the rule (3.104). Graph 11 does not contain poles in ε owing to the compensating zero in the chain with link indices $2\alpha, 4\alpha$. We also note the appearance of the constant $\pi^2 \sim \psi'(1) \sim \zeta(2)$, which did not occur in the examples studied earlier.

Using the data of Table 4.5 and the general rule (3.118) (now $r_\gamma = 1$), we can find the constants $Z_{1,2}$ through u^3 and Z_3 through u^2:

$$Z_1 = 1 - u^2/360\varepsilon + u^3[1/243\varepsilon - 1/324\varepsilon^2] + ...,$$

$$Z_2 = 1 + u^2/24\varepsilon + u^3[5/108\varepsilon^2 - 11/27\varepsilon - \pi^2/72\varepsilon] + ..., \qquad (4.173)$$

$$Z_3 = 1 + 5u/3\varepsilon + u^2[25/9\varepsilon^2 - 75/8\varepsilon - 15\pi^2/16\varepsilon] + ...$$

Then, using the general expressions (1.113) or, more simply, expressions like (4.3) and (4.5) (here with $k = 4$, since $g_B = g\mu^{4\varepsilon}$), we find all the needed RG functions. They are conveniently expressed in terms of $u \equiv g/64\pi^2$, giving

$$\gamma_\varphi = \gamma_1/2 = u^2/90 - 2u^3/81 + ...,$$

$$\gamma_\tau = \gamma_2 - \gamma_1 = -16u^2/45 + u^3 \left[400/81 + \pi^2/6\right] + ...,$$

$$\beta \equiv \tilde{\mathcal{D}}_\mu u = u(-4\epsilon - \gamma_g) = -4\epsilon u + 20u^2/3 - u^3[1124/15 + 15\pi^2/2] + ...$$
$$(4.174)$$

From these RG functions we can find the coordinate of the IR-stable fixed point u_* and the corresponding critical anomalous dimensions (we recall that $d = 3 - 2\varepsilon$):

$$u_* = 3\varepsilon/5 + 27\varepsilon^2[1124/7500 + 3\pi^2/200] + ...,$$

$$\gamma_\varphi^* = \varepsilon^2/250 + \varepsilon^3[2279/46875 + 27\pi^2/5000] + ..., \qquad (4.175)$$

$$\gamma_\tau^* = -16\varepsilon^2/125 - \varepsilon^3[30928/46875 + 171\pi^2/1250] + ...$$

With these results, (1.162) and (1.165) can be used to easily find all the other exponents. They characterize the tricritical behavior described by the model (4.165) of the system in dimension $d < 3$. The logarithmic corrections to the Landau theory can be found directly in dimension $d = 3$ using the RG functions (4.174) and the general rules of Secs. 1.43 and 1.44.

4.22   The $\varphi^4 + \varphi^6$ model

Let us consider a generalization of the model (4.165) to include the φ^4 interaction:

$$S(\varphi) = -(\partial\varphi)^2/2 - \tau_0\varphi^2/2 - \lambda_0\varphi^4/24 - g_0\varphi^6/6!, \qquad (4.176)$$

$$S_B(\varphi) = -(\partial\varphi)^2/2 - \tau\varphi^2/2 - \lambda\mu^{2\varepsilon}\varphi^4/24 - g\mu^{4\varepsilon}\varphi^6/6!. \tag{4.177}$$

The monomials M in (3.88) can now also contain positive integer powers of the renormalized φ^4 charge λ with $d_\lambda = 1$, and so additional counterterms $L\Gamma_n$ from the functions $\Gamma_{0,4}$ with odd (for $\varepsilon = 0$) dimensions can appear: $L\Gamma_0 = \lambda^3\mu^{-2\varepsilon}c + \lambda\tau\mu^{-2\varepsilon}c$, $L\Gamma_4 = \lambda\mu^{2\varepsilon}c$, where c are various dimensionless coefficients depending only on g and ε. The counterterm $L\Gamma_6$ does not change, and $c\lambda^2$ is added to the ordinary contributions $c\tau + cp^2$ in $L\Gamma_2$ of the simple φ^6 model. The fact that the additional counterterms are polynomials in λ implies that they are generated only by graphs with a definite number of φ^4 vertices for any number of φ^6 vertices. For example, the counterterms of $L\Gamma_4$ that are multiples of λ are generated by graphs of Γ_4 with a single φ^4 vertex, while graphs of Γ_4 with two or more such vertices will not be SD graphs owing to their dimension.

Adding all these counterterms to the functional (4.177) (and omitting the vacuum counterterm), we arrive at the following renormalized action:

$$S_R(\varphi) = -Z_1(\partial\varphi)^2/2 - (\tau Z_2 + \lambda^2 Z_5)\varphi^2/2 - Z_4\lambda\mu^{2\varepsilon}\varphi^4/24 - Z_3 g\mu^{4\varepsilon}\varphi^6/6!. \tag{4.178}$$

Here the constants $Z_{1,2,3}$ are the same as in (4.167), and the new $Z_{4,5}$ are expressed in terms of the corresponding normalized functions $\overline{\Gamma}$ ($\Gamma_2 = \overline{\Gamma}_2$, $\Gamma_4 = -\lambda\mu^{2\varepsilon}\overline{\Gamma}_4$) as

$$Z_4 = 1 - [L_3\overline{\Gamma}_4]\Big|_\lambda, \qquad \lambda^2 Z_5 = [L_3\overline{\Gamma}_2]\Big|_{\lambda^2}, \tag{4.179}$$

where L_3 is the usual operator selecting superficial divergences (poles in ε in $R'\gamma$) of logarithmic graphs, and the symbol $|_{...}$ indicates that in $\Gamma_{2,4}$ only graphs with the required number (one for Γ_4 and two for Γ_2) of φ^4 vertices are selected. Only graphs without contracted lines are important (Sec. 4.21).

In lowest-order perturbation theory the constants (4.179) are determined by the graphs

$$\Gamma_2 = \frac{1}{6} \; \bigcirc \quad + ... \; ; \; \Gamma_4 = \times + \frac{2}{3} \; \bigcirc\hspace{-0.3em}\times + ... \tag{4.180}$$

and are easy to calculate (as in Sec. 4.21), which gives

$$Z_4 = 1 + 2u/3\varepsilon + ..., \qquad Z_5 = 1/384\pi^2\varepsilon + ..., \tag{4.181}$$

where the ellipsis denotes corrections of higher order in $u \equiv g/64\pi^2$.

The action (4.178) is obtained from (4.176) by standard multiplicative renormalization, differing from (4.168) by only the change of the constant $Z_\tau \to Z'_\tau$ and the additional renormalization λ:

$$\lambda_0 = \lambda\mu^{2\varepsilon}Z_\lambda, \quad \tau_0 = \tau Z_\tau + \lambda^2\overline{Z} \equiv \tau Z'_\tau,$$

$$Z_\lambda = Z_4 Z_1^{-2}, \quad Z_\tau = Z_2 Z_1^{-1}, \quad \overline{Z} = Z_5 Z_1^{-1}. \tag{4.182}$$

4.23 RG analysis of the tricritical asymptote in the $\varphi^4 + \varphi^6$ model

As explained in Secs. 1.14 and 1.16, the model (4.176) describes tricritical behavior in systems with one-component order parameter φ and symmetry $\varphi \to -\varphi$. The values $\tau_0 = \lambda_0 = 0$ correspond directly to the tricritical point, and for the tricritical asymptote $\tau_0 \to 0$, $\lambda_0 \to 0$, $g_0 =$const, so that the parameters τ_0 and λ_0 differ essentially from g_0 in the statement of the problem. The relative smallness of τ_0 and λ_0 is arbitrary, since it is determined by the choice of trajectory by which the tricritical point is approached, i.e., by the experimental conditions. It was shown in Sec. 1.16 that depending on the relative smallness of τ_0 and λ_0, the corresponding asymptote can be either purely tricritical (i.e., the φ^4 interaction is insignificant; for brevity, we shall refer to this as case A), or modified critical (the φ^6 interaction is insignificant; case B), or combined tricritical, when both interactions are important (case C).

These conclusions were arrived at in Sec. 1.16 by comparing only the canonical dimensions. Now we wish to see how they agree with the RG analysis and how they are refined by the inclusion of the anomalies generated by the renormalization.

In renormalization using the exact scheme with cutoff Λ, the regular contributions (Sec. 1.19) would lead to some renormalization of the parameters τ and λ, in particular, to a shift of the location of the tricritical point like the shift of T_c in the φ^4 model (Sec. 1.20); the IR-irrelevant contributions of the action can play the same role. Discarding them in the construction of the model and using the formal scheme for the calculations (Secs. 1.20 and 3.16), we shall neglect all these unimportant effects. Then in the renormalized variables the values $\tau = \lambda = 0$ correspond to the tricritical point, and we are interested in the asymptote $\tau \to 0$, $\lambda \to 0$, $g =$const in dimension $d = 3 - 2\varepsilon$ (we assume that $g \sim \varepsilon \neq 0$, $\mu =$const, so that the relation between the bare and renormalized parameters is one-to-one and nonsingular).

Passing to the RG analysis, from the general rule (1.108) and the renormalization formulas (4.168) and (4.182) we find the form of the RG operator $\mathcal{D}_{\mathrm{RG}} = \widetilde{\mathcal{D}}_\mu$ for the model (4.178):

$$\mathcal{D}_{\mathrm{RG}} = \mathcal{D}_\mu + \beta\partial_u - (2\varepsilon + \gamma_\lambda)\mathcal{D}_\lambda - (\tau\gamma_\tau + \lambda^2\overline{\gamma})\partial_\tau \qquad (4.183)$$

with the usual definition (1.113) of the RG functions $\beta \equiv \beta_u = \widetilde{\mathcal{D}}_\mu u$ and γ_a for the corresponding constants Z_a (now $a = g, \lambda, \tau, \varphi$) and the new "cross" RG function,

$$\overline{\gamma}(u) = Z_\tau^{-1}(\widetilde{\mathcal{D}}_\mu - 4\varepsilon - 2\gamma_\lambda)\overline{Z}, \qquad \widetilde{\mathcal{D}}_\mu = \mathcal{D}_{\mathrm{RG}}. \qquad (4.184)$$

In the lowest order in u from (4.173), (4.181), and (4.182) we find the needed constants Z and from them the RG functions (on any function of u we have

$\widetilde{\mathcal{D}}_\mu = \beta\partial_u = -4\varepsilon\mathcal{D}_u + ...$, and the corrections in lowest order are unimportant):

$$\gamma_\lambda(u) = -8u/3 + ..., \qquad \overline\gamma(u) = -1/96\pi^2 + ... \qquad (4.185)$$

The other RG functions are known from (4.174) with higher accuracy.

To be specific, let us consider the RG equation (1.150) for the renormalized correlator $D_R = D_R(p, u, \lambda, \tau, \mu)$ and, as usual, change over to dimensionless quantities, taking

$$D_R = p^{-2}\Phi(s, u, \lambda', \tau'), \quad s \equiv p\mu^{-1}, \quad \lambda' \equiv \lambda\mu^{-1}, \quad \tau' \equiv \tau\mu^{-2}. \qquad (4.186)$$

For convenience, we shall distinguish the dimensionless parameters we have introduced λ', τ' from the original ones by only a prime, and use the traditional notation s for the scaling variable.

Substituting (4.186) into (1.150) with the RG operator (4.183), we obtain the following RG equation for Φ:

$$[-\mathcal{D}_s + \beta\partial_u - (1 + 2\varepsilon + \gamma_\lambda)\mathcal{D}_{\lambda'} - (2 + \gamma_\tau)\mathcal{D}_{\tau'} - \overline\gamma\lambda'^2\partial_{\tau'} + 2\gamma_\varphi]\Phi = 0. \qquad (4.187)$$

It differs from (1.142) by the presence of the cross term $\sim \lambda'^2\partial_{\tau'}$, but after going to $F = \ln\Phi$ it becomes the usual expression (1.131) and satisfies the conditions of Statement 3 in Sec. 1.29. Therefore, the general solution of equation (4.187) can be written in the form of (1.148):

$$\Phi(s, u, \lambda', \tau') = \Phi(1, \overline u, \overline\lambda', \overline\tau')\exp\left[2\int_u^{\overline u}dx\,\frac{\gamma_\varphi(x)}{\beta(x)}\right]. \qquad (4.188)$$

The invariant charge $\overline u(s, u)$ is determined by (1.144), and the two other invariant variables by (1.140), namely,

$$\mathcal{D}_s\overline\lambda' = -[1 + 2\varepsilon + \gamma_\lambda(\overline u)]\overline\lambda', \quad \overline\lambda'|_{s=1} = \lambda', \qquad (4.189a)$$

$$\mathcal{D}_s\overline\tau' = -[2 + \gamma_\tau(\overline u)]\overline\tau' - \overline\gamma(\overline u)\overline\lambda'^2, \quad \overline\tau'|_{s=1} = \tau'. \qquad (4.189b)$$

It is easy to write the general solution of these equations: for (4.189a) it is given by an expression like (1.147), and then for the function $\overline\lambda'(\overline u)$ already found, the variable substitution $s \to \overline u$ in (4.189b) gives an easily solved linear inhomogeneous equation for $\overline\tau'$.

We shall not present these solutions, because we are interested only in the critical regime $s \to 0$, $\tau' \to 0$, $\lambda' \to 0$, $u =$const. For sufficiently small $s = p\mu^{-1}$ we can assume that $\overline u(s, u) \simeq u_*$ and replace all the RG functions $\gamma(\overline u)$ in the equations by their values $\gamma^* \equiv \gamma(u_*)$ at the IR-stable fixed point, which leads only to the loss of unimportant normalization factors (Sec. 1.33). In this approximation from (1.157), (4.186), (4.188), and (4.189) we find that

$$D_R \simeq p^{-2}s^{2\gamma_\varphi^*}\Phi(1, u_*, \overline\lambda', \overline\tau'), \qquad (4.190a)$$

$$\overline{\lambda}' = \lambda' s^{-\Delta_\lambda}, \quad \overline{\tau}' = (\tau' + a\lambda'^2)s^{-\Delta_\tau} - a\lambda'^2 s^{-2\Delta_\lambda}, \tag{4.190b}$$

$$\Delta_\lambda \equiv 1 + 2\varepsilon + \gamma_\lambda^*, \quad \Delta_\tau = 2 + \gamma_\tau^*, \quad a = -\overline{\gamma}^*/b, \quad b \equiv 2\Delta_\lambda - \Delta_\tau. \tag{4.190c}$$

From (4.175) and (4.185) we find $\gamma_\lambda^* = -8\varepsilon/5 + ...,$ $\overline{\gamma}^* = -1/96\pi^2 + ...,$ and $\gamma_\tau^* \simeq 0$, and after substituting these values into (4.190c) we obtain

$$\Delta_\lambda = 1 + 2\varepsilon/5 + ..., \quad b \equiv 2\Delta_\lambda - \Delta_\tau = 4\varepsilon/5 + ..., \quad a = 5/384\pi^2\varepsilon + ... \tag{4.191}$$

The equations (4.190) describe combined tricritical scaling, realized for $\overline{u} \simeq u_*$ (which is ensured by the smallness of $s = p/\mu$) and invariant variables (4.190b) of finite order of magnitude (Sec. 1.34). The latter is ensured by the parameters $\lambda' = \lambda/\mu$ and $\tau' = \tau/\mu^2$ being appropriately small. The exact statement of the corresponding asymptotic regime is the following:

$$p \to 0, \quad \lambda \sim p^{\Delta_\lambda} \to 0, \quad \tau + a\lambda^2 \sim p^{\Delta_\tau} \to 0. \tag{4.192}$$

It follows from (4.190) that the variables λ and $\tau + a\lambda^2$ have critical dimensions Δ_λ and Δ_τ, respectively, while the parameter τ itself is a mixture of two contributions with different dimensions.

Of course, this does not prevent us from studying various trajectories of the type $\tau \to 0$, $\lambda \sim \tau^\alpha \to 0$ in the phase space of the parameters λ and τ, corresponding to different approaches to the tricritical point. In Sec. 1.16 at the level of the canonical dimensions it was shown that for $d = 3$ the case $\alpha > 1/2$ corresponds to case A (see above), the case $\alpha < 1/2$ to case B, and $\alpha = 1/2$ to case C. For $d = 3 - 2\varepsilon$ the canonical limiting value will be $\alpha_0 = 1/2 + \varepsilon$. The results of the RG analysis for $d = 3 - 2\varepsilon$ given above support these conclusions, and refine the limiting value $\alpha_0 = 1/2$ by the inclusion of all corrections (canonical and anomalous) of order ε and higher. The agreement with the statements of Sec. 1.16 is most transparent in the RG equation for objects like the free energy (see below), which do not contain momentum variables, since in Sec. 1.16 we discussed only the relationship between λ and τ without reference to the momentum scale. Nevertheless, the required correspondence can also be seen in the formulation (4.190).

Let us consider the trajectory $\lambda \to 0$, $\tau \sim \lambda^{2-\kappa} \to 0$ in the λ, τ plane, choosing λ as the fundamental variable, since it has definite critical dimension (see above). Since $\alpha = 1/(2 - \kappa)$, the limiting value $\alpha_0 = 1/2$ in the zeroth-order (in ε) approximation corresponds to $\kappa_0 = 0$, case A to $\kappa > 0$, and case B to $\kappa < 0$. We shall substitute $\tau = c\lambda^{2-\kappa}$ (c is a positive constant) into (4.190b) for $\overline{\tau}'$ and assume that the rate of falloff of λ is consistent with the momentum according to the second equation in (4.192) [otherwise, the argument $\overline{\lambda}'$ of the function Φ in (4.190a) will tend either to zero or infinity, which leads directly to case A or B]. So, let us make the substitution $\tau = c\lambda^{2-\kappa}$ with $\lambda \sim s^{\Delta_\lambda} \to 0$ in (4.190b) for $\overline{\tau}'$. It is then easily checked, taking into account the signs of (4.191), that the exact limiting value κ_0 will now be

$$2 - 1/\alpha_0 \equiv \kappa_0 = (2\Delta_\lambda - \Delta_\tau)/\Delta_\lambda = 4\varepsilon/5 + ... \tag{4.193}$$

If $\kappa > \kappa_0$, then $\bar{\tau}' \to +\infty$ for $s \to 0$, i.e., case A in the language of the solution (4.190) (additional explanations follow). If $\kappa < \kappa_0$, then $\bar{\tau}'$ vanishes for some finite value $\overline{\lambda}' > 0$, which corresponds not to tricritical but simply critical behavior governed completely by the φ^4 interaction. Finally, when $\kappa = \kappa_0$ the quantity $\bar{\tau}'$ varies within finite limits as s varies from one to zero, always remaining positive for sufficiently large amplitude coefficient c in the law $\tau = c\lambda^{2-\kappa}$, which is case C. If the amplitude c is small, $\bar{\tau}'$ changes sign and we have case B.

It is instructive to discuss the other form of the RG equation (4.187), making the substitution $\tau' \to u_2 = \lambda'^2/\tau' = \lambda^2/\tau$ in it and treating the dimensionless parameter u_2 as the second charge along with the original one $u \equiv u_1$. After this substitution, (4.187) takes the form of the RG equation of the two-charge model for the new function $\Phi = \Phi(s, u \equiv u_1, u_2, \lambda')$:

$$[-\mathcal{D}_s + \beta_1\partial_{u_1} + \beta_2\partial_{u_2} - (1 + 2\varepsilon + \gamma_\lambda)\mathcal{D}_{\lambda'} + 2\gamma_\varphi]\Phi = 0,$$

$$u_1 \equiv u, \quad \beta_1 \equiv \beta, \quad \beta_2 = u_2[\gamma_\tau - 4\varepsilon - 2\gamma_\lambda + \overline{\gamma}u_2].$$

(4.194)

All the RG funcitons γ and $\beta = \beta_1$ depend only on u_1, only β_2 depends on the second charge u_2, and the u_2 dependence indicated in (4.194) is exact. The general solution of (4.194) is given by (4.188) with the replacement $\bar{\tau}' \to \bar{u}_2$, the basic invariant charge $\bar{u} \equiv \bar{u}_1$ is unchanged, and the second charge is determined by an equation like (1.140): $\mathcal{D}_s\bar{u}_2 = \beta_2(\bar{u}_1, \bar{u}_2)$. After the substitution $\bar{u}_1 \to u_{1*}$ (see above), this equation is easy to solve:

$$\beta_2(u_{1*}, u_2) = -bu_2(1 + au_2),$$ (4.195)

$$\bar{u}_2(s, u) = \overline{\lambda}'^2/\bar{\tau}' = u_2/[(1 + au_2)s^b - au_2]$$ (4.196)

with constants a and b from (4.191). The first equation in (4.196) is exact, because any function of first integrals is also a first integral, and any of these is uniquely determined by its initial data (Sec. 1.29).

It is known from (4.191) that the parameters a and b are positive. It then follows that the trivial fixed point $u_{2*} = 0$ of the β function (4.195) is UV-attractive, and the IR-stable point $u_{2*} \sim \varepsilon$ is located in the unphysical region $u_2 < 0$ (for our definitions $u_2 = \lambda^2/\tau$, $\tau = c\lambda^{2-\kappa}$ with positive c and λ).

However, the behavior of the invariant charge (4.196) in this case is not simply determined by the fixed points, because we are interested not in the usual case (Sec. 1.31) $s \to 0$, $u_2 = \text{const}$, but in the asymptote at which simultaneously with $s \to 0$ in (4.196) the initial value $u_2 = \lambda^2/\tau$ changes in the regime $\tau = c\lambda^{2-\kappa}$, $\lambda \sim s^{\Delta_\lambda} \to 0$, i.e., $u_2 \sim s^{\kappa\Delta_\lambda}$.

This situation is analogous to that discussed in Sec. 4.20. It does not require special analysis, since the first equation in (4.196) allows the behavior of $\bar{u}_2$ in the regime studied to be related to the behavior of $\bar{\tau}'$ already discussed above, namely, $\bar{u}_2 \to 0$ in case A ($\kappa > \kappa_0$) and $\bar{u}_2 \to +\infty$ for some finite s_0 in case B ($\kappa < \kappa_0$). We note that in this case the arguments about the behavior of $\bar{u}_2$ remain reliable also for large $\bar{u}_2$, since the dependence of β_2 on u_2 given in (4.195) is exact. Therefore, the usual arguments regarding the unreliability of the knowledge of the β function at large charges do not apply to this case.

This instructive example clearly shows that the behavior of the invariant charge $\bar{u}(s, u)$ can be highly nontrivial and not subject to the rules of Sec. 1.31 if at the asymptote of interest the "initial data" u change simultaneously with $s \to 0$. For example, in case A the invariant charge $\bar{u}_2$ in the $s \to 0$ limit reaches the UV-stable fixed point $u_{2*} = 0$, even though we are studying the IR asymptote.

Let us conclude with a brief discussion of the RG equation for a purely thermodynamical object without any momentum variables, the specific free energy $\mathcal{F} = \mu^d \Phi(u_1, u_2, \lambda')$. In the usual manner, for Φ we find that

$$[\beta_1 \partial_{u_1} + \beta_2 \partial_{u_2} - (1 + 2\varepsilon + \gamma_\lambda)\mathcal{D}_{\lambda'} + d]\Phi = (*) \qquad (4.197)$$

with RG functions from (4.194) and an unimportant inhomogeneous term $(*)$ on the right-hand side generated by the vacuum counterterm (Sec. 4.4). Up to the inhomogeneous term, (4.197) can be brought to the canonical form (1.142) by division by $(1 + 2\varepsilon + \gamma_\lambda)$. The natural scaling variable in (4.197) is now $\lambda' \equiv \lambda/\mu$. For the new invariant charges $\bar{u}(\lambda', u)$, as before, for $\lambda' \to 0$ we have $\bar{u}_1 \to u_{1*}$, and after the substitution $\bar{u}_1 \simeq u_{1*}$ the solution for $\bar{u}_2$ differs from (4.196) by only the replacement $s \to \lambda'$, $b \to b/\Delta_\lambda \equiv b'$, i.e.,

$$\bar{u}_2(\lambda', u_2) = u_2/[(1 + au_2)\lambda'^{b'} - au_2]. \qquad (4.198)$$

It is in this form that we can most clearly see the relation to the analysis of Sec. 1.16, since the trajectory $\tau \sim \lambda^{2-\kappa}$ corresponds to the dependence $u_2 \sim \lambda'^\kappa$, now expressed directly in terms of the scaling variable in (4.198). The qualitative behavior of the invariant charge (4.198) for $u_2 \sim \lambda'^\kappa$, $\lambda' \to 0$ in the three different cases remains the same as for the charge (4.196).

Therefore, the results of the RG analysis of the asymptote $\lambda \sim \tau^\alpha$ fully agree with the conclusions of Sec. 1.16, and only refine the limiting value of the exponent $\alpha_0 = 1/(2 - \kappa_0)$ at which combined tricritical behavior involving both interactions should be observed. In case A the behavior is purely tricritical, and φ^4 gives only corrections to scaling. In case B the φ^6 interaction is unimportant. In this case the expressions obtained using the model (4.176) do not give specific predictions, but only show that the φ^4 interaction becomes "infinitely stronger" than the φ^6 one. Therefore, the φ^6 term should be simply discarded, and the corresponding modified critical behavior is determined by the equations of Sec. 4.20. For $d = 3 - 2\varepsilon$ the criterion $\alpha < \alpha_0 = 1/(2 - \kappa_0)$ for realizing case B is compatible at small ε with the known (Sec. 4.20) condition for the modified φ^4 interaction to be IR-relevant, $\alpha < (4 - d)/2$.

The accuracy of calculating the RG functions (4.185) and the corresponding limiting value (4.193) was improved by an order of magnitude in [110], with generalization to an O_n-symmetric model with arbitrary n. There it was found that $(d = 3 - 2\varepsilon)$

$$\alpha_0 = \frac{1}{(2 - \kappa_0)} = \frac{1}{2} + \frac{(6-n)(2\varepsilon)}{2(3n+22)} + \frac{(n+4)^2(2\varepsilon)^2}{16(3n+22)^3} \times$$

$$\times \left[\pi^2(n^3 + 8n^2 - 496n - 2888) - 8(19n^2 + 508n + 2428) \right] + \ldots \tag{4.199}$$

The leading order of the $1/n$ expansion in this model was studied in Ref. [111].

4.24 Renormalization of the φ^3 model in dimension $d = 6 - 2\varepsilon$

In real problems, one often uses not the simple φ^3 model with one-component field $\varphi(x)$, but multicomponent generalizations of it like the Potts model. Up to the structure factors r_γ (Sec. 3.20), their renormalization constants are determined by the divergences of the graphs of the simple φ^3 model. It is therefore meaningful to study it separately to see the counterterm contributions of the individual graphs, which we shall do in the present section. The obvious instability of this model (the absence of an absolute energy minimum for the Landau functional) does not hinder us from using the standard RG technique for analyzing the perturbation series. An obvious difference between the φ^3 model and those studied earlier is the absence of the $\varphi \to -\varphi$ symmetry, making odd functions Γ_n forbidden for zero external field h. The divergences of the function Γ_1 can be eliminated in two ways: by a shift of the field $\widehat{\varphi}(x)$ itself, or of the external field h, if one exists. The first method leads to violation of the rule (1.74), and so it is undesirable, because it was used to derive the general RG equations in Sec. 1.24. We shall therefore choose the second variant, in which (1.74) is preserved, but in return it becomes necessary to have an external field in the unrenormalized and basic action functionals:

$$S(\varphi) = -(\partial\varphi)^2/2 - \tau_0\varphi^2/ - g_0\varphi^3/6 + h_0\varphi, \tag{4.200}$$

$$S_{\mathrm{B}}(\varphi) = -(\partial\varphi)^2/2 - \tau\varphi^2/2 - g_{\mathrm{B}}\varphi^3/6 + h\varphi, \quad g_{\mathrm{B}} \equiv g\mu^\varepsilon. \tag{4.201}$$

The canonical dimensions of all quantities, including the 1-irreducible functions Γ_n defined as usual (Sec. 3.2), are given in Table 4.6.

Table 4.6 Canonical dimensions d_F of various quantities F in the φ^3 model in dimension $d = 6 - 2\varepsilon$.

F	$\varphi(x)$	τ, τ_0	h, h_0	g_0, g_{B}	g	μ	Γ_n
d_F	$2 - \epsilon$	2	$4 - \epsilon$	ϵ	0	1	$6 - 2n + \epsilon(n-2)$

When analyzing the UV divergences we can study the functions Γ_n of the basic model (4.201) with $h = 0$, since the external field cannot enter into the counterterms (Sec. 3.13). In this case the Γ_n will be the coefficients of the expansion of the functional $\Gamma(\varphi)$ from (2.111) in φ at the origin rather than at the stationarity point. All the graphs Γ_n with their symmetry coefficients needed for the two-loop calculation of the counterterms are the following:

$$\Gamma_0 = \frac{1}{12} \; \text{⬡} \; + \ldots \; ; \; \Gamma_1 = \frac{1}{2} \; \text{◯} \; + \frac{1}{4} \; \text{⬭} \; + \ldots \; ;$$

$$\Gamma_2 = -p^2 - \tau + \frac{1}{2} \; \text{◯} \; + \frac{1}{2} \; \text{⬭} \; + \frac{1}{2} \; \text{⬭} \; + \ldots \; ; \qquad (4.202)$$

$$\Gamma_3 = \text{⋏} + \text{△} + 3 \; \text{△} + \frac{3}{2} \; \text{△} + \frac{1}{2} \; \text{△} + \ldots \; .$$

We see from the dimensions (see Table 4.6) that in this model the graphs of the functions Γ_n with $n = 0, 1, 2, 3$ are SD graphs according to the general rule (3.88). The normalized functions $\overline{\Gamma}_n$ with integer dimensions $d[\overline{\Gamma}_n] = 6 - 2n$ analogous to (3.91) can now be defined as $\overline{\Gamma}_n = (-g_B)^{2-n}\Gamma_n$, and their graphs $\overline{\gamma}$ have two factors $-g_B$ per loop for any n. This is all obvious from the dimensions. The structure of the primitive divergences, i.e., the form of the counterterms in the MS scheme, is also obvious: $c\tau^3$ in $\overline{\Gamma}_0$, $c\tau^2$ in $\overline{\Gamma}_1$, $c\tau + cp^2$ in $\overline{\Gamma}_2$, and c in $\overline{\Gamma}_3$, where all the c are dimensionless constants. Then, taking into account the relation between $\overline{\Gamma}_n$ and Γ_n and the expressions analogous to (3.93) and (3.97), we find the explicit form of the counterterms and the renormalized action corresponding to the basic theory (4.201):

$$S_R(\varphi) = \Delta S_{\text{vac}} - Z_1(\partial\varphi)^2/2 - Z_2\tau\varphi^2/2 - Z_3 g_B\varphi^3/6 + Z_4\tau^2\varphi/2g_B + h\varphi \tag{4.203}$$

with vacuum counterterm $\Delta S_{\text{vac}} = -VZ_0\tau^3/6g_B^2$ analogous to (3.106) and constants

$$Z_1 = 1 + [L_2\overline{\Gamma}_2], \quad Z_2 = 1 + [L_3\partial_\tau\overline{\Gamma}_2], \quad Z_3 = 1 - [L_3\overline{\Gamma}_3],$$

$$Z_4 = [L_3\partial_\tau^2\overline{\Gamma}_1], \quad Z_0 = [L_3\partial_\tau^3\overline{\Gamma}_0], \quad \overline{\Gamma}_n \equiv (-g_B)^{2-n}\Gamma_n. \tag{4.204}$$

The operators $L_{2,3}$ of the MS scheme entering here are the same as in Sec. 3.18, and the graphs of the derivatives $\partial_\tau\overline{\Gamma}_2$, $\partial_\tau^2\overline{\Gamma}_1$, and $\partial_\tau^3\overline{\Gamma}_0$ are identified with the vertex graphs (Sec. 3.19). Equations (4.204) are the analogs of (3.107) and (3.114). After performing the differentiations with respect to τ in them it is possible, as usual, to pass to massless graphs.

Up to the vacuum counterterm, the action (4.203) is obtained from (4.200) by multiplicative renormalization:

$$\hat{\varphi} = \hat{\varphi}_R Z_\varphi, \quad \tau_0 = \tau Z_\tau, \quad g_0 = g\mu^\varepsilon Z_g, \quad h_0 = h Z_h, \quad Z_1 = Z_\varphi^2,$$

$$Z_2 = Z_\tau Z_\varphi^2, \quad Z_3 = Z_g Z_\varphi^3, \quad h Z_h = Z_\varphi^{-1}[h + \tau^2 Z_4/2g_B]. \tag{4.205}$$

All the constants Z in the MS scheme depend only on g^2. The expansions begin at unity for all except Z_0 and Z_4. The results for the counterterm contributions of the individual graphs to the various constants in the action (4.203) are given in Table 4.7, the exact analog of Table 3.1 in Sec. 3.20.

Table 4.7 Contributions of individual graphs to the renormalization constants in the φ^3 model (4.203), $d = 6 - 2\varepsilon$, $u \equiv g^2/64\pi^3$.

No.	Graph γ	$[L_a\bar{\gamma}] = u^l\xi_\gamma$		Symmetry coefficient c_γ in constants				
		u^l	ξ_γ	Z_0	Z_1	Z_2	Z_3	Z_4
1		u	$-\dfrac{1}{6\varepsilon}$	0	$1/2$	0	0	0
2		u	$\dfrac{1}{2\varepsilon}$	0	0	-1	-1	1
3		u^2	$\dfrac{1 - 2\varepsilon/3}{12\varepsilon^2}$	0	$1/2$	0	0	0
4		u^2	$-\dfrac{1 - 11\varepsilon/6}{72\varepsilon^2}$	0	$1/2$	0	0	0
5		u^2	$-\dfrac{1 - \varepsilon/2}{8\varepsilon^2}$	-3	0	-3	-3	3
6		u^2	$\dfrac{1 - 7\varepsilon/6}{24\varepsilon^2}$	$-3/2$	0	$-3/2$	$-3/2$	$3/2$
7		u^2	$\dfrac{1}{4\varepsilon}$	$-1/2$	0	$-1/2$	$-1/2$	$1/2$

Let us briefly explain the procedure for calculating the quantities in Table 4.7. The coefficients c_γ in the various constants (4.204) are easily found from the coefficients in (4.202) taking into account the rule (3.112).

The quantities $[L_a\overline{\gamma}] = [K_a R'\overline{\gamma}]$ for graphs 1 to 6 are calculated, like for the φ^4 model in Sec. 3.21, from massless graphs with the simple flow of a single momentum, and the R' operator is determined by the general rules of Sec. 3.8. In our dimension $d = 6 - 2\varepsilon$ the presence of two massless lines with the same integration momentum no longer leads to an IR divergence (at least three are needed). This allows us to perform the calculation with horizontal momentum flow for graph 5 also, after which it reduces to simple chains.

The analogs of (3.122) and (3.128) for an arbitrary graph $\overline{\gamma} \in \overline{\Gamma}_n$ with l loops and $n = 2, 3$ are

$$\alpha = 2 - \varepsilon, \quad a = \|2 - \varepsilon\|/4\pi^{3-\varepsilon},$$

$$\overline{\gamma} = u^l t^{\varepsilon l}[p^2 H(\alpha)]^{3-n} H(\lambda_{nl}) \cdot q, \quad u \equiv g^2/64\pi^3, \quad (4.206)$$

$$t^{\varepsilon} \equiv \|2 - \varepsilon\|^3 (4\pi\mu^2/p^2)^{\varepsilon}, \quad \lambda_{nl} = 6 - n - \varepsilon(l+1).$$

As for the φ^4 model (Sec. 3.21), all the simple factors (a, g, and powers of p, μ, 2, and π) are grouped into the combination $u^l t^{\varepsilon l}[p^2 H(\alpha)]^{3-n}$. The coefficient q in (4.206) is the product of only the nontrivial factors H arising when a coordinate graph is reduced to a simple line with known (from power counting) index λ_{nl}, and the factor $H(\lambda_{nl})$ arises from its Fourier transform (3.125).

The elementary integration formulas of the simple chains (3.124) are sufficient for calculating the quantities q in graphs 1, 2, 4, 5, and 6 with horizontal momentum flow. In the end we find $q_1 = 1$, $q_2 = H(\alpha, \alpha, d - 2\alpha)$, $q_4 = H(\alpha, \alpha, 2\alpha, 3d/2 - 4\alpha)$, $q_5 = H(\alpha, \alpha, d - 2\alpha, \alpha, \alpha, 3\alpha - d/2, 2d - 5\alpha)$, and $q_6 = H(\alpha, \alpha, d - 2\alpha, \alpha, \alpha, 2\alpha, 3d/2 - 4\alpha)$. For graph 3, q is found from (3.135). Only graph 7 is complicated to calculate in coordinate space. It reduces to the graph (3.134) with indices α on the four outer edges and $\alpha - 1$ on the diagonal. It is easier to calculate it with masses using formula (3.77) exactly like graph 13 in Table 3.1.

The two-loop calculation in the MS scheme for the φ^3 model was first done in [112]. We note that the two-loop result quoted in [60] with reference to [112] is incorrect.

Using the data of Table 4.7 and the general rule (3.118) (in the present case with $r_\gamma = 1$ and without the unit contributions for $Z_{0,4}$), all the constants Z are easily calculated:

$$Z_1 = 1 - u/12\varepsilon + u^2(30 - 13\varepsilon)/864\varepsilon^2 + ...,$$

$$Z_3 = 1 - u/2\varepsilon + u(30 - 23\varepsilon)/96\varepsilon^2 + ..., \quad (4.207)$$

$$Z_2 = Z_3, \quad Z_4 = 1 - Z_3, \quad Z_0 = Z_3 - 1 + u/2\varepsilon. \quad (4.208)$$

Relations (4.208) are consequences of the fact that the coefficients c_γ for different Z in Table 4.7 coincide and are exact relations for the simple φ^3

model. In fact, for the action (4.201) with $h = 0$ we easily find the following expression from an equation like (3.192) with derivative with respect to τ and a Schwinger equation of the type (3.187) integrated over x with all renormalized quantities replaced by basic quantities:

$$g_B \partial_\tau [W(A) - \ln C_R] = \int dx [\tau \delta W(A)/\delta A(x) - A(x)] \qquad (4.209)$$

for the generating functional $W(A)$ of the connected Green functions of the basic theory. When the Legendre transform $\Gamma(\alpha)$ is done (Sec. 2.5), (4.209) takes the form $g_B \partial_\tau [\Gamma(\alpha) - \ln C_R] = \int dx [\tau \alpha(x) + \delta \Gamma(\alpha)/\delta \alpha(x)]$. Comparing the coefficients of the powers of α on the two sides of this equation, we obtain several relations between the basic functions Γ_n. Their inclusion in (4.204) leads to (4.208).

In general, these equations are violated for multicomponent generalizations of the φ^3 model owing to the appearance of additional structure factors r_γ in the graphs. The data of Table 4.7 can be used to calculate the constants Z of any generalization of the simple φ^3 model taking into account only these additional factors r_γ.

4.25 RG equations for the φ^3 model including vacuum loops

The RG functions β and γ are defined in the usual manner through the constants Z (Sec. 1.25). Instead of (1.114), from the corresponding equations (4.205) we now obtain

$$\beta_g \equiv \widetilde{\mathcal{D}}_\mu g = -\varepsilon g - g\gamma_g, \qquad \widetilde{\mathcal{D}}_\mu \tau = -\tau \gamma_\tau, \qquad (4.210)$$

and from the equation $Z_2 = Z_3$ we find for the constants in (4.205)

$$\gamma_\tau = \gamma_g + \gamma_\varphi. \qquad (4.211)$$

The terms involving ∂_g and ∂_τ in the RG operator (1.108) are the usual ones [as in (1.122)]. Only the contribution $[\widetilde{\mathcal{D}}_\mu h]\partial_h$ is nontrivial owing to the nontriviality of the renormalization h in (4.205). To calculate $\widetilde{\mathcal{D}}_\mu h$ it is necessary to operate with $\widehat{\mathcal{D}}_\mu$ on both sides of the equation $h_0 = hZ_\varphi^{-1} + Z_\varphi^{-1} Z_4 \tau^2/2g_B$ following from (4.205), taking into account the definitions of all the RG functions and the relations (4.205), (4.208) between the constants Z. Finally, we obtain an explicit expression for $\widetilde{\mathcal{D}}_\mu h$, the substitution of which into (1.108) gives

$$\mathcal{D}_{RG} = \mathcal{D}_\mu + \beta_g \partial_g - \gamma_\tau \mathcal{D}_\tau + [h\gamma_\varphi + (3\gamma_\varphi + \gamma_g)\tau^2/2g_B]\partial_h. \qquad (4.212)$$

For completeness, let us consider the RG equation for the functional $W_R = W_R(A; g, \tau, h, \mu)$ taking into account vacuum loops, which can be written in the form (4.12) or (4.13). We immediately note that the functionals W_R and W'_R for the action (4.203) depend on A and h only through the sum $A + h$.

It is easy to check that the addition of the divergent part of $\ln C_R$ in (4.13c) to the vacuum counterterm in (4.203) is equivalent to the replacement $Z_0 \to Z_0 - u/2\varepsilon = Z_3 - 1$, where the last equation follows from (4.208). Taking this into account along with the definition of the RG functions and all the relations between the constants Z, it is easy to calculate the RG function ξ' explicitly in (4.13). Finally, we obtain the RG equation

$$[\mathcal{D}_{RG} + \gamma_\varphi \mathcal{D}_A]W'_R = -V\tau^3(\gamma_g + 3\gamma_\varphi)/6g_B^2. \tag{4.213}$$

The functional W'_R also satisfies the equation

$$(g_B\partial_\tau - \tau\partial_h)W'_R = -\int dx[A(x) + h], \tag{4.214}$$

which is obtained using the technique described in Sec. 3.29 and relations like (3.187) and (3.192) with derivatives with respect to $e = \tau, h$. Equation (4.214) is specific to the φ^3 model and is the renormalized equivalent of (4.209) with $h \neq 0$. The general solution of (4.214) has the form

$$W'_R = \overline{W}'_R(A + h', g, \mu) - \int dx[\tau(A(x) + h)/g_B + \tau^3/3g_B^2]. \tag{4.215}$$

The second term is the particular solution of the inhomogeneous equation and the first — the general solution of the homogeneous equation — is an unknown function $\overline{W}'_R$ of its first integrals g, μ, A, and $h' = h + \tau^2/2g_B$; the latter two enter into $\overline{W}'_R$ only as the sum (see above). It is important here that the variables τ and h enter into $\overline{W}'_R$ only as the combination h'.

When the substitution $\mu, g, \tau, h \to \mu, g, \tau, h'$ is performed in the operator (4.212), the cross term vanishes and we obtain

$$\mathcal{D}_{RG} = \mathcal{D}_\mu + \beta_g\partial_g - \gamma_\tau\mathcal{D}_\tau + \gamma_\varphi\mathcal{D}_{h'}, \quad h' = h + \tau^2/2g_B. \tag{4.216}$$

It is easy to check that the second term in (4.215), the particular solution of (4.214), is also the particular solution of the inhomogeneous RG equation (4.213). Therefore, upon substitution of (4.215) into (4.213) for $\overline{W}'_R$ we obtain the homogeneous equation

$$[\mathcal{D}_\mu + \beta_g\partial_g + \gamma_\varphi(\mathcal{D}_{h'} + \mathcal{D}_A)]\overline{W}'_R(A + h', g, \mu) = 0. \tag{4.217}$$

The contribution involving $\mathcal{D}_\tau$ from (4.216) vanishes because in the new variables the functional $\overline{W}'_R$ does not depend explicitly on τ.

It follows from (4.215) and (4.217) that all the renormalized connected Green functions W_{nR}, except for the vacuum loops W_{0R} and the tadpole W_{1R}, actually depend not on the two thermodynamic parameters τ and h, but only on their combination h', and satisfy RG equations like (1.107):

$$[\mathcal{D}_\mu + \beta_g\partial_g + \gamma_\varphi\mathcal{D}_{h'} + n\gamma_\varphi]W_{nR}(...; g, h', \mu) = 0. \tag{4.218}$$

The grouping of τ and h together to form a single parameter h' is easily explained directly from the form of the basic action (4.201). By making the shift $\varphi \to \varphi + \text{const}$ in this functional (changing only the connected functions $W_{0,1}$), it is possible to eliminate either the linear or the quadratic contribution, and it is easily checked that the coefficient of the remaining contribution depends only on the combination h'. The form of the inhomogeneity in (4.215) corresponds to eliminating $\tau\varphi^2$.

Therefore, h' is the only real thermodynamical parameter. Its sign has a simple meaning. It is easy to check that for $h' > 0$ the system (4.201) is metastable, while for $h' < 0$ it is unstable. The point $h' = 0$ at which the local energy minimum vanishes (we recall that the action differs by a sign from the Landau functional; see Sec. 1.12) is somewhat reminiscent of the critical point in φ^4-type models: in both cases this point is where the regimes change, although, of course, the regimes themselves are different. It can therefore be expected that singularities of the Green functions will appear at the IR asymptote $h' \to 0$, $p \to 0$. These singularities can be studied using the standard RG technique if the β function in (4.217) has the corresponding IR-stable fixed point. In fact, as we shall verify directly, there is no such point for the simple φ^3 model in dimension $d = 6 - 2\varepsilon < 6$.

Let us calculate the corresponding RG functions using the constants (4.207). Since all the Z depend simply on $u = g^2/64\pi^3$, it is more convenient to use this quantity rather than g as the renormalized charge. For its β function from (4.210) we have $(d = 6 - 2\varepsilon)$

$$\beta_u \equiv \widetilde{\mathcal{D}}_\mu u = 2g\beta_g/64\pi^3 = -2\varepsilon u - 2u\gamma_g, \quad \beta_g\partial_g = \beta_u\partial_u. \tag{4.219}$$

The RG functions are conveniently calculated using relations like (4.3) and (4.5); for $Z = 1 + A(u)/\varepsilon + \ldots$ we have $\gamma = \widetilde{\mathcal{D}}_\mu \ln Z = -\mathcal{D}_g A = -2\mathcal{D}_u A$. In this manner, from (4.207) we find that $\gamma_1 = 2\gamma_\varphi$ and $\gamma_3 = \gamma_g + 3\gamma_\varphi$ [these follow from (4.205)], from which we obtain all the needed RG functions in the two-loop approximation:

$$\gamma_\varphi = u/12 + 13u^2/432 + \ldots, \quad \gamma_g = 3u/4 + 125u^2/144 + \ldots,$$
$$\beta_u = -2\varepsilon u - 3u^2/2 - 125u^3/72 + \ldots. \tag{4.220}$$

The behavior of the β function is shown in Fig. 4.2 by the solid line for $\varepsilon > 0$ and the dashed line for $\varepsilon < 0$.

For $\varepsilon > 0$ the IR-stable point $u_* \sim \varepsilon$ is located in the unphysical (for real g) region $u < 0$. Therefore, the RG technique in this case does not give any information about the IR asymptote, and predicts only UV-asymptotic freedom for the model with the normal sign $u > 0$. The problem with $u > 0$ can be studied formally in dimension $d > 6$. For it the RG technique predicts IR-asymptotic freedom and UV scaling with nontrivial dimensions, which can be calculated in the form of $6 + 2\varepsilon$ expansions. However, none of this has anything to do with the theory of critical behavior. Normal IR scaling in

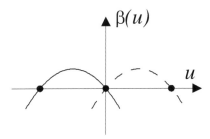

Figure 4.2   Behavior of the β function (4.220) for $d < 6$ (solid line) and $d > 6$ (dashed line).

dimension $d < 6$ can be obtained only in the φ^3 model with purely imaginary constant g (then $u < 0$) or in multicomponent generalizations of it with $u > 0$, owing to the possible change of sign of the coefficient of u^2 in the β function (4.220) due to the appearance of additional structure factors in the graphs. In these cases (4.218) gives the usual formulas of critical scaling for the IR asymptote $h' \to 0$, $p \to 0$ and the $6 - 2\varepsilon$ expansions of the corresponding critical exponents. This is the case in particular for the Potts model, used to describe percolation [113].

4.26   The $2 + \varepsilon$ expansion in the nonlinear σ model: multiplicative renormalizability of low-temperature perturbation theory

The unrenormalized full Green functions of the σ model with n-component ($n > 1$) field $\varphi = \{\varphi_a\}$ can be defined by the generating functional (Sec. 2.26)

$$\overline{G}(A) = \int D\varphi\, \delta(\varphi^2 - 1) \exp[-(\partial\varphi)^2/2t_0 + A\varphi], \qquad (4.221)$$

where t_0 is the absolute temperature. The factor ensuring normalization $\overline{G}(0) = 1$ in (4.221) and below will always be assumed to be included in $D\varphi$ (we do not consider vacuum loops).

Calculation of the integral (4.221) by the stationary phase method leads to low-temperture perturbation theory in powers of the parameter t_0, which in this case plays the role of the bare charge. Here we shall always take the σ model to be a model with a given perturbation theory (others are possible); more details are given below.

Since the field φ (owing to the relation $\varphi^2 = 1$) and the exponent in (4.221) are dimensionless, we have $d[t_0] = 2 - d$, and so the logarithmic dimension for this perturbation theory is $d^* = 2$ and we shall always take $d = 2 + 2\varepsilon$. In [114], [115] the Ward identities expressing the O_n symmetry were used to show that the σ model with a given perturbation theory is multiplicatively renormalizable, i.e., all its UV divergences (poles in ε) are eliminated by field and charge renormalization: $\widehat{\varphi} = \widehat{\varphi}_R \overline{Z}_\varphi$, $t_0 = t_B \overline{Z}_t$, where $t_B = t\mu^{-2\varepsilon}$ is the basic charge and μ is the renormalization mass (in the notation of [114], $\overline{Z}_\varphi^2 \equiv Z$, $\overline{Z}_t \equiv Z_1$). The generating functional of renormalized Green functions corresponding to (4.221) is

$$\overline{G}_R(A) = \int D\varphi \; \delta(\overline{Z}_1 \varphi^2 - 1) \exp[-\overline{Z}_2 (\partial\varphi)^2 / 2t_B + A\varphi] \qquad (4.222)$$

with the constants $\overline{Z}_1 = \overline{Z}_\varphi^2$, $\overline{Z}_2 = (\overline{Z}_t)^{-1} \overline{Z}_\varphi^2$, and $t_B = t\mu^{-2\varepsilon}$.

The general theory of renormalization described in Ch. 3 pertains to the case of polynomial action without constraints. Therefore, to understand the proof of [114] and [115] it is convenient to reproduce the constraint using the O_n-scalar auxiliary field $\psi(x)$ (Sec. 2.27), passing to the local theory of two fields with basic action

$$S_B = [-(\partial\varphi)^2 + \psi(\varphi^2 - 1)]/2t_B, \qquad t_B = t\mu^{-2\varepsilon}. \qquad (4.223)$$

The low-temperature perturbation theory for the action (4.223) is obtained by the stationary phase method. Contributions like the free action for ψ appear in (4.223) after shifting to the stationarity point. The general theory of Ch. 3, in particular, the criterion (3.88), is applicable to the model (4.223). For $\varepsilon = 0$ from (4.223) we find that $d_\varphi = 0$, $d_\psi = 2$, and so in (3.88) we have $d[\Gamma_n] = 2 - 2n_\psi$, so that the only SD graphs are the graphs with $n_\psi = 0, 1$ for any n_φ. They correspond to local counterterms with two derivatives ∂ or a single ψ and any number of fields φ. At first glance this implies nonrenormalizability, because the number of such counterterms is infinite. However, this is not the case. When the locality and the O_n symmetry of the counterterms are taken into account (in the low-temperature solution the symmetry is broken spontaneously, but it must be preserved in the renormalized action S_R; see Sec. 1.28), the set of all contributions with a single ψ in S_R must have the form $\psi f(\varphi^2) \equiv \int dx \psi(x) f(\varphi^2(x))$ with some function $f(a)$ in the perturbation theory beginning with $(a - 1)/2t_B$. Since ψ is a Lagrange multiplier, such a contribution to S_R corresponds to the constraint $f(\varphi^2(x)) = 0$, equivalent to $\varphi^2(x) = c$, where c is the root of the equation $f(c) = 0$. In perturbation theory this root exists and is unique. Therefore, S_R actually does not change when $\psi f(\varphi^2)$ is replaced by $\psi(\varphi^2 - c)$ or $\psi(Z\varphi^2 - 1)$ with $Z = c^{-1}$, i.e., the inclusion of all the counterterms involving ψ is equivalent to the multiplicative renormalization of the field φ in this term of (4.223).

There remain the counterterms with two ∂ instead of ψ. In their analysis we can use the constraint $\varphi^2(x) = c$, assuming the integration over ψ has already been performed. It is easily verified that for $\varphi^2(x) = c$ the only independent local O_n-scalar construction with two ∂ and any number of φ is the form $(\partial\varphi)^2 \equiv \partial\varphi_a \cdot \partial\varphi_a$ present in (4.223). Therefore, the inclusion of these counterterms is equivalent to the introduction of a factor of Z for this contribution, which completes the proof.

The form in which (4.221) is written in [114] is not standard (Sec. 1.16), since the free part of the action contains a dimensional overall coefficient. We therefore perform the dilatation $\varphi \to t_0^{1/2}\varphi$, introducing

$$G(A) = \int D\varphi \, \delta\left[t_0\varphi^2 - 1\right] \exp\left[-(\partial\varphi)^2/2 + t_0^{-1/2}h_0\varphi + A\varphi\right]. \quad (4.224)$$

We have added the contribution of the bare uniform external field h_0 in the form of an additional term $t_0^{-1}h_0\varphi$ in (4.221) in order to preserve the usual (as in [114]) meaning and dimension of the parameter h_0. In the model (4.224) the field φ already has the standard (Sec. 1.15) canonical dimension $d_\varphi = d/2 - 1 = \varepsilon$, but now $d[h_0] = 2$ instead of obeying the shadow relation with d_φ.

The renormalized analog of (4.224) for $d = 2 + 2\varepsilon$ is

$$G_R(A) = \int D\varphi \, \delta\left[Z_1 t_B\varphi^2 - 1\right] \exp\left[-Z_2(\partial\varphi)^2/2 + t_B^{-1/2}h\varphi + A\varphi\right], \quad (4.225)$$

which corresponds to multiplicative renormalization:

$$\widehat{\varphi} = \widehat{\varphi}_R Z_\varphi, \quad t_0 = t_B Z_t = t\mu^{-2\varepsilon}Z_t, \quad h_0 = hZ_h,$$

$$\quad (4.226)$$

$$Z_h = Z_\varphi^{-1}Z_t^{1/2}, \quad Z_1 = Z_t Z_\varphi^2, \quad Z_2 = Z_\varphi^2.$$

These constants are related to $\overline{Z}$ from (4.222) as $Z_t = \overline{Z}_t$, $Z_\varphi = \overline{Z}_\varphi(\overline{Z}_t)^{-1/2}$. It is easily checked that for $h = 0$ the relation $G_R(A) = \overline{G}_R(t_B^{-1/2}A)$ is satisfied, so that the UV-finiteness of the functional (4.222) and the parameter t_B lead to UV finiteness of the functional (4.225) with $h = 0$, which is not spoiled by a subsequent UV-finite shift $A \to A + t_B^{-1/2}h$.

Let $h_a = he_a$, where $h \equiv |h| > 0$. If we pass from φ to the fields $\sigma \equiv \varphi^{\parallel}$ (projection on e) and $\pi \equiv \varphi^\perp$ (the projection onto the subspace orthogonal to e, with the components of π in this basis denoted as π^i, $i = 1, ..., n-1$) in the usual way, the renormalized action in (4.225) takes the form

$$S_R(\pi, \sigma) = -Z_2\left[(\partial\pi)^2 + (\partial\sigma)^2\right]/2 + t_B^{-1/2}h\sigma. \quad (4.227)$$

The calculation of the integral (4.225) with $D\varphi \equiv D\pi D\sigma$ by the stationary phase method corresponds to simply integrating over σ using the functional δ function. It is defined as the formal product of local δ symbols over all points x (Sec. 2.27) and up to an insignificant factor is equivalent to the expression $J(\pi)\delta[\sigma - \sigma(\pi)] \equiv J(\pi)\prod_x \delta[\sigma(x) - \sigma(x; \pi)]$, where

$$\sigma(x; \pi) = \left[Z_1^{-1}t_B^{-1} - \pi^2(x)\right]^{1/2}, \quad (4.228)$$

and $J(\pi) = \prod_x \sigma^{-1}(x; \pi)$ is the Jacobian of the substitution. Such constructions are interpreted by (4.122), in our case

$$J(\pi) = \exp\left\{-\frac{1}{2}\delta(0)\int dx \ln\left[Z_1^{-1}t_B^{-1} - \pi^2(x)\right]\right\}. \quad (4.229)$$

The integral (4.225) is finally reduced to the form

$$G_R(A) = \int D\pi\, J(\pi) \exp[S_R(\pi) + A^i\pi^i + A^0\sigma(\pi)] \qquad (4.230)$$

with action S_R obtained by substituting (4.228) into (4.227):

$$S_R(\pi) = -\frac{1}{2}Z_2(\partial\pi)^2 - \frac{Z_1 Z_2 t_B (\partial\pi^2)^2}{8(1 - Z_1 t_B \pi^2)} + Z_1^{-1/2} t_B^{-1} h \left[1 - Z_1 t_B \pi^2\right]^{1/2}. \quad (4.231)$$

The quantities appearing here, including the constants Z, are expanded in series in t. The calculation in finite order in t requires only finite segments of these series, and in this approximation S_R has an ordinary polynomial structure.

The addition to the action generated by (4.229) contains $\delta(0) \sim \Lambda^d$, where Λ is the UV cutoff and serves only to cancel the analogous UV divergences in the graphs which are powers of Λ. If the calculations are performed using formal dimensional regularization (Sec. 1.20), all such contributions must simply be ignored, setting $J = 1$.

4.27 Calculation of the constants Z and the RG functions in the one-loop approximation

After expanding in t and discarding the unimportant constant $Z_1^{-1/2} t_B^{-1} h$, only contributions with positive powers of t remain in (4.231). The free action is the zeroth approximation $S_{0R} = -[(\partial\pi)^2 + h\pi^2]/2$, corresponding to the correlator $[k^2 + h]^{-1}\delta_{is}$ of the field π with $h > 0$ instead of τ. For the one-loop calculation it is sufficient to restrict ourselves to the following approximation in (4.231):

$$S_R \cong -Z_2(\partial\pi)^2/2 - Z_1^{1/2} h\pi^2/2 - t_B \left[(\partial\pi^2)^2 + h\pi^4\right]/8, \qquad (4.232)$$

including in Z only corrections of first order in t, which must cancel the divergences of the one-loop graphs. To find them it is sufficient to calculate the renormalized function $\Gamma_{\pi\pi} = -D_\pi^{-1}$ with momentum flow p. From (4.232) we find that

$$\Gamma_{\pi\pi} = -Z_2 p^2 - Z_1^{1/2} h + \Sigma_1, \qquad (4.233)$$

where Σ_1 are the contributions of the one-loop self-energy graphs for the interaction in (4.232). They are conveniently written as

$$\Sigma_1 = \frac{h}{2}\; \raisebox{-0.5em}{\includegraphics{}} \;+\; h\; \raisebox{-0.2em}{\includegraphics{}} \;+\; \raisebox{-0.2em}{\includegraphics{}} \;+\; \raisebox{-0.2em}{\includegraphics{}}\;. \qquad (4.234)$$

The dashed lines denote the interaction line $\delta(x - x')$, which is understood to include the factor $-t_B$, the closed loops of solid π lines generate a factor of $n-1$ from the summation over the indices, and the vertical line crossing the ends of these lines denotes differentiation with respect to the corresponding coordinate x. The following analytic expression is associated with the graphs (4.234):

$$\Sigma_1 = -\frac{t_B}{(2\pi)^d} \left\{ \left[h\frac{(n+1)}{2} + p^2 \right] \int \frac{dk}{k^2 + h} + \int dk \frac{k^2}{k^2 + h} \right\}. \quad (4.235)$$

Substituting $t_B = t\mu^{-2\varepsilon}$, $d = 2 + 2\varepsilon$, and calculating the d-dimensional integrals using the expressions of Sec. 3.15, we find that

$$\Sigma_1 = u\|1 - \varepsilon\|(h/4\pi\mu^2)^\varepsilon[(n-1)h + 2p^2]/4\varepsilon, \quad u \equiv t/2\pi. \quad (4.236)$$

Requiring that the poles in ε in (4.233) cancel with Σ_1 from (4.236), we find the first corrections of order u/ε in the renormalization constants: $Z_1 = 1 + u(n-1)/2\varepsilon + ...$, $Z_2 = 1 + u/2\varepsilon + ...$ From this using (4.226) we can find all the other constants Z, but it is simpler to directly calculate the anomalous dimensions $\gamma_i = \tilde{\mathcal{D}}_\mu \ln Z_i$ using relations like (4.3) following from (4.226) and taking $u = t/2\pi$ instead of t as the renormalized charge. When calculating γ using expressions like (4.5) it is necessary to use $\mathcal{D}_g = \mathcal{D}_u$ and $k = -2$ in them according to the definition $t_B = t\mu^{-2\varepsilon}$.

Two independent RG functions γ_φ and $\gamma_t = \gamma_u$ correspond to the two independent constants Z. Using them and (4.226) we find that

$$\beta \equiv \tilde{\mathcal{D}}_\mu u = 2\varepsilon u - u\gamma_u, \qquad \gamma_h = -\gamma_\varphi + \gamma_u/2, \quad (4.237)$$

and the RG operator (1.108) is now

$$\mathcal{D}_{RG} = \mathcal{D}_\mu + \beta\partial_u - \gamma_h \mathcal{D}_h \quad (4.238)$$

(in [114] the variable u is again denoted as t, and the independent RG functions are taken to be $W \equiv \beta$ and $\zeta \equiv \gamma_\varphi + \gamma_u$).

From the expressions obtained above for Z it is possible to find all the RG functions only in the one-loop approximation, but we shall give the two independent RG functions γ_φ and β [γ_u and γ_h are expressed in terms of them by (4.237)] with the highest, four-loop, accuracy obtained so far (the two-loop contributions were found in [114], the three-loop contributions in [116], and the four-loop contributions in [117]):

$$\gamma_\varphi = \frac{1}{2}u + \frac{1}{24}(n-2)u^2 \left\{ -12 + 3u(2n-5) + \right.$$

$$(4.239)$$

$$\left. + u^2[-3n^2 + 2n + 14 + 3(3-n)(n+5)\zeta(3)] + ... \right\},$$

$$\beta = 2\varepsilon u - u\gamma_u = 2\varepsilon u - \frac{(n-2)u^2}{12} \left\{ 12 + 12u + \right.$$

$$(4.240)$$

$$\left. + 3u^2(n+2) + u^3[-n^2 + 22n - 34 + 18(n-3)\zeta(3)] + ... \right\}.$$

It was shown in [114] that for $n = 2$ the first terms of the series for γ_φ and β are the exact answers, i.e., $\gamma_\varphi = u/2$, $\gamma_u = 0$, and $\beta = 2\varepsilon u$. This is explained by the fact that for $n = 2$ we can use the parametrization $\pi \sim \sin\Theta$, $\sigma \sim \cos\Theta$, for which $(\partial\varphi)^2 \sim (\partial\Theta)^2$ and $D\pi \sim D\Theta$. In this case all the Green functions of π and σ in zero external field (and from them the RG functions) can be calculated exactly, because the integral (4.230) reduces to a Gaussian after expansion in the sources.

4.28 The Goldstone and critical asymptotes. $2 + \varepsilon$ expansion of the critical exponents

The multiplicative renormalizability of the model leads to the standard RG equations (Sec. 1.24) with the operator (4.238), in particular, the equation (1.150) for the renormalized correlator D (we shall drop the subscript R) of any of the fields π, σ. Changing to canonically dimensionless variables in the usual manner,

$$D(p, u, h, \mu) = p^{-2}\Phi(s, u, z), \quad s \equiv p\mu^{-1}, \quad z \equiv h\mu^{-2}, \tag{4.241}$$

from (1.150) and (4.238) we obtain the RG equation for Φ:

$$[-\mathcal{D}_s + \beta\partial_u - (2 + \gamma_h)\mathcal{D}_z + 2\gamma_\varphi]\Phi(s, u, z) = 0. \tag{4.242}$$

In the special case $n = 2$, owing to the triviality of the RG functions (Sec. 4.27), (4.242) leads to scaling with known, u-dependent exponents. In what follows we shall assume that $n > 2$.

The β function (4.240) for $n > 2$ and $\varepsilon > 0$ behaves as shown by the dashed line in Fig. 4.2. It has IR-stable point $u_* = 0$ and UV-stable point $u_* \sim \varepsilon$. In contrast to models of the φ^4 type with fixed (when studying the asymptotes) charge $u \sim g$, in the present case $u = t/2\pi$ is a thermodynamical variable — the renormalized temperature, and $u_* \sim \varepsilon$ has the meaning of the critical temperature, because the RG equations for $u = u_*$ predict scaling with nontrivial exponents, i.e., singularities of thermodynamical quantities. The arguments of Sec. 1.27 remain in force, and show that $0 < u < u_*$ for any value of the bare temperature t_0, i.e., only the low-temperature phase can be studied in this perturbation theory.

Let us return to (4.242). Its solution is described by expressions analogous to (1.152) and (1.153):

$$\Phi(s, u, z) = \Phi(1, \overline{u}, \overline{z}) \exp\left[\int_u^{\overline{u}} dx \frac{2\gamma_\varphi(x)}{\beta(x)}\right], \tag{4.243a}$$

$$\overline{z} = zs^{-2} \exp\left[-\int_u^{\overline{u}} dx \frac{\gamma_h(x)}{\beta(x)}\right]. \tag{4.243b}$$

The invariant charge $\bar{u} = \bar{u}(s, u)$ appearing here is defined by the usual rule (1.144) and in the one-loop approximation differs from (1.154) by only the replacement $\varepsilon \to -\varepsilon$, i.e., $\bar{u}(s, u) = uu_* / [u_* s^{-2\varepsilon} + u(1 - s^{-2\varepsilon})]$ with $u_* \sim \varepsilon$.

As in Sec. 1.33, the equations (4.243) can be used to analyze the $s \to 0$, $z \to 0$, $u = $ const asymptote. Formerly it corresponded to the critical asymptote, and now it is the Goldstone one: $p \to 0$, $h \to 0$, $u = $const. In this regime $\bar{u}(s, u) \to u_* = 0$, and all the anomalous dimensions $\gamma^* \equiv \gamma(u_*)$ vanish at the point $u_* = 0$. However, it would be incorrect to refer to (1.165) and conclude that the Goldstone asymptote corresponds to ordinary scaling with canonical dimensions, because the case $u_* = 0$ is special.

Let us consider this case in more detail, beginning with refinement of the rate of falloff of $\bar{u}$ for $s = p\mu^{-1} \to 0$. It follows from (1.170) that $\bar{u} \sim s^{\omega}$ with $\omega = \beta'(0) = 2\varepsilon$ according to (4.240), i.e., $\bar{u} \sim s^{2\varepsilon} \to 0$. Both integrals in (4.243) for $\bar{u} \to 0$ can be assumed constants up to unimportant corrections. Therefore, from (4.241) and (4.243) we have

$$D \cong c_1 p^{-2} \Phi_0(1, \bar{u}, c_2 h p^{-2}), \tag{4.244}$$

where $c_{1,2}$ are normalization factors generated by the exponents in (4.243), and $\Phi_0(1, u, z)$ is the leading term of the $u \to 0$ asymptote of the function $\Phi(1, u, z)$ from (4.241), determined from the lowest-order perturbative contribution for D.

The representation (4.244) is valid for both correlators D_π and D_σ, but the form of Φ_0 is different for them. For D_π it is determined by the bare contribution $(p^2 + h)^{-1}$, which corresponds to $\Phi_0(1, u, z) = (1 + z)^{-1}$, and upon substitution into (4.244) we find the asymptote

$$D_\pi \cong c_1 p^{-2} (1 + c_2 h p^{-2})^{-1} = c_1 (p^2 + c_2 h)^{-1} \tag{4.245}$$

in agreement with (4.116). The correlator $D_\sigma = \langle \sigma\sigma \rangle$ for the field (4.228) in lowest-order perturbation theory is equal to the connected part of the unperturbed expectation value $(t_B/4)\langle \pi^2(x)\pi^2(x') \rangle$. Going to momentum space, we obtain $D_\sigma = \pi u \mu^{-2\varepsilon} \Pi(p, h)$, where Π is a simple loop with unperturbed π lines, i.e., (2.187) with the replacement $\xi \to h$ and an additional factor of $n - 1$ from summation over the indices of π. From power counting we have $\Pi = p^{-2+2\varepsilon} f_0(hp^{-2})$, where f_0 is the scaling function of the simple loop. Therefore, from the definition of Φ in (4.241) we have $\Phi_0(1, u, z) = \pi u f_0(z)$, and upon substitution into (4.244) using $\bar{u}(s, u) \sim s^{2\varepsilon}$ (see above) we obtain

$$D_\sigma \sim p^{-2} \bar{u} f_0(c_2 h p^{-2}) \sim p^{-2+2\varepsilon} f_0(c_2 h p^{-2}). \tag{4.246}$$

Addition of the factor c_2 to the argument of f_0 corresponds to replacing the bare π lines of the loop by their Goldstone asymptote (4.245), as required. The resulting expressions agree with the results of the general analysis of the leading Goldstone singularities in Sec. 4.15. If desired, the RG method can also be used to find the corrections to the leading contributions with any degree of accuracy.

Let us now discuss the critical regime, which in the present case is controlled by the UV-stable fixed point $u_* \sim \varepsilon$ interpreted as the critical temperature, and corresponds in (4.241) to the asymptote $\tau \equiv u_* - u \to 0$ for any (not necessarily small, as for the φ^4 model) values of s and z. This statement of the problem is possible only because now u is a thermodynamical variable (the temperature) that can be varied in an experiment, whereas $u = \text{const}$ in standard models of the φ^4 type. In this case the approach of the invariant charge to the critical regime $\bar{u}(s, u) \simeq u_*$ is ensured not by going to the small-s asymptote, as usual, but simply by the external conditions: $\bar{u}(u, s) \to u_*$ for $u \to u_*$ (i.e., for $T \to T_c$) independently of the value of $s = p\mu^{-1}$.

The desired equation of critical scaling can be obtained most simply by the method described at the end of Sec. 1.33, by studying the RG equation (1.150) with the operator (4.238) in the vicinity of $u_* \sim \varepsilon$. If we simply set $u = u_*$, we obtain the equation of critical scaling directly at the point $T = T_c$. To determine the critical dimension $\Delta_\tau \equiv 1/\nu$ of the parameter $\tau \equiv u - u_* \sim T_c - T$ (now it is canonically dimensionless), it is necessary to keep the term linear in τ in the β function. In this approximation, (1.150) with the operator (4.238) takes the form

$$[\mathcal{D}_\mu + \beta'(u_*)\mathcal{D}_\tau - \gamma_h^* \mathcal{D}_h + 2\gamma_\varphi^*]D = 0. \qquad (4.247)$$

Subtracting it from the canonical scaling equation $[\mathcal{D}_\mu - \mathcal{D}_x + 2\mathcal{D}_h - 2d_\varphi]D = 0$ with $d_\varphi = d/2 - 1 = \varepsilon$, written for the function in coordinate space [the form of (4.247) is independent of the representation], we obtain the desired equation of critical scaling $[-\mathcal{D}_x + \Delta_\tau \mathcal{D}_\tau + \Delta_h \mathcal{D}_h - 2\Delta_\varphi]D = 0$ with the critical dimensions

$$\Delta_\varphi = d_\varphi + \gamma_\varphi^*, \qquad \Delta_h = 2 + \gamma_h^*, \qquad \Delta_\tau \equiv 1/\nu = -\beta'(u_*). \qquad (4.248)$$

Equation (4.240) for the β function leads to the exact relation $\gamma_u^* = 2\varepsilon$ for $u_* \sim \varepsilon$, and so from the second equation in (4.237) we have $\gamma_h^* = \varepsilon - \gamma_\varphi^*$. Taking this into account, from (4.248) we find $\Delta_\varphi + \Delta_h = d_\varphi + 2 + \varepsilon = 2 + 2\varepsilon = d$, i.e., the critical dimensions of φ and h (in contrast to the canonical ones) obey the shadow relation, as required, since only two critical exponents can be independent. If, as usual, we take these to be $\eta = 2\gamma_\varphi^*$ and $\Delta_\tau = 1/\nu$, from the RG functions (4.239) and (4.240) we can find their $2 + \varepsilon$ expansions through order ε^4. The results are written most compactly in terms of the variable $z \equiv 2\varepsilon/(n - 2) \equiv (d - 2)/(n - 2)$:

$$u_* = z - z^2 + \frac{(6 - n)}{4} z^3 + \frac{z^4}{12}[n^2 - 7n + 4 + 18(3 - n)\zeta(3)] + ..., \qquad (4.249)$$

$$\begin{aligned}
\eta &= z + \frac{(n - 1)z^2}{4}\Big\{ -4 + 2nz + \\
&\quad + z^2[6 - 2n - n^2 + (3 - n)(n + 4)\zeta(3)] + ...\Big\},
\end{aligned}$$

$$1/\nu = z(n-2)\left\{1 + \frac{z}{4}\left[4 + 2z(n-2) + \right.\right.$$
$$\left.\left. + z^2[-n^2 + 14n - 30 - 18(3-n)\zeta(3)] + ...\right]\right\}, \tag{4.250}$$

$$\nu = \frac{1}{(n-2)z}\left\{1 - z + \frac{(4-n)z^2}{2} + \right.$$
$$\left. + \frac{z^3}{4}[n^2 - 10n + 18 + 18(3-n)\zeta(3)] + ...\right\},$$

where $\eta = 2\gamma_\varphi^*$ and $1/\nu = \Delta_\tau$. We note that there is a misprint in the result for ν given in Ch. 29 of [45].

The generalization of γ_φ^* to the case of arbitrary irreducible tensor power φ^l of the basic field, the traceless part of the tensor product of l operators $\varphi_a(x)$ [for two factors this is $\varphi_a(x)\varphi_b(x) - \delta_{ab}\varphi^2(x)/n$, and so on] is also given in [45]. All these operators are multiplicatively renormalizable and have the following critical anomalous dimensions [45]:

$$\gamma^*[\varphi^l] = \frac{l^2 z}{2} + \frac{l(n+l-2)z^2}{8}\left\{-4 + 2nz + \right.$$
$$\left. + z^2\left[6 - 2n - n^2 + \zeta(3)[2(7-2n) - l(n+l-2)(n-2)]\right] + ...\right\}. \tag{4.251}$$

The combination $l(l+n-2) \equiv a_l$ appears as a common factor in the full critical dimension $\Delta[\varphi^l]$ $(= -\zeta_l(u_*)$ in the notation of [45]), obtained by the addition of $d[\varphi^l] = ld_\varphi = l\varepsilon = lz(n-2)/2$ to (4.251). The same combination also enters into the coefficient of $\zeta(3)$ in (4.251), and so it is easiest to write $\Delta[\varphi^l]$ in terms of a_l. For completeness, we also give the RG function $l\gamma_u/2 + \gamma[\varphi^l] \equiv -\zeta_l(u)$ from [45] ($\gamma[\varphi^l]$ is the RG function of the anomalous dimension of the operator φ^l), the value of which at the fixed point coincides with $\Delta[\varphi^l]$ owing to the equation $\gamma_u^* = 2\varepsilon$:

$$-\zeta_l(u) = \frac{ua_l}{2}\left\{1 + \frac{(n-2)u^2}{12}\left[9 + u[4(5-n) + 3(2-a_l)\zeta(3)] + ...\right]\right\}, \tag{4.252}$$

where $a_l \equiv l(n+l-2)$. The special case $l = 1$ in (4.251) and (4.252) corresponds to the field φ itself.

Explicit expressions for the renormalized correlators and the equation of state are also given in [114] with two-loop accuracy.

The principal corrections to scaling in this model are generated by composite operators with $d_F^* = 4$ (for action operators $d_F^* = 2$), namely, $((\partial\varphi)^2)^2$, $(\partial_i\varphi \cdot \partial_k\varphi)^2$, and $(\partial^2\varphi)^2$, with which two more nontrivial operators with the field σ in the denominator mix. The corresponding 5×5 matrix of renormalization constants in the one-loop approximation was calculated in [115], and used in [118] to determine the corresponding exponents ω (Sec. 4.8). The smallest one (i.e., the principal correction exponent) is

$$\omega = 2 - 2\varepsilon n/(n-2) + ... = 2 - nz + ... \tag{4.253}$$

4.29 The $1/n$ expansion of the critical exponents of the O_n φ^4 and σ models

The technique for constructing $1/n$ expansions for these two models has been described in detail in Ch. 2. It was shown (Sec. 2.27) that they have identical critical behavior, because in the $1/n$ expansion the corresponding action functionals differ only by IR-irrelevant terms. A complete calculation (correlators, the equation of state, the critical exponents) in leading (zeroth) order in $1/n$ was also performed in Ch. 2. In particular, from Table 2.2 in Sec. 2.24 we have

$$\eta_0 = 0, \quad \nu_0 = 1/(d-2), \quad \omega_0 = 4 - d. \tag{4.254}$$

Here and below we use η_k to denote the coefficient of $1/n^k$ in the $1/n$ expansion of η, and similarly for the other exponents. In (4.254) and below we give the results for the two independent principal exponents η and ν, related by (1.162), (1.164), and (2.203) to the critical dimensions of the fields, and also the results for the principal correction exponent ω, given by (2.204) for the σ model and (1.156) for the φ^4 model. All the other principal exponents are expressed in terms of η and ν using (1.165).

In first order in $1/n$ all three exponents are known [the notation of (3.68) is used everywhere]:

$$\eta_1 = -4H(2 - d/2, d/2 - 2, d/2 - 1)/\|d/2 + 1\|, \tag{4.255a}$$

$$\nu_1/\eta_1 = -2(d-1)/(d-2)(4-d), \tag{4.255b}$$

$$\omega_1/\eta_1 = -2(d-1)^2. \tag{4.255c}$$

Only η and ν are known in second order:

$$\eta_2/\eta_1^2 = \frac{d^2 - 3d + 4}{4-d} R_0 + \frac{1}{d} + \frac{1}{d-2} + \frac{9}{4-d} + \frac{4}{(4-d)^2} - 2 - d, \tag{4.256a}$$

$$\nu_2/\eta_1^2 = \frac{d}{(d-2)(4-d)^2} \left\{ \frac{2(d^4 - 6d^3 + 5d^2 + 12d - 8)}{d(d-2)} R_0 + \right.$$

$$+ \frac{3d(11 - 4d)}{2} R_2 + \frac{2d(d-3)^2}{4-d} [R_1^2 + R_3 - 6R_2] + \tag{4.256b}$$

$$\left. + \frac{16}{4-d} - \frac{8}{d-2} + \frac{12}{(d-2)^2} - \frac{16}{d^2(d-2)} - d^2 + d - 34 \right\},$$

where

$$R_0 \equiv \psi(d-2) + \psi(2-d/2) - \psi(2) - \psi(d/2-2),$$

$$R_1 \equiv \psi(d-3) + \psi(2-d/2) - \psi(1) - \psi(d/2-1),$$

$$R_2 \equiv \psi'(d/2-1) - \psi'(1), \quad R_3 \equiv \psi'(d-3) - \psi'(2-d/2) - R_2.$$

Only the exponent η is known in order $1/n^3$. The expression for η_3 in arbitrary dimension d is very awkward and will be given later in Sec. 4.40. Here we give η_3 together with the other known coefficients of the $1/n$ expansions only for dimension $d=3$:

$$\eta_1 = 8/3\pi^2, \quad \nu_1/\eta_1 = -4, \quad \omega_1/\eta_1 = -8,$$

$$\eta_2/\eta_1^2 = -8/3, \quad \nu_2/\eta_1^2 = 56/3 - 9\pi^2/2, \qquad (4.257)$$

$$\eta_3/\eta_1^3 = -797/18 - 61\pi^2/24 + 9\pi^2\ln 2/2 + 27\psi''(1/2)/8$$

(numerically, $\eta_1 \simeq 0.27$, $\eta_2 \simeq -0.195$, and $\eta_3 \simeq -1.88$).

The quantities (4.255) were calculated in [119], and η_2 for $d=3$ in [120]. For arbitrary dimension, η_2 was first calculated by K. Symanzik (unpublished), and then by the authors of [89] and [121]. The value of ν_2 for $d=3$ was calculated along with other quantities in a series of four articles concluding with [122], and for arbitrary dimension in [89]. The value of η_3 for arbitrary dimension (and for $d=3$ as a special case) was found in [123].

In addition to the exponents, the scaling functions of the correlators [124], the equations of state [99], and the critical dimensions of some composite operators have been calculated for this model in first order in $1/n$. These will be discussed in more detail in Sec. 4.34, where the results of the calculations are given.

Similar calculations of the exponents for more complicated generalizations of the nonlinear σ model like those studied in Sec. 2.28 never go beyond first order in $1/n$. In higher orders only the values of $\eta_{2,3}$ and ν_2 in the simplest σ model, which we have already discussed, are known. The relative dearth of results compared to the case of ε expansions is just due to technical difficulties, like the rapid growth in the number of graphs and their complexity with increasing order in $1/n$; already in the lowest orders in $1/n$ graphs with a large number of loops are encountered.

Let us conclude by discussing the relation between the $1/n$ and ε expansions. It is possible to construct $4-\varepsilon$ expansions by the RG method only in the φ^4 model and $2+\varepsilon$ expansions only in the σ model. The $1/n$ expansions for these models are the same. Since these two models are known to be equivalent in the critical region, all the universal characteristics of the leading critical singularities (the principal critical exponents and the normalized scaling functions) must be the same in them. This means that the $4-\varepsilon$,

$2 + \varepsilon$, and $1/n$ expansions of a given physical quantity obtained by different methods are different versions of the expansions of some common object depending on d and n (of course, assuming the absence of exponentially small contributions, which are completely lost in power expansions). This statement (or, rather, hypothesis, in view of the last remark) is confirmed by a direct check that the coefficients of the double expansions in ε and $1/n$ obtained by the different methods coincide. This is done either by expanding in $1/n$ the first few coefficients of the ε expansions known for any n, or by expanding in the corresponding parameter 2ε ($4 - d$ or $d - 2$) the first few coefficients of the $1/n$ expansions known for any d. The coefficients of the double expansions found in this way always coincide, which, first, confirms the hypothesis that exponentially small contributions are absent and thereby that the models are critically equivalent, and, second, serves as a method of checking the calculations. The possibility of performing such a check is one of the reasons (in addition to aesthetic considerations) that it is desirable to calculate the coefficients of the $1/n$ expansions for arbitrary dimension d.

The statements that the models are equivalent and that the different variants of the expansions are consistent pertain not only to the leading critical singularities, but also to the principal (of smallest magnitude) correction critical exponent ω. In the σ model, ω is determined by the dimension of the IR-irrelevant operator $\psi^2(x)$, the addition of which transforms the σ model into the φ^4 model (Sec. 2.27), i.e., the "σ model + $\psi^2(x)$ corrections = the φ^4 model." The complete models on the two sides of this symbolic equation include the leading correction terms, and the equality implies that the models coincide through the corrections we have discussed. This means that the exponent ω calculated in the σ model from (2.204) in terms of the critical dimension of the operator $\psi^2(x)$ must coincide with the analogous exponent of the φ^4 model, which is determined by (1.156) in terms of the β function, and also with the principal correction exponent ω of the σ model itself in the $2 + \varepsilon$ expansion (Sec. 4.28).

4.30 Calculation of $1/n$ expansions of the exponents in terms of the RG functions of the φ^4 model

The results given in Sec. 4.29 were obtained by various methods. One of them, the most direct, is the use of the standard RG technique for the O_n φ^4 model with subsequent resummation of the series of Sec. 4.2 in $u \equiv g/16\pi^2$ for the constants Z and the RG functions γ. For this it is necessary to substitute $u = \lambda/n$ in them, taking λ to be a new independent parameter (a charge) with $\beta_\lambda = -\lambda(2\varepsilon + \gamma_g)$ according to (4.2), and then to systematically select all contributions of the same order in $1/n$. In zeroth order the problem is trivial, because in this approximation $\gamma_\varphi = 0$ and γ_τ and γ_g are determined by the first terms of the corresponding series in u. In fact, in this order only those graphs contribute for which the number of factors n in the structure coefficients r_γ of Table 3.2 is equal to the number of loops. These are only graphs 1, 3, and 6 of Table 3.1 and their analogs in higher orders. Owing to the second equation in (3.104), none of these graphs except for graph 1 has poles of first order in ε, and so they do not contribute to the RG functions [see the discussion following (4.4)].

Taking this into account, the expressions of Sec. 4.2 with the substitution $u = \lambda/n$ in zeroth order in $1/n$ gives $\gamma_\varphi = 0$, $\gamma_\tau = -\lambda/3$, and $\beta_\lambda = \lambda(-2\varepsilon + \lambda/3)$, from which $\lambda_* = 6\varepsilon$, $\gamma_\varphi^* = 0$, $\gamma_\tau^* = -2\varepsilon$, and $\omega = 2\varepsilon$, which is equivalent to (4.254) for $d = 4 - 2\varepsilon$ (we recall that in the φ^4 model, $\eta = 2\gamma_\varphi^*$ and $1/\nu = \Delta_\tau = 2 + \gamma_\tau^*$).

In next (first) order in $1/n$ infinite summations are already needed in all the RG functions. The constants Z and the RG functions in this approximation were calculated in [125], and the values (4.255) were obtained using them (a simpler calculation will be given in Sec. 4.34). However, already in the next order $1/n^2$ it is practically impossible to obtain anything in this way owing to the extreme complexity of the calculations. We note that in this scheme for calculating the $1/n$ expansions, the renormalization is understood to be the usual procedure for the φ^4 model, namely, eliminating the poles in $\varepsilon \sim 4 - d$. In other schemes the renormalization procedure is different.

4.31 The analog of dimensional regularization and nonmultiplicative renormalization of the massless σ model

Calculations of the $1/n$ expansions of exponents are considerably simplified if they are performed on the basis of the massless (= critical) σ model (2.205), first, because the model contains no unnecessary parameters and, second, because all the lines of the graphs are massless (massless graphs are much easier to calculate than massive ones). However, to carry out the program it is necessary first of all to clearly formulate the procedure for regularizing and renormalizing the model (2.205).

For the convenience of the reader, here we reproduce the action (2.205) with relabeling of the principal field $\varphi \to \phi$, so that $\varphi \equiv \phi, \psi$ denotes the set of all fields:

$$S(\varphi) = -\frac{1}{2}(\partial\phi)^2 - \frac{1}{2}\psi L\psi + \frac{1}{2}\psi\phi^2 + \frac{1}{2}\psi L\psi, \qquad L \equiv -\frac{1}{2}\,\bigcirc\,. \qquad (4.258)$$

The first two contributions are assumed to be the free part, and the others the interaction. The last term cancels all graphs with insertions of simple loops (2.173) in the line ψ (Sec. 2.20). The lines of the graphs of the model (4.258) are associated with the bare propagators (correlators) $D_\phi = (-\partial^2)^{-1}$, $D_\psi = L^{-1}$. In coordinate space

$$D_\phi = Ax^{-2\alpha}, \qquad D_\psi = Bx^{-2\beta}, \qquad (4.259)$$

where α and β are the canonical dimensions of the fields (2.206),

$$\alpha \equiv d_\phi = d/2 - 1, \quad \beta \equiv d_\psi = 2, \tag{4.260}$$

and A and B are known amplitude factors:

$$A = H(1)/4\pi^{d/2}, \quad B = -32H(d/2 - 2, 2 - d/2)/nH^2(1). \tag{4.261}$$

In the notation of (4.259) and below we have omitted the index factor δ_{ab} of the propagator ϕ, the summation over whose indices in closed loops of ϕ lines gives a factor n. The expression for D_ϕ is known from (3.121), and $D_\psi = L^{-1}$ is calculated using the Fourier transform of arbitrary degree (3.72) with respect to the kernel $L(x) = -nD_\phi^2(x)/2$ known from the definition (4.258).

For the graphs of the model (4.258) to be meaningful the model must be regularized, because for fields with canonical dimensions (4.260) the vertex $\psi\phi^2$ in (4.258) is logarithmic for any d, and so any graph with vertex subgraphs diverges. We stress the fact that the variability of d in this case does not ensure regularization, in contrast to the $4 - \varepsilon$ or $2 + \varepsilon$ schemes. Of course, any model can be regularized by a cutoff parameter Λ, but this is undesirable because calculations of graphs with a cutoff are extremely complicated and nearly impossible to perform when there are many loops. It would therefore be best to have a different regularization of the dimensional type that would not spoil the masslessness of the lines and would preserve the simple power-law form (4.259) of the corresponding propagators, which is very important for the calculations.

A regularization of the needed type can be performed by making an arbitrary small shift of the canonical dimensions (4.260), violating the condition for the vertex to be logarithmic $2\alpha + \beta = d$. In practice, it is sufficient [126] to regularize only the propagator of the auxiliary field ψ by a small ε shift $\beta \to \beta - \varepsilon$ of its canonical dimension, leaving the dimension and propagator of the principal field ϕ unchanged:

$$\beta \to \beta - \varepsilon, \quad D_\psi = Bx^{-2\beta} \to D_{\psi,\text{reg}} = Bx^{-2\beta + 2\varepsilon}, \quad \varepsilon > 0. \tag{4.262}$$

Such a regularization is effected by the replacement $L = D_\psi^{-1} \to L_\varepsilon \equiv D_{\psi,\text{reg}}^{-1}$ of the kernel of the quadratic form in the free action for ψ, i.e., in the second term in (4.258). In coordinate space

$$L(x) = -nD_\phi^2(x)/2 \longrightarrow L_\varepsilon(x) = C_\varepsilon L(x) x^{-2\varepsilon}, \tag{4.263}$$

where C_ε is an additional amplitude factor which becomes unity for $\varepsilon \to 0$ (its explicit form is easily found from the definitions, but we do not need it).

To compensate for the shift d_ψ, the regularization (4.263) must be accompanied by the introduction of an additional factor μ^ε in the interaction $\psi\phi^2$ in (4.258) (μ is the renormalization mass), and to preserve the cancellations of the insertions of simple loops by the last term in (4.258), it should be multiplied by $\mu^{2\varepsilon}$, because a factor μ^ε now appears at each of the two vertices of a loop. The kernel L is left unchanged, because the lines of the principal field in the loop are not regularized. Therefore, the regularized functional (4.258) should be written as

$$S_B(\varphi) = -\frac{1}{2}(\partial\phi)^2 - \frac{1}{2}\psi L_\varepsilon \psi + \frac{1}{2}\mu^\varepsilon \psi\phi^2 + \frac{1}{2}\mu^{2\varepsilon}\psi L\psi \qquad (4.264)$$

and treated as the basic action of the regularized massless σ model. The regularized graphs differ from the original ones only by the shift $\beta = 2 \to 2-\varepsilon$ of the "indices" of the ψ lines and additional factors of μ^ε at all vertices. In the formal scheme (Sec. 1.20) all the regularized graphs are well defined, and, as usual, UV divergences appear in them in the form of poles in ε. It should be stressed that the parameter ε now has no relation to the space dimension d, which remains completely arbitrary.

Let us now consider the renormalization of the model (4.264). According to the general rule (3.88) taking into account the values of d_φ in (4.260), superficial divergences can occur only in graphs of the 1-irreducible functions Γ for which

$$d[\Gamma] = d - (d/2 - 1)n_\phi - 2n_\psi = 2m, \qquad m = 0, 1, ..., \qquad (4.265)$$

where $n_{\phi,\psi}$ is the number of corresponding fields in Γ [in the present model the monomials M in (3.88) are scalars constructed only from the external momenta, and so $d_M = 0, 2, 4, ...$]. It follows from the criterion (4.265) that in the general case of arbitrary dimension there are only logarithmic superficial diverges in the vertex graphs $\Gamma_{\phi\phi\psi}$ and quadratic ones in the graphs $\Gamma_{\phi\phi}$. They generate counterterms of the same type as the first and third contributions in (4.264), so that the renormalized action has the form

$$S_R(\varphi) = -\frac{1}{2}Z_1(\partial\phi)^2 - \frac{1}{2}\psi L_\varepsilon \psi + \frac{1}{2}Z_2\mu^\varepsilon \psi\phi^2 + \frac{1}{2}\mu^{2\varepsilon}\psi L\psi \qquad (4.266)$$

and contains only two independent renormalization constants Z.

These arguments do not pertain to the case of so-called *exceptional dimensions*, which are defined by the same criterion (4.265), viewed as an equation for d. The range natural for the present problem (Sec. 2.19) $2 < d < 4$ contains only the following series ($m = n_\psi = 0$, $n_\phi = 2s$) of exceptional dimensions:

$$d_s = 2s/(s-1), \qquad s = 3, 4, 5, ... \qquad (4.267)$$

For $d = d_s$ an additional logarithmic divergence arises in 1-irreducible functions of the type ϕ^{2s}, requiring the introduction of the corresponding O_n-scalar local counterterm. In particular, the real dimension $d = 3$ is an exceptional one; here an additional logarithmic divergence appears in the six-tail graph involving the principal field. We note that the regularization (4.264) violates the constraint $\phi^2 =$const [which takes the form $\phi^2 = 0$ in the formal model (4.258) without dimensional parameters], and so the counterterms $\sim \phi^{2s} \equiv (\phi^2)^s$ cannot be eliminated by simply applying the constraint. The constraint is restored only in the renormalized Green functions after the regularization is removed [126].

All the counterterms and constants Z are constructed as series in $1/n$. Their form depends on the chosen subtraction operator (Sec. 3.10), in particular, in the standard MS scheme,

$$Z = 1 + \sum_{k=1}^{\infty} n^{-k} \sum_{m=1}^{k} \varepsilon^{-m} c_{km}(d). \tag{4.268}$$

The renormalized Green functions are defined as sums of the basic graphs with selected (by the subtraction scheme) R operator according to the general rules of Ch. 3. The constants Z in (4.266) can be expressed in terms of the counterterms of the basic graphs by relations like (3.107).

A special feature of the massless model under consideration is the fact that the renormalization is nonmultiplicative: there is no unrenormalized action without Z and μ related to the functional (4.266) by the standard expression of multiplicative renormalization (1.84), and this is even more true for exceptional dimensions (4.267) requiring the introduction of new counterterms $\sim \phi^{2s}$.

When the constants $Z_{1,2}$ analogous to (4.266) are included in the formal action (4.258), the renormalization appears multiplicative (the dilatation of two fields), but without regularization all the constructions are meaningless. The regularization spoils the exact cancellation of the two nonlocal contributions in (4.258). In (4.266) they are not renormalized in accordance with the general theory (Sec. 3.13), which leads to the renormalization being nonmultiplicative. The renormalization becomes multiplicative if two additional parameters are introduced into the model after inserting independent coefficients in front of the nonlocal contributions in the action (4.264). This is discussed in more detail in Sec. 4.32.

In conclusion, let us return to the problem of exceptional dimensions (4.267). The need to introduce a new counterterm for $d = d_s$ naturally gives rise to the question of whether or not the renormalized Green functions and their various characteristics, in particular, the critical exponents, are continuous in d at the points $d = d_s$. Let us explain by an example. The critical dimensions of fields for $d = 3$ can be found from the renormalized correlators calculated by two methods: either directly in $d = 3$ with an additional counterterm for subgraphs of the ϕ^6 type, or from the action (4.266) for an arbitrary nonexceptional dimension $d \neq 3$, followed by taking the limit $d \to 3$. The question is whether the results obtained in these two ways coincide.

The answer we propose is the following. The first procedure ($d = d_s$, $\varepsilon \to 0$) is the rigorous definition of the renormalized Green functions themselves for $d = d_s$, because for the second procedure (first $d \neq d_s$, $\varepsilon \to 0$, then $d \to d_s$) the limit $d = d_s$ may not exist owing to singularities in d at the points (4.267). However, these singularities can be removed by a suitable transformation of the UV-finite (in the sense of there being no poles in ε) renormalization, which has no effect (Sec. 1.18) on the

universal characteristics of the critical behavior — the exponents and normalized scaling functions. Such objects must therefore be continuous in d
at the points (4.267) and can be calculated by any method, including continuation in d from nonexceptional dimensions on the basis of the action
(4.266) without additional counterterms. As justification for this it can be
noted that the universal characteristics (which are the only thing of objective
value) must be independent of the method by which the model is regularized
and renormalized and are certainly continuous in d for cutoff regularization
(Sec. 1.19). We also note that UV divergences from subgraphs of the type ϕ^{2s}
for $d = d_s$ in the simplest Green functions (propagators and vertices) appear
only in high orders in $1/n$ that are presently inaccessible, so that for now
this problem is academic. Actually, all calculations are performed for arbitrary nonexceptional dimension, and the results have no singularities in d in
exceptional dimensions.

4.32   Critical scaling. Calculation of the critical dimensions from the Green functions

When discussing the Green functions of the model (4.266), in what follows
we shall assume, unless stated otherwise, that the $\varepsilon \to 0$ limit has already
been taken in them (we recall that for us ε is only an auxiliary regularizer, in
contrast to the $4 - \varepsilon$ or $2 + \varepsilon$ schemes).

The renormalized Green functions of the massless model (4.266) depend
only on the external momenta (or coordinates) and the renormalization
mass μ, and do not contain a free parameter like the renormalized
charge g, in contrast to usual models of the φ^4 type. Physically they describe
the IR-asymptote of the Green functions of the O_n φ^4 model or, equivalently,
the σ model in the $1/n$ expansion, so that they themselves, and not only
their IR asymptotes, must possess the property of critical scale invariance
(critical scaling). This implies, in particular, that the exact field correlators
must have the form (4.259) but with different amplitudes and the exact critical dimensions of the fields instead of the canonical ones as the exponents α
and β.

The presence of scaling in the renormalized functions is obvious from physical considerations, but rather complicated to prove directly within the model
(4.266) itself owing to its special features like the nonmultiplicative nature of
the renormalization and the absence of the parameter g. It is not possible to
use the standard method (Sec. 1.24) to obtain the RG equation, from which
critical scaling automatically follows at the IR asymptote (Sec. 1.33). It is also
impossible to obtain any formulas expressing the desired critical dimensions
in terms of the constants Z in (4.266). The reason is that formulas of this
type in ordinary models always contain not only the constants Z themselves,
but also their derivatives with respect to g, and here we have no such parameter. In this respect the model (4.266) is similar to the φ^4 model studied only

in the regime of critical scaling, i.e., at the single point $g = g_*$, and not for arbitrary g. Such a formulation of the φ^4 model would have similar problems.

The proof of scaling in the model (4.266) was given in [126] on the basis of an "extended" model in which additional independent coefficients u and v are introduced for the two nonlocal contributions in (4.264) and (4.266). The model then becomes multiplicatively renormalizable, and the standard technique of Ch. 1 is applicable to it. In this language the problem is that the RG operator contains, among other things, derivatives with respect to u and v, so that the RG equation for the Green functions of the extended model does not automatically become the equation for the Green functions of the original model (4.266) for $u = v = 1$. It was shown in [126] that for some subtraction scheme (not MS) the contributions involving derivatives of the Green functions with respect to u and v vanish at the point $u = v = 1$, which

leads to the equation of critical scaling for the Green functions of the model (4.266) itself. Since the scaling is a universal property (independent of what sort of renormalization is used), it will occur for any subtraction scheme, including the MS one.

In this manner we prove the existence of critical scaling in the model (4.266) and obtain explicit expressions, analogous to those of Sec. 1.25, for the desired anomalous dimensions in terms of the renormalization constants Z of the extended model and their derivatives with respect to u and v at the point $u = v = 1$. The derivatives do not drop out of these expressions, and so to use them in practice it is necessary to calculate the constants Z of the extended model, and, moreover, not in the most convenient (not MS) subtraction scheme. No infinite summations of graphs with the insertions (2.173) in ψ lines are required, because it is sufficient to calculate the constants Z only to first order in the deviations of u and v from unity, since the results involve only the first derivatives of the Z at the point $u = v = 1$.

Working with the extended model is useful for proving the existence of scaling, but if scaling has already somehow been established it is easier to calculate the desired critical dimensions not in terms of the constants Z of the extended model, but directly in terms of the Green functions of the original model (4.266).

For example, let $\Gamma_{\mathrm{R}} = \Gamma_{\mathrm{R}}^{(0)}\overline{\Gamma}_{\mathrm{R}}$ be a renormalized 1-irreducible Green function depending on a single external momentum p (the inverse propagator or a vertex with a single zero external momentum) with definite critical dimension $\Delta[\Gamma] = d[\Gamma] + \gamma^*[\Gamma]$ (canonical plus anomalous) and $\Gamma_{\mathrm{R}}^{(0)}$ be its zeroth-order (in $1/n$) approximation with critical dimension $d[\Gamma]$. By construction, the $1/n$ expansion of the function $\overline{\Gamma}$ begins at unity and for it $d[\overline{\Gamma}] = 0$ and $\Delta[\overline{\Gamma}] = \gamma^*[\Gamma]$, so that

$$\overline{\Gamma}_{\mathrm{R}}(p,\mu) = A(p/\mu)^{\gamma^*[\Gamma]}, \qquad (\Gamma_{\mathrm{R}} = \Gamma_{\mathrm{R}}^{(0)}\overline{\Gamma}_{\mathrm{R}}). \tag{4.269}$$

If we manage to calculate the function $\overline{\Gamma}_{\mathrm{R}}$ in some order in $1/n$ (from the basic graphs with the R operator), we can directly verify the validity of the representation (4.269) in this order and find the parameters A and $\gamma^*[\Gamma]$ entering into it. If scaling has been proved, there is no need to worry about it, and to determine the parameters themselves it is sufficient to calculate the coefficients of the zeroth and first powers of the logarithm in the expansion of $\overline{\Gamma}_R$ in powers of $\ln(p/\mu)$.

A representation analogous to (4.269) can also be written down for Green functions of the type (3.158) with an arbitrary system $\widehat{F}_R = \{\widehat{F}_{iR}\}$ of composite operators that mix in renormalization (Sec. 3.28). The final result of the calculation will be the matrix γ_F^* of their anomalous dimensions in (4.43). To completely describe the renormalization of this system of operators in the presence of mixing it is desirable to also calculate the corresponding matrix Q from (3.176). It should be noted that the nonmultiplicative nature of the renormalization of the model (4.266) makes the concept of unrenormalized field $\widehat{\varphi}$ and all the expressions of Ch. 3 involving it meaningless, including the standard expressions of multiplicative renormalization (3.183) for composite operators that mix. Only the expressions involving renormalized and basic objects remain meaningful, in particular, the fundamental relation (3.176) determining the functional form of the renormalized operators.

The direct method of calculating dimensions described above is applicable not only to the simple σ model, but also to various generalizations of it like those considered in Sec. 2.28. Theoretically, it permits the calculation of the critical dimensions in any order in $1/n$, but in practice it has never been possible to go beyond first order using this method owing purely to technical difficulties. Nevertheless, the method is universal, in contrast to the special technique used to calculate higher orders of η and ν in the σ model (more on this below).

4.33 Calculation of the dimensions of fields and composite operators using counterterms of graphs in first order in $1/n$

Let us discuss in more detail the method of calculating anomalous dimensions γ^* in lowest (first) order in $1/n$. In ordinary models of the φ^4 type it is preferable to calculate γ^* using not the Green functions, but the renormalization constants Z, since they are simpler objects not involving momentum variables. This is impossible in the model (4.266), and so we are forced to deal with the Green functions. However, in first order in $1/n$ this calculation turns out not to be more complicated than the usual calculations in terms of Z, since to obtain the results it is sufficient to find the residues of the poles in ε (divergent subgraphs and higher-order poles in ε do not occur in lowest order) of all the individual graphs, which involves an amount of work equal to that to calculate the constants Z.

Let us explain this first for the simple example of a function like (4.269). Its $1/n$ expansion begins with unity, and so

$$\overline{\Gamma}_R = 1 + \frac{1}{n}\gamma_1^*[\Gamma]\ln(p/\mu) + \frac{1}{n}A_1 + O(1/n^2) \qquad (4.270)$$

with the usual (Sec.4.29) meaning of the subscript 1. For an individual basic graph $\gamma \in \Gamma$ of first order in $1/n$ we have

$$\overline{\gamma}_B = (\mu/p)^{2\varepsilon m_\gamma}[v_\gamma/\varepsilon + ...], \qquad (4.271)$$

where the constant v_γ is the residue at the ε pole, the ellipsis denotes the nonpole contribution, and $m_\gamma = 1, 2, ...$ is the regularizer multiplicity, the number of internal ψ lines in γ containing regularizers ε. All the vertices γ contain factors of μ^ε, but some can go into the zeroth approximation $\Gamma^{(0)}$, while the remaining ones make the factors $x^{2\varepsilon} \sim p^{-2\varepsilon}$ of the regularized ψ lines (4.262) dimensionless. The counterterm $L\overline{\gamma}_B$ of a graph (4.271) is the quantity v_γ/ε (MS scheme), and $R\overline{\gamma}_B = \overline{\gamma}_B - L\overline{\gamma}_B$, so that

$$\overline{\gamma}_R = \lim_{\varepsilon\to 0} R\overline{\gamma}_B = \lim_{\varepsilon\to 0}[(\mu/p)^{2\varepsilon m_\gamma} - 1]v_\gamma/\varepsilon + ... = -2m_\gamma v_\gamma \ln(p/\mu) + ..., \quad (4.272)$$

where the ellipsis denotes an unimportant constant.

Therefore, the logarithms are generated by only the pole contributions of the basic graphs γ, while the coefficient of the logarithm in (4.270) is determined by the residues v_γ at the poles. Combining (4.272) with the symmetry coefficients c_γ and comparing with (4.270), in first order in $1/n$ we find the anomalous dimension $\gamma^*[\Gamma] \simeq \gamma_1^*[\Gamma]/n$:

$$\gamma^*[\Gamma] = -2\sum_\gamma c_\gamma m_\gamma v_\gamma. \qquad (4.273)$$

This is the final result for a function like (4.269). The sum in (4.273) differs from the analogous one in the counterterm $L\overline{\Gamma}_B = \varepsilon^{-1}\sum c_\gamma v_\gamma$ by additional factors of m_γ, and so one sum cannot be reconstructed from the other. However, in calculating the counterterms it is still necessary to calculate the residues v_γ of the individual graphs, and the regularizer multiplicity m_γ for each is easily found from the form of the graph, so that calculating (4.273) is just as complicated as calculating the counterterm.

Let us now consider an arbitrary system of local monomials that mix under renormalization $F_i \equiv F_i(x; \varphi)$. We shall use the standard notation of Ch. 3, always understanding φ to be the set of all the fields of the model; here, $\varphi \equiv \psi, \phi$. The number of fields n_φ is then a multiple index, and expressions like $n_\varphi d_\varphi$ are sums over all types of fields. Owing to the absence of dimensional parameters, in our model the only operators that can mix under renormalization are the ones with canonical dimensions $d_i \equiv d[F_i]$ differing only by quantities of order ε, i.e., operators, which without regularization, have the same value of d_F.

Turning now to the derivation of the matrix analog of (4.273), we first note that it is important to choose a convenient initial object. The function $\overline{\Gamma}_R$ in (4.269) was convenient because its $1/n$ expansion begins with unity, so that the desired γ^* is uniquely determined by the coefficient of the logarithm. A convenient object analogous to $\overline{\Gamma}$ for composite operators is the generating functional $\Gamma_i(x; \varphi)$ of 1-irreducible Green functions with a single operator $F \equiv F_i(x; \varphi)$ and any number of simple fields φ, which is defined by (3.160) followed by selection of the 1-irreducible part. In massless perturbation theory in $1/n$ we have

$$\Gamma_{i\mathrm{B}}(x; \varphi) = F_i(x; \varphi) + \sum_{\gamma} c_{\gamma} \gamma_{i\mathrm{B}}(x; \varphi) + ... \qquad (4.274)$$

The summation runs over all the basic graphs γ of order $1/n$, c_{γ} are their symmetry coefficients in the functional (3.160), and the ellipsis denotes terms of higher order in $1/n$ that we do not need. In general, the functional (3.160) contains, in addition to the F_i, other 1-irreducible contributions of zeroth order obtained by combining pairs to form the contracted lines of the fields involved in F_i. We have no such terms in (4.274), because contributions with contracted massless lines are assumed to be zero (Sec. 3.19).

When (4.274) is substituted into the definition of the renormalized functional $\Gamma_{i\mathrm{R}}(x; \varphi) = R\Gamma_{i\mathrm{B}}(x; \varphi)$, the R operator gives unity on the contribution F_i and $1 - L$ on the contributions γ (there are no divergent subgraphs), so that

$$\Gamma_{i\mathrm{R}}(x; \varphi) = F_i(x; \varphi) + \sum_{\gamma} c_{\gamma}[1 - L]\gamma_{i\mathrm{B}}(x; \varphi) + ... \qquad (4.275)$$

The operator counterterm $L\gamma_{i\mathrm{B}}(x; \varphi)$ of any graph is a linear combination of monomials F (Sec. 3.28). For the graphs of first order in $1/n$ in the MS scheme we have

$$L\gamma_{i\mathrm{B}}(x; \varphi) = \frac{1}{\varepsilon} \sum_{k} \mu^{d_i - d_k} v_{\gamma}^{ik} F_k(x; \varphi), \qquad (4.276)$$

where the numbers v_{γ}^{ik} are the residues at the ε poles, and the factor $\mu^{d_i - d_k}$ cancels the possible small (of order ε) difference of the canonical dimensions of the operators in the regularized theory [it is impossible to take the $\varepsilon = 0$ limit in (4.276)].

As before, the logarithms of interest in the answers will be generated only by the pole (in ε) contributions of the graphs. In the graphs $\gamma_{i\mathrm{B}}(x; \varphi)$ themselves these contributions differ from the counterterms (4.276) by only additional factors of $(\mu x)^{2\varepsilon m_{\gamma}}$, where m_{γ} are the regularizer multiplicities, as in (4.273). Let us briefly explain the origin of the factors $(\mu x)^{\varepsilon}$. The calculations of the operator counterterms are performed in momentum space (Sec. 3.26). In general, along with a pole $1/\varepsilon$ there are several factors $(\mu/p)^{\varepsilon}$ and some polynomial in the momenta. In coordinate space the momenta in the polynomial become partials ∂ and are included in the operators F_k, whose coefficients retain only the coordinate image of the factors $(\mu/p)^\varepsilon$. The choice of the momentum p in them is arbitrary, since a change of p generates only additions of order ε which cancel the pole in ε. It is convenient to choose p to be the momentum flowing into the vertex of the composite operator (Sec. 3.26). Then in coordinate space p becomes differentiation $\partial \equiv \partial/\partial x$ of the operators $F_k(x; \varphi)$ with respect to x: $(\mu/p)^\varepsilon \to (\mu/\partial)^\varepsilon$. The last equation can be rewritten as $(\mu x)^\varepsilon \cdot (x\partial)^{-\varepsilon}$ and then the second factor discarded, as it, in contrast to the first, has zero critical dimension and therefore does not generate interesting logarithms that affect scale transformations (see below).

From this and (4.276) we then have

$$[1 - L]\gamma_{i\mathrm{B}}(x; \varphi) = \frac{1}{\varepsilon} \sum_k \mu^{d_i - d_k} [(\mu x)^{2\varepsilon m_\gamma} - 1] v_\gamma^{ik} F_k(x; \varphi) + ..., \qquad (4.277)$$

which, after substitution into (4.275) and taking the limit $\varepsilon = 0$, gives

$$\Gamma_{i\mathrm{R}}(x; \varphi) = F_i(x; \varphi) + 2\ln(\mu x) \sum_k u^{ik} F_k(x; \varphi) + ..., \qquad (4.278)$$

where the dots denote unimportant contributions not containing $\ln(\mu x)$ and

$$u^{ik} = \sum_\gamma c_\gamma m_\gamma v_\gamma^{ik}. \qquad (4.279)$$

With our accuracy ($1/n$ in u), (4.278) can be rewritten compactly as

$$\Gamma_{\mathrm{R}}(x; \varphi) = (\mu x)^{2u} F(x; \varphi) + ..., \qquad (4.280)$$

where Γ and F are understood as vectors in the labels i, k and the objects acting on them as matrices in these indices.

We still need to relate the exponent in (4.280) to the desired matrix of critical dimensions Δ_F of the type (4.43). It follows from the equation of critical scaling (4.42) (now without contributions involving $\mathcal{D}_e$) for renormalized composite operators and its analog for simple fields that critical scaling is expressed in terms of the renormalized functional $\Gamma_{\mathrm{R}}(x; \varphi)$ by an equation analogous to (1.45):

$$\Gamma_{\mathrm{R}}(\lambda^{-1}x; \varphi_\lambda) = \lambda^{\Delta_F} \Gamma_{\mathrm{R}}(x; \varphi), \qquad \varphi_\lambda(x) \equiv \lambda^{\Delta_\varphi} \varphi(\lambda x), \qquad (4.281)$$

in which $\lambda > 0$ is the arbitrary parameter of the scale transformation and $\Delta_{\varphi, F}$ are the critical dimensions of the corresponding quantities (a matrix for F). It is easily verified directly that simple monomials $F(x, \varphi)$ then transform as

$$F_i(\lambda^{-1}x; \varphi_\lambda) = \lambda^{d_F + n_\varphi^{(i)} \gamma_\varphi^*} F_i(x; \varphi), \qquad \Delta_\varphi = d_\varphi + \gamma_\varphi^*, \qquad (4.282)$$

where $d_{\varphi, F}$ are the canonical dimensions (we assume $\varepsilon = 0$, so that d_F is the same for all F_i), and $n_\varphi^{(i)}$ is the number of factors φ in the monomial F_i

(a multiple index for a set of fields, in which case expressions like $n_\varphi^{(i)} \gamma_\varphi^*$ are understood as sums over all types of fields). To explain (4.282) we note that $d_F - n_\varphi^{(i)} d_\varphi$ is equal to the number of symbols ∂ in F_i.

Let us perform the scale transformation (4.281) in (4.280) and then use (4.282) and again (4.280) (here the noncommutativity of the matrices does not present a problem, since the only zeroth-order contribution in the exponent is a multiple of the unit matrix d_F, and the noncommutativity of the contributions of order $1/n$ can be neglected with our accuracy). Finally, we obtain

$$\Gamma_R(\lambda^{-1}x; \varphi_\lambda) = \lambda^{d_F + n_\varphi \gamma_\varphi^* - 2u} \Gamma_R(x; \varphi), \qquad (4.283)$$

which when compared to (4.281) gives the desired relation between the matrices (4.279) and (4.43):

$$\Delta_F = d_F + \gamma_F^* = d_F + n_\varphi \gamma_\varphi^* - 2u. \qquad (4.284)$$

Here n_φ is understood as a diagonal matrix with elements $n_\varphi^{ik} = \delta^{ik} \cdot n_\varphi^{(k)}$ (for convenience, we write the matrix indices as superscripts), and $n_\varphi \gamma_\varphi^*$ is the sum of contributions from all types of fields. In detailed notation we have

$$(\gamma_F^*)^{ik} = -2u^{ik} + \delta^{ik} \sum_\varphi n_\varphi^{(k)} \gamma_\varphi^*. \qquad (4.285)$$

This is the final answer and the analog of (4.273). We recall that $n_\varphi^{(k)}$ is the number of factors φ in the monomial F_k (in general, a multiple index), γ_φ^* are the anomalous dimensions of the fields themselves, and the matrix u is expressed in terms of the counterterm contributions of graphs of order $1/n$ by (4.279).

To complete the description of the renormalization it is desirable to also calculate the matrix Q in (3.176) determining the functional form $[F_i(x; \varphi)]_R$ of the renormalized operators. The matrix Q is nonuniversal (it depends on the choice of subtraction scheme), but it still contains some objective information: it tells which operators mix under renormalization. In our case, from the definition (3.176) and the equations (4.274) and (4.276) we have

$$[F_i(x; \varphi)]_R = F_i(x; \varphi) - \sum_k q^{ik} F_k(x; \varphi), \qquad (4.286)$$

$$q^{ik} = \frac{1}{\varepsilon} \sum_\gamma c_\gamma v_\gamma^{ik}, \qquad Q^{ik} = \delta^{ik} - q^{ik} \qquad (4.287)$$

with all quantities having the same meaning as in (4.279).

We recall that the renormalized composite operators whose critical dimensions were discussed above are determined from the functional (4.286) by the second equation in (3.184), while the first equation becomes meaningless in our model. The critical dimensions themselves are the eigenvalues of the matrix Δ_F, and it is the corresponding basis operators (4.44) that have these dimensions, as discussed in Sec. 4.7.

Let us make one final remark. Since only divergent graphs contribute to the answer, in (4.279) and (4.287) it is necessary to sum only over the SD graphs (Sec. 3.25) of the functional (3.160), and not over all its 1-irreducible graphs γ of first order.

4.34 Examples

Let us give some examples of calculations in first order in $1/n$, beginning with the dimensions of the fields $\varphi = \phi, \psi$ themselves. They can be found from the two 1-irreducible functions $\Gamma_{\varphi\varphi} = -D_\varphi^{-1}$ of the fields ϕ, ψ, but it is more convenient to use $\Gamma_{\phi\phi}$ and the vertex $\Gamma_{\phi\phi\psi}$. It follows from the formulas for critical scaling (Sec. 1.33) that up to a sign the anomalous dimension of an arbitrary Green function is equal to the sum of the anomalous dimensions of the fields entering into it. For full or connected functions the field dimensions enter with a plus sign, and for 1-irreducible ones they enter with a minus sign. Therefore,

$$\gamma^*[\Gamma] = -\sum_\varphi n_\varphi \gamma_\varphi^*, \quad \gamma^*[\Gamma_{\phi\phi}] = -2\gamma_\phi^*, \quad \gamma^*[\Gamma_{\phi\phi\psi}] = -2\gamma_\phi^* - \gamma_\psi^*. \quad (4.288)$$

We have given the general expression and the special cases we shall need.

In first order in $1/n$ the functions Γ of the model (4.264) that we are interested in are represented by the graphs (4.289) with the indicated coefficients (all are unity):

$$\Gamma_{\phi\phi} = -p^2 + \quad \raisebox{-0.5ex}{\includegraphics[height=1em]{g1}} \quad ; \; \Gamma_{\phi\phi\psi} = \quad \raisebox{-0.5ex}{\includegraphics[height=1em]{g2}} + \quad \raisebox{-0.5ex}{\includegraphics[height=1em]{g3}} + \quad \raisebox{-0.5ex}{\includegraphics[height=1em]{g4}} . \quad (4.289)$$

The lines of the graphs are associated with the propagators (4.259) with amplitudes known from (4.261) and indices (terminology of Sec. 3.21) $\alpha = d/2-1$ for ϕ lines and $\beta - \varepsilon$ with $\beta = 2$ for regularized ψ lines. All vertices carry a factor of μ^ε. The zeroth approximation $\Gamma^{(0)}$ is $-p^2$ for $\Gamma_{\phi\phi}$ and μ^ε for $\Gamma_{\phi\phi\psi}$.

Let us begin with the calculation of γ_ϕ^*. The only graph of $\Gamma_{\phi\phi}$ in coordinate space is a simple product of lines, and so the entire calculation amounts to a single Fourier transform according to the rule (3.72), giving (using the notation of Sec. 3.15 everywhere)

$$\Gamma \equiv \Gamma_{\phi\phi} = -p^2 + \mu^{2\varepsilon} AB\pi^{d/2} H(\alpha + \beta - \varepsilon)(2/p)^{d-2\alpha-2\beta+2\varepsilon}. \quad (4.290)$$

A pole in ε is contained in the function $H(\alpha + \beta - \varepsilon) = H(d/2 + 1 - \varepsilon) = \| - 1 + \varepsilon \| / \| d/2 + 1 - \varepsilon \|$. Cancelling it with the corresponding counterterm $\sim p^2/\varepsilon$ and then passing to the limit $\varepsilon = 0$, for the renormalized function (4.290) we obtain the expression $\Gamma_R = -p^2[1 - 2a \ln(p/\mu)]$ with coefficient

$a = \pi^{d/2}AB/4\|d/2+1\|$. Comparison of this result with (4.270) and (4.288) shows that $a = \gamma^*_\phi$ in our approximation. Substituting the known values into a (see above), we obtain the result (4.255a) for the exponent $\eta = 2\gamma^*_\phi$ in order $1/n$.

Let us now consider the vertex function (4.289). Its nontrivial graphs $\gamma_{1,2}$ (numbered in the order they occur) diverge logarithmically. Their countert- erms are easily calculated with vertical momentum flow using (3.123)–(3.125). The amplitude factors in γ_1 are grouped together to form the combination $A^2B \equiv u$, and in γ_2 to form $nA^4B^2 = nu^2$, where

$$u \equiv A^2B = -\frac{2H(d/2-2,2-d/2)}{\pi^d n} = \frac{\|d/2+1\|\|d/2-1\|}{2\pi^d} \cdot \frac{\eta_1}{n}. \qquad (4.291)$$

After separating out these amplitudes and the single factor of μ^ε, we are left with the canonically dimensionless graphs $\overline{\gamma}_{1,2}$ with unit amplitudes of all the lines. The calculation can be simplified using the following very useful formula for the pole contribution (in the Fourier transform) defining the counterterm. It is the formula for an arbitrary *logarithmic n-hedron* ($\forall n \geq 2$) with links having indices $\alpha_1...\alpha_n$ and all vertices and lines having unit amplitudes:

$$\sum_{i=1}^{n}\alpha_i = \frac{(n-1)d}{2} - \Delta \quad \Rightarrow \quad \textcircled{n} = \frac{\pi^{(n-1)d/2}H(\alpha_1...\alpha_n)}{\Delta\|d/2\|} + ..., \qquad (4.292)$$

where Δ is the total small regularizer and the dots stand for the finite contri- butions. The logarithmicity condition is the first equation in (4.292), and the total regularizer Δ is the sum of the individual regularizers Δ_i contained in the indices α_i of all or some of the links. The Δ_i are assumed to be the quan- tities of the same order and sign and the ratios Δ_i/Δ are finite, and so the contributions Δ_i in the arguments of $H(...)$ in (4.292) can be neglected when isolating the pole. At an n-hedron there is understood to be a factor $\mu^{2\Delta}$ ensuring that the entire expression is dimensionless. In dimensionless graphs $\overline{\gamma}$ such factors are formed automatically from the required number of μ^ε at the vertices, and the regularizers ε of all the ψ lines play the role of the Δ_i.

Equation (4.292) is valid only for graphs with a simple pole, i.e., with- out subdivergences. In this regard only links with index $\alpha = d/2 + m$, $m = 0, 1, 2, ...$ (up to regularizers), are dangerous in the n-hedron. There must be no such links in (4.292). Finally, when using (4.292) there is no need to worry about IR divergences and the choice of momentum flow (Sec. 3.19); it can be assumed, for example, that all the external momenta are zero and that the graph is regularized by masses in some lines.

The first vertex graph γ_1 in (4.289) is a logarithmic triangle with $\Delta = \varepsilon$, and so its counterterm $L\overline{\gamma}_1$ is found immediately from the rule (4.292) taking into account (4.291): $L\overline{\gamma}_1 = u\pi^d H(\alpha,\alpha,\beta)/\varepsilon\|d/2\|$. The second ver- tex graph γ_2 reduces to a logarithmic square with $\Delta = 2\varepsilon$ by single in- tegration according to the rule (3.124) of the leftmost chain, and we find $L\overline{\gamma}_2 = nu^2\pi^{2d}H(\alpha,\alpha,\beta,\beta,\beta,\alpha,3\alpha-d/2)/2\varepsilon\|d/2\|$. The first three arguments of H come from integration of the chain with links α, α, and the last four from the resulting square. Substituting in the known quantities, we obtain the following final expressions for the counterterms $L\overline{\gamma}$ of the vertex graphs (4.289) in standard units of η_1/n:

$$
L\left[\cdots \!\!\!\!<\!\!\!| \right] = -\frac{d}{2(4-d)\varepsilon} \; ; \; L\left[\cdots \!\!\!\!<\!\!\!\!| \right] = \frac{d(3-d)}{2(4-d)\varepsilon} \; . \tag{4.293}
$$

The residues $v_\gamma \equiv v_{1,2}$ in (4.273) for the given function Γ are thereby calculated, and from (4.289) we find $c_1 = c_2 = 1$ for the symmetry coefficients and $m_1 = 1$, $m_2 = 2$ for the regularizer multiplicities (number of ψ lines). Substituting all this into (4.273), we obtain $\gamma^*[\Gamma_{\phi\phi\psi}] = -2v_1 - 4v_2 = d(2d-5)/(4-d)$ in the standard units, which allows us to find γ_ψ^* from the last equation in (4.288) for the already known $\gamma_\phi^* = \eta/2$. In the end we find that

$$
\gamma_\phi^* = \frac{1}{2} \cdot \frac{\eta_1}{n}, \qquad \gamma_\psi^* = -\frac{2(d-1)(d-2)}{(4-d)} \cdot \frac{\eta_1}{n} \tag{4.294}
$$

with η_1 from (4.255a). The result (4.294) for γ_ψ^* is equivalent to (4.255b) for v_1. To make the comparison easier, let us give in one place the basic definitions and relations between the exponents in the perturbation theory in $1/n$:

$$
\Delta_\varphi = d_\varphi + \gamma_\varphi^*, \quad \varphi \equiv \phi, \psi; \quad d_\phi = d/2 - 1, \quad d_\psi = 2;
$$
$$
\eta = 2\gamma_\phi^*; \quad 1/\nu \equiv \Delta_\tau = d - \Delta_\psi = d - 2 - \gamma_\psi^*, \quad \omega = \Delta[\psi^2] - d. \tag{4.295}
$$

The critical dimension Δ_F of a given physical quantity F is an objective (unique) concept, but its splitting into canonical and anomalous contributions $\Delta_F = d_F + \gamma_F^*$ is not unique, because in general it depends on the choice of perturbation theory (if d_F is defined in the usual manner as the zeroth approximation of Δ_F in a given perturbation theory). For example, for $\Delta_\tau = d_\tau + \gamma_\tau^*$ in the usual perturbation theory of the φ^4 model we have $d_\tau = 2$, $\gamma_\tau^* = O(4-d)$, whereas in the $1/n$ perturbation theory it follows from (4.295) that $d_\tau = d - 2$, $\gamma_\tau^* = O(1/n)$.

Let us now consider several examples involving composite operators. For our dimensions d_φ, coincidence of the d_F for two operators (the necessary condition for mixing under renormalization) is possible only in two cases: (1) when the operators are related by a simple regrouping of the symbols ∂ or the O_n indices, for example, $\phi\partial^2\phi$ and $\partial^2\phi^2$ or $\phi_a\phi_b$ and $\delta_{ab}\phi^2$; (2) when they are related by the replacement of one factor of ψ by two ∂, for example, ψ^2 and $\partial^2\psi$.

For some operators it is immediately clear that they do not mix with anything, and so the renormalization is multiplicative. This will be the case, for example, for the family of O_n irreducible symmetric tensors $F = \phi^l$ discussed in Sec. 4.28 (where $\phi_a \to \varphi_a$). (Here the first of the $F = \phi^l$ is the field ϕ_a

itself, the second is $\phi_a\phi_b - \delta_{ab}\phi^2/n$, and so on, and ϕ^l is the conventional notation.) Owing to the irreducibility and absence of symbols ∂, these operators do not mix with anything. Of course, this does not prevent us from using the expressions of Sec. 4.33 for the calculations, assuming that all matrices are trivial (1×1).

Let us present the calculation of $\gamma^*[\phi^l]$. The following graphs are SD graphs of the functional (3.160) with operator $F(x;\varphi) = \phi^l$ in first order in $1/n$ that generate counterterms:

$$\Gamma_{\mathrm{B}}(x;\varphi) \;=\; F(x;\varphi) + \frac{1}{2}\; \vartriangleleft + \frac{1}{2}\; \vartriangleleft\!\!\rceil . \tag{4.296}$$

They are similar to the vertex graphs (4.289), but now the external lines are associated with the corresponding (to the type of line) factors $\varphi \equiv \phi, \psi$ (in Sec. 3.26 they were shown as tails, but the graphical notation (4.296) can also be used, if it is interpreted appropriately). The large dot in (4.296) stands for the vertex of a composite operator, which corresponds to a standard vertex factor of the type (2.21), i.e., the derivative $\delta^k F/\delta\varphi^k$, where k is the number (in general, a multiple index) of lines attached to a given vertex [in our model $k \equiv k_\phi, k_\psi$ and $(\delta/\delta\varphi)^k \equiv (\delta/\delta\phi)^{k_\phi}(\delta/\delta\psi)^{k_\psi}$]. The difference between the coefficients in (4.289) and (4.296) is due to the fact that the coefficients in the Green functions are obtained from the coefficients of the analogous graphs in the generating functional by multiplying by $n!$, where the number n (in general, a multiple index) is the multiplicity of external lines of the same type (Sec. 2.3).

For our operator the second graph in (4.296) actually does not contribute, since the ϕ lines contain the symbol δ_{ab} and the inclusion of the vertex F in a closed cycle of ϕ lines generates the contraction of a pair of indices in the vertex factor, which gives zero since F is irreducible.

Therefore, only the contribution of the first graph in (4.296) remains. Obviously, the index structure of the operator ϕ^l is reproduced in its counterterm, and this allows us to pass to the usual graph (4.289) with an additional combinatorial factor $l(l-1)$ from the vertex factor $\delta^2\phi^l/\delta\phi\delta\phi$. For an ordinary graph with unity at the vertex F the residue at the pole v_γ is known from (4.293). For it $m_\gamma = 1$ and $c_\gamma = 1/2$ in (4.296), which allows us to find the 1×1 matrix u directly from (4.279) and then the desired quantity $\gamma_F^* = \gamma^*[\phi^l]$ from the rule (4.285) with $n_\varphi\gamma_\varphi^* = l\gamma_\phi^* = l/2$ in the standard units. Finally, we obtain the following expression for the critical dimension of this operator:

$$\Delta[\phi^l] = l(d/2 - 1) + \frac{l(ld + 4 - 2d)}{2(4 - d)} \cdot \frac{\eta_1}{n} + O(1/n^2). \tag{4.297}$$

This answer agrees (Sec. 4.29) with the expression known from (4.251) for the anomalous dimenson of the given operator in the $2 + \varepsilon$ expansion (there $\phi \to \varphi$). The result (4.297) was obtained in [127] by a more complicated method using the Wilson operator expansion. The results of a similar calculation for a special class of operators of the type ϕ^l with additional symbols ∂ are also given there.

Let us now consider the family of operators $F = \psi^l(x)$. From power counting we can expect the counterterms of ψ^l to contain a set of other operators obtained from the original one by successive replacements $\psi \to \partial\partial$, in particular, two new independent operators $\psi^{l-2}\partial^2\psi$ and $\partial^2(\psi^{l-1})$ for a single replacement. Therefore, in general the calculation of the counterterms of ψ^l is difficult owing to the large (and growing with l) dimension of the matrices. However, in first order in $1/n$ everything simplifies, because in this order, as will be shown below, there is no $\psi \to \partial\partial$ mixing and the renormalization of ψ^l is simply multiplicative.

Turning now to the calculations, let us consider the SD graphs of the functional (3.160) for $F = \psi^l$ in the model (4.264):

$$\text{(a)} = \frac{1}{2} \; ; \quad \text{(b)} = \frac{1}{2} \; + \frac{1}{4} \; + \frac{1}{2} \; . \tag{4.298}$$

Here we give all the SD graphs of first order in $1/n$ with their symmetry coefficients. The notation is the same as in (4.296), and the large dot denotes the standard vertex factor, in the present case, $\delta^2\psi^l/\delta\psi\delta\psi = l(l-1)\psi^{l-2}$. The graphs of (4.298b) diverge logarithmically and generate multiplicative renormalization of ψ^l. The $\psi \to \partial\partial$ mixing in first order arises from only the single graph (4.298a), which from power counting is seen to diverge quadratically and corresponds to a counterterm $\sim \psi^{l-2}\partial^2\psi$. However, this graph actually does not generate a counterterm, since it does not contain a pole in ε due to its "accidental" cancellation by the zero of the residue. Let us explain in more detail. As in (4.296), the counterterm of the functional graph (4.298a) is determined by an ordinary graph of the same form which, according to the rule (3.134), reduces to a simple line with index $3\alpha + 2\beta - 2\varepsilon - d = d/2 + 1 - 2\varepsilon$ and a numerical coefficient Π. With our indices on the links in the graph (4.298a) this coefficient Π can be calculated exactly from (3.135) and turns out to be a multiple of ε (i.e., without regularization $\Pi = 0$). Therefore, in this case the pole in ε in the Fourier transform of the resulting line (the presence of which is predicted by power counting) is cancelled by the zero of the coefficient Π. The graph (4.298a) thus does not generate a counterterm, i.e., $\psi \to \partial\partial$ mixing is absent in first order (but it can arise in higher orders from new graphs).

The multiplicative renormalization of $F = \psi^l$ in first order in $1/n$ is generated by the functional SD graphs (4.298b). Their counterterms, as before, are determined by the ordinary graphs $\gamma_{1,2,3}$ (numbered in order) of the same form with an additional coefficient $l(l-1)$ from the vertex F. The counterterms $L\bar{\gamma}$ of the ordinary graphs (now $\bar{\gamma} = \gamma$) with unity at the vertex F are easily calculated by the method described above (reduction to a logarithmic n-hedron by the integration of simple chains), and in the standard units η_1/n turn out to be

$$L\bar{\gamma}_1 = \frac{d(3-d)}{4\varepsilon}; \quad L\bar{\gamma}_2 = \frac{d(3-d)(6-d)}{2(4-d)\varepsilon}; \quad L\bar{\gamma}_3 = -\frac{2d(3-d)^2}{3(4-d)\varepsilon}. \quad (4.299)$$

The residues of these poles with the additional coefficient $l(l-1)$ are $v_\gamma \equiv v_{1,2,3}$ in (4.279), and the coefficients c_γ and regularizer multiplicities m_γ of the graphs (4.298b) are known ($m_1 = m_2 = 2$, $m_3 = 3$). Using these data, from (4.279) we find the 1×1 matrix u, and from it using (4.285) with $n_\varphi \gamma_\varphi^* = l\gamma_\psi^*$ in the present case and the value of γ_ψ^* known from (4.294) we find the anomalous dimension of the operator in question. Finally, we obtain the following expression for the critical dimension:

$$\Delta[\psi^l] = 2l + \frac{l(d-1)[ld(d-3) - d^2 + d + 4]}{(4-d)} \cdot \frac{\eta_1}{n} + O(1/n^2). \quad (4.300)$$

From this in the special case $l = 2$ we obtain the result (4.255c) for the correction exponent $\omega = \Delta[\psi^2] - d$ [and, conversely, from the results in (4.294) and (4.255c) and the general form of the l dependence we could obtain the result (4.300) without calculation]. Here it should also be noted that when $\partial^2\psi$ is mixed with ψ^2 in higher orders the quantity $\Delta[\psi^2]$ in the expression for ω should be understood as the dimension not of ψ^2 itself, but of the operator (4.44) associated with it. This does not affect the justification of (2.204), since the contribution $\partial^2\psi$ vanishes inside the integral over x in the action functional.

The result (4.300) was obtained in [128]. The next most complicated scalar operators involving ψ and two ∂ (with contracted indices) were studied there. We can take $\psi^l\partial^2\psi$ and $\partial^2\psi^{l+1}$ to be the two independent operators of this type; the operator with two ∂ on different ψ reduces to these. The renormalization of $\partial^2\psi^{l+1}$ is determined by the renormalization of ψ^{l+1}, and so it is sufficient to restrict ourselves to the operator $F \equiv \psi^l\partial^2\psi$. Under renormalization it can mix with the operator with fewer ψ ($\psi \to \partial\partial$ mixing) generated by the graph (4.298a), the two operators with the same number of ψ (F itself and $\partial^2\psi^{l+1}$) from the graphs (4.298b), and the operator ψ^{l+2} ($\partial\partial \to \psi$ mixing) from the graphs

$$\frac{1}{2} \quad \cdots \quad + \frac{1}{2} \quad \cdots \quad + \quad \cdots \quad + \frac{1}{2} \quad \cdots \quad + \frac{1}{2} \quad \cdots \quad . \quad (4.301)$$

Also in this case $\psi \to \partial\partial$ mixing is absent: the graph (4.298a) for the new operator differs from that studied earlier by only the possible appearance of the operator ∂^2 acting on one of the ψ lines at the vertex of the composite operator, which increases its index by unity (i.e., $2 \to 3$ neglecting regularization). Instead of $\beta, \gamma = 2, 2$ in (3.135), we now obtain $\beta, \gamma = 2, 3$, but again in this case (and in general for any $\beta, \gamma = m, n$ with integer $n, m \geq 2$) Eq. (3.135) for the coefficient Π has a zero. We note that this proof does not generalize directly to operators with more than two ∂, and so the question of $\psi \to \partial\partial$ mixing remains open for them even in lowest order.

We do not have this problem, and since the matrices are triangular the new critical dimension associated with $F = \psi^l \partial^2 \psi$ is uniquely determined by the corresponding diagonal element. The others are needed only to calculate the coefficients of the admixtures $\partial^2 \psi^{l+1}$ and ψ^{l+2} in the corresponding exact operator (4.44) having the given dimension. Only the 2×2 block generated by the graphs (4.298b) was calculated explicitly in [128] (it involves the needed diagonal element), and the graphs (4.301) were not calculated. The following expression was obtained for the critical dimension of the operator associated with $\psi^l \partial^2 \psi$ ($l > 0$):

$$
\Delta[\psi^l \partial^2 \psi] = 2l + 4 + \frac{(d-1)}{6(4-d)} \Big\{ (d-3) \big[6dl^2 + 2l(d-8) +
$$

$$
(4.302)
$$

$$
+ d^2 - 2d - 24 \big] - 12(l+1)(d-2) \Big\} \cdot \frac{\eta_1}{n} + O(1/n^2).
$$

Here we conclude our discussion of the calculational technique for the model (4.266), only adding that the general statements of Secs. 4.11 and 3.29 remain valid for it, allowing the UV finiteness of operators like $\delta S_R / \delta \varphi$, $\varphi \delta S_R / \delta \varphi$, and so on to be proved without calculation, and their critical dimensions to be found.

This technique also works, with slight changes, for various generalizations of the simple σ model, including the CP^{n-1} and matrix σ models (Sec. 2.28). These two models contain, in addition to an auxiliary scalar field ψ, a vector field B with $d_B = 1$. The anomalous dimensions of all the fields [129] and all (two for CP^{n-1} and four for the matrix model) the principal correction exponents ω [130] are presently known, but only in first order in $1/n$. In the CP^{n-1} model this calculation requires the computation of the counterterms of another 23 graphs in addition to the 6 graphs of the simple σ model discussed above, while in the matrix model there are 14 more graphs. The results for the individual graphs are given as a table in [130], and can be used in analogous calculations for related models, just like the data of Table 3.1 above.

4.35 Calculation of the principal exponents using the self-consistency equations for the correlators

The procedure for calculating the $1/n$ expansions of the exponents in the σ model can be simplified considerably by using self-consistency equations of a different type. The simplest are the self-consistency equations for the exact correlators ("dressed lines") D, obtained from the usual Dyson equations (2.51). The self-energy Σ is represented by graphs with exact tadpoles $\alpha = \langle \hat{\varphi} \rangle$ but bare lines and vertices. It is well known (see, for example, [53]) that Σ is not changed if the bare lines Δ are replaced by dressed lines D, at the same time discarding all graphs with self-energy insertions, as their summation leads to the replacement $\Delta \to D$. We then obtain the Dyson equation in the form

$$D^{-1} = \Delta^{-1} - \Sigma(\alpha, D), \qquad (4.303)$$

which can be understood as a self-consistency equation for D. If we add to (4.303) the stationarity equation (1.51) with the same rearrangement $\Delta \to D$ in the graphs, we obtain a system of two equations determining the two unknowns α, D. This is true for any model; when there are several fields (4.303) becomes a matrix equation. The formal apparatus for the rigorous derivation of such skeleton equations and their generalizations with additional dressing of the vertices is the technique of higher Legendre functional transforms [53] (in this terminology, the transform studied in Sec. 1.11 is the first one; dressing lines requires the second one, dressing triple vertices the third one, and so on).

When calculating the principal exponents (the critical dimensions of the fields), (4.303) is written out directly for the critical (massless) theory. Then $\alpha = 0$ is used and the correlators D are sought in the form of functions with power-law IR asymptote. Here an important role is played by the idea of *discarding bare quantities*: the inverse bare correlator, being a simple polynomial in the momentum, cannot affect the procedure of making the desired fractional powers of the IR asymptote on the two sides of (4.303) consistent. It then follows that when calculating the exponents in the critical theory it is possible to take $\alpha = 0$ in (4.303) and discard the contribution of the bare quantity Δ^{-1}, which gives

$$D^{-1} = -\Sigma(0, D). \qquad (4.304)$$

For the σ model (4.258) in the simplest Hartree–Fock approximation, (4.304) for two fields $\varphi \equiv \phi, \psi$ takes the form

$$D_\phi^{-1} = - \;\overset{\frown}{\underset{\bullet\;\;\;\bullet}{}}\;; \quad D_\psi^{-1} = -\frac{n}{2} \; \overset{}{\underset{}{\bigcirc}} \; . \qquad (4.305)$$

The lines are now associated with the exact correlators (propagators) D_φ, and the vertices with unity [regularization of the type (4.262) is not required in the self-consistency equations]. In (4.305) and below the factors of n from closed loops are always explicitly isolated.

The results (4.255) for η_1 and ν_1 can be obtained very simply using (4.305) (see below). To raise the accuracy to the next order, it is necessary to add to the right-hand sides of (4.305) all the skeleton graphs of the next highest order in $1/n$, namely,

$$D_\phi^{-1} = - \;\overset{\frown}{\underset{\bullet\;\;\;\bullet}{}}\; - \;\overset{}{\underset{}{\diamondsuit}}\; - \; n \;\overset{}{\underset{}{\triangledown}}\; , \qquad (4.306a)$$

$$D_\psi^{-1} = -\frac{n}{2} \; \text{[diagram]} \; - \frac{n}{2} \; \text{[diagram]} \; - \frac{n^2}{2} \; \text{[diagram]} \; . \tag{4.306b}$$

This accuracy is sufficient for obtaining the results (4.256).

The integral over the momentum k circulating in the loops of (4.305) can be arbitrarily split into regions of high (hard) and low (soft) k, and the external momentum p at the IR asymptote assumed small. Clearly, the region of large k can give only regular contributions (Sec. 1.19) at small p, so that the procedure of making the fractional powers self-consistent involves only small k, in which the exact correlators associated with the lines of the loops in (4.305) can be replaced by their IR asymptote. This is also true for the higher-order graphs in (4.306), but with the caveat that this is accurate up to vertex renormalization. Let us explain. The contribution from hard integration momenta inside some vertex subgraph will also be regular at soft external (to the subgraph) momenta, i.e., constant in leading order. Replacement of the vertex subgraph by a constant is equivalent to "shrinking" it (like in the construction of the counterterms in Sec. 3.17), which leads to one-loop graphs of the type (4.305), i.e., effectively to a change (renormalization) of the vertex factors in the one-loop contributions (4.306). After this renormalization is taken into account, the region of hard momenta can be completely ignored in the graphs of (4.306), and the desired IR-asymptote of the exact correlators substituted as the lines.

This explains the physical meaning of vertex renormalization, which we shall need later, and at the same time justifies the self-consistency equations for the original microscopic lattice model (Sec. 2.26), in which not the correlators themselves, but only their IR asymptote has a scaling form (which is the usual case). This distinguishes the exact microscopic model from the renormalized field model used in Sec. 4.31, which is simplified to the extent that its exact correlators already represent the IR asymptote and have the form (4.259). It also follows from the procedure of justifying (4.304) that these equations can be used not only at the critical point itself, but also in its vicinity (small $\tau \geq 0$ and $h = 0$ to ensure $\alpha = 0$), where the exact correlators are not simple powers, but expressions like (1.161).

Returning to the equations (4.305) right at the critical point, we shall seek their solution in the form (4.259) with unknown amplitudes and exponents, representing them as

$$\alpha = d/2 - 1 + \eta/2, \qquad \beta = 2 - \eta - \kappa. \tag{4.307}$$

Here $\alpha = \Delta_\phi$ and $\beta = \Delta_\psi$ are the desired critical dimensions of the fields, and $\kappa = \gamma^*[\Gamma_{\phi\phi\psi}]$ is the anomalous dimension of the vertex owing to the last equation in (4.288). We can find the exponent ν from η and κ using (4.295) and (4.307):

$$1/\nu = d - \beta = d - 2 + \eta + \kappa. \tag{4.308}$$

We shall work in coordinate space, where the right-hand sides of (4.305) are simple products of the lines (4.259), and the coordinate representation of the inverse propagators is easily found using the Fourier transform (3.72):

$$D(x) = Ax^{-2\alpha} \Longleftrightarrow D^{-1}(x) = A^{-1}p(\alpha)x^{2\alpha - 2d},$$
$$\tag{4.309}$$
$$p(z) \equiv \pi^{-d}H(z - d/2, d/2 - z) = \pi^{-d}\|z\|d - z\|/\|d/2 - z\|z - d/2\|.$$

Finally, (4.305) with the lines (4.259) takes the form

$$A^{-1}p(\alpha)x^{2\alpha - 2d} + ABx^{-2\alpha - 2\beta} = 0,$$
$$\tag{4.310}$$
$$B^{-1}p(\beta)x^{2\beta - 2d} + nA^2x^{-4\alpha}/2 = 0.$$

The condition for cancellation of the powers of x is $2\alpha + \beta = d$, equivalent to $\kappa = 0$ in (4.307) and corresponding to logarithmic behavior of the vertex $\psi\phi^2$. When this condition is satisfied, from (4.310) we obtain

$$p(\alpha) + u = 0, \quad 2p(\beta)/n + u = 0, \quad u \equiv A^2B, \quad 2\alpha + \beta = d. \tag{4.311}$$

Eliminating the parameter u, we arrive at the equation for the exponent η in (4.307):

$$p(\alpha) = 2p(\beta)/n, \quad \alpha = d/2 - 1 + \eta/2, \quad \beta = 2 - \eta. \tag{4.312}$$

For the function p known from (4.309), it is easily checked that when $\eta \to 0$ the quantity $p(\beta) = p(2) + O(\eta)$ is finite, while $p(\alpha) = p(d/2 - 1 + \eta/2) = -\eta H(1)\|d/2 + 1\|/2\pi^d + O(\eta^2)$ has a first-order zero in η. This allows us to find the first coefficients of the $1/n$ expansions for η and u from (4.311) and (4.312):

$$\eta_1 = -4\pi^d p(2)/H(1)\|d/2 + 1\|, \quad u_1 = -2p(2). \tag{4.313}$$

The resulting expression for η_1 is the same as (4.255a).

The next coefficients of the $1/n$ expansion of η could also be found by iterating (4.312). However, these would not be exact results, but only the Hartree–Fock contributions to the higher η_k, since the original equations (4.305) themselves are approximate and guarantee only $1/n$ accuracy.

To calculate η_2 it is necessary to turn to (4.306). Substituting the lines (4.259) with the exponents (4.307) into the graphs of (4.306), we isolate all the amplitude factors and the known (from power counting) powers of x, denoting the remaining numerical coefficients (values of the graphs) $\Sigma_{1,2}$ for the two higher-order graphs in (4.306a) and $\Pi_{1,2}$ in (4.306b). In this notation the system (4.306) takes the form

$$p(\alpha) + x^{2\kappa}u + x^{4\kappa}u^2\Sigma_1 + x^{6\kappa}nu^3\Sigma_2 = 0, \tag{4.314a}$$

$$2p(\beta)/n + x^{2\kappa}u + x^{4\kappa}u^2\Pi_1 + x^{6\kappa}nu^3\Pi_2 = 0. \qquad (4.314b)$$

Here only the contributions of order $1/n$ and $1/n^2$ are reliable (we recall that $u \sim 1/n$), and it is only them that we shall follow, neglecting all higher-order corrections.

At first glance it appears that consistency of the powers of x in (4.314) requires simply setting $\kappa = 0$, as in (4.310). However, this cannot be done owing to the presence in Σ_i and Π_i of poles in κ generated by the logarithmic divergences of their vertex (of the type $\psi\phi^2$) subgraphs:

$$\Sigma_i(\eta, \kappa) = K_i(\eta)/\kappa + \Sigma_i'(\eta) + ..., \quad \Pi_i(\eta, \kappa) = K_i(\eta)/\kappa + \Pi_i'(\eta) + ..., \quad (4.315)$$

where the ellipsis denotes corrections of order κ corresponding to unimportant contributions of order $1/n^3$ in (4.314). The explicit calculation of (4.315) will be given later in Sec. 4.37; here we only note that the coincidence of the residues K_i in Σ_i and Π_i is a consequence of the renormalizability of the model.

The parameter κ determining the deviation from logarithmicity of the $\psi\phi^2$ interaction vertex now plays the same role as the regularizer ε in (4.264). As usual, the poles in the regularizer can be eliminated by a renormalization procedure, in this case, vertex renormalization. This amounts to replacing the unit vertex factor by a renormalization constant $Z_g = 1 + a/\kappa + ...$ with residue $a \sim 1/n$ and unimportant higher-order corrections. In $1/n$ perturbation theory the corrections from Z_g need to be taken into account only in the one-loop contributions of (4.314) in the form of additional factors of $Z_g^2 = 1 + 2a/\kappa + ...$ from the two vertices, and can be neglected in the higher-order graphs. Using (4.315), this gives

$$p(\alpha) + x^{2\kappa}(1 + 2a/\kappa)u + x^{4\kappa}u^2 K_1/\kappa +$$
$$+x^{6\kappa}nu^3 K_2/\kappa + x^{4\kappa}u^2\Sigma_1' + x^{6\kappa}nu^3\Sigma_2' = 0 \qquad (4.316)$$

and similarly for (4.314b). For our accuracy $(1/n^2)$ the factors of x^κ in front of Σ_i' in (4.316) should simply be neglected, and in the other contributions it is sufficient to limit ourselves to the approximation $x^{m\kappa} \simeq 1 + m\kappa \ln x$. The condition for the κ poles in (4.316) to cancel has the form $2au + u^2 K_1 + nu^3 K_2 = 0$ and serves as the definition of the constant a in Z_g. For this choice of a, only terms of the factors x^κ that are a multiple of $\ln x$ remain in (4.316) with our accuracy. The condition for them to cancel is

$$\kappa = -uK_1 - 2nu^2 K_2, \qquad (4.317)$$

allowing κ_1 to be found from the residues of the poles in (4.315) and then ν_1 from (4.308). Jumping ahead (the calculations are given in Sec. 4.37), let us give the result thus obtained for the anomalous dimension of the vertex $\kappa = \gamma^*[\Gamma_{\phi\phi\psi}]$ (it has already been calculated by a different method in Sec. 4.34):

$$\kappa_1/\eta_1 = d(2d - 5)/(4 - d), \tag{4.318}$$

which corresponds to the result (4.255b) for ν_1.

For the above choice of parameters a and κ, the poles and logarithms in the two equations (4.314) cancel simultaneously, and we arrive at the following renormalized version of these equations:

$$p(\alpha) + u + u^2\Sigma_1' + nu^3\Sigma_2' = 0, \tag{4.319a}$$

$$2p(\beta)/n + u + u^2\Pi_1' + nu^3\Pi_2' = 0 \tag{4.319b}$$

with accuracy $1/n^2$. By explicitly calculating the "finite" parts Σ_i' and Π_i' of the higher-order graphs in (4.306) (it is sufficient to do this for $\eta = 0$ with our accuracy), from (4.319) we can find the second coefficients η_2 and u_2 of the $1/n$ expansions, which leads to the result (4.256a). The details of the calculation are explained in Sec. 4.37.

If we were able to improve the accuracy by one more order by adding the next graphs to (4.306), we would be able to find κ_2, η_3, and u_3, and then ν_2 from (4.308). In practice, this is impossible owing to technical difficulties (there are 32 new, more complicated graphs), but η_3 and ν_2 can be found by other methods (Secs. 4.39 and 4.40).

4.36   The technique for calculating massless graphs

In this section we shall give several new formulas and describe technical tricks useful for calculating massless graphs. We shall use the notation and formulas of Sec. 3.15, and also the following graphical notation:

$$\underset{\alpha}{\overset{x \qquad y}{\bullet\!\!-\!\!\!-\!\!\bullet}} = |x - y|^{-2\alpha}, \quad \alpha \equiv \text{line index}, \quad \alpha' \equiv \frac{d}{2} - \alpha \quad \forall\alpha;$$

$$\underset{\alpha}{\overset{x\, i \qquad y}{\bullet\!\!+\!\!\!-\!\!\bullet}} = \frac{\partial}{\partial x_i}|x - y|^{-2\alpha} = -\frac{\partial}{\partial y_i}|x - y|^{-2\alpha} = -\underset{\alpha}{\overset{x \qquad i\, y}{\bullet\!\!-\!\!\!-\!\!+\!\!\bullet}} \; ;$$

$$\underset{\alpha}{\overset{x \quad i \quad y}{\bullet\!\!\!\longrightarrow\!\!\!\bullet}} = (y - x)_i|x - y|^{-2\alpha}, \tag{4.320}$$

$$\underset{\alpha}{\overset{x \quad ik \quad y}{\bullet\!\!\!\longrightarrow\!\!\!\bullet}} = (y - x)_i(y - x)_k|x - y|^{-2\alpha}$$

and so on. For vector indices i, k we shall often use numerical notation $1, 2, ...,$ understood as $i_1, i_2, ...$ The coordinate arguments x, y of the line ends will usually be omitted.

The formulas for transforming derivatives into "arrows" are

$$\overset{1}{\underset{\alpha}{\vdash\!\!\!-\!\!\!\bullet}} = 2\alpha \overset{1}{\underset{\alpha+1}{\triangle}} \; ; \; \text{contraction} \; \overset{1\,1}{\underset{\alpha}{\bullet\!\!\#\!\!-\!\!\bullet}} = 4\alpha(1-\alpha') \underset{\alpha+1}{\bullet\!\!-\!\!-\!\!\bullet} ;$$

$$\overset{1\,2}{\underset{\alpha}{\bullet\!\!\#\!\!-\!\!\bullet}} = 2\alpha[\, -\delta_{12} \underset{\alpha+1}{\bullet\!\!-\!\!-\!\!\bullet} + 2(\alpha+1) \overset{12}{\underset{\alpha+2}{\triangle}} \,];$$

$$\overset{123}{\underset{\alpha}{\bullet\!\!\#\!\!-\!\!\bullet}} = -4\alpha(\alpha+1)[\, \delta_{12} \overset{3}{\underset{\alpha+2}{\triangle}} + \delta_{13} \overset{2}{\underset{\alpha+2}{\triangle}} + \delta_{23} \overset{1}{\underset{\alpha+2}{\triangle}} -$$

$$- 2(\alpha+2) \overset{123}{\underset{\alpha+3}{\triangle}} \,]; \; \overset{1234}{\underset{\alpha}{\bullet\!\!\#\!\!-\!\!\bullet}} = 4\alpha(\alpha+1)\Big\{ [\delta_{12}\delta_{34} + 2\text{ more}] \times$$

$$\times \underset{\alpha+2}{\bullet\!\!-\!\!-\!\!\bullet} - 2(\alpha+2)[\, \delta_{12} \overset{34}{\underset{\alpha+3}{\triangle}} + 5\text{ more}] + 4(\alpha+2)(\alpha+3) \overset{1234}{\underset{\alpha+4}{\triangle}} \Big\}.$$

$$(4.321)$$

Expressions like "+ 2 more" denote sums over all versions of the given structure fully symmetrized in the indices, for example, $\delta_{12}\delta_{34} + 2$ more $= \delta_{12}\delta_{34} + \delta_{13}\delta_{24} + \delta_{14}\delta_{23}$.

The formulas for the inverse transformations are

$$\overset{1}{\underset{\alpha}{\triangle}} = \frac{1}{2(\alpha-1)} \overset{1}{\underset{\alpha-1}{\vdash\!\!\!-\!\!\!\bullet}} \; ; \; \overset{12}{\underset{\alpha}{\triangle}} = \frac{1}{4(\alpha-1)(\alpha-2)} \Big[\overset{1\,2}{\underset{\alpha-2}{\bullet\!\!\#\!\!-\!\!\bullet}} +$$

$$+ 2(\alpha-2)\,\delta_{12} \underset{\alpha-1}{\bullet\!\!-\!\!-\!\!\bullet} \Big]; \; \overset{123}{\underset{\alpha}{\triangle}} = \frac{1}{8(\alpha-1)(\alpha-2)(\alpha-3)} \times$$

$$\times \Big\{ \overset{123}{\underset{\alpha-3}{\bullet\!\!\#\!\!-\!\!\bullet}} + 2(\alpha-3)[\, \delta_{12} \overset{3}{\underset{\alpha-2}{\bullet\!\!-\!\!\bullet}} + \delta_{13} \overset{2}{\underset{\alpha-2}{\bullet\!\!-\!\!\bullet}} + \delta_{23} \overset{1}{\underset{\alpha-2}{\bullet\!\!-\!\!\bullet}} \,] \Big\} ;$$

$$\overset{1234}{\underset{\alpha}{\triangle}} = \frac{1}{16(\alpha-1)(\alpha-2)(\alpha-3)(\alpha-4)} \Big\{ \overset{1234}{\underset{\alpha-4}{\bullet\!\!\#\!\!-\!\!\bullet}} + 2(\alpha-4)\times$$

$$\times [\delta_{12} \overset{34}{\underset{\alpha-3}{\bullet\!\!\#\!\!-\!\!\bullet}} + 5\text{ more}] + 4(\alpha-3)(\alpha-4)[\delta_{12}\delta_{34} + 2\text{ more}] \underset{\alpha-2}{\bullet\!\!-\!\!-\!\!\bullet} \Big\}.$$

$$(4.322)$$

The operation of "suspending an arrow," equivalent to differentiating with respect to momentum, is convenient in coordinate space for reducing graphs with linear, quadratic, etc. divergences to logarithmic graphs. In the end one often obtains an n-hedron of the type (4.292) with additional ∂ symbols on the links. The condition for this object to be logarithmic is $\sum \alpha_i + \frac{1}{2} \times$

(number of ∂) = $(n-1)d/2$ up to regularizers, and its pole part is obtained by multiplying (4.292) by additional "structure factors" corresponding to the ∂ symbols:

$$\partial_1\partial_2 \Rightarrow \delta_{12}[-4/d], \quad \partial_1\partial_2\partial_3\partial_4 \Rightarrow (\delta_{12}\delta_{34} + 2 \text{ more})[16/d(d+2)],$$

$$\partial_1...\partial_{2m} \Rightarrow (\delta_{12}\delta_{34}...\delta_{2m-1,2m} + ...)[(-4)^m/d(d+2)...(d+2m-2)] \tag{4.323}$$

(the sum over permutations contains $(2m-1)!!$ terms). It is assumed that all the ∂ are oriented in the same direction (clockwise or counterclockwise), and their distribution over links is unimportant. A logarithmic n-hedron has no poles for an odd number of ∂.

When discussing graphs, we shall always have in mind (unless stated otherwise) arbitrary scalar graphs in coordinate space with local pointlike vertices and unit amplitudes of all vertices and lines. The expression for a graph γ is then a function of the coordinates $x \equiv x_1, x_2, ...$ of all its outer vertices. If the coordinate origin is located at one of them so that $x = 0$ is assigned to it, that vertex will be termed the *base*. The *index* (ind) *of a polyhedron* is the sum of the indices of its links, and the *index of a vertex* is the sum of the indices of the lines converging at it. A vertex with ind = d is termed *unique* (this term was introduced in [89]), and one with ind = $d+m$ (here and below $m = 0, 1, 2, ...$) is termed unique in the broad sense. The value ind = 0 is considered a unique value of a line index (the line vanishes), ind = $-m$ is unique in the broad sense. For an n-hedron these values are ind = $(n-2)d/2$ and $(n-2)d/2 - m$, respectively. An integer power of $(x-y)^2$ corresponds to a line that is unique in the broad sense, so that such a line can also be eliminated using the relation $(x-y)^2 = x^2 + y^2 - 2xy$, or, graphically,

$$\underset{-1}{\overset{0\quad\quad 0}{\triangle}} = \underset{0}{\overset{-1\quad\quad 0}{\triangle}} + \underset{0}{\overset{0\quad\quad -1}{\triangle}} + 2\underset{0\quad 0}{\overset{i\quad\quad i}{\triangle}} \tag{4.324}$$

with arbitrary choice of the third vertex.

Any graph with two outer vertices reduces to a simple line:

$$\blacktriangleleft\!\!\!\blacktriangleright = \Pi(\alpha) \underset{\lambda}{\bullet\!\!-\!\!\bullet} , \tag{4.325}$$

where α is the set of indices of the original graph, and the index λ of the resulting line is determined from power counting [$\lambda = \sum \alpha_i - (d/2)\times$ the number of integration vertices]. The numerical coefficient Π in (4.325) will be termed the *value of the graph*, and the graph itself is termed *integrable* if Π can be calculated explicitly in terms of known special functions (usually gamma functions and their derivatives). All simple chains are integrated according to the rule (3.124), in particular,

$$\underset{a\quad b}{\bullet\!\!-\!\!\bullet\!\!-\!\!\bullet} \; = \; \pi^{d/2}H(a,b,c) \; \underset{a+b\,-\,d/2}{\bullet\!\!-\!\!-\!\!\bullet} \,, \quad c = d - a - b \,, \tag{4.326}$$

and for a chain with n links with indices $\alpha_1...\alpha_n$ we will have

$$\Pi = \pi^{(n-1)d/2}H(\alpha_1, \alpha_2, ..., \alpha_n, nd/2 - \sum \alpha_i).$$

A very useful trick is *conformal transformation with the given base of an arbitrary graph*: $\gamma(x) \rightarrow \gamma(x') \equiv \gamma'(x)$. Here $x = x_1, x_2, ...$ is the set of coordinates of all the outer vertices different from the base (see above), and the transformation is the conformal inversion $x \rightarrow x' = x/x^2$ of each of these d-dimensional vectors. Accompanying it by the analogous change of variable $y \rightarrow y'$ of the coordinates of all the integration vertices and using the equation $|x' - y'|^2 = |x - y|^2/x^2 y^2$, it is easy to verify that the transformed graph $\gamma'(x)$ is related to the original one $\gamma(x)$ as follows. In γ' any vertex v is connected to the base by a line with index

$$\alpha = \left\{ \begin{array}{l} d - \mathrm{ind}[v], \;\; \text{if } v \text{ is an integration vertex,} \\ -\mathrm{ind}[v], \;\;\; \text{if } v \text{ is an outer vertex} \end{array} \right\} \tag{4.327}$$

($\mathrm{ind}[v]$ is the index of the vertex v in the original graph), and the indices of all the other lines are not changed (i.e., they are identical in γ and γ').

The rule (4.327) has an important consequence: *after a conformal transformation, all unique integration vertices are disconnected from the base.* This can be used, in particular, to easily prove the following expression [131] (we recall the notation $\alpha' \equiv d/2 - \alpha \;\; \forall \alpha$):

$$\underset{a+b+c=d}{\underset{a\quad b}{\diagup\!\!\!\diagdown}^{c}} \; = \; \pi^{d/2}H(a,b,c) \; \underset{c'}{\overset{b'\diagup\!\!\!\diagdown a'}{\triangle}} \,, \tag{4.328}$$

relating a unique vertex to a unique triangle, which together with (4.326) actually forms the basis of all the calculations. Equation (4.328) is proved by the following transformation chain:

$$\underset{a\quad b}{\diagup\!\!\!\diagdown^{c}} \; \rightarrow \; \underset{a\quad b}{{}^{-a}\triangle^{-b}} \; \rightarrow \; {}^{-a}\triangle^{-b}_{\;c'} \; \rightarrow \; \underset{c'}{{}^{b'}\triangle^{a'}} \,.$$

The first step is conformal transformation from the base at the upper vertex, the second is integration of a simple chain ($c' = a + b - d/2$), and the third is another conformal transformation returning to the original $\gamma(x)$. The analogous procedure for a vertex that is unique in the broad sense leads, after the first step, to a vertical line with index $-m$, which can be eliminated using (4.324). Finally, for $a + b + c = d + 1$ we obtain

$$\begin{picture}(0,0)\end{picture} = \pi^{d/2} H(a,b,c)\left\{ a'b' \begin{picture}(0,0)\end{picture} + a'c' \begin{picture}(0,0)\end{picture} + b'c' \begin{picture}(0,0)\end{picture} \right\}, \qquad (4.329)$$

and in general,

$$\begin{picture}(0,0)\end{picture}_{a+b+c=d+m} = \frac{\pi^{d/2} m!}{\|a\|\|b\|\|c\|} \sum_{\substack{0 \le m_i \le m \\ m_1+m_2+m_3=m}} C(m_1,m_2,m_3) \begin{picture}(0,0)\end{picture}, \qquad (4.330)$$

with coefficients

$$C(m_1,m_2,m_3) = \|a' + m - m_1\| \|b' + m - m_2\| \|c' + m - m_3\| / m_1! m_2! m_3!.$$

The following expressions for the inverse transformations are also useful:

$$\begin{picture}(0,0)\end{picture} = \frac{\pi^{-d/2} H(a,b,c)}{(a'-1)(b'-1)(c'-1)} \times \qquad (4.331)$$

$$\times \left\{ (a'-1) \begin{picture}(0,0)\end{picture} + (b'-1) \begin{picture}(0,0)\end{picture} + (c'-1) \begin{picture}(0,0)\end{picture} \right\}$$

for $a + b + c = d/2 - 1$, and in general,

$$\begin{picture}(0,0)\end{picture}_{a+b+c=d/2-m} = \frac{\pi^{-d/2} m!}{\|a\|\|b\|\|c\|} \sum_{\substack{0 \le m_i \le m \\ m_1+m_2+m_3=m}} C(m_1,m_2,m_3) \begin{picture}(0,0)\end{picture}, \qquad (4.332)$$

with coefficients

$$C(m_1,m_2,m_3) = \|a' - m + m_1\| \|b' - m + m_2\| \|c' - m + m_3\| / m_1! m_2! m_3!.$$

Let us discuss the example of the following graph frequently encountered in calculations:

$$\begin{picture}(0,0)\end{picture} = \Pi \begin{picture}(0,0)\end{picture} \quad ; \qquad \left. \begin{aligned} v_1 &\equiv \alpha_1 + \alpha_2 + \alpha_5, \\ v_2 &\equiv \alpha_3 + \alpha_4 + \alpha_5, \\ t_1 &\equiv \alpha_1 + \alpha_4 + \alpha_5, \\ t_2 &\equiv \alpha_2 + \alpha_3 + \alpha_5, \\ \beta &\equiv \alpha_1 + \alpha_2 + \alpha_3 + \alpha_4 + \alpha_5 . \end{aligned} \right\} \qquad (4.333)$$

The values $\alpha = -m$ for lines, $v = d + m$ for vertices, $t = d/2 - m$ for triangles, and $\beta = 3d/2 + m$ for the "complete" index β introduced in (4.333) are unique. The presence of any uniqueness directly leads to integrability of the graph [for

v and t, reduction to chains by means of (4.328), and for β we explain below], for example,

$$v_2 = d \Rightarrow \Pi = \pi^d H(\alpha_3, \alpha_4, \alpha_5, \alpha_1 + \alpha_3', \alpha_2 + \alpha_4', \alpha_1' + \alpha_2' - \alpha_5), \quad (4.334a)$$

$$t_1 = d/2 \Rightarrow \Pi = \pi^d H(\alpha_2, \alpha_3, \alpha_1 + \alpha_3', \alpha_4 + \alpha_2', \alpha_2' + \alpha_3', \alpha_2 + \alpha_3 - \alpha_5'), \quad (4.334b)$$

$$\beta = 3d/2 \Rightarrow \Pi = \pi^d H(\alpha_1, \alpha_2, \alpha_3, \alpha_4, \alpha_2' + \alpha_3', \alpha_1' + \alpha_4'). \quad (4.334c)$$

For reference, we give the analogous formulas for uniqueness in the broad sense for the simplest case of $m = 1$:

$$v_2 = d + 1 \Rightarrow$$

$$\Pi = \pi^d H(\alpha_3, \alpha_4, \alpha_5, \alpha_1 + \alpha_3' + 1, \alpha_2 + \alpha_4' + 1, \alpha_1' + \alpha_2' - \alpha_5) \times$$

$$\times [\alpha_3' \alpha_4' (\alpha_1' + \alpha_2' - \alpha_5 - 1)(\alpha_1 + \alpha_2 - \alpha_5') +$$

$$+ \alpha_3' \alpha_5' (\alpha_2 + \alpha_4')(\alpha_4 - \alpha_2 - 1) + \alpha_4' \alpha_5' (\alpha_1 + \alpha_3')(\alpha_3 - \alpha_1 - 1)], \quad (4.335a)$$

$$t_1 = d/2 - 1 \Rightarrow$$

$$\Pi = \pi^d H(\alpha_2, \alpha_3, \alpha_2' + \alpha_4 + 1, \alpha_3' + \alpha_1 + 1, \alpha_2 + \alpha_3 - \alpha_5', \alpha_2' + \alpha_3' + 1) \times$$

$$\times [\alpha_1 \alpha_4 (\alpha_2' + \alpha_3' - \alpha_5)(\alpha_2 + \alpha_3 - \alpha_5' - 1) +$$

$$+ \alpha_4 \alpha_5 (\alpha_3' + \alpha_1)(\alpha_3 - \alpha_1 - 1) + \alpha_1 \alpha_5 (\alpha_2' + \alpha_4)(\alpha_2 - \alpha_4 - 1)], \quad (4.335b)$$

$$\beta = 3d/2 + 1 \Rightarrow$$

$$\Pi = \pi^d H(\alpha_1, \alpha_2, \alpha_3, \alpha_4, \alpha_2' + \alpha_3' + 1, \alpha_1' + \alpha_4' + 1) \times$$

$$\times \{\alpha_3' \alpha_4' \alpha_5 (\alpha_5' - 1) + (\alpha_3 - \alpha_4' - 1)[\alpha_3'(\alpha_1' + \alpha_4')(\alpha_4 - \alpha_1' - 1) +$$

$$+ \alpha_4' (\alpha_2' + \alpha_3')(\alpha_3 - \alpha_2' - 1)] \}. \quad (4.335c)$$

There exists a discrete group of transformations allowing the indices in the graph (4.333) to be changed and the following equation to be used in calculations:

$$\Pi(\alpha) = C(\alpha) \cdot \Pi(\alpha^*), \quad (4.336)$$

where α is the original set of indices, α^* is the transformed set, and $C(\alpha)$ is the transfer coefficient. First, these are the above-mentioned *conformal transformations* from the base at the left ($\rightarrow$) or right ($\leftarrow$) outer vertex; for them $C(\alpha) = 1$. Another transformation is *passage to momentum space*:

rewriting the definition of Π in (4.333) for the graph in momentum space, the resulting expression can be interpreted as the coordinate-space one for the graph with the indices changed. Next, there are several *insertion operations*, which are constructed as follows. Equation (4.326) is used to split a selected line of a graph into two "halves" by the insertion of a point. The index of the half adjacent to the integration vertex is chosen such that the vertex becomes unique (and then the index of the second half is uniquely determined). Then, using the rule (4.328) to integrate the resulting unique vertex, we arrive at the same graph with different indices. For the graph (4.333) there are six such transformations, and we shall denote them as $\nearrow$, $\uparrow$, $\nwarrow$ for the upper vertex (the slope of the arrow indicates the line chosen for the insertion) and $\searrow$, $\downarrow$, $\swarrow$ for the lower one. Two more transformations of this type are obtained by attaching to the graph (4.333) an additional external line on the left ($\rightarrow$) or right ($\leftarrow$) with index such that a unique vertex is obtained, which is then integrated using the rule (4.328).

These 11 operations were introduced in [89] and form a complete basis of the group in question, i.e., products of these operators exhaust all the presently known transformations (4.336) of the graph (4.333) (there are 1440 of them altogether [132]). An important contribution to understanding the structure of a symmetry subgroup of this group [transformations with $C(\alpha) = 1$ in (4.336)] was made later in [133], where it was noted that the symmetry of any graph of the type (4.325) is manifested most clearly after passage to the corresponding logarithmic vacuum graph. This is done as follows. We (mentally) regularize all the lines of the original graph (4.325) by a small shift $\alpha_i \rightarrow \alpha_i + \Delta_i$ of the indices of its lines while adding the required powers of the renormalization mass that compensate for the change of dimension. We then "close" the two sides of (4.325) by a "reference line" with index $d/2 - \lambda$, so as to obtain a logarithmic 2-hedron (4.292) on the right-hand side. In the resulting expression we then select the pole in the total regularizer Δ. Taking into account the fact that for the 2-hedron at the residue of (4.292) we have $H(\lambda, d/2 - \lambda) = 1$ owing to (3.73), we find that

$$\overset{\frown}{\smile} = \Pi(\alpha) \overset{\lambda}{\underset{d/2-\lambda}{\bigcirc}} = \Pi(\alpha)\pi^{d/2}/\Delta\|d/2\| + ..., \tag{4.337}$$

where the ellipsis stands for the nonpole contribution. It is understood that the lines contain regularizers, but they do not need to be taken into account in the residue of the pole.

Therefore, the residue at the pole in the regularizer of the resulting logarithmic vacuum graph turns out to be proportional to the desired quantity Π with coefficient independent of the graph indices. This leads to the main statement [133]: since all the vertices of a vacuum graph are on an equal footing (all except an arbitrarily chosen one are integrated over), the quantity Π should not change for any other choice of reference line of the vacuum graph on the left-hand side of (4.337).

As applied to our graph (4.333) this implies that

$$\cong \quad = \quad , \quad \sum_{i=1}^{6} \alpha_i = 3d/2, \qquad (4.338)$$

where the logarithmicity condition uniquely determines the index $\alpha_6 = 3d/2 - \beta$ of the added reference line.

The vacuum graph in (4.338) is a tetrahedron with symmetry group containing $4! = 24$ elements, represented as the six ways of choosing the reference line times the four elements of the symmetry group of the graph (4.333) (left–right, upper–lower). The natural parameters of the tetrahedron are the indices of its six links and four vertices, which is equivalent to a set of ten parameters in (4.333). In this language, uniqueness of a triangle is equivalent to uniqueness of the opposite vertex of the tetrahedron, and uniqueness of the complete index β is equivalent to uniqueness of the reference line.

The original graph (4.333) corresponds to choosing line 6 as the reference line, and the nontrivial *transformations of the tetrahedron* are the other ways of choosing the reference line (1,2,3,4,5 in Table 4.8). The quantity Π is independent of this choice, and so for these transformations we have $C(\alpha) = 1$ in (4.336).

Let us make one final remark. The transformations we have discussed are most simply formulated in terms of ten other parameters a_i determined by the equations $\alpha_i = d/4 + a_i$ for the six links of the tetrahedron and by the sums of the corresponding triplets a_i for its four vertices. In these variables all the transformations reduce to permutations and reflections in the set of the ten parameters a_i. This simplifies the study of the general structure of this group and its invariants, which are polynomials in the set of the a_i; this is very useful for some calculations using the $4 - \varepsilon$ expansion [132]. However, in calculations in arbitrary space dimension it is simpler to use the usual parameters (4.333).

Here we give a table of transformations of the graph indices (4.333) useful for calculations; the table will be explained later. It has been taken from [89] with the addition of the five nontrivial transformations of the tetrahedron defined above, numbered by the choice of reference line. It is clear from the symmetry of the tetrahedron that two more conformal transformations (in the vertex number) can be added to the two introduced above, and four more insertion transformations can be added to the eight above. We do not list these in the table as it would become too large, and all the new transformations can be obtained as products of the ones introduced above. This is also true for the "tetrahedron transformations" themselves: for example, transformation No. 2 can be obtained by the insertion $\downarrow$, followed by $\searrow$, and then the conformal transformation $\leftarrow$.

The starred indices of the transformed graph on the right-hand side of (4.336) are given in the first column of Table 4.8 in the notation of (4.333),

Table 4.8 Transformations of the indices in (4.336) for the graph (4.333). The unique values of the indices are $\alpha = -m$, $v = d + m$, $t = d/2 - m$, $\beta = 3d/2 + m$, and the antiunique values are $\alpha = d/2 + m$ (UV), $v = d/2 - m$ (IR), $t = d + m$ (UV), and $\beta = d - m$ (IR). ms — momentum space, coef — coefficient.

		TRANSFORMATIONS						
		Conformal		Tetrahedron symmetry				
	ms	$\rightarrow$	$\leftarrow$	1	2	3	4	5
α_1^*	$\dfrac{d}{2} - \alpha_2$	$d - v_1$	α_1	$\dfrac{3d}{2} - \beta$	α_1	α_5	α_1	α_1
α_2^*	$\dfrac{d}{2} - \alpha_3$	α_2	$d - v_1$	α_2	$\dfrac{3d}{2} - \beta$	α_2	α_5	α_4
α_3^*	$\dfrac{d}{2} - \alpha_4$	α_3	$d - v_2$	α_5	α_3	$\dfrac{3d}{2} - \beta$	α_3	α_3
α_4^*	$\dfrac{d}{2} - \alpha_1$	$d - v_2$	α_4	α_4	α_5	α_4	$\dfrac{3d}{2} - \beta$	α_2
α_5^*	$\dfrac{d}{2} - \alpha_5$	α_5	α_5	α_3	α_4	α_1	α_2	$\dfrac{3d}{2} - \beta$
v_1^*	$\dfrac{3d}{2} - t_2$	$d - \alpha_1$	$d - \alpha_2$	$\dfrac{3d}{2} - t_1$	$\dfrac{3d}{2} - t_2$	v_1	v_1	$\dfrac{3d}{2} - t_2$
v_2^*	$\dfrac{3d}{2} - t_1$	$d - \alpha_4$	$d - \alpha_3$	v_2	v_2	$\dfrac{3d}{2} - t_2$	$\dfrac{3d}{2} - t_1$	$\dfrac{3d}{2} - t_1$
t_1^*	$\dfrac{3d}{2} - v_1$	$2d - \beta$	t_1	$\dfrac{3d}{2} - v_1$	t_1	t_1	$\dfrac{3d}{2} - v_2$	$\dfrac{3d}{2} - v_2$
t_2^*	$\dfrac{3d}{2} - v_2$	t_2	$2d - \beta$	t_2	$\dfrac{3d}{2} - v_1$	$\dfrac{3d}{2} - v_2$	t_2	$\dfrac{3d}{2} - v_1$
β^*	$\dfrac{5d}{2} - \beta$	$2d - t_1$	$2d - t_2$	$\dfrac{3d}{2} - \alpha_1$	$\dfrac{3d}{2} - \alpha_2$	$\dfrac{3d}{2} - \alpha_3$	$\dfrac{3d}{2} - \alpha_4$	$\dfrac{3d}{2} - \alpha_5$
coef	$C[\text{ms}]$	1	1	1	1	1	1	1

Table 4.8 *(Continued)*

	Point insertion transformations							
	↗	↑	↘	↘	↓	↙	→	←
α_1^*	$v_1 - \dfrac{d}{2}$	$\dfrac{d}{2} - \alpha_2$	$\dfrac{d}{2} - \alpha_5$	α_1	α_1	$t_1 - \dfrac{d}{2}$	$\dfrac{d}{2} - \alpha_4$	α_1
α_2^*	$\dfrac{d}{2} - \alpha_5$	$\dfrac{d}{2} - \alpha_1$	$v_1 - \dfrac{d}{2}$	$t_2 - \dfrac{d}{2}$	α_2	α_2	α_2	$\dfrac{d}{2} - \alpha_3$
α_3^*	$t_2 - \dfrac{d}{2}$	α_3	α_3	$\dfrac{d}{2} - \alpha_5$	$\dfrac{d}{2} - \alpha_4$	$v_2 - \dfrac{d}{2}$	α_3	$\dfrac{d}{2} - \alpha_2$
α_4^*	α_4	α_4	$t_1 - \dfrac{d}{2}$	$v_2 - \dfrac{d}{2}$	$\dfrac{d}{2} - \alpha_3$	$\dfrac{d}{2} - \alpha_5$	$\dfrac{d}{2} - \alpha_1$	α_4
α_5^*	$\dfrac{d}{2} - \alpha_2$	$v_1 - \dfrac{d}{2}$	$\dfrac{d}{2} - \alpha_1$	$\dfrac{d}{2} - \alpha_3$	$v_2 - \dfrac{d}{2}$	$\dfrac{d}{2} - \alpha_4$	$t_1 - \dfrac{d}{2}$	$t_2 - \dfrac{d}{2}$
v_1^*	$\dfrac{d}{2} + \alpha_1$	$\dfrac{d}{2} + \alpha_5$	$\dfrac{d}{2} + \alpha_2$	v_1	$\beta - \dfrac{d}{2}$	v_1	v_1	v_1
v_2^*	v_2	$\beta - \dfrac{d}{2}$	v_2	$\dfrac{d}{2} + \alpha_4$	$\dfrac{d}{2} + \alpha_5$	$\dfrac{d}{2} + \alpha_3$	v_2	v_2
t_1^*	t_1	t_1	$\dfrac{d}{2} + \alpha_4$	t_1	t_1	$\dfrac{d}{2} + \alpha_1$	$\dfrac{d}{2} + \alpha_5$	$\beta - \dfrac{d}{2}$
t_2^*	$\dfrac{d}{2} + \alpha_3$	t_2	t_2	$\dfrac{d}{2} + \alpha_2$	t_2	t_2	$\beta - \dfrac{d}{2}$	$\dfrac{d}{2} + \alpha_5$
β^*	β	$\dfrac{d}{2} + v_2$	β	β	$\dfrac{d}{2} + v_1$	β	$\dfrac{d}{2} + t_2$	$\dfrac{d}{2} + t_1$
coef	$C[\uparrow]$	$C[\uparrow]$	$C[\uparrow]$	$C[\downarrow]$	$C[\downarrow]$	$C[\downarrow]$	$C[\rightarrow]$	$C[\leftarrow]$

and their expressions in terms of the indices of the original graph on the left-hand side of (4.336) are given in the other columns. The transfer coefficient $C(\alpha)$ in (4.336) is also given for all the transformations (last line). There are five such coefficients, and they are expressed in terms of the indices of the original graph as

$$C[\uparrow] = H(\alpha_1, \alpha_2, \alpha_5, d - v_1); \quad C[\downarrow] = H(\alpha_3, \alpha_4, \alpha_5, d - v_2);$$

$$C[\rightarrow] = H(\alpha_1, \alpha_4, \alpha_6, t_2 - d/2); \quad C[\leftarrow] = H(\alpha_2, \alpha_3, \alpha_6, t_1 - d/2); \quad (4.339)$$

$$C[\text{ms}] = H(\alpha_1, \alpha_2, \alpha_3, \alpha_4, \alpha_5, \alpha_6), \quad \alpha_6 \equiv 3d/2 - \beta.$$

In the table caption along with the unique values of the indices of the graph in (4.333) we give the *antiunique values* [89] corresponding to various divergences of the subgraphs. For an arbitrary graph in coordinate space, a UV divergence of a given subgraph γ is associated with the procedure of collapsing all its vertices and the lines connecting them to a point, and an IR divergence is associated with the procedure of taking all the vertices of γ to infinity (and an infinite distance from each other) while "stretching" the corresponding lines. The criterion for the corresponding (UV or IR) divergence to be present in a given subgraph γ with N vertices is stated as

$$\delta_{\text{UV}}[\gamma] = \sum \alpha_i - (N-1)d/2 = m, \quad \delta_{\text{IR}}[\gamma] = \sum \alpha_i - Nd/2 = -m \quad (4.340)$$

with the usual $m = 0, 1, 2, \ldots$. In the first expression in (4.340) the summation runs over the indices of all the lines pulled to a point ($N - 1$ there because all the vertices are pulled to a single one), and in the second it runs over the indices of all the lines stretched in the process of separating the vertices. The exact meaning of the term "associated" used above is the following. When the criterion (4.340) is satisfied for a subgraph γ, there is a (UV or IR) divergence generated by the corresponding (to the procedure of collapsing or separating all the vertices of γ) region in the integral over the coordinates of the internal vertices of the graph. It follows, in particular, from (4.340) that the antiunique values of the indices in the graph (4.333) are $\alpha - d/2 + m$ (UV), $v = d/2 - m$ (IR), $t = d + m$ (UV), and $\beta = d - m$ (IR), with the type of divergence given in parentheses.

Since antiuniqueness implies divergence, it can never be exact (in contrast to uniqueness), and it always presupposes the presence of some regularization by a small shift Δ of the indices. It is easily checked that all the transformations listed in Table 4.8 preserve the total number of uniquenesses and antiuniquenesses in the graph (4.333). However, a uniqueness can change into an antiuniqueness and vice versa, which is accompanied by the appearance of a zero or a pole in the transfer coefficient $C(\alpha)$ in (4.336) via the regularizer Δ. This can be used, for example, to reduce the problem of computing the constant and the contribution of order Δ in the original finite graph to the simpler problem of calculating the pole and the constant in the transformed graph with an antiuniqueness. In general, the number of terms of the Taylor expansion in the regularizers of a given finite graph that can be found by standard techniques is determined by the number of uniquenesses in the graph without regularizers. Let us explain. The presence of a uniqueness means that a graph with arbitrary regularizers Δ_i in all the lines is integrated exactly in satisfying the constraint on the regularizers $\sum c_i \Delta_i = 0$ that ensures the uniqueness. The contributions linear in Δ_i are then determined up to the term $\sum c_i \Delta_i$ with unknown coefficient. This coefficient is found if there is another uniqueness ensuring integrability when another constraint $\sum c'_i \Delta_i = 0$ is satisfied. Therefore, two uniquenesses allow all the contributions linear in Δ_i to be found, while in the quadratic contributions only the coefficient of the product of constraints $\sum c'_i \Delta_i \cdot \sum c_k \Delta_k$ remains unknown. It, in turn, is found when a third uniqueness is present, and so on. To make maximal use of this trick for a graph containing an antiuniqueness, it is necessary to try to change the latter to a uniqueness by a suitable transformation after isolating the poles explicitly.

Another very useful trick is to use various recursion relations obtained as follows [89]: (1) by acting with the operator $x \partial_x$ on the coordinate x of an outer vertex of a graph; (2) by multiplying an integration vertex with coordinate y by a constant $d = \partial_y y$ and then integrating by parts with respect to y; (3) by inserting a point (see above) into one of the lines in order to obtain a vertex with index $d + 1$, then using (4.329) followed by the inverse transformation (4.328) after splitting the line with index $+1$ in order to obtain unique triangles. The second trick for a vertex with arbitrary indices and upper base leads to (4.341a) and the third to (4.341b):

$$\underset{a}{\overset{|c}{\diagup\!\!\diagdown}}_{b} = \frac{1}{d-a-b-2c}\left\{ a\left[\overset{-}{\underset{+}{\diagdown\!\!\diagup}} - \underset{+}{\overset{-}{\diagup\!\!\diagdown}} \right] + b\left[\underset{+}{\overset{-}{\diagup\!\!\diagdown}} - \underset{+}{\overset{-}{\diagdown\!\!\diagup}} \right] \right\} = \qquad (4.341a)$$

$$= \frac{c(c'-1)}{(a-1)(b-1)} \;\overset{|+}{\underset{-\,\triangle\,-}{}}_{+} + \frac{a+b-1-d/2}{a-1} \;\underset{+}{\overset{|}{\triangle}}_{-} + \frac{a+b-1-d/2}{b-1} \;\underset{+}{\overset{|}{-\triangle}} . \qquad (4.341b)$$

Here and below in similar expressions the symbols $\pm$ on the right-hand side indicate only additions of ± 1 to the indices of the original graph. One of the three lines of the vertex in (4.341) turns out to be distinguished, and we shall call it the *basic line*; of course, its choice is arbitrary.

For the graph (4.333) the above tricks lead to the following complete (up to left–right and upper–lower symmetry) system of recursion relations:

$$\diamondsuit = \begin{cases} \dfrac{1}{k_1}\left\{ \alpha_2\left[\overset{-}{\underset{}{\diamondsuit}}\overset{+}{} - \overset{}{\diamondsuit}\overset{+}{} \right] + \alpha_3\left[\underset{-}{\diamondsuit}\underset{+}{} - \overset{}{\diamondsuit}\underset{+}{} \right] \right\}, & (4.342a) \\[2.2em] \dfrac{1}{k_2}\left\{ \alpha_2\left[\underset{-}{\diamondsuit}\overset{+}{} - \overset{}{\diamondsuit}\overset{+}{} \right] + \alpha_5\left[\overset{-}{\diamondsuit}\underset{+}{} - \overset{}{\diamondsuit}\underset{+}{} \right] \right\}, & (4.342b) \\[2.2em] \dfrac{1}{k_3}\left\{ \alpha_1\left[\overset{+}{\underset{-}{\diamondsuit}} - \overset{+}{\underset{-}{\diamondsuit}} \right] + \alpha_2\left[\overset{+}{\underset{-}{\diamondsuit}} - \overset{+}{\underset{-}{\diamondsuit}} \right] \right\}, & (4.342c) \end{cases}$$

$$
\langle\diamond\rangle = \begin{cases}
c_{11}\,\langle\diamond\rangle + c_{12}\,\langle\diamond\rangle + c_{13}\,\langle\diamond\rangle\,, & \text{(4.342d)} \\[2em]
c_{21}\,\langle\diamond\rangle + c_{22}\,\langle\diamond\rangle + c_{23}\,\langle\diamond\rangle\,, & \text{(4.342e)} \\[2em]
c_{31}\,\langle\diamond\rangle + c_{32}\,\langle\diamond\rangle + c_{33}\,\langle\diamond\rangle\,, & \text{(4.342f)}
\end{cases}
$$

where $k_1 \equiv 2(d-\beta) + \alpha_2 + \alpha_3$, $k_2 \equiv d - \alpha_1 - v_1$, $k_3 \equiv d - \alpha_5 - v_1$, and

$$
\begin{aligned}
c_{11} &= \alpha_1(\alpha_1' - 1)/(\alpha_2 - 1)(\alpha_5 - 1), \\
c_{12} &= (\alpha_2 + \alpha_5 - 1 - d/2)/(\alpha_5 - 1), \\
c_{13} &= (\alpha_2 + \alpha_5 - 1 - d/2)/(\alpha_2 - 1), \\
c_{21} &= \alpha_5(\alpha_5' - 1)/(\alpha_1 - 1)(\alpha_2 - 1), \\
c_{22} &= (\alpha_1 + \alpha_2 - 1 - d/2)/(\alpha_1 - 1), \\
c_{23} &= (\alpha_1 + \alpha_2 - 1 - d/2)/(\alpha_2 - 1), \\
c_{31} &= (\beta - 1 - d)(3d/2 - \beta)/(\alpha_1 - 1)(\alpha_4 - 1), \\
c_{32} &= (\alpha_1 + \alpha_4 - 1 - d/2)/(\alpha_1 - 1), \\
c_{33} &= (\alpha_1 + \alpha_4 - 1 - d/2)/(\alpha_4 - 1).
\end{aligned}
$$

The first of these is obtained using the operator $x\partial_x$, the next two using (4.341a) with two topologically inequivalent choices of the basic line, and the last three using (4.341b). The figures in (4.342) are understood as the values of the corresponding graphs Π without an overall power of x. To correctly reproduce the latter it would be necessary to add external lines (connecting outer vertices) with suitable index ± 1 to the graphs with unequal number ± 1.

Let us give some examples using these techniques. The first is the derivation of (3.135) for the graph (4.333) with index $\alpha \equiv d/2 - 1$ on all the links of one of the triangles (we shall take the right-hand one). We denote the value of this graph as $V(\beta, \gamma)$, where β and γ are the other arbitrary indices. The result (3.135) for $V(\beta, \gamma)$ was first obtained in [90] by summing infinite series involving Gegenbauer polynomials. Here we give an easier derivation [89]: applying the transformation $\uparrow$ and then $\diagup$ to the original graph (see Table 4.8), we obtain the graph (4.333) with indices $v = d - 1$ of the two vertices and $t_1 = d/2 + 1$ of the left-hand triangle. This graph has no uniqueness, and so it cannot be integrated directly. However, if we use the recursion relation (4.342a) for it, all the graphs on the right-hand side turn out to be integrable owing to uniqueness of one of the vertices or the left-hand triangle.

Another method is to transform the original graph to momentum space and then use (4.341a) for the lower vertex with vertical basic line. We then obtain (dropping all inessential indices and coefficients, and taking $\alpha \equiv d/2 - 1$)

$$(4.343)$$

All the graphs on the right-hand side are explicitly integrable (they are simple chains).

It is easy to obtain the generalization of the V formula to the case of the graph (4.333) with indices $\alpha_2 = \alpha_5 = \alpha$, $\alpha_3 = \alpha - m$ and arbitrary $\alpha_{1,4}$. For this we take the second graph in (4.343), a multiple of the usual $V(m = 0)$, and use (4.342a):

$$(4.344)$$

All these graphs are known except for the second one on the right-hand side (we shall call it graph 121), and so it is thereby calculated. Another use of (4.342a) for graph 121 then allows us to find the analogous graph 131, and so on. Finally, the momentum-space transformation reduces graph $1m1$ to the desired generalization of the V graph.

Let us give an example of a more complicated graph which we shall need later on (Sec. 4.39). We consider a graph of the type Π_2 in (4.306b) in the zeroth-order approximation for the indices α, β with replacement of the index $\beta = 2$ of one of the horizontal lines by $2 - \lambda$ with arbitrary $\lambda \neq -m$ (this is the condition for UV divergences to be absent in the subgraphs of the $\psi\phi^2$ vertex). This graph is calculated using the following transformation chain (everywhere $\alpha \equiv d/2 - 1$, $\lambda \neq -m$ is arbitrary, and $\lambda' \equiv d/2 - \lambda$):

$$(4.345)$$

The first step is to integrate the lower left-hand unique vertex with indices $\alpha, \alpha, 2$, the second is to insert a point (4.345) into the upper horizontal line, and the third is to integrate the upper right-hand unique vertex. For the resulting graph (the last one on the top) we then use (4.341a) for the upper left-hand vertex, choosing 1 as the basic line. This leads to the sum of graphs 1 to 4 on the right-hand side of (4.345) (the coefficients and indices of the basic graph have been dropped). In each of them one of the lines vanishes (addition of -1 to the index $+1$), and they all turn out to be integrable. After chain integration and the $\nearrow$ transformation of Table 4.8, graphs 1 and 3 reduce to $V(1, \lambda')$, graph 2 is the contraction of a simple line and a graph of the type (4.333), which is reduced to $V(1, 2-\lambda)$ by the $\nearrow$ transformation, and graph 4 reduces to $V(\alpha, 2-\lambda)$ after chain integration and the $\nearrow$ transformation. The final result is the following (we recall that $\alpha \equiv d/2 - 1$, $\lambda \neq -m$ is arbitrary, and $\lambda' \equiv d/2 - \lambda$):

$$\Pi\left\{ \begin{array}{c} \overset{2-\lambda}{\underset{2}{\overset{\alpha\quad\quad\alpha}{\underset{\alpha\quad\quad\alpha}{\square}}}} \end{array} \right\} = \frac{\pi^{2d} H(\alpha, 2\alpha)}{(1-\alpha)(1-\lambda)^2 (1-\lambda')^2} \left\{ H(\alpha, 2\alpha-1, 2-\lambda, 2-\lambda') \times \right.$$

$$\left. \times \left[\frac{(1-\lambda)(1-\lambda')}{(1-\alpha)} [2B(\lambda) - B(2-\lambda) - B(2-\lambda')] + 1 \right] - 1 \right\},$$

$$(4.346)$$

where $B(z) \equiv \psi(z) + \psi(d/2 - z)$. This function naturally arises when $H(z+\Delta)$ is expanded in the small parameter Δ [owing to (3.69) and the definition $H(z) \equiv \|z'\|/\|z\|$; everywhere $z' \equiv d/2 - z$]:

$$H(z + \Delta) = H(z) \cdot \exp[-\Delta B(z) - \Delta^2 C(z)/2 - \Delta^3 D(z)/6 - \ldots],$$

$$(4.347)$$

$$B(z) \equiv \psi(z) + \psi(z'), \quad C(z) \equiv \psi'(z) - \psi'(z'), \quad D(z) \equiv \psi''(z) + \psi''(z'),$$

and so on. Such an expansion must be used, in particular, to calculate $V(1, \beta)$, since an ambiguity arises in the general expression (3.135) for $V(\beta, \gamma) = V(\gamma, \beta)$ for $\gamma \to 1$ and must be resolved by introducing a regularizer Δ. For reference we give the results thus obtained for the special cases in (3.135) ($\alpha \equiv d/2 - 1$, arbitrary index β):

$$V(1, \beta) = \frac{\pi^d H(\alpha, 2\alpha)}{(1-\beta)^2} \left\{ -H(1,1) + (1-\beta)H(\beta, 2-\beta) \left[B(\beta) - \right. \right.$$

$$(4.348)$$

$$\left. \left. -B(2-\beta) + \frac{1}{1-\beta} \right] \right\}, \quad V(1,1) = 3\pi^d H(1, 2\alpha) C(\alpha)$$

with the functions B and C from (4.347). With this we conclude our brief review of the techniques for calculating massless graphs. We can summarize as follows. The basis for integrability is always some uniqueness, and graphs with a sufficient number of uniquenessess can be integrated directly. Then there are graphs one step from uniqueness, i.e., graphs which become integrable upon a single application of recursion relations like (4.341) and (4.342), and so on. In particular, this class contain the V graph and all graphs obtained from it by the transformation group discussed above. Then come graphs that are integrable after two applications of (4.341), and so on. In the end, the class of integrable graphs turns out to be rather large, although it is often very difficult to establish the integrability of a particular graph owing to lack of a unique algorithm for choosing the needed transformation chain. For example, the result (4.346) was obtained in [89] only for the special case $\lambda = \alpha$ and in a very complicated way, because at that time the more convenient transformation chain (4.345) had not been found.

Finally, a few words about the history of this problem. The basic relation (4.328) for a unique vertex has long been known and widely used already in early studies on the conformal bootstrap method (see Sec. 4.40 and references therein). All the tricks used in this section, except the idea [133] of passing to vacuum graphs using (4.337), were stated in [89] as brief explanations of the methods used there to calculate the graphs for η_2 and ν_2 in arbitrary spatial dimension. The technique of [89] was later used by Kazakov (see [134] and references therein) and generally referred to as the uniqueness method for calculating the counterterms of 4- and 5-loop graphs in the φ^4 model. Such calculations were performed earlier in [65] and [66], but there some of the integrals were evaluated numerically. The author of [134] attempted to prove that the uniqueness method can be used to calculate all the needed integrals explicitly in the form of combinations of zeta functions $\zeta(n)$ with rational coefficients. This turned out to be fairly simple for all the 4-loop graphs and nearly all the 5-loop graphs. The problem at 5 loops reduced to the graph (4.333) in dimension $d = 4 - 2\varepsilon$ with indices $\alpha_i = 1 + \Delta_i$ on all five lines, which required calculation through order ε^4 in the expansion in the set of all the small parameters ε, $\Delta_i \sim \varepsilon$. The technique described above allowed the answer to be found only through order ε^3. The needed order-of-magnitude increase in the accuracy was achieved in [134] by a new technique, which the author referred to as the method of functional equations. It essentially is the following. One of the indices of the graph is assumed to be the variable α, and standard methods are used to obtain a recursion relation for the desired $f(\alpha)$ of the type $f(\alpha+1) = f(\alpha)+$known contributions (this is the "functional equation"). If the uniqueness of the solution, i.e., the absence of nontrivial solutions of the homogeneous recursion relation, is then somehow justified, then for $f(\alpha)$ we obtain a series representation, which can sometimes be summed. This is how the order-ε^4 contribution was calculated in [134] (see above), which made it possible to explicitly find the most difficult coefficient of $\zeta(7)$ in the 5-loop RG functions of the φ^4 model. The results [134] for the graph in question were then completed in [132] by the fuller inclusion of the structure of the transformation group for this graph.

The technique described in this section is now widely used in various calculations. One of its several advantages is the elegance with which the calculations are done — they are all represented graphically as a transformation chain without using any integral signs, and the coefficients that arise are successively accumulated to form the answer.

4.37   Calculation of η_2

The technique of the preceding section can be used to calculate the needed first terms of the expansions (4.315) for the four higher-order graphs in (4.306) even for arbitrary η, although to calculate ν_1 and η_2 it is sufficient to find them only in the zeroth-order approximation $\eta = 0$. Poles in κ are generated by logarithmic UV divergences [$m = 0$ in (4.340)] of the $\psi\phi^2$ vertex subgraphs. There are two such subgraphs in each of our graphs, and their UV divergences come from two nonoverlapping regions of integration corresponding to pulling all the inner vertices to either the right- or left-hand outer vertices. The parameter κ in (4.307) plays the role of the regularizer Δ for these UV divergences. For $\kappa = 0$ the principal $\psi\phi^2$ vertex with indices α, α, β is unique, which forms the basis of all the calculations. The second important feature is that the pole contribution of a logarithmic UV divergence from a given subgraph γ is not changed if the point of attachment of any line of the graph external to γ is shifted in any manner inside γ. To prove this it is sufficient to note that the difference of the original and new expressions contains an additional "small" factor $\sim y_1 - y_2$ (the difference of the coordinates of the points where the line is attached), which cancels the UV divergence when all the vertices of the subgraph γ are pulled to a point.

These two considerations form the basis of the calculational procedure, which we shall explain for the example of the most complicated graph Π_2 in (4.306b). We rewrite the original expression for Π_2 in the form $\Pi_2 \equiv A = (A - B - C) + (B + C)$, namely,

$$\Pi_2 = \left\{ \text{<graph>} - \text{<graph>} - \text{<graph>} \right\} + \{\text{last 2 graphs}\}. \quad (4.349)$$

The graphs B and C each contain only one of the two poles and are chosen such that the two vertex poles of the original graph A cancel in the combination $A - B - C$ (see the remark above). This makes it possible to calculate the combination for $\kappa = 0$ using the uniqueness of the α, α, β vertex, which gives one of the contributions to the desired quantity Π_2' in the expansion (4.315).

However, this procedure, like any cancellation of infinities, requires some care. We shall proceed as follows. Noting that the divergence arises only at the very last stage of the integration (i.e., when all the inner vertices are pulled close to a given outer one), we isolate any one of the inner vertices, which must be common to the three graphs in the combination $A - B - C$ (this is where one must be careful), and first we consider the integral over the coordinates of all the other vertices. This integral converges for all three graphs, and so the order of integration can be changed in it and the contributions A, B, and C can be considered separately. For our graphs (4.349) these integrals are easily done for $\kappa = 0$ using (4.328) if the lower left vertex is assumed to be fixed in all the graphs. The factor arising in the integration turns out to be the same for all three graphs, and the expression $A - B - C$ leads to the combination

$$\underbrace{\overset{d/2\;d/2}{\frown}}_{d/2-\beta} - \underbrace{\overset{\alpha\;\;d/2}{\frown}}_{\alpha} - \underbrace{\overset{d/2\;\;\alpha}{\frown}}_{\alpha}, \tag{4.350}$$

in which α and β are the indices (4.307) with $\kappa = 0$. Only the integration over the coordinate of the vertex fixed in the first stage remains in the combination (4.350), and this combination can be treated only as a whole (the integral of a sum of contributions), because each individual term diverges. However, since we know that (4.350) is finite as a whole, any other auxiliary regularization that makes the individual terms meaningful can be used to calculate it. It is simplest to do this by a small ε shift of the indices of the two links of the integrated chains; the shift must, of course, be the same in all three terms of (4.350). Integrating the chains and verifying that all the ε poles in the sum of graphs in (4.350) cancel, in the $\varepsilon \to 0$ limit we obtain the desired final result, the contribution to Π_2' from the combination $A - B - C$ in (4.349).

The similarity of the trick of taking $A \to A - B - C$ to the standard R' operation eliminating subgraph divergences (Sec. 3.17) should be noted. However, there is a very important difference. The subtracted terms B and C do not explicitly involve the regularizer κ, and this allows us to set it equal to zero in the entire combination, which cannot be done in the usual expressions of Sec. 3.17 with the R' operator (κ is the analog of ε).

We still need to calculate the contribution $(B+C)$ to (4.349). It contains poles and so it must be calculated using κ. This is possible because the graphs are simpler than the original one. Graph C can be integrated directly because it reduces to chains. Graph B with $\kappa \neq 0$ cannot be integrated directly, and so we again resort to a subtraction, taking $B = (B - D) + D$, namely,

$$B = \left\{ \text{⬭} - \text{⬭} \right\} + \{\text{last graph}\} \tag{4.351}$$

and we assume that the lower right-hand vertex is fixed in the first stage in the combination $B - D$. This UV-finite combination is again easily calculated for $\kappa = 0$, and the graph D itself in (4.351) is integrated with $\kappa \neq 0$, because it reduces to chains. Selecting the pole (in κ) and finite contributions in the results, we obtain the desired quantities (4.315) for the graph Π_2. For all the other graphs (Π_1 and $\Sigma_{1,2}$) the calculation is basically the same but simpler, and in the end we obtain

$$K_1 = 2\pi^d H(\alpha, \alpha, \beta)/\|d/2\|;$$
$$K_2 = \pi^{2d} H(\alpha, \alpha, \alpha, \beta, \beta, \beta, \alpha + \beta')/\|d/2\|; \tag{4.352}$$

$$\Sigma_1' = K_1[B(\beta) - B(\alpha)]/2; \quad \Sigma_2' = 2K_2[B(\beta) - B(\alpha)];$$
$$\Pi_1' = 2\Sigma_1'; \quad \Pi_2' = K_2[4B(\beta) - 3B(\alpha) - B(\alpha + \beta')] \tag{4.353}$$

with α and β from (4.312), $\beta' \equiv d/2 - \beta$, and the function $B(z)$ from (4.347). Using the residues (4.352) with $\eta = 0$, from (4.317) we obtain the result (4.318), and using Σ_i' and Π_i' from (4.319) we obtain (4.256a) for η_2. These equations can also be used to find u_2.

Let us make a final remark regarding this calculational scheme. The difficulties with calculating the graph B (see above) naturally suggest that for the subtractions in (4.349) it would be more convenient to use not B, but an explicitly integrable graph that is symmetric to C. However, in this new combination $A - B - C$ it is not possible to calculate the integrals over the coordinates of all the vertices except for a single particular one with the condition that this particular vertex is the same for all three graphs. If different vertices are chosen in the three graphs, it is not surprising that we obtain a finite but incorrect result, as this is equivalent to an illegal change of the order of integration in divergent integrals.

4.38   Generalization of the self-consistency equations to the case of correction exponents

Up to now we have used the self-consistency equations (4.304) as a method of calculating the leading terms of the IR asymptote of the correlators directly in the critical theory, and so we have sought power-law solutions like (4.259). The same equations can also be used to analyze various perturbations of simple power-law solutions. Such perturbations may be related, first, to IR-irrelevant corrections to scaling right at the critical point, and, second, to small deviations in $\tau \equiv T - T_c \geq 0$ from this point (Sec. 4.35). The numerical parameters for perturbations of the first type are the coefficients a of the IR-irrelevant operators F in the corrections to the action (Sec. 4.8), while for perturbations of the second type the parameter is $\tau \geq 0$ characterizing the deviation from the critical point for $h = 0$ [this condition is necessary to preserve the form of the equations (4.304)].

We denote the entire set of these parameters by e and assume that they are chosen such that each has definite critical dimension Δ_e. For parameters of the first group $\Delta_e < 0$ and the quantities $\omega = -\Delta_e$ are the corresponding correction exponents (4.52). For the second group $\Delta_e > 0$. Each of the parameters e is a "spatially uniform source" for some composite operator F with definite critical dimension Δ_F, related to Δ_e by the shadow relation $\Delta_F + \Delta_e = d$. Since e is uniform, all operators of the type $\partial F'$ with exterior derivatives drop out of the class of possible F. Moreover, we shall restrict ourselves to scalar parameters, thereby excluding possible perturbations by vector, tensor, etc. operators. Finally, we note that in the $1/n$ expansion the perturbation of a correlator by a given composite operator F may be manifested not immediately, but only in higher order in $1/n$. For example, addition to the action of an operator like $F = \phi \partial^{2m} \phi$ perturbs the correlator D_ϕ already in leading order in $1/n$, but operators F with four or more fields ϕ

do not affect D_ϕ in this order, because graphs of D_ϕ with the insertion of such F must contain a contracted massless line equal to zero. An operator with four ϕ can perturb D_φ only in the next-to-leading order, one with six ϕ in the next highest order, and so on. The same arguments hold for operators with several fields ψ: in leading order only operators with two ψ, in particular, the operator ψ^2 determining the leading correction exponent ω in (4.295), can perturb the correlators. All these arguments may be useful for interpreting the results.

It then follows that a small perturbation of a purely power-law solution for the correlators can be written as

$$D_\varphi(x; e) \underset{e \to 0}{=} D_\varphi(x; 0)\{1 + \textstyle\sum_e c_e \cdot e \cdot x^{\Delta_e} + ...\} \qquad (4.354)$$

with coefficients c_e not containing quantities with nonzero critical dimensions. In the correction (4.354) we have explicitly isolated only the primary perturbations linear in the corresponding parameters e, and the ellipsis denotes all possible secondary perturbations (products of the primary ones) and, in general, contributions containing fractional powers of e. For the parameter τ the presence of such contributions is obvious from (4.105): the powers of τ given there correspond in the correction factor (4.354) to contributions $[\tau \cdot x^{\Delta_\tau}]^{n + \Delta_F / \Delta_\tau} \sim x^{\Delta_F + n \Delta_\tau}$ made dimensionless by the replacement $\tau \to \tau \cdot x^{\Delta_\tau}$, with arbitrary $n = 0, 1, ...$ and dimensions Δ_F of all composite operators contributing to the asymptote (4.105). Contributions with $n = 0$ give corrections $\sim x^{\Delta_F}$ in (4.354), the exponents of which satisfy shadow relations with the exponents $\Delta_e = d - \Delta_F$ of the contributions (4.354) linear in the corresponding parameters e.

We shall seek a perturbative solution of (4.304) of the form

$$D_\phi = A \cdot x^{-2\alpha}[1 + A' x^{2\lambda}], \qquad D_\psi = B \cdot x^{-2\beta}[1 + B' x^{2\lambda}], \qquad (4.355)$$

assuming that the correction contributions are small and doing all calculations only through first order in the correction amplitudes A', B'. In this approximation all the effects of the various perturbations in (4.354) are additive and independent, which means that we can limit ourselves to only a single perturbation in (4.355). In this case the idea of self-consistency expresses the requirement that the perturbation in question be consistent with the basic solution, and this requirement must determine the possible values of the exponent 2λ in (4.355). It is clear that there must be many solutions and that different values of 2λ correspond to different perturbations in (4.354), but not to all of them. In the linear approximation (4.355) the secondary perturbations (products of primary ones) are certainly lost, and so the various solutions for the parameter 2λ obtained from the self-consistency equations must correspond either to exponents of the primary perturbations (4.354), or, possibly (see above), to values Δ_F related to them by the shadow relation:

$$\{2\lambda\} = \{\Delta_e = d - \Delta_F, \Delta_F\}. \qquad (4.356)$$

It is possible to identify the various solutions as the dimension of a particular parameter or of a particular operator only in the zeroth-order (in $1/n$) approximation. We can therefore expect difficulties to arise, not only in the interpretation, but also in the formulation of the equations themselves in the case of perturbations that are degenerate in zeroth order, i.e., when there are several operators F with different Δ_F which coincide in zeroth order. This will always be the case for any system of operators of high dimensions that mix nontrivially under renormalization, and it should be stated at the outset that for such systems the method of calculating dimensions that we have described has never been applied, and it is unclear in general how it can be generalized to this case. There is probably some generalization, but the question remains open. Finally, let us just add that the exclusion of operators like ∂F (see above) significantly limits the number of operators and, consequently, the mixing possibilities. For example, in the class of scalars with two ϕ and any number of ∂ there remains only the family of operators $F = \phi \partial^{2m} \phi$, each of which is uniquely identified by its canonical dimension. However, the mixing problem $\psi \to \partial \partial$ (Sec. 4.34) remains.

Specifically, we shall be interested only in the following two solutions associated with the exponents ν and ω:

$$2\lambda = 1/\nu \equiv \Delta_\tau = d - \Delta_\psi = \Delta[\phi^2] = d - 2 + O(1/n), \quad \lambda_0 = d/2 - 1; \quad (4.357\mathrm{a})$$

$$2\lambda = -\omega = d - \Delta[\psi^2] = d - 4 + O(1/n), \quad \lambda_0 = d/2 - 2. \quad (4.357\mathrm{b})$$

We have also given the interpretation of these values of 2λ in terms of (4.356), and the equation $\Delta[\phi^2] = d - \Delta_\psi$ is a consequence of one of the Schwinger equations of the type (3.187) for the model (4.266) (the equation with the derivative with respect to ψ leads to $[\phi^2]_\mathrm{R} \sim A_\psi$, and the dimension of the source A_ψ obeys the shadow relation with the dimension of the field ψ, and owing to (4.295) coincides with the dimension of the parameter τ). Therefore, in the $1/n$ expansion $\Delta[\phi^2] = \Delta_\tau$, while in the $4 - \varepsilon$ scheme the same operator in the massless theory has dimension obeying the shadow relation with τ: $\Delta[\phi^2] = d - \Delta_\tau$. This example shows that the critical dimension of a given composite operator, in contrast to the dimensions of parameters [see the remark following (4.295)] is not an objective quantity, because it depends on the choice of perturbation theory. This is not surprising, as a symbol like ϕ^2 corresponds in different schemes to different random quantities $\widehat{F} = \widehat{\phi}^2$ owing to the difference of the random field $\widehat{\phi}$ itself in them.

The exponents (4.357) are uniquely identified by their zeroth-order approximations λ_0, i.e., there are no mixing problems for them (owing to the exclusion of $\partial^2 \psi$; see above). It is also clear that corrections involving the exponents (4.357) can be primary perturbations in (4.354) [contributions with $e = \tau, a$, where a is the coefficient of ψ^2 in the addition to the action (4.266)], so that the corresponding values should be present in the set of possible solutions for the correction exponent 2λ in (4.355).

Turning now to the calculations, first we give the inversion formula for a correlator of the type (4.355) through first order in A', which is easily obtained using (3.72):

$$D(x) = Ax^{-2\alpha} [1 + A'x^{2\lambda}] \Rightarrow$$
$$D^{-1}(x) = A^{-1}p(\alpha)x^{2\alpha-2d}[1 - A'q(\alpha, \lambda)x^{2\lambda}] \tag{4.358}$$

with the function $p(\alpha)$ from (4.309) and the new function

$$q(z, \lambda) \equiv H(z - \lambda, z + \lambda - d/2, d - z, d/2 - z). \tag{4.359}$$

In this section we shall study (4.304) in the simple Hartree–Fock approximation (4.305). Substituting the lines (4.355) into the graphs (4.305) and using (4.358) for the inverse correlators on the left-hand sides, in the linear approximation after discarding the overall power of x we obtain the system

$$p(\alpha)[1 - A'q(\alpha, \lambda)x^{2\lambda}] + u[1 + (A' + B')x^{2\lambda}] = 0,$$
$$2p(\beta)[1 - B'q(\beta, \lambda)x^{2\lambda}] + nu[1 + 2A'x^{2\lambda}] = 0 \tag{4.360}$$

with the relation for the basic exponents and the constant u from (4.311). Assuming that Eqs. (4.311), which ensure the cancellation of the main contributions in (4.360), are satisfied, the condition for the cancellation of the correction contributions can be written as the system of equations

$$q(\alpha, \lambda)A' + A' + B' = 0, \quad q(\beta, \lambda)B' + 2A' = 0. \tag{4.361}$$

Its solvability condition is the equation

$$[1 + q(\alpha, \lambda)]q(\beta, \lambda) = 2, \tag{4.362}$$

satisfaction of which gives the ratio B'/A' from (4.361), with arbitrary common coefficient. For the basic exponents α and β already known from (4.312), Eq. (4.362) is the desired equation determining the possible values of the correction exponent 2λ in the simplest approximation (4.305).

Only the solution (4.357a) was studied in Ref. 89. Here we shall discuss (4.362) in more detail. It is easily verified from the definitions (4.359) and (3.68) that for any z and λ we have

$$q(z, 0) = 1, \quad q(z, \lambda) = q(z, \lambda'), \quad \lambda' \equiv d/2 - \lambda. \tag{4.363}$$

From the first it follows that $\lambda = 0$ is one of the solutions of (4.362), and the second shows that this equation and, therefore, the set of its solutions are invariant with respect to the shadow transformation:

$$2\lambda \rightarrow 2\lambda' = d - 2\lambda. \tag{4.364}$$

We immediately note that this symmetry is preserved even when the higher-order graphs in (4.306) are included in the correction equations

(Sec. 4.39). It follows that together with any solution 2λ, the shadow value $d - 2\lambda$ is also a solution in accordance with the general structure of the set of solutions given in (4.356).

This property and the asymptote (4.106) are related. The latter involves powers of τ with exponents 1, $1 - \alpha$, $2 - \alpha$, and $2 - \alpha + \omega\nu$, where $\alpha = 2 - d\nu$ is the exponent of the specific heat. The passage to the corrections (4.354) is effected by the replacement $\tau \to \tau \cdot x^{\Delta_\tau}$ with $\Delta_\tau = 1/\nu$, so that these powers of τ correspond to $x^{2\lambda}$ contributions in (4.354) with exponents $2\lambda = 1/\nu$, $(1 - \alpha)/\nu = d - 1/\nu$, $(2 - \alpha)/\nu = d$, and $(2 - \alpha + \omega\nu)/\nu = d + \omega$. The first is the exponent (4.357a), the second obeys the shadow relation with it, the third obeys the shadow relation with the known (see above) solution $2\lambda = 0$, and the last one obeys the shadow relation with (4.357b). Therefore, all the fractional powers of τ in (4.106) arise naturally along with the analytic contribution $\sim \tau$ in the solution of the self-consistency equations for the correction exponent 2λ.

Let us briefly discuss the set of all solutions of (4.362) for the exponent 2λ within the $1/n$ expansion. The functions $H(z) = \|d/2 - z\|/\|z\|$ (defined in Sec. 3.15) entering into (4.359) have poles at the points $z = d/2 + m$ and zeros at $z = -m$, where here and below $m = 0, 1, 2, ...$. It then follows that in our zeroth-order approximation $\alpha_0 = d/2 - 1$ and $\beta_0 = 2$, the factor $H(d-\alpha)$ in $q(\alpha, \lambda)$ has a pole in the deviation $\alpha - \alpha_0 \equiv \delta\alpha$, and so satisfaction of (4.362) requires the presence of a compensating zero either in the same function $q(\alpha, \lambda) \equiv q_1$ or in $q(\beta, \lambda) \equiv q_2$. The possible values of the zeroth-order approximation $2\lambda_0$ are thus determined: the first case corresponds to values $2\lambda_0 = 2 - 2m$ and values obeying the shadow relation with them (series A), and the second to values $2\lambda_0 = 4 + 2m$ and values obeying the shadow with them (series B). For series A we still need to distinguish the cases $m = 0, 1$ and $m \geq 2$. In the first the function q_1 has a single pole $\sim 1/\delta\alpha$ and a single zero $\sim (\delta\alpha + \delta\lambda)$, which allows us to find the ratio λ_1/α_1 of the corresponding first coefficients of the $1/n$ expansion (for all finite factors H without zeros and poles it is sufficient to limit ourselves to the known zeroth-order approximation, and for series A all the factors H in q_2 are finite). For the second case ($m \geq 2$ in series A) the function q_1 contains another pole $\sim 1/(\delta\alpha - \delta\lambda)$, and so the zero in $\delta\alpha + \delta\lambda$ must be of second order in $1/n$. It then immediately follows that $\lambda_1 = -\alpha_1$, and then we can also find the ratio $(\alpha_2 + \lambda_2)/\alpha_1^2$. We recall that $2\alpha_k = \eta_k$ for all k owing to (4.307), but in the present case η_2 should be understood not as the exact expression (4.256a), but only the Hartree–Fock contribution in $\eta_2 = 2\alpha_2$ obtained from (4.312).

Therefore, for the solutions of series A ($2\lambda_0 = 2 - 2m$ and the shadow values) with $m = 0, 1$ the exponent λ can be found from (4.362) through first order in $1/n$ and for $m \geq 2$ through second order. The solution with $m = 1$ is, in fact, exact ($2\lambda_0 = 2\lambda = 0$; see above), and the solution with $m = 0$ is the shadow value of the exponent (4.357a) determining the exponent ν, so that the result (4.255b) for ν_1 can be obtained from λ_1 for this solution. It should be emphasized that obtaining the same accuracy in the calculation of ν in terms of the exponent β using (4.308) would require the inclusion of higher-order graphs (Sec. 4.35), i.e., the calculation of ν as a correction exponent is technically simpler.

The remaining solutions with $m \geq 2$ in series A are, judging from the zeroth-order approximation $2\lambda_0 = 2-2m$, the shadow values of the dimensions Δ_F of the operators $F = \phi\partial^{2m}\phi$. However, for them there is the $\psi \to \partial\partial$ mixing problem (Sec. 4.34), and so no unique interpretation is possible, even though the solutions themselves for λ can be found through order $1/n^2$.

Let us now discuss the solutions of series B with $2\lambda_0 = 4 + 2m$ and their shadow values, which includes the exponent (4.357b) as the shadow of the solution with $2\lambda_0 = 4$. For all the solutions of this series the coefficient $q_1 \equiv q(\alpha, \lambda)$ in (4.362) has a pole $\sim 1/\delta\alpha$, and $q_2 \equiv q(\beta, \lambda)$ has a zero $\sim (\delta\lambda - \delta\beta) = \delta\lambda + 2\delta\alpha$ [using (4.311) relating α and β], which allows the ratio λ_1/α_1 to be found from (4.362).

However, this result for λ_1 will actually be incorrect, because owing to the anomalous smallness of the coefficient $q_2 \sim 1/n$ (in contrast to $q_2 \sim 1$ for series A) in the term $q_2 B'$ of the second equation in (4.361), the corrections from the higher-order graphs in (4.306) not included in (4.361) become comparable in magnitude. This is true only for the coefficient of B' in the second equation in (4.361), i.e., the corrections from ψ lines in the higher-order graphs in (4.306b). The inclusion of these corrections makes it possible to find the correct value of λ_1 for the solutions of series B, in particular, the value (4.255c) for ω_1. More details are given in the following section.

4.39 Calculation of ν_2 and ω_1

Let us consider the equations for the correction exponents taking into account the higher-order graphs $\Sigma_{1,2}$ in (4.306a) and $\Pi_{1,2}$ in (4.306b). When the lines (4.355) are substituted into them, the Hartree–Fock contributions in (4.316) are modified as in (4.360), and in the linear approximation each of the four higher-order graphs generates correction terms, for example,

$$\Sigma_i \to \Sigma_i + x^{2\lambda}[A'\Sigma_{iA} + B'\Sigma_{iB}] \tag{4.365}$$

and similarly for Π_i. Quantities like Σ_{iA} are represented as a sum of graphs obtained from the original graph Σ_i by all possible insertions of the correction factor $x^{2\lambda}$ into ϕ lines (which is equivalent to decreasing their indices by λ), while for Σ_{iB} the insertions are made into ψ lines. For example,

$$\Sigma_{1A} = 2 \,\text{—}\!\!\triangleleft\!\!\bigcirc\!\!\triangleright\text{—} \; + \; \text{—}\!\!\triangleleft\!\!\diamondsuit\!\!\triangleright\text{—} \;, \quad \Sigma_{1B} = 2 \,\text{—}\!\!\triangleleft\!\!\bigcirc\!\!\triangleright\text{—} \;, \tag{4.366}$$

where the cross denotes a perturbation of a line leading to decrease of its index by λ, and topologically equivalent versions of perturbation insertions are included by means of coefficients. The four higher-order graphs in (4.306) generate 13 correction graphs like (4.366).

Each of the four higher-order graphs in (4.306) has two identical poles in the "vertex exponent" κ from the left- and right-hand vertex subgraphs (Sec. 4.35). A perturbation $\sim x^{2\lambda}$ of any line inside a given vertex subgraph eliminates its UV divergence and the corresponding pole in κ, and so, for example, in the graph of Σ_{1B} and in the first graph of Σ_{1A} in (4.366) only a single pole from the right-hand vertex subgraph remains, while none remains in the second graph of Σ_{1A}. An elementary calculation of the number of poles remaining in the correction graphs shows that they are correlated with the poles of the basic graphs (4.315) as follows:

$$\Sigma_{iA} = K_i/\kappa + \Sigma'_{iA} + ..., \quad \Sigma_{iB} = K_i/\kappa + \Sigma'_{iB} + ...,$$

$$\Pi_{iA} = 2K_i/\kappa + \Pi'_{iA} + ..., \quad \Pi_{iB} = \Pi'_{iB} + ... \tag{4.367}$$

Here K_i are the residues of the poles in (4.315) known from (4.352), quantities like Σ'_{iA}, as in (4.315), denote κ-independent, finite parts of the graphs (in this case, correction graphs), and the ellipsis denotes unimportant corrections of order κ.

Adding all the correction contributions to (4.314), it is easily verified that owing to the relations between the residues in (4.315) and (4.367), all the poles in κ and contributions that are multiples of $\ln x$ in the basic and correction equations cancel simultaneously for the choice of vertex renormalization constant Z and exponent κ given in Sec. 4.35. Finally, we obtain the following renormalized system of correction equations:

$$A'[-p(\alpha)q(\alpha,\lambda) + u + u^2\Sigma'_{1A} + nu^3\Sigma'_{2A}] + B'[u + u^2\Sigma'_{1B} + nu^3\Sigma'_{2B}] = 0,$$

$$A'[2u + u^2\Pi'_{1A} + nu^3\Pi'_{2A}] + B'[-2p(\beta)q(\beta,\lambda)/n + u^2\Pi'_{1B} + nu^3\Pi'_{2B}] = 0. \tag{4.368}$$

The requirement that the determinant of this system be zero for the principal exponents α, β, and u known from (4.317) and (4.319) is the desired equation determining the correction exponent λ.

This equation can be solved by iterating in $1/n$ with arbitrary zeroth-order approximation. For the solution (4.357a) all the coefficients of A' and B' in (4.368) have the normal order of magnitude (leading contributions $\sim 1/n$ plus corrections $\sim 1/n^2$). It is sufficient to calculate the contributions of all the higher-order graphs in zeroth order in $1/n$, and then the "correction" ϕ line vanishes owing to the equation $\lambda_0 = \alpha_0 = d/2 - 1$. In this approximation all the needed graphs are fairly easy to integrate using the tricks described in Sec. 4.36. The most difficult is the graph Π_{2B}, the result for which can be obtained by substituting $\lambda = d/2 - 1$ into (4.346) (in [89] this graph is calculated in a more complicated manner). The final result of the calculation is the value (4.256b) for ν_2 in arbitrary dimension d obtained in [89].

When ω_1 is calculated in terms of the correction exponent (4.357b), the quantity $p(\beta)q(\beta,\lambda)/n$ in (4.368) has, as noted in Sec. 4.38, anomalous small value $\sim 1/n^2$ (rather than $1/n$, as in the case of ν), whereas the contributions of the higher-order graphs in Π'_{1B} and Π'_{2B} to this coefficient preserve the order of magnitude $\sim 1/n^2$ normal for them. Therefore, all these contributions of the same order must be included already when calculating the first coefficient $\omega_1 = -2\lambda_1$. The contributions of all other higher-order graphs to (4.368) can be neglected, but it is impossible to ignore the contribution κ_1 in the argument $\beta = 2 - \eta - \kappa$ of the function $q(\beta, \lambda)$ known from (4.318). The values of the higher-order graphs

$$\Pi_{1B} = \text{--}\!\!\!\!\raisebox{-0.3em}{\includegraphics}\!\!\!\!\text{--} \; , \quad \Pi_{2B} = 2\,\text{--}\!\!\!\!\raisebox{-0.3em}{\includegraphics}\!\!\!\!\text{--} \tag{4.369}$$

needed for this calculation have no UV divergences (so that $\Pi_{iB} = \Pi'_{iB}$) and can be calculated in zeroth order in $1/n$ ($\alpha = d/2 - 1$, $\beta = 2$, $\lambda = d/2 - 2$). The value of Π_{2B} can be obtained from (4.346) by the substitution $\lambda = d/2 - 2$. The result (4.346) is symmetric under the transformation (4.364), and the graph Π_{1B} in (4.369) has the same symmetry. This allows us to replace $\lambda = d/2 - 2$ by $\lambda = 2$, making the calculation of the graphs in (4.369) trivial, because for $\lambda = 2$ perturbation of a ψ line causes it to vanish, and the graphs of (4.369) then reduce to simple chains. This calculation of ω_1 was performed in the diploma study of J. R. Honkonen (Leningrad State University, 1980, unpublished).

Increasing the accuracy by one order, i.e., calculating ν_3 or ω_2 by this method, requires including the next highest-order graphs in (4.306). This cannot be done in practice owing to technical difficulties. It is also not possible to generalize such a technique (except for the calculation of η_1) to more complicated models of the CP^{n-1} type (Sec. 2.28), and so for them all the results are limited to first order in $1/n$ and are obtained more simply by the method described in Sec. 4.34. The technique of self-consistency equations for the propagators has been successfully used in the Gross-Neveu quantum-field model [135] and analogs of it. These are models containing fermions, which in the classical version correspond to anticommuting Grassmannian variables. They therefore do not bear any direct relation to the theory of the critical behavior of real classical systems. However, they are interesting for elementary particle physics and can serve as a good illustration of the possibilities of this technique (see Secs. 4.44 to 4.49 for more details about the Gross–Neveu model).

4.40 Calculation of η_3 in the σ model by the conformal bootstrap technique

The 1982 calculation of η_3 in the σ model [123] remained the only result in order $1/n^3$ for a decade, and only in 1993 did a second appear: the analogous calculation of η_3 for the Gross–Neveu model [136], [137]. These results were obtained using the self-consistency skeleton equations of the conformal bootstrap technique with not only dressed lines, but also dressed triple vertices of the $\psi\phi^2$ type. The known conformal invariance of the massless σ

model (Sec. 2.17) along with other symmetries determining the explicit form of the dressed lines and triple vertices up to parameters (Sec. 2.18) were essential in the calculation. The method is appicable to any massless model with conformal symmetry and only triple (no higher) bare vertices, including our model (4.258). To avoid confusion, it should be made clear that by conformal symmetry of a model we always mean the conformal invariance of all the Green functions of its fields (Sec. 2.18), but this symmetry does not carry over automatically to scale-invariant Green functions with composite operators.

Let us explain the conformal bootstrap technique for the simple example of the massless φ^3 model with one scalar field $\varphi(x)$ (Sec. 4.24). It is well known (see, for example, [53]) that in models of this type the full vertex (a 1-irreducible three-tail object) can be represented as the sum of the bare vertex and all 3-irreducible skeleton graphs with dressed lines and vertices and the standard symmetry coefficients. Graphs that do not contain nontrivial 2- and 3- subgraphs (i.e., self-energy and vertex subgraphs) are termed 3-irreducible. The corresponding equation is written graphically in (4.370), and the representation for the self-energy Σ in the Dyson equation (2.51) following from it is given in (4.371):

$$\hspace{3cm} , \qquad (4.370)$$

$$\Sigma = \hspace{5cm} . \qquad (4.371)$$

Here and below the symmetry coefficients, in particular, the $1/2$ with the graphs of Σ, will be implicit. A shaded block with three outgoing lines (a circle or triangle, whichever is most convenient) denotes a full vertex, a point denotes a bare vertex, and all the internal lines of the graphs are dressed and the external ones are amputated. The second equation in (4.370) is the definition of the shaded square (4-block) in the form of a known infinite sum of skeleton graphs with standard coefficients.

In general, a vertex equation like (4.370) together with the Dyson equation (2.51) and the stationarity equation (1.51) in the needed form make up a system of three equations determining three unknowns - the vertex Γ, the propagator D, and the expectation value $\langle \hat{\varphi} \rangle \equiv \alpha$. We set $\alpha = 0$ directly in the critical (massless) theory, and so only the propagator and vertex equations remain. The conformal bootstrap technique amounts to discarding all the bare contributions, which leads to "self-consistency" equations for D and Γ:

$$\sqsubset = \sqsubset\!\!\!\boxed{} \quad , \quad D^{-1} = -\Sigma = \text{⟨▨▨▷} - \text{⟨▨▷} \quad . \qquad (4.372)$$

Their solutions are sought in the form of conformally invariant structures of the type (2.147), namely,

$$D(x) = Ax^{-2\alpha}, \qquad \Gamma(x_1, x_2, x_3) = C \cdot r_{12}^{-2a_{12}} r_{13}^{-2a_{13}} r_{23}^{-2a_{23}} \qquad (4.373)$$

with $r_{ik} \equiv |x_i - x_k|$ and unknown amplitudes and exponents. In (4.373) we give the general form of the 1-irreducible vertex Γ corresponding to the full function (2.147b) with three arbitrary field dimensions α_i in terms of which the three exponents in Γ are expressed. Graphically,

$$D = A \, \bullet\!\!-\!\!\bullet \, ; \, \Gamma \sim \begin{array}{c} \text{(triangle diagram)} \end{array} \qquad \left.\begin{array}{l} \alpha_1 + a_{12} + a_{13} = d, \\ \alpha_2 + a_{12} + a_{23} = d, \\ \alpha_3 + a_{13} + a_{23} = d. \end{array}\right\} \qquad (4.374)$$

The index α of the line D is the dimension of the corresponding field, the vertex Γ corresponds only to the conformal triangle itself in (4.374), and the indices of its three sides a_{ik} are uniquely determined by the requirement that all the vertices be unique (Sec. 4.36) after external lines with specified indices α_i are attached to them. In the special case of the simple φ^3 model we have

$$D = A \, \bullet\!\!-\!\!\bullet \, ; \quad \Gamma \equiv \begin{array}{c} \text{(triangle diagram)} \end{array} = C \cdot \begin{array}{c} \text{(triangle diagram)} \end{array}, \quad 2a + \alpha = d. \qquad (4.375)$$

When these quantities are substituted into (4.372) the general structure of all the expressions is self-reproduced, and in the end we obtain simple algebraic equations for the numerical parameters contained in (4.375).

Let us explain this, beginning with the vertex equation (4.372). The 4-block entering into it is understood as the sum of irreducible skeleton graphs known from (4.370). Substitution of the elements (4.375) into them leads to ordinary graphs with power-law lines and local triple vertices. *Definition.* We shall term an arbitrary 1-irreducible graph conformal if all its internal vertices are unique. By construction, the right-hand side of the vertex equation (4.372) is the sum of conformal vertex graphs. The basis of the conformal bootstrap method is the following simple statement: *every conformal vertex graph coincides exactly up to a coefficient with a simple conformal triangle of the type (4.374), specified by the indices of three external lines* (the latter are uniquely reconstructed by the requirement that the three external vertices be unique after external lines are attached to them). The proof of this statement is elementary. Making a conformal transformation of the original graph with the base at one of its three outer vertices, we obtain a graph in which all the inner vertices are disconnected from the base (Sec. 4.36). In it only a single block of the type (4.325) joining the two other outer vertices is nontrivial. It reduces to a simple line, and so the vertex graph itself reduces to a triangle. Another conformal transformation leads to a vertex triangle of the type (4.374).

It then follows that any graph on the right-hand side of the vertex equation (4.372) and, therefore, the sum of all these graphs coincide up to a numerical factor with the original vertex (4.375) on the left-hand side. We therefore have

$$ \text{[graph]} = V \; \text{[graph]} \;, \quad V = V(\alpha, u), \quad u \equiv C^2 A^3, \tag{4.376} $$

which is the definition of the *vertex function V*. It depends on the single combination of amplitudes $u = C^2 A^3$ and on the index α. The dependence on the spatial dimension d is not written out explicitly but is everywhere understood. In this notation the vertex equation (4.372) reduces to the algebraic equation $V(\alpha, u) = 1$, which gives one equation for the two unknowns α and u. The function $V(\alpha, u)$ itself is assumed to be known from its definition (4.376) in terms of the sum of graphs in (4.370). Of course, in practice it can only be calculated approximately in terms of the first graphs.

A second relation between the two unknowns α and u can be obtained from the propagator self-consistency equation in (4.372). However, it contains an indeterminacy of the type $0 \cdot \infty$, which must first be resolved. Actually, on the one hand, the two graphs in (4.372) for Σ must cancel each other out owing to the vertex equation (4.372) [in other words, according to the first equation in (4.371) the self-energy Σ is proportional to the bare vertex, which is set to zero in the vertex equation (4.372)]. On the other hand, the substitution of the elements (4.375) into any of the skeleton graphs of (4.372) for Σ gives a conformal graph with UV divergences from the two logarithmic vertex subgraphs corresponding to pulling all the inner vertices to one of the outer ones. We note that the representation (4.371) itself for Σ is exact; the $0 \cdot \infty$ indeterminacy in it arises only when the bare vertex is discarded and the conformal structures (4.375) substituted. It therefore must be resolved before taking the limit of zero bare vertex.

In order to do this, (4.372) is rewritten as

$$ x_i D^{-1}(x) = -x_i \Sigma(x) \equiv -\Sigma_i(x). \tag{4.377} $$

Graphically [using the notation of (4.320) with the base at the left-hand vertex],

$$ \Sigma_i = \text{[graph]} - \text{[graph]} \;. \tag{4.378} $$

Using the obvious equations [from the definition of a line with an arrow in (4.320)],

$$(4.379)$$

we can bring (4.378) to the form

$$(4.380)$$

Actually, when (4.379) is substituted into (4.378), owing to the exact vertex equation (4.370) pairs of the first and second graphs in (4.379) add to form the combination

$$= 0 , \qquad (4.381)$$

equal to zero owing to an equation of the type $x_i \delta(x) = 0$ following from the assumed locality of the bare vertex and the absence of any ∂ in it (the simple φ^3 interaction). Only the contributions of the last graphs in (4.379) remain, and they transform the representation (4.378) into (4.380). It has an advantage over (4.378) in that it does not contain any UV divergences when the structures (4.375) are substituted, owing to the introduction of additional factors of x_i (lines with an arrow) eliminating the logarithmic UV divergences which arise from pulling in both the left- and the right-hand vertex subgraphs [the same mechanism as for the graphs (4.369)].

From (4.309), (4.377), and (4.380) we obtain the desired form of the propagator equation without the $0 \cdot \infty$ indeterminacy (a line with an arrow is understood to carry a free vector index):

$$(4.382)$$

Equation (4.382) together with the vertex equation (4.372) can already be used for specific calculations if we limit ourselves to some approximation for the 4-block in them. However, it is much more convenient to use the modification of (4.382) proposed in [138], which allows us to restrict ouselves to the calculation of the vertex function V using some ε regularization. For this it is necessary to introduce a regularized vertex Γ_ε and a regularized vertex function V_ε:

$$\Gamma_\varepsilon \equiv \text{(graph)} = C \cdot \text{(triangle)} \; ; \quad \text{(graph)} = V_\varepsilon \; \text{(graph)} . \qquad (4.383)$$

This regularization preserves the uniqueness of the vertices of the conformal triangle, and so all the vertex graphs remain conformal. In the graphs defining V_ε only the vertex triangle directly adjacent to the regularized external line is subject to regularization; the indices of all the other lines and all the amplitude factors are not changed.

The quantities (4.383) are used to simplify the right-hand side of (4.382) as follows. Since there are no divergences in the graphs of (4.382), they are regular in the indices of their external lines near their canonical values, and so the following transformation chain is valid:

$$\text{(graph)} = \lim_{\varepsilon \to 0} \text{(graph)} =$$
$$= \lim_{\varepsilon \to 0} \left\{ \text{(graph)} - \text{(graph)} \right\} =$$
$$= \lim_{\varepsilon \to 0} \left\{ V \text{(graph)} - V_\varepsilon \text{(graph)} \right\} = \qquad (4.384)$$
$$= \lim_{\varepsilon \to 0} \left\{ [V - V_\varepsilon] \text{(graph)} - V \text{(graph)} \right\} .$$

In the derivation we have used the definitions of V and V_ε in (4.376) and (4.383).

The vertex equation (4.372) gives $V = 1$, and so the last graph of (4.384) cancels one of the graphs when substituted into (4.382), and in the end we find that

$$A^{-1} p(\alpha) \; \text{(graph)}_{d-\alpha} = \lim_{\varepsilon \to 0} [V_\varepsilon - V] \; \text{(graph)} . \qquad (4.385)$$

Substituting the triangles (4.375) and (4.383) into this, we obtain a single graph, and in it we need to calculate only the residue at the ε pole generated by pulling all the inner vertices to the left-hand outer one. This calculation is easily performed using the tricks of Sec. 4.36. The residue for the graph with unit amplitudes reduces to the graph on the left-hand side of (4.385) with the coefficient

$$S(\alpha) = \pi^{2d} H(\alpha, \alpha, a, d/2 + a - \alpha)/\|d/2\|, \qquad (2a + \alpha = d), \qquad (4.386)$$

and all the amplitudes and the symmetry factor of $1/2$ (which so far we have omitted) combine to form the factor $u/2$ on the right-hand side. Finally, the complete system of conformal bootstrap equations for the φ^3 model takes the form

$$V(\alpha, u; \varepsilon)|_{\varepsilon=0} = 1, \tag{4.387a}$$

$$2p(\alpha) = uS(\alpha)\partial V(\alpha, u; \varepsilon)/\partial \varepsilon|_{\varepsilon=0} \tag{4.387b}$$

with the functions p from (4.309), S from (4.386), and $V_\varepsilon \equiv V(\alpha, u; \varepsilon)$ from (4.383).

To make practical use of (4.387) we need to perform a diagrammatic calculation of only the regularized vertex function V_ε only through terms of first order in the regularizer ε. By iterating (4.387) in the φ^3 model we can find the coefficients of the $6 + 2\varepsilon$ expansion of the critical dimension of the field $\Delta_\varphi \equiv \alpha = d/2 - 1 + \eta/2$ (Sec. 4.25).

The technique explained for the example of the φ^3 model is easily generalized to the σ model (4.258), allowing the coefficients of the $1/n$ expansions of the critical dimensions $\alpha \equiv \Delta_\phi$ and $\beta \equiv \Delta_\psi$ to be calculated by iterating the conformal bootstrap equations. Solutions for the correlators (dressed lines) are sought in the form (4.259) and the $\psi\phi^2$ vertex is represented by the conformal triangle (4.388). Its regularization is the following (the link indices a and b are expressed in terms of α and β by the uniqueness conditions $2a + \beta = d$, $a + b + \alpha = d$, and in the second triangle we show only the additions from the regularizers):

$$\tag{4.388}$$

Owing to the presence of two fields it is necessary to introduce two independent regularizers ε and ε', and the regularized vertex function $V_{\varepsilon,\varepsilon'}$ is determined not by (4.383), but by

$$\tag{4.389}$$

As usual, in the graphs determining $V_{\varepsilon,\varepsilon'}$ only the conformal triangles right next to the regularized external lines are regularized (now there are two: one with ε and the other with ε'). The indices of the other vertex triangles and all the internal lines are not changed. To calculate the dimensions α, β through $1/n^3$ in α (i.e., η_3) and $1/n^2$ in β (i.e., ν_2), it is necessary to include the following nine graphs in the definition of the vertex function (for compactness, we indicate the vertex conformal triangles simply by blackened circles):

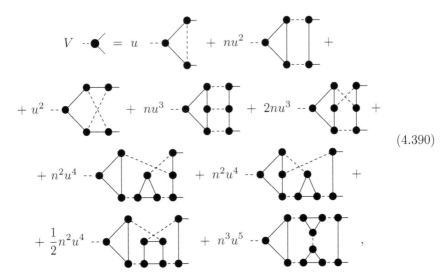

$$(4.390)$$

where $u \equiv C^2 A^2 B$ [A and B are the amplitudes of the lines (4.259) and C is the amplitude of the vertex]. In the preceding order ($1/n^2$ in α and $1/n$ in β) only the first two graphs would be needed. The complete system of conformal bootstrap equations analogous to (4.387) for determining the two exponents α, β and the amplitude combination $u \equiv C^2 A^2 B$ in terms of the regularized vertex function $V_{\varepsilon,\varepsilon'} \equiv V(\alpha, \beta, u; \varepsilon, \varepsilon')$ then looks as follows in this model:

$$V(\alpha, \beta, u; \epsilon, \epsilon')|_{\epsilon=\epsilon'=0} = 1,$$

$$p(\alpha) = uS(\alpha, \beta)\partial V(\alpha, \beta, u; \epsilon, \epsilon')/\partial \epsilon'|_{\epsilon=\epsilon'=0}, \qquad (4.391)$$

$$2p(\beta)/n = uS(\alpha, \beta)\partial V(\alpha, \beta, u; \epsilon, \epsilon')/\partial \epsilon|_{\epsilon=\epsilon'=0}$$

with the usual functions p from (4.309) and the quantity analogous to (4.386):

$$S(\alpha, \beta) = \pi^{2d} H(\alpha, \alpha, a, a, \beta, b, d/2 + b - \beta)/\|d/2\|. \qquad (4.392)$$

The details of the calculation of η_3 using (4.391) can be found in [123]. The result for arbitrary spatial dimension $d = 2\mu$ is (in the order of exclusion here we use the notation of [123] $\mu \equiv d/2$)

$$\eta_3/\eta_1^3 = 3\mu^2\alpha(4\mu - 5)I(\mu)S_3/2\beta^2 +$$

$$+2\mu^2\alpha(2\mu - 3)^2[3S_0 S_1 - S_2 - S_0^3]/3\beta^3 + \frac{1}{2}\left\{70 + 26\mu + \right.$$

$$+8\mu^2 - 177/\beta + 67/\beta^2 + 58/\beta^3 - 16/\beta^4 + 9/\alpha + 1/\alpha^2 + 1/\mu^2 +$$

$$+B[66 + 14\mu + 4\mu^2 - 187/\beta + 102/\beta^2 + 16/\beta^3 + 2/\alpha + 3/\mu] +$$

$$+B^2[20 - 50/\beta + 32/\beta^2] + S_3[-45 - 10\mu + 7\mu^2 + 127/\beta - \tag{4.393}$$

$$-64/\beta^2 - 48/\beta^3 + 32/\beta^4] + S_4[14 + 8\mu + 8\mu^2 - 30/\beta] +$$

$$+BS_3[-45 - 13\mu - 2\mu^2 + 136/\beta - 108/\beta^2 + 32/\beta^3]\bigg\},$$

where $\mu \equiv d/2$, $\alpha \equiv \mu-1$, $\beta \equiv 2-\mu$, $B \equiv B(2-\mu)-B(2)$, $S_0 \equiv B(2-\mu)-B(1)$, $S_1 \equiv C(2-\mu) - C(1)$, $S_2 \equiv D(2-\mu) - D(1)$, $S_3 \equiv C(\mu-1) = -C(1)$, and $S_4 \equiv C(2-\mu)$ with the functions B, C, and D from (4.347).

The result (4.393) contains the quantity $I(\mu \equiv d/2)$, which cannot be expressed explicitly in terms of gamma functions and their derivatives and is determined by the relation $\Pi(\mu, \Delta) = \Pi(\mu, 0)[1 + \Delta I(\mu) + O(\Delta^2)]$, where $\Pi(\mu, \Delta)$ is the value of the graph (4.333) with unit amplitudes and link indices $\alpha_1 = \alpha_4 = 1$, $\alpha_2 = \alpha_3 = \mu - 1$, $\alpha_5 = \mu - 1 + \Delta$ in dimension $d = 2\mu$. For $\Delta = 0$ this quantity is $V(1, 1)$ from (4.348), but already the next contribution linear in Δ in arbitrary dimension cannot be calculated explicitly using the standard technique. It is possible to find only the particular value $I(\mu = 3/2) = 3\psi''(1/2)/2\pi^2 + 2\ln 2$, which allows the value of η_3/η_1^3 given in (4.257) to be obtained from (4.393) for $d = 3$ $[\psi''(1/2) = -14\zeta(3)]$ along with the first coefficients of the ε expansions $I(\mu = 1 + \varepsilon) = -2/3\varepsilon + 0 + O(\varepsilon)$, $I(\mu = 2 - \varepsilon) = 0 + O(\varepsilon)$, which are needed to check the consistency of the result (4.393) with the $2 + \varepsilon$ and $4 - \varepsilon$ expansions [see Sec. 4.49 for more details about $I(\mu = 1 + \varepsilon)$].

In (4.393) we have corrected a misprint in the exponent $\beta \equiv 2 - \mu$ in the denominator of the second contribution in the result of [123] for arbitrary dimension.

The obvious advantages of the conformal bootstrap method are, first, the smaller number of graphs [the calculation of η_3 by the technique of Sec. 4.34 would require 36 graphs, whereas there are only 9 in (4.390)] and, second, the absence of UV divergences in the graphs so that there is no need for renormalization. However, the method is applicable only to models with guaranteed conformal invariance in the critical regime (i.e., at the fixed point of the renormalization group) and only with triple bare vertices. At present (1997) it has been successfully applied in practice only to the simple σ model discussed above, and to the Gross–Neveu model, which will be studied in detail later on (Sec. 4.44). From the purely technical point of view the calculation of η_3 apparently represents the limit of the possibilities offered by this (or any other) method — the next order $1/n^4$ is clearly inaccessible unless there is a major advance in the calculational technique.

4.41   Conformal invariance in the critical regime

In this section we supplement the material of Sec. 2.17 using the results of Ch. 3 on the theory of the renormalization of composite operators. The problem is stated as follows. We are given a multiplicatively renormalizable local model for an arbitrary system of fields φ (not necessarily scalar) with non-trivial fixed point of the renormalization group g_* at which there is critical scaling, and, consequently, critical scale invariance of the Green functions of the fields in the massless model. We wish to know if these scale-invariant Green functions are also conformally invariant.

The analysis starts with the Ward identities (2.130) for scale and conformal transformations with canonical dimensions of the fields in the renormalized massless model (i.e., $\Delta = d_\varphi$, $S \equiv S_R$, $W \equiv W_R$ in all the expressions of Secs. 2.15 and 2.16), namely,

$$\partial_i J_i^\alpha(x) + [\delta S_R/\delta \varphi(x)] T^\alpha \varphi(x) = -N^\alpha(x), \qquad (4.394)$$

where J_i^α are the canonical currents defined by the expressions of Sec. 2.15 in terms of the given functional $S_R(\varphi)$, T^α are the group generators, and $N^\alpha(x)$ is the corresponding *symmetry breaking operator*. For our groups in the general case of fields with spin and arbitrary dimension Δ the generators have the form (Sec. 2.15)

$$T^\alpha \equiv T = x\partial + \Delta \qquad \text{(scale),} \qquad (4.395a)$$

$$T^\alpha \equiv T^k = x^2 \partial_k - 2x_k(x\partial) - 2x_k \Delta + 2\Sigma^{ks} x_s \quad \text{(conformal),} \qquad (4.395b)$$

where Σ^{ks}, the spin part of the rotation generator, are matrices acting on the indices of the fields and sources, which we shall not indicate explicitly. We recall that in the Euclidean formalism there is no difference between upper and lower indices, and we shall arrange them in the formulas only on the basis of uniformity and convenience of the notation.

The dimension Δ in the generators (4.395) is arbitrary. In the original identities (4.394) we considered canonical transformations with $\Delta = d_\varphi$ in the generators (4.395) and in the corresponding currents. The identity (4.394) leads to the local relation (2.129) with the addition of $N^\alpha(x)$ to the current divergence; integrating it over x using the definition (2.122) for $W = W_R$ in our case we obtain

$$\mathcal{D}_A^\alpha W_R(A) = \int dx \langle\!\langle N^\alpha(x) \rangle\!\rangle, \quad \mathcal{D}_A^\alpha \equiv \int dx\, A(x) T^\alpha \delta/\delta A(x) \qquad (4.396)$$

[the contribution involving $\delta(0)$ which is unimportant in dimensional regularization has been dropped]. The equality (4.396) without N^α would imply the presence of the corresponding symmetry of the connected (and therefore all other) Green functions. However, canonical scale and conformal symmetries are always violated by regularization (Sec. 2.15), i.e., the contribution $N^\alpha(x)$ in (4.396) must be present in our case.

To analyze the consequences of (4.396) in the critical regime it is essential (see below) to express $N^\alpha(x)$ in terms of renormalized composite operators and then set $g = g_*$ in the coefficients. The scale operator $N^\alpha \equiv N$ has canonical dimension $d[N] = d$ and its expansion in renormalized quantities can contain, among other things, UV-finite (Sec. 3.29) composite operators:

$$\varphi(x)\delta S_R(\varphi)/\delta\varphi(x) = [\varphi(x)\delta S_B(\varphi)/\delta\varphi(x)]_R. \qquad (4.397)$$

Their contribution to the scaling equation (4.396) is, owing to (3.188), equivalent to a shift of the canonical dimension d_φ in the generators (4.395a). If *a priori* we were sure of the presence of critical scale invariance with dimensions $\Delta_\varphi = d_\varphi + \gamma_\varphi^*$ (which is assumed), then the shift $d_\varphi \to \Delta_\varphi$ would be the only effect of breaking the corresponding canonical symmetry. This means that the scale operator $N^\alpha \equiv N$ expressed in terms of renormalized operators must have the following representation for $g = g_*$:

$$N(x) = \sum_\varphi \gamma_\varphi^* \varphi(x)\delta S_R(\varphi)/\delta\varphi(x) + \sum \partial F, \qquad (g = g_*) \qquad (4.398)$$

with summation over all types of fields in the first term and any operator of the type ∂F in the second. The presence of the exterior derivative is the necessary condition for the contribution of this operator to (4.396) to vanish upon integration over x. We note that the operator (4.397) is defined precisely for $g = g_*$ by the right-hand side of this equation. The left-hand side involves constants Z that have a singularity for $g = g_*$ (Sec. 1.27). Such singularities are also present implicitly in unrenormalized local monomials. This is why deciphering the meaning of any composite operator for $g = g_*$ requires expressing it in terms of renormalized quantities.

Comparison of the two generators (4.395) shows that critical conformal invariance with dimensions $\Delta_\varphi = d_\varphi + \gamma_\varphi^*$ will occur when and only when the conformal operator $N^\alpha \equiv N^k$ has the following representation for $g = g_*$:

$$N^k(x) = -2x_k \sum_\varphi \gamma_\varphi^* \varphi(x)\delta S_R(\varphi)/\delta\varphi(x) + \sum \partial F, \qquad (g = g_*). \qquad (4.399)$$

The validity of (4.398) is guaranteed by the presence of (the assumed) critical scale invariance (it usually follows directly from the RG equation for $g = g_*$; see Sec. 1.33), and proof of the associated conformal symmetry is equivalent to proof of the representation (4.399). In any specific model the operators N and N^k can be constructed explicitly from the definition (4.394); after analyzing the renormalization of all the monomials entering into them it is possible to directly verify the validity of the representation (4.399). However, the answer can sometimes be obtained without calculation on the basis of general arguments, as was done in Sec. 2.17 for purely scalar theories. We shall now describe a broader class of models for which it can be proved that critical conformal symmetry is an automatic consequence of scale invariance.

Let $S_R(\varphi) = \int dx \mathcal{L}(x)$, where $\mathcal{L}(x) = \mathcal{L}_0(x) + \mathcal{V}(x)$ is the renormalized Lagrangian density, $\mathcal{L}_0$ is its free part, and $\mathcal{V}$ is the interaction. In this section we shall take $\mathcal{L}_0$ to include all contributions to $\mathcal{L}$ quadratic in the fields (in the usual terminology, all counterterms, including quadratic ones, are included in the interaction), and $\mathcal{V}$ to be the higher powers. We assume that (1) the action $S_R(\varphi)$ is local and possesses the usual (translational and rotational) symmetries; (2) the quadratic part of the action $S_0(\varphi) = \int dx \mathcal{L}_0(x)$ is scale and conformally invariant under transformations with canonical dimensions d_φ; (3) the interaction $\mathcal{V}(x)$ contains only the fields $\varphi(x)$ themselves and does not contain their derivatives $\partial \varphi(x)$. Clearly, all this is true of many multiplicatively renormalizable massless models.

With these assumptions it is easy to prove the presence of the following simple relation between the scale (N) and conformal (N^k) breaking operators in the Ward identities:

$$N^k(x) = -2x_k N(x). \tag{4.400}$$

For the proof it is sufficient to note that with our assumptions, the operators N and N^k are generated only by the interaction $\mathcal{V}$, whose contributions to the left-hand side of (4.394) are easily traced using the explicit expressions for the currents given below (2.119), namely, the contribution of $\mathcal{V}$ to the derivative $\partial \mathcal{L}/\partial(\partial_i \varphi_a) \equiv \phi_{ia}$ is zero, it is $-\mathcal{V}\delta_{ik}$ in the energy-momentum tensor ϑ_i^k, $-\mathcal{V}x_i$ in the scale current J_i, $(-x^2\delta_{ik} + 2x_i x_k)\mathcal{V}$ in the conformal current J_i^k, and $\partial \mathcal{V}/\partial \varphi_a$ in the variational derivative $\delta S_R(\varphi)/\delta \varphi_a(x) \equiv u_{aR}(x;\varphi)$ (in the expressions of this section we shall omit indices like a, understanding that they are present as needed). Substituting these expressions into the left-hand sides of the corresponding identities (4.394) with $\Delta = d_\varphi$ in the generators (4.395), we easily obtain the following explicit expressions for the breaking operators:

$$N(x) = \partial_i[-\mathcal{V}x_i] + [\partial \mathcal{V}/\partial \varphi]T\varphi = -d\mathcal{V} + d_\varphi[\partial \mathcal{V}/\partial \varphi]\varphi, \tag{4.401}$$

$$N^k(x) = \partial_i[(-x^2\delta_{ik} + 2x_i x_k)\mathcal{V}] + [\partial \mathcal{V}/\partial \varphi]T^k\varphi =$$
$$= -2x_k N(x) + 2x_s[\partial \mathcal{V}/\partial \varphi]\Sigma^{ks}\varphi. \tag{4.402}$$

In the derivation we have used the explicit form of the generators (4.395) with $\Delta = d_\varphi$ and the assumed (see above) absence of derivatives $\partial \varphi$ in the interaction $\mathcal{V}$. If $\mathcal{V}$ contains only a single monomial, from (4.401) it follows that $N \sim \mathcal{V}$; in general, N is a linear combination of all the monomials making up $\mathcal{V}$ with coefficients of order $\varepsilon \sim (d^* - d)$, since in the logarithmic dimension $d = d^*$ the interaction is also canonically scale-invariant owing to the fact that all charges are dimensionless (Sec. 1.16).

The last term on the right-hand side of (4.402) is a multiple of the variation of the interaction $\mathcal{V}$ for a purely spin rotation, and with our assumptions (locality, rotational invariance, and the absence of derivatives in the interaction) it must vanish owing to the corresponding invariance of $\mathcal{V}$. This leads to the desired relation (4.400).

Let us now consider its consequences. In general, there are several independent operators like (4.397) (as many as there are fields in the system φ), and they can conveniently be used as some of the basis elements of the complete system (including all independent monomials with the symmetry of $\mathcal{L}$ and $d_F^* = d^*$, where d_F^* is the value of d_F in the logarithmic dimension $d = d^*$) of renormalized composite operators in which $N(x)$ is expanded. In general, this expansion contains contributions of three types, denoted A, B, and C:

$$N(x) = \sum_\varphi a_\varphi \varphi(x) \delta S_\mathrm{R}(\varphi)/\delta\varphi(x) + \sum b[F(x)]_\mathrm{R} + \sum c\partial[F(x)]_\mathrm{R}. \quad (4.403)$$

The first sum (A) contains all the operators (4.397), and the other two contain all the other independent operators with the needed symmetry and dimension, either without exterior derivatives (contribution B) or with them (contribution C). [Example: in the simple φ^4 model there is a single operator (4.397), one contribution $[\varphi^4]_\mathrm{R}$ of type B, and one $\partial^2[\varphi^2]_\mathrm{R}$ of type C]. The various coefficients a, b, and c in (4.403) depend on the charge (or charges) g, on the parameter $\varepsilon \sim d^* - d$, and on the renormalization mass μ through the factor μ^ε, but certainly do not contain any explicit dependence on x in the case of operators of the type (4.401) that we are interested in. We also note that coefficients like a and b must be UV-finite owing to the obvious UV-finiteness of the left-hand side of (4.396).

Critical scale invariance leads to the representation (4.398), which compared with (4.403), implies

$$a_\varphi(g_*) = \gamma_\varphi^*, \quad b(g_*) = 0, \quad (4.404)$$

while the values of $c(g_*)$ are not determined. In specific models the coefficients of the operators in the expansion (4.403) are found from the matrix of renormalization constants of the closed system (Sec. 3.28) containing all the monomials of $\mathcal{L}$. Coefficients like a and b (but not c, like Z^{43} in Sec. 4.10) can be expressed in terms of the renormalization constants of the action (Sec. 3.30), the coefficients b turn out to be multiples of β functions, and the a_φ are the anomalous dimensions of the fields γ_φ, which ensures that (4.404) are satisfied.

Let us now turn to the operator N^k. Since external factors of x_k do not affect the renormalization, it follows from (4.400) that a representation for $N^k(x)$ analogous to (4.403) is obtained simply by multiplying both sides of (4.403) by $-2x_k$. Equations (4.404) then guarantee that terms like A in the expansion of N^k have the needed form (4.399), and the contributions to N^k generated by terms of type B in (4.403) vanish at $g = g_*$. Contributions of type C with two or more exterior derivatives ∂ are also unimportant: after multiplication by x_k they become operators like ∂F in (4.399) that vanish upon integration over x in the identity (4.396).

It then follows that the breaking of critical conformal symmetry in the presence of scale invariance can be generated only by *dangerous contributions* of the type $\partial_k \mathcal{O}_k(x)$ in (4.403) with one and only one exterior derivative ∂. In the integral over x the derivative is "cancelled" by the factor x_k, so that this operator does not contribute to the scale Ward identity (4.396), but does contribute to the conformal one. Therefore, in general the conformal Ward identity (4.396) for $g = g_*$ has the form

$$\mathcal{D}_A^k W_{\mathrm{R}}(A) = c(g_*) \int dx \langle\langle \mathcal{O}_k(x) \rangle\rangle, \qquad (g = g_*) \tag{4.405}$$

with some local operator $\mathcal{O}_k$ without explicit x dependence [because there is none in (4.403)] and some numerical factor $c(g_*)$. If there are several monomials in $\mathcal{O}_k$, the coefficient of any one of them can be taken as $c(g)$. All quantities in (4.405) are UV-finite owing to the obvious UV-finiteness of the left-hand side, and Δ in the generators $T^\alpha = T^k$ (4.395b) entering into the definition (4.396) of the conformal operator $\mathcal{D}_A^\alpha \equiv \mathcal{D}_A^k$ are now understood as the exact critical dimensions Δ_φ.

The expression on the left-hand side of (4.405) has definite critical dimension -1 (which coincides with the canonical one), because the functional $W_{\mathrm{R}}(A)$ operated on by $\mathcal{D}_A^\alpha$ can be assumed dimensionless (Sec. 4.11), and the conformal operator $\mathcal{D}_A^k$ lowers the dimension by unity. In the massless model we are considering all the coefficients in (4.403) and, therefore, also $c(g_*)$ in (4.405) can only contain factors μ^ε with zero critical dimension (Sec. 4.7). Therefore, for $c(g_*) \neq 0$ the operator $\mathcal{O}_k$ in (4.405) must have completely defined critical dimension, namely,

$$\Delta[\mathcal{O}_k] - d \quad 1, \quad d^*[\mathcal{O}_k] = d^* - 1, \tag{4.406}$$

where $d^*[F]$ is the value of $d[F] \equiv d_F$ in the logarithmic dimension d^*. The second equation in (4.406) follows from the first in the zeroth-order approximation in $\varepsilon \sim d^* - d$ and is useful for the preliminary selection of possible candidates for monomials in $\mathcal{O}_k$. If such monomials exist and a linear combination $\mathcal{O}_k$ can be constructed from them without an external derivative and with given critical dimension (4.406), then in (4.405) it is possible that $c(g_*) \neq 0$, i.e., the conformal symmetry can then be broken. However, if the operator $\mathcal{O}_k$ with the required properties does not exist, then necessarily $c(g_*) = 0$ in (4.405) and in this case conformal symmetry is guaranteed.

It is sometimes obvious from elementary arguments that the expansion (4.403) cannot contain dangerous contributions $\partial_k \mathcal{O}_k$ with the required properties. This will occur, for example, in any model with a single scalar field, because in this case a scalar operator with an exterior derivative can be constructed only in the form $\partial^2 F$. This is essentially the proof of Sec. 2.17 in the language we are using now. Another important class is that of massless models with spinor (or spinor and scalar) fields without derivatives in the interaction. The simplest example is the U_N-symmetric Gross–Neveu model, which we shall study in Sec. 4.44. This is also a good example for understanding the features of the spinor fields and models with charge symmetry, in which dangerous contributions can be generated also by several combinations of operators of the type (4.397).

4.42 Generalization to composite operators

Let $F(y)$ be a renormalized local composite operator with definite canonical and critical (for $g = g_*$) dimension (in detailed notation $F(y) = \sum_i c_i(\varepsilon) \cdot \mu^{\varepsilon n_i} \cdot [M_i(y; \varphi)]_{\mathrm{R}}$ is a linear combination of renormalized local monomials M_i; see Sec. 4.7). Let $\langle\!\langle F(y) \rangle\!\rangle$ be the corresponding functional (2.107a), i.e., the expectation value of $F(\varphi)$ with weight $\exp[S(\varphi) + A\varphi]$ and renormalized action $S = S_{\mathrm{R}}$ in the exponent (in the present section we shall omit the label R on operators and the action in order to simplify the expressions). We recall (Sec. 2.13) that the quantity $\langle\!\langle F(y) \rangle\!\rangle \equiv W_F(y; A)$ is the generating functional of renormalized connected Green functions with a single operator F and any number of simple fields φ, and this functional for $g = g_*$ has definite critical dimension Δ_F in the following sense:

$$W_F(\lambda^{-1}y; A_\lambda) = \lambda^{\Delta_F} W(y; A), \qquad A_\lambda(x) \equiv \lambda^{d-\Delta_\varphi} A(\lambda x). \qquad (4.407)$$

The invariance of the corresponding Green functions under critical [i.e., with $\Delta = \Delta_\varphi$ in the generators (4.395)] scale or conformal transformations is expressed as

$$[\mathcal{D}_A^\alpha + T_F^\alpha]\langle\!\langle F(y) \rangle\!\rangle = 0 \qquad (4.408)$$

with the operator $\mathcal{D}_A^\alpha$ from (4.396) and a given (by the dimension and spin) generator T_F^α acting on the argument y of the operator F [we want to reserve x as the argument in (4.394)]. From the Schwinger equation,

$$0 = \int D\varphi \; \delta\{F(y; \varphi) \cdot T^\alpha \varphi(x) \exp[S(\varphi) + A\varphi]\}/\delta\varphi(x) \qquad (4.409)$$

in the usual manner (Sec. 3.29) discarding the contribution involving $\delta(0)$ after integration over x we obtain

$$\mathcal{D}_A^\alpha \langle\!\langle F(y) \rangle\!\rangle + \langle\!\langle F(y)\delta^\alpha S + \delta^\alpha F(y) \rangle\!\rangle = 0. \qquad (4.410)$$

Here and below, $\delta^\alpha \phi$ for any functional $\phi(\varphi)$ will denote the quantity

$$\delta^\alpha \phi \equiv \int dx [\delta\phi(\varphi)/\delta\varphi(x)] T^\alpha \varphi(x), \qquad (4.411)$$

i.e., the coefficient of the corresponding infinitesimal parameter ω_α [see (2.213)] in the variation of ϕ generated by variation of its functional argument φ. Rewriting (4.410) as

$$[\mathcal{D}_A^\alpha + T_F^\alpha]\langle\!\langle F(y) \rangle\!\rangle = \langle\!\langle T_F^\alpha F(y) - \delta^\alpha F(y) - F(y)\delta^\alpha S \rangle\!\rangle \qquad (4.412)$$

and comparing with (4.408), we conclude that proof of the presence of critical symmetry with the generators T^α is equivalent to proof that the right-hand side of (4.412) vanishes for $g = g_*$. The statement of the problem is the same as in Sec. 4.41. Assuming critical scale invariance of the functional $\langle\langle F(y)\rangle\rangle$, we want to find out whether or not the corresponding conformal symmetry is present. The question is meaningful only when the given symmetry is present for the Green functions of the simple fields. We shall therefore assume that all the assumptions of Sec. 4.41 about the model are satisfied and that there are no dangerous contributions of the type $\partial_k \mathcal{O}_k$ in the expansion (4.403) for $g = g_*$. When the functional $\langle\langle F \rangle\rangle$ is conformally invariant, we shall term the operator F *conformal*.

As before, the answer to the question is obtained by analyzing the corresponding Ward identities (4.394). Multiplying both sides of (4.394) by $F(y)$ and integrating over x, we obtain

$$F(y)\delta^\alpha S|_{\text{can}} = -\int dx\, F(y)N^\alpha(x), \qquad (4.413)$$

where the subscript can recall that we are considering transformations with the canonical dimensions.

For the scale equation the operator $N^\alpha \equiv N$ in (4.413) must be written in the form (4.403). The contributions of type A in (4.403) are moved to the left-hand side of (4.413) and the canonical dimensions are rearranged to form the critical ones. Then $\delta^\alpha S|_{\text{can}}$ becomes the full variation $\delta^\alpha S$ entering into (4.412). The contributions of type C in (4.403) do not contain any dangerous terms by assumption (see above), and all the others (with two and more exterior derivatives) are unimportant, because they vanish upon integration over x in both the scale and the conformal equation (4.413). Therefore, after the rearrangement $\delta^\alpha S|_{\text{can}} \to \delta^\alpha S$ only contributions of type B remain on the right-hand side of (4.413) (for example, the single operator $[\varphi^4]_{\text{R}}$ in the φ^4 model), so that this equation can be rewritten as

$$F(y)\delta^\alpha S = -\int dx\, F(y)N^\alpha(x)|_{\text{B}},$$

$$\qquad (4.414)$$

$$N|_{\text{B}} = \sum_i b_i[M_i]_{\text{R}}, \qquad N^k|_{\text{B}} = -2x_k N|_{\text{B}},$$

where $N^\alpha|_{\text{B}}$ are the contributions to $N^\alpha = \{N, N^k\}$ of only terms of type B in (4.403). We have given the explicit form of these operators above; in general, $N|_{\text{B}}$ is a linear combination of renormalized local monomials $M_i \equiv M_i(x; \varphi)$ without exterior derivatives with coefficients $b_i = b_i(g; \mu, \varepsilon)$ depending on g, ε, and μ (through μ^ε). The conformal operator is related to the scale operator by (4.400). The operator $F(y)$ in question is also understood as a linear combination $F = \sum c_a[M_a]_{\text{R}}$ of renormalized local monomials with coefficients $c_a = c_a(\varepsilon, \mu)$ depending only on ε and μ (through μ^ε). All the coefficients b and c are UV-finite, i.e., they have no poles in ε.

We have mentioned all this because the subtraction operator (3.34) of the standard MS scheme does not commute with multiplication by ε-dependent coefficients, and so it is necessary to precisely specify the order of operations when working with polynomials. By convention (Ch. 3), all the renormalization operators of the MS scheme for composite operators are initially defined on simple monomials and are generalized to polynomials by linearity, i.e., the symbol $[\sum c_i M_i]_R$ is always understood as $\sum c_i [M_i]_R$ and analogously for all the other renormalization operators.

The right-hand side of the first equation in (4.414) involves the double sum $\sum c_a b_i \cdot [M_a]_R \cdot [M_i]_R$. All its coefficients $b_i = b_i(g; \mu, \varepsilon)$ vanish at the point g_* owing to (4.404). However, this does not imply that the entire expression vanishes, since it involves products of renormalized operators, which are not UV-finite objects (Sec. 3.25). It follows from the general rule (3.159) and the definition (2.107) that inside the average $\langle\langle ... \rangle\rangle$ for any two local monomials with different arguments x, y we have $\langle\langle \{[M_a]_R \cdot [M_i]_R\} \rangle\rangle = \langle\langle \{...\} \rangle\rangle_R + \langle\langle \text{p.c.} \{...\} \rangle\rangle$. The first term is the UV-finite "fully renormalized" contribution (the result of applying the R operator to the functional $\langle\langle M_a \cdot M_i \rangle\rangle$ of the basic theory), and the second is the contribution of the local [a multiple of the function $\delta(x - y)$ or its derivative] *pair counterterm operator* containing all the UV-divergences. Written out in detail, p.c.$\{[M_a(y; \varphi)]_R \cdot [M_i(x; \varphi)]_R\} = LP\{M_a(y; \varphi) \cdot M_i(x; \varphi) \exp V(\varphi)\}$ in the notation of (3.159). In the MS scheme the splitting into the fully renormalized expectation value and the pair counterterms is unique: for simple monomials (but not polynomials!) the first term contains the entire UV-finite part, and the second contains only the poles in ε. We note that this classification is possible only away from the point $g = g_*$, because at $g = g_* \sim \varepsilon$ any difference between the UV-finite and infinite contributions is lost.

When the first expression in (4.414) is substituted into (4.412) we obtain $\sum c_a b_i \langle\langle \{[M_a]_R \cdot [M_i]_R\} \rangle\rangle = \sum c_a b_i [\langle\langle \{...\} \rangle\rangle_R + \langle\langle \text{p.c.} \{...\} \rangle\rangle]$. The contributions of fully renormalized expectation values vanish at the point g_* owing to the vanishing of the coefficients b_i. However, the pair counterterms, like the ordinary renormalization constants Z, may contain singularities at the point $g = g_*$ (Sec. 1.27) that cancel the zero of the coefficients b, so that in general these contributions do not vanish at the point g_*.

The general form of the pair counterterms can be written in a different way. It is convenient for us to use the following representation of the scale counterterms in (4.414) (the conformal ones are obtained simply by multiplying by $-2x_k$):

$$\text{p.c.}\{F(y) \cdot N(x)|_B\} = [P^{(0)}(y) + P_k^{(1)}(y)\partial/\partial x_k + ...]\delta(x - y), \qquad (4.415)$$

where $P^{(s)}(y)$ are local composite operators from the field $\varphi(y)$ and its derivatives without explicit y dependence (they are understood to carry all the indices of F, if any), and the ellipsis stands for contributions involving two or more derivatives with respect to the argument x which are unimportant for us. The canonical dimension of the counterterms (4.415) is equal to the sum of the dimensions of the operators on the left-hand side. The scale operator $N|_B$ in (4.415) has dimension d equal to the dimension of the δ function on the right-hand side, and so $d[P^{(s)}] = d_F - s$. In general, $P^{(s)}$ can contain all monomials M with $d_M^* = d_F^* - s$ and the given index structure and symmetry; their coefficients are functions of g, ε, and μ (through μ^{ε}) and, in general, contain both UV-divergent and UV-finite contributions [the latter are present only when the coefficients of the factors on the left-hand side of (4.415) involve ε].

Substituting (4.415) into (4.414) taking into account (4.400) for $g = g_*$ (so that only the contributions of pair counterterms remain), we obtain

$$F(y)\delta^{\alpha}S = \left\{ \begin{array}{ll} -P^{(0)}(y) & \text{(scale)} \\[2mm] 2y_k P^{(0)}(y) - 2P_k^{(1)}(y) & \text{(conformal)} \end{array} \right\}. \qquad (4.416)$$

The "normal" contributions involving $P^{(0)}$ have standard correlation of the type (4.400), and the contribution $P^{(1)}$ is dangerous in the terminology of Sec. 4.41.

Let us consider the operator on the right-hand side of (4.412) at $g = g_*$. It is written as the sum of the canonical contribution

$$[T_F^{\alpha}F(y) - \delta^{\alpha}F(y)]|_{\text{can}} \qquad (4.417)$$

and all the anomalies generated by the renormalization, namely,

$$\gamma_F^* F(y) - \gamma_{\varphi}^* \int dx[\delta F(y)/\delta\varphi(x)]\varphi(x) + P^{(0)}(y) \qquad (4.418)$$

for the scale equation and

$$-2y_k\gamma_F^* F(y) + \gamma_{\varphi}^* \int dx[\delta F(y)/\delta\varphi(x)]2x_k\varphi(x) - 2y_k P^{(0)}(y) + 2P_k^{(1)}(y) \qquad (4.419)$$

for the conformal one [we have used the explicit form of the generators (4.395) with the exact dimensions and relation (4.416)]. We shall refer to an operator as a *canonically scale (conformal) operator* if the corresponding quantity (4.417) vanishes. This means that a canonical transformation (scale or conformal) of the argument φ generates the correct (with total dimension) transformation of the functional $F(\varphi)$ as a whole. Any local monomial and any linear combination of such monomials that does not contain dimensional parameters in the coefficients is a canonically scale operator, while not every such object is a conformal operator, as the differentiation operator usually spoils the conformal symmetry. For example, in the class of scalar operators of the φ^4 model, any monomial constructed from the factors φ and $\partial^2\varphi$ is a canonically conformal operator (the Laplacian ∂^2 does not violate canonical conformal symmetry when and only when it acts on an object with canonical dimension of the field d_{φ}). Then monomials like $\partial^2\varphi^2$ and $\varphi\partial^4\varphi$ ($\partial^4 \equiv \partial^2\partial^2$) are not conformal.

In the standard ε-expansion scheme, the necessary condition for the right-hand side of (4.412) to vanish at $g = g_*$ is that it equal zero in zeroth order in ε. The renormalized operator F with definite critical dimension is usually some linear combination of renormalized monomials with factors of μ^{ε} in the coefficients (Sec. 4.7). For finite ε these factors even violate canonical scale invariance, but for $\varepsilon = 0$ the latter is restored. Therefore, critical scale invariance (assumed to exist) guarantees the vanishing of, first, the corresponding expression (4.417) in zeroth order in ε, and, second, the sum of contributions of order ε from this expression and the anomalies (4.418). We immediately note that the analogous sum in the conformal equations, if it contained no dangerous contributions, would be obtained from the scale equation by simply multiplying by $-2y_k$, and so it would automatically vanish. Therefore, there can be only two reasons for the violation of conformal invariance when scale invariance is present: (1) the absence of canonical conformal invariance in zeroth order in ε (i.e., in the logarithmic theory), and (2) dangerous contributions in the conformal relations.

If the first occurs, the answer is clear: it is impossible to speak of exact conformal invariance. We shall therefore assume that the operator in question is canonically conformal for $\varepsilon = 0$, and discuss the second reason — the presence of dangerous contributions. The contributions $P^{(1)}$ in (4.419) are such, but they are not the only ones. If the operator F contains derivatives $\partial\varphi$, $\partial\partial\varphi$, and so on (as is usual), the analogous dangerous contributions are generated also by the conformal variation $\delta^k F$, in particular, by the second term in (4.419) and corrections of order ε in the conformal expression (4.417). Let us explain this for the example of the canonically conformal operator $F = \varphi\partial^2\varphi$. In its variation $\delta^k F(y) \equiv \int dx[\delta F(y)/\delta\varphi(x)]T^k\varphi(x) = T^k\varphi(y) \cdot \partial^2\varphi(y) + \varphi(y) \cdot \partial^2 T^k\varphi(y)$, the derivatives ∂^2 act not only on the factors $\varphi(y)$ (giving "normal" contributions), but also on the coordinate arguments contained explicitly in the generators (4.395), here y. For us it is not the entire variation $\delta^k\varphi = T^k\varphi$ that is important, but only the part of it generated by canonical contributions of order ε in (4.417) and the anomaly γ^*_φ in the exact critical dimension. The argument y enters into this part of $\delta^k\varphi$ only linearly [see (4.395)], and so it always vanishes upon differentiation. It then follows that the dangerous contributions generated by $\delta^k F$ have exactly the same general structure as the contributions of the counterterms $P^{(1)}_k$ in (4.419): they are local operators that do not explicitly involve the argument y and that carry the additional (compared to F) vector index k and have canonical dimension $d_F - 1$.

Collecting all the dangerous contributions, the general expression (4.412) at $g = g_*$ for an operator which is canonically conformal at $\varepsilon = 0$ can be written in a form analogous to (4.405):

$$[\mathcal{D}^k_A + T^k_F]\langle\langle F(y)\rangle\rangle = c(g_*)\langle\langle\mathcal{O}_k(y)\rangle\rangle, \quad (g = g_*), \qquad (4.420)$$

where $\mathcal{O}_k$ is a dangerous operator and $c = c(g; \mu, \varepsilon)$ is a scalar numerical factor. If F is not a scalar and carries indices, they are all also included in $\mathcal{O}_k$.

The left-hand side of (4.420) is UV-finite and has definite canonical and critical dimensions [in the sense of (4.407) for the functional $\langle\langle F\rangle\rangle$, the operators $\mathcal{D}_A^k$ and T_F^k lower the dimension by unity]. As in (4.405), the coefficient $c(g_*)$ in (4.420) has zero critical dimension. Therefore, for $c(g_*) \neq 0$, i.e., when the critical conformal symmetry is violated, it follows from (4.420) that

$$\Delta[\mathcal{O}_k] = \Delta_F - 1, \quad d^*[\mathcal{O}_k] = d_F^* - 1. \tag{4.421}$$

If for a given F an operator $\mathcal{O}_k$ with the required properties does not exist, then the coefficient $c(g_*)$ in (4.420) must vanish, i.e., the given operator F is conformal.

Therefore, violation of critical conformal invariance for an operator with definite critical dimension which is canonically conformal for $\varepsilon = 0$ is possible only when a linear combination $\mathcal{O}_k$ with the known dimension (4.421) can be constructed from the given (by the general structure and the value of d_F^*) finite system of renormalized monomials. Otherwise, the coefficients $c(g; \mu, \varepsilon)$ of each monomial in the original exact expression for the right-hand side of (4.412) must vanish at $g = g_*$. We note that we can always expect the conformal symmetry to be violated for operators of the type $F = \partial_k [F_k']_\mathrm{R}$ with exterior derivative and determined value of Δ_F, because for them in (4.420) it is always possible to choose $\mathcal{O}_k = [F_k']_\mathrm{R}$, i.e., an operator $\mathcal{O}_k$ with the required properties certainly exists for such F.

4.43   Examples

In analyzing the consequences of (4.420) it is convenient to use the following terminology. Operators whose critical dimensions differ by an integer are assumed to belong to the same family, inside which the operators with higher dimension are considered to be the descendants of operators with lower dimension (for example, ∂F is a descendant of F, and F is an ancestor of ∂F). Since the dimensions of all operators are positive (in the ε scheme), each family has a "founder" — the operator with the lowest dimension. In our case it follows from (4.421) that the operator $\mathcal{O}_k$ is the closest ancestor of F. Owing to the absence of ancestors, for any founder the right-hand side of (4.420) must vanish $[c(g_*) = 0]$, i.e., each founder is conformal, of course, if it is canonically conformal in the logarithmic theory.

Therefore, the property of being conformal is not so rare — it can be violated owing to anomalies only when there is a closest ancestor with an additional vector index.

Let us explain this for the example of the scalar operators of the massless φ^4 model. The independent monomials are classified according to the value of $d_F^* \equiv d_F|_{\varepsilon=0}$ (equal to an integer), and according to this criterion are grouped into the sets φ; φ^2; φ^3, $\partial^2 \varphi$; φ^4, $\varphi \partial^2 \varphi$, $\partial^2 \varphi^2$, and so on. The multiplicatively renormalizable operators φ and φ^2 are certainly conformal as they are the founders of families with independent dimensions $\Delta[\varphi] \equiv \Delta_\varphi$ and $\Delta[\varphi^2] = d - \Delta_\tau$ (Sec. 4.9). In the following sets with $d_F^* \geq 3$ it is no longer simply renormalized monomials that have definite critical dimensions, but certain linear combinations of them calculated using the corresponding matrices of renormalization constants (Sec. 4.7). Not all of them are non-trivial in the sense that they generate new critical dimensions, because they include, first, explicit descendants of operators of lower dimension (for example, $\partial^2\varphi$ or $\partial^2\varphi^2$), and, second, *special operators* of the form $F = V \cdot U$ with $U \equiv \delta S/\delta\varphi$ and arbitrary UV-finite coefficient V with the lowest (compared to F) dimension. Owing to (3.191), such an operator is UV-finite (we recall that now $S \equiv S_R$) and has definite dimension equal to the simple sum of the dimensions of the cofactors $\Delta[V]$ and $\Delta[U] = \Delta_A = d - \Delta_\varphi$, of course, if V is taken to be an operator with definite dimension. A typical criterion for an operator to be like this follows from the same expression (3.191): the Green functions $\langle F(y)\varphi(x_1)\varphi(x_2)...\rangle$ should contain some coordinate δ functions $\delta(y-x_i)$. Therefore, special operators, if they are conformal, correspond to special conformal structures involving δ functions, discussed at the end of Sec. 2.18. By this criterion (the presence of δ functions), operators involving factors ∂U, $\partial^2 U$, ... are then also special, since the derivatives can always be moved onto the second cofactor or the operator as a whole (for example, $V \cdot \partial U = \partial[VU] - \partial V \cdot U$, $V\partial^2 U = \partial^2[VU] + \partial^2 V \cdot U - 2\partial[\partial V \cdot U]$, where all the terms on the right-hand side have the same definite critical dimensions). We also note that for special operators like $V \cdot U$ the pair counterterms (4.415) can be expressed in terms of the renormalization constants of the operators V and $N|_B$ by analyzing the UV divergences in the Schwinger equation analogous to (4.409) with derivative $\delta/\delta\varphi(y)$ and factor $V(y) \cdot N|_B(x)$ in the preexponent.

After separating out the special operators and the explicit descendants in a set with a given d_F^*, there remain the independent nontrivial operators that can generate new critical dimensions. They become the founders of new families which are then (see above) necessarily conformal when there is canonical conformal invariance for $\varepsilon = 0$. There are few such operators, as can be checked by analyzing the first few sets. There are none at all in the set φ^3, $\partial^2\varphi$ with $d_F^* = 3$, since a complete basis of operators with definite critical dimensions is formed by the special operator $U \equiv \delta S/\delta\varphi$ and the explicit descendant $\partial^2\varphi$. In the next set with $d_F^* = 4$ complete basis contains three operators: the descendant $\partial^2[\varphi^2]_R$, the special operator $\varphi \cdot U$, and the single nontrivial operator (4.66) in the critical regime ($\tau = 0$, $g = g_*$). The latter is associated with φ^4 and has critical dimension $d + \omega$ with the new index ω (Sec. 4.10). It is therefore the founder of a new family, like the special operator $\varphi \cdot U$ with critical dimension $\Delta[\varphi \cdot U] = d$ (Sec. 4.11). The special operator is constructed from the canonically conformal monomials $\varphi\partial^2\varphi$ ($\equiv F_4$) and φ^4 ($\equiv F_5$), and the operator (4.66) also contains an admixture of the canonically nonconformal monomial $\partial^2\varphi^2$ ($\equiv F_3$) [in the parentheses we give the notation of Sec. 4.9].

However, the coefficient of the monomial F_3 in the operator (4.66) for $g = g_*$ is a quantity of order ε, since the expansion of the RG function $\gamma_F^{43}(g)$ begins at g^3 or higher (see Secs. 4.10 and 3.30). Therefore, for $\varepsilon = 0$ the two operators in question are canonically conformal, and so they are critically conformal for any ε as they are the founders of families. The third basis operator $\partial^2[\varphi^2]_\mathrm{R}$ of this system is certainly not conformal, since it is not even canonically conformal for $\varepsilon = 0$.

We can use this example to explain the general recipe stated in the text following (4.421). From the structure and the canonical dimension, the only candidate for $\mathcal{O}_k$ in (4.420) for scalars with $d_F^* = 4$ is the operator $\mathcal{O}_k = \partial_k[\varphi^2]_\mathrm{R}$. However, in the total dimension [the rule (4.421)] it is suitable only for the certainly nonconformal operator $F = \partial^2[\varphi^2]_\mathrm{R}$. It then follows that for the two other operators with $d_F^* = 4$ the coefficient $c \equiv c(g; \mu, \varepsilon)$ of a given monomial on the right-hand side of (4.412) must have a zero at $g = g_*$ (and the vanishing of all the other contributions is guaranteed by the critical scale invariance).

These arguments allow the presence of critical conformal symmetry to be proved without any explicit (and awkward) calculation of the pair counterterms (4.415). However, for reference we note that some information about these counterterms for the operators we have discussed can be obtained by selecting the UV-divergent contributions in equations like (4.409) with the operator $\delta/\delta\varphi(y)$ and a suitable preexponent, and also their analogs with derivative ∂_g instead of $\delta/\delta\varphi$ (similar to the calculations of Sec. 4.11).

Let us conclude with another example: the set of scalars with $d_F^* = 6$. The following complete set of seven monomials can be taken as the independent ones: φ^6, $\varphi^3\partial^2\varphi$, $\partial^2\varphi^4$, $\partial^2\varphi \cdot \partial^2\varphi$, $\partial^4\varphi^2$, $\varphi\partial^4\varphi$, and $\partial^2[\varphi\partial^2\varphi]$. The complete basis of renormalized operators with definite critical dimensions is constructed from the corresponding renormalized monomials. It contains, first, the three descendents of $\partial^2 F$ of the operators with $d_F^* = 4$ discussed earlier and, second, three more independent special operators $U \cdot U$, $\partial^2\varphi \cdot U$, and $\varphi \cdot \partial^2 U$ ($\partial\varphi \cdot \partial U$ is dependent) with dimensions $\Delta_F = 2d - 2\Delta_\varphi$, $d+2$, and $d+2$, respectively. Of these six operators, only the special ones $U \cdot U$ and $\partial^2\varphi \cdot U$ are canonically conformal for $\varepsilon = 0$. Verifying that they are exactly conformal requires the analysis of vector operators with $d_F^* = 5$, which we recommend to the reader as a useful exercise. However, independently of the answer to this question, there is enough space remaining in the complete basis for only a single nontrivial operator. It is apparently associated with φ^6 and in turn generates a new independent index, and so it must be conformal.

Generalizing all these observations, it can be assumed that any set of scalars with a given d_F^* contains no more than one nontrivial operator. Each of them generates a new family (if we exclude the possibility of an accidental coincidence of dimensions for any ε) and is therefore a conformal operator, of course, when canonical conformal invariance is present in the logarithmic theory. Special operators and the descendents, most of which are nonconformal, form the bulk of the operators.

4.44 The chiral phase transition in the Gross–Neveu model

The U_N-symmetric Gross–Neveu model [139] describes a system of N copies of complex d-dimensional Dirac spinor fields $\varphi \equiv \{\psi_a, \overline{\psi}_a, a = 1, ..., N\}$. Like the σ model (Sec. 4.26), it is logarithmic in dimension $d^* = 2$, and its renormalized action in dimension $d = 2 + 2\varepsilon$ has the form

$$S_R(\varphi) = Z\overline{\psi}\widehat{\partial}\psi + Z' g_B(\overline{\psi}\psi)^2/2, \qquad g_B = g\mu^{-2\varepsilon}, \qquad (4.422)$$

where the integration over x and the required summations over the "isotopic" index a and the Dirac indices of the spinors ψ_a, $\overline{\psi}_a$ are understood. The form $\widehat{\partial} \equiv \gamma_i \partial_i$ involves Dirac γ matrices obeying the commutation relation

$$\gamma_i \gamma_k + \gamma_k \gamma_i = 2\delta_{ik} \qquad (4.423)$$

in the Euclidean formalism that we are using.

Let us first give some information about the d-dimensional spinors ψ, $\overline{\psi}$. They describe a field with spin $1/2$. The general rules of Sec. 2.15 for including spin contributions Σ^{ks} in generators of rotations T^{ks} and conformal transformations T^k (the addition of Σ^{ks} to T^{ks} and $2x_s\Sigma^{ks}$ to T^k) hold also for spinors if they are written in universal notation with the single field $\varphi \equiv \{\psi, \overline{\psi}\}$ and with suitable choice of block (2×2) matrix Σ^{ks}, namely, $\Sigma^{ks}\varphi \equiv \Sigma^{ks}\{\psi, \overline{\psi}\} = \{\sigma_{ks}\psi, -\overline{\psi}\sigma_{ks}\}$, where $\sigma_{ks} \equiv (\gamma_s\gamma_k - \gamma_k\gamma_s)/4$ is the spin part of the rotation generator for ψ (we again recall that in the Euclidean formalism there is no difference between upper and lower indices). This means that when the expressions for ψ and $\overline{\psi}$ are written separately, the matrix σ_{ks} plays the role of Σ^{ks} for ψ, acting from the left on ψ, while for $\overline{\psi}$ it is the matrix $-\sigma_{ks}$ acting from the right (moving a matrix from left to right is equivalent to its transposition). The definition (2.114) of the conformal inversion operator r with $r^2 = 1$ is modified for spinors by multiplication by the matrix $\widehat{n}(x)$ for ψ from the left and by $-\widehat{n}(x)$ for $\overline{\psi}$ from the right, where $\widehat{n}(x) \equiv n_i\gamma_i$, $n_i \equiv x_i/|x|$. Special conformal transformations are defined by the standard expression $rt_\omega r$ in terms of the inversion r and ordinary translations t_ω by the vector ω, and are given by the expressions of Sec. 2.15 with additional multiplication by $Q^{1/2}(1 + \widehat{x}\widehat{\omega})$ for ψ from the left and by $Q^{1/2}(1 + \widehat{\omega}\widehat{x})$ for $\overline{\psi}$ from the right.

The spinors ψ and $\overline{\psi}$ are the elements of a complex linear space in which the matrices γ act. The dimension of this space, i.e., the value of tr1 for the unit operator acting on it, for even integer d and then odd d, following it, is usually given by the rule tr1 $= 2^{d/2}$ for even d [this is the minimum dimension for which matrices with the properties (4.423) can be constructed]; in particular, tr1 $= 2, 2, 4$ for $d = 2, 3, 4$, respectively. This expression for tr1 has no unique or even simply natural extension to arbitrary values of d (in contrast to a vector field, for which always tr1 $= \delta_{ii} = d$). Therefore, in calculations involving spinors with variable d the quantity tr1 must be retained intact without explicit expression in terms of d.

Furthermore, according to the general principles of quantum theory, fields with half-integer spin correspond to particles with Fermi statistics. Therefore, at the classical level, in particular, in the action functional and other functional constructions, the quantities ψ and $\overline{\psi}$ and the sources $\overline{A}$ and A conjugate to them must be assumed not to be ordinary functions but Grassmann variables. This means that all factors ψ, $\overline{\psi}$, A, and $\overline{A}$ anticommute both with each other and with $\delta/\delta\psi$, $\delta/\delta\overline{\psi}$, $\delta/\delta A$, and $\delta/\delta\overline{A}$. This makes it necessary to distinguish between left $(\vec{\delta}/\delta...)$ and right $(\overleftarrow{\delta}/\delta...)$ derivatives. The first, by definition, act on the expression from the left, and the second act on the expression from the right according to the usual rules of differentiation also taking into account the anticommutativity. In the Feynman graphs the Fermi nature of the fields leads to only one modification: the appearance of an additional factor of -1 from each closed loop of fermion lines. The general rules of functional integration remain valid also for Grassmann variables with only slight changes due to the anticommutativity of the fields. In particular, the basic formula of Gaussian integration for Grassmann variables ψ, $\overline{\psi}$, A, $\overline{A}$, namely,

$$\int D\psi D\overline{\psi}\exp[-\overline{\psi}L\psi + \overline{A}\psi + \overline{\psi}A] = \text{const}\cdot\det L\cdot\exp[\overline{A}L^{-1}A] \qquad (4.424)$$

differs from the analogous expression of Sec. 2.1 only by the replacement $L \to L^{-1}$ inside the determinant (this leads to the rule of a factor of -1 for a closed fermion loop). All the other general expressions of Ch. 2 generalize to Grassmann fields, only now we must pay attention to the ordering of anticommuting factors and the choice of derivative (left or right).

Let us return to the model (4.422). The corresponding unrenormalized (S) and basic (S_B) actions are obtained from S_R in the obvious manner. The Green functions are given by the usual expressions of Sec. 1.13, but now with Grassmann fields and sources. This statement of the problem, used in most studies of this model, corresponds to an abstract classical statistical physics with Grassmann variables. The charge g in (4.422) then has the meaning of the temperature (in classical statistical physics, the original $S = -\text{energy}/kT$, and the parameter kT is eliminated from the free part by rescaling the fields and then appears as a factor in front of the interaction). Since classical Grassmann systems do not exist in nature, a realistic prototype of this problem is the analogous quantum system of massless fermions with zero Fermi energy and local pair interaction. In the quantum problem it would be necessary to work with temperature (Matsubara) Green functions, for which there are other functional integral representations (see, for example, [53]); the temperature enters into them in a different way, and the interaction strength is characterized by a second independent parameter g.

Phase transitions of various types are possible in this system. Here we are interested in the *chiral transition* of the second kind, expressed as the spontaneous appearance of an anomalous scalar expectation value $\langle\overline{\psi}(x)\psi(x)\rangle$, i.e., of a fermion mass, since masslessness corresponds to $\langle\overline{\psi}(x)\psi(x)\rangle = 0$

(in this case one usually speaks of spontaneous breakdown of the discrete γ_5 symmetry, but for the d-dimensional problem this terminology is not really appropriate owing to the absence of a natural definition of the γ_5 matrix in arbitrary dimension). For fermions it is also possible to have a transition to the superconducting phase, expressed as the spontaneous appearance of anomalous correlators $\langle \psi\psi \rangle$ and $\langle \overline{\psi}\,\overline{\psi} \rangle$ forbidden by global gauge symmetry. The detailed analysis of this quantum system is therefore a separate, complex problem that we shall not tackle, limiting ourselves to only the chiral phase transition in the classical model (4.422) and identifying the charge $g > 0$ in (4.422) as the temperature.

Since the anomalous expectation value $\langle \overline{\psi}\psi \rangle$ is bilinear in the fields, the actual existence of this transition is not manifested at the level of the loopless approximation, i.e., directly in the form of the action (4.422) (in quantum field theory this situation is usually referred to as dynamical symmetry breaking). It is most easily found in the leading order of the $1/n$ expansion ($n \equiv N$tr1 everywhere), which is constructed according to the standard scheme (Sec. 2.19) by introducing an auxiliary scalar field $\sigma(x)$ and going over to the theory of three fields $\varphi \equiv \psi, \overline{\psi}, \sigma$ with unrenormalized action

$$S(\varphi) = \overline{\psi}\hat{\partial}\psi - \sigma^2/2g_0 + \overline{\psi}\psi\sigma \tag{4.425}$$

and the usual condition $g_0 \sim 1/n$ in the $1/n$ expansion. Integration over σ gives back the original (unrenormalized) model (4.422), and integration over ψ and $\overline{\psi}$ using the rule (4.424) leads to an integral over σ with exponent $S(\sigma) = N$tr$\ln(\hat{\partial} + \sigma) - \sigma^2/2g_0$. For $g_0 \sim 1/n$ the two contributions are of order n, and so calculation of the integral over σ by the saddle-point method (Sec. 2.19) automatically gives the desired $1/n$ expansion.

It is easily verified that for all negative and for small positive values of the bare temperature g_0 the equation for a spatially uniform stationarity point σ_0 ($= \langle \sigma \rangle$ in leading order in $1/n$) has only the trivial solution $\sigma_0 = 0$. For increasing g_0 at some critical value g_{0c} the normal solution becomes unstable [this is manifested in the violation of the positive-definiteness of the correlator $\langle F(x)F(y) \rangle$ for the composite operator $F(x) = \overline{\psi}(x)\psi(x)$], but then a stable, nontrivial solution degenerate in sign appears: $\sigma_0 = \pm$const $\neq 0$. We see from (4.425) that after the shift to the stationarity point $\sigma \to \sigma + \sigma_0$, the fermion field propagator in this approximation turns out to be $\langle \psi\overline{\psi} \rangle = -[\hat{\partial} + m_{\text{eff}}]^{-1}$ with effective mass $m_{\text{eff}} = \sigma_0 = \langle \sigma \rangle$, which is the analog of the spontaneous magnetization in the σ model.

The standard RG analysis (more details are given below) shows that near the transition point there is ordinary critical scaling with definite critical dimensions $\Delta_F \equiv \Delta[F]$ of the various quantities F. Of particular interest are the dimensions of the field ψ (unless indicated otherwise, the dimensions of ψ and $\overline{\psi}$ are always assumed to be the same) and the renormalized parameters: the "temperature" $\tau \equiv g - g_c$ and the fermion mass. The traditional notation is $\Delta_\tau \equiv 1/\nu$ and $2\gamma_\psi^* = \eta$. For the masses we distinguish the "primary" mass m, a parameter introduced into the original action (4.422) or (4.425) by the replacement $\widehat{\partial} \to \widehat{\partial} + m$ (in the renormalized version $\widehat{\partial} \to \widehat{\partial} + m_0$ with $m_0 = mZ_m$), and the full "effective" mass m_{eff} formed (sometimes spontaneously) in the fermion propagator. In the vicinity of the critical point $\tau = 0$, $m = 0$ the quantity m_{eff} is a function of m and τ. In the critical region all these parameters are small and have definite dimensions; the dimensions of m and m_{eff} are different and always obey the shadow relation: $\Delta[m] + \Delta[m_{\text{eff}}] = d$ (explanations follow).

All the critical dimensions can be calculated as $2 + \varepsilon$ expansions using the Gross–Neveu model (4.422), as $1/n$ expansions using the model (4.425), and as $4 - \varepsilon$ expansions using the "partner" of the GN model proposed in [140] with the same IR behavior, obtained by adding to the action (4.425) terms $\sim (\partial\sigma)^2$ and σ^4 that are IR-irrelevant in the $1/n$ expansion (see below). This partner model and the GN model are related to each other just as are the O_n φ^4 and the σ model, or the CP^{n-1} model and scalar electrodynamics, or the matrix σ model and scalar chromodynamics (Sec. 2.28).

The GN model is exactly analogous to the nonlinear σ model in its properties. The parameter g has the same meaning (the temperature) in both, and the behavior of the β function with UV fixed point $g_* = g_c$ (the renormalized critical temperature) is the same. The passage to the critical regime is also ensured by the external condition $g \to g_c \equiv g_*$, and not by passing to the small-momentum asymptote for fixed $g \neq g_c$, as in "normal" models of the φ^4 type (Sec. 4.28). In this model, the $2 + \varepsilon$ expansions of the various critical exponents can be calculated using the following expressions:

$$d_\psi = (d-1)/2, \quad d_m = 1, \quad \Delta_\psi = d_\psi + \gamma_\psi^*, \quad \eta \equiv 2\gamma_\psi^*,$$

$$\Delta_m = d_m + \gamma_m^* = d - \Delta[\overline{\psi}\psi], \quad \Delta[m_{\text{eff}}] = \Delta[\overline{\psi}\psi], \quad (4.426)$$

$$\Delta_\tau \equiv \Delta[g - g_c] \equiv 1/\nu = -\beta'(g_*)$$

with the usual notation $\gamma_F^* \equiv \gamma_F(g_*)$ for any F. Equation (4.426) for Δ_τ is justified just like the analogous expression (4.248). For m_{eff} the leading singular contribution in the massless model (4.422) is determined by $\langle \overline{\psi}(x)\psi(x)\rangle$. The operator $\overline{\psi}\psi$ in this model is renormalized multiplicatively and therefore has definite dimension $\Delta[\overline{\psi}\psi]$. It obeys the shadow relation with the dimension of the primary mass m in the massive generalization of the model (4.422), and so the quantity $\Delta[\overline{\psi}\psi]$ can be calculated also in terms of the anomalous dimension γ_m^* using the corresponding renormalization constant Z_m (the calculation is given in the following section).

The $1/n$ expansions of the same critical dimensions can be calculated using the model (4.425) or, more conveniently, using its maximally simplified, completely massless version [the analog of the model (2.205), discussed in detail in Sec. 4.31]:

$$S(\varphi) \; = \; \overline{\psi}\widehat{\partial}\psi + \overline{\psi}\psi\sigma - \frac{1}{2}\sigma L\sigma + \frac{1}{2}\sigma L\sigma \;, \quad L = \;\bigcirc \;. \qquad (4.427)$$

The dimensional-like regularization procedure described in Sec. 4.31 and the proof of renormalizability in the $1/n$ expansion for arbitrary dimension d generalize directly to the model (4.427). The dimensions of the various quantities in the $1/n$ expansion are given by

$$d_\psi = (d-1)/2, \quad d_\sigma = 1, \quad \Delta_\varphi = d_\varphi + \gamma_\varphi^*, \quad \varphi \equiv \psi, \sigma,$$

$$\Delta_m = d - \Delta_\sigma = \Delta[\overline{\psi}\psi], \quad \Delta[m_{\mathrm{eff}}] = \Delta_\sigma, \quad \Delta_\tau \equiv 1/\nu = d - \Delta[\sigma^2]. \tag{4.428}$$

Let us explain these expressions. The canonical dimension $d_\sigma = 1$ in the $1/n$ expansion is determined from the requirement that the interaction $\overline{\psi}\psi\sigma$ be dimensionless. The contribution quadratic in σ in (4.425) is now IR-relevant and analogous to the contribution in (2.199) linear in the auxiliary field. In calculations using dimensional-like regularization the value of this contribution at the critical point g_{0c} itself can be discarded [like the contribution $h_c\psi$ in (2.201)], leaving only the correction proportional to $\tau\sigma^2$. This leads to (4.428) for Δ_τ, which is the analog of (2.203). The leading singular contribution to m_{eff} comes from $\langle\sigma\rangle$, which also leads to (4.428) for $\Delta[m_{\mathrm{eff}}]$. The expression for Δ_m can be explained as follows. The addition of the primary mass m in the replacement $\widehat{\partial} \to \widehat{\partial} + m$ in (4.425) can be eliminated by the shift $\sigma \to \sigma - m$, which transforms the mass insertion $m\overline{\psi}\psi$ into the contribution $h\sigma$ with $h = m/g_0 \simeq m/g_{0c}$. It should be added that the dimensions of the parameter $h \sim m$ and the field σ obey the shadow relation. The second equation for Δ_m in (4.428) is proved exactly like its analog in (4.357), using the Schwinger equation (3.187) with $\delta/\delta\sigma$ for the renormalized model (4.427) (we obtain $[\overline{\psi}\psi]_{\mathrm{R}} = -A_\sigma$, and the dimension of the source A_σ obeys the shadow relation with Δ_σ). To avoid confusion, we note that there is no contradiction in the difference between the expressions for Δ_m in (4.426) and (4.428), since the same operator $F = \overline{\psi}\psi$ in the $1/n$ and $2+\varepsilon$ schemes has different critical dimensions [see the discussion of the analogous problem for the operator ϕ^2 in (4.357)].

The $4-\varepsilon$ expansions of the critical dimensions are constructed according to the model [140], whose unrenormalized action with dimensional regularization (allowing the contribution σ^2 to be discarded directly at the critical point itself) can be written as

$$S(\varphi) = \overline{\psi}[\widehat{\partial} + g_0\sigma]\psi - (\partial\sigma)^2/2 - \tau_0\sigma^2/2 - \lambda_0\sigma^4/24. \tag{4.429}$$

According to an estimate from the canonical dimensions (4.428), the contributions $(\partial\sigma)^2$ and σ^4 in (4.429) are IR-irrelevant and can be discarded, which gives back the action (4.425) up to a change of notation and thereby proves (Sec. 2.27) that the critical behavior is identical in the models (4.422) and (4.429) within the $1/n$ expansion. In ordinary perturbation theory in the charges g, λ, the quantity d_σ in the model (4.429) is determined by the contribution $(\partial\sigma)^2$, i.e., $d_\sigma = d/2 - 1$. For this estimate of the dimensions the two interactions in (4.429) are logarithmic at $d^* = 4$, and the model is multiplicatively renormalizable for $d = 4 - 2\varepsilon$. The one-loop calculation [140] shows that its β functions have IR-attractive fixed point $\{g_*, \lambda_*\} \sim \varepsilon$. This allows the $4 - \varepsilon$ expansions of the critical dimensions of the basic quantities $F = \psi, \sigma$, and τ to be calculated using the usual expressions

$$\Delta_F = d_F + \gamma_F^*, \quad d_\psi = (d-1)/2, \quad d_\sigma = (d-2)/2, \quad d_\tau = 2 \qquad (4.430)$$

in terms of the coresponding RG functions γ_F.

Let us conclude by giving the results of the two-loop calculation [141] of the dimensions (4.426) for $d = 2 + 2\varepsilon$ (explanations are given in the following section):

$$\gamma_\psi^* \equiv \eta/2 = \varepsilon^2(n-1)/(n-2)^2 + ...,$$

$$\gamma_m^* = 2(n-1)[\varepsilon/(n-2) + \varepsilon^2/(n-2)^2 + ...], \qquad (4.431)$$

$$\Delta_\tau \equiv 1/\nu = 2\varepsilon - 4\varepsilon^2/(n-2) + ...$$

and of the one-loop calculation [140] of the dimensions (4.430) in the renormalized model (4.429) for $d = 4 - 2\varepsilon$:

$$\gamma_\psi^* \equiv \eta/2 = \varepsilon/(n+6) + ..., \qquad \gamma_\sigma^* = \varepsilon n/(n+6) + ...,$$

$$\Delta_\tau \equiv 1/\nu = 2 + \gamma_\tau^* = 2 - \varepsilon[5n + 6 + (n^2 + 132n + 36)^{1/2}]/3(n+6) + ..., \qquad (4.432)$$

where the ellipsis denotes corrections of higher order in ε. We have added to (4.431) the result for γ_m^* (see Sec. 4.45), which was not calculated in [141], and we have expressed all quantities in terms of the universal parameter $n \equiv N\mathrm{tr}1$ (the value $\mathrm{tr}1 = 2$ was used in [141], and $\mathrm{tr}1 = 4$ in [140], which makes comparison with the $1/n$ expansions difficult). The results on the $1/n$ expansions of the exponents will be given in Sec. 4.48.

4.45   Two-loop calculation of the RG functions of the GN model in dimension $2 + 2\varepsilon$

This calculation is standard, and so we shall describe it only briefly. To be general, we study the massive analog of the model (4.422) in dimension $d = 2 + 2\varepsilon$ taking into account vacuum loops:

$$S_\mathrm{R}(\varphi) = Z_0 m^2 \mu^{2\varepsilon} + \overline{\psi}[Z_1\widehat{\partial} + Z_2 m]\psi + Z_3 g_\mathrm{B}(\overline{\psi}\psi)^2/2, \qquad g_\mathrm{B} = g\mu^{-2\varepsilon}. \quad (4.433)$$

All these contributions include the integration $\int dx...$, and so the factor $V \equiv \int dx$ in the vacuum counterterm with Z_0 has been dropped. All constants

$Z = Z(g; \varepsilon)$ are dimensionless and in the MS scheme that we shall always use are represented by series of the type (1.104) without the unit contribution for Z_0. The model (4.433) is multiplicatively renormalizable (explanations are given below) according to the general rule (1.84) with the constants $Z_\psi = Z_{\overline{\psi}}$ for the fields $\varphi \equiv \psi, \overline{\psi}$ and $m_0 = mZ_m$, $g_0 = g_B Z_g$ for the parameters, with $Z_1 = Z_\psi^2$, $Z_2 = Z_m Z_\psi^2$, and $Z_3 = Z_g Z_\psi^4$. In the calculations it is more convenient technically to work with the graphs of a model like (4.425) with three fields $\varphi \equiv \psi, \overline{\psi}, \sigma$, taking the basic action to be the functional

$$\widetilde{S}_B(\varphi) = \overline{\psi}[\hat{\partial} + m]\psi - \sigma^2/2 + \lambda_B\overline{\psi}\psi\sigma, \qquad \lambda_B \equiv g_B^{1/2} = \lambda\mu^{-\varepsilon} \qquad (4.434)$$

$(\lambda_B^2 = g_B, \lambda^2 = g)$. Adding all the needed counterterms, we find

$$\widetilde{S}_R(\varphi) = \widetilde{Z}_0 m^2 \mu^{2\varepsilon} + \overline{\psi}[\widetilde{Z}_1 \hat{\partial} + \widetilde{Z}_2 m]\psi - \widetilde{Z}_3 \sigma^2/2 +$$
$$+ \widetilde{Z}_4 \lambda_B \overline{\psi}\psi\sigma + \widetilde{Z}_5 \lambda_B^2 (\overline{\psi}\psi)^2/2 + \widetilde{Z}_6 m\mu^\varepsilon \sigma \qquad (4.435)$$

with dimensionless constants $\widetilde{Z}_i = \widetilde{Z}_i(\lambda, \varepsilon)$; the unit contribution is absent in $\widetilde{Z}_{0,5,6}$. The last two terms in (4.435) make the renormalization of this model nonmultiplicative. In the purely fermionic sector (i.e., after integration over σ in the integral with weight $\exp \widetilde{S}_R(\varphi)$ and only fermionic sources), the model (4.435) becomes the multiplicatively renormalizable model (4.433) with constants

$$Z_1 = \widetilde{Z}_1, \qquad\qquad Z_2 = \widetilde{Z}_2 + \lambda\widetilde{Z}_3^{-1}\widetilde{Z}_4\widetilde{Z}_6,$$
$$Z_3 = \widetilde{Z}_5 + \widetilde{Z}_3^{-1}\widetilde{Z}_4^2, \quad Z_0 = \widetilde{Z}_0 + \frac{1}{2}\widetilde{Z}_3^{-1}\widetilde{Z}_6^2. \qquad (4.436)$$

These expressions can be used to calculate the desired constants Z of the model (4.433) using the counterterms of the superficially divergent graphs of the basic theory (4.434). As usual, it is technically simplest to perform the calculations using the massless model, but this requires first getting rid of the constants $\widetilde{Z}_{0,2,6}$ related to the mass. This can be done by the usual method (Sec. 3.19), expressing them in terms of the derivatives with respect to the mass of suitable 1-irreducible Green functions Γ of the model (4.434) and then setting $m = 0$. However, in this case there is a simpler way. The constants $\widetilde{Z}_{0,2,6}$ of the model (4.435) can be expressed purely algebraically in terms of other $\widetilde{Z}$ using a relation like (3.194) with derivative with respect to σ and (3.199) with derivative with respect to the parameter $e = m$. In this case $\ln C_R = -\text{tr} \ln(\hat{\partial} + m) + \text{const}$ and the UV-divergent part of $\ln C_R = V n m^2 \mu^{2\varepsilon}/8\pi\varepsilon$ with the factors $V = \int dx$ and $n \equiv N\text{tr}\, 1$, the latter coming from the summation over isotopic and spinor indices. Finally, we obtain the following two relations:

$$\widetilde{Z}_4 \lambda_B \overline{\psi}\psi - \widetilde{Z}_3\sigma + \widetilde{Z}_6 m\mu^\varepsilon = [\lambda_B\overline{\psi}\psi - \sigma]_R = \lambda_B[\overline{\psi}\psi]_R - \sigma, \qquad (4.437)$$

$$2\tilde{Z}_0 m\mu^{2\varepsilon} + \tilde{Z}_2\overline{\psi}\psi + \tilde{Z}_6\mu^\varepsilon\sigma + nm\mu^{2\varepsilon}/4\pi\varepsilon = [\overline{\psi}\psi]_{\mathrm{R}}, \qquad (4.438)$$

containing information about the renormalization of the set of mixing operators $1 \equiv F_1$, $\sigma \equiv F_2$, $\overline{\psi}\psi \equiv F_3$ in the model (4.435). In writing the second equation in (4.437) we have used the relation $[\varphi]_{\mathrm{R}} = \varphi$ for all φ (Sec. 3.24), and in deriving (4.438) from (3.199) we have taken into account the absence of operators like ∂F with exterior derivative in this system. In the present case this allows us to remove the integration over x in (3.199) [see the analogous derivation of (3.200) in Sec. 3.30]. Comparing the expressions for $[\overline{\psi}\psi]_{\mathrm{R}}$ obtained from (4.437) and (4.438) and equating the coefficients of the three independent operators $F_{1,2,3}$, we find that

$$\tilde{Z}_0 + n/8\pi\varepsilon = (1 - \tilde{Z}_3)/2\lambda^2, \quad \tilde{Z}_2 = \tilde{Z}_4, \quad \tilde{Z}_6 = (1 - \tilde{Z}_3)/\lambda, \qquad (4.439)$$

expressing the three "undesirable" constants $\tilde{Z}_{0,2,6}$ in terms of other constants $\tilde{Z}$ which can now be calculated directly in the massless model (4.434), as required. Here it should be noted that the nonmultiplicative nature of the renormalization does not render expressions like (3.176) for a system of composite operators meaningless, but it does not allow passage from the mixing matrix Q to the matrices $Z_a = Z_F^{-1}$ and then to $\gamma_F = -\gamma_a$ using (3.181) and (4.31) owing to the indeterminacy of the Z_φ in this situation.

The counterterms in (4.435) are expressed by the general rule (3.37) in terms of 1-irreducible Green functions Γ of the basic theory (4.434) containing superficially divergent graphs. In the special case of the massless model we have

$$\begin{aligned} \tilde{S}_{\mathrm{R}}(\varphi) = \tilde{S}_{\mathrm{B}}(\varphi) - L\Gamma_{\mathrm{B}}(\varphi) = \tilde{S}_{\mathrm{B}}(\varphi) - L\Big[\overline{\psi}\Gamma_{\overline{\psi}\psi}\psi + \\ + \frac{1}{2}\sigma\Gamma_{\sigma\sigma}\sigma + \overline{\psi}\Gamma_{\overline{\psi}\psi\sigma}\psi\sigma + \frac{1}{4}\overline{\psi}\,\overline{\psi}\Gamma_{\overline{\psi}\,\overline{\psi}\psi\psi}\psi\psi\Big], \end{aligned} \qquad (4.440)$$

where L is the counterterm operator (Sec. 3.8) and $\Gamma_{\dots}$ are the corresponding 1-irreducible (over all fields $\varphi \equiv \overline{\psi}, \psi, \sigma$) Green functions of the massless basic model (4.434). Defining the dimensionless "normalized" functions $\overline{\Gamma}_{\dots}$ as

$$\begin{aligned} \Gamma_{\overline{\psi}\psi} = -\hat{\partial}\cdot\overline{\Gamma}_{\overline{\psi}\psi}, \qquad \Gamma_{\sigma\sigma} = \overline{\Gamma}_{\sigma\sigma}, \\ \Gamma_{\overline{\psi}\psi\sigma} = \lambda_{\mathrm{B}}\overline{\Gamma}_{\overline{\psi}\psi\sigma}, \qquad \Gamma_{\overline{\psi}\,\overline{\psi}\psi\psi} = 2\lambda_{\mathrm{B}}^2\overline{\Gamma}_{\overline{\psi}\,\overline{\psi}\psi\psi}, \end{aligned} \qquad (4.441)$$

by comparing (4.435) and (4.440) we obtain the following equations:

$$\tilde{Z}_1 = 1 + L\overline{\Gamma}_{\overline{\psi}\psi}, \quad \tilde{Z}_3 = 1 + L\overline{\Gamma}_{\sigma\sigma}, \quad \tilde{Z}_4 = 1 - L\overline{\Gamma}_{\overline{\psi}\psi\sigma}, \quad \tilde{Z}_5 = -L\overline{\Gamma}_{\overline{\psi}\,\overline{\psi}\psi\psi}, \quad (4.442)$$

expressing the desired constants $\tilde{Z}$ in terms of the counterterms of the normalized graphs of the massless theory (4.434).

For the two-loop calculation we need the following graphs:

$$\Gamma_{\bar\psi\psi} = i\widehat{p} + \text{} - n \,\text{} + \dots, \tag{4.443a}$$

$$\Gamma_{\sigma\sigma} = -1 - n \,\text{} - n \,\text{} + \dots, \tag{4.443b}$$

$$\Gamma_{\bar\psi\psi\sigma} = \lambda_{\mathrm B} + \text{} + \text{} + \text{} + \tag{4.443c}$$

$$+ \text{} + \text{} - n \,\text{} + \dots,$$

$$\Gamma_{\bar\psi\,\bar\psi\psi\psi} = 2\left\{ \text{} + \text{} + \begin{array}{c}\text{the same with}\\\text{insertions}\end{array} + \text{} + 5 \text{ more} + \dots \right\}. \tag{4.443d}$$

We have also given the bare contributions in momentum space $(\widehat\partial \to i\widehat p)$, to which we need to return in the last stage of calculating the counterterms. As usual, all the preceding steps are conveniently performed in coordinate space.

In the graphs (4.443) the ψ lines correspond to the propagators $\langle\psi\bar\psi\rangle = -\widehat\partial^{-1}$, the σ lines $\delta(x - x')$ generate "contractions" of graphs, the factor $\lambda_{\mathrm B} = \lambda\mu^{-\varepsilon}$ is understood to be present at each vertex, and all the symmetry coefficients and factors of $-n = -N\mathrm{tr}\,1$ (where the minus sign is due to the fermionic nature of the fields) from summation over indices in ψ cycles are isolated explicitly. Equation (4.443a) does not contain the one-loop graph, because the σ-line contraction of the ends of the massless ψ line generates a massless subgraph of the type (3.116) equal to zero. For the same reason there are no graphs with insertions of such subgraphs into internal ψ lines. By "the same with insertions" in (4.443d) we mean the first two graphs with all possible insertions of one-loop subgraphs (4.443b) and (4.443c). The last graph in (4.443d) plus "5 more" is the sum over the $3! = 6$ permutations of the points where the ends of the three σ lines are attached to the lower ψ line (the analogous symmetrization then arises automatically on the upper ψ line). Graphs like (4.443) can also be used in the model (4.433), with the interaction represented by a σ line by analogy with the representation (2.148). However, we must bear in mind that the concepts of 1-irreducibility are different in the models (4.433) and (4.434), because irreducibility with respect to such artificial "interaction lines" is not required, in contrast to the σ lines of the model (4.434). Therefore, the function $\Gamma_{\bar\psi\,\bar\psi\psi\psi}$ of the model (4.433) would also contain graphs that are 1-reducible in the σ lines, which are not present in (4.443d).

As usual, we present the results of the calculation in the form of a table.

Table 4.9 Contributions of graphs to the renormalization constants $\widetilde{Z}$ in the massless model (4.435). $u \equiv g/4\pi$, $g \equiv \lambda^2$, $n \equiv N\,\text{tr}\,1$, $d = 2 + 2\varepsilon$.

No.	Graph γ	Counterterm $L\overline{\gamma}$	Coefficient in constants			
			$\widetilde{Z}_1$	$\widetilde{Z}_3$	$\widetilde{Z}_4$	$\widetilde{Z}_5$
1		u/ε	0	$-n$	0	0
2		u/ε	0	0	-1	0
3		$-u^2/2\varepsilon$	1	0	0	0
4		$-u^2/2\varepsilon$	$-n$	0	0	0
5		$-u^2/\varepsilon^2$	0	$-n$	0	0
6		$-u^2/2\varepsilon^2$	0	0	-1	0
7		$-u^2/2\varepsilon^2$	0	0	n	0
8		$-u^2/2\varepsilon^2$	0	0	-2	0
9	+ 5 more	$-u^2/\varepsilon$	0	0	0	-1

Regarding these results, it should first be mentioned that writing the four-fermion counterterms as $(\overline{\psi}\psi)^2$ in all the expressions is based on the assumption that the graphs in (4.443d) have exactly these divergences, i.e., the index structure $L\Gamma_{\overline{\psi}\,\overline{\psi}\psi\psi} \sim 1 \otimes 1$ (the cofactors in $A \otimes B$ correspond to the two solid ψ lines of the graphs in (4.443d) and counterterms of the form $\overline{\psi}A\psi \cdot \overline{\psi}B\psi$).

This does not follow from dimensional analysis; since the γ matrices are dimensionless, counterterms with the structure $\gamma_i \otimes \gamma_i$, $\gamma_i\gamma_k \otimes \gamma_i\gamma_k$, etc. are allowed by the dimensions. Such additional structures in the counterterms of individual graphs do actually arise. For example, for each of the two one-loop graphs (4.443d) the divergences have the structure $\gamma_i \otimes \gamma_i$, because in these graphs each of the two solid lines carries only one factor of γ. However, these "nonstandard" divergences of the two one-loop graphs in (4.443d) cancel each other out, as also will occur for "graphs with insertions," since the set of all insertions does not spoil this symmetry (summation over permutations of the ends of σ lines), on which the cancellations are based. In the two-loop graphs (line 9 in Table 4.9), each of the solid ψ lines carries two γ matrices, and so the individual graphs can have divergences with the structure $1 \otimes 1$ or $\gamma_i\gamma_k \otimes \gamma_i\gamma_k$. In the summation over all the $3! = 6$ permutations of the ends of the σ lines the divergences of the second type cancel out and only the needed counterterms with the structure $1 \otimes 1$ remain. The statement that the model (4.433) is multiplicatively renormalizable is based on the assumption that this mechanism of the mutual cancellation of "unnecessary" divergences will also operate in higher orders (although, as far as we know, for arbitrary dimension $d = 2 + 2\varepsilon$ this has not been rigorously proven in general). This explains the absence from Table 4.9 of the one-loop graphs in (4.443d), and also the grouping of the graphs in lines 6 and 9 (without this grouping there would be four-fermion subdivergences in the individual graphs of line 6, whereas they are absent in the sum of the two graphs).

The rest of the calculation is performed following the standard scheme in coordinate space with passage to momentum space at the final stage. The massless fermion propagator $\langle \psi\bar{\psi} \rangle = -\widehat{\partial}^{-1} = -\widehat{\partial} \cdot \partial^{-2}$ in coordinate space is obtained by acting with the operator $\widehat{\partial} = \gamma_i\partial_i$ on the scalar line (3.121), which gives

$$< \psi\bar{\psi} > = a \cdot \gamma_i \;\overset{i}{\underset{\alpha}{\bullet\!+\!\bullet}}\; = 2\alpha a \cdot \gamma_i \;\overset{i}{\underset{\alpha+1}{\frown}}\; = \frac{\|d/2\|}{2\pi^{d/2}} \cdot \gamma_i \;\overset{i}{\underset{d/2}{\frown}}\; \quad (4.444)$$

in the graphical notation of (4.320) and with the parameters α and a from (3.121). In arbitrary dimension d the trace of the product of an odd number of γ matrices is zero, while for an even number it is uniquely determined by the condition of cyclical invariance and (4.423), which gives

$$\text{tr}\,[\gamma_1\gamma_2] = \text{tr}\,1 \cdot \delta_{12},$$
$$\text{tr}\,[\gamma_1\gamma_2\gamma_3\gamma_4] = \text{tr}\,1 \cdot [\delta_{12}\delta_{34} + \delta_{14}\delta_{23} - \delta_{13}\delta_{24}]$$
$$(4.445)$$

and so on with $1 \equiv i_1, \dots$. Additional useful reference formulas involving fermion lines will be given below in Sec. 4.48; the formulas given above are sufficient for calculating the counterterms of all the graphs of Table 4.9. This is elementary to do in the case of the simple flow of a single momentum (Sec. 3.21), as the graphs reduce to simple lines. For the graphs of $\Gamma_{\bar\psi\psi}$ (4.443a) this line is a spinor line, and the isolation from it of the factor $-\widehat{\partial}$

in going to the normalized graph [see (4.441)] is also conveniently done in coordinate space (Sec. 4.36). Then the passage to momentum space in the final stage is done for a simple scalar line using the rule (3.125). With the grouping in line 6, the only divergent subgraphs in our graphs are those in lines 1 and 2. The corresponding subdivergences are eliminated by the standard operator R' in $L = KR'$ (Sec. 3.21).

Multiplying the counterterms of the individual graphs by the coefficients from the corresponding column of Table 4.9 and combining the results, we find the needed constants $\widetilde{Z}$ of the massless model (4.435) in the two-loop approximation ($u \equiv g/4\pi$, $g = \lambda^2$, $d = 2 + 2\varepsilon$):

$$\widetilde{Z}_1 = 1 + u^2(n-1)/2\epsilon + ...; \quad \widetilde{Z}_3 = 1 - un/\epsilon + u^2 n/\epsilon^2 + ...;$$

$$\widetilde{Z}_4 = 1 - u/\epsilon + u^2(3-n)/2\epsilon^2 + ...; \quad \widetilde{Z}_5 = u^2/\epsilon + ... \ . \tag{4.446}$$

From these using (4.436) and (4.439) we find the constants Z of the original GN model (4.433):

$$Z_0 = un(n-1)/8\pi\epsilon^2 + ...; \quad Z_1 = 1 + u^2(n-1)/2\epsilon + ...;$$

$$Z_2 = 1 + u(n-1)/\epsilon + u^2(n-1)(2n-3)/2\epsilon^2 + ...; \tag{4.447}$$

$$Z_3 = 1 + u(n-2)/\epsilon + u^2/\epsilon + u^2(n-2)^2/\epsilon^2 + ... \ .$$

Using the convenient charge variable $u \equiv g/4\pi$, from (4.447) following the standard scheme (Sec. 4.2) we find the RG functions $\gamma_{\psi,m,g}$ and $\beta \equiv \beta_u = \widetilde{\mathcal{D}}_\mu u = 2\varepsilon u - u\gamma_g$ (for completeness we give them with the addition of the three-loop contributions calculated in [142]–[144]):

$$\gamma_\psi(u) = u^2(n-1) - u^3(n-1)(n-2) + ...;$$

$$\gamma_m(u) = 2u(n-1) - 2u^2(n-1) - 2u^3(n-1)(2n-3) + ...;$$

$$\beta(u) = u[2\epsilon - \gamma_g(u)] = 2\epsilon u - 2u^2(n-2) + 4u^3(n-2) + 2u^4(n-2)(n-7) + ...,$$
$$\tag{4.448}$$

and also the "vacuum" RG function $\gamma_0(u)$, defined analogously to (4.18) by the expression

$$\widetilde{\mathcal{D}}_\mu[\Delta S_{\text{vac}} + \text{UV-div. part of } \ln C_{\text{R}}] = Vm^2\mu^{2\epsilon}\gamma_0(u), \tag{4.449}$$

with $\Delta S_{\text{vac}} = VZ_0 m^2\mu^{2\varepsilon}$ and the UV-divergent part of $\ln C_{\text{R}}$ equal to $Vnm^2\mu^{2\varepsilon}/8\pi\varepsilon$:

$$\gamma_0(u) = m^{-2}\mu^{-2\epsilon}\widetilde{\mathcal{D}}_\mu[m^2\mu^{2\epsilon}(Z_0 + n/8\pi\epsilon)] =$$

$$= [2\epsilon - 2\gamma_m + \beta\partial_u](Z_0 + n/8\pi\epsilon) = n/4\pi + 0 + O(u^2). \tag{4.450}$$

From (4.448) we easily find the values (4.431) of the critical exponents at the UV-stable (like in the σ model) fixed point

$$u_* = \varepsilon/(n-2) + 2\varepsilon^2/(n-2)^2 + \varepsilon^3(n+1)/(n-2)^3 + \ldots$$

of the β function (4.448).

4.46 The multiplicatively renormalizable two-charge GN model with σ field

Since the renormalization is nonmultiplicative, the field σ in the model (4.435) does not have definite critical dimension, in contrast to the field ψ, which has dimension owing to the possibility of passing to the multiplicatively renormalizable model (4.433). Thus the fields σ in the models (4.427) and (4.435) cannot be said to be the same, which makes it difficult to compare the $2 + \varepsilon$ and $1/n$ expansions. Moreover, we cannot use the model (4.435) as the basis for analyzing the critical conformal invariance of the Green functions of all the fields $\varphi = \psi, \overline{\psi}, \sigma$ (Sec. 4.41).

The multiplicative nature of the renormalization is spoiled by the last two counterterms in the action (4.435), and so it can be restored by introducing into the basic functional (4.434) the contributions of a uniform external field $h\sigma$ and $(\overline{\psi}\psi)^2$ as a second independent interaction. Such a two-charge theory in dimension $d = 2 + 2\varepsilon$ corresponds to the following unrenormalized ($\overline{S}$), basic ($\overline{S}_{\mathrm{B}}$), and renormalized ($\overline{S}_{\mathrm{R}}$) action functionals:

$$\overline{S}(\varphi) = \overline{\psi}[\hat{\partial} + \overline{m}_0]\psi - \sigma^2/2 + \lambda_0\overline{\psi}\psi\sigma + g_{20}(\overline{\psi}\psi)^2/2 + h_0\sigma, \qquad (4.451)$$

$$\overline{S}_{\mathrm{B}}(\varphi) = \overline{\psi}[\hat{\partial} + \overline{m}]\psi - \sigma^2/2 + \lambda_{\mathrm{B}}\overline{\psi}\psi\sigma + g_{2\mathrm{B}}(\overline{\psi}\psi)^2/2 + h\sigma, \qquad (4.452)$$

$$\begin{aligned}
\overline{S}_{\mathrm{R}}(\varphi) &= \overline{Z}_0\overline{m}^2\mu^{2\epsilon} + \overline{\psi}[\overline{Z}_1\hat{\partial} + \overline{Z}_2\overline{m}]\psi - \overline{Z}_3\sigma^2/2 + \\
&\quad + \overline{Z}_4\lambda_{\mathrm{B}}\overline{\psi}\psi\sigma + \overline{Z}_5 g_{2\mathrm{B}}(\overline{\psi}\psi)^2/2 + \overline{Z}_6\mu^\epsilon\overline{m}\sigma + h\sigma
\end{aligned} \qquad (4.453)$$

with independent bare parameters $\overline{m}_0$, h_0, $\lambda_0 \equiv g_{10}^{1/2}$, g_{20} and their basic and renormalized analogs ($\lambda_{\mathrm{B}} = \lambda\mu^{-\varepsilon}$, $g_{i\mathrm{B}} = g_i\mu^{-2\varepsilon}$). All the constants $\overline{Z}$ in (4.453) are functions of the two dimensionless renormalized charges $g \equiv \{g_1 \equiv \lambda^2, g_2\}$, and the contribution $h\sigma$ is not renormalized according to the general rule (see Sec. 3.13 and the analysis of the φ^3 model in Sec. 4.24). The renormalized action (4.453) is clearly related to the unrenormalized one (4.451) by the standard relation of multiplicative renormalization (1.84) with the constants $\overline{Z}_\varphi \equiv \{\overline{Z}_\psi = \overline{Z}_{\overline{\psi}}, \overline{Z}_\sigma\}$ for the fields and $\overline{m}_0 = \overline{m}\overline{Z}_m$, $\lambda_0 = \lambda_{\mathrm{B}}\overline{Z}_\lambda$, $g_{i0} = g_{i\mathrm{B}}\overline{Z}_{g_i}$ ($\overline{Z}_{g_1} = \overline{Z}_\lambda^2$), $h_0 = \overline{Z}_\sigma^{-1}[h + \overline{Z}_6\overline{m}\mu^\epsilon]$ for the parameters, and these constants $\overline{Z}$ for the fields and parameters are uniquely expressed in terms of the constants $\overline{Z}$ in (4.453). On the other hand, it is easy to show that all the constants $\overline{Z}_i$ can be expressed in terms of the four constants Z of the massive GN model (4.433). Actually, if we limit ourselves to the fermionic sector, after performing the integration over σ in the basic theory (4.452) we arrive at the "effective GN model" (4.433) with parameters

$$m = \overline{m} + h\lambda_B, \qquad g = g_1 + g_2 \qquad (4.454)$$

and with the constant $Vh^2/2 + \ln[\overline{C}_R/C_R]$ added to its action, where the last term comes from the difference between the normalization factors $C_R \sim \det(\hat{\partial}+m)^{-1}$ in functionals of the type (2.30) for the models in question. The UV divergences of this effective GN model are eliminated by the counterterms (4.433) with the parameters (4.454) and the addition of the known (Sec. 4.45) constant $-$UV-div. part of $\ln[\overline{C}_R/C_R] = Vn\mu^{2\varepsilon}(m^2 - \overline{m}^2)/8\pi\varepsilon$ to the vacuum counterterm in (4.433) to eliminate the divergenes of the extra added constant. Owing to the UV-finiteness in the same subtraction scheme (MS) the set of counterterms of this effective GN model must be the same as the set of counterterms of the original model (4.453) with the field σ eliminated (by integration). It then follows that

$$\overline{Z}_0\overline{m}^2\mu^{2\epsilon} + \overline{\psi}[\overline{Z}_1\hat{\partial} + \overline{Z}_2\overline{m}]\psi + \frac{1}{2}\overline{Z}_5 g_{2B}(\overline{\psi}\psi)^2 +$$

$$+\frac{1}{2}\overline{Z}_3^{-1}[\overline{Z}_4\lambda_B\overline{\psi}\psi + \overline{Z}_6\overline{m}\mu^\epsilon + h]^2 = \frac{1}{2}h^2 + \frac{n\mu^{2\epsilon}}{8\pi\epsilon}(m^2 - \overline{m}^2) + \qquad (4.455)$$

$$+Z_0 m^2\mu^{2\epsilon} + \overline{\psi}[Z_1\hat{\partial} + Z_2 m]\psi + \frac{1}{2}Z_3 g_B(\overline{\psi}\psi)^2$$

[as in (4.433) the factor $V = \int dx$ in the vacuum contributions has been omitted].

Equating the coefficients of the independent structures on both sides of (4.455), after simple algebra we obtain the following seven equations:

$$\overline{Z}_1 = Z_1, \quad \overline{Z}_3^{-1} = 1 + 2g_1(Z_0 + n/8\pi\epsilon), \quad \overline{Z}_2 = \overline{Z}_4 = Z_2\overline{Z}_3,$$
$$\qquad (4.456)$$
$$\lambda\overline{Z}_6 = 1 - \overline{Z}_3 = 2g_1(\overline{Z}_0 + n/8\pi\epsilon), \quad g_1\overline{Z}_3^{-1}\overline{Z}_4^2 + g_2\overline{Z}_5 = gZ_3$$

($g_1 \equiv \lambda^2$), making it possible to express all seven constants $\overline{Z} = \overline{Z}(g_1, g_2)$ in terms of the four constants $Z = Z(g)$ of the model (4.433) with $g = g_1 + g_2$. We note that (4.437) and (4.438) and their consequences (4.439) remain valid for the model (4.453) with the simple replacement $\tilde{Z}_i \to \overline{Z}_i$, but the three relations like (4.439) will now be automatic consequences of (4.456).

From these using expressions like (4.5) it is easy to find the relation between the RG functions of the models (4.453) and (4.433). Omitting the straightforward calculations, we give only the results for the anomalous dimensions of the fields $\overline{\gamma}_\varphi = \tilde{\mathcal{D}}_\mu \ln \overline{Z}_\varphi$ and the β functions $\overline{\beta}_i = \tilde{\mathcal{D}}_\mu g_i$ of the two charges of the model (4.453) in terms of the RG function of the model (4.433):

$\overline{\gamma}_\psi = \gamma_\psi$, $\overline{\gamma}_\sigma = -g_1\gamma_0$, $\overline{\beta}_1 = 2g_1[\varepsilon - \gamma_m + g_1\gamma_0]$, $\overline{\beta}_1 + \overline{\beta}_2 = \beta_g$. It is convenient to make the variable substitution $\{g_1, g_2\} \to \{g_1, u\}$, taking the second
independent variable to be not g_2, but the charge $u \equiv g/4\pi = (g_1 + g_2)/4\pi$
of the effective GN model, which is the argument of its RG functions (4.448).
In terms of the new variables u and g_1 the above relations take the form

$$\overline{\gamma}_\psi(u, g_1) = \gamma_\psi(u), \qquad \overline{\gamma}_\sigma(u, g_1) = -g_1\gamma_0(u), \qquad (4.457)$$

$$\overline{\beta}_u(u, g_1) = \beta(u), \qquad \overline{\beta}_{g_1}(u, g_1) = 2g_1[\epsilon - \gamma_m(u) + g_1\gamma_0(u)] \qquad (4.458)$$

with the known RG functions (4.448) and (4.450). Within the ε expansion
the β functions (4.458) have nontrivial fixed point $u = u_*$, $g_1 = g_{1*}$ with the
former (Sec. 4.45) value $u_* = g_*/4\pi$ and

$$g_{1*} = \frac{\gamma_m(u_*) - \epsilon}{\gamma_0(u_*)} = \frac{4\pi\epsilon(3n - 2)}{n(n - 2)} + O(\epsilon^2). \qquad (4.459)$$

For $n > 2$ this point is located in the physical region $g_1 \equiv \lambda^2 > 0$ (which
is important for comparison with $1/n$ expansions) and is IR-attractive in the
variable g_1 [since $\gamma_0(u_*) > 0$], whereas in the variable u we have a fixed point
of the UV type usual for the GN model. Therefore, reaching the critical regime
is assured only when the GN charge u is a parameter like the temperature and
the asymptote $u \to u_*$ is specified by the external conditions (the experimental
setup). In contrast to u, the second charge g_1 can be arbitrary, and the
corresponding invariant charge will tend to g_{1*} at the IR asymptote in the
momenta and the mass parameters $\overline{m}$, h (Sec. 1.34). Therefore, the Green
functions of all three fields $\varphi = \psi, \overline{\psi}, \sigma$ will possess critical scale invariance
in the massless ($\overline{m} = 0$, $h = 0$) model (4.453) at the asymptote $u \to u_*$
in the GN charge $u = (g_1 + g_2)/4\pi$ and $p \to 0$ in all the momenta for any
value of the second charge $g_1 \equiv \lambda^2$, or for any values of the momenta right
at the fixed point $u = u_*$, $g_1 = g_{1*}$ in the two charges. The corresponding
critical anomalous dimensions of the fields γ_φ^* are the values of the known RG
functions (4.457) at the fixed point. Then for the ψ field we obtain the usual
dimension of the GN model, and for the σ field (absent in the GN model),
taking into account its canonical dimension $d_\sigma = d/2 = 1 + \varepsilon$ in the functional
(4.453), from (4.457) and (4.458) we find that

$$\Delta_\sigma = d_\sigma + \overline{\gamma}_\sigma^* = 1 + 2\varepsilon - \gamma_m^* = d - \Delta_m. \qquad (4.460)$$

Therefore, Δ_σ in the model (4.453) obeys the shadow relation with the mass
dimension Δ_m in the model (4.433) and therefore coincides with the critical
dimension $\Delta[\overline{\psi}\psi]$ of the composite operator $[\overline{\psi}\psi]_R$ in the GN model. Equation
(4.460) can be used to calculate the $2 + \varepsilon$ expansion of the same quantity Δ_σ,
the $1/n$ expansion of which is calculated using the model (4.427) in arbitrary
dimension d, and the $4 - \varepsilon$ expansion using the model (4.429). Naturally, all
these expansions turn out to be mutually consistent.

In conclusion, we note that for the purely fermionic Green functions of the model (4.453) the critical regime is associated not with a single point (as for the set of all the Green functions), but with an entire plane in its four-dimensional parameter space determined by the criticality conditions $m = 0$, $g = g_*$ for the parameters (4.454) of the "effective" GN model. Moreover, although right now we are interested in only massless critical regimes, passage to the massive analogs is necessary for obtaining the desired relations (4.456) between the renormalization constants of these models.

4.47   Proof of critical conformal invariance

In this section we shall show that in the critical regime the Green functions of the fields of both the purely fermionic model (4.422) and the model (4.453) with the field σ are conformally invariant with the usual critical conformal dimensions of the fields [145] (i.e., the same as the scale dimensions and identical for ψ and $\overline{\psi}$). We recall that by "critical regime" we here mean the massless model studied right at the fixed point $g = g_*$ for the charges, independently of what type of fixed point this is.

Let us first consider the model (4.422). It satisfies all the requirements listed in Sec. 4.41 (multiplicative renormalizability, the presence of a fixed point g_* and critical scaling with definite dimensions of the fields, canonical conformal invariance of the free part of the action, and no derivatives in the interaction), and so all the general relations and conclusions of Sec. 4.41 are valid for it, including the basic relation (4.400). The anticommutativity of the variables makes it necessary to carefully interpret the formulas. In particular, the expression $d_\varphi[\partial \mathcal{V}/\partial \varphi]\varphi$ in (4.401) must now be understood as $d_{\overline{\psi}} \cdot \overline{\psi}[\vec{\partial}\mathcal{V}/\partial\overline{\psi}] + d_\psi \cdot [\vec{\partial}\mathcal{V}/\partial\psi]\psi$, which for the interaction $\mathcal{V}$ from (4.422) with the usual assumption that the dimensions are equal $d_\psi = d_{\overline{\psi}} = (d-1)/2$ gives

$$N = (4d_\psi - d)\mathcal{V} = (d-2)\mathcal{V}, \quad \mathcal{V} \equiv Z' g_{\mathrm{B}}(\overline{\psi}\psi)^2/2. \tag{4.461}$$

Here it should be noted that the formal rules of Sec. 1.15 for the action (4.422) with independent fields ψ and $\overline{\psi}$ can be used to find only the sum of their canonical dimensions, not each one separately. The same will occur with the anomalies in renormalization, because from the graphs one always calculates only the product $Z_\psi \cdot Z_{\overline{\psi}}$. However, in real situations the fields ψ and $\overline{\psi}$ are not independent, but are related by some conjugation operator, and so it is natural to require that their dimensions be equal. We also note that the quadratic part of the action (4.422) is conformally invariant only under transformations with identical dimensions $d_\psi = d_{\overline{\psi}}$.

Let us now consider the representation (4.403) for the operator (4.461). In this case the right-hand side of (4.403) may involve all the independent U_N-symmetric scalars with $d_F^* = 2$. The basis of this set consists of (assuming inner closure of the GN model; see Sec. 4.45) three renormalized monomials

$[\overline{\psi}\hat{\partial}\psi]_R$, $\partial[\overline{\psi}\gamma\psi]_R$, and $[(\overline{\psi}\psi)^2]_R$, and the two fields ψ and $\overline{\psi}$ correspond to two independent operators of the type (4.397), namely,

$$\overline{\psi}\left[\overrightarrow{\delta S_R/\delta\overline{\psi}}\right] = Z\overline{\psi}\hat{\partial}\psi + g_B Z'(\overline{\psi}\psi)^2 = [\overline{\psi}\hat{\partial}\psi]_R + g_B[(\overline{\psi}\psi)^2]_R,$$

(4.462)

$$\left[\delta S_R/\delta\psi\right]\psi = -Z\partial\overline{\psi}\cdot\gamma\psi + g_B Z'(\overline{\psi}\psi)^2 = -[\partial\overline{\psi}\cdot\gamma\psi]_R + g_B[(\overline{\psi}\psi)^2]_R.$$

They generate two contributions of type A in the expansion (4.403), and the third independent basis element can be taken to be the operator $[(\overline{\psi}\psi)^2]_R$ of type B, so that in this case the expansion (4.403) does not contain any separate terms of type C with external derivatives. However, the problem of dangerous contributions $\sim \partial_k \mathcal{O}_k$ is still present, because their role is now taken over by the difference of the two operators (4.462), equal to

$$Z\partial(\overline{\psi}\gamma\psi) = \partial[\overline{\psi}\gamma\psi]_R.$$

(4.463)

Therefore, the operator $[\overline{\psi}\gamma_k\psi]_R$ with canonical dimension $d - 1$ is a candidate for the role of $\mathcal{O}_k$ in (4.405). It also satisfies the criterion (4.406), because it is a multiplicatively renormalizable (it does not mix with anything) canonical conserved current for the group of global gauge transformations $\psi(x) \to \exp(i\omega)\psi(x)$, $\overline{\psi}(x) \to \exp(-i\omega)\overline{\psi}(x)$, and so (Sec. 3.31) it does not have an anomalous dimension.

The dangerous contribution (4.463) is present in the expansion (4.403) for $g = g_*$ when and only when the coefficients $a(g_*)$ of the two operators (4.462) do not coincide, which owing to the first relation in (4.404) would imply $\gamma^*[\psi] \neq \gamma^*[\overline{\psi}]$. Conformal symmetry is conserved here in the broad sense, and the only violation is the "splitting" of the conformal critical dimensions of the fields ψ and $\overline{\psi}$. We note that such a splitting by the addition of $\pm$const to the anomalous dimensions would not be manifested either in the scale Ward identity or in the RG equations for Green functions with equal numbers of the fields ψ and $\overline{\psi}$, since these quantities involve only the sum of their anomalous dimensions.

However, this beautiful but speculative splitting of the dimensions is actually impossible, as we shall now show. The simplest argument is to use the charge symmetry of the theory, which, crudely speaking, implies that the fields ψ and $\overline{\psi}$ are on a completely equal footing. This can be shown for the simple example of the complex φ^4 model with the fields $\varphi, \overset{+}{\varphi}$ and the interaction $\mathcal{V} \sim (\overset{+}{\varphi}\varphi)^2$, where the analog of the dangerous contribution (4.463) will be the divergence of the gauge current $\partial\{\overset{+}{\varphi}\cdot\partial\varphi - \partial\overset{+}{\varphi}\cdot\varphi\}$. This operator is odd under the discrete charge conjugation transformation $\varphi \leftrightarrow \overset{+}{\varphi}$, and so in renormalization it cannot mix with the even operator $N \sim \mathcal{V} \sim (\overset{+}{\varphi}\varphi)^2$. For Dirac spinors charge conjugation corresponds to the transformation $\{\psi, \overline{\psi}\} \to \{\psi' = C\overline{\psi}, \overline{\psi}' = -\psi C^{-1}\}$ with some matrix C chosen so as to ensure the charge invariance of the action (4.422) taking into account the anticommutativity of the fields [in the free part this corresponds to the condition $C^{-1}\gamma_i C = -\gamma_i^{\mathrm{T}}$ for a specific representation of the γ matrices, and the interaction in (4.422) is automatically invariant owing to the equation $\overline{\psi}'\psi' = \overline{\psi}\psi$ for all C]. Under this transformation the operators (4.462) transform into each other, and so their difference is C-odd and in renormalization cannot mix with the C-even operator (4.461). All this is the same as in the scalar model. However, there is one fine point. In the scalar case the charge conjugation transformation $\varphi \leftrightarrow \overset{+}{\varphi}$ is well defined in arbitrary dimension d, whereas in the spinor case the explicit representation of the matrix C in terms of the Dirac matrices γ_i is easily constructed only in even integer dimension d and has no natural extension to arbitrary d (this is the same problem as for the matrix γ_5, which is defined in even integer dimension as the product of all the matrices γ_i up to a coefficient). Therefore, in calculations involving the variable d [$d = 2 + 2\varepsilon$ in the model (4.422) or the $1/n$ expansion in arbitrary dimension d] the argument that the dangerous contribution (4.463) is forbidden by the charge symmetry is of doubtful validity.

However, it can be argued that this contribution is forbidden without using charge symmetry. For a spinor field (like for a scalar field and in contrast to a vector field) the usual symmetries (translational plus rotational plus scale) without conformal symmetry fix the form of the massless propagator up to the normalization (we do not consider the exceptional structures mentioned in Sec. 2.18):

$$\langle \psi(y)\overline{\psi}(z) \rangle = A \cdot \widehat{n} \cdot |x|^{-2\alpha}, \tag{4.464}$$

where $x \equiv y - z$, $n_i \equiv x_i/|x|$, $2\alpha = \Delta[\psi] + \Delta[\overline{\psi}]$ is the sum of the critical dimensions of the fields, and A is an amplitude factor. Performing the conformal inversion of the fields (Sec. 4.44) with arbitrary (unknown) dimensions $\Delta[\psi]$ and $\Delta[\overline{\psi}]$, it is easily verified that the expression on the right-hand side of (4.464) will be conformally invariant when and only when $\Delta[\psi] = \Delta[\overline{\psi}]$. Since conformal invariance in the broad sense (with unknown $\Delta[\psi]$ and $\Delta[\overline{\psi}]$) has already been established, this argument demonstrates the need for the conformal critical dimensions to be equal, i.e., the impossibility that they are split.

Therefore, the Green functions of the fields ψ and $\overline{\psi}$ of the model (4.422) in the critical regime are conformally invariant in the usual sense, i.e., they are invariant under transformations with critical conformal dimensions of the fields ψ and $\overline{\psi}$ that are equal and the same as the scale dimensions.

In order to use the conformal symmetry in calculating the $1/n$ expansions of the critical exponents in the model (4.427), it is necessary to generalize the proof to the Green functions of all the fields $\varphi = \psi, \overline{\psi}, \sigma$. We cannot do this in the model (4.427) itself, because its renormalization is very specific (Sec. 4.31), and our proof (Sec. 4.41) presupposes the standard scheme of multiplicative renormalization with charge (or charges) g, fixed point g_*, and some sort of ε expansion. All these requirements are satisfied by the massless two-charge model (4.453), and we shall return to it to prove the critical conformal symmetry.

Here it should be stressed that the critical symmetry is an objective property of the Green functions and therefore independent of the choice of perturbation theory ($2 + 2\varepsilon$, $4 - 2\varepsilon$, or $1/n$). Therefore, if it is proved using some particular scheme, it can be considered proved for any other. Of course, here we are speaking only of the various types of expansion for models with identical critical behavior (which in our case is known) and of Green functions of quantities that are identical up to normalization and that must have definite critical dimensions according to which they are identified.

Let us consider the massless ($\overline{m} = 0$, $h = 0$) two-charge model (4.453). The additional term $\sim \sigma^2$ in its free action is also canonically conformal (with $d_\sigma = d/2$), and so this model satisfies all the requirements listed in Sec. 4.41. In the expansion (4.403), the operators $[\sigma^2]_R$ and $[\overline{\psi}\psi\sigma]_R$ are now added to the three fermionic monomials (see the beginning of the present section). This does not lead to the appearance of new dangerous linear combinations with a single external derivative different from (4.463), and so the proof of critical conformal invariance given above for the purely fermionic model (4.422) carries over to the model (4.453) without change: the Green functions of all three fields $\varphi = \psi$, $\overline{\psi}$, σ of the model (4.453) in the critical regime are conformally invariant. This statement is also valid for the Green functions of the fields of the renormalized models (4.427) and (4.429), since they have the same critical behavior (Sec. 4.44) and the auxiliary field σ has the same meaning in them (it is a quantity with critical dimension Δ_σ which is definite and therefore the same for these models).

For what follows (Sec. 4.48) it is also important to note that all these models contain a scalar composite operator F with critical dimension $\Delta_F = d - \Delta_\tau = d - 1/\nu$, which is critically conformal as it is the founder of a new family (associated with Δ_τ) (Sec. 4.43). The explicit form of this operator depends on the model. In the models (4.427) and (4.429) $F = [\sigma^2]_R$, and in the model (4.453) F is a linear combination of five renormalized monomials with $d_F^* = d^* = 2$ associated with one of its two exponents ω for the system of β functions (4.458). Let us explain in more detail. Generalizing the arguments of Sec. 4.11 to the massless model with arbitrary fields φ and charges $g = \{g_i\}$, it is easily verified that the family of operators $F = \{F_i\}$, defined by the relations

$$\left[\left(\sum_k \omega_{ki}\partial_k - \sum_\varphi \partial_i\gamma_\varphi \cdot \varphi\delta/\delta\varphi \right) S_B(\varphi) \right]_R \Bigg|_{g=g_*} = \int dx\, F_i(x) \qquad (4.465)$$

with $\partial_i \equiv \partial/\partial g_i$ and $\omega_{ik} \equiv \partial\beta_i/\partial g_k$, satisfies equations of critical scaling of the type (4.42) with the matrix of critical dimensions $\Delta_F = d + \omega^\top$. The eigenvalues of the matrices ω and $\omega^\top$ are the exponents ω_a, and the corresponding basis operators (4.44) have dimensions $d + \omega_a$, $a = 1, 2, \ldots$ and are the analogs of the operator (4.66) for the φ^4 model. In our model (4.453) the complete set of scalars with $d_F^* = d^* = 2$ consists of five monomials (see above), and from them are constructed three operators of the type (4.397) with $\Delta_F = d$

for the three fields $\varphi = \psi, \overline{\psi}, \sigma$ and two operators of the type (4.66) with
$\Delta_F = d + \omega_a$ corresponding to the two charges u, g_1 of the system of β functions (4.458). In this case the matrix ω is triangular and its eigenvalues ω_a are simply the corresponding diagonal elements, namely, $d\beta(u)/du|_* = -\Delta_\tau$ for the charge u [see (4.426)] and $\partial\overline{\beta}_{g_1}(u, g_1)/\partial g_1|_* = 2\varepsilon - 2\gamma_m^*$ for the charge g_1. The operator of interest to us is associated with the first of these, and owing to the triangularity of the matrix ω it is simply an operator of the type (4.465) corresponding to the charge u.

The presence of a conformal operator with critical dimension $\Delta_F = d - \Delta_\tau$ is the necessary condition for it to be possible to calculate this dimension by the conformal bootstrap method.

4.48 $1/n$ expansion of the critical exponents of the GN model

Calculations of $1/n$ expansions are most easily carried out using the completely massless model (4.427) with the needed regularization and renormalization (Sec. 4.31). Here it is possible to use any of the technical methods described above for the σ model, including the most powerful one: the conformal bootstrap method taking into account the proven (Sec. 4.47) critical conformal invariance of the GN model. Let us introduce the following parametrization of the three independent exponents in (4.428):

$$\gamma_\psi^* \equiv \eta/2, \quad \gamma_\sigma^* \equiv -\eta - \kappa, \quad \Delta_\tau \equiv 1/\nu \equiv 2\lambda \qquad (4.466)$$

and list all the presently (1996) available results for the coefficients of expansions of the type $z = z_0 + z_1/n + z_2/n^2 + ...$, $n \equiv N\mathrm{tr}\,1$, of the three independent exponents $z \equiv \eta, \kappa, \lambda$:

$$\eta_0 = 0, \quad \kappa_0 = 0, \quad \lambda_0 = \mu - 1,$$

$$\eta_1 = -2\|2\mu - 1\|/\|1 - \mu\|\mu - 1\|\mu\|\mu + 1\|,$$

$$\kappa_1/\eta_1 = \mu/\alpha, \quad \lambda_1/\eta_1 = 1 - 2\mu,$$

$$\eta_2/\eta_1^2 = (2\mu - 1)B_1/\alpha + 1/2\mu - 1/2\alpha - 1/2\alpha^2,$$

$$\kappa_2/\eta_1^2 = (\mu/\alpha)[(2\mu - 1)B_1/\alpha - 3\mu C(1) - 2\mu - 1 - 6/\alpha - 4/\alpha^2],$$

$$\lambda_2/\eta_1^2 = (\mu/2\alpha)\Big\{8/\eta_1\beta^2 + 4\mu(3 - 2\mu)[C(2 - \mu) - B_2^2]/\beta +$$

$$+ B_2[-8\alpha^2 - 16\alpha - 6 - 8/\alpha - 10/\beta + 4/\beta^2 - 2/\mu] +$$

$$+ \mu C(1)[6\alpha - 29 + 22/\beta] + 8\alpha^2 - 10 + 14/\alpha - 5/\alpha^2 +$$

$$+42/\beta - 4/\beta^2 + 4/\mu - 1/\mu^2 \Big\}, \tag{4.467}$$

$$\eta_3/\eta_1^3 = (1/4)\Big\{2(2\mu - 1)^2[C(1) - C(1 - \mu)]/\alpha^2 - 6\mu^2 I(\mu)C(1)/\alpha+$$

$$+C(1)[8 - \mu + 16/\alpha + 9/\alpha^2] - 12\mu^2 B_1 C(1)/\alpha + 6(2\mu - 1)^2 B_1^2/\alpha^2+$$

$$+B_1[6/\mu - 6 - 4\mu - 28/\alpha - 40/\alpha^2 - 14/\alpha^3] + 2/\mu^2 - 6-$$

$$-4\mu - 9/\alpha + 3/\alpha^2 + 12/\alpha^3 + 5/\alpha^4 \Big\},$$

where $\mu \equiv d/2$, $\alpha \equiv \mu - 1$, $\beta \equiv 2 - \mu$, $B_1 \equiv B(1 - \mu) - B(1)$, $B_2 \equiv B(2 - \mu) - B(1)$, $B(z)$ and $C(z)$ are the notation of (4.347), and $I(\mu)$ in η_3 is the same quantity as in (4.393) and cannot be expressed explicitly in terms of gamma functions and their derivatives. The contributions of first order in $1/n$ have been calculated by many authors using different methods (sometimes even with direct verification of scaling, for which there is really no need, because the arguments of Sec. 2.25 establishing the existence of scaling remain valid for the GN model). The value of η_2 was calculated in [135] using the propagator self-consistency equations, η_3 and κ_2 were obtained in [136], and λ_2 was found in [146] by the conformal bootstrap method and independently in [137] and [147]. In the double expansion in ε and $1/n$ all the results (4.467) agree with the results for the $2 + \varepsilon$ and $4 - \varepsilon$ expansions known from (4.431) and (4.432).

Let us explain the most complicated calculations [136], [146], paying special attention to the details of working with spinor objects. Spinor propagators of the type (4.444) are conveniently represented using the following graphical notation [see also (4.320)]:

$$(\widehat{y} - \widehat{x}) \cdot |y - x|^{-2a} = \gamma_i \quad \underset{a}{\overset{i}{\longrightarrow}} \quad \equiv \quad \underset{a}{\bullet\!\!\longrightarrow\!\!\bullet} \ , \tag{4.468}$$

where the letter on a line is the index. Closed loops of spinor lines generate the trace of a product of γ matrices which is calculated using the rules (4.445). Chains with spinor and ordinary links are integrated using the expressions [notation of (3.68)]

$$\underset{a+1 \quad b}{\bullet\!\!\longrightarrow\!\!\bullet\!\!-\!\!\bullet} \quad = \quad \frac{\pi^{d/2} H(a, b, d - 1 - a - b)}{a(d - 1 - a - b)} \quad \underset{a+b+1-d/2}{\bullet\!\!\longrightarrow\!\!\bullet} , \tag{4.469a}$$

$$\underset{a+1 \quad b+1}{\bullet\!\!\longrightarrow\!\!\bullet\!\!\longrightarrow\!\!\bullet} \quad = \quad - \frac{\pi^{d/2} H(a, b, d - 1 - a - b)}{ab} \quad \underset{a+b+1-d/2}{\bullet\!\!\longrightarrow\!\!\bullet} , \tag{4.469b}$$

and the presence of the unit matrix with spinor indices on the right-hand side of (4.469b) is understood. The analog of the basic rule for integrating a unique vertex (4.328) will now be

$$
\int_{\substack{a+1 \quad b+1 \\ a+b+c=d-1}}^{c} = \frac{\pi^{d/2} H(a,b,c)}{ab} \; \stackrel{b'}{\underset{c'}{\triangle}}\!\!^{a'} \;, \tag{4.470}
$$

where $z' \equiv d/2 - z$ for all z, and the uniqueness condition is given below the graph.

In real calculations it is convenient to use the following parametrization of the propagator indices:

$$
G_\psi = A \;\underset{\alpha+1}{\longrightarrow}\; , \; G_\sigma = B \;\underset{\beta-2\Delta}{\bullet\!\!-\!\!\bullet}\; , \quad
\begin{aligned}
\alpha &= d/2 - 1 + \eta/2 \;, \\
\beta &= 1 - \eta \;, \; \Delta = \kappa/2 \;.
\end{aligned} \tag{4.471}
$$

In the notation (4.466) the quantity $2\Delta = \kappa$ has the meaning of the critical dimension of the basic vertex $\overline{\psi}\psi\sigma$ (Sec. 4.35), and for $\Delta = 0$ and any η the local vertex $\overline{\psi}\psi\sigma$ is unique owing to the equation $2\alpha + \beta = d - 1$ for the indices (4.471). The scalar propagator G_σ is inverted using the rule (4.309), and its analog for the spinor propagator is conveniently written as

$$
[A \;\underset{\alpha+1}{\bullet\!\!-\!\!\bullet}\;]^{-1} = A^{-1}\, p_\psi(\alpha) \;\underset{d-\alpha}{\bullet\!\!-\!\!\bullet}\; , \; p_\psi(z) = zp(z)/(d/2-z) \tag{4.472}
$$

with the function $p(z)$ from (4.309). The full basic vertex $\langle \overline{\psi}\psi\sigma \rangle$ and the scalar vertex $\langle F\sigma\sigma \rangle$ with composite operator $F = \sigma^2$ in the notation of (4.471) and $\Delta_F \equiv 2\lambda$ [the shadow value of 2λ from (4.466)] are represented up to amplitude factors by the following conformal triangles:

$$
\underset{\quad}{\otimes} \; \sim \;
\overset{\beta-2\Delta}{\underset{\underset{\beta-\Delta}{\alpha+1}}{\alpha+1+\Delta \,\triangle\, \alpha+1+\Delta}}\!{}^{\alpha+1} \; , \quad
\underset{\quad}{\otimes} \; \sim \;
\overset{2\lambda}{\underset{\underset{\beta'+\lambda+2\Delta}{\beta-2\Delta}}{\lambda' \,\triangle\, \lambda'}}\!{}^{\beta-2\Delta} \; , \tag{4.473}
$$

where $z' \equiv d/2 - z$ for all z. The regularizers ε, ε' are introduced into the basic vertex using the rule (4.388). The conformal bootstrap equations analogous to (4.391) are written in terms of the regularized vertex function $V \equiv V(\alpha, \Delta, u; \varepsilon, \varepsilon')$ as

$$
V\,|_{\varepsilon=\varepsilon'=0} = 1,
$$

$$
p_\psi(\alpha) = us[\partial V/\partial \varepsilon']\,|_{\varepsilon=\varepsilon'=0} \;, \tag{4.474}
$$

$$
p(\beta - 2\Delta)/n = us[\partial V/\partial \varepsilon]\,|_{\varepsilon=\varepsilon'=0}
$$

with the functions $p_\psi(z)$ from (4.472) and $p(z)$ from (4.309), the standard amplitude combination $u \equiv A^2 C^2 B$ [A and B are the amplitudes of the propagators (4.471), and C is the amplitude factor at the basic vertex (4.473)], and the quantity

$$s = \pi^{2d} H(\alpha, \alpha, \beta - 2\Delta, \alpha + \Delta, \alpha + \Delta, \beta - \Delta, d/2 + \Delta)/\alpha^2(\alpha + \Delta)^2 \|d/2\|.$$

(4.475)

The following graphs must be included in the calculation of the dimension Δ_ψ through order $1/n^3$ and simultaneously Δ_σ through order $1/n^2$ (i.e., η_3 and κ_2) in the expression analogous to (4.390) for the regularized vertex function V in (4.474):

(4.476)

The notation is the same as in (4.390), i.e., the full vertices in the graphs are shown as dark circles. The summation over all indices in closed loops of ψ lines gives factors of $-N \operatorname{tr} 1 \equiv -n$ (the minus sign arises because the field is fermionic). In (4.476) they are isolated explicitly along with the amplitude combinations u and the symmetry coefficients. Only the very first (lowest-order) graph in (4.476) is needed to calculate η_2 and κ_1; the others are needed to calculate η_3 and κ_2 (details are given in [136]). With this accuracy in $1/n$, (4.476) contains fewer graphs than the analogous expression (4.390) owing to the exclusion of graphs with odd cycles of ψ lines (such a cycle contains the trace of a product of an odd number of γ matrices, which is zero). The gain from the decreased number of graphs is greater than the loss due to the greater complexity of their lines (they now are spinor lines), and so on the whole calculations of higher orders in the GN model turn out to be easier than in the σ model.

This is what made it possible to also calculate the "correction" exponent $\Delta_\tau \equiv 1/\nu = d - \Delta[\sigma^2]$ in the GN model in order $1/n^2$ [146]. At the present time (1996) this is the only example of a calculation in higher than first order in $1/n$ of a critical dimension of a composite operator [like Δ_τ in (4.428), the exponent ω in (4.295) for the simple σ model is known only in first order in $1/n$]. All the other known results in higher orders pertain only to the dimensions of the fields themselves. It is possible that this is the reason that in the result (4.467) for λ_2 the empirical rule $z_k = A\eta_1^k$ with coefficient A not containing simple gamma functions is violated: from (4.467) we have $\lambda_2 = A\eta_1^2 + B\eta_1$ in the same notation.

The dimension $\Delta[\sigma^2]$ was calculated in [146] by the conformal bootstrap method for the vertex $\langle F\sigma\sigma \rangle$ with operator $F = \sigma^2$. The technique described in Sec. 4.40 generalizes directly to a vertex of the type $\langle F\varphi\varphi \rangle$ if, first, it is conformal and, second, the operator F contains only two fields φ. The former is necessary for the vertex to be representable by a suitable conformal triangle, and the second is required to preserve the general structure of vertex equations of the type (4.370) and (4.372) and the possibility of resumming all the graphs in them to obtain skeleton graphs with dressed lines and triple (but not higher) vertices. The operator $F = \sigma^2$ in the GN model satisfies all these requirements, and so the vertex $\langle F\sigma\sigma\rangle$ (which we shall refer to as the correction vertex) satisfies conformal bootstrap equations analogous to (4.372) and (4.376):

$$\text{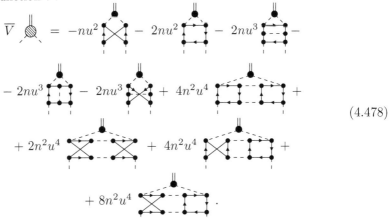} \quad = \quad = \overline{V} \quad , \quad \overline{V} = 1 . \tag{4.477}$$

The expression involving the shaded rectangle is the sum of irreducible skeleton graphs with dressed basic vertices and a single (upper) correction vertex. The second equation in (4.477) is the definition, analogous to (4.376), of the corresponding vertex function $\overline{V}$. It depends on all the parameters, including the desired critical dimension $\Delta_F \equiv 2\lambda$; the other dimensions and the quantity u are assumed to be already known with the needed accuracy from the solution of (4.474). We note that irreducibility implies, among other things, the absence of nontrivial vertex subgraphs not only for the basic vertex but also for the correction vertex. Therefore, the shaded rectangle in (4.477) is a Bethe–Salpeter kernel with no horizontal 2-sections in the σ lines (however, the analogous 2-sections in ψ lines are allowed). We also note that the exact vertex equation differs from the first equation in (4.477) by only the addition of the contribution of the bare vertex $\langle F\sigma\sigma\rangle$ to the right-hand side.

The following graphs must be included in calculating the dimension $\Delta_F \equiv \Delta[\sigma^2] \equiv 2\lambda$ through order $1/n^2$ in the definition (4.477) of the vertex function $\overline{V}$:

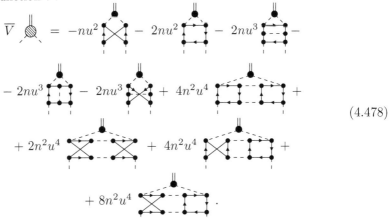

The upper dark circle in the graphs of (4.478) corresponds to the correction conformal triangle (4.473) with unknown value of the exponent 2λ (which we want to find), and all the others correspond to the basic triangles (4.473). The first two graphs in (4.478) are lower-order ones, the following seven are higher-order, and no regularization by parameters ε and ε' is needed.

The algebraic vertex equation $\overline{V}(2\lambda) = 1$ must have many solutions corresponding to different operators F. A particular solution can be constructed by iterating in $1/n$ starting from a given zeroth-order approximation. The solution $2\lambda = \Delta[\sigma^2]$ corresponds to the zeroth-order approximation $2\lambda_0 = 2$ (the canonical dimension of the operator $F = \sigma^2$ in the $1/n$ expansion). If the triangles (4.473) are substituted into the graphs (4.478) and a conformal transformation is done with the base at the uppermost vertex (Sec. 4.36), it is easy to check that any of the graphs in (4.478) and, consequently, the function $\overline{V}(2\lambda)$ as a whole are invariant with respect to the shadow transformation (4.364). This implies that the set of all solutions for the desired dimension 2λ has the general structure (4.356), in particular, the value of the exponent $\Delta_\tau \equiv 1/\nu = d - \Delta[\sigma^2]$ in (4.428) obeying the shadow relation with $\Delta[\sigma^2]$ must be one of the possible solutions of the vertex equation $\overline{V}(2\lambda) = 1$ for $F = \sigma^2$. This is what we must construct starting from the zeroth-order approximation $\lambda_0 = d/2 - 1$. We note that passage from $\lambda_0 = 1$ to $\lambda_0 = d/2 - 1$ leads to important simplifications in the calculation of the graphs; the details can be found in [146].

4.49   Use of the $1/n$ expansions of exponents to calculate RG functions

The results for the coefficients of the $1/n$ expansions of the critical exponents (4.467) can be used to calculate the RG functions (4.448). The basic idea is simple: the coefficients of the double expansions of the critical exponents in ε and $1/n$ obtained by various methods must coincide. Here it is important that in the MS scheme all the nontrivial coefficients of the series (4.448) are independent of ε and are polynomials in n of known order. For example, for the β function (4.448) we have

$$\beta(u) = 2\varepsilon u - 2u^2(n - 2) + \sum_{l=2}^{\infty} u^{l+1} P_{l-1}(n), \qquad (4.479)$$

where l is the number of loops and $P_k(n)$ is a polynomial in n of order k.

Let us explain the basic idea for the example of the exponent $1/\nu \equiv 2\lambda$, which is expressed in terms of the β function (4.448) by the relation $2\lambda = -\beta'(u_*)$ known from (4.426). Since $u_* \sim 1/n$, the coefficient λ_0 in the $1/n$ expansion of λ is determined by only the first two terms in (4.479), the coefficient λ_1 gives information about all contributions of the type $u^2(un)^m$ in (4.479), the coefficient λ_2 gives information about contributions of the type $u^3(un)^m$, and so on. Therefore, λ_1 can be used to find the coefficient of the highest power of n^{l-1} in all the polynomials in (4.479), λ_1 and λ_2 to find the coefficients of n^{l-1} and n^{l-2}, and so on.

Let us consider a specific example. We assume that we know only the general structure $u^3(an + b)$ of the 2-loop contribution to (4.479), but not the values of the coefficients a and b. We can then use the 2-loop β function $\beta(u) = 2\varepsilon u - 2u^2(n-2) + u^3(an+b)$ to find the value of $2\lambda = -\beta'(u_*) = 2\varepsilon + \varepsilon^2 f(n) + O(\varepsilon^3)$, where the function $f(n)$ and, consequently, all the coefficients of its $1/n$ expansion $f(n) = f_1/n + f_2/n^2 + \dots$ are uniquely expressed in terms of the desired parameters a and b. On the other hand, the same quantities $f_{1,2}$ can be found by expanding the functions $\lambda_{1,2}(d)$ known from (4.467) in $2\varepsilon = d - 2$ with the needed accuracy (ε^2). Equating the expressions for $f_{1,2}$ obtained by the two methods, we can express the desired coefficients a, b in terms of the known quantities $\lambda_{1,2}$, i.e., we can find the complete 2-loop contribution to (4.479).

If we also knew the next coefficient λ_3 of the $1/n$ expansion, then by the same method we would be able to find the complete 3-loop contribution to the β function, and so on. Then, assuming that the β function is known with the required accuracy, the coefficients of the $1/n$ expansions of the exponents η and κ in (4.467) could be used to obtain information about the RG functions γ_ψ and γ_m in (4.448) by the same method.

From the purely practical point of view the specific example given above is not the best one, because the calculation of the coefficients λ_1 and λ_2 of the $1/n$ expansions is technically more complicated than the direct calculation of the 2-loop contribution to the β function. However, there are additional considerations that make this technique more powerful. For example, noting that the factor $n - 2$ is common to all the polynomials in (4.479) [142], we can write $a(n - 2)$ instead of $an + b$, and this allows us to find the complete 2-loop contribution using λ_1 and the complete 3-loop one using λ_1 and λ_2, i.e., to raise the accuracy of the calculation by an order of magnitude. It is even more effective to use the results (4.467) to calculate the RG functions $\gamma_\psi(u)$ and $\gamma_m(u)$ in (4.448) with the factor $n - 1$ isolated from these two RG functions [142], [148]. In [148] this calculation was performed with 5-loop accuracy using the results (4.467) and gave

$$\gamma_\psi(u) = u^2(n-1) - u^3(n-1)(n-2) + u^4(n-1)[n^2 - 7n + 7]+$$

$$+u^5[(n-1)/3]\left\{3n^3[-2\zeta(3) - 1] + 2n^2[-3\zeta(3) + 17]+\right.$$

$$\left.+12n[-11\zeta(3) - 10] + c_1\right\},$$

$$\gamma_m(u) = 2u(n-1) - 2u^2(n-1) - 2u^3(n-1)(2n-3)+ \qquad (4.480)$$

$$+2u^4[(n-1)/3]\left\{n^2[-6\zeta(3) + 5] + n[90\zeta(3) + 23] + c_2\right\}+$$

$$+u^5[(n-1)/3]\left\{3n^3[-6\zeta(3) - 5] + n^2[-81\zeta(4) + 264\zeta(3)+\right.$$

$$\left.+169] + nc_3 + c_4\right\},$$

$$\beta(u) = 2\varepsilon u - 2u^2(n-2) + 4u^3(n-2) + 2u^4(n-2)(n-7)+$$

$$+4u^5[(n-2)/3]\left\{-n^2 - n[66\zeta(3) + 19] + c_5\right\}+$$

$$+u^6[(n-2)/3]\left\{3n^3[-2\zeta(3)+1]+n^2[261\zeta(4)-168\zeta(3)-83]+\right.$$

$$\left.+nc_6+c_7\right\},$$

where $c_1,...,c_7$ are the remaining unknown coefficients. We note that the result in [148] contains a misprint in the 3-loop contribution in γ_m (the coefficient $n-2$ is given instead of the correct $n-1$).

The results of the analogous 5-loop calculation of the RG functions (4.239) and (4.240) of the nonlinear σ model using the information from the $1/n$ expansions of the exponents are also given in [148]:

$$\gamma_\varphi(u) = ... + u^5[(n-2)/192]\left\{6n^3[3\zeta(4)+4\zeta(3)+2]-\right.$$

$$\left.-n^2[54\zeta(3)+5]-n[315\zeta(4)+1764\zeta(3)+307]+c_1\right\},$$

$$\tag{4.481}$$

$$\beta(u) = ... + u^6[(n-2)/96]\left\{3n^3[2\zeta(3)-1]+\right.$$

$$\left.+n^2[81\zeta(4)+144\zeta(3)-11]+nc_2+c_3\right\},$$

where the ellipsis denotes the 4-loop expressions known from (4.239) and (4.240), and $c_{1,2,3}$ are the remaining unknown coefficients in the 5-loop contributions.

For the 5-loop calculation of γ_ψ in (4.480) and (4.481) it is necessary to know the ε expansion of the function $I(\mu)$, $\mu \equiv d/2 = 1 + \varepsilon$, entering into (4.393) and (4.467) through order ε^2. This expansion has the form

$$I(1+\varepsilon) = -2/3\varepsilon+0+0+2\zeta(3)\varepsilon^2/3-\zeta(4)\varepsilon^3+13\zeta(5)\varepsilon^4/3+.... .$$

$$\tag{4.482}$$

The first two terms were obtained in [89] using the technique described in Sec. 4.36, and the higher ones were calculated in [149] by summing infinite series in Gegenbauer polynomials. The following recursion relation for the function $I(\mu)$ is derived in an appendix to [148]:

$$I(\mu) = \frac{1}{\alpha^2 C(1)}\left[B_2+\frac{1}{6\alpha}\right]+\left[1+\frac{1}{\alpha^2 C(1)}\right]\left[\frac{1}{\alpha}+I(\mu+1)\right], \tag{4.483}$$

where all the notation is the same as in (4.467). For $\mu = 1 + \varepsilon$ the coefficient of $I(\mu+1) = I(2+\varepsilon) = O(\varepsilon)$ [89] in (4.483) is a quantity of order ε^3, and so from (4.483) it is possible to find $I(1+\varepsilon)$ through order ε^3 without summing any infinite series.

Equation (4.483) was obtained in [148] using a new technical trick proposed in [150], which allows the value of an arbitrary massless graph in dimension d

to be expressed in terms of a finite sum of analogous graphs with transformed line indices in dimension $d-2$. As applied to a graph like (4.333) with arbitrary link indices, this trick leads to the relation

$$\{\text{graph (4.333) in dimension } d\} = [\pi^2/(d-3)(d-4)] \times$$

$$\times \{\text{the same graph in dimension d-2}\}$$

with integrand containing the additional factor

$$Q = -r_6^4 r_5^2 + r_6^2[-r_5^4 + r_5^2(r_1^2 + r_2^2 + r_3^2 + r_4^2) + \tag{4.484}$$

$$+r_1^2 r_3^2 + r_2^2 r_4^2 - r_1^2 r_2^2 - r_3^2 r_4^2] + r_5^2[r_1^2 r_3^2 + r_2^2 r_4^2 - r_1^2 r_4^2 - r_2^2 r_3^2] +$$

$$+r_1^2 r_2^2 r_3^2 + r_1^2 r_2^2 r_4^2 + r_1^2 r_3^2 r_4^2 + r_2^2 r_3^2 r_4^2 - r_1^4 r_3^2 - r_1^2 r_3^4 - r_2^4 r_4^2 - r_2^2 r_4^4,$$

in which r_i^2 is the squared length of the corresponding link of the tetrahedron in the notation of (4.338). The expression on the right-hand side of (4.484) is the sum of graphs with various additions of three -1 to the indices of the original graph, and the factors r_6^2 correspond to the unimportant external line in the graph (4.333). If the original graph contains many lines with indices $+1$, many terms on the right-hand side of (4.484) either vanish or can be integrated exactly, so that useful information can be obtained from (4.484). The situation is the same regarding the graph $\Pi(\mu, \Delta)$ in terms of which the function $I(\mu)$ is defined [see the text following (4.393) in Sec. 4.40]. The transformations studied in Sec. 4.36 can be used to reduce $\Pi(\mu, \Delta)$ to the graph (4.333) with link indices $\alpha_1 = \alpha_2 = \alpha_3 = \alpha_4 = 1$, $\alpha_5 = \mu - 1 - \Delta$, following which (4.484) can be used.

In concluding our discussion of the GN model let us make two remarks. First, everything we have said about the GN model in the $2 + \varepsilon$ scheme was based on the assumption that it is multiplicatively renormalizable in noninteger dimension $d = 2 + 2\varepsilon$. This is a nontrivial assumption, because the γ-matrix algebra in noninteger dimension is infinite-dimensional, and so there are an infinite number of independent four-fermion vertices which, in principle, could be generated by the basic interaction $(\bar{\psi}\psi)^2$ of the GN model as counterterms. Let us explain this in more detail. In noninteger dimension d only the basic commutation relation (4.423) can be used for the γ matrices. It allows any product of γ matrices (in noninteger dimension it would be more correct to say γ symbols, but we shall use the traditional term) to be reduced to a linear combination of fully antisymmetrized (As) products

$$\gamma_{i_1 \ldots i_n}^{(n)} \equiv \text{As}[\gamma_{i_1} \cdot \ldots \cdot \gamma_{i_n}]. \tag{4.485}$$

In any integer dimension d all structures (4.485) with $n > d$ vanish, but in noninteger dimension they must all be considered nonzero, independent quantities forming a complete basis in the space of matrices acting on the spinors ψ. Therefore, in noninteger dimension it is possible to construct an infinite number of independent U_N-symmetric four-fermion interactions:

$$V_n = \overline{\psi}\gamma^{(n)}_{i_1...i_n}\psi \cdot \overline{\psi}\gamma^{(n)}_{i_1...i_n}\psi \qquad (4.486)$$

(repeated indices $i_1...i_n$ are summed over). All the vertices (4.486) are logarithmic in dimension $d = 2$, so that in the $2 + \varepsilon$ scheme they can mix under renormalization, i.e., any of them can generate others as counterterms. In integer dimension $d = 2$ only the first three interactions (4.486) are nonzero. The vertex V_0 corresponds to the GN model, V_1 to the Thirring model, and V_2 to the pseudoscalar analog of the GN model. The GN model in dimension $d = 2$ with cutoff regularization possesses a specific U_{2N} symmetry that guarantees that it is multiplicatively renormalizable. However, for $d = 2 + 2\varepsilon$ this argument becomes invalid, and so there is no guarantee that the GN interaction V_0 does not generate higher-order vertices (4.486) as counterterms. Direct calculation shows [142], [144] that through three loops this does not occur. The analysis of the renormalization of the GN model in Secs. 4.45 and 4.46 was based on the assumption that the GN interaction V_0 does not generate higher-order vertices (4.486) even in higher orders of perturbation theory. However, this is only an assumption, because there is no general proof that the GN model is multiplicatively renormalizable in the $2 + \varepsilon$ scheme.

If the GN interaction V_0 generates higher-order vertices (4.486) in higher orders of perturbation theory, the GN model in the $2 + \varepsilon$ scheme cannot be considered internally closed. In this situation only the complete four-fermion model [151] including all the vertices (4.486) with independent coefficients (charges) will be multiplicatively renormalizable. Then it is no longer possible to speak of the RG functions of the GN model in the $2 + \varepsilon$ scheme, but the concept of the $2 + \varepsilon$ expansions of the critical exponents of the GN model remains meaningful, since this critical regime corresponds to one of the fixed points in the infinite-dimensional charge space of the full model [152]. All these problems are generated by the specific nature of the four-fermion interaction: in models with triple vertices of the type $\overline{\psi}\psi\sigma$ in (4.427) it is impossible for counterterms with higher-order ($n > d$) structures (4.485) to appear (many antisymmetrized indices do not contract with anything). Therefore, in the $1/n$ expansion the internal closure of the GN model is guaranteed.

Let us make a final remark regarding the chiral phase transition in the GN model. Here we are dealing with an ordinary second-order transition with scalar order parameter $\varphi(x) = \langle\overline{\psi}(x)\psi(x)\rangle$ and the symmetry $\varphi \to -\varphi$. The question arises of why this transition is not described by the simple φ^4 model in accordance with the Landau theory. The answer is that the standard Landau theory pertains only to microscopic models with short-range interaction, which in field language corresponds to the primary microscopic fields being massive. In our case the fields $\overline{\psi}$, ψ are massless. This particular example shows that in microscopic models with long-range interaction the standard arguments of the Landau theory are not applicable, and the critical behavior of the system is determined by the microscopic model itself and not by its "phenomenological image."

Chapter 5

Critical Dynamics

The term dynamics in the broad sense refers to any problem involving time. Stochastic dynamics is the study of fluctuating quantities and their statistical characteristics. In the real world stochasticity is generated by causes internal to a system: random collisions between molecules owing to thermal Brownian motion, the spontaneous appearance and interaction of vortices in a turbulent flow, and so on. At present it is difficult or impossible to rigorously describe such processes mathematically. Therefore, the stochastic dynamics are usually modeled phenomenologically by introducing random "noise" into the dynamical equations — either random forces or other random parameters with a simple (usually Gaussian) distribution. The Gaussian distribution is specified by the noise correlator, which is selected on the basis of physical considerations. In particular, models of critical dynamics have been constructed in this manner in order to study the critical singularities of relaxation times and various kinetic coefficients. Critical dynamics is based on stochastic equations of a particular form, the Langevin equations. Here the noise corresponds to thermal fluctuations, and the noise correlator is uniquely determined by the requirement that the dynamics and statics be mutually consistent.

In this chapter we shall first give some general information about the equations of stochastic dynamics and their reduction to quantum-field models. Then we will discuss models of critical dynamics in more detail for some specific examples. Other stochastic problems of the non-Langevin type (turbulence and so on) will be studied in the following chapter.

5.1 Standard form of the equations of stochastic dynamics

As usual, we shall write out all the general expressions in universal notation (Sec. 1.9), i.e., $\varphi(x)$ will denote the entire set of fields of the problem, $x \equiv \{t, \mathbf{x}\}$, where t is the time and $\mathbf{x}$ are all the other variables — the spatial coordinates and indices, if the latter are not indicated explicitly. In this notation the standard problem of stochastic dynamics is stated as

$$\partial_t \varphi(x) = U(x; \varphi) + \eta(x), \quad \langle \widehat{\eta}(x)\widehat{\eta}(x') \rangle = D(x, x'), \tag{5.1}$$

where $\varphi(x)$ is the desired field (or system of fields), $U(x; \varphi)$ is a given t-local functional not containing time derivatives of φ (i.e., a functional depending only on the fields φ and their spatial derivatives at a single instant of time t), $\widehat{\eta}(x)$ is a random external force (the conventional terminology), and $\eta(x)$ in the equation for φ is any particular realization of the force. We assume that $\widehat{\eta}$ obeys a Gaussian distribution with zero mean $\langle \widehat{\eta}(x) \rangle = 0$ and given correlator D. All the terms in the first equation of (5.1) depend on the single argument $x \equiv t, \mathbf{x}$, which will henceforth be omitted.

The problem (5.1) is augmented by the retardation condition and is studied only on the semiaxis $t \geq 0$ with given initial data for φ at $t = 0$ (the Cauchy problem), or on the entire time axis with given (usually simply zero) asymptote for $t \to -\infty$. The second version is technically simpler, and everywhere below, unless stated otherwise, we shall study (5.1) on the entire time axis with $\varphi \to 0$ for $t \to -\infty$ and also for $|\mathbf{x}| \to \infty$ in any plane $t = \mathrm{const}$. This will be termed the *standard* formulation of the problem.

These conditions ensure the existence and uniqueness of the solution $\varphi(x; \eta)$ for any specific realization $\eta \in E$ (Sec. 1.13). The random quantity $\varphi(x; \widehat{\eta}) \equiv \widehat{\varphi}(x)$ is thereby determined. The calculated objects are the *correlation functions* (1.33) of the field $\widehat{\varphi}$ and the *response functions* describing the response to an external force, i.e., the quantities

$$\langle \delta^m [\widehat{\varphi}(x_1) \ldots \widehat{\varphi}(x_n)] / \delta \widehat{\eta}(x_1') \ldots \delta \widehat{\eta}(x_m') \rangle. \qquad (5.2)$$

All these objects will be referred to as Green functions. The simplest of the quantities (5.2) is the linear response function

$$\langle \delta \widehat{\varphi}(x) / \delta \widehat{\eta}(x') \rangle. \qquad (5.3)$$

The symbol $\langle ... \rangle$ denotes the average over the Gaussian noise distribution $\widehat{\eta}(x)$, which in functional language is the average over configurations $\eta(x)$ with weight $\exp[-\eta D^{-1} \eta / 2]$. We note that in the definitions (5.2) and (5.3) variational differentiation with respect to the random force $\widehat{\eta}(x)$ can be replaced by differentiation with respect to an analogous nonrandom force $f(x)$ introduced additively into $U(x; \varphi)$, since $\widehat{\varphi}$ will then depend only on the sum $f(x) + \widehat{\eta}(x)$. We also note that for the standard formulation of the problem and translationally invariant U and D, all the Green functions will also be translationally invariant, i.e., they will depend only on differences of time and spatial coordinates, and functions at a single time will then not depend on a time t common to all arguments x_i. This standard formulation of the problem is preferable to the Cauchy problem, where translational invariance is restored only at the asymptote $t \to +\infty$, when the system "forgets" the initial data.

The response functions are retarded functions, which follows from the natural requirement of causality: the solution φ of (5.1) at time t cannot depend on the value of the force η at future times. It then follows that the functions (5.2) are nonzero only when any one of the times t' is bounded above by one of the times t; in particular, the function (5.3) is nonzero only for $t \geq t'$.

A special case of (5.1) is the *stochastic Langevin equation* describing the simple dynamics of a system with given static action $S^{\mathrm{stat}}(\varphi)$, a functional of the time-independent field $\varphi(\mathbf{x})$ (we recall that in the earlier chapters we dealt only with static objects, and so we wrote simply S and x instead of S^{stat} and $\mathbf{x}$):

$$\partial_t \varphi(x) = \alpha [\delta S^{\text{stat}}(\varphi)/\delta \varphi(\mathbf{x})]\big|_{\varphi(\mathbf{x}) \to \varphi(x)} + \eta(x), \tag{5.4a}$$

$$\langle \widehat{\eta}(x)\widehat{\eta}(x')\rangle = 2\alpha\delta(x - x'). \tag{5.4b}$$

The *Onsager kinetic coeffifcient* α appearing here is a φ-independent symmetric linear operator acting only on the argument $\mathbf{x}$ (it is usually simply a numerical coefficient or a matrix in the φ indices, or the analogous object with an additional Laplace operator). In (5.4b) for the correlator $\delta(x - x') \equiv \delta(t - t')\delta(\mathbf{x} - \mathbf{x}')$, and the operator α acts on one of the two arguments $\mathbf{x}$ or $\mathbf{x}'$ (it does not matter which because the δ function is symmetric). The specific choice of the correlator (5.4b) is determined by the requirement that the dynamics and statics be mutually consistent (see Sec. 5.8 for more details).

The general problem (5.1) differs from (5.4) by the arbitrariness of the correlator D and the functional U, which may not reduce to the variational derivative of some functional.

Let us conclude by giving a few examples. The simplest models A and B of critical dynamics in the terminology of [153] are described by (5.4) with the static action of the φ^4 model for the order-parameter field φ and Onsager coefficient $\alpha = \text{const}$ for model A or $\alpha = \text{const} \cdot \partial^2$ for model B.

The simplest example of problem (5.4) is the equation describing Brownian diffusion (random walk of a particle):

$$\partial_t r_i(t) = \eta_i(t), \qquad \langle \widehat{\eta}_i(t)\widehat{\eta}_k(t')\rangle = 2\alpha\delta_{ik}\delta(t - t'). \tag{5.5}$$

In this example $\varphi(x) \equiv r_i(t)$ are the d-dimensional particle coordinates at time t, the functional U is zero, and the constant α is the diffusion coefficient.

The final example is the *stochastic Navier–Stokes equation* for an incompressible fluid (a gas), on which one of the traditional approaches to the theory of developed turbulence is based [154]:

$$\nabla_t \varphi_i(x) = \nu\partial^2 \varphi_i(x) - \partial_i p(x) + \eta_i(x), \qquad \nabla_t \equiv \partial_t + (\varphi\partial). \tag{5.6}$$

Here $\varphi_i(x)$ is the vector velocity field satisfying the transversality condition $\partial\varphi \equiv \partial_k\varphi_k = 0$ (incompressibility), ν is the viscosity coefficient, p is the pressure per unit mass, and the random force $\widehat{\eta}$ is also assumed to be transverse. The operator ∇_t is a Galilean-covariant derivative, and (5.6) takes the form (5.1) after moving the contribution of the nonlinear term $(\varphi\partial)\varphi_i$ to the right-hand side. The purely longitudinal contribution of the pressure $\partial_i p$ in (5.6) cancels the longitudinal part of the quantity $(\varphi\partial)\varphi_i$. Therefore, the pressure can be eliminated by discarding the contribution $\partial_i p$ and placing the transverse projection operator in front of $(\varphi\partial)\varphi_i$. In this form (5.6) becomes the equation for a single field φ. It falls within the general framework of the problem (5.1), but does not reduce to the Langevin equation (5.4a).

5.2   Iterative solution of the stochastic equations

In general, the functional U on the right-hand side of (5.1) contains the contribution of the nonrandom force f, the "free" part $L\varphi$ linear in φ and involving some operator L, and the nonlinear contribution $n(\varphi)$:

$$U(\varphi) = L\varphi + n(\varphi) + f. \tag{5.7}$$

The linear problem (5.1) can be solved exactly, and the nonlinearity $n(\varphi)$ can be taken into account perturbatively by iterating (5.1). For this it is convenient to rewrite it in integral form:

$$\varphi = \Delta_{12}[f + \eta + n(\varphi)], \qquad \Delta_{12} \equiv (\partial_t - L)^{-1}, \tag{5.8}$$

where $\Delta_{12} \equiv \Delta_{12}(x, x')$ is the retarded Green function of the linear operator $\partial_t - L$ [by retarded we mean that $\Delta_{12}(x, x') = 0$ for $t < t'$ in the arguments $x \equiv t, \mathbf{x}$ and $x' \equiv t', \mathbf{x}'$].

As a simple example, let us consider the case of a local quadratic nonlinearity:

$$n(x; \varphi) = v\varphi^2(x)/2, \tag{5.9}$$

where v is the vertex factor playing the role of the coupling constant (in general, v is a three-index object with the indices of the field φ, if they exist). Equation (5.8) with the nonlinearity (5.9) and $f = 0$ (the generalization to $f \neq 0$ is trivial) can be represented graphically as

$$\text{\raisebox{-0.5ex}{\includegraphics{}}} \quad = \quad \bullet\!\!-\!\!\!\times \quad + \quad \frac{1}{2} \; \bullet\!\!-\!\!\!\big\langle \quad , \tag{5.10}$$

where φ is represented by the wavy external line (tail), η by the cross, and $\Delta_{12}(x, x')$ by the straight line with marked end corresponding to the argument x'. The point where the three graphical elements (straight lines and tails) are joined is associated with the vertex factor v.

Iterating (5.10) leads to representation of $\widehat{\varphi}(x)$ in the form of an infinite sum of tree graphs:

$$\widehat{\varphi} \equiv \text{\raisebox{-0.5ex}{\includegraphics{}}} \quad = \quad \bullet\!\!-\!\!\!\times \quad + \quad \frac{1}{2} \; \bullet\!\!-\!\!\!\big\langle \quad + \; ... \tag{5.11}$$

with root at the point x and crosses $\widehat{\eta}$ on the ends of all the branches.

Correlation functions like (1.33) are obtained by multiplying together the tree graphs in (5.11) for all the factors of $\widehat{\varphi}$ and then averaging over $\widehat{\eta}$, which corresponds graphically to contracting pairs of crosses to form correlators D in all possible ways. This operation leads to the appearance of a new graphical element:

$$\Delta_{11} \equiv < \widehat{\varphi} \; \widehat{\varphi} >_0 = \Delta_{12} \; D \; \Delta_{12}^{\mathrm{T}} = \; \text{•}\!\!-\!\!\sigma\sigma\sigma\!\!-\!\!\text{•} \; \equiv \; \text{•}\!\!-\!\!-\!\!\text{•} \qquad (5.12)$$

(the "spring" denotes D), interpreted as the unperturbed correlator $\langle \widehat{\varphi}\widehat{\varphi} \rangle_0$, which we shall denote as a simple line without marks. All the crosses $\widehat{\eta}$ should be contracted to form correlators D, and graphs with an odd number of crosses give zero when the averaging is done.

Finally, the correlation functions $\widehat{\varphi}$ are represented as Feynman-type graphs with triple interaction vertices and two types of line: Δ_{12} and Δ_{11}. Graphical representations of the response functions are constructed from (5.11) and the definition (5.2) in exactly the same way, and variational differentiation with respect to $\widehat{\eta}$ corresponds graphically to removing the cross in all possible ways with the appearance of the free argument x' on the end of the corresponding line. As an example, let us present the first few graphs of the correlator $\langle \widehat{\varphi}\widehat{\varphi} \rangle$ and the response fucntion (5.3) obtained in this way:

$$< \widehat{\varphi}\widehat{\varphi} > = \quad \text{——} \quad + \frac{1}{2} \; \text{⚬̣} \; + \frac{1}{2} \; \text{⚬̣} \; +$$

$$+ \frac{1}{2} \; \text{⌢} \; + \; \text{⌢} \; + \; \text{⌢} \; + \; \dots \; , \qquad (5.13)$$

$$< \delta\widehat{\varphi}/\delta\widehat{\eta} > = \text{——} + \frac{1}{2} \; \text{⚬̣} \; + \; \text{⌢} \; + \; \dots \; .$$

This graphical technique was first introduced and analyzed in detail for the Navier–Stokes equation (5.6) by Wyld [155], and for the general problem (5.1) later on [156].

5.3 Reduction of the stochastic problem to a quantum field model

The graphs (5.13) can be identified as Feynman graphs if we admit the existence of a second field $\widehat{\varphi}'$ in addition to $\widehat{\varphi}$ with zero bare correlator $\langle \widehat{\varphi}'\widehat{\varphi}' \rangle_0 = 0$, and if we interpret the line Δ_{12} as a bare correlator and the response function (5.3) as the exact correlator $\langle \widehat{\varphi}\widehat{\varphi}' \rangle$. The triple vertex corresponds to the interaction $V = \varphi'v\varphi\varphi/2$. Clearly, (5.13) involves all the graphs of this quantum-field model without contracted lines $\Delta_{12} = \langle \widehat{\varphi}\widehat{\varphi}' \rangle_0$ and no others, so that the complete proof of the equivalence would require only showing that the symmetry coefficients coincide. We shall not do this, since there exists a general proof of the equivalence of any stochastic problem (5.1) to a certain quantum-field model with twice the number of fields [157], [158].

Turning now to the exposition of this proof (following [159]), we use $\overline{\varphi} \equiv \overline{\varphi}(x; \eta)$ to denote the solution of (5.1) and $G(a)$ the generating functional of the correlation functions of the field $\hat{\varphi}$. Clearly, for fixed η we will have $G(a; \eta) = \exp(a\overline{\varphi})$, and after averaging we obtain

$$G(a) = \frac{\int D\eta \exp[-\eta D^{-1}\eta/2 + a\overline{\varphi}]}{\int D\eta \exp[-\eta D^{-1}\eta/2]}, \tag{5.14}$$

where $a(x)$ is the source and $a\overline{\varphi}$ is a standard linear form of the type (1.48).

In (5.14) we use the identity

$$\exp(a\overline{\varphi}) = \int D\varphi \delta(\varphi - \overline{\varphi}) \exp(a\varphi) \tag{5.15}$$

with functional δ function $\delta(\varphi - \overline{\varphi}) \equiv \prod_x \delta[\varphi(x) - \overline{\varphi}(x)]$. Since $\overline{\varphi}$ is the unique solution of (5.1) [the initial or asymptotic conditions ensuring uniqueness are understood to have been included in the definition of the region of integration over φ in (5.15)], the two equations in (5.16a) are equivalent and so (5.16b) is valid. The latter is the analog of an identity well known for finite-dimensional objects:

$$\varphi = \overline{\varphi} \quad \Leftrightarrow \quad Q(\varphi, \eta) \equiv -\partial_t \varphi + U(\varphi) + \eta = 0, \tag{5.16a}$$

$$\delta(\varphi - \overline{\varphi}) = \det M \cdot \delta[Q(\varphi, \eta)], \quad M \equiv \delta Q/\delta\varphi. \tag{5.16b}$$

We recall that Q and φ carry the arguments x, so that $M = M(x, x') = \delta Q(x)/\delta\varphi(x')$ and $\det M$ is understood as the determinant of the linear integral operator with kernel $M(x, x')$ (Sec. 2.1).

Let us now use an identity of the type (2.213), representing the δ function on the right-hand side of (5.16b) as an integral over an auxiliary field φ' of the same nature as Q and φ (i.e., with the same set of indices, transversality properties when indices are present, and so on):

$$\delta[Q(\varphi, \eta)] = \int D\varphi' \exp[\varphi' Q(\varphi, \eta)], \tag{5.17}$$

where all the normalization factors are assumed to be included in $D\varphi'$. Using (5.15) to (5.17) in (5.14), we arrive at an integral over η, φ, and φ'. Substituting into it the explicit expression (5.16a) for Q and performing the Gaussian integration over η, we find

$$G(a) = \int \int D\varphi D\varphi' \det M \exp[\varphi' D\varphi'/2 + \varphi'(-\partial_t \varphi + U(\varphi)) + a\varphi]. \tag{5.18}$$

Let us study $\det M \equiv \det[\delta Q/\delta\varphi]$ in more detail. From the representation (5.7) and the definitions of Q in (5.16a) and Δ_{12} in (5.8) we have

$$M = M_0 + M_1, \quad M_0 \equiv -\partial_t + L = -\Delta_{12}^{-1}, \quad M_1 \equiv \delta n(\varphi)/\delta\varphi, \tag{5.19}$$

and so $\det M = \det M_0 \det[1 - \Delta_{12}M_1]$. The first factor is independent of the fields and is therefore an unimportant constant, and for the second the rule (2.7) gives

$$\ln \det[1 - \Delta_{12}M_1] = -\mathrm{tr}\,[\Delta_{12}M_1 + \Delta_{12}M_1\Delta_{12}M_1/2 + ...] =$$

$$= - \bigcirc\!\!\!\!\bullet \; - \frac{1}{2} \bigcirc\!\!\!\!\bullet\!\!\!\!\bullet \; - \frac{1}{3} \bigcirc\!\!\!\!\bullet\!\!\!\!\bullet\!\!\!\!\bullet \; - \; ... = - \bigcirc\!\!\!\!\bullet \;, \qquad (5.20)$$

since all multiple closed loops with retarded [i.e., a multiple of $\theta(t - t')$] lines $\Delta_{12}(x, x')$ give zero. Here it is important that the vertex factors M_1 of these loops do not contain, by assumption (Sec. 5.1), derivatives ∂_t, which could lead to the contraction of a loop to a point owing to the transformation of a θ function into a δ function upon operation by ∂_t.

The single loop (5.20) corresponds to the expression $\int\int dx dx' \Delta_{12}(x, x')M_1(x', x)$ with kernel $M_1(x', x) = \delta(t' - t)\widetilde{M}_1(x', x)$, because the quantity $M_1 \equiv M_1(x', x) \equiv \delta n(x'; \varphi)/\delta\varphi(x)$ defined in (5.19) necessarily contains a factor $\delta(t' - t)$ owing to the assumed (Sec. 5.1) t-locality of the functional (5.7). It then follows that the single loop (5.20) contains the function $\Delta_{12}(x, x')$ with coincident times $t = t'$. This quantity is undefined, because Δ_{12} involves the discontinuous function $\theta(t - t')$. More precisely, the definition $(\partial_t - L)\Delta_{12}(x, x') = \delta(x - x')$ (the operator L acts on the variable $\mathbf{x}$) and the retardation condition lead to the representation

$$\Delta_{12}(x, x') = \theta(t - t')R(x, x'), \quad R(x, x')\,|_{t=t'} = \delta(\mathbf{x} - \mathbf{x}'), \qquad (5.21)$$

where $R(x, x')$ is the solution of the homogeneous equation $(\partial_t - L)R(x, x') = 0$ in the variable x satisfying the condition in (5.21). Equation (5.21) uniquely determines the limits of $\Delta_{12}(x, x')$ for $t \to t' \pm 0$:

$$\Delta_{12}(x, x')\,|_{t=t'+0} = \delta(\mathbf{x} - \mathbf{x}'), \quad \Delta_{12}(x, x')\,|_{t=t'-0} = 0, \qquad (5.22)$$

while the value of Δ_{12} itself at $t = t'$ remains undefined.

It is usual to make the definition of such quantities precise by taking half the sum of the limits:

$$\theta(t - t')\,|_{t=t'} = \tfrac{1}{2}, \quad \Delta_{12}(x, x')\,|_{t=t'} = \tfrac{1}{2}\delta(\mathbf{x} - \mathbf{x}'). \qquad (5.23)$$

Then we have

$$\det M = \mathrm{const} \cdot \exp[-\tfrac{1}{2}\int dx\,\widetilde{M}_1(x, x)], \qquad (5.24)$$

where $\widetilde{M}_1$ is the coefficient of the time δ function in the kernel M_1 (see above). This is the definition of $\det M$ which is usually used.

In this book we shall adopt a different point of view [159], namely, the following *convention*: in (5.20) for $\det M$ and in all Feynman graphs, a single loop with a contracted line Δ_{12} is defined to be zero, i.e., the second limit in

(5.22). Then $\det M$ in the integral (5.18) becomes an unimportant constant that can be discarded.

This is the simplest and most natural definition. Actually, as noted at the beginning of this section, graphs with contracted lines Δ_{12} do not directly arise when (5.1) is iterated. However, they appear, along with other graphs, in the construction of Feynman graphs using the standard rules (Ch. 2) for the integral (5.18), and the only role of the factor $\det M$ in (5.18) is to exactly cancel these unnecessary graphs (of course, this occurs only when the contracted lines Δ_{12} in the graphs and in $\det M$ are defined in the same way). The simplest (but not, of course, only) way of ensuring these cancellations is to define the contracted line Δ_{12} to be zero in all these objects. This is what we shall do in this book.

Then $\det M$ in (5.18) can be discarded as it is an unimportant constant, and the final result is stated as follows. *Any stochastic problem (5.1) is completely equivalent to a quantum-field model with twice the number of fields* $\phi \equiv \varphi, \varphi'$ *and the action functional*

$$S(\phi) = \varphi' D\varphi'/2 + \varphi'[-\partial_t\varphi + U(\varphi)]. \qquad (5.25)$$

As usual, the required integrations over x and summations over indices are understood.

This equivalence implies that the Green functions of the stochastic problem are the ordinary Green functions of the quantum-field model (5.25), i.e., they are determined by the generating functional

$$G(A) = \int D\phi \, \exp\,[S(\phi) + A\phi] \qquad (5.26)$$

with $D\phi \equiv D\varphi D\varphi'$ and $A\phi \equiv a\varphi + a'\varphi'$. The normalization factor ensuring that $G(0) = 1$ is included in $D\phi$.

The source $a' = a'(x)$ of the auxiliary field φ' in (5.26) plays the role of a nonrandom external force, and so variational differentiation with respect to the set of sources $a, a' \equiv A$ reproduces both the correlation functions of the field $\hat{\varphi}$ of the type (1.33) and all the response functions (5.2). Therefore, in what follows we shall not introduce the term f in (5.7); the role of f will be played by the value of the source a' in the final expressions. Then from (5.7) and (5.25) we have

$$S(\phi) = \varphi' D\varphi'/2 + \varphi'[-\partial_t\varphi + L\varphi + n(\varphi)]. \qquad (5.27)$$

The contributions quadratic in the fields form the free action $S_0(\phi)$, and the contribution of the nonlinear term $\varphi'n(\varphi)$ is the interaction $V(\phi)$ in the notation of (2.13). The Feynman graphs are determined from the action (5.27) according to the standard rules of Ch. 2. The lines in the graphs are the elements of the 2×2 matrix $\Delta = K^{-1}$ of the bare correlators, and in matrix notation we assume that $\varphi \equiv \phi_1$, $\varphi' \equiv \phi_2$. The matrix K is found by a symmetrization of the type (2.8) of the contributions to the action (5.27) quadratic in ϕ:

$$S_0(\phi) = -\frac{1}{2}\begin{pmatrix} \varphi \\ \varphi' \end{pmatrix}\begin{pmatrix} 0, & (\partial_t - L)^\top \\ \partial_t - L, & -D \end{pmatrix}\begin{pmatrix} \varphi \\ \varphi' \end{pmatrix} \equiv -\frac{1}{2}\phi K\phi. \quad (5.28)$$

For such a block matrix with symmetric operator $P = P^\top$ and any Q we have

$$\begin{vmatrix} 0, & Q^\top \\ Q, & P \end{vmatrix}^{-1} = \begin{vmatrix} -Q^{-1}PQ^{-1\top}, & Q^{-1} \\ Q^{-1\top}, & 0 \end{vmatrix}, \quad (5.29)$$

which for $\Delta = K^{-1}$ with K from (5.28) gives

$$\Delta_{12} = \Delta_{21}^\top = (\partial_t - L)^{-1}, \quad \Delta_{11} = \Delta_{12}D\Delta_{21}, \quad \Delta_{22} = 0,$$

$$\Delta_{ik}(x,x') \equiv \langle \widehat{\phi}_i(x)\widehat{\phi}_k(x')\rangle_0, \quad \phi_1 \equiv \varphi, \phi_2 \equiv \varphi' \quad (5.30)$$

in accordance with the definitions introduced above. Since the function Δ_{12} is taken to be retarded, its transpose $\Delta_{21} = \Delta_{12}^\top$ will be advanced, and the symmetric correlator $\Delta_{11} = \Delta_{11}^\top$ will contain both retarded and advanced contributions. Since $\Delta_{22} = 0$, *the field $\varphi' \equiv \phi_2$ can be contracted in the graphs only with the field $\varphi \equiv \phi_1$.*

The interaction in (5.27) corresponds to a vertex with a single field φ' and two or more fields φ [if $n(\varphi)$ is a monomial in φ].

It is understood that all possible boundary or asymptotic conditions on φ of the problem (5.1), along with the conditions ensuring the choice of the required (retarded) Green function $(\partial_t - L)^{-1}$ (see the remarks of Sec. 2.1), are included in the definition of the region of functional integration over ϕ in (5.26).

Reduction of the stochastic problem (5.1) to the quantum-field model (5.25) makes it possible to use all the technical apparatus and terminology of Ch. 2, which we shall do in what follows without additional explanation.

5.4 Some consequences of retardation

The general properties of the lines and vertices of models like (5.25) listed at the end of the preceding section lead to several simple but important consequences.

Statement 1. Any 1-irreducible or S-matrix (Sec. 2.2) graph with external lines of only the basic field φ gives zero contribution because it contains a closed loop of advanced functions Δ_{21}.

Proof. Let us consider any external vertex of such a graph. It contains a single field φ' that, by convention, is not associated with an external line (which are associated only with φ by assumption), and so it must be contracted with a field φ from some other vertex to form the advanced line Δ_{21}. If we move along this line to the second vertex, there we also find a field φ'

that also must be contracted with a φ from another vertex. We then move to that vertex, and so on. Clearly, in this construction we must sometime return to one of the vertices we have already visited, so that we thereby obtain a closed loop of advanced functions Δ_{21}, which is equal to zero. This construction can also be used to prove the following.

Statement 2. Any graph of a connected Green function of only the fields φ' (without φ) and also all vacuum loops in the model (5.25) give zero contribution. Actually, graphs of connected functions of the fields φ' are reduced to the S-matrix graphs of the fields φ studied above by amputating bare external lines $\langle \varphi' \varphi \rangle$, and the above construction generalizes directly to vacuum loops.

For equal-time Green functions, the analogous result is obtained not only by returning to one of the vertices already visited in this construction, but also by going to any other external vertex, because it has, by condition, the same time as the initial one, and this is equivalent to the appearance of a closed loop of time θ functions. Therefore, we have zero contribution from all graphs for which this construction can even be started, which is ensured by the presence of at least one field φ for 1-irreducible and S-matrix functions or at least one field φ' in connected functions. From this we find:

Statement 3. All nontrivial (nonbare) graphs of equal-time 1-irreducible or S-matrix functions with at least one field φ or connected functions with at least one field φ' give zero contribution. In particular, always $\langle \widehat{\varphi}' \rangle = 0$, and so the response function $\langle \widehat{\varphi} \widehat{\varphi}' \rangle$ coincides with its connected part, i.e., with the correlator.

The stipulation regarding trivial bare contributions pertains only to functions of the type $\langle \widehat{\varphi} \widehat{\varphi}' \rangle$: for a connected function $\langle \widehat{\varphi} \widehat{\varphi}' \rangle$ the bare contribution is Δ_{12}, which corresponds to the bare contribution $-\Delta_{12}^{-1} = -\partial_t + L$ for the 1-irreducible function $\langle \widehat{\varphi}' \widehat{\varphi} \rangle_{1-\mathrm{ir}}$, and only these contributions remain for coincident times in the corresponding exact Green functions.

It then follows that for the response function (5.3), i.e., the exact connected function $\langle \widehat{\varphi} \widehat{\varphi}' \rangle$, the limiting values for $t \to t' \pm 0$ are the same as for its bare contribution (5.22):

$$\langle \widehat{\varphi}(x) \widehat{\varphi}'(x') \rangle = \begin{cases} \delta(\mathbf{x} - \mathbf{x}') & \text{for } t = t' + 0, \\ 0 & \text{for } t < t'. \end{cases} \tag{5.31}$$

Instead of the limit $t = t' - 0$ we have quoted the exact retardation condition of the response function (Sec. 5.1).

Note. It follows from the form of the limit for $t = t' + 0$ in (5.22) that the quantity $\Delta_{12}(x, x')$ defined as the retarded Green function of the linear asymptotic problem (Sec. 5.1) is at the same time the Green function of the Cauchy problem with initial data at time t', i.e., the convolution of Δ_{12} with the initial data is the desired solution of the Cauchy problem.

5.5 Stability criterion for a system in stochastic dynamics

In statics, the stability criterion for a system in the Landau theory is the requirement that an absolute energy minimum, i.e., a maximum of the action functional, exist (Sec. 1.13); in the exact theory the action $S(\varphi)$ in this formulation must be replaced by the functional $\Gamma(\varphi)$ (Sec. 2.9), for which the action is the zeroth approximation in the number of loops.

However, this criterion pertains only to real fields φ and is not needed for the various auxiliary fields introduced artificially as Lagrange multipliers and so on (see Secs. 2.19 and 2.27).

It is thus already clear that the static stability criterion is inapplicable to the model (5.25) containing an unphysical auxiliary field φ' [entering into (5.17) as a Lagrange multiplier], without which the quantum-field reformulation of the problem (5.1) is impossible.

In stochastic dynamics it is fairly easy to formulate the stability criterion of a system only for small perturbations (which is analogous to having a local energy minimum). This is done using the response function, requiring that the latter vanish in the limit $t - t' \to \infty$:

$$\langle \widehat{\varphi}(x) \widehat{\varphi}'(x') \rangle \to 0 \quad \text{for } t - t' \to +\infty. \tag{5.32}$$

Physically, this corresponds to requiring that all small perturbations be damped out, because the response function, by definition, is the coefficient of the linear response of the system to perturbation by an external force.

For the procedure of iterating (5.1) (and thereby also the diagrammatic technique) to be correct it is necessary to have stability in both the exact and the free (linear) problem. The latter is ensured by the condition

$$\Delta_{12}(x, x') \to 0 \text{ for } t - t' \to +\infty \quad \Leftrightarrow \quad \text{Re } L < 0, \tag{5.33}$$

where Re L is the real part of the linear operator L in (5.7) and (5.8).

All of the above is valid both for the standard formulation (Sec. 5.1), and for the Cauchy problem with initial data. In the first case, owing to the translational invariance of the Green functions, quantities like (5.3) possessing two arguments depend only on the difference $x - x'$. The Fourier transform of such functions for d-dimensional $\mathbf{x}$ will always be defined by relations analogous to (1.35):

$$F(x, x') = (2\pi)^{-d-1} \int d\omega \int dk \, \widetilde{F}(\omega, k) \exp[-i\omega(t - t') + ik(\mathbf{x} - \mathbf{x}')], \tag{5.34a}$$

$$\widetilde{F}(\omega, k) = \int dt \int d\mathbf{x} \, F(x, x') \exp[i\omega(t - t') - ik(\mathbf{x} - \mathbf{x}')], \tag{5.34b}$$

where ω is the frequency and k is the wave vector (the momentum in the terminology of the preceding chapters). As in (1.35), the integral over $x \equiv t, \mathbf{x}$ in (5.34b) can be replaced by an integral over x' or over the difference $x - x'$.

The stability criterion (5.32) for translationally invariant response function is expressed in terms of its Fourier transform $\widetilde{F}(\omega, k)$ by the requirement of analyticity in ω in the upper half plane, which for the bare function Δ_{12} is equivalent to the condition on L in (5.33).

5.6 Equations for equal-time correlation functions of the field $\widehat{\varphi}$

In this section we shall isolate the variable t and write it explicitly as an index, treating it as a fixed parameter, in contrast to the variable $\mathbf{x}$ (coordinates and indices) assumed to be the argument of the functions.

Let us consider the family of equal-time correlation functions of the basic field $\widehat{\varphi}(x) \equiv \widehat{\varphi}(t, \mathbf{x}) \equiv \widehat{\varphi}_t(\mathbf{x})$ and their generating functional $g_t(a)$, where $a \equiv a(\mathbf{x})$ is the time-independent source for the field $\widehat{\varphi}_t(\mathbf{x})$. The following *statement* is valid. If the correlator D in (5.1) has the form

$$D(x, x') = \delta(t - t')\overline{D}(\mathbf{x}, \mathbf{x}') \tag{5.35}$$

with arbitrary symmetric kernel $\overline{D}$, the functional $g_t(a)$ satisfies the equation

$$\partial_t g_t(a) = a\left[U(\varphi)\big|_{\varphi=\delta/\delta a} + \tfrac{1}{2}\overline{D}a\right]g_t(a). \tag{5.36}$$

The expression in the square brackets in (5.36) and the coefficient a in front of it each have the single argument $\mathbf{x}$ and are contracted with each other (integration over coordinates and summation over indices).

Equation (5.36) is satisfied both for the standard formulation (the asymptotic problem on the entire time axis) and for the Cauchy problem. In the first case, (5.35) leads to time translational invariance of all the Green functions and, therefore, t-independence of the equal-time functions and their generating functional $g_t(a)$, and so both sides of (5.36) should vanish in this case. The time translational invariance is violated, in particular, when the kernel $\overline{D}$ in (5.35) contains explicit t dependence. All the expressions of the present section can be generalized directly to this case (which is rather exotic) by the replacement $\overline{D}(\mathbf{x}, \mathbf{x}') \rightarrow \overline{D}_t(\mathbf{x}, \mathbf{x}') \equiv \overline{D}(t; \mathbf{x}, \mathbf{x}')$.

Proof [159]. From (5.14) we have

$$g_t(a) = \text{const} \int D\eta \exp[-\eta D^{-1}\eta/2 + a\overline{\varphi}_t], \tag{5.37}$$

where $\overline{\varphi}_t$ is the solution of (5.1) at time t functionally dependent on η, $a\overline{\varphi}_t \equiv \int d\mathbf{x}\, a(\mathbf{x})\overline{\varphi}_t(\mathbf{x})$, and, owing to (5.35),

$$\eta D^{-1}\eta = \int dt \iint d\mathbf{x}d\mathbf{x}'\eta_t(\mathbf{x})(\overline{D})^{-1}(\mathbf{x}, \mathbf{x}')\eta_t(\mathbf{x}'). \tag{5.38}$$

Differentiating (5.37) with respect to t, using (5.1) we find that

$$\partial_t g_t(a) = \langle\!\langle a\partial_t\overline{\varphi}_t\rangle\!\rangle = \langle\!\langle a[U(\overline{\varphi}_t) + \eta_t]\rangle\!\rangle, \tag{5.39}$$

where here and below $\langle\langle F \rangle\rangle$ denotes the expression on the right-hand side of (5.37) with the additional factor of F inside the integral.

Now we shall use the Schwinger equation (Sec. 2.11) for the integral (5.37): $0 = \int D\eta \, \delta \exp[...]/\delta \eta_{t'}(\mathbf{x}')$. Taking into account (5.38), we then find that

$$\langle\langle -\int d\mathbf{x}''(\overline{D})^{-1}(\mathbf{x}',\mathbf{x}'')\eta_{t'}(\mathbf{x}'') + \int d\mathbf{x} \, a(\mathbf{x})\delta\overline{\varphi}_t(\mathbf{x})/\delta\eta_{t'}(\mathbf{x}')\rangle\rangle = 0. \qquad (5.40)$$

We want to set $t = t'$ here. The variational derivative $\delta\overline{\varphi}_t(\mathbf{x})/\delta\eta_{t'}(\mathbf{x}')$ is a special case of the response function for a system with fixed value of η. Therefore, it possesses the usual properties of a response function, in particular, a discontinuity at $t = t'$ with the limits in (5.31) for $t \to t' \pm 0$. In this case it is necessary (see below for the justification) to make the definition more precise as in (5.23):

$$\delta\overline{\varphi}_t(\mathbf{x})/\delta\eta_{t'}(\mathbf{x}')\,|_{t=t'} = \tfrac{1}{2}\delta(\mathbf{x}-\mathbf{x}'). \qquad (5.41)$$

Substituting (5.41) into (5.40) with $t = t'$ and convoluting the result with the kernel $\overline{D}(\mathbf{x},\mathbf{x}')$ in the argument $\mathbf{x}'$, we find that

$$\langle\langle -\eta_t + \overline{D}a/2 \rangle\rangle = 0, \qquad (5.42)$$

or, in detailed notation, $\langle\langle -\eta_t(\mathbf{x}) + \int d\mathbf{x}'\overline{D}(\mathbf{x},\mathbf{x}')a(\mathbf{x}')/2 \rangle\rangle = 0$.

Equation (5.42) allows us to replace η_t in (5.39) by the expression $\overline{D}a/2$. If we also take into account the fact that multiplication by $\overline{\varphi}_t$ inside the integral (5.37) is equivalent to differentiating the integral with respect to a, then from (5.39) we immediately obtain the desired equation (5.36).

In the case of the Cauchy problem (5.1) with given initial value $\varphi_0(\mathbf{x})$ at $t = 0$, (5.36) is supplemented by the initial data $g_0(a) = \exp(a\varphi_0)$, and the t dependence of $g_t(a)$ vanishes only at the asymptote $t \to +\infty$. We note that choosing the initial value $g_0(a)$ in a form different from $\exp(a\varphi_0)$ allows us to reproduce the Cauchy problem with stochastic initial data. For the standard formulation (Sec. 5.1) the functional $g_t(a)$ is independent of t, which is natural, since the asymptotic problem can be understood as the Cauchy problem with initial data at $t = -\infty$, so that stationarity is established by any finite time t.

Let us now explain the choice of the definition (5.41). Substituting into (5.36) the functional U in the form (5.7), we can seek a stationary (time-independent) solution for $g(a)$ by iterating in powers of the nonlinearity $n(\varphi)$. The result can be represented using some new (non-Feynman) diagrammatic technique. The same result must be obtained directly from the graphs of Sec. 5.2 for the correlation functions if all the external times in them are set equal to each other and the integration over the times of all internal vertices is performed explicitly. It can be shown that the expressions for $g(a)$ obtained by these two methods really do coincide, which actually justifies the use in the general case of the seemingly arbitrary definition (5.23) used to derive (5.36). This can also be checked for the particular example of the exactly solvable free problem.

5.7 The Fokker–Planck equation for the equal-time distribution function of the field $\widehat{\varphi}$

The statement (5.36) can be reformulated as an equation for the equal-time distribution function $P_t(\varphi)$ of configurations $\varphi(\mathbf{x})$ of the random field $\widehat{\varphi}_t(\mathbf{x})$, defining it as

$$g_t(a) = \int D\varphi \, P_t(\varphi) \exp(a\varphi), \quad \varphi \equiv \varphi(\mathbf{x}), \quad a \equiv a(\mathbf{x}). \tag{5.43}$$

Clearly, the action of the operator $\delta/\delta a$ on the integral (5.43) is equivalent to multiplying the integrand by φ, and multiplication by a is equivalent to differentiating the exponential in (5.43) with respect to φ and becomes the operator $-\delta/\delta\varphi$ after the derivative is moved onto $P_t(\varphi)$ by integrating by parts. These substitutions $\delta/\delta a \to \varphi$, $a \to -\delta/\delta\varphi$ allow (5.36) to be immediately rewritten as the following *Fokker–Planck equation* for the distribution function:

$$\left\{ \partial_t + \frac{\delta}{\delta\varphi} \left[U(\varphi) - \frac{1}{2}\overline{D}\frac{\delta}{\delta\varphi} \right] \right\} P_t(\varphi) = 0. \tag{5.44}$$

In the stationary case (the standard problem), $P_t(\varphi)$ does not depend on t, and then the contribution of ∂_t in (5.44) vanishes.

As a very simple example, let us consider the problem (5.5). In this case $\varphi(\mathbf{x}) = \{r_i\}$ is the $\mathbf{x}$-independent coordinate vector of the diffusing particle, $U = 0$, $\overline{D} = 2\alpha\delta_{ik}$ according to (5.5) and (5.1), and so (5.44) is the usual diffusion equation:

$$[\partial_t - \alpha\partial^2]P_t(\mathbf{x}) = 0, \tag{5.45}$$

where ∂^2 is the Laplace operator, α is the diffusion coefficient, and we have replaced the argument r of the function P_t by the more usual $\mathbf{x}$. The standard formulation of the problem for (5.45) is not meaningful owing to the absence of stationary solutions compatible with the natural normalization condition

$$\int d\mathbf{x} P_t(\mathbf{x}) = 1. \tag{5.46}$$

For this system and its various generalizations we always have the Cauchy problem with initial data,

$$P_0(\mathbf{x}) = \delta(\mathbf{x}), \tag{5.47}$$

which corresponds to the diffusing particle starting at the coordinate origin at $t = 0$. Equation (5.45) leads to stationarity of the integral in (5.46), and so the normalization condition (5.46) is guaranteed to be satisfied for any $t \geq 0$ when it is satisfied at $t = 0$. We also note that for the solution of (5.45) with the initial data (5.47) the quantity

$$\theta(t - t')P_{t-t'}(\mathbf{x} - \mathbf{x}') \tag{5.48}$$

is the retarded Green function of the linear operator in (5.45).

5.8 Relation between dynamics and statics for the stochastic Langevin equation

The following *statement* is valid. Let the correlator in (5.1) have the form (5.35), and U be written as

$$U(\varphi) = \frac{1}{2}\overline{D}\frac{\delta S^{\text{stat}}(\varphi)}{\delta\varphi}, \qquad (5.49)$$

where $S^{\text{stat}}(\varphi)$ is a functional of $\varphi(\mathbf{x})$ not involving the time (a static functional; the replacement $\varphi(\mathbf{x}) \to \varphi(x)$ indicated explicitly in (5.4a) in going from static to dynamical objects is understood in (5.49) and analogous expressions). Then (5.44) has the stationary solution

$$P_t(\varphi) = \text{const} \cdot \exp S^{\text{stat}}(\varphi), \qquad (5.50)$$

which is unique in perturbation theory.

In fact, when (5.49) and (5.50) are substituted into (5.44), the contribution of ∂_t vanishes, the linear operator $\overline{D}/2$ is divided by a common factor, and the equation is satisfied owing to the equality

$$\left[\frac{\delta S^{\text{stat}}(\varphi)}{\delta\varphi(\mathbf{x})} - \frac{\delta}{\delta\varphi(\mathbf{x})}\right]\exp S^{\text{stat}}(\varphi) = 0. \qquad (5.51)$$

In perturbation theory, the equal-time correlation functions, and therefore also the corresponding distribution funciton $P(\varphi)$ in (5.43), are determined uniquely (Sec. 5.1). The constant in (5.50) is fixed by the normalization condition $g_t(0) = 1$ for the integral in (5.43) following from the definition of $g_t(a)$.

Clearly, a stationary solution can exist only for the standard formulation and not for the Cauchy problem with initial data different from the functional (5.50).

Consequence: For the stochastic Langevin problem (5.4) the functional (5.50) is the stationary solution of the corresponding Fokker–Planck equation (5.44). In fact, the necessary condition (5.35) for the correlator (5.4b) is satisfied and $\overline{D} = 2\alpha$, which for (5.4a) corresponds to the condition (5.49).

The meaning of these statements is clear. For the problem (5.4) in the standard formulation, the equal-time dynamical correlation functions are independent of time and coincide with the static functions for the equilibrium distribution (5.50) with static action given in (5.4a). This amounts to the condition that the dynamics and statics to be consistent for the problem (5.4). It, in fact, is what determines the choice of the noise correlator in the form (5.4b). In the case of the Cauchy problem the equilibrium distribution (5.50) is realized, of course, only at the asymptote $t \to +\infty$, when the dependence on the time and the initial data disappears (assuming that the system is stable).

The above proof of the consistency of the dynamics and statics can also be generalized to equal-time correlation functions involving any local composite operators $F(\varphi)$ constructed only from the fields φ. [For the proof it is necessary to make the replacement $a\overline{\varphi}_t \to a\overline{\varphi}_t + bF(\overline{\varphi}_t)$ in (5.37) and $a\varphi \to a\varphi + bF(\varphi)$ in (5.43), where $b \equiv \{b(\mathbf{x})\}$ are additional sources of composite operators. This leads to the replacement $a \to a + b\delta F(\varphi)/\delta\varphi|_{\varphi=\delta/\delta a}$ of the two factors of a in (5.36), while the Fokker–Planck equation (5.44) and its stationary solution (5.50) remain unchanged.]

As already mentioned, even when the condition (5.35) is satisfied, the general problem (5.1) does not always reduce to (5.4), because the static functional $(\overline{D})^{-1}U(\varphi)$ cannot always be represented, as in (5.4), as a variational derivative of something (similar to the way that not every vector field is the gradient of something). One counter-example is the Navier–Stokes equation (5.6). Of course, also in this case it is possible to construct a stationary solution $P(\varphi)$ of (5.44) by iterating in the nonlinear term, so that $\ln P(\varphi)$ is the equilibrium static action for the given dynamical problem. However, it will be an object as complex as the Green functions themselves, and so this procedure is not followed in practice.

Equation (5.4) describes the very simple, purely relaxational dynamics of a system with given statics: the maximum of the action $S^{\text{stat}}(\varphi)$, i.e., the energy minimum in the Landau theory, corresponds to some equilibrium configuration $\varphi_0 = \text{const}$ (Sec. 1.13). The regular contribution to the right-hand side of (5.4a) can be viewed as the "force" returning the system to the equilibrium state, while the contribution of the random force $\eta(x)$ can be viewed as thermal fluctuations. In the simplest linear approximation neglecting noise, (5.4a) gives

$$\partial_t \delta\widetilde{\varphi}(t, k) = -\widetilde{\alpha}(k) \cdot \Delta^{-1}(k) \cdot \delta\widetilde{\varphi}(t, k), \tag{5.52}$$

where $\delta\widetilde{\varphi}(t, k)$ is the Fourier component of the fluctuation $\delta\varphi(x) \equiv \varphi(x) - \varphi_0$ with momentum (wave vector) k, $\widetilde{\alpha}(k)$ is the Fourier transform of the Onsager coefficient α in (5.4), and $\Delta^{-1} = -\delta^2 S^{\text{stat}}(\varphi)/\delta\varphi^2|_{\varphi=\varphi_0}$ is the inverse correlator of the field φ in the Landau approximation (Sec. 1.12). If φ is the field of the order parameter, the value of $\Delta(k = 0)$ is interpreted as the static susceptibility (Sec. 1.9). It diverges at the critical point, which causes the coefficient on the right-hand side of (5.52) to be anomalously small for small k. *Definition:* A *soft mode* is any quantity whose large-scale fluctuations relax slowly, i.e., in momentum space the relaxation rate of a fluctuation with a given wave vector k tends to zero for $k \to 0$.

It then follows that the field of the order parameter right at the critical point is always a soft mode. The density $\rho(x) \equiv \rho(t, \mathbf{x})$ of any conserved quantity is also a soft mode, independently of how close the system is to T_c, because the conservation law

$$\partial_t \int d\mathbf{x}\, \rho(x) = 0 \tag{5.53}$$

for the Fourier transform of ρ gives $\partial_t \widetilde{\rho}(t, k = 0) = 0$, which corresponds to the definition of a soft mode. Examples of conserved quantities are any generator of a continuous symmetry group of the system (for example, all three components of the total spin for the isotropic Heisenberg ferromagnet), the total energy, the momentum, the mass of a liquid or gas, and so on. Already from these examples it is clear that in some cases (the density of a gas, the spin for an isotropic ferromagnet) the order-parameter field itself is the density of a conserved quantity, i.e., it satisfies a condition like (5.53). Then it must be a soft mode even far from T_c, which can be ensured only by the Onsager coefficient $\widetilde{\alpha}(k)$ in (5.52) being small at small k, i.e., the condition $\widetilde{\alpha}(0) = 0$. This justifies the following simple selection rule for the Onsager coefficient:

$$\partial_t \int d\mathbf{x}\, \varphi(x) = 0 \quad \Rightarrow \quad \widetilde{\alpha}(k) = \text{const} \cdot k^2, \qquad (5.54a)$$

$$\partial_t \int d\mathbf{x}\, \varphi(x) \neq 0 \quad \Rightarrow \quad \widetilde{\alpha}(k) = \text{const}. \qquad (5.54b)$$

It should be noted that in general $\widetilde{\alpha}(k)$ is assumed to be a function of k^2 regular at the origin, and in the critical region (small k) it can always be approximated by only the leading term of its asymptote for $k \to 0$. If the order parameter [in general, the field component on the left-hand side of (5.4a)] is not conserved, i.e., if it does not satisfy a condition like (5.53), then it is always assumed that $\widetilde{\alpha}(k) \simeq \widetilde{\alpha}(0) = \text{const}$. When it is conserved, necessarily (see above) $\widetilde{\alpha}(0) = 0$, and so the leading term is the next term in the expansion in k^2, which corresponds to the rule (5.54). For a multicomponent field φ the coefficient α is in general a matrix in the indices, and then the rule (5.54) holds for its rows.

5.9 General principles for constructing models of critical dynamics. The intermode interaction

In critical statics we dealt only with the field of the order parameter. In dynamics the situation is much more complicated, because often when constructing a realistic model it is necessary to take into account the interaction of the order-parameter field with other soft modes. A typical example is the inclusion of the motion of the medium when describing the dynamics of the liquid–gas transition. The justification of a specific dynamical model is a separate problem to which a large literature is devoted; references can be found in [153] and [160]. These questions lie outside the scope of the present book; we shall discuss only some general principles for the construction of models of critical dynamics, and then illustrate them by examples. From now on we shall use the term dynamical equations to mean only the standard formulation of the problem on the entire time axis with translationally invariant Green functions (Sec. 5.1).

The analysis begins with the isolation of all the soft modes important for the dynamics of the order parameter. The set of these modes is in gerneral a multicomponent field $\varphi \equiv \{\varphi_a\} = \{\psi, ...\}$. Here and below ψ will denote the field of the order parameter, and the ellipsis denotes the other important soft modes, if any. It is assumed that the interaction between the soft modes can be described by hydrodynamical equations like (5.1), where the noise is effectively reproduced by hard (small-scale and rapidly oscillating) contributions. In critical dynamics the functional U on the right-hand side of (5.1) always contains the relaxation (dissipative) contribution indicated explicitly in (5.4a) and determined by a given static action for the system of fields φ, plus in general some nondissipative (time-reversible) interactions that do not affect the equilibrium distribution function (5.50). Such interactions are referred to as intermode couplings or *"streaming terms,"* and the entire formalism itself as *mode-coupling theory.* The first example of streaming terms was studied in [161], and the general theory was constructed by Kadanoff and Swift [162] and Kawasaki [163]. These studies introduced a generalization of the problem (5.4) to the case of a system of fields $\varphi \equiv \{\varphi_a\}$ (repeated indices are always summed over):

$$\partial_t \varphi_a = (\alpha_{ab} + \beta_{ab})\frac{\delta S^{\text{stat}}(\varphi)}{\delta \varphi_b} + \eta_a, \quad \langle \widehat{\eta}_a(x)\widehat{\eta}_b(x')\rangle = 2\alpha_{ab}\delta(x - x'), \quad (5.55)$$

where $\alpha_{ab} = \alpha_{ba}^{\top}$ are the Onsager coefficients and β_{ab} are new *streaming coefficients* with the properties

$$\beta_{ab} = -\beta_{ba}^{\top}, \quad \delta\beta_{ab}(\mathbf{x}; \varphi)/\delta\varphi_a(\mathbf{x}) = 0 \qquad (5.56)$$

(the transpose pertains to possible operators like ∂ in the coefficients α_{ab} and β_{ab}, i.e., as a whole, $\alpha = \alpha^{\top}$, $\beta = -\beta^{\top}$).

Let us explain these properties and the general form of the coefficients β_{ab} in more detail. Like α_{ab}, these are static (i.e., not explicitly involving t and ∂_t) local and translationally invariant (i.e., not containing explicit $\mathbf{x}$ dependence) linear operators on the fields $\varphi(\mathbf{x})$ and other objects of a similar nature [like in (5.55)]. However, in these coefficients, in contrast to α_{ab}, it is possible to have a local dependence on the fields $\varphi(\mathbf{x})$ and their derivatives (i.e., only spatial derivatives), and thereby an implicit $\mathbf{x}$ dependence through the field. Upon substitution into dynamical constructions the replacement $\varphi(\mathbf{x}) \to \varphi(x)$ in all static objects is understood, which introduces an implicit (through the field) dependence on the time t. This holds also for the coefficients β, so that both types of notation are suitable for them: the static one $\beta_{ab} = \beta_{ab}(\mathbf{x}; \varphi)$ and the dynamical one $\beta_{ab} = \beta_{ab}(x; \varphi)$, with field $\varphi = \varphi(\mathbf{x})$ in the first case and $\varphi = \varphi(x)$ in the second. Which one is meant will always be clear from the context. In the second equation in (5.56) we have used the static notation; the dynamical analog with the replacement $\mathbf{x} \to x$ and $\varphi(\mathbf{x}) \to \varphi(x)$ follows automatically from it. The variational derivative in (5.56) is a multiple of $\delta(0)$, but (5.56) itself must be satisfied not owing to the rule $\delta(0) = 0$, postulated only in dimensional regularization, but as a consequence of the general structure and symmetry properties of the coefficients β (see the examples in Sec. 5.12 below).

The Fokker–Planck equation (5.44) for the problem (5.55) has the form

$$\left\{\partial_t + \frac{\delta}{\delta\varphi_a}\left[(\alpha_{ab} + \beta_{ab})\frac{\delta S^{\text{stat}}}{\delta\varphi_b} - \alpha_{ab}\frac{\delta}{\delta\varphi_b}\right]\right\} P_t(\varphi) = 0. \qquad (5.57)$$

Its stationary solution is the same functional (5.50) as at $\beta = 0$ owing to the equation

$$\int d\mathbf{x} \frac{\delta}{\delta \varphi_a(\mathbf{x})} \left[\beta_{ab}(\mathbf{x}; \varphi) \frac{\delta S^{\text{stat}}(\varphi)}{\delta \varphi_b(\mathbf{x})} \exp S^{\text{stat}}(\varphi) \right] = 0 \qquad (5.58)$$

[we have explicitly indicated the integration over $\mathbf{x}$ understood in (5.44)], the validity of which is ensured by the two conditions (5.56).

Therefore, the intermode interaction in (5.55) has no effect on the static equilibrium distribution specified by the functional $S^{\text{stat}}(\varphi)$. In this sense dynamics is less universal than statics, since a single static action can correspond to many different dynamical models with different streaming terms. In specific models (Sec. 5.12) the coefficients β_{ab} are determined from the form of the symmetry transformations, the generators of which are additional soft modes.

It should also be noted that in statics we were dealing only with the field of the order parameter ψ and the corresponding action $S^{\text{stat}}(\psi)$, while in dynamics the functional $S^{\text{stat}}(\varphi)$ in the presence of additional soft modes is a geneneralization of $S^{\text{stat}}(\psi)$. Of course, this generalization should not distort the equilibrium distribution for ψ known from $S^{\text{stat}}(\psi)$; only renormalization of the parameters in it that does not affect the universal characteristics of the critical statics is permitted. This requirement significantly limits the possible form of the interaction of the order-parameter field ψ with the other soft modes in $S^{\text{stat}}(\varphi)$ and is certainly satisfied in the absence of such an interaction.

In addition to the basic conditions (5.56) ensuring that the intermode interaction does not affect the statics, in specific models this interaction always satisfies the condition of t-reversibility. This means that the dynamical equation (5.55) with purely intermode right-hand side (i.e., without noise and the contribution involving α) is invariant under time reversal. The latter corresponds to the field transformation

$$\varphi(x) \rightarrow \varphi_T(x) = I\varphi(Tx), \quad I^2 = 1, \quad Tx \equiv T\{t, \mathbf{x}\} = \{-t, \mathbf{x}\}. \qquad (5.59)$$

The matrix $I = I^{-1}$ representing the discrete T-reflection operator is usually but not always simply diagonal, and then its diagonal elements $\varepsilon_a = \pm 1$ determine the t-parity of the corresponding field component φ_a. Under T-reflection (5.59) all the times and the operator ∂_t in the dynamical equation (5.55) change sign, and in specific models the functional $S^{\text{stat}}(\varphi)$ on fields $\varphi(\mathbf{x})$ is always invariant under the replacement $\varphi(\mathbf{x}) \rightarrow I\varphi(\mathbf{x})$, and so its variational derivative is transformed by the transposed matrix $I^\top$. T-invariance (reversibility) of the intermode interaction implies that it transforms exactly like the derivative $\partial_t \varphi$ on the left-hand side of (5.55), i.e., by the matrix $-I$, while the dissipative contribution with coefficients α_{ab} is transformed by the matrix I, so that it changes sign relative to $\partial_t \varphi$. This is all ensured by the following properties of the matrices α and $\beta = \beta(\mathbf{x}; \varphi)$ (static version) in (5.55):

$$I \alpha I^\top = \alpha, \quad I \beta(\mathbf{x}; I\varphi) I^\top = -\beta(\mathbf{x}; \varphi). \qquad (5.60)$$

These properties are always satisfied in actual models (see Sec. 5.12).

Let us conclude by giving the action functional (5.25) corresponding to the problem (5.55):

$$S(\phi) = \varphi' \alpha \varphi' + \varphi' \left[-\partial_t \varphi + (\alpha + \beta) \frac{\delta S^{\text{stat}}}{\delta \varphi} \right] \qquad (5.61)$$

with the needed integrations and summations understood.

5.10   Response to an external field

A stationary external field $h(\mathbf{x})$ gives the contribution $h\varphi = \int d\mathbf{x}\, h(\mathbf{x})\varphi(\mathbf{x})$ to the static action. To determine the dynamical response functions in an external field h, the latter must be assumed to also depend on the time. We make the replacement

$$S^{\text{stat}}(\varphi) \to S^{\text{stat}}(\varphi) + \int d\mathbf{x}\, h(x)\varphi(\mathbf{x}), \qquad (5.62)$$

which in the action (5.61) corresponds to the replacement

$$S(\phi) \to S(\phi) + h\overline{\varphi}, \qquad \overline{\varphi} \equiv \varphi'(\alpha + \beta) = (\alpha - \beta)\varphi' \qquad (5.63)$$

(we recall that the matrix α is symmetric and β is antisymmetric). The functions describing the dynamical response of the quantity $\langle \widehat{\varphi} \rangle$ to a field h are defined as

$$R(x; x_1' \ldots x_n') = \frac{\delta^n}{\delta h(x_1') \ldots \delta h(x_n')} \langle \widehat{\varphi}(x) \rangle \big|_{h=0} =$$

$$= \int D\phi\, \varphi(x)\overline{\varphi}(x_1') \ldots \overline{\varphi}(x_n') \exp S(\phi). \qquad (5.64)$$

Similarly, it is possible to introduce more complicated functions describing the response of any correlation function $\langle \widehat{\varphi}(x_1) \ldots \widehat{\varphi}(x_n) \rangle$ to an external field.

In the absence of streaming terms ($\beta = 0$) the field $\overline{\varphi}(x)$ conjugate to $h(x)$ in (5.63) differs from $\varphi'(x)$ by only a coefficient α independent of the fields ϕ, and then the quantities (5.64) have a simple relation to the functions introduced in Sec. 5.1 describing the response to an external force $\langle \widehat{\varphi}(x)\widehat{\varphi}'(x_1') \ldots \widehat{\varphi}'(x_n') \rangle$. However, in general for $\beta(\varphi) \neq 0$ the field $\overline{\varphi}$ in (5.63) is a composite operator, and then these two types of response function differ significantly.

It is obvious from the definitions that the functions (5.2) and (5.64) possess identical retardation properties (Sec. 5.1). The following important relation is also valid for the functions (5.64):

$$\int \ldots \int dt_1' \ldots dt_n'\, R(x; x_1' \ldots x_n') = \langle \widehat{\varphi}(\mathbf{x})\widehat{\varphi}(\mathbf{x}_1') \ldots \widehat{\varphi}(\mathbf{x}_n') \rangle_{\text{stat}}. \qquad (5.65)$$

For the proof we consider the special case of a stationary external field $h(x) = h'(\mathbf{x})$. Its addition to (5.62) causes the static action to acquire the additional term $h'\varphi \equiv \int d\mathbf{x}\, h'(\mathbf{x})\varphi(\mathbf{x})$. For the standard problem the dynamical expectation value of a single field $\langle\widehat{\varphi}(x)\rangle$ always coincides with the static one (Sec. 5.8), and so for any $h'(\mathbf{x})$

$$\langle\widehat{\varphi}(x)\rangle = \int D\phi\, \varphi(x)\exp[S(\phi) + h\overline{\varphi}] = \int D\varphi\, \varphi(\mathbf{x})\exp[S^{\text{stat}}(\varphi) + h'\varphi]$$
(5.66)

(the needed normalization factors are assumed to be included in $D\phi$ and $D\varphi$). Clearly, the coefficients of the Taylor expansion in $h'(\mathbf{x})$ of the static functional integral in (5.66) are the static correlation functions on the right-hand side of (5.65). On the other hand, the same coefficients for the dynamical functional integral in (5.66) with the linear form $h\overline{\varphi} = \int d\mathbf{x}\, h'(\mathbf{x})\overline{\varphi}(x) = \int d\mathbf{x}\, h'(\mathbf{x}) \int dt\, \overline{\varphi}(t,\mathbf{x})$ are the left-hand sides of (5.65) according to the definition (5.64), which is what we wanted to prove.

Equation (5.65) implies that the dynamical response functions (5.64) with zero external frequencies are the same as the static correlation functions. This statement turns out to be very useful (like the Ward identities) in proving various relations between the renormalization constants in actual models [158].

5.11   The fluctuation–dissipation theorem

The fluctuation–dissipation (FD) theorem is the following equation (see, for example, [164]) relating the simplest response function (5.64) $R(x;x')$ to the dynamical correlator $\langle\widehat{\varphi}(x)\widehat{\varphi}(x')\rangle \equiv C(x,x')$ (possible indices are assumed to be included in the argument x):

$$R(x;x') = -\theta(t - t')\partial_t C(x,x'),$$
(5.67)

or, equivalently (taking into account the retardation of R and the symmetry of C),

$$R(x;x') - R(x';x) = -\partial_t C(x,x').$$
(5.68)

To obtain (5.68) from (5.67) it is necessary to subtract from (5.67) the analogous equation with the interchange $x \leftrightarrow x'$ followed by the replacement $\partial_{t'} \to -\partial_t$, which is possible owing to the translational invariance of the Green functions; the reverse passage from (5.68) to (5.67) is obvious in view of the fact that R is retarded. In the momentum–frequency representation (5.34), Eq. (5.68) takes the form

$$R(\omega, k) - R^{\top}(-\omega, -k) = i\omega C(\omega, k),$$
(5.69)

where the transpose pertains only to matrix indices.

Proof. For $h \neq 0$ taking into account (5.63) we have

$$\langle\widehat{\varphi}(x)\rangle = \int D\phi\, \varphi(x)\exp[S(\phi) + \varphi'(\alpha + \beta)h],$$
(5.70)

where $S(\phi)$ is the functional (5.61). According to the definition (5.64), the response function $R(x;x')$ is the variational derivative of (5.70) with respect to $h(x')$ at $h = 0$. If we differentiate the integral in (5.70) itself, we obtain for R a representation like (5.64). The same quantity can be represented in a different manner if we differentiate with respect to h after performing the Gaussian integration over φ' in (5.70). We then find that

$$-2R(x; x') = c \int D\varphi \, \varphi(x) A(x') \exp \widetilde{S}(\varphi) \equiv \langle\!\langle \varphi(x) A(x') \rangle\!\rangle, \qquad (5.71)$$

where $c = \det[-\alpha/\pi]^{-1/2}$ is an unimportant constant, $\langle\!\langle ... \rangle\!\rangle$ is convenient notation for integration over φ with weight $c \exp \widetilde{S}(\varphi)$, and

$$\widetilde{S} = -F\alpha^{-1}F/4, \qquad F \equiv -\partial_t\varphi + (\alpha + \beta)[\delta S^{\text{stat}}/\delta\varphi], \qquad (5.72a)$$

$$A = F\alpha^{-1}(\alpha + \beta) = (1 - \beta\alpha^{-1})F \equiv -\partial_t\varphi + B, \qquad (5.72b)$$

$$B \equiv \alpha[\delta S^{\text{stat}}/\delta\varphi] + \beta\alpha^{-1}(\partial_t\varphi - \beta[\delta S^{\text{stat}}/\delta\varphi]). \qquad (5.72c)$$

The quantities φ, F, A, and B should be understood as vectors in the indices of φ, the quantities α, α^{-1}, and β are matrices acting on them from the right or left (moving a matrix from left to right or vice versa is accompanied by transpose of the matrix and $\alpha^\top = \alpha$, $\beta^\top = -\beta$), $\widetilde{S}(\varphi)$ in (5.72a) is a quadratic form with kernel α^{-1}, and the required arguments x and integrations over them are everywhere understood.

It has been shown (Sec. 5.9) that under T-reflection (5.59) the quantities $\partial_t\varphi$ and $\beta[\delta S^{\text{stat}}/\delta\varphi]$ are transformed by the matrix $-I$, while φ and $\alpha[\delta S^{\text{stat}}/\delta\varphi]$ are transformed by I. Quantities of the first type will be termed I-odd, and those of the second I-even. Using (5.60), it is easily verified that all the contributions in (5.72c) for B are I-even [when F from (5.72a) is substituted into (5.72b), the two I-odd contributions cancel each other out]. In the notation of (5.71) and (5.72) we have

$$-2R(x; x') = \langle\!\langle \varphi(x)[-\partial_{t'}\varphi(x') + B(x')] \rangle\!\rangle, \qquad (5.73)$$

where the contributions involving $\partial_{t'}\varphi(x')$ and $B(x')$ have different I-parity.

Let us now show that the action $\widetilde{S}(\varphi)$ in (5.71) is T-even, i.e., invariant under T-reflection (5.59). In the quadratic form $F\alpha^{-1}F$ in (5.72a) for $\widetilde{S}$ the "vector" F is the sum of I-even and I-odd contributions. It follows from the first equation in (5.60) that in the form $F\alpha^{-1}F$ products of F contributions with the same I-parity give a T-even contribution to $\widetilde{S}$. Therefore, a T-odd contribution can be generated only by a cross term in the form $F\alpha^{-1}F$ that is a multiple of the expression

$$\left[-\partial_t\varphi + \beta\frac{\delta S^{\text{stat}}}{\delta\varphi}\right]\alpha^{-1}\left[\alpha\frac{\delta S^{\text{stat}}}{\delta\varphi}\right] = \frac{\delta S^{\text{stat}}}{\delta\varphi}\left[-\partial_t\varphi + \beta\frac{\delta S^{\text{stat}}}{\delta\varphi}\right]. \qquad (5.74)$$

Both contributions to the right-hand side are zero, the second owing to antisymmetry of the matrix β, while the first reduces to the integral of a total derivative. In fact, in detailed notation [see (5.4)] this contribution has the form $-\int dx \partial_t \varphi(x) \delta S^{\text{stat}}(\varphi)/\delta\varphi(\mathbf{x})|_* = -\int dt \partial_t L(t)$, where $L(t) \equiv S^{\text{stat}}(\varphi)|_*$, and $|_*$ everywhere denotes the replacement $\varphi(\mathbf{x}) \to \varphi(x)$.

Therefore, the action $\widetilde{S}(\varphi)$ is T-invariant, and so all expectation values $\langle\!\langle ... \rangle\!\rangle$ with weight $\exp\widetilde{S}(\varphi)$ are also invariant, in particular,

$$\langle\!\langle\varphi(x)[-\partial_{t'}\varphi(x') + B(x')]\rangle\!\rangle = I\langle\!\langle\varphi(\overline{x})[\partial_{\overline{t}'}\varphi(\overline{x}') + B(\overline{x}')]\rangle\!\rangle I^{\top}, \qquad (5.75)$$

where $\overline{x} \equiv Tx$, $\overline{x}' \equiv Tx'$, and we have used the known (see above) I-parity properties of all quantities [(5.75) is a matrix equation].

Let us now take into account the retardation of the response function, from which it follows that the right-hand side of (5.73) vanishes for $t < t'$, i.e.,

$$\langle\!\langle\varphi(x)\partial_{t'}\varphi(x')\rangle\!\rangle = \langle\!\langle\varphi(x)B(x')\rangle\!\rangle \quad \text{for } t < t'. \qquad (5.76)$$

We take $t > t'$ in (5.75). Then on the right-hand side we will have $\overline{t} < \overline{t}'$ and we can use (5.76) for it. This allows the entire right-hand side of (5.75) to be replaced by twice its first term, which coincides with the contribution of the first term on the left-hand side of (5.75) owing to T-invariance. Therefore, for $t > t'$ we have $\langle\!\langle\varphi(x)[-\partial_{t'}\varphi(x') + B(x')]\rangle\!\rangle = -2\langle\!\langle\varphi(x)\partial_{t'}\varphi(x')\rangle\!\rangle$, which taking into account the retardation, time translational invariance, and (5.73) is equivalent to the desired equation (5.67).

The following generalization of the FD theorem (5.67) is proved in exactly the same way:

$$\langle M(x) \cdot \overline{\varphi}(x') \cdot N(x')\rangle = \theta(t - t')\langle M(x) \cdot \partial_{t'}\varphi(x') \cdot N(x')\rangle, \qquad (5.77)$$

where $\overline{\varphi} = \varphi'(\alpha + \beta)$ from (5.63), and M and N are any local composite operators with any tensor structure in the field indices (free or with contractions) possessing definite I-parity and constructed only from the fields φ (without φ') and their derivatives. The definite I-parity is needed only to ensure that the right-hand sides of the relations analogous to (5.73) and (5.75) for the objects in (5.77) are proportional to each other. Equation (5.67) is a special case of (5.77) with vector $M = \varphi$ and $N = 1$. Another example is (5.270) in Sec. 5.26, where M is a scalar and N is a nontrivial vector contracted in the indices with $\overline{\varphi}(x')$ on the left-hand side and with $\varphi(x')$ on the right-hand side of (5.77).

5.12 Examples of actual models of critical dynamics

In this section we shall present the standard list [153] of specific dynamical models. They all correspond to the static O_n φ^4 model with various values of n for the field of the order parameter. Their physical meaning is discussed in Sec. 5.13. In this section we shall denote the field of the order parameter by ψ (and use φ to denote the set of all fields), and so the static action of the type (1.63) for $\psi(\mathbf{x})$ will be written as

$$S^{\text{stat}}(\psi) = -(\partial\psi)^2/2 - \tau\psi^2/2 - v_1(\psi^2)^2/24 \qquad (5.78)$$

with implicit integration over the d-dimensional coordinate $\mathbf{x}$ and the required summations over the O_n indices of ψ when present (Sec. 1.14), as in the other static functionals. The charge g from (1.63) has been denoted by v_1 in (5.78), and the additional interaction parameters in various models will be denoted by v_2, v_3, ... for uniformity. They are all real constants, like the parameters $\lambda > 0$ in the various Onsager coefficients α, which will be introduced below.

All the models are described by the standard equations (5.55), and so when formulating them we shall indicate only the set of fields φ, the form of $S^{\text{stat}}(\varphi)$, and the coefficients α_{ab} and β_{ab}, and also the form of the matrix I in (5.59). Unless stated otherwise, all fields should be assumed real, and ψ is an n-component field if (5.78) is viewed as an O_n-symmetric model. We shall only give one of the values (the second one can be deduced from the symmetry) for the coefficients $\alpha_{ab} = \alpha_{ba}^{\top}$ and $\beta_{ab} = -\beta_{ba}^{\top}$, and for the off-diagonal coefficients we shall also use the notation $\alpha_{ab} = \alpha[\varphi_a, \varphi_b]$ and $\beta_{ab} = \beta[\varphi_a, \varphi_b]$ with a particular identification of the components φ_a and φ_b. Use of the notation α_a always implies that for a given component φ_a the corresponding row of the matrix α is diagonal (i.e., $\alpha_{ab} = \alpha_a \delta_{ab}$). All coefficients α_{ab} and β_{ab} not indicated explicitly, except those obtained by the replacement $ab \to ba$, are to be assumed zero. The term "conserved" for a field always means simply that it is the density of a conserved quantity (Sec. 5.8).

Let us turn to the specific models [153]. The first four models A–D are purely relaxational ($\beta_{ab} = 0$); A and B are the simplest ones, and C and D are more complicated versions of them.

Model A. In this model $\varphi = \psi$ (only the field of the order parameter) with the static action (5.78) and $\alpha_{\psi} = \lambda_{\psi}$ (here and below $\lambda_a = \text{const} > 0$ for all a).

Model B. The same as model A, but with $\alpha_{\psi} = -\lambda_{\psi}\partial^2$.

Model C. In this model $\varphi = \{\psi, m\}$, where m is an additional one-component conserved field, and

$$S^{\text{stat}}(\varphi) = S^{\text{stat}}(\psi) - m^2/2 - v_2 m\psi^2/2, \quad \alpha_{\psi} = \lambda_{\psi}, \quad \alpha_m = -\lambda_m\partial^2. \quad (5.79)$$

Model D. The same as model C, but with $\alpha_{\psi} = -\lambda_{\psi}\partial^2$.

In models B and D the field ψ is conserved, while it is not in A and C. The matrix I in the T-reflection operator (5.59) is defined as $I\psi = \varepsilon_{\psi}\psi$ for models A and B and $I\{\psi, m\} = \{\varepsilon_{\psi}\psi, m\}$ for models C and D, where the t-parity $\varepsilon_{\psi} = \pm 1$ is not fixed.

All the following models contain the intermode interaction.

Model H. In this model $\varphi = \{\psi, j\}$, where ψ is a one-component scalar and $j = \{j_i\}$ is a transverse ($\partial_i j_i = 0$) vector field. Both fields are conserved, and

$$S^{\text{stat}}(\varphi) = S^{\text{stat}}(\psi) - j^2/2, \quad \alpha_{\psi} = -\lambda_{\psi}\partial^2, \quad \alpha_j = -\lambda_j\partial^2,$$

$$\beta[\psi, j_i] = v_2 \partial_i \psi, \quad I\{\psi, j\} = \{\psi, -j\}. \quad (5.80)$$

On the right-hand side of (5.55) for j the transverse part is selected by the corresponding projector $P^\perp$ (in momentum space $P_{is}^\perp = \delta_{is} - k_i k_s/k^2$). We note that in the case of substitution into equations like (5.57) this projector can be omitted, since in the variational derivative of any functional with respect to a transverse field the transverse part is selected automatically [as in the functional (5.61) with transverse j']. The noise $\hat{\eta}$ in the equation for j and its correlator is also assumed to be transverse.

Model F. In this model $\varphi = \{\psi, \overset{+}{\psi}, m\}$ with complex N-component fields ψ and $\overset{+}{\psi}$, m is a conserved real field, and

$$S^{\text{stat}}(\varphi) = -\partial \overset{+}{\psi} \cdot \partial \psi - \tau \overset{+}{\psi} \psi - v_1 (\overset{+}{\psi} \psi)^2/4 - m^2/2 - v_2 m \overset{+}{\psi}\psi,$$

$$\beta[\psi, \overset{+}{\psi}] = iv_3, \qquad \beta[\psi, m] = iv_4\psi, \qquad \beta[\overset{+}{\psi}, m] = -iv_4\overset{+}{\psi}, \qquad (5.81)$$

$$\alpha[\psi, \overset{+}{\psi}] = \lambda_\psi, \quad \alpha_m = -\lambda_m \partial^2, \quad I\{\psi, \overset{+}{\psi}, m\} = \{i\overset{+}{\psi}, -i\psi, m\}$$

(see the conventions preceding the description of the model).

Model E. This is a special case of model F with $v_2 = v_3 = 0$.

Model G. In this model $\varphi = \{\psi, m\}$ with real fields $\psi = \{\psi_a, a = 1, 2, 3\}$, $m = \{m_a, a = 1, 2, 3\}$, and

$$S^{\text{stat}}(\varphi) = S^{\text{stat}}(\psi) - m^2/2, \quad \alpha_\psi = \lambda_\psi, \quad \beta[\psi_a, m_b] = v_2 \varepsilon_{abc}\psi_c,$$

$$\alpha_m = -\lambda_m \partial^2, \quad \beta[m_a, m_b] = v_2 \varepsilon_{abc} m_c, \quad I\{\psi, m\} = \{-\psi, -m\}, \qquad (5.82)$$

where ε_{abc} is a fully antisymmetric tensor with $\varepsilon_{123} = 1$. We note that the intermode interaction generated by the coefficients $\beta[m_a, m_b]$ contributes only when an external field h conjugate to m is introduced following the general rule (5.63).

Model J. In this model $\varphi = \{\psi\} \equiv \{\psi_a, a = 1, 2, 3\}$ with S^{stat} from (5.78) and

$$\alpha_\psi = -\lambda_\psi \partial^2, \quad \beta[\psi_a, \psi_b] = v_2 \varepsilon_{abc}\psi_c, \quad I\{\psi\} = \{-\psi\}. \qquad (5.83)$$

This exhausts the list [153] of the most thoroughly studied models; however, it would obviously be easy to come up with more. The physical interpretation of these models will be discussed in Sec. 5.13; here we only note that model H is used to describe the critical dynamics of the liquid–gas transition and the lamination point in binary mixtures. All the other models are used for various types of magnet, and model F is also used to describe the transition to the superfluid phase of He$_4$.

It is easily checked that the general relations (5.56) and (5.60) are satisfied for all the models listed above. The second equation in (5.56) is satisfied owing to cancellations of contributions when the index a is summed over, and in model H it is satisfied owing to the equation $\partial_i \delta(0) = 0$. We note that this follows from the symmetry under spatial rotations, and not from the formal rule $\delta(0) = 0$ in dimensional regularization [see the text following (5.56)].

5.13   The physical interpretation of models A–J

Let us briefly discuss (see [153] for more details) some of the possible applications of these models and the physical interpretation of the quantities entering into them. For models A to D we shall limit ourselves to magnetic systems, although these models also have other applications [153].

Possible candidates for conserved quantities in magnetic systems are the total spin energy $E = \int d\mathbf{x}\, E(x)$ and the generators of spin rotations — the components of the total spin $S_a = \int d\mathbf{x}\, S_a(x)$. In reality, the spins exchange energy not only between each other, but also with other degrees of freedom, for example, with lattice vibrations (phonons). If the first process is considerably faster than the second, then E can be assumed to be conserved, and $E(x)$ is an important soft mode describing temperature fluctuations. The second limiting case is rapid energy exchange between the spins and other degrees of freedom with high thermal conductivity. Then the spins are essentially located in a heat bath with uniform temperature T of the entire sample. The first situation corresponds to a model of type C with $m(x) = E(x)$, and the second to a model without energy conservation, i.e., without $E(x)$ and with the single parameter $\tau = T - T_c$.

The generators S_a are conserved in the presence of the corresponding rotational symmetry (spin rotations; see Sec. 1.14). In general, a Heisenberg magnet with three-component spin vector is characterized by the values of the three exchange integrals J_a in the corresponding generalization of the Hamiltonian (1.16). For $J_1 = J_2 = J_3$ (an isotropic magnet) all three components of S_a are conserved, for $J_1 = J_2 = J_\perp \neq J_3$ (a uniaxial or planar magnet) there is only axial symmetry with conservation of S_3, and for three different J_a (an anisotropic magnet) there are no conserved generators S_a. For a nonisotropic magnet the spins are always ordered in the direction (or directions) corresponding to maximum absolute value of J_a, with the case $J_{\max} > 0$ corresponding to ferromagnetic ordering and $J_{\max} < 0$ to antiferromagnetic ordering. An anisotropic magnetic therefore always has a single "light" axis and is equivalent to an Ising magnet [153], while for $J_1 = J_2 \equiv J_\perp \neq J_3$ there are two possible cases: $|J_3| > |J_\perp|$ corresponding to ordering along the light 3-axis (a uniaxial magnet), and $|J_3| < |J_\perp|$ corresponding to ordering in the "light" 1,2-plane (a planar magnet). Instead of a real two-component order parameter $\psi_{1,2}$, for a planar magnet it is usual to use the complex pair $\psi = \psi_1 + i\psi_2$, $\overset{+}{\psi} = \psi_1 - i\psi_2$ and to distinguish between symmetric and asymmetric magnets. In the first case there is an additional symmetry under rotation by 180° about any axis in the 1,2-plane (for example, $\{S_1, S_2, S_3\} \rightarrow \{S_1, -S_2, -S_3\}$), while in the second there is no such symmetry.

The order parameter ψ for a ferromagnet is the magnetization $\psi(x) = \langle \widehat{S}(x) \rangle$, which is conserved when the corresponding symmetry is present. For the simplest antiferromagnet with two sublattices A and B the order parameter is, by definition, the difference between the magnetizations of the sublattices, i.e., the field $\psi(x) = \langle \widehat{S}^A(x) - \widehat{S}^B(x) \rangle$ (staggered magnetization). This quantity is not associated with any symmetry and so is never conserved, in contrast to the total magnetization $\langle \widehat{S}^A(x) + \widehat{S}^B(x) \rangle = \langle \widehat{S}(x) \rangle \equiv m(x)$, which is conserved in the presence of spin isotropy and thus plays the role of an additional soft mode.

The information about the fields and magnet models with various symmetries [153] is summarized in Table 5.1.

Table 5.1 Dynamical models without energy conservation for magnets (F–ferro-, AF–antiferro-) with various symmetries: isotropic, uniaxial, planar (symmetric and asymmetric), and anisotropic. All fields have argument x, the 3-axis is always the distinguished one, ψ is the field of the order parameter, and m are additional soft modes.

System		Model	Cons. field	Noncons. field
Anisotr. F		$A(n=1)$	–	$\psi = \langle \widehat{S}_3 \rangle$
Anisotr. AF		$A(n=1)$	–	$\psi = \langle \widehat{S}_3^A - \widehat{S}_3^B \rangle$
Uniaxial F		$B(n=1)$	$\psi = \langle \widehat{S}_3 \rangle$	–
Uniaxial AF		$A(n=1)$	$m = \langle \widehat{S}_3 \rangle$ (nonint.)	$\psi = \langle \widehat{S}_3^A - \widehat{S}_3^B \rangle$
Plan. F	sym.	E	$m = \langle \widehat{S}_3 \rangle$	$\psi_a = \langle \widehat{S}_a \rangle$, $a = 1, 2$ or $\psi = \psi_1 + i\psi_2$, $\overset{+}{\psi} = \psi_1 - i\psi_2$
	asym.	F		
Plan. AF	sym.	E	$m = \langle \widehat{S}_3 \rangle$	$\psi_a = \langle \widehat{S}_a^A - \widehat{S}_a^B \rangle$, $a = 1, 2$ or $\psi = \psi_1 + i\psi_2$, $\overset{+}{\psi} = \psi_1 - i\psi_2$
	asym.	F		
Isotr. F		J	$\psi_a = \langle \widehat{S}_a \rangle$, $a = 1, 2, 3$	–
Isotr. AF		G	$m_a = \langle \widehat{S}_a \rangle$, $a = 1, 2, 3$	$\psi = \langle \widehat{S}_a^A - \widehat{S}_a^B \rangle$, $a = 1, 2, 3$

The intermode coupling between the field components $\{\psi, m\} \equiv \varphi$ for the models listed in Table 5.1 corresponds physically to including the Larmor spin precession [153]. For intermode coupling of this type the coefficients β_{ab} in the corresponding equations (5.55) are nonzero only for conserved components

$\varphi_b(x) = \langle \widehat{S}_b(x) \rangle$, and then

$$\beta_{ab} \equiv \beta[\varphi_a, \varphi_b] \sim \delta_b \varphi_a, \qquad (5.84)$$

where $\delta_b F$ is the variation of F under an infinitesimal transformation with the conserved generator $\int dx\, \varphi_b(x)$ $(= S_b$ in this case). In the literature on critical dynamics, the expression on the right-hand side of (5.84) is often referred to as the Poisson bracket of $\varphi_a(x)$ and $\int dx\, \varphi_b(x)$. For the uniaxial antiferromagnet (see Table 5.1) the order parameter ψ is invariant under rotations about the 3-axis. Therefore, there is no intermode coupling between ψ and m, and the dynamics of the system are described by the simple model A (without energy conservation).

A special case is that of the intermode interaction between ψ and $\overset{+}{\psi}$ in model F. It is not associated with any rotational symmetry and is formally equivalent to imaginary additions $\pm i v_3$ to the real Onsager coefficient $\alpha[\psi, \overset{+}{\psi}] = \lambda_\psi$ in (5.55) for ψ and $\overset{+}{\psi}$. However, it is actually impossible to add an imaginary quantity to α; the intermode interaction must be treated separately, because complexity of the matrix α_{ab} for the pair ψ, $\overset{+}{\psi}$ would spoil its symmetry, which is not admissible owing to (5.4b) (the correlator is always symmetric).

The additional intermode coupling of ψ and $\overset{+}{\psi}$ in model F is introduced only because such contributions arise as counterterms in renormalization (details in Sec. 5.20) if $v_2 \neq 0$ in the action (5.81). For the symmetric planar magnet the additional symmetry under the transformation $\{\psi, \overset{+}{\psi}, m\} \to \{\overset{+}{\psi}, \psi, -m\}$ (see above) forbids a contribution $m|\psi|^2$ in the static action and thus ensures reality of the renormalization of the Onsager coefficient. Therefore, in model E simultaneously with $v_2 = 0$ in (5.81) we set $v_3 = 0$.

Up to now we have discussed models of magnets in a heat bath, i.e., for constant temperature T throughout the sample. The generalization to the case of conserved energy E will be discussed only for the simplest models A and B. For them we introduce the additional field $m(x) = \langle \widehat{E}(x) \rangle -$ const describing temperature fluctuations. The static action has the form (5.79), and the dynamics are described by the purely relaxational equations (5.4) without streaming terms ($\beta_{ab} = 0$). Then model A becomes model C and B becomes D.

Model F is also used to describe the critical dynamics of the transition from the normal to the superfluid phase in He$_4$. In this case ψ, $\overset{+}{\psi}$ is a complex order parameter (the expectation values of the bosonic field operators $\widehat{\psi}$ and $\widehat{\psi}^+$), the conserved field $m(x)$ is a linear combination of the energy and density that effectively describes the temperature fluctuations, and the intermode coupling of ψ and $\overset{+}{\psi}$ with m (outwardly exactly the same as for the magnet) corresponds physically to using the exact Josephson relation [153] between the phase of the complex order parameter and the chemical potential.

Of the models listed in Sec. 5.12, only model H has nothing to do with magnets. This model is used to describe the critical dynamics of the liquid–gas transition, and also the critical lamination point in binary mixtures. For the liquid–gas transition the role of ψ in model H is played by a linear combination of the quantities in (1.31) (see Sec. 5.23 for more details), and the field j is interpreted as the transverse part of the momentum density vector. Both fields are conserved. The transversality of j is related to the fact that the longitudinal (sound) modes are IR-irrelevant for the critical dynamics [153], so that when describing the motion of the medium the compressibility can be neglected, even though it diverges at the critical point. The intermode interaction (5.80) between ψ and j satisfies (5.84), because the momentum is the generator of spatial translations. It is easily verified that the inclusion of the intermode interaction (5.80) in (5.55) for ψ is equivalent to replacing the simple derivative ∂_t by the Galilean-covariant derivative:

$$\partial_t \to \nabla_t = \partial_t + v_2(j\partial) \equiv \partial_t + (v\partial). \tag{5.85}$$

This replacement is the standard method of including the motion of the medium. The vector $v = v_2 j$ is the velocity of the medium, and in this case it is transverse, which corresponds to incompressibility. Introducing the intermode coupling by the replacement (5.85) in the relaxation equation of the type (5.4) for ψ and interpreting $\partial\psi$ as $\beta[\psi, j]$ (up to coefficients), from the known derivative $\delta S^{\text{stat}}/\delta j \sim j$ we can find the j dependence of S^{stat}. Then, taking into account the antisymmetry of β, we can find the contribution of the intermode coupling of ψ and j in the equation for j. The standard dynamical equations of model H are thus derived from rather simple arguments.

In order to ensure the exact Galilean invariance of the model, the replacement (5.85) should have been made also in the equation for j, which would have led to the introduction of an additional intermode interaction between the various components of the vector j. However, this additional interaction turns out to be IR-irrelevant, and so it is not usually taken into account. For the same reason it would actually be correct to discard the derivative $\partial_t j$ in the equation for j (see Sec. 5.24 for more details). Therefore, the traditional formulation of this model (Sec. 5.12) from the viewpoint of the general principles for the construction of fluctuation models (Sec. 1.16) is not completely correct, because the IR-irrelevant contribution is retained in it. Model H will be studied in detail in Secs. 5.23 to 5.25.

5.14 Canonical dimensions in dynamics

The action functional (5.61) for all dynamical models is invariant under two independent scale transformations: in one all quantities F and the time (frequency) are dilated, and in the other the quantities F and the coordinates (momenta) are. Therefore, for any F it is possible to introduce two independent canonical dimensions: the momentum dimension d_F^p and the frequency dimension d_F^ω (or $d^p[F]$ and $d^\omega[F]$ for complicated F). They are found (Sec. 1.15) from the requirement that each term in the action functional be dimensionless (the momentum and frequency dimensions should vanish separately). In this way it is easy to check that the momentum dimensions $d^p[F]$ of all quantities F (fields and parameters) entering into the static action $S^{\text{stat}}(\varphi)$ coincide with the usual static dimensions $d^{\text{stat}}[F]$, and that their frequency dimensions are zero:

$$d^p[F] = d^{\text{stat}}[F], \quad d^\omega[F] = 0 \ \text{ for all } F \text{ in } S^{\text{stat}}(\varphi). \tag{5.86}$$

This is true, in particular, for all the components of the basic field φ_a, which, taking into account the rule (1.69) generalized to the case $x = t, \mathbf{x}$ and the required zero dimension of all the contributions in the action (5.61), allows us to find the dimensions of the auxiliary fields φ_a' and the coefficients α_{ab}, β_{ab}. The results for the problem with d-dimensional $\mathbf{x}$ are summarized in Table 5.2.

Table 5.2 Momentum and frequency canonical dimensions of various quantities F in the dynamical action (5.61).

F	p	$\mathbf{x}$	ω	t	$\varphi_a(x)$	$\varphi_a'(x)$	$\alpha_{ab}, \ \beta_{ab}$
d_F^p	1	–1	0	0	$d^{\text{stat}}[\varphi_a]$	$d - d^{\text{stat}}[\varphi_a]$	$d^{\text{stat}}[\varphi_a] + d^{\text{stat}}[\varphi_b] - d$
d_F^ω	0	0	1	–1	0	0	1

In most dynamical models (all except models D and H in the above list), the highest power of ∂ in the operator L from (5.7) is identical for all components φ_a. This means that for $T = T_c$ the operators ∂ and ∂_t in the linearized dynamical equations occur in the combination $\partial_t + \text{const} \cdot \partial^{d_\omega}$, identical for all components φ_a ($d_\omega = 2$ for models A, C, F, E, G and $d_\omega = 4$ for models B and J). Then the action (5.61) will also be invariant under some "total" scale transformation, in which all the times and coordinates (or frequencies and momenta) are dilated simultaneously and consistently ($t \sim |\mathbf{x}|^{d_\omega}$). The corresponding total canonical dimensions $d_F \equiv d[F]$ of any F are then defined as

$$d_F = d_F^p + d_\omega \cdot d_F^\omega, \quad (\omega \sim p^{d_\omega}). \tag{5.87}$$

The presence of a uniquely defined total frequency dimension d_ω is natural and even necessary if we are discussing critical scaling. The latter corresponds to an asymptote of the type (1.14) with simultaneous and mutually consistent approach of all IR-relevant parameters e to zero. In dynamics these include ω and p, and if IR scaling is a general property of the model in question, then the relative smallness $\omega \sim p^{d_\omega}$ at the IR asymptote must be completely defined and the same for any object in the model. The renormalization group will then only specify the exponents, leading to replacement of the canonical dimension d_ω by the critical one $\Delta_\omega = d_\omega + O(\varepsilon)$ ($\Delta_\omega \equiv z$ is the traditional notation for this critical exponent in dynamics), but in zeroth order in ε the quantity $\Delta_\omega \simeq d_\omega$ must be fixed *ab initio*.

In this respect models D and H represent a special case, because for them the combination $\partial_t + \text{const} \cdot \partial^4$ enters into the equation for the fundamental order-parameter field ψ, while the combination $\partial_t + \text{const} \cdot \partial^2$ enters into the equation for the second field. In this situation the choice between the two possibilities ($\omega \sim p^4$ or $\omega \sim p^2$) is not unique and is actually just part of the statement of the problem. All real calculations for these models correspond to the condition $\omega \sim p^4$, i.e., to nontrivial dynamics for ψ. But then in the second equation the contribution $\partial_t \sim \omega \sim p^4$ turns out to be IR-irrelevant compared to $\partial^2 \sim p^2$. It must simply be discarded as an IR-irrelevant correction, which corresponds to neglecting retardation for the second field. These models will be studied in more detail in Secs. 5.19 and 5.25.

It is the total dimension (5.87) that plays the same role in dynamics as the ordinary (momentum) dimension in statics when analyzing the UV divergences of graphs and the form of the corresponding counterterms.

5.15 Analysis of the UV divergences and counterterms in dynamics

We shall discuss only correctly constructed dynamical models with uniquely defined total frequency dimension d_ω and all other quantities (5.87). For a correctly constructed model the free part of the action must not contain any parameters with nonzero total dimension, which would remain finite at the IR asymptote. It is only in this case that the simple rules of Sec. 1.16 are valid, allowing us to judge the IR-relevance of the various terms of the interaction from the dimensions of the corresponding coefficients. In going from static formulations to dynamical ones, the term "dimension" should always be understood as "total dimension." In particular, the logarithmic dimension of space d^* for any monomial in the interaction is determined by the condition that the total dimension of its coefficient be zero. In a correctly constructed model all the included interaction terms must become logarithmic simultaneously, and all the IR-irrelevant contributions to the action, including its free part, must be discarded. We note that this is not only desirable as a means of simplifying the model, but is also necessary for revealing the correlation between the IR and UV singularities (Sec. 1.23), which in the end makes it possible to study the IR asymptote using the RG technique (see the discussion of model H in Sec. 5.25 for more details).

All the models listed in Sec. 5.12 are correctly constructed in the above sense, except for models D and H, which require simplification. All of them, except for model J, are logarithmic for $d = 4$ and will therefore be studied in dimension $d = 4 - 2\varepsilon$ (model J is logarithmic for $d = 6$). As in statics, it is simplest to perform the calculations within the formal scheme of dimensional regularization without the cutoff Λ (Sec. 1.20), using the MS scheme for the renormalization. This is what we shall always do in what follows.

For a correctly constructed dynamical model all the main statements of the general theory of renormalization (Ch. 3) remain valid, if the term "dimension" is now understood as the total canonical dimension (5.87). The canonical dimensions of 1-irreducible functions Γ with n_ϕ (a multiple index) fields $\phi = \varphi, \varphi'$ are given as in (3.3) by the equations

$$d_\Gamma^p = d - \sum_\phi n_\phi d_\phi^p, \qquad d_\Gamma^\omega = 1 - \sum_\phi n_\phi d_\phi^\omega,$$

$$d_\Gamma = d + d_\omega - \sum_\phi n_\phi d_\phi, \quad d_\phi = d_\phi^p + d_\omega d_\phi^\omega \qquad (5.88)$$

with summation over all components of ϕ in a given function Γ. The *formal UV exponent* (Sec. 3.2) is

$$d_\Gamma^* \equiv d_\Gamma \mid_{\varepsilon=0} . \qquad (5.89)$$

However, the presence of superficial divergences must be judged from the *real UV exponent* δ_Γ (Sec. 3.2), which in dynamical models is often smaller than the formal exponent owing to the shifting of ∂ symbols onto the external lines of the auxiliary fields φ' in the graphs of Γ (and onto the fields φ' in the corresponding counterterms). According to the general rules (Sec. 3.16), the graphs of a given function Γ contain superficial divergences and generate the corresponding counterterms only when δ_Γ is a nonnegative integer. The combined total dimension of the ∂ and ∂_t symbols in the counterterm (plus those of external fields h when they are present) is equal to (5.89), where it is assumed that $d[\partial] = 1$ and $d[\partial_t] = d_\omega$. We note that the number of symbols ∂_t in a counterterm cannot be determined from the frequency dimension d_Γ^ω (5.88), because the theory always involves parameters with zero total dimension but nonzero frequency dimension, multiplication by which can change d_Γ^ω without changing the total dimension d_Γ.

When determining the form of the counterterms we should take into account, as in statics, the symmetry properties of the problem and also the fact that in dynamics all graphs of vacuum loops and the functions Γ involving only the basic fields φ (without φ') are zero because they contain closed loops of retarded functions (Sec. 5.4). They therefore naturally do not generate counterterms.

In models of critical dynamics, very important information about the renormalization constants can also be obtained from the FD theorem (Sec. 5.11) and the general principles of the correspondence between dynamics and statics (Secs. 5.8 and 5.9). If the unrenormalized and basic action functionals of a model have the standard form (5.61), then the equal-time dynamical Green functions of the basic fields φ are determined by the distribution (5.50) with the corresponding (unrenormalized or basic) functional S^{stat} in the exponent. If the basic functions and the renormalization scheme are specified (for us the latter is the MS scheme), the corresponding renormalized Green functions are thereby uniquely determined, and so they are given by the same distribution (5.50) with known functional $S_{\mathrm{R}}^{\text{stat}}$ in the exponent. This means that the functional $\exp S_{\mathrm{R}}^{\text{stat}}$ must necessarily be a stationary solution of the Fokker–Planck equation (5.44) for the renormalized dynamical model when the renormalization schemes are the same (MS) in dynamics and statics. From this we can obtain information about the dynamical renormalized functional S_{R}.

For all the models discussed in Sec. 5.12, the functional S^{stat} is renormalized multiplicatively. If we were sure that the renormalization were multiplicative also in the corresponding dynamical model, then from the above arguments it would follow directly that the dynamical and static constants Z in the MS scheme coincide for all quantities entering into S^{stat}:

$$Z_F = Z_F^{\text{stat}} \quad \text{for all } F \text{ in } S^{\text{stat}}(\varphi) \tag{5.90}$$

(in the general case they differ by only some UV-finite renormalization). Usually, in dynamical models the standard dimensional analysis allows easy proof of only the fact that all the counterterms can be reproduced by the introduction of various constants Z for the terms of the basic action. However, the number of such constants usually exceeds the number of fields and parameters, and so whether the renormalization is multiplicative is *a priori* not obvious. In this situation it is the above statements about the relation to renormalized statics which allow us to demonstrate the presence of relations between dynamical renormalization constants Z which ensure that the model is multiplicatively renormalizable (another method is detailed analysis of the various response functions [158]).

Let us explain in more detail. We assume that the unrenormalized and basic action functionals have the standard form (5.61), and that the renormalized action S_{R} can be written as

$$S_{\mathrm{R}}(\phi) = \varphi' \alpha' \varphi' + \varphi'[-Z\partial_t \varphi + A(\varphi)], \tag{5.91}$$

where A is a vector and α' and Z are matrices in the indices of the fields φ and φ'. It is usually obvious from power counting that α' and A contain the same contributions as the basic action, but with additional coefficients Z. In order to obtain a Fokker–Planck equation of the type (5.44) for the renormalized theory it is necessary to first bring the functional (5.91) to the canonical form (5.25). This can be done by dilatation of the auxiliary fields $\varphi' \to \varphi' Z^{-1} = Z^{-1\top}\varphi'$, which [as for any dilatation of the coefficients α and β in (5.61)] obviously does not affect the equal-time distribution of the fields φ. Then (5.91) takes the form $\varphi' Z^{-1}\alpha' Z^{-1\top}\varphi' + \varphi'[-\partial_t\varphi + Z^{-1}A(\varphi)]$, from which it follows that the usual Fokker Planck equation (5.44) with $U(\varphi) = Z^{-1}A(\varphi)$ and $\overline{D} = 2Z^{-1}\alpha' Z^{-1\top}$ is satisfied for the renormalized theory. We know (see above) that this equation must have the functional $P_{\mathrm{R}}(\varphi) \sim \exp S_{\mathrm{R}}^{\text{stat}}(\varphi)$ as its stationary solution, i.e., the equation

$$\frac{\delta}{\delta\varphi}\left\{\left[Z^{-1}A(\varphi) - Z^{-1}\alpha' Z^{-1\top}\frac{\delta}{\delta\varphi}\right]\exp S_{\mathrm{R}}^{\text{stat}}(\varphi)\right\} = 0 \tag{5.92}$$

must be satisfied. In detailed notation,

$$\int d\mathbf{x} \frac{\delta}{\delta\varphi_a(\mathbf{x})} \left[B_a(\varphi) \exp S_R^{\text{stat}}(\varphi) \right] = 0 \tag{5.93}$$

(we recall that repeated indices are always summed over), where

$$B(\varphi) \equiv Z^{-1}A(\varphi) - Z^{-1}\alpha' Z^{-1\top} \delta S_R^{\text{stat}}(\varphi)/\delta\varphi. \tag{5.94}$$

The vector B is local, i.e., it depends only on the fields φ and their derivatives at a single point $\mathbf{x}$, the same as over which the integration in (5.93) runs. Therefore, when the variational differentiation is performed in (5.93) all contributions from derivatives of B will be multiples of $\delta(0)$, while contributions from derivatives of the exponent will not contain this factor. Since these two types of contribution are clearly independent, (5.93) reduces to two equations:

$$\int d\mathbf{x} \, \delta B_a(\varphi)/\delta\varphi_a = 0, \tag{5.95a}$$

$$\int d\mathbf{x} \, B_a(\varphi) \cdot \delta S_R^{\text{stat}}(\varphi)/\delta\varphi_a = 0. \tag{5.95b}$$

These equations can be used for specific models to obtain information about the renormalization constants in S_R, allowing the multiplicative nature of the renormalization of the dynamical model and, consequently, (5.90) to be proved. In this way it is fairly easy to prove the multiplicative renormalizability of all the models listed in Sec. 5.12 except for models D and H, which require preliminary simplification.

In conclusion, let us discuss the renormalization of contributions involving the external field h in multiplicatively renormalizable dynamical models. In statics, h enters into the unrenormalized action in the form φh_0, and into the basic and renormalized actions as φh, from which we obtain (1.85) for the renormalization constant $h_0 = hZ_h$: $Z_h^{\text{stat}} \cdot Z_\varphi^{\text{stat}} = 1$. In dynamics the contribution of h to the basic action of the type (5.61) has the form $\varphi'(\alpha+\beta)h$, and only for $\beta = 0$, i.e., in models A, B, C, and D without streaming terms, is the contribution $\varphi'\alpha h$ not renormalized, as in statics (we recall that the coefficient α does not depend on the fields, in contrast to β). From this for these models [taking into account (5.90) for $F = \varphi, h$] we obtain

$$Z_{\varphi'} Z_\alpha = (Z_h^{\text{stat}})^{-1} = Z_\varphi^{\text{stat}} = Z_\varphi, \tag{5.96}$$

relating the renormalization constants of the auxiliary fields φ' and the corresponding Onsager coefficients α.

When there are streaming terms with coefficients β depending on the fields φ, the arguments of Sec. 3.13 become meaningless and in general various nontrivial coefficients Z appear in front of the terms $\varphi'\alpha h$ and $\varphi'\beta h$ in the renormalized action.

5.16 Models A and B

Since this is the first and simplest example, we shall study it fairly thoroughly, in order not to have to repeat the arguments later. Everywhere in what follows we shall assume that the standard MS scheme is used to calculate the renormalization constants.

The unrenormalized action (5.61) of model A (Sec. 5.12) is written as

$$S(\phi) = \lambda_0 \psi' \psi' + \psi'[-\partial_t \psi + \lambda_0 \delta S^{\text{stat}}(\psi)/\delta \psi] =$$
$$= \lambda_0 \psi' \psi' + \psi'[-\partial_t \psi + \lambda_0 (\partial^2 \psi - \tau_0 \psi - g_0 \psi^3/6 + h_0)] \tag{5.97}$$

with unrenormalized static action of the type (5.78), to which is added the contribution of the external field h_0:

$$S^{\text{stat}}(\psi) = -(\partial \psi)^2/2 - \tau_0 \psi^2/2 - g_0 \psi^4/24 + \psi h_0. \tag{5.98}$$

The fields $\phi = \psi, \psi'$ are assumed to be n-component fields and the interaction isotropic. The required summations over the field indices and integration over d-dimensional $\mathbf{x}$ in the static functional and over $x = t, \mathbf{x}$ in the dynamical functional is everywhere understood. In (5.97) λ_0 is the unrenormalized Onsager coefficient. We see from (5.97) that in this model $\partial_t \sim \partial^2$, i.e., $d_\omega = 2$ in the definition (5.87). The canonical dimensions of the fields ϕ and the parameters $e_0 \equiv \lambda_0, \tau_0, g_0, h_0$ of model A are given in Table 5.3 along with their renormalized analogs.

Table 5.3 Canonical dimensions of the fields and parameters of model A.

F	p, ∂	ω, ∂_t	ψ	ψ', h, h_0	λ, λ_0	τ, τ_0	g_0	g	μ
d_F^p	1	0	$d/2-1$	$d/2+1$	-2	2	$4-d$	0	1
d_F^ω	0	1	0	0	1	0	0	0	0
d_F	1	2	$d/2-1$	$d/2+1$	0	2	$4-d$	0	1

We see from Table 5.3 that this model is logarithmic in dimension $d^* = 4$ [found from the condition that the total dimension of the coefficient $\lambda_0 g_0$ of the interaction in (5.97) is zero]. Therefore, henceforth we shall assume that $d = 4 - 2\varepsilon$ and shall write the basic action corresponding to (5.97) as

$$S_{\text{B}}(\phi) = \lambda \psi' \psi' + \psi'[-\partial_t \psi + \lambda(\partial^2 \psi - \tau \psi - g_{\text{B}} \psi^3/6 + h)], \tag{5.99}$$

where $g_{\text{B}} = g\mu^{2\varepsilon}$ is the basic charge, μ is the renormalization mass, and $\{\lambda, \tau, g, h\} \equiv e$ are the renormalized parameters.

The counterterms of the basic theory (5.99) needed to eliminate the UV-divergences (poles in ε) are determined from the real UV exponent δ_Γ of the graphs of the 1-irreducible functions Γ. In this model it coincides with the formal exponent (5.89), and from the results in Table 5.3 we find

$$\delta_\Gamma = d_\Gamma^* = 6 - n_\psi - 3n_{\psi'}, \qquad (5.100)$$

where $n_{\psi,\psi'}$ are the numbers of fields ψ, ψ' in the given function Γ. In determining the form of the counterterms it is also necessary to take into account the facts that, first, all functions Γ with $n_{\psi'} = 0$ vanish (Sec. 5.4) and, second, that the counterterms cannot depend on the external field h (Sec. 3.13) and therefore must possess the symmetry $\psi, \psi' \to -\psi, -\psi'$ of the model with $h = 0$. They can therefore be generated only by functions Γ with $n_{\psi'} \neq 0$ and even value of $n_\psi + n_{\psi'}$. It follows from these arguments and (5.100) that the counterterms of the model (5.99) are generated only by 1-irreducible functions Γ of the type $\langle\psi'\psi\rangle$ ($d_\Gamma^* = 2$, possible counterterms $\psi'\partial_t\psi$, $\psi'\partial^2\psi$, $\psi'\tau\psi$) and $\langle\psi'\psi\psi\psi\rangle$ ($d_\Gamma^* = 0$, counterterm $\psi'\psi^3$) and $\langle\psi'\psi'\rangle$ ($d_\Gamma^* = 0$, counterterm $\psi'\psi'$). Therefore, all the counterterms have the form of the terms (5.99), and so they can be reproduced by introducing into (5.99) the corresponding renormalization constants Z:

$$S_R(\phi) = Z_1\lambda\psi'\psi' + \psi'[-Z_2\partial_t\psi + \lambda(Z_3\partial^2\psi - Z_4\tau\psi - Z_5g_B\psi^3/6 + h)]. \qquad (5.101)$$

All the constants Z are completely dimensionless (in both momentum and frequency) and cannot contain h and μ (MS scheme), and so they can depend only on the completely dimensionless renormalized charge g.

The model is multiplicatively renormalizable, because the functional (5.101) can be obtained from (5.97) by the standard procedure of multiplicative renormalization: $S_R(\phi, e, \mu) = S(Z_\phi\phi, e_0(e, \mu))$, where $Z_\phi\phi \equiv \{Z_\psi\psi, Z_{\psi'}\psi'\}$ is the field renormalization and

$$\lambda_0 = \lambda Z_\lambda, \quad \tau_0 = \tau Z_\tau, \quad g_0 = g\mu^{2\varepsilon}Z_g, \quad h_0 = hZ_h \qquad (5.102)$$

is the parameter renormalization, where in the action (5.101)

$$Z_1 = Z_\lambda Z_{\psi'}^2, \quad Z_2 = Z_{\psi'}Z_\psi, \quad Z_3 = Z_{\psi'}Z_\lambda Z_\psi,$$

$$Z_4 = Z_{\psi'}Z_\lambda Z_\tau Z_\psi, \quad Z_5 = Z_{\psi'}Z_\lambda Z_g Z_\psi^3, \quad 1 = Z_{\psi'}Z_\lambda Z_h. \qquad (5.103)$$

The divergences of the graphs of Γ in the model (5.99) can be used to calculate the constants Z in (5.101), and (5.103) can then be used to find the field and parameter renormalization constants.

However, such a complete calculation is actually unnecessary, because (5.90) and (5.96) follow from the multiplicative renormalizability of the static and dynamical models; in this case,

$$Z_F = Z_F^{\text{stat}} \quad \text{for all } F = \psi, \tau, g, h, \quad Z_h^{\text{stat}} \cdot Z_\psi^{\text{stat}} = 1, \quad Z_{\psi'}Z_\lambda = Z_\psi. \qquad (5.104)$$

Therefore, in dynamics it is sufficient to calculate only a single new renormalization constant Z_λ, which can be done either using Z_1 in (5.101) and the graphs of the 1-irreducible function $\langle \psi' \psi' \rangle$, or using Z_2 and the graphs of $\langle \psi' \psi \rangle$ (the constants $Z_{3,4,5}$ are purely static). As usual, it is simplest to perform the calculations in the massless theory, because the constants Z are independent of τ in the MS scheme. The details of the three-loop calculation can be found in [165]; some remarks concerning it are given below.

The RG functions of anomalous dimension γ_a of any quantity a (a field or parameter) and the β functions of the charges g (of which we have one) are determined from the corresponding renormalization constants Z_a by the general relations (1.113). The RG equation for renormalized connected Green functions W_{nR} has the standard form (1.107), (1.108). In our model all the RG functions are known from statics (Sec. 4.2), except for

$$\gamma_\lambda = \widetilde{\mathcal{D}}_\mu \ln Z_\lambda = \beta_g \partial_g \ln Z_\lambda, \quad \gamma_{\psi'} = \gamma_\psi - \gamma_\lambda \tag{5.105}$$

[the second equation follows from the last relation in (5.104)], and the RG equation (1.107) can be written as follows taking into account the renormalization formulas (5.102) and the definitions (1.113):

$$\left[\mathcal{D}_{\mathrm{RG}} + \sum_\phi n_\phi \gamma_\phi \right] W_{nR} = 0, \quad \mathcal{D}_{\mathrm{RG}} = \widetilde{\mathcal{D}}_\mu = \mathcal{D}_\mu + \beta_g \partial_g - \sum_a \gamma_a \mathcal{D}_a \tag{5.106}$$

with summation (in the present case) over the parameters $a = \lambda, \tau, h$. Its general solution can be studied just as in statics (Ch. 1), by changing over to functions with completely dimensionless (zero momentum and frequency dimension) arguments. The presence of an IR-attractive fixed point $g_* \sim \varepsilon$ for the β function in (5.106) (known from statics) guarantees IR scaling for the asymptote of small ω, p, τ, and h at fixed μ and λ. If we are interested only in the equations of critical scaling for the asymptote $W_{nR}|_{\mathrm{IR}}$ of the function W_{nR}, then they are most simply obtained by the method of Sec. 1.33, combining the RG equation (5.106) with $g = g_*$ [then the contribution involving the β function vanishes and $\gamma_a(g) \to \gamma_a(g_*) \equiv \gamma_a^*$] with two (momentum and frequency) scale equations of the type (1.6). For the functions W_{nR} in coordinate space (where the dimension of W_{nR} is equal to the simple sum of the dimensions of the fields entering into it) these two equations can be written in the general case as

$$\left[\mathcal{D}_\mu - \mathcal{D}_{\mathbf{x}} + \sum_e d_e^p \mathcal{D}_e - \sum_\phi n_\phi d_\phi^p \right] W_{nR}(t, \mathbf{x}, e, \mu) = 0, \tag{5.107a}$$

$$\left[-\mathcal{D}_t + \sum_e d_e^\omega \mathcal{D}_e - \sum_\phi n_\phi d_\phi^\omega \right] W_{nR}(t, \mathbf{x}, e, \mu) = 0 \tag{5.107b}$$

(t is the set of all times, $\mathbf{x}$ is the set of all coordinates, and e is the set of all renormalized parameters). In our model $\phi = \psi, \psi'$, $e = \lambda$, τ, g, h, and all the dimensions are known from Table 5.3. To obtain the equation of critical scaling for $W_{nR}|_{IR}$ we need to eliminate the operators $\mathcal{D}_\mu$ and $\mathcal{D}_\lambda$ from the system of three equations (5.107a), (5.107b), and (5.106) with $g = g_*$, because the parameters μ and λ are assumed to be fixed at the asymptote in question. This procedure always leads to an equation of critical scaling of the following form:

$$\left[-\mathcal{D}_{\mathbf{x}} - \Delta_\omega \mathcal{D}_t + \sum_a \Delta_a \mathcal{D}_a - \sum_\phi n_\phi \Delta_\phi\right] W_{nR}(t, \mathbf{x}, e, \mu)\big|_{IR} = 0 \qquad (5.108)$$

$(a = \tau, h$ for model A), in which

$$\Delta_\omega = -\Delta_t = -[d_\lambda^p + \gamma_\lambda^*]/d_\lambda^\omega \equiv z \qquad (5.109)$$

is the frequency critical dimension ($\Delta_\omega \equiv z$ is the traditional notation for this critical exponent), and the other Δ_F are the critical dimensions of the corresponding quantities F, defined as

$$\Delta_F = d_F^p + \Delta_\omega d_F^\omega + \gamma_F^*. \qquad (5.110)$$

For convenience we have given the general expression in (5.110); for our $F = \tau, h, \psi, \psi'$ the contribution involving Δ_ω in (5.110) vanishes because d_F^ω is zero. The critical dimensions Δ_F of the quantities $F = \tau, h, \psi$ occurring in statics coincide with their static values $\Delta_F = d_F^p + \gamma_F^*$. Only the following dimensions are new for the dynamics in model A:

$$z \equiv \Delta_\omega = 2 - \gamma_\lambda^*, \qquad \Delta_{\psi'} = d_{\psi'}^p + \gamma_\psi^* - \gamma_\lambda^*, \qquad (5.111)$$

where the second equation follows from (5.110) and (5.105) and the first from (5.109) and the data of Table 5.3. We note that the critical dimension $\Delta_\psi + \Delta_{\psi'} = d + 2\gamma_\psi^* - \gamma_\lambda^*$ of the renormalized response function $\langle\psi\psi'\rangle$ differs from the dimension d on the right-hand side of the normalization condition (5.31). However, there is no contradiction, because critical scaling corresponds to large $t, \mathbf{x}$ with the possible subsequent limit $\lambda t/\mathbf{x}^2 \to 0$ inside this region, where the contribution of the δ function from (5.31) vanishes. We also note that in converting the critical dimensions of coordinate space to those in momentum–frequency space we must take into account the fact that the symbol $d\mathbf{x}$ in the Fourier integrals has critical dimension $-d$, while the symbol dt has dimension $\Delta_t = -\Delta_\omega$.

Therefore, the only new anomalous dimension in dynamics is γ_λ^*. One-loop graphs do not contribute to Z_λ, and so the ε expansion of γ_λ^* begins at ε^2. At present we know the order-ε^2 and ε^3 contributions. The results of calculating γ_λ^* are usually written as

$$\gamma_\lambda^* = -R \cdot \eta, \qquad z = 2 + R\eta, \qquad (5.112)$$

where $\eta = 2\gamma_\psi^*$ is the known static exponent (Sec. 1.37). The answer is given for the coefficient R in (5.112), which in the two lowest orders (1 and ε) turns out to be independent of the number of field components n. The result for $d = 4 - 2\varepsilon$ has the form

$$R = 0.7621(1 - 2\varepsilon \cdot 0.1885 + \ldots). \tag{5.113}$$

The first coefficient $6\ln(4/3) - 1 \simeq 0.7621$ was calculated in [166] and the second in [165]. Earlier in [167] the second term was calculated by a rather complicated method different from that we have discussed here (which had not been developed at that time). The calculation of [167] is basically correct, but near the end there is an error in the calculations, so that instead of 0.1885 the value 1.687 was obtained [167]. This is the result that was reproduced in the review [153], where it was also erroneously stated that it agrees with the known [166] lowest (zeroth) order of the $1/n$ expansion for R in arbitrary dimension d [for η see (4.255a)]:

$$R = \frac{4}{4 - d} \left\{ \frac{d \cdot B(d/2 - 1, d/2 - 1)}{8 \int\limits_0^{1/2} dx[x(2 - x)]^{d/2 - 2}} - 1 \right\} + O(1/n), \tag{5.114}$$

where $B(a, b) \equiv \|a\|\|b\|/\|a + b\|$ (in the review [153] the result of [166] contains a misprint). Actually, (5.113) agrees with (5.114).

The first term of the $2 + \varepsilon$ expansion has also been calculated for R [168]. The expansion begins with the contribution of order ε [for η see (4.250)]:

$$R = [1 - \ln(4/3)](d - 2) + \ldots. \tag{5.115}$$

This exhausts the presently available information on the dynamical exponent (5.112) of this model.

Let us briefly explain the calculational technique. Graphical representations of the functions Γ are constructed according to the usual Feynman rules (Sec. 5.3). The lines of the graphs for the model (5.99) in momentum–frequency space correspond to the propagators (we omit the factor δ_{ab} carrying the indices):

$$\langle \psi\psi' \rangle = \langle \psi'\psi \rangle^\top = \frac{1}{-i\omega + \varepsilon_k}, \quad \langle \psi\psi \rangle = \frac{D_k}{|-i\omega + \varepsilon_k|^2} \tag{5.116}$$

with $\varepsilon_k = \lambda(k^2 + \tau)$, $D_k = 2\lambda$ in the present case, and the vertices $\psi'\psi\psi\psi$ correspond to factors $-\lambda g_{\mathrm{B}} = -\lambda g\mu^{2\varepsilon}$. The calculations can also be performed in t, k space, where

$$\langle \psi\psi' \rangle = \theta(t - t') \exp[-\varepsilon_k(t - t')],$$

$$\langle \psi\psi \rangle = \frac{D_k}{2\varepsilon_k} \exp[-\varepsilon_k|t - t'|]. \tag{5.117}$$

Then for each graph we consider all possible time versions, i.e., variants of ordering the corresponding time vertices t_i. For a particular version the integration over the times is performed following simple rules, beginning with the smallest time, then the next smallest, and so on. The number of terms is equal to the number of possible time versions taking into account the retardation properties of the lines $\langle \psi \psi' \rangle$. This is the usual calculational method, because it is technically simpler than integration over circulating internal frequencies. It is useful to note that in dynamical models the sum of contributions from graphs with different arrangements of the slashes at the line ends (Sec. 5.2) and different time versions for each is usually much simpler than the individual terms.

On the whole, the calculations in dynamics are more complicated than in statics. The calculation of the $\sim \varepsilon^3$ contribution to γ_λ^* performed in [165] and [167] for model A is the only (as far as we know) three-loop calculation in critical dynamics; all the others do not go beyond two loops (if they do not reduce to the static case). The three-loop calculation [165] was performed using the standard scheme described in this section for the massless model in coordinate space $t, \mathbf{x}$ for the propagators. At two-loop order it is necessary to calculate three graphs, and in three-loop order there are four more. In [165] four renormalization constants in (5.101) (without Z_4) were calculated and (5.104) was verified explicitly in the dynamical model. The results are presented as a table of contributions of individual graphs to each of the renormalization constants (like Table 3.1 in Sec. 3.20), so that they can be used for various generalizations of model A with a more complicated index structure of the fields. We note that the dynamical constants Z already contain contributions like $\ln(4/3)$ at two loops (which is unusual in statics), while in three loops various dilogarithms appear (the Spence integral).

Model B. This model differs from model A by the replacement $\lambda_0 \to -\lambda_0 \partial^2$ in (5.97) and, accordingly, $\lambda \to -\lambda \partial^2$ in (5.99). Therefore, now $\partial_t \sim \partial^4$, i.e., $d_\omega = 4$ and $d_\lambda^p = -4$, $d_\lambda^\omega = 1$, $d_\lambda = 0$, and all the other dimensions in Table 5.3 are unchanged. The model is logarithmic in dimension $d^* = 4$, and instead of (5.100) we obtain $d_\Gamma^* = 8 - n_\psi - 3n_{\psi'}$. However, the real exponent δ_Γ will now be smaller than the formal one, since for the interaction $\psi' \partial^2 \psi^3 \sim \partial^2 \psi' \cdot \psi^3$ a "factor" ∂^2 will be present on each external line ψ' of the function Γ, and so $\delta_\Gamma = d_\Gamma^* - 2n_{\psi'}$. An analysis of the counterterms as in model A shows that for model B they can be reproduced by constants Z that are the same as in (5.101), with all the expressions (5.102)–(5.104) left unchanged. Here $Z_1 = Z_2 = 1$ owing to the absence of the corresponding divergences because the real exponent is decreased. It follows from these expressions and (5.103) that $Z_{\psi'} = Z_\psi^{-1}$ and $Z_\lambda = Z_\psi^2$, i.e., in this case all the dynamical constants Z are expressed in terms of the static ones. The final expressions for the critical dimensions are obtained from the general relations (5.109) and (5.110) using the new (see above) values of d_λ^p and d_λ^ω:

$$z \equiv \Delta_\omega = 4 - 2\gamma_\psi^* = 4 - \eta, \qquad \Delta_{\psi'} = d_{\psi'}^p - \gamma_\psi^*. \qquad (5.118)$$

We note that in this case the critical dimension $\Delta_\psi + \Delta_{\psi'}$ of the response function $\langle \psi\psi' \rangle$ is canonical $(= d)$, in contrast to the case in model A.

5.17 Model C (slow heat conduction): statics

Let us begin by analyzing the statics of model C, taking the massless version (5.79) plus the external field contributions $\varphi h = \psi h_\psi + m h_m$ as the unrenormalized action and the corresponding basic static action for the fields $\varphi = \psi, m$ in dimension $d = 4 - 2\varepsilon$:

$$S^{\text{stat}}(\varphi) = -(\partial\psi)^2/2 - g_{10}\psi^4/24 + g_{20}m\psi^2/2 - m^2/2 + \varphi h_0, \qquad (5.119)$$

$$S_{\text{B}}^{\text{stat}}(\varphi) = -(\partial\psi)^2/2 - g_{1\text{B}}\psi^4/24 + g_{2\text{B}}m\psi^2/2 - m^2/2 + \varphi h, \qquad (5.120)$$

where g_{10} and g_{20} are the bare charges and $g_{1\text{B}} = g_1\mu^{2\varepsilon}$ and $g_{2\text{B}} = g_2\mu^\varepsilon$ are the corresponding basic charges. The standard analysis of the divergences (taking into account the fact that they are independent of h) shows that they can be completely eliminated by introducing four constants $\widetilde{Z}$ into (5.120):

$$S_{\text{R}}^{\text{stat}}(\varphi) = -\widetilde{Z}_1(\partial\psi)^2/2 - \widetilde{Z}_2 g_{1\text{B}}\psi^4/24 + \widetilde{Z}_3 g_{2\text{B}}m\psi^2/2 - \widetilde{Z}_4 m^2/2 + \varphi h.$$
$$(5.121)$$

This corresponds to multiplicative renormalization $S_{\text{R}}^{\text{stat}}(\varphi, e, \mu) = S^{\text{stat}}(\widetilde{Z}_\varphi\varphi, e_0)$ of the model (5.119) with $\widetilde{Z}_\varphi\varphi \equiv \widetilde{Z}_\psi\psi$, $\widetilde{Z}_m m$ for the fields and $g_{i0} = g_{i\text{B}}\widetilde{Z}_{g_i}$ $(i = 1, 2)$, $h_0 = h\widetilde{Z}_h$, $\widetilde{Z}_h = \widetilde{Z}_\varphi^{-1}$ $(\varphi = \psi, m)$ for the parameters $e = g_1, g_2, h_\psi, h_m$.

If we had added the "massive contribution" $-\tau\psi^2/2$ to (5.120), this would have led to the appearance of three more constants $\widetilde{Z}$ in (5.121): one from the renormalization of this contribution, one from the additional vacuum counterterm $\sim \tau^2\mu^{-2\varepsilon}$, and one from the counterterm $\sim m\tau\mu^{-\varepsilon}$ corresponding to shifting the external field h_m. However, these three new constants $\widetilde{Z}$ are actually not independent and must be expressed in terms of the four constants in (5.121), because the contribution $-\tau\psi^2/2$ in the basic action can be eliminated by a shift $m \to m+$ const, giving the massless functional (5.120) with the replacement $h_m \to h_m - \tau/g_{2\text{B}}$ (plus an additive constant). The shift of m affects only the vacuum loops and the expectation value $\langle m \rangle$. It then follows that all the other connected Green functions for the massive generalization of the model (5.120) will actually depend not on the parameters τ and h_m separately, but only on the particular combination $h_m - \tau/g_{2\text{B}}$, and only this combination will have a definite critical dimension (the shadow dimension of m). In this respect the situation is analogous to that in the φ^3 model studied in detail in Ch. 4. For these reasons it is not advisable to introduce a massive contribution into (5.120), since it leads only to unnecessary complications without giving any new information.

The four constants $\widetilde{Z}$ in (5.121) can be expressed in terms of the four known renormalization constants of the simple φ^4 model from (3.105). This statement is proved using the scheme of Sec. 4.46, and so we shall discuss it only briefly. Let us consider the generating functional (2.30) with the source $\varphi A = \psi A_\psi$ of only the field ψ for the model (5.120) and perform the Gaussian integration over the field m in it. It is easy to check that in the exponent of the remaining integral over ψ we obtain the basic action of the simple φ^4 model with parameters

$$\tau = -g_{2B}h_m, \quad g = g_1 - 3g_2^2, \quad g_B = g\mu^{2\epsilon} \tag{5.122}$$

and the additional additive constant $\ln[C/C_\tau] + \int dx h_m^2/2$, where C is the normalization factor in (2.30) associated with the field ψ for the original model (5.120), and C_τ is the analogous factor for the φ^4 model with the parameters (5.122). These factors are determined by the general rule (2.16) with $K = -\partial^2$ for C and $K = -\partial^2 + \tau$ for C_τ, and the quantity $-\ln[C/C_\tau]$ coincides with the right-hand side of (3.233).

The UV divergences of the resulting φ^4 model with the parameters (5.122) are eliminated by the known renormalization constants in (3.105) along with the subtraction of the UV-divergent part of the contribution $\ln[C/C_\tau]$, which is equivalent to adding (3.236) to the action (3.105). As is well known (Sec. 4.4), this addition leads to the replacement $Z_0 \to \overline{Z}_0 = Z_0 - ng/32\pi^2\varepsilon$ in the vacuum counterterm (3.106).

On the other hand, the same renormalized action for the field ψ should be obtained if we perform the integration over m in the functional (2.30) with the action (5.121) in the exponent (here an additive constant $\ln[C/C_\tau]$ also appears). Equating the expressions for $S_R^{stat}(\psi)$ obtained by these two methods (the contribution $\ln[C/C_\tau]$ cancels), we obtain the equation

$$-Z_1(\partial\psi)^2/2 - Z_2\tau\psi^2/2 - Z_3 g_B\psi^4/24 + \overline{Z}_0\tau^2/2g_B + h_m^2/2 =$$

$$= -\widetilde{Z}_1(\partial\psi)^2/2 - \widetilde{Z}_2 g_{1B}\psi^4/24 + \widetilde{Z}_4^{-1}(h_m + \widetilde{Z}_3 g_{2B}\psi^2/2)^2/2,$$

from which using (5.122) we find that

$$Z_1 = \widetilde{Z}_1, \quad gZ_3 = g_1\widetilde{Z}_2 - 3g_2^2\widetilde{Z}_4^{-1}\widetilde{Z}_3^2,$$

$$Z_2 = \widetilde{Z}_4^{-1}\widetilde{Z}_3, \quad \overline{Z}_0 g_2^2 = g(\widetilde{Z}_4^{-1} - 1). \tag{5.123}$$

The constants (5.121) depending on the two charges $g_{1,2}$ are thereby expressed in terms of the known constants $Z_{1,2,3}$ and $\overline{Z}_0$ of the simple φ^4 model, depending only on the single charge $g = g_1 - 3g_2^2$.

Using expressions like (1.120), it is easy to show that the relations (5.123) between the renormalization constants lead to the following relations between the corresponding RG functions (1.113):

$$\widetilde{\gamma}_\psi = \gamma_\psi, \quad \widetilde{\gamma}_m = -g_2^2 \gamma_0/2,$$

$$\widetilde{\beta}_{g_1} = g_1(-2\varepsilon - \widetilde{\gamma}_{g_1}) = g_1(-2\varepsilon - \gamma_g) + 3g_2^2(\gamma_g - 2\gamma_\tau + g_2^2\gamma_0),$$

$$\widetilde{\beta}_{g_2} = g_2(-\varepsilon - \widetilde{\gamma}_{g_2}) = g_2(-\varepsilon - \gamma_\tau + g_2^2\gamma_0/2),$$

in which $\gamma_{\psi,\tau,g}$ are the known (Sec. 4.2) RG functions of the simple φ^4 model ($\gamma_\psi \equiv \gamma_\varphi$ in the notation of Sec. 4.2), and γ_0 is its vacuum RG function (4.19). In the model (5.121) it is convenient to make a change of variable, going from the charges $g_{1,2}$ to the pair g, g_2 with the new RG function $\widetilde{\beta}_g \equiv \mathcal{D}_\mu g = \widetilde{\beta}_{g_1} - 6g_2\widetilde{\beta}_{g_2}$ for the charge g from (5.122). In these variables the RG functions of the model (5.121) take the form

$$\widetilde{\beta}_g(g, g_2) = \beta_g(g), \quad \widetilde{\beta}_{g_2}(g, g_2) = g_2[-\varepsilon - \gamma_\tau(g) + g_2^2\gamma_0(g)/2],$$

$$\widetilde{\gamma}_\psi(g, g_2) = \gamma_\psi(g), \quad \widetilde{\gamma}_m(g, g_2) = -g_2^2\gamma_0(g)/2,$$

$$(5.124)$$

i.e., the β function of the charge g in the model (5.121) coincides with the usual β function of the simple φ^4 model.

In the ε expansion the β functions (5.124) have four fixed points on the g, g_2^2 plane, of which the following two may be IR-attractive:

$$\text{No. 1:} \ \{g_* \sim \varepsilon, \ g_{2*}^2 = 0\}, \quad \text{No. 2:} \ \{g_* \sim \varepsilon, \ g_{2*}^2 = 2(\varepsilon + \gamma_\tau^*)/\gamma_0^*\}, \quad (5.125)$$

where $g_* \sim \varepsilon$ is the coordinate of the IR-attractive point of the simple φ^4 model and $\gamma_a^* \equiv \gamma_a(g_*)$. The points (5.125) correspond to two possible critical regimes: in the first the interaction of the fields ψ and m is switched off, and in the second it is important. The choice between them is determined by the signs of the correction exponents ω_a (Sec. 1.42). Since the matrix $\partial\beta_i/\partial g_k$ (1.189) for the pair of charges g, g_2 with the β functions (5.124) is triangular, its eigenvalues ω_a are simply the diagonal elements $\omega_g = d\beta_g(g)/dg$ for $g = g_*$ and $\omega_{g_2} = -\varepsilon - \gamma_\tau^* + 3g_{2*}^2\gamma_0^*/2$. The first of these is positive (Sec. 1.33), and the second takes the value $-(\varepsilon + \gamma_\tau^*)$ at point No. 1 and $2(\varepsilon + \gamma_\tau^*)$ at point No. 2, so that one of these two points must be IR-attractive. The choice is determined by the sign of the quantity $\varepsilon + \gamma_\tau^*$, which is expressed by (1.19) and (1.40) in terms of the traditional critical exponents of the φ^4 model:

$$\varepsilon + \gamma_\tau^* = \alpha/2\nu, \quad (5.126)$$

where $\alpha = 2 - d\nu$ is the specific heat exponent and $1/\nu \equiv \Delta_\tau = 2 + \gamma_\tau^*$ is the critical dimension of the temperature τ. Regime No. 1 is realized for $\alpha < 0$ and regime No. 2 for $\alpha > 0$.

It follows from (5.124) that the anomalous dimension of the field ψ in the two regimes is the same as in the φ^4 model ($\widetilde{\gamma}_\psi^* = \gamma_\psi^* = \eta/2$), and for the field m we have $\widetilde{\gamma}_m^* = 0$ at point No. 1 and $\widetilde{\gamma}_m^* = -(\varepsilon + \gamma_\tau^*) = -\alpha/2\nu$ at point No. 2. Therefore, the critical dimension $\widetilde{\Delta}_m = d_m + \widetilde{\gamma}_m^*$ of the field m in regime No. 1 coincides with its canonical dimension $d_m = d/2 = 2 - \varepsilon$. In regime No. 2 we have $\widetilde{\Delta}_m = 2 - \varepsilon - (\varepsilon + \gamma_\tau^*) = d - \Delta_\tau$, i.e., in this case the quantity $\widetilde{\Delta}_m$ is the shadow value of Δ_τ, and so it coincides with the dimension of the composite operator φ^2 in the φ^4 model (Sec. 4.9).

The choice between the two regimes is determined by the sign of the specific heat exponent α in the φ^4 model. In the $4 - \varepsilon$ expansion it is known through contributions of order ε^5, and the explicit expression for the first three terms of the series is given in Sec. 1.37. We see from this expression that α changes sign at some boundary value $n_c = n_c(d)$: for $n > n_c$ we have $\alpha < 0$, i.e., regime No. 1 is realized, while for $n < n_c$ we have regime No. 2. For $d = 4 - 2\varepsilon$ in the first two orders we obtain $n_c(d) = 4 - 8\varepsilon + \ldots$ from the expression for α. According to some estimates, $n_c(3) \simeq 1.8$ [153], and if this is true then for the real dimension $d = 3$ the nontrivial critical regime is realized only for a one-component field ($n = 1$), and for any $n \geq 2$ the interaction between ψ and m is unimportant in the critical regime.

5.18   Model C: dynamics

We write the unrenormalized action of the dynamical model C (Sec. 5.12) as

$$S(\phi) = \lambda_{0\psi}\psi'\psi' + \psi'[-\partial_t\psi + \lambda_{0\psi}H_\psi] - \lambda_{0m}m'\partial^2 m' + m'[-\partial_t m - \lambda_{0m}\partial^2 H_m], \tag{5.127}$$

where $\lambda_{0\varphi}$ with $\varphi = \psi, m$ are the bare Onsager coefficients and

$$H_\varphi = \delta S^{\text{stat}}(\varphi)/\delta\varphi \tag{5.128}$$

is the compact notation for the variational derivatives of the static action, in this case the functional (5.119). The basic action $S_{\text{B}}(\varphi)$ is obtained from (5.127) by replacing the bare parameters $\lambda_{0\varphi}$ by their renormalized analogs λ_φ, and $S^{\text{stat}}(\varphi)$ in (5.128) by the basic functional (5.120). In this model we have $\partial_t \sim \lambda\partial^2$ for both fields, and so

$$d_\omega = 2, \quad d^p[\lambda_\varphi] = -2, \quad d^\omega[\lambda_\varphi] = 1, \quad d[\lambda_\varphi] = 0, \quad \varphi \equiv \psi, m. \tag{5.129}$$

In what follows, instead of $\lambda_{\psi,m}$ we shall use the notation

$$\lambda_\psi \equiv \lambda, \quad \lambda_m \equiv u\lambda \tag{5.130}$$

and similarly for the bare parameters. The fully dimensionless parameter $u \equiv \lambda_m/\lambda_\psi$ enters into the dynamical renormalization constants, and so it will play the role of the third charge (in addition to the static charges $g_{1,2}$) of the dynamical model.

For $d = 4$ the total dimensions (5.87) of the fields ψ, ψ', m, and m' are 1, 3, 2, and 2, respectively, and so for the formal exponent of divergence (5.89) we obtain $d_\Gamma^* = 6 - n_\psi - 3n_{\psi'} - 2n_m - 2n_{m'}$. However, the real exponent δ_Γ will be smaller than the formal one, because it follows from the form of the vertex $m'\partial^2\psi^2 \sim \partial^2 m' \cdot \psi^2$ in (5.127) that the operator ∂^2 is singled out on each external line of the field m' in the graphs of the 1-irreducible functions Γ and in the corresponding counterterms, so that $\delta_\Gamma = d_\Gamma^* - 2n_{m'}$. It then follows that the graphs of the 1-irreducible function $\langle m'm'\rangle$ do not generate counterterms ($\delta_\Gamma = -2$), i.e., the contribution $\sim m'\partial^2 m'$ in the basic action is not renormalized. For the same reason the graphs of $\langle m'm\rangle$ can generate only counterterms of the type $m'\partial^2 m$, and cannot generate counterterms $m'\partial_t m$ of the same dimension. Owing to the absence of intermode coupling, contributions with external field h also are not renormalized (Sec. 5.15), and the parameters h cannot enter into the counterterms, which must therefore possess the $\psi, \psi' \to -\psi, -\psi'$ symmetry of the problem with $h = 0$. Taking into account these arguments and the value of δ_Γ, it is easily verified that all possible counterterms of the dynamical model have the form of the terms in the basic action, and so they are reproduced by the introduction of the needed renormalization constants. In detailed notation using (5.120) and (5.130) we have

$$S_R(\phi) = Z_1\lambda\psi'\psi' + \psi'[-Z_2\partial_t\psi + \lambda(Z_3\partial^2\psi - Z_4 g_{1B}\psi^3/6 + Z_5 g_{2B}m\psi +$$

$$+ h_\psi)] - \lambda u m'\partial^2 m' + m'[-\partial_t m - \lambda u\partial^2(-Z_6 m + Z_7 g_{2B}\psi^2/2 + h_m)].$$
$$(5.131)$$

If the renormalization were multiplicative, we would have

$$Z_1 = Z_\lambda Z_{\psi'}^2, \quad Z_2 = Z_{\psi'}Z_\psi, \quad Z_3 = Z_{\psi'}Z_\lambda Z_\psi, \quad Z_4 = Z_{\psi'}Z_\lambda Z_\psi^3 Z_{g_1},$$

$$Z_5 = Z_{\psi'}Z_\lambda Z_m Z_\psi Z_{g_2}, \quad 1 = Z_\lambda Z_u Z_{m'}^2 = Z_{m'}Z_m,$$

$$Z_6 = Z_{m'}Z_\lambda Z_u Z_m, \quad Z_7 = Z_{m'}Z_\lambda Z_u Z_\psi^2 Z_{g_2}.$$
$$(5.132)$$

It is easily checked that these relations can be satisfied when and only when there is a single relation between the constants (5.131):

$$Z_1 Z_7 = Z_2 Z_5, \qquad (5.133)$$

and so the multiplicative nature of the renormalization (5.131) is not obvious.

However, the renormalization is in fact multiplicative, as can be checked using (5.93)–(5.95). Equating (5.91) and (5.131), we see that in this case Z and α' in (5.91) are diagonal matrices with the elements $Z_\psi = Z_2$, $Z_m = 1$, $\alpha'_\psi = \lambda Z_1$, $\alpha'_m = -\lambda u\partial^2$, and for the "vector" A_a with $a = \psi, m$ from (5.131) we find $A_\psi = \lambda(Z_3\partial^2\psi - Z_4 g_{1B}\psi^3/6 + Z_5 g_{2B}m\psi + h_\psi)$ and $A_m = -\lambda u\partial^2(-Z_6 m + Z_7 g_{2B}\psi^2/2 + h_m)$. Substituting these quantities and S_R^{stat} from (5.121) into the definition (5.94) and then into (5.95), we easily verify that the latter equations can be satisfied only for $B_\psi = B_m = 0$. From this [taking into account the explicit form of $A_{\psi,m}$ and the definition (5.94)] follows the multiplicative renormalizability and, consequently, the validity of (5.132) and (5.133), the equations (5.90) for $F = \psi, m, g_1, g_2, h_\psi, h_m$, and also (5.96), which in this case take the form

$$Z_{\psi'} Z_\lambda = Z_\psi, \quad Z_{m'} Z_\lambda Z_u = Z_m. \tag{5.134}$$

The first of these is equivalent to the relation $Z_1 = Z_2$ for the constants in (5.131), and the second follows automatically from the (already proven) relations (5.132).

Of the four renormalization constants of the quantities ψ', m', λ, and u absent in statics, the only essentially new one is Z_λ. All the others are expressed in terms of Z_λ and the static constants by relations following from (5.132) and (5.134):

$$Z_{\psi'} = Z_\psi Z_\lambda^{-1}, \quad Z_{m'} = Z_m^{-1}, \quad Z_u = Z_m^2 Z_\lambda^{-1}. \tag{5.135}$$

From this for the anomalous dimensions (1.113) we obtain

$$\gamma_{\psi'} = \gamma_\psi - \gamma_\lambda, \quad \gamma_{m'} = -\gamma_m, \quad \gamma_u = 2\gamma_m - \gamma_\lambda. \tag{5.136}$$

The operator $\mathcal{D}_{\mathrm{RG}}$ in the standard RG equation (5.106) now has the form

$$\mathcal{D}_{\mathrm{RG}} = \mathcal{D}_\mu + \sum_g \beta_g \partial_g - \sum_a \gamma_a \mathcal{D}_a \tag{5.137}$$

with summation over $a = u, \lambda, h$ and with the static β functions for the charges $g_{1,2}$ or g, g_2 in the notation of (5.124). Calculations (the two-loop one was performed in [169]) show that the constant Z_λ depends not only on the static charges, but also on the parameter u introduced in (5.130). In the one-loop approximation

$$Z_\lambda = 1 + g_2^2/16\pi^2 \epsilon(1 + u) + \dots. \tag{5.138}$$

Therefore, the parameter u in dynamics plays the role of a third charge, and the corresponding contribution in (5.137) determines its β function:

$$-\gamma_u \mathcal{D}_u \equiv \beta_u \partial_u, \quad \beta_u = -u\gamma_u = u(\gamma_\lambda - 2\gamma_m). \tag{5.139}$$

In the one-loop approximation, from (5.138), (5.124) (in Sec. 5.17 the static RG functions were denoted as $\widetilde{\beta}$ and $\widetilde{\gamma}$), and the explicit expression (4.21) for γ_0 we obtain

$$\beta_u(g, g_2, u) = \frac{ug_2^2}{16\pi^2} \left[n - \frac{2}{1 + u} \right]. \tag{5.140}$$

The coordinates of the fixed points of the three-charge (g, g_2, u) operator (5.137) for the static charges g, g_2 are known from (5.125). When regime No. 1 is realized [$n > n_c(d)$, $\alpha < 0$, $g_{2*} = 0$], the interaction of the fields ψ and m in the critical regime is switched off, and so the dynamics of the field ψ in model C will be the same as in model A, and the dynamics of m will be free.

If regime No. 2 is realized $[n < n_c(d),\ \alpha > 0,\ g_{2*} \neq 0]$, the β function (5.140) is nontrivial and has three possible fixed points:

$$u_* = 0 \quad (A), \quad u_* = \infty \quad (B), \quad u_* = 2/n - 1 + O(\varepsilon) \quad (C) \qquad (5.141)$$

(we shall refer to these as regimes 2A, 2B, and 2C). The matrix ω (1.189) in the variables g, g_2, and u is triangular, and so the IR-stability of the points (5.141) is determined by the sign of the derivative $\partial\beta_u/\partial u \equiv \omega_u$ [in order to find and study the stability of the point $u_* = \infty$, it is necessary to make the change of variable $u \to u' \equiv 1/u$ in (5.139)]. It follows from (5.140) that point A will be IR-attractive for $n > 2$, point C for $n < 2$, and point B (in the variables g, g_2, and $u' = 1/u$) will be unstable for any $n > 0$. From the estimate of $n_c(d)$ determining the boundary of the region $\alpha > 0$ (Sec. 5.17), it appears equally likely that in the real dimension $d = 3$ only the value $n = 1$ will ensure the static regime No. 2 with $g_{2*} \neq 0$. For $n = 1$ in dynamics the point C with $u_* \neq 0$ will be IR-stable. The value of γ_λ^* at this point is expressed in terms of the static exponents owing to the zero value of the β functions (5.124) and (5.139) and the relation (5.124) for $\gamma_m \equiv \tilde{\gamma}_m$:

$$\gamma_\lambda^* \Big|_C = 2\gamma_m^* = -g_{2*}^2 \gamma_0^* = -2(\varepsilon + \gamma_\tau^*) = -\alpha/\nu. \qquad (5.142)$$

From this, using the general rule (5.109) and taking into account (5.129) we find the frequency critical dimension (the exponent z):

$$z \equiv \Delta_\omega = 2 - \gamma_\lambda^* = 2 + \alpha/\nu. \qquad (5.143)$$

Therefore, in the nontrivial critical regime 2C with $g_{2*} \neq 0$, $u_* \neq 0$ all the dynamical exponents are expressed by (5.136) and (5.142) in terms of the known static exponents of the φ^4 model.

5.19 Model D

Model D differs from model C by the replacement $\lambda_\psi \to -\lambda_\psi \partial^2$ in (5.127) and in the corresponding basic action. Then the equation for ψ involves the combination $\partial_t + \lambda_\psi \partial^4$ and in the equation for m we obtain $\partial_t - \lambda_m \partial^2$, so that the choice of the total frequency dimension d_ω is not unique. In principle, we can consider both versions: $\omega \sim p^4$ and $\omega \sim p^2$ (up to corrections of order ε and higher in the exponents). For each of these two versions certain terms of the basic action will be IR-irrelevant. Only the case $\omega \sim p^4$ (i.e., $d_\omega = 4$) has been discussed in the literature; it corresponds to nontrivial dynamics for the field of the order parameter ψ. In this case the contribution $\partial_t \sim p^4$ in the equation for m turns out to be unimportant compared to the contribution $\lambda_m \partial^2 \sim p^2$ and must be dropped when analyzing the leading terms of the IR asymptote of interest $p \to 0$, $\omega \sim p^4 \to 0$. In the end we obtain a dynamical model with the basic action

$$S_{\mathrm{B}}(\phi) = -\lambda_\psi \psi' \partial^2 \psi' + \psi'[-\partial_t \psi - \lambda_\psi \partial^2 H_\psi^{\mathrm{B}}] - \lambda_m m' \partial^2 m' + m'[-\lambda_m \partial^2 H_m^{\mathrm{B}}], \tag{5.144}$$

where H_φ^{B} are the variational derivatives (5.128) of the basic action (5.120). Instead of (5.129) we now have $d^p[\lambda_\psi] = -4$, $d^p[\lambda_m] = -2$, and $d^\omega[\lambda_\psi] = d^\omega[\lambda_m] = 1$, from which for the total dimensions (5.87) with $d_\omega = 4$ we obtain

$$d[\lambda_\psi] = 0, \quad d[\lambda_m] = 2. \tag{5.145}$$

The second equation in (5.145) implies that the action (5.144) is not "correctly constructed" (Sec. 5.15), because its free part contains the parameter λ_m, which remains finite at the IR asymptote in question and has nonzero total dimension $d[\lambda_m] = 2$. Because of this we can say nothing about the IR-relevance of the various interaction terms in (5.144) simply on the basis of the dimensions of their coefficients (Sec. 1.16).

The parameter λ_m in the free part of (5.144) can be eliminated by a field dilatation $m \to \lambda_m^{-1/2} m$, $m' \to \lambda_m^{-1/2} m'$. Then the interaction (5.144) takes the form of the sum of three terms $u_1 V_1 + u_2 V_2 + u_3 V_3$ with the monomials $V_1 = \psi' \partial^2 \psi^3 / 6$, $V_2 = -\psi' \partial^2 (m\psi)$, $V_3 = -m' \partial^2 \psi^2 / 2$ and the coefficients $u_1 = \lambda_\psi g_{1\mathrm{B}}$, $u_2 = \lambda_\psi \lambda_m^{-1/2} g_{2\mathrm{B}}$, $u_3 = \lambda_m^{1/2} g_{2\mathrm{B}}$. For $d = 4$ these coefficients have total dimension 0, -1, and 1, respectively, i.e., the first interaction is logarithmic, the second is less IR-relevant, and the third is more IR-relevant. If we proceed formally according to the rules of Sec. 1.16, we have to keep only the most important interaction V_3 and discard the "weaker" ones $V_{1,2}$.

However, in the present case it is impossible to do this if we are interested in the dynamics of the fields ψ, ψ' (which is consistent also with the statement of the problem $\omega \sim p^4$). In fact, if only the interaction V_3 is kept, all the graphs of the Green functions of the fields ψ and ψ' simply vanish, because they cannot be constructed without the vertices $V_{1,2}$, and each vertex V_3 in these graphs is accompanied by at least one vertex V_2. Since the dimensions -1 and 1 of the coefficients u_2 and u_3 at the vertices V_2 and V_3 cancel each other out, it is clear that the most important graphs will be those with identical numbers of vertices V_2 and V_3 for an arbitrary number of vertices V_1. All these graphs are logarithmic in dimension $d = 4$, and all the others are IR-irrelevant compared to them owing to the presence of additional vertices V_2 and the corresponding coefficients u_2.

The important graphs are selected automatically if we perform the Gaussian integration over the fields m, m' in the functional integral (5.26) with weight $\exp S_{\mathrm{B}}(\phi)$ and $S_{\mathrm{B}}(\phi)$ from (5.144). This gives the addition

$$-2\lambda_\psi^2 \lambda_m^{-1} g_{2\mathrm{B}}^2 (\psi \partial^2 \psi') \partial^{-2} (\psi \partial^2 \psi') - \lambda_\psi g_{2\mathrm{B}}^2 (\psi \partial^2 \psi')(\psi^2 + 2h_m/g_{2\mathrm{B}}) \tag{5.146}$$

to the action functional of the fields ψ, ψ' of model B. The nonlocal first term in (5.146) is generated by the two V_2 vertices and is an IR-irrelevant correction that must be discarded. The local term proportional to $\psi^3\partial^2\psi'$ is generated by the interference of V_2 and V_3. It has the same form as the interaction V_1, and so it generates only a change of coefficient of the latter, while the contribution involving h_m in (5.146) reproduces the mass insertion $\sim \tau\psi^2$ in model B. It then follows that the dynamics of the fields ψ, ψ' in model D turn out to be exactly the same as in model B (Sec. 5.16), i.e., in this case the interaction of the fields ψ and m is irrelevant in the critical regime.

5.20 Models F and E

We shall take the unrenormalized static action of the model F to be the massless version of (5.81) with the additional contribution of the external field h_m:

$$S^{\text{stat}}(\varphi) = -\partial\overset{+}{\psi}\cdot\partial\psi - g_{10}(\overset{+}{\psi}\psi)^2/6 + g_{20}m\overset{+}{\psi}\psi - m^2/2 + mh_{0m} \qquad (5.147)$$

with real one-component field m, complex N-component field ψ, $\overset{+}{\psi}$, and U_N-symmetric interaction. The model (5.147) is equivalent to the real O_n-symmetric model (5.119) with $n = 2N$ and with the same normalization for the charges $g_{1,2}$ [this is why in (5.147) we have the coefficient $1/6$ instead of the natural symmetry coefficient $1/4$]. The real problem for superfluid He_4 corresponds to $N = 1$, i.e., $n = 2$. Owing to the presence of intermode coupling, relations like (5.96) are not satisfied for models F and E, and so we have not inserted the entire set of external fields h_φ into (5.147).

We write the unrenormalized dynamical action of model F as

$$S(\phi) = 2\lambda_{0\psi}\overset{+}{\psi'}\psi' + \overset{+}{\psi'}\left[-\partial_t\psi + \lambda_{0\psi}(1+ib_0)\overset{+}{H}_\psi + i\lambda_{0\psi}g_{30}\psi H_m\right] +$$

$$+\psi'[-\partial_t\overset{+}{\psi} + \lambda_{0\psi}(1-ib_0)H_\psi - i\lambda_{0\psi}g_{30}\overset{+}{\psi}H_m] - \lambda_{0m}m'\partial^2 m' + \qquad (5.148)$$

$$+m'[-\partial_t m - \lambda_{0m}\partial^2 H_m + i\lambda_{0\psi}g_{30}(\overset{+}{\psi}\overset{+}{H}_\psi - \psi H_\psi)],$$

where $\lambda_{0\varphi}$ are positive parameters with dimensions (5.129), b_0 and g_{30} are real parameters corresponding to the intermode coupling, H_φ with $\varphi = \psi, m$ are the derivatives (5.128) of the action (5.147), and $\overset{+}{H}_\psi$ is the analogous derivative with respect to $\overset{+}{\psi}$. Henceforth, we shall use the notation (5.130) for the $\lambda_{0\varphi}$ and their renormalized analogs. Then the basic action corresponding to (5.148) written out in detail has the form

$$S_B(\phi) = 2\lambda\overset{+}{\psi'}\psi' + \overset{+}{\psi'}\left[-\partial_t\psi + \lambda(1+ib)(\partial^2\psi - g_{1B}\psi(\overset{+}{\psi}\psi)/3 +\right.$$

$$+g_{2B}m\psi) + i\lambda g_{3B}\psi(-m + g_{2B}\overset{+}{\psi}\psi + h_m)] + \text{H.c.} - \lambda um'\partial^2 m' + \qquad (5.149)$$

$$+m'[-\partial_t m - \lambda u\partial^2(-m + g_{2B}\overset{+}{\psi}\psi + h_m) + i\lambda g_{3B}\partial(\overset{+}{\psi}\partial\psi - \psi\partial\overset{+}{\psi})],$$

where $g_{1\mathrm{B}} = g_1\mu^{2\varepsilon}$, $g_{2\mathrm{B}} = g_2\mu^{\varepsilon}$, $g_{3\mathrm{B}} = g_3\mu^{\varepsilon}$ are the basic charges ($d = 4 - 2\varepsilon$), $g_{1,2,3}$ are the renormalized charges, and H.c. denotes the addition of the complex conjugate of the preceding expression.

The total dimensions of the fields ψ, $\overset{+}{\psi}$, ψ', $\overset{+}{\psi}'$, m, m', and the parameter h_m for $d = 4$ are 1, 1, 3, 3, 2, 2, and 2, respectively, and so for the formal exponent (5.89) we obtain $d_\Gamma^* = 6 - n_\psi - n_{\overset{+}{\psi}} - 3n_{\psi'} - 3n_{\underset{\psi'}{+}} - 2n_m - 2n_{m'}$. However, the real exponent δ_Γ is smaller than the formal one, because from the form of the vertices involving m' in (5.149) it follows that at least one ∂ is singled out on each external line m' (and on the field m' in the corresponding counterterms) in the graphs of the 1-irreducible functions Γ. Therefore, $\delta_\Gamma = d_\Gamma^* - n_{m'}$. Taking this into account, it is easy to verify that all the needed counterterms have the form of the various terms in the action (5.149), where the contribution $m'\partial_t m$ is not renormalized, as in model C, and the contribution $m'\partial^2 m'$ now can be renormalized [in model C two ∂ symbols are singled out on each external line m', while two or one are in (5.149)].

As usual, (5.95) can be used to prove that the renormalization is multiplicative. There are similar terms in the functional (5.149), and so it makes no sense to reproduce the counterterms by introducing independent constants Z in each term of (5.149). However, it follows from the above discussion about the structure of the counterterms that the renormalized action can in any case be represented as

$$S_{\mathrm{R}}(\phi) = 2Z_1\lambda\,\overset{+}{\psi}'\psi'+ \overset{+}{\psi}'[-Z_2\partial_t\psi+A_\psi]+\text{H.c.}-Z_3\lambda um'\partial^2 m'+m'[-\partial_t m+A_m], \tag{5.150}$$

where A_φ are linear combinations

$$A_\psi = \sum_{a=1}^{4} a_a M_a, \qquad A_m = \sum_{a=5}^{8} a_a M_a \tag{5.151}$$

of known monomials $\{M_a\} \equiv \partial^2\psi$, $\psi(\overset{+}{\psi}\,\psi)$, $m\psi$, ψh_m, $\partial^2 m$, $\partial^2(\overset{+}{\psi}\,\psi)$, $\partial^2 h_m$, $i\partial(\overset{+}{\psi}\,\partial\psi - \psi\partial\overset{+}{\psi})$ (numbered in this order). It follows from the formal reality of the action (5.149), which must be preserved also after renormalization, that the constants $Z_{1,3}$ and the coefficients a_a in A_m must be real, and Z_2 and the coefficients a_a in A_ψ may be complex.

Equating (5.150) and (5.91), we can write relations (5.95) for our system of fields $\varphi = \psi, \overset{+}{\psi}, m$ of the following form:

$$\int d\mathbf{x}\,[\delta B_\psi/\delta\psi + \text{H.c.} + \delta B_m/\delta m] = 0, \tag{5.152}$$

$$\int d\mathbf{x}\,[B_\psi H_\psi^{\mathrm{R}} + \text{H.c.} + B_m H_m^{\mathrm{R}}] = 0, \tag{5.153}$$

where H_φ^{R} are the quantities (5.128) with the replacement $S^{\text{stat}}(\varphi) \to S_{\mathrm{R}}^{\text{stat}}(\varphi)$

and

$$B_\psi = Z_2^{-1} A_\psi - \lambda Z_1 |Z_2|^{-2} \overset{+}{H}{}^R_\psi \equiv \sum_{a=1}^4 b_a M_a,$$

$$B_m = A_m + Z_3 \lambda u \partial^2 H^R_m \equiv \sum_{a=5}^8 b_a M_a \tag{5.154}$$

with the same M_a and the same reality properties of the coefficients b_a as for the a_a in (5.151). For the H^R_φ we have

$$H^R_\psi \equiv \delta S^{\text{stat}}_R / \delta\psi = c_1 \partial^2 \overset{+}{\psi} + c_2 (\overset{+}{\psi} \psi) \overset{+}{\psi} + c_3 m \overset{+}{\psi},$$

$$H^R_m \equiv \delta S^{\text{stat}}_R / \delta m = c_4 m + c_3 \overset{+}{\psi}\psi + h_m \tag{5.155}$$

with known [from the form of $S^{\text{stat}}_R(\varphi)$] real coefficients c_a.

Substituting (5.155) and (5.154) with the specific (see above) monomials M_a into (5.153), it is easy to check that it can be satisfied only for $b_5 = b_6 = b_7 = 0$, arbitrary real b_8, and purely imaginary $b_1, ..., b_4$ satisfying the relations

$$c_1 b_4 = i b_8, \quad c_1 b_2 - c_2 b_1 = i c_3 b_8, \quad c_1 b_3 - c_3 b_1 = i c_4 b_8. \tag{5.156}$$

These relations leave two of the five nonzero coefficients b arbitrary, for example, real b_8 and purely imaginary b_1. Introducing the parameterization $b_1 = i\alpha c_1$, $b_8 = \beta c_1$ with arbitrary real α and β, for the solution with $\alpha \neq 0$, $\beta = 0$ from (5.154)–(5.156) we obtain

$$B_\psi = i\alpha[c_1 \partial^2 \psi + c_2 \psi(\overset{+}{\psi} \psi) + c_3 m\psi] = i\alpha \overset{+}{H}{}^R_\psi, \quad B_m = 0, \tag{5.157}$$

and for the solution with $\alpha = 0$, $\beta \neq 0$ we similarly have

$$B_\psi = i\beta[c_3 \psi(\overset{+}{\psi} \psi) + c_4 m\psi + \psi h_m] = i\beta \psi H^R_m,$$

$$B_m = i\beta c_1 \partial(\overset{+}{\psi} \partial\psi - \psi\partial\overset{+}{\psi}) = i\beta[\psi\overset{+}{H}{}^R_\psi - \overset{+}{\psi} H^R_\psi]. \tag{5.158}$$

The general solution for B_φ is the sum of (5.157) and (5.158) [(5.152) is satisfied automatically for it], and using the known B_φ in (5.154) we find the A_φ in (5.150):

$$A_\psi = \lambda Z_1 \overset{+}{Z}{}_2^{-1} \overset{+}{H}{}^R_\psi + Z_2[i\alpha \overset{+}{H}{}^R_\psi + i\beta \psi H^R_m],$$

$$A_m = -Z_3 \lambda u \partial^2 H^R_m + i\beta[\psi\overset{+}{H}{}^R_\psi - \overset{+}{\psi} H^R_\psi]. \tag{5.159}$$

This form of A_φ demonstrates that the renormalization is multiplicative: it is easily checked that the functional (5.150) with A_φ from (5.159) is obtained from the unrenormalized action (5.148) by the standard multiplicative renormalization of all the fields and parameters ($e_0 = e Z_e$ for $e = \lambda, u, b, h_m$ and $g_{i0} = g_{iB} Z_{g_i}$ for the three charges g, where all these Z are real), in agreement with the static case, i.e., with satisfaction of the general relations (5.90) for $F = \psi, m, h_m, g_1, g_2$. Here the constants Z and the parameters α and β in (5.159) are expressed in terms of the multiplicative renormalization constants by the relations $Z_1 = Z_\lambda |Z_{\psi'}|^2$, $Z_2 = \overset{+}{Z}_{\psi'} Z_\psi$, $Z_3 = Z_\lambda Z_u Z_{m'}^2$, $\alpha = \lambda b Z_\lambda Z_b Z_\psi^{-2}$, $\beta = \lambda g_{3B} Z_\lambda Z_{g_3} Z_m^{-1}$ and

$$Z_{m'} Z_m = 1 \tag{5.160}$$

owing to the absence of renormalization of the contribution $m' \partial_t m$ in (5.150). Therefore, the only new constants in dynamics are the renormalization constants of the parameters λ, u, b, g_3, and the field ψ', and only $Z_{\psi'}$ can be complex.

The external field h_m in the static functionals must be assumed to be static (independent of t), and only in this case is the above proof of multiplicative renormalizability valid. However, in dynamical objects like (5.149) h_m can be assumed to be an arbitrary function of $x = t, \mathbf{x}$, and the proof generalizes directly to this case: once we have verified (from power counting and the value of the real exponent δ_Γ) that all the counterterms of the dynamical model (5.149) with arbitrary h_m have the form of the terms (5.149) and therefore do not contain $\partial_t h_m$, the dynamical model with arbitrary h_m must be renormalized by the same counterterms (or constants Z) as for static h_m.

There is another relation between the dynamical constants Z of model F, namely,

$$Z_\lambda Z_{g_3} = Z_m. \tag{5.161}$$

For the proof, we consider the field transformation $\phi \to \phi_\alpha$:

$$\psi_\alpha = e^{i\alpha} \psi, \quad \psi'_\alpha = e^{i\alpha} \psi', \quad m_\alpha = m, \quad m'_\alpha = m', \tag{5.162}$$

where $\alpha = \alpha(t)$ is an arbitrary real function depending only on the time [the transformations of $\overset{+}{\psi}$ and $\overset{+}{\psi}'$ are obtained from (5.162) by complex conjugation]. We see from (5.149) and (5.162) that the transformation $\phi \to \phi_a$ in (5.149) is equivalent to the shift $h_m \to h_m - \partial_t \alpha / \lambda g_{3B}$ of the external field h_m. For the unrenormalized dynamical functional we similarly have

$$S(\phi_\alpha, h_{0m}) = S(\phi, h_{0m} - \partial_t \alpha / \lambda_0 g_{30}) \tag{5.163}$$

(we show only the dependence on the important quantities). Since the multiplicative nature of the renormalization has already been proven, it can be stated that both terms of the expression $h_{0m} - \partial_t \alpha / \lambda_0 g_{30}$ are renormalized by the same factor Z which, using the fact that $Z_{h_m} = Z_m^{-1}$ (these are the static constants), is equivalent to (5.161). We note that the transformation (5.163) in general takes us outside the class of static h_m, but this is not important owing to the possibility of transforming to arbitrary h_m, as discussed above. It would be possible to stay within the class of static h_m if in (5.163) we restricted outselves to functions $\alpha(t) = at$ with arbitrary real coefficient a, which is sufficient for proving (5.161).

Explicit calculations show [158] that the dynamical renormalization constants contain powers of the ratio g_3^2/u, so that the RG functions will also contain "generational contributions" with u in the denominators. The variables used in [158] are more convenient for the RG analysis:

$$f = g_3^2/8\pi^2 u, \qquad w = 1/u \qquad (5.164)$$

and similarly for the bare quantities. It follows from (5.164) that $Z_w = Z_u^{-1}$, $Z_f = Z_{g_3}^2 Z_w$, so that (5.161) in the variables (5.164) takes the form $Z_f = Z_m^2 Z_\lambda^{-2} Z_w$ and we obtain the equation

$$\gamma_f = 2\gamma_m + \gamma_w - 2\gamma_\lambda \qquad (5.165)$$

for the corresponding anomalous dimensions $\gamma_a \equiv \widetilde{\mathcal{D}}_\mu \ln Z_a$. The dynamical constants Z of model F are series in powers of the charges g_1, g_2^2, and f. The coefficients of these series depend on the parameters w and b, and so in the RG equations they also act as charges. The β functions of the charges $g_{1,2}$ are static (Sec. 5.17), and for the three new ones we have

$$\beta_f = f[-2\varepsilon - \gamma_f], \quad \beta_w = -w\gamma_w \,, \quad \beta_b = -b\gamma_b \qquad (5.166)$$

with γ_f from (5.165).

Model E. This model is a special case of model F with $g_2 = b = 0$. It possesses an additional symmetry under the discrete transformation

$$\psi \overset{+}{\leftrightarrow} \psi, \ \psi' \overset{+}{\leftrightarrow} \psi', \ m, m', h_m \leftrightarrow -m, -m', -h_m. \qquad (5.167)$$

The contributions involving g_2 and b in (5.149) are odd under the transformation (5.167) and all the others are even, and so when there are no odd contributions to the basic action they will not be generated by renormalization (symmetry conservation). All the general relations of model F are satisfied for model E, and, in addition (as is known from statics),

$$Z_m = 1 \quad \Rightarrow \quad \gamma_m = 0. \qquad (5.168)$$

For the dynamical β functions of model E, from (5.165), (5.166), and (5.168) we have

$$\beta_f = f[-2\varepsilon - \gamma_w + 2\gamma_\lambda], \qquad \beta_w = -w\gamma_w. \qquad (5.169)$$

We note that for model E the transformation (5.162) of the fields ψ, ψ' in the basic action, accompanied by the transformation $m_\alpha = m - \partial_t \alpha/\lambda g_{3\mathrm{B}}$, $m'_\alpha = m'$ of the fields m, m', is equivalent to the shift $A_{m'} \to A_{m'} + \partial_t^2 \alpha/\lambda g_{3\mathrm{B}}$ of the source $A_{m'}$ of the field m' in a generating functional of the type (5.26). This allows us to write the corresponding Ward identities (Sec. 2.16), from which, in particular, (5.161) also follows. However, this trick does not generalize to model F, in contrast to the relations (5.163) used above.

Let us now briefly discuss the possible critical regimes, beginning with model E. The two-loop calculation for this model in the MS scheme was performed in [158]. In the one-loop approximation it is found that $(n = 2N)$

$$\beta_f = f[-2\varepsilon + f/(1+w) + fn/4], \quad \beta_w = w[fn/4 - f/(1+w)], \quad (5.170)$$

and the rather awkward explicit expressions for the two-loop β functions (5.169) are given in Eqs. 3.26–3.30 of [158] (in the notation of [158], $\gamma_\lambda \equiv -\eta_\Gamma$ and $\gamma_w \equiv \eta_\lambda - \eta_\Gamma$, where the values of η_Γ and η_λ are given for arbitrary $n \equiv 2N$). The results of the RG analysis [158] reduce to the following. The fixed point with $f_* = 0$ corresponding to model A is IR-unstable, because for it the exponent ω_f associated with the charge f is negative. Variants with $w_* = \infty$, $w_* = 0$, and $w_* = \text{const} \neq 0$ are possible for points with $f_* \neq 0$. The point with $w_* = \infty$ is always unstable, and one of the other two points is IR-stable: for small $n < n_c(d)$ it is the point with $w_* \neq 0$ (the dynamical scaling fixed point), and for $n > n_c(d)$ it is the point with $w_* = 0$ (the weak scaling fixed point). For the boundary value $n_c(d)$ the study [158] found $(d = 4 - 2\varepsilon)$

$$n_c(d) = 4 - 2\varepsilon[19\ln(4/3) - 11/3] + O(\varepsilon^2). \quad (5.171)$$

For the dynamical fixed point with $f_* \neq 0$, $w_* \neq 0$, (5.169) gives $\gamma_w^* = 0$, $\gamma_\lambda^* = \varepsilon$, and so for the exponent $z \equiv \Delta_\omega$ the general equation (5.109) using (5.129) gives the exact expression

$$z = 2 - \gamma_\lambda^* = 2 - \varepsilon = d/2 \quad (f_* \neq 0, \; w_* \neq 0, \; n < n_c(d)). \quad (5.172)$$

For the second regime $w_* = 0$, $\gamma_w^* \neq 0$, and so from (5.169) we have $\gamma_\lambda^* = \varepsilon + \gamma_w^*/2$, and then in this case the exponent z is nontrivial:

$$z = 2 - \gamma_\lambda^* = d/2 - \gamma_w^*/2 \quad (f_* \neq 0, \; w_* = 0, \; n > n_c(d)). \quad (5.173)$$

The case of He$_4$ corresponds to $N = 1$, i.e., $n = 2$. For $n = 2$, $d = 4 - 2\varepsilon$ the exponents ω associated with the charges f and w are [158]

$$\omega_f = 2\varepsilon - 0.230 \cdot (2\varepsilon)^2 + ..., \quad \omega_w = 2\varepsilon/4 - 0.139 \cdot (2\varepsilon)^2 + ... \quad (5.174)$$

for the point (5.172) and

$$\omega_f = 2\varepsilon - 0.126 \cdot (2\varepsilon)^2 + ..., \quad \omega_w = -2\varepsilon/3 + 0.214 \cdot (2\varepsilon)^2 + ... \quad (5.175)$$

for the point (5.173). At small ε the dynamical regime (5.172) is IR-stable for $n = 2$. However, as noted in [158], owing to the numerical smallness of the coefficients of the ε expansions (5.174), (5.175) of ω_w we cannot discard the possibility that in the real dimension $d = 3$ ($2\varepsilon = 1$) the sign of the exact ω_w is different from that of the contribution of lowest order in ε (this is the same problem as for the sign of the specific heat exponent α). In this case the regime (5.173) will be realized for $n = 2$, $d = 3$.

Let us now turn to model F, which is considered more realistic for He$_4$. It contains two static charges with the β functions (5.124) and three dynamical charges with the β functions (5.166) and γ_f from (5.165). Depending on the sign of the specific heat exponent α, in statics one of the two regimes (5.125) is realized, and for them $\gamma_m^* = -\widetilde{\alpha}/2\nu$, where $\widetilde{\alpha} \equiv \max(0,\alpha)$ (Sec. 5.17). For dynamical fixed points with $f_* \neq 0$, $w_* \neq 0$ and any b_*, from (5.165) and (5.166) we have $\gamma_w^* = 0$, $\gamma_\lambda^* = \varepsilon + \gamma_m^* = \varepsilon - \widetilde{\alpha}/2\nu$, and so

$$z = 2 - \gamma_\lambda^* = d/2 + \widetilde{\alpha}/2\nu, \quad \widetilde{\alpha} \equiv \max(0,\alpha), \quad (f_* \neq 0, \ w_* \neq 0). \qquad (5.176)$$

This exponent z differs from (5.172) only for $\alpha > 0$, when $g_{2*} \neq 0$. Judging from the lowest order of the ε expansion for α, it is this variant which is realized for $n = 2$. However, from the experimental data for He$_4$ ($n = 2$, $d = 3$) it is known that $\alpha \simeq -0.02$ [153], and in this case the main candidates for the role of the IR-stable points of model F are the fixed points (5.172), (5.173) of model E with $g_{2*} = b_* = 0$ for the additional charges. Positivity of the exponent ω associated with g_2 is then ensured by the inequality $\alpha < 0$ (Sec. 5.17), and for the exponent ω_b it can be checked by explicit calculation [158]: for $g_{2*} = b_* = 0$ we obtain

$$\omega_b = 3(2\varepsilon)/4 - 0.299 \cdot (2\varepsilon)^2 + ... \qquad (5.177)$$

for the regime (5.172) and

$$\omega_b = 2(2\varepsilon)/3 - 0.159 \cdot (2\varepsilon)^2 + ... \qquad (5.178)$$

for the regime (5.173).

Therefore, for He$_4$ ($n = 2$) in the real dimension $d = 3$ the IR-attractive fixed points of model F are the points (5.172) or (5.173) corresponding to model E with $g_{2*} = b_* = 0$ for the additional charges. Judging from the lowest order of the ε expansion of ω_w, the dynamical regime (5.172) with exactly known exponent $z = d/2$ will be stable. However, it is not impossible that for real dimension $d = 3$ the other regime (5.173) with nontrivial exponent z will be stable.

More detailed information about the two-loop calculations in model F and comparison of their results with the experimental data for He$_4$ can be found in [170].

5.21   Model G

This model describes two real 3-component fields $\varphi \equiv \{\psi_a, m_a, a = 1, 2, 3\}$ with isotropic static action without interaction between ψ and m. We shall take $S^{\mathrm{stat}}(\varphi)$ to be the massless functional

$$S^{\mathrm{stat}}(\varphi) = -(\partial\psi)^2/2 - g_{10}\psi^4/24 - m^2/2 + mh_0 \qquad (5.179)$$

with the additional contribution of the bare external field $h_0 \equiv h_{0m}$ (without $h_{0\psi}$). Then the unrenormalized dynamical action for this model will be the functional (Sec. 5.12)

$$S(\phi) = \lambda_{0\psi}\psi'_a\psi'_a + \psi'_a[-\partial_t\psi_a + \lambda_{0\psi}H_{\psi_a} + \lambda_{0\psi}g_{20}\epsilon_{abc}\psi_b H_{m_c}]$$

$$-\lambda_{0m}m'_a\partial^2 m'_a + m'_a[-\partial_t m_a - \lambda_{0m}\partial^2 H_{m_a} + \lambda_{0\psi}g_{20}\epsilon_{abc}(\psi_b H_{\psi_c} + m_b H_{m_c})] \tag{5.180}$$

with the derivatives (5.128) of the functional (5.179) with respect to the fields $\varphi \equiv \{\psi_a, m_a\}$ and the dimensions (5.129) of the parameters λ. All the interactions in (5.180) are logarithmic for $d = 4$. In the notation (5.130) the basic action corresponding to the functional (5.180) will have the following form in detailed notation:

$$S_{\mathrm{B}}(\phi) = \lambda\psi'_a\psi'_a + \psi'_a[-\partial_t\psi_a + \lambda(\partial^2\psi_a - \tfrac{1}{6}g_{1\mathrm{B}}\psi_a\psi^2)+$$

$$+\lambda g_{2\mathrm{B}}\epsilon_{abc}\psi_b(-m_c + h_c)] - \lambda u m'_a\partial^2 m'_a + m'_a[-\partial_t m_a- \tag{5.181}$$

$$-\lambda u\partial^2(-m_a + h_a) + \lambda g_{2\mathrm{B}}\varepsilon_{abc}\psi_b\partial^2\psi_c + \lambda g_{2\mathrm{B}}\varepsilon_{abc}m_b h_c)],$$

where $g_{1\mathrm{B}} = g_1\mu^{2\varepsilon}$, $g_{2\mathrm{B}} = g_2\mu^\varepsilon$ are the basic charges ($d = 4 - 2\varepsilon$). In this model there are two different intermode interactions (ψ with m and m with m), but the coupling constants associated with them in the unrenormalized and basic actions are assumed identical in accordance with the general rule (5.84). The second intermode interaction [last term in (5.181)] contributes only for $h \neq 0$.

The renormalization of the model (5.181) is analyzed following the general scheme already described in detail for model F, and so we shall discuss it only briefly, and focus on the problems specific to model G. For $d = 4$ the total dimensions (5.87) of the fields ψ, ψ', m, m' and the parameter h are, respectively, 1, 3, 2, 2, and 2, and from this for the formal exponent (5.89) ($d_\omega = 2$, $d = 4$) we obtain $d^*_\Gamma = 6 - n_\psi - 3n_{\psi'} - 2n_m - 2n_{m'}$. As in model F, the real exponent δ_Γ is smaller than the formal one, because at the triple vertex involving m' in (5.181) a factor ∂ acting on the field m' is also singled out: inside the integral over x

$$m'_a\epsilon_{abc}\psi_b\partial^2\psi_c = m'_a\epsilon_{abc}\partial_i(\psi_b\partial_i\psi_c) = -\partial_i m'_a \cdot \epsilon_{abc}\psi_b\partial_i\psi_c. \tag{5.182}$$

Therefore, $\delta_\Gamma = d^*_\Gamma - n_{m'}$, and the standard analysis of the divergences of the functions Γ on the basis of their dimensions and the real exponent shows that all the needed counterterms of model G have the form of the various terms in (5.181), with the contribution $m'\partial_t m$ not renormalized, as in model F. Using this information and the general relations (5.95), it can then be shown that the renormalized action $S_{\mathrm{R}}(\phi)$ is obtained from the unrenormalized action by the standard renormalization of the static action and by additional multiplicative renormalization of the auxiliary fields ψ', m', the constants λ_φ, and the coefficients of the two intermode interactions.

If these coefficients in the basic action were independent (different charges), then these arguments would constitute a proof of multiplicative renormalizability of the model consistent with the static case. However, in (5.181) the coefficients were taken to be equal, and so these arguments imply only the multiplicative renormalizability of model G without an external field h, when the contribution of the second intermode interaction vanishes.

To avoid this difficulty, let us consider a generalization of model G in which for the second intermode interaction [last term in (5.181)] instead of g_{2B} we use the coefficient $g_{3B} = g_3\mu^\varepsilon$ with a third independent renormalized charge g_3. This model will already certainly be multiplicatively renormalizable, and all its renormalization constants, except for Z_{g_3}, will depend only on the original charges $g_{1,2}$ because they can be calculated in the model without the external field.

Let us now consider the special case $h = h(t)$ for this generalized renormalized model [see the text following (5.160) in Sec. 5.20]. Owing to its multiplicative renormalizability, it can be stated that

$$S_R(\phi, h) = S_R(\phi, 0) + Z_{\psi'}Z_\psi Z_\lambda Z_{g_2}\lambda g_{2B}\psi'_a\epsilon_{abc}\psi_b h_c + Z_\lambda Z_{g_3}\lambda g_{3B}m'_a\varepsilon_{abc}m_b h_c.$$
(5.183)

In deriving (5.183) from (5.181) we have used the relations

$$Z_m = Z_h^{-1} = 1, \quad Z_{m'} = 1,$$
(5.184)

where the first follow from statics and the last one from the first ones and the absence of renormalization of the contribution $m'\partial_t m$ in (5.181).

The contribution $S_R(\phi, 0)$ for the generalized model is the same as for the original one, and is obtained from the unrenormalized action (5.180) with $h = 0$ by multiplicative renormalization consistent with the static case.

Let us now consider a rotation $\phi \to U\phi$ of the fields $\phi = \psi, \psi', m, m'$ by an orthogonal ($U^\top = U^{-1}$) UV-finite matrix $U(t)$ that depends only on time. It follows from the multiplicative renormalizability of the model (5.181) with $h = 0$ and (5.184) that under such a rotation,

$$S_R(U\phi, 0) = S_R(\phi, 0) - Z_{\psi'}Z_\psi\psi'_a(U^\top\partial_t U)_{ab}\psi_b - m'_a(U^\top\partial_t U)_{ab}m_b.$$
(5.185)

For $h = h(t)$ the functional (5.183) ensures UV-finiteness of the Green functions for any $g_{2,3}$, in particular, also for $g_3 = g_2$, but we do not yet know whether or not the constants Z_{g_2} and Z_{g_3} are the same. To find out, we set $g_3 = g_2$ in (5.183) and compare the resulting expression to the functional (5.185), which also ensures UV-finiteness of the Green functions for any UV-finite orthogonal matrix $U(t)$. If we associate $U(t)$ with the renormalized external field $h(t)$ via the relation

$$-(U^\top\partial_t U)_{ab} = \lambda g_{2B}\varepsilon_{abc}h_c$$
(5.186)

[or, explicitly, $h_c = -\varepsilon_{abc}(U^\top \partial_t U)_{ab}/2\lambda_{2B}$], it can be stated that up to renormalization constants the functional (5.185) has exactly the same form as the functional (5.183) with $g_2 = g_3$. Furthermore, since these two functionals ensure UV-finiteness of the Green functions, they must differ in general by only a UV-finite renormalization. In our MS scheme they simply coincide [since UV-finiteness of a ratio of any two constants of the form (1.104) is possible only when they are equal]. We therefore have

$$1 = Z_\lambda Z_{g_2} = Z_\lambda Z_{g_3}\Big|_{g_3=g_2} . \qquad (5.187)$$

The second equation demonstrates the multiplicative renormalizability of the original model G with identical constants $g_3 = g_2$ for the two intermode interactions, and the first implies that $\gamma_\lambda + \gamma_{g_2} = 0$ for the corresponding anomalous dimensions. Then for the β function of the dynamical charge g_2 we obtain

$$\beta_{g_2} = g_2(-\varepsilon - \gamma_{g_2}) = g_2(-\varepsilon + \gamma_\lambda). \qquad (5.188)$$

Therefore, at the nontrivial fixed point with $g_{2*} \neq 0$ we have $\gamma_\lambda^* = \varepsilon$, which leads to the same result (5.172) as in model E for the dynamical exponent: $z \equiv \Delta_\omega = d/2$.

5.22   Model J

In this model there is only one real field $\varphi \equiv \{\psi_a, a = 1, 2, 3\}$ with the usual isotropic static action of the φ^4 model. We shall assume the action is massless but contains the contribution of an external field h. Then the unrenormalized dynamical action for this model (Sec. 5.12) will be the functional

$$S(\phi) = -\lambda_0 \psi_a' \partial^2 \psi_a' + \psi_a'[-\partial_t \psi_a - \lambda_0 \partial^2 H_{\psi_a} + \lambda_0 g_{20} \varepsilon_{abc} \psi_b H_{\psi_c}] \qquad (5.189)$$

with the usual notation (5.128): $H_{\psi_a} \equiv \delta S^{\text{stat}}/\delta \psi_a = \partial^2 \psi_a - g_{10} \psi_a \psi^2/6 + h_a$. In this model $\partial_t \sim \partial^4$, and so $d_\omega = 4$ in (5.87). The model (5.189) differs in its dimensions from those studied earlier, and so for convenience we give a table of the dimensions of its fields and parameters, which are easily found from the general rules of Sec. 5.14.

We see that the intermode coupling becomes logarithmic for $d = 6$, while the static ψ^4 interaction is logarithmic for $d = 4$. For $d > 2$ the latter is IR-irrelevant compared to the intermode interaction and should be discarded when analyzing the leading terms of the IR asymptote (Sec. 1.16). This distinguishes model J from those studied earlier.

Therefore, the standard RG technique allows model J to be studied in dimension $d = 6 - 2\varepsilon$ when the IR-irrelevant static ψ^4 interaction is discarded, i.e., with free statics for the field ψ. Then the basic action will be the

Table 5.4 Canonical dimensions of the fields and parameters of model J in arbitrary spatial dimension d.

F	$\psi(x)$	ψ', h_0	λ_0	g_{10}	g_{20}
d_F^p	$d/2 - 1$	$d/2 + 1$	-4	$4 - d$	$3 - d/2$
d_F^ω	0	0	1	0	0
d_F	$d/2 - 1$	$d/2 + 1$	0	$4 - d$	$3 - d/2$

functional,

$$S_B(\phi) = -\lambda \psi'_a \partial^2 \psi'_a +$$
$$+ \psi'_a [-\partial_t \psi_a - \lambda \partial^2 (\partial^2 \psi_a + h_a) + \lambda g_{2B} \epsilon_{abc} \psi_b (\partial^2 \psi_c + h_c)] \tag{5.190}$$

with the single basic charge $g_{2B} = g_2 \mu^\varepsilon$ for $d = 6 - 2\varepsilon$.

For $d = 6$ the total dimensions of the fields ψ, ψ' and the parameter h are 2, 4, and 4, respectively, and so for the exponent (5.89) using $d_\omega = 4$, $d = 6$ we obtain $d_\Gamma^* = 10 - 2n_\psi - 4n_{\psi'}$ and $\delta_\Gamma = d_\Gamma^* - n_{\psi'}$ for the same reasons as for model G. Then, as usual, it follows that all the needed counterterms have the form of the various terms in (5.190) and that the contribution $\psi' \partial_t \psi$ is not renormalized. Taking into account this information, using (5.95) it is then easy to prove the multiplicative renormalizability of the dynamical model, with

$$Z_\psi = Z_h = 1, \quad Z_{\psi'} = 1. \tag{5.191}$$

The first equations are known from (free) statics, and the last one is a consequence of the first and the absence of renormalization of the contribution $\psi' \partial_t \psi$ in (5.190).

Then, just as for model G, the transformation $\phi \to U\phi$ with arbitrary orthogonal matrix $U(t)$ depending only on time can be used to easily show that the coefficient λg_{2B} of the intermode coupling in (5.190) is not renormalized. This leads to the first equation in (5.187) and its consequence $\gamma_\lambda^* = \varepsilon$ for the nontrivial fixed point with $g_{2*} \neq 0$. However, now $\varepsilon = (6 - d)/2$ [in contrast to $\varepsilon = (4 - d)/2$ for model G] and different (see Table 5.4) dimensions of the parameter λ, and so from the general relation (5.109) for model J we obtain [164]

$$z \equiv \Delta_\omega = 4 - \gamma_\lambda^* = 4 - \varepsilon = d/2 + 1. \tag{5.192}$$

In addition to model J, the authors of [164] also studied a modification of it with $-\lambda_0\partial^2 \to \lambda_0$ in (5.189). In this case the intermode interaction becomes IR-irrelevant and everything reduces to model A.

5.23   Model H: determination of dynamical variables

Model H (Sec. 5.12) serves to describe the dynamics in the vicinity of the critical point corresponding to the liquid–gas transition and other systems of the same universality class, in particular, the critical point of lamination in binary mixtures. For definiteness, we shall discuss the critical point of the liquid–gas transition, and in this section we discuss the choice of variables of the dynamical fluctuation model.

The fundamental microscopic theory of a fluid that we take as our starting point will be the nonequilibrium statistical physics of a system of interacting classical particles. In this theory the random fields $\widehat{F}(x) = \widehat{H}(x)$, $\widehat{p}_i(x)$, and $\widehat{n}(x)$, the volume densities of the conserved energy, momentum, and particle number, are correctly defined. Their averages $\langle ... \rangle$ over the equilibrium distribution are known thermodynamical quantities: $\langle\widehat{H}(x)\rangle = U$ is the internal energy per unit volume, $\langle\widehat{n}(x)\rangle = \rho$ is the particle number density, and $\langle\widehat{p}_i(x)\rangle = 0$ is the momentum density. According to the general principles of the Landau theory (Sec. 1.12), deviations $\Delta\widehat{F}(x) = \widehat{F}(x) - F_c$ of the random fields $\widehat{F}(x)$ from their equilibrium expectation values at the critical point $\langle\widehat{F}(x)\rangle_c \equiv F_c$ can be assumed to be "small" in the critical region, so that their higher orders can be neglected relative to the lower ones (and in thermodynamics they can be treated as differentials). Then using the original fields $\widehat{H}(x) - U_c \equiv \Delta\widehat{U}(x)$ and $\widehat{n}(x) - \rho_c \equiv \Delta\widehat{\rho}(x)$ and the thermodynamic relation $\Delta U = T\Delta S + \mu\Delta\rho \simeq T_c\Delta S + \mu_c\Delta\rho$, we can determine the analogous random field for the entropy (per unit volume) $\Delta\widehat{S}(x) = T_c^{-1}[\Delta\widehat{U}(x) - \mu_c\Delta\widehat{\rho}(x)]$, and in place of the momentum $\widehat{p}_i(x)$ we can introduce the random velocity field with the same accuracy: $\widehat{v}_i(x) = \widehat{p}_i(x)/M\rho_c$, where M is the particle mass. Although the field v, in contrast to the momentum p, is not exactly conserved, it follows from the definitions that the difference between these two objects can be neglected in the critical region, i.e., all the corrections associated with this difference must be IR-irrelevant. Of course, this statement is confirmed by the following rigorous analysis.

The fields $\Delta\widehat{S}(x)$, $\Delta\widehat{\rho}(x)$, and $\widehat{v}_i(x)$ form a complete basis of the variables involved in the dynamics. We note that in their construction in terms of the initial densities of conserved quantities we have used only linear relations with coefficients that are finite at the critical point, and have neglected the corrections to these coefficients because they are certainly IR-irrelevant. The quantities $\Delta\widehat{F}(x)$ thus determined cannot be identified with thermodynamical fluctuations (we shall discuss this in more detail at the end of this section).

The equilibrium expectation values $\langle\Delta\widehat{\rho}(x)\rangle$ and $\langle\Delta\widehat{S}(x)\rangle$ are known from statics; see (1.31). Each contains a leading singular contribution $\sim \partial_1 p^{(s)}$

and a less singular one $\sim \partial_2 p^{(s)}$ with critical dimensions $d - \Delta_h = \Delta_\varphi$ and $d - \Delta_\tau = \Delta[\varphi^2]$, respectively, where Δ_F are the critical dimensions of the simple $(n = 1)$ φ^4 model ($\Delta_h \equiv \Delta_1$, $\Delta_\tau \equiv \Delta_2$ in the notation of Sec. 1.8). It follows from (1.31) that the linear combinations of $\Delta\widehat{\rho}(x)$ and $\Delta\widehat{S}(x)$ that are singled out are the ones whose equilibrium expectation values have the definite critical dimensions of $\partial_1 p^{(s)}$ and $\partial_2 p^{(s)}$, namely,

$$\psi(x) = \text{const}[\Delta\rho(x) - b\Delta S(x)], \tag{5.193a}$$

$$m(x) = \text{const}[\Delta S(x) - a\Delta\rho(x)], \tag{5.193b}$$

where a and b are the coefficients from (1.28). The field ψ is the order parameter of the φ^4 model in the exact sense of the term (now $\varphi \to \psi$), and the field m is the composite operator ψ^2 in this model. If we wish to consider these quantities independent, they must be identified with the corresponding fields of model C (Sec. 5.17). The proportionality $m \sim \psi^2$ will then be a consequence of one of the Schwinger equations (Sec. 2.11) for this model. When we referred to just $\Delta\widehat{\rho}(x)$ as the order parameter in statics (Sec. 1.9), we added the stipulation "up to leading singular contributions." With the same accuracy, either $\Delta\widehat{S}(x)$ or indeed any linear combination of $\Delta\widehat{S}(x)$ and $\Delta\widehat{\rho}(x)$ in whose expectation value the leading singular contribution $\partial_1 p^{(s)}$ does not cancel could have been taken as the order parameter. It is the expectation value of the random quantity (5.193a), which therefore should be taken to be the fundamental field of the ψ^4 model, i.e., its order parameter in the exact sense of the term.

The definition (5.193a) of the order parameter ψ is not the same as the usual definition [153], [171] based on the equations of the classical hydrodynamics of an ideal fluid (i.e., neglecting heat conduction and viscosity). According to the equations of hydrodynamics [172], the particle number density $\rho(x)$ and entropy density $S(x)$ (but not the energy density) satisfy continuity equations of the same type, $\partial_t F + \partial_i(v_i F) = 0$, $F = \rho, S$. From this for the difference $\Delta F(x) \equiv F(x) - F_c$ we have

$$\partial_t \Delta F + F_c \partial_i v_i + \partial_i(v_i \Delta F) = 0, \quad F = \rho, \ S. \tag{5.194}$$

From the viewpoint of hydrodynamics, a special role is played by the combination of $\Delta\rho$ and ΔS which in the linear approximation (in the set of "small" variables $\Delta\rho$, ΔS, and v_i) is split off from the longitudinal ($\sim \partial_i v_i$) part of the velocity field associated with sound oscillations. It follows from (5.194) that this combination has the form

$$\text{const}[S_c^{-1}\Delta S(x) - \rho_c^{-1}\Delta\rho(x)], \tag{5.195}$$

i.e., with the same accuracy it is proportional to ΔF for $F = S/\rho$, the entropy per particle (or unit mass). It is the combination (5.195) that is usually taken as the order-parameter field ψ in model H, and the usual ψ^4 model is used to describe the statics of this field.

This description can only be approximate, because the field (5.195) is actually a mixture of two fields (5.193) having different critical dimensions. We shall therefore take the quantities in (5.193) to be the fundamental fields, where each of these has definite critical dimension. The latter is more important than the decoupling from $v_\parallel$, which actually does not play much of a role (see below). The statics of the fields (5.193) are described by model C (Sec. 5.17) with one-component field ψ, which automatically ensures the correct critical dimensions (Δ_φ and $\Delta[\varphi^2]$ in the terminology of the φ^4 model) for these fields. The needed canonical dimensions ($d/2 - 1$ for ψ and $d/2$ for m) are ensured by choosing the normalization factors in (5.193) to be suitable powers of the cutoff parameter $\Lambda = r_{\min}^{-1}$, where $r_{\min}$ is the minimum characteristic scale of the problem: the interatomic separation.

The nonuniversal coefficients a and b in (5.193) are determined by (1.28). The meaning of a is $a = -d\mu/dT$ at a finite point of the coexistence curve $\Delta\mu + a\Delta T \equiv e_1 = 0$. As explained in Sec. 1.8, linear continuation of this curve into the region $T > T_c$ neglecting corrections corresponds to the line $\rho = \text{const}$ and (simultaneously) $S = \text{const}$, and so we can take $a = -d\mu/dT|_{\rho=\text{const}} = -d\mu/dT|_{S=\text{const}}$ (all derivatives are taken in the limit $T \to T_c + 0$). Using the relations of Sec. 1.8, it is easily checked that with the same accuracy $a = dS/d\rho|_{p=\text{const}}$. It is this expression for the coefficient a in terms of the derivative at constant pressure p that is usually used in defining the "second diffusion mode" (5.193b). We recall that the coefficients a and b for the grand canonical ensemble can be calculated theoretically but cannot be measured experimentally owing to the natural arbitrariness in the definition of the entropy (see the discussion at the end of Sec. 1.8).

The hydrodynamical equations of an ideal fluid for the fields ψ and m are obtained from their definitions (5.193) and from the equations (5.194):

$$\partial_t \psi = -\psi_0 \partial_i v_i - \partial_i(v_i \psi), \qquad \partial_t m = -m_0 \partial_i v_i - \partial_i(v_i m), \qquad (5.196)$$

where ψ_0 and m_0 are constants. Adding the contributions of damping and noise and interpreting the terms involving v as streaming terms, it is easy to bring the equations (5.196) to the form of the full fluctuation model of all three fields ψ, m, and v. We shall do this in the next section. There we shall show that the "coupling" of ψ and $v_\parallel$ in (5.196) [which was not present for the field (5.195)] actually does not present any problem, because it leads only to a renormalization of the Onsager coefficient in the stochastic equation for ψ. The main argument in favor of choosing (5.195) as the field of the order parameter thereby becomes meaningless. It is therefore preferable to choose the variables (5.193), because it is these combinations which have definite critical dimensions.

In conclusion, we again state that fluctuation fields like $\Delta\widehat{F}(x)$ for $F = S, \rho$ must not be identified with thermodynamical fluctuations $\delta\widehat{F}(x) = \widehat{F}(x) - \langle\widehat{F}(x)\rangle$ of the corresponding quantities. By definition, for fluctuations we have $\langle\delta\widehat{F}(x)\rangle = 0$ for any value of the external parameters T and μ, whereas the expectation values $\langle \Delta \widehat{F}(x) \rangle$ for fluctuation fields are nontrivial quantities like (1.31). There is another important difference. In the general theory of thermodynamical fluctuations [51] all quantities F are assumed equally likely, and for any of them (including T and μ) it is possible to define the corresponding quantity $\delta \widehat{F}(x)$ as a linear combination of the "primary" quantities $\delta \widehat{U}(x)$ and $\delta \widehat{\rho}(x)$. The coefficients of these linear combinations are calculated from the equations of equilibrium thermodynamics and in general contain critical singularities. However, when constructing the fields of fluctuation models it is not possible to have singularities in the coefficients, because they must be generated later on within the model itself owing to interactions between its fields, rather than introduced by hand in the definition of the fields. Therefore, the only possible candidates for fluctuation fields are the quantities $\Delta F = \alpha \Delta U + \beta \Delta \rho$, for which at least one of the coefficients α, β remains finite at the critical point, where the small corrections $\alpha - \alpha_c$, $\beta - \beta_c$ in these coefficients can (as they are IR-irrelevant), and even must (in order not to introduce any critical singularities) be neglected. In this approach the equality of the roles of the various thermodynamical quantities characteristic of fluctuation theory are lost. For example, for the pressure p in (1.29) the quantity Δp, when the correction from $p^{(s)}$ is neglected, contains only contributions $\Delta p = S_c \Delta T + \rho_c \Delta \mu$ that are regular in T and μ at the critical point, whereas for $\Delta \rho$ and ΔS the leading terms are the singular components (1.31). Therefore, Δp cannot be used as the order-parameter field, while $\Delta \rho$ and ΔS can be. All these general considerations have been taken into account above in defining the fields of the dynamical fluctuation model.

5.24 Model H: IR irrelevance of the sound modes in the regime $\omega \sim p^4$

Model H [153], [171] involves only the order-parameter field ψ and the transverse part $v_\perp$ of the velocity field, because the interaction of these fields with the "sound modes" m and $v_\parallel$ is unimportant for their dynamics. In this section we present a formal proof of this statement, based on the standard (Secs. 5.15 and 1.16) analysis of the canonical dimensions.

We begin by describing the complete fluctuation model for the system of fields ψ, m, and v. As already mentioned in Sec. 5.23, the statics of the fields ψ and m are naturally described by a model like C (Sec. 5.17), and the inclusion of the velocity v reduces to the addition of the kinetic energy contribution $\sim v^2$. Therefore, the unrenormalized static action will be the functional

$$S^{\text{stat}}(\varphi) = -(\partial \psi)^2/2 - \tau_0 \psi^2/2 - g_{10} \psi^4/24 - m^2/2 + g_{20} m \psi^2/2 - a v^2/2.$$

$$(5.197)$$

The parameter $a = M \rho_c / k T_c$ has been introduced into (5.197) so as to preserve the usual physical meaning and dimensions of the velocity field v.

We recall (Sec. 1.8) that in the description of the critical point of the liquid–gas transition using the ψ^4 model, the analogs of the usual variables h and τ for a magnet are the quantities e_1 and e_2 defined in (1.28). Using the functional (5.197) without the contribution of the external field $h\psi$, we set $h \equiv e_1 = 0$, which for $T > T_c$ approximately corresponds to the critical isochore (Sec. 1.8). On this line the second variable e_2 from (1.28) is proportional to $\Delta T \equiv T - T_c \equiv \tau$, i.e., the coefficient of ψ^2 in (5.197) can, as usual, be assumed to be the bare value of this variable.

In an ideal fluid the hydrodynamics of the fields ψ and m are described by (5.196) and the hydrodynamical equation [172] for the velocity: $\partial_t(\overline{\rho}v_i) + \partial_k T_{ik} = 0$, where $\overline{\rho} = M\rho$ is the mass density of the medium and T_{ik} is the stress tensor, containing, among other things, the contribution of the pressure $p\delta_{ik}$ and the "kinetic term" $\overline{\rho}v_i v_k$. In our problem the form of T_{ik} is uniquely determined by the general principles for constructing models of critical dynamics (Sec. 5.9). In fact, if the contributions involving v in (5.196) are interpreted as intermode coupling of the fields ψ and m with v, then from the rules of Sec. 5.9 we easily find the corresponding contributions in the equation for v:

$$\partial_t v_i + (v_k \partial_k) v_i = a^{-1}(\psi_0 + \psi)\partial_i H_\psi + a^{-1}(m_0 + m)\partial_i H_m, \qquad (5.198)$$

where H_φ are the derivatives (5.128) of the functional (5.197). The term $(v\partial)v$ on the left-hand side of (5.198) ensuring Galilean invariance is generated by the kinetic contribution $\overline{\rho}v_i v_k$ to T_{ik}. If the velocity field v were purely transverse (incompressibility), the contribution $(v\partial)v$ (rather, its "transverse" part) would not violate the consistency with statics (5.197), because it could be interpreted as the intermode interaction between the components of v. The additions from the longitudinal part of v in $(v\partial)v$ violate the consistency with statics (5.197), and in this sense are bad. However, it will be shown later that the contribution $(v\partial)v$ is actually IR-irrelevant, i.e., it can be discarded in analyzing the critical asymptote, and all the problems associated with it vanish.[*]

It is necessary to add to (5.196) and (5.198) the correlated, according to (5.4), contributions of dissipation and noise. For brevity we write

$$\{\psi, m, v_\parallel, v_\perp\} \equiv \{\varphi_a, a = 1, 2, 3, 4\} \qquad (5.199)$$

and discuss the possible general form of the 4×4 matrix of Onsager coefficients α_{ab}. It is determined taking into account the following considerations:

---

[*]In principle, it is easy to modify the model so as to ensure complete consistency with statics. For this it is necessary to change the static action (5.197) by replacing the approximate expression $\rho_c v^2$ by the exact one ρv^2, while appropriately modifying the contributions of the intermode interaction on the right-hand side of (5.198) (most easily done using the variables ΔS, $\Delta \rho$), and to replace the contribution $(v\partial)v$ on the left-hand side of (5.198) by adding to the right-hand side the intermode interaction between the components of v with the coefficients $\beta[v_i, v_k] \sim \rho^{-1}(\partial_k v_i - \partial_i v_k)$. Then all the discrepancies from the simple equation (5.198) will be only IR-irrelevant corrections, which can be neglected in the critical region. Therefore, in what follows we shall start from (5.196) and (5.198).

(1) the matrix α must be symmetric and positive-definite; (2) the transverse vector field $v_\perp \equiv \varphi_4$ cannot mix with the other fields, and so $\alpha_{4a} = 0$ for all $a = 1, 2, 3$; (3) since all our fields are generated by conserved quantities (Sec. 5.23), all nontrivial elements α_{ab} must contain at least one factor ∂ (Sec. 5.8). From this, taking into account the index structure and the positive-definiteness of the matrix α, it follows that all four diagonal elements α_{aa} and the off-diagonal element α_{12} must be proportional to ∂^2, while the elements α_{a3} with $a = 1, 2$ are multiples of (at least) $\partial^2 \partial_i$. Let us explain the latter: the form $\alpha_{a3} \sim \partial_i$ (i.e., $\sim p_i$ in momentum space) allowed by the index structure would contradict the Cauchy–Bunyakov inequality $\alpha_{a3}^2 \leq \alpha_{aa}\alpha_{33} \sim p^4$ at the asymptote $p \to 0$, and this inequality follows from the positive-definiteness of the matrix α.

Summarizing the above arguments, we can introduce the following parameterization of the elements of the matrix α in the notation (5.199):

$$\alpha_{aa} = -\lambda_{aa}\partial^2 \text{ for all } a, \ \alpha_{12} = -\lambda_{12}\partial^2, \ \alpha_{a3} = -\lambda_{a3}\partial^2\partial_i, \ a = 1, 2, \ (5.200)$$

where all the λ are numerical coefficients.

There is also another important consideration regarding the possible form of the matrix α. If the original variables ΔS and $\Delta \rho$ were used instead of the fields ψ and m, we would expect contributions from dissipation in the equations for the entropy (heat conduction) and velocity (viscosity), but there should not be any in the equation for the density, because the continuity equation (5.194) for ρ (in contrast to S) is exact and does not contain contributions from dissipation even for a nonideal fluid. This means that in these variables we would have $\alpha_{\rho\rho} = 0$, and so also $\alpha_{\rho S} = \alpha_{\rho v} = 0$ because the matrix α is positive-definite. Then only the element α_{SS}, two elements (longitudinal and transverse) of the type α_{vv}, and one element of the type α_{Sv} can be nonzero. The passage to the variables ψ, m corresponds to a linear transformation of the fields $\varphi \to L\varphi$ under which $\alpha \to L\alpha L^\top$, as is easily checked using (5.4). Therefore, from the above arguments about the structure of the matrix α in the variables S, ρ, and v we obtain definite relations for the elements of the new matrix α in the variables ψ, m, and v. It is easy to check that in the notation (5.199) these relations have the form

$$\alpha_{11}/\alpha_{12} = \alpha_{12}/\alpha_{22} = \alpha_{13}/\alpha_{23} = L_{11}/L_{21}, \quad (5.201)$$

where L_{11} is the coefficient of ΔS in (5.193a) and L_{21} is the coefficient in (5.193b). The relations (5.201) are equivalent to the analogous ones for the coefficients λ in (5.200). It follows, in particular, from (5.201) that the three elements α_{11}, α_{12}, and α_{22} are proportional to each other and can vanish only simultaneously.

Adding to (5.196) and (5.198) the contributions of dissipation and noise with the Onsager coefficients (5.200), from the general rules of Sec. 5.8 we obtain the following action functional (5.61) for the fields $\phi = \varphi, \varphi'$ with

$\varphi = \psi, m, v_\parallel$, and $v_\perp$:

$$S(\phi) = -\lambda_{11}\psi'\partial^2\psi' - \lambda_{22}m'\partial^2m' - \lambda_{33}v'_\parallel\partial^2v'_\parallel - \lambda_{44}v'_\perp\partial^2v'_\perp - 2\lambda_{12}\psi'\partial^2m' -$$

$$-2\lambda_{13}\psi'\partial^2(\partial v'_\parallel) - 2\lambda_{23}m'\partial^2(\partial v'_\parallel) + \psi'[-\partial_t\psi - \lambda_{11}\partial^2 H_\psi - \lambda_{12}\partial^2 H_m +$$

$$+a\lambda_{13}\partial^2(\partial v_\parallel) - \partial(v\psi) - \psi_0(\partial v_\parallel)] + m'[-\partial_t m - \lambda_{22}\partial^2 H_m - \lambda_{12}\partial^2 H_\psi +$$

$$+a\lambda_{23}\partial^2(\partial v_\parallel) - m_0(\partial v_\parallel) - \partial(vm)] + v'_\parallel[-\partial_t v_\parallel + a\lambda_{33}\partial^2 v_\parallel - \lambda_{13}\partial(\partial^2 H_\psi) -$$

$$-\lambda_{23}\partial(\partial^2 H_m) - (v\partial)v + a^{-1}\psi\partial H_\psi + a^{-1}\psi_0\partial H_\psi + a^{-1}m\partial H_m +$$

$$+a^{-1}m_0\partial H_m] + v'_\perp[-\partial_t v_\perp + a\lambda_{44}\partial^2 v_\perp - (v\partial)v + a^{-1}\psi\partial H_\psi + a^{-1}m\partial H_m].$$

$$(5.202)$$

We have written out this lengthy complete expression for $S(\phi)$, and we are now going to show that most of the terms in it are actually IR-irrelevant and can be discarded. In the end this procedure for simplifying the model (5.202) leads to the complete elimination of the sound modes m, $v_\parallel$ and to the usual formulation (Sec. 5.12) of model H.

Let us prove this statement. To judge the relative IR-relevance of the various terms in (5.202), it is necessary first to exactly define the concept of the total dimension (5.87), i.e., to fix the frequency dimension d_ω. We see from (5.197) and (5.202) that for $T = T_c$ the dynamical equation for ψ involves the combination $\partial_t + \text{const} \cdot \partial^4$, while the equation for the fields m and v involves $\partial_t + \text{const} \cdot \partial^2$. Therefore, the choice of d_ω is not unique and in this case should be understood simply as one of the elements of the exact formulation of the asymptotic problem (Sec. 5.14).

The common choice for model H is $\omega \sim p^4$ ($d_\omega = 4$), corresponding to nontrivial dynamics for the field of the order parameter ψ. This is what we shall use, along with the spatial dimension $d = 4 - 2\varepsilon$, because the two static interactions in (5.197) and also the intermode interaction of ψ with v in (5.202) are logarithmic for $d = 4$. The other necessary information about the total dimensions (5.87) of the quantities in (5.202) is given in Table 5.5 for $d_\omega = 4$ and $d = 4$.

Table 5.5 Total dimensions d_F of the quantities F in the functional (5.202) for $d = 4$, $d_\omega = 4$.

F	ψ,ψ_0	m,m_0	v	H_ψ	H_m	a	λ_{11}	$\lambda_{12},\lambda_{13}$	$\lambda_{22},\lambda_{23}$	$\lambda_{33},\lambda_{44}$
d_F	1	2	3	3	2	-2	0	1	2	4

We see from Table 5.5 that the numerical coefficients of the various terms in (5.202) have different total dimensions, and so some of them can certainly be discarded as they are IR-irrelevant. However, this procedure must be performed with great care, and so we shall discuss it in some detail.[*]

We perform the following dilatation of the auxiliary fields in (5.202):

$$\{\psi', m', v'_\parallel, v'_\perp\} \rightarrow \{\psi', m_0^{-1}m', m_0^{-1}av'_\parallel, av'_\perp\}, \qquad (5.203)$$

and then we shift the new fields m' and $v'_\parallel$:

$$m' \rightarrow m' - \psi_0\psi', \qquad v'_\parallel \rightarrow v'_\parallel + (\psi_0 m_0^{-1}\lambda_{22} - \lambda_{12})\partial\psi'. \qquad (5.204)$$

It is easy to check that after these two transformations the action functional (5.202) takes the form

$$S(\phi) = S_0(\phi) + V(\phi), \qquad (5.205)$$

where $V(\phi)$ is the sum of the various terms whose coefficients have negative dimensions and

$$S_0(\phi) = -\lambda'_{11}\psi'\partial^2\psi' - a^2\lambda_{44}v'_\perp\partial^2 v'_\perp + \psi'[-\partial_t\psi - \lambda'_{11}\partial^2 H_\psi -$$

$$-\partial(v\psi)] + m'[-m_0^{-1}\lambda_{22}\partial^2 H_m - \partial v_\parallel] + v'_\parallel[\partial H_m] +$$

$$+v'_\perp[a^2\lambda_{44}\partial^2 v_\perp + \psi\partial H_\psi + m\partial H_m], \qquad (5.206)$$

where $\lambda'_{11} \equiv \lambda_{11} - 2\alpha\lambda_{12} + \alpha^2\lambda_{22}$, $\quad \alpha \equiv \psi_0 m_0^{-1}$.

Let us consider only the leading singular contributions in the regime $\omega \sim p^4$ of the Green functions of the fields ψ, ψ', $v_\perp$, and $v'_\perp$ in the model (5.205). It is easily verified that these Green functions are correctly determined also in the model (5.206) (see below for details), and so we can ask whether or not the contribution $V(\phi)$ in (5.205) is IR-relevant. This question is answered very simply using the standard dimensional arguments of Sec. 1.16. From Table 5.5 we see that all the coefficients in the functional (5.206) have zero total dimension, and the functional $V(\phi)$, as already mentioned, consists of terms whose coefficients have negative dimensions. It then follows directly that the contribution $V(\phi)$ in (5.205) is IR-irrelevant compared to $S_0(\phi)$ and can be discarded, because it gives only corrections to the leading singularities in this regime.

After the dilatation (5.203) the functional (5.202) still contains two terms whose coefficients have positive dimensions, namely, $-\lambda_{12}\psi'\partial^2 H_m$ and $-\psi_0\psi'(\partial v_\parallel)$, which are formally even more important than the terms of $S_0(\phi)$. These dangerous terms are eliminated by making the additional shift (5.204).

---

[*]Here we are speaking of the dimensions of only the coefficients that remain finite at the IR asymptote in question (then, for example, the contributions $\sim (\partial\psi)^2$ and $\sim \tau\psi^2$ are assumed to be "codimensional"); it is also important that the fields φ are not mixtures, but quantities with definite critical dimensions.

Then only contributions whose coefficients have zero and negative dimensions remain, with the first grouped together to form $S_0(\phi)$ and the second to form $V(\phi)$. We also note that the coupling of the fields ψ and $v_\parallel$ ($\psi_0 \neq 0$) leads only to the renormalization $\lambda_{11} \to \lambda'_{11}$ of the Onsager coefficient for the field ψ indicated in (5.206).

Let us now study the simplified model (5.206) in more detail. It is easily seen from the form of the functional (5.206) that this is a theory with constraints and so some of the fields can be eliminated. In fact, integration over $v'_\parallel$ and m' in the generating functional (5.26) with the action (5.206) and without sources for the sound modes generates two functional δ functions corresponding to the constraints $\partial H_m = 0$ and $\partial v_\parallel = 0$. When the standard requirement that the fields fall off for $|\mathbf{x}| \to \infty$ (Sec. 5.1) and the explicit form of the derivative (5.128) of the functional (5.197) with respect to m are taken into account, these constraints are equivalent to the following simpler relations:

$$H_m = -m + g_{20}\psi^2/2 = 0, \qquad v_\parallel = 0. \tag{5.207}$$

The constraint $\partial H_m = 0 \Leftrightarrow H_m = 0$ is generated by integration over $v'_\parallel$, after which all the other contributions involving H_m in (5.206) can be dropped. Then the subsequent integration over m' gives the second constraint (5.207). From the mathematical point of view the fields $v'_\parallel$ and m' in the model (5.206) enter as Lagrange multipliers.

The constraints (5.207) correspond to functional δ functions that "perform" the integration over the fields $v_\parallel$ and m: the field $v_\parallel$ is simply set equal to zero, and m is expressed in terms of ψ using the first equation in (5.207). Then m is thereby eliminated also from the variational derivative H_ψ of the action (5.197). It is easy to check that this operation is equivalent to eliminating m in the functional (5.197) itself and leads to the standard static action of the simple ψ^4 model with renormalization $g_{10} \to g_{10} - 3g_{20}^2$ of the coupling constant of the ψ^4 interaction.

The proof that it is possible to replace the full model (5.202) by the simplified model (5.206) in this regime would be more transparent if we worked directly in the language of the stochastic equations (5.55) for the system of fields (5.199).[*] The relative IR-relevance of the various contributions in these equations can also be judged simply from the dimensions of the corresponding numerical coefficients. In the equations for m and $v_\parallel$ it is necessary to keep only the most important contributions (terms involving m_0) and the first corrections to them, i.e., the contributions with coefficients one unit lower in dimension (compared to the leading terms). The other contributions, including $\partial_t m$ and $\partial_t v_\parallel$, can be neglected. Then from the equations for m and $v_\parallel$ we can find ∂H_m and $\partial v_\parallel$ and substitute them into the equations for ψ and $v_\perp$. Only the contributions with zero and negative dimensions of the coefficients

---

[*]This conclusion was reached in [241]. We warn the reader that Eqs. (19) and (25) of that paper contain misprints: in (19) a factor a should appear before λ_{33}, and in (25) $v_\parallel$ should be replaced by $v_\perp$.

will be left in these equations. Discarding the contributions with negative dimensions as they are IR-irrelevant, we arrive at the model (5.206). The renormalization $\lambda_{11} \to \lambda'_{11}$ of the Onsager coefficient indicated in (5.206) occurs, which in the language of the stochastic equations corresponds to the renormalization $\eta_1 \to \eta'_1 \equiv \eta_1 - \alpha\eta_2$ of the random force η_1.

The only technical inconvenience in this language is the problem of evaluating the relative IR-relevance of the contributions involving random forces η_a, particularly when the nondiagonal nature of their correlators is taken into account. In principle, this problem is easily solved by isolating the factors $\lambda_{aa}^{1/2}$ from η_a (i.e., by the substitution $\eta_a = \lambda_{aa}^{1/2}\eta'_a$), which should be viewed as "numerical coefficients" on an equal footing with the others. However, here it is difficult not to resort to the language of the action functional and the arguments of Sec. 1.16. This is why we have preferred to use this language from the start, to some degree sacrificing simplicity in favor of uniformity and accuracy.

The final conclusion is the following. When analyzing the leading terms of the IR asymptote of the Green functions of the fields ψ, ψ', $v_\perp$, and $v'_\perp$ in the regime $\omega \sim p^4$, we can completely neglect their interaction with the sound modes $v_\parallel$ and m. This leads to the usual model H (Sec. 5.12) with transverse velocity field v, where the contribution $\partial_t v$ in the equation for the velocity in this regime must also be considered IR-irrelevant. This model will be studied in detail in the following section.

5.25   Model H: renormalization and RG analysis in the regime $\omega \sim p^4$

Neglecting the interaction with the sound modes (Sec. 5.24), we can use the following functional as the unrenormalized static action of the fields $\varphi = \{\psi, v = v_\perp\}$:

$$S^{\text{stat}}(\varphi) = -(\partial\psi)^2/2 - \tau_0\psi^2/2 - g_{10}\psi^4/24 - av^2/2 = S^{\text{stat}}(\psi) - av^2/2. \tag{5.208}$$

The coefficient a will not be renormalized, and so it can be considered to be a UV-finite renormalized parameter, while the others are bare coefficients.

For the basic and renormalized static action we have

$$S_{\text{B}}^{\text{stat}}(\varphi) = S_{\text{B}}^{\text{stat}}(\psi) - av^2/2, \quad S_{\text{R}}^{\text{stat}}(\varphi) = S_{\text{R}}^{\text{stat}}(\psi) - av^2/2, \tag{5.209}$$

where $S_{\text{B}}^{\text{stat}}(\psi)$ and $S_{\text{R}}^{\text{stat}}(\psi)$ are known functionals of the simple ψ^4 model.

The unrenormalized dynamical action of model H consistent with the statics (5.208) will be written as

$$S(\phi) = -\lambda_0\psi'\partial^2\psi' + \psi'[-\partial_t\psi - \lambda_0\partial^2 H_\psi - v\partial\psi]-$$
$$-\lambda_0^{-1}g_{20}^{-1}v'\partial^2v' + v'[-a\partial_t v + \lambda_0^{-1}g_{20}^{-1}\partial^2 v + \psi\partial H_\psi], \tag{5.210}$$

where H_ψ is the derivative (5.128) of the functional (5.208), and the constants λ_0 and g_{20} are related to the parameters in (5.206) as $\lambda_0 = \lambda'_{11}$, $\lambda_0^{-1} g_{20}^{-1} = a^2 \lambda_{44}$. Here λ_0 and $\eta_0 \equiv (a\lambda_0 g_{20})^{-1}$ are treated as the bare heat conduction and viscosity coefficients, g_{20} is the bare charge of the intermode interaction, and g_{10} in (5.208) is the static charge. The vector fields v and v' are taken to be transverse, and so no additional insertion of the transverse projector in front of the term $\psi \partial H_\psi$ in (5.210) is needed.

In detailed notation, the basic action corresponding to (5.210) has the form ($d = 4 - 2\varepsilon$)

$$S_B(\phi) = -\lambda\psi'\partial^2\psi' + \psi'[-\partial_t\psi - \lambda\partial^2(\partial^2\psi - \tau\psi - g_{1B}\psi^3/6) - v\partial\psi] -$$

$$-\lambda^{-1}g_{2B}^{-1}v'\partial^2 v' + v'[-a\partial_t v + \lambda^{-1}g_{2B}^{-1}\partial^2 v + \psi\partial(\partial^2\psi)],$$

(5.211)

where $g_{iB} = g_i\mu^{2\varepsilon}$ with $i = 1, 2$ are the basic charges, and λ, τ, a, and $g_{1,2}$ are renormalized parameters. Since the field v' is transverse, only the contribution $\partial^2\psi$ remains in the term $v'\psi\partial H_\psi^B$ from the derivative (5.128) of the static basic action. The canonical dimensions of the fields and parameters of model H are given in Table 5.6 for $d = 4 - 2\varepsilon$, $d_\omega = 4$.

Table 5.6   Canonical dimensions of the fields and parameters of model H for $d = 4 - 2\varepsilon$, $d_\omega = 4$.

F	ψ	ψ'	v, v'	a	λ, λ_0	g_{10}, g_{20}	g_1, g_2
d_F^p	$1 - \epsilon$	$3 - \varepsilon$	-1	$6 - 2\varepsilon$	-4	2ε	0
d_F^ω	0	0	1	-2	1	0	0
d_F	$1 - \epsilon$	$3 - \epsilon$	3	$-2 - 2\epsilon$	0	2ϵ	0

The renormalization is performed by applying the R operator of the MS scheme to the graphs of the basic theory (5.211), with the contribution $-av'\partial_t v$ understood as the insertion of an IR-irrelevant composite operator, and the renormalized action written as a series in powers of a (Sec. 3.26). From the general rule (3.157) we have

$$S_R(\phi) = S_R^{(0)}(\phi) - a[v'\partial_t v]_R + ..., \qquad (5.212)$$

where $S_R^{(0)}(\phi)$ are the contributions of zeroth order in a, $[v'\partial_t v]_R$ is the renormalized composite operator, and the dots stand for corrections of order a^2 and higher from the counterterms of graphs with insertions of two or more such operators.

We shall refer to the dynamical theory with $a = 0$ as the H_0 model. It is its renormalization that we wish to study, as the contribution of the operator $v' \partial_t v$ gives only corrections to the leading IR-singularities of the Green functions. We have retained this contribution only because we want to use (5.95) to obtain additional information about the renormalized action of the H_0 model $S_R^{(0)}(\phi)$ [the derivation of (5.95) is possible only when the action contains contributions involving $\partial_t \varphi$ for all the fields]. Moreover, it is also necessary to add the term $-av' \partial_t v$ and then take the limit $a \to 0$ to correctly define the graphs of the H_0 model in those cases where the direct substitution of propagators with $a = 0$ into them leads to an ambiguity.

Let us consider the renormalization of the H_0 model (5.211) with $a = 0$, i.e., the contribution $S_R^{(0)}(\phi)$ in (5.212). According to the data of Table 5.6, for the formal exponent of divergence (5.89) of the 1-irreducible graphs of Γ of the model (5.211) with $a = 0$ we obtain

$$d_\Gamma^* = 8 - n_\psi - 3n_{\psi'} - 3n_v - 3n_{v'}. \tag{5.213}$$

However, the real exponent δ_Γ is smaller than the formal one, namely,

$$\delta_\Gamma = d_\Gamma^* - n_{\psi'} - n_{v'}, \tag{5.214}$$

because at least one factor ∂ acting on the fields $\varphi' \equiv \psi', v'$ is isolated at all the vertices of the action (5.211) (and therefore in the counterterms). Actually, this is obvious for the static interaction $\sim \psi' \partial^2 \psi^3 = \partial^2 \psi' \cdot \psi^3$, and for the two intermode-coupling vertices taking into account the transversality of the fields v, v' inside the integral $\int dx...$ we have

$$-\psi' v \partial \psi \equiv -\psi' v_i \partial_i \psi = \partial_i \psi' \cdot v_i \psi$$

and

$$v' \psi \partial (\partial^2 \psi) \equiv v'_i \psi \partial_i \partial_k \partial_k \psi = -v'_i \partial_i \psi \cdot \partial_k \partial_k \psi = \partial_k v'_i \cdot \partial_i \psi \cdot \partial_k \psi + v'_i \partial_k \partial_i \psi \cdot \partial_k \psi,$$

where the last term vanishes owing to the equation $v'_i \partial_k \partial_i \psi \cdot \partial_k \psi = v'_i \partial_i (\partial_k \psi \cdot \partial_k \psi)/2..$ The interaction functionals corresponding to the three vertices in (5.211) can therefore be written as

$$\lambda g_{1B} \psi' \partial^2 \psi^3 / 6 = \lambda g_{1B} \partial^2 \psi' \cdot \psi^3 / 6,$$
$$-\psi' v \partial \psi = \partial_i \psi' \cdot v_i \psi, \quad v' \psi \partial (\partial^2 \psi) = \partial_k v'_i \cdot \partial_i \psi \cdot \partial_k \psi, \tag{5.215}$$

which proves (5.214).

From the exponents (5.213) and (5.214) and the general rules of Sec. 5.15 it is easy to show that all the counterterms needed to eliminate the UV divergences of the H_0 model (5.211) with $a = 0$ have the form of the terms of the basic action. The contribution $\psi' \partial_t \psi$ is not renormalized, because owing to (5.214) the fields φ' must enter into the counterterms only in the form $\partial \varphi'$.

Additional information about the renormalization comes from the Galilean invariance of the H_0 model (5.211), which is ensured by the fact that the symbol ∂_t enters into it only through the covariant derivative $\nabla_t = \partial_t + v\partial$, and the fields v in the other contributions enter only with ∂. The same must also be true for the renormalized action $S_R^{(0)}(\phi)$, because the symmetry is preserved by the MS renormalization. Therefore, it follows from the absence of the counterterm $\sim \psi'\partial_t\psi$ that there must also not be any counterterm $\sim \psi'v\partial\psi$. All other counterterms are allowed by the symmetry and the exponents (5.213) and (5.214), and so the renormalized dynamical action of the H_0 model must have the form

$$S_R^{(0)}(\phi) = -Z_1\lambda\psi'\partial^2\psi' + \psi'[-\partial_t\psi - \lambda\partial^2(Z_2\partial^2\psi - Z_3\tau\psi - Z_4 g_{1B}\psi^3/6) -$$

$$-v\partial\psi] - Z_5\lambda^{-1}g_{2B}^{-1}v'\partial^2v' + v'[Z_6\lambda^{-1}g_{2B}^{-1}\partial^2v + Z_7\psi\partial(\partial^2\psi)]$$

$$(5.216)$$

with constants Z_1–Z_7.

We want to show that the action (5.216) is obtained from the H_0 model (5.210) by the standard procedure of multiplicative renormalization of the fields and parameters consistent with the statics for the field ψ. This is not obvious *a priori*, because the functional (5.216) contains too many constants Z. It is easily verified that for the renormalization in (5.216) to be multiplicative it is necessary (and sufficient) that the constants obey the following relation:

$$Z_1 Z_6 Z_7 = Z_2 Z_5. \qquad (5.217)$$

Similar problems also occurred in the models studied earlier, and they were always solved fairly simply using (5.95), which express the conditions for the renormalized dynamics and statics to be consistent. This is how we shall proceed in the present case, but here matters are complicated by the absence in the H_0 model of the contribution involving $\partial_t v$, without which it is impossible to obtain the required Fokker–Planck equations in the usual manner (Sec. 5.7). We are therefore forced to study the full model H (5.211) and its renormalization including terms linear in a. The renormalized composite operator $[v'\partial_t v]_R$ in (5.212) is given by (3.176):

$$v'\partial_t v \equiv F_1(\phi), \qquad [F_i(\phi)]_R = \sum_k Q_{ik}F_k(\phi), \qquad (5.218)$$

where $F_k(\phi)$ with $k = 1, 2, 3, \ldots$ are the various composite operators which can mix with $v'\partial_t v \equiv F_1(\phi)$ in renormalization, and Q_{ik} are the elements of the corresponding mixing matrix (3.176).

In order to use (5.95) for the functional (5.212), it is necessary to first verify that the latter admits the representation (5.91) through terms linear in a. For this it is necessary to have information about the operators $F_k(\phi)$ in (5.218). Of course, the problem would be simpler if we knew that the operator $F_1 \equiv v'\partial_t v$ is renormalized without mixing or is not renormalized at all.

However, so far that has not been studied. Therefore, to determine the form of the operators $F_k(\phi)$ that can mix with $F_1(\phi) \equiv v'\partial_t v$ in renormalization, we can be guided only by the following general considerations.

1. Since the action involves operators inside the integral $\int dx...$, we can neglect operators like ∂F, $\partial_t F$ with exterior derivatives, as they are "equivalent" to zero.

2. All 1-irreducible graphs with the insertion of a single operator F_1 and external lines of only the fields $\varphi = \psi, v$ are equal to zero because they contain closed loops of retarded lines (Sec. 5.4). Such graphs therefore do not generate operator counterterms, and so any of them must contain at least one factor $\varphi' = \psi', v'$.

3. It follows from the form of the vertices (5.215) that any factor φ' in the counterterms must enter together with ∂, i.e., in the form $\partial\varphi'$.

4. All counterterms must possess the $\psi, \psi' \to -\psi, -\psi'$ symmetry of the model (5.211), i.e., the total number of factors ψ and ψ' in them must be even.

If instead of $v'\partial_t v$ we had taken F_1 to be the Galilean-invariant operator $v'\nabla_t v$ with $\nabla_t = \partial_t + v\partial$ [which is completely natural, judging from the form of (5.198)], we would also be able to state that all its operator counterterms must also be Galilean-invariant, i.e., they must contain the derivative ∂_t only in the combination $\nabla_t = \partial_t + v\partial$ and other factors of v only behind a ∂. This would simplify the analysis, as the number of possible counterterms would be decreased. However, this simplification is unimportant for what follows, and we shall assume that $F_1 = v'\partial_t v$ and use only the above statements 1–4.

From these statements (taking into account the dimensions $d_F^* = 1, 3, 3, 3$ for the fields ψ, ψ', v, and v' and 1, 4, and 2 for the symbols ∂, ∂_t, and τ) it follows that the counterterms F_1 can only be operators of the form $\partial\varphi' \cdot \partial\varphi'$ with additional factors of dimension 2 (possible variants are τ, two ψ, two ∂, ψ and ∂), or operators of the form $\partial\varphi'$ with additional factors of dimension 6. Among them there cannot be any counterterms with two or more ∂_t, and the only one allowed with a single ∂_t is $\sim \partial\psi' \cdot \partial_t\partial\psi \sim \psi' \cdot \partial^2\partial_t\psi$. The operator $F_1 = v'\partial_t v$ cannot be a counterterm itself owing to the fact that it does not contain a ∂ acting on v', and so in (5.218) $Q_{11} = 1$ and $Q_{i1} = 0$ for all $i \neq 1$.

This brief qualitative analysis shows which operators F_k can mix with $F_1 = v'\partial_t v$ in renormalization. The only important thing for us right now is the fact that the coefficient of F_1 is not renormalized ($Q_{11} = 1$) and the functional (5.212) through terms linear in a can be written in the form (5.91) after the expressions (5.218) for $[F_1(\phi)]_R$ is substituted into it:

$$S_R(\phi) = \varphi'\alpha'\varphi' + \psi'[-Z\partial_t\psi + A_\psi(\varphi)] + v'[-a\partial_t v + A_v(\varphi)]. \quad (5.219)$$

The quantities $\{\alpha', Z, A\} \equiv X$ appearing here are written as $X^{(0)} + X^{(1)}$, where $X^{(0)}$ is the zeroth approximation known from (5.216) and $X^{(1)}$ is the correction of order a. Here $Z = 1 + Z^{(1)}$ with $Z^{(1)} = \text{const} \cdot \partial^2$, and the matrix α' for the fields $\varphi = \psi$, v is diagonal in zeroth order in a, while in first order it may contain the cross term $\alpha'_{\psi v}$ and, which is more exotic, explicit dependence on the field ψ.

We have not yet encountered this problem, because it was always assumed that the matrix of Onsager coefficients α_{ab} (in contrast to the intermode coupling coefficients β_{ab}) does not contain any dependence on the fields φ. In the language of the stochastic problem (5.4), allowing α to depend on φ is, of course, rather strange. However, if this dependence appears only in the correction terms of some iteration procedure (here expansion in powers of the parameter a), then the problem with $\alpha = \alpha(\varphi)$ can be formulated correctly. In the language of the functional integral (5.26) the generalization to the case of an action of the type (5.61) with $\alpha = \alpha(\varphi)$ does not generate any fundamental problems at all.

However, the question arises of whether the Fokker–Planck equation (5.57) needed to obtain (5.95) remains valid in this case. The answer is yes. Analyzing the derivation of (5.57), it can be shown that it remains valid even when the coefficients α depend on φ. This is done most simply by using the alternative derivation of (5.36) given in the appendix of [159], from which (5.57) follows.

Therefore, (5.57) and its consequence (5.95) are valid also for the functional (5.219) through contributions linear in a. The rest is simple. Using the functional (5.219) to construct the corresponding quantities (5.94), substituting them into (5.95b) and selecting the contributions of zeroth order in a it is easily seen that the functional (5.219) for $a = 0$ must have the form

$$S_R^{(0)}(\phi) = \psi'\alpha_\psi\psi' + v'\alpha_v v' + \psi'[-\partial_t\psi + \alpha_\psi H_\psi^R - \beta v\partial\psi] + v'[-\alpha_v v + \beta\psi\partial H_\psi^R],$$
(5.220)

where β is an arbitrary coefficient and H_ψ^R is the derivative (5.128) of the renormalized functional (5.209). Identifying (5.216) with (5.220), we conclude that in (5.220) $\alpha_\psi = -Z_1\lambda\partial^2$, $\alpha_v = -Z_5\lambda^{-1}g_{2B}^{-1}\partial^2$, $\beta = 1$, and in (5.216) $Z_5 = Z_6$, $Z_1 Z_7 = Z_2$ in agreement with (5.217).[*]

The representation (5.220) obtained for the functional (5.216) shows that the renormalization of model H_0 is multiplicative and consistent with statics, that the fields v and v' are not renormalized, and that the renormalization of ψ' is determined by that of ψ:

$$Z_v = Z_{v'} = 1, \quad Z_{\psi'} = Z_\psi^{-1}, \quad Z_F = Z_F^{\text{stat}} \quad \text{for all } F = \psi, \tau, g_1. \qquad (5.221)$$

---

[*]A more detailed proof can be found in [242].

For the constants Z in (5.216) we have

$$Z_1 = Z_\lambda Z_{\psi'}^2, \quad Z_2 = Z_\lambda, \quad Z_3 = Z_\lambda Z_\tau, \quad Z_4 = Z_\lambda Z_{g_1} Z_\psi^2,$$

$$Z_5 = Z_6 = Z_\lambda^{-1} Z_{g_2}^{-1}, \quad Z_7 = Z_\psi^2. \tag{5.222}$$

The only important elements in the derivation of (5.220) are the fact that equations (5.95) can be written down for the functional (5.212) and the explicit form of the contributions of zeroth order in a in these equations. The form of these contributions is determined only by the term $S_R^{(0)}(\phi)$ and the coefficient $-aQ_{11} = -a$ for $F_1 \equiv v'\partial_t v$ in (5.212). If we had considered also contributions linear in a in (5.95), we would have obtained some information about the renormalization of the composite operator F_1 in (5.218). The point is that the contributions of the admixtures to $F_1 \equiv v'\partial_t v$ in (5.212) must not violate the consistency with renormalized statics, i.e., they should reduce to only a change of the coefficients α_{ab} and β_{ab} plus a possible dilatation of the field ψ' by a factor of the type Z from (5.219). This, of course, would allow us to significantly decrease the number of unknown coefficients in (5.218). However, it is impossible to obtain complete information about the renormalization of the operator F_1 in this way, although the critical dimension associated with it can be found exactly using the equations $Q_{11} = 1$, $Q_{i1} = 0$ for all $i \neq 1$ (see above) for the mixing matrix Q in (5.218) and relations (5.221) for the fields v and v'.

Returning to model H_0, from (5.221) and (5.222) we conclude that its only new dynamical renormalization constants are Z_λ and Z_{g_2}. They can be calculated in terms of suitable constants (5.222) from the divergences of the 1-irreducible graphs of the basic model (5.211) in the limit $a \to 0$. The ψ, ψ'-field lines in these graphs correspond to the propagators (5.116) and (5.117) with $\varepsilon_k = \lambda k^2(k^2 + \tau)$, $D_k = 2\lambda k^2$, and the v, v'-field lines in the ω, k representation correspond to the propagators

$$\langle vv'\rangle = \langle v'v\rangle^\top = \frac{\lambda g_{2B}}{-i\omega\delta + k^2}P^\perp, \quad \langle vv\rangle = \frac{2\lambda g_{2B}k^2}{|-i\omega\delta + k^2|^2}P^\perp \tag{5.223}$$

with the parameter $\delta \equiv a\lambda g_{2B} \to 0$ (with $\delta^{-1} = \eta$ interpreted as the viscosity coefficient) and the transverse projector $P^\perp$ in the field indices: $P_{is}^\perp(k) = \delta_{is} - k_i k_s/k^2$. In the t, k representation the propagators (5.223) correspond to

$$\langle vv'\rangle = P^\perp \delta^{-1}\lambda g_{2B}\theta(t - t')\exp[-\delta^{-1}k^2(t - t')],$$

$$\langle vv\rangle = P^\perp \delta^{-1}\lambda g_{2B}\exp[-\delta^{-1}k^2|t - t'|]. \tag{5.224}$$

In the limit $\delta = 0$ Eqs. (5.223) take the form

$$\langle vv'\rangle = P^\perp \lambda g_{2B}k^{-2}, \quad \langle vv\rangle = 2P^\perp \lambda g_{2B}k^{-2}, \tag{5.225}$$

which in the t, k representation formally corresponds to

$$\langle vv' \rangle = P^{\perp} \lambda g_{2B} k^{-2} \delta(t - t'), \quad \langle vv \rangle = 2 P^{\perp} \lambda g_{2B} k^{-2} \delta(t - t'). \quad (5.226)$$

Although we are interested in the limiting H_0 model with $\delta = 0$, we cannot just directly substitute the lines (5.225) or (5.226) into all the graphs, because in some of them we would obtain an ambiguity that must be resolved by substituting (5.223) or (5.224) and then taking the limit $\delta \to 0$. Such limits always exist and are uniquely determined.

The constants Z_λ and Z_{g_2} can be calculated, for example, from the UV divergences of the graphs of the 1-irreducible response functions $\Gamma^B_{\varphi'\varphi}(\omega, p)$ of the basic theory (5.211) with $a = 0$:

$$\Gamma^B_{\psi'\psi} = i\omega - \varepsilon_p + \Sigma_{\psi'\psi}, \quad \Gamma^B_{v'v} = P^{\perp}[-\lambda^{-1} g_{2B}^{-1} p^2 + \Sigma_{v'v}], \quad (5.227)$$

where $\varepsilon_p \equiv \lambda p^2 (p^2 + \tau)$ and $\Sigma_{\varphi'\varphi}$ are the sums of the corresponding 1-irreducible graphs. The one-loop contributions of the intermode interaction are represented by the graphs

$$\Sigma_{\psi'\psi} = \;\;+\!\!\!\underset{}{\overbrace{}}\!\!\!+ \;\; + \;\;+\!\!\!\underset{}{\overbrace{}}\!\!\!\times \;\; , \quad \Sigma_{v'v} = \;-\!\!\bigcirc\!\!\cdot\cdot \;. \quad (5.228)$$

The propagators of the fields ψ, ψ' are shown here as solid lines, those of the fields v, v' as dashed lines, and the line ends with slashes correspond to the fields $\varphi' = \psi'$, v' in accordance with the notation of Sec. 5.2. When the propagators (5.223) for the fields v, v' are substituted into (5.228), we obtain the following expressions for the contributions of the intermode interaction to $\Sigma_{\varphi'\varphi}(\omega, p)$:

$$\Sigma_{\psi'\psi} = -\lambda g_{2B}(p^2 + \tau) p_i p_s \int \frac{dk}{(2\pi)^d} \frac{P^{\perp}_{is}(k)}{[q^2 + \tau][-i\omega\delta + k^2 + \delta\varepsilon_q]}, \quad (5.229)$$

$$\Sigma_{v'v} = \frac{1}{(d-1)} \int \frac{dk}{(2\pi)^d} \frac{[k_i P^{\perp}_{is}(p) k_s][p^2 - 2(pk)]}{[q^2 + \tau][-i\omega + \varepsilon_k + \varepsilon_q]}, \quad (5.230)$$

where $q \equiv p - k$, d is the spatial dimension and ε_k is the usual [as in (5.227)] notation. We note that Σ for the vector field denotes the scalar coefficient of the transverse projector in (5.227), and that a factor of p^2 must be separated out of all quantities $\Sigma_{\varphi'\varphi}$.

For $d = 4$ the integral in (5.230) contains a logarithmic UV divergence leading to a pole in ε in dimensional regularization with $d = 4 - 2\varepsilon$. However, the integral (5.229) converges for $d = 4$ and $\delta \neq 0$. The UV divergence and the corresponding pole in ε are manifested in it only for $\delta = 0$, i.e., only after the limit to the H_0 model is taken. This particular example gives a good illustration of the general statement that it is not only possible but even necessary to discard the IR-irrelevant contributions in the action if we wish to use the standard RG technique based on correlations between IR and UV singularities (see Secs. 1.16 and 1.23).

From the dynamical renormalization constants Z_a with $a = \lambda$, g_2 we find the corresponding anomalous dimensions $\gamma_a = \tilde{\mathcal{D}}_\mu \ln Z_a$ and the β function of the second charge (1.117) [for g_1 it is purely static and known from (4.8)]. The two-loop calculation was performed in [171]. Its results in our notation can be written as follows if interpreted in the MS scheme ($d = 4 - 2\varepsilon$):

$$\gamma_\lambda(g) = 3u_2/2 + 0.160 \cdot u_2^2 + u_1^2/6 + ...,$$

$$\gamma_{g_2}(g) = -19u_2/12 - 0.206 \cdot u_2^2 - u_1^2/6 + ..., \quad \beta_{g_2}(g) = g_2[-2\varepsilon - \gamma_{g_2}(g)],$$
$$(5.231)$$

where $u_i \equiv g_i/16\pi^2$. The two β functions of the charges $g \equiv g_{1,2}$ have an IR-attractive fixed point with coordinates

$$u_{1*} = 2\varepsilon/3 + ..., \quad u_{2*} = 12(2\varepsilon)/19 - 0.0635 \cdot (2\varepsilon)^2 + ... \quad (5.232)$$

[the quantity u_{1*} is known from (4.8), and u_{2*} is found from (5.231)], in which

$$\gamma_\lambda^* = 18(2\varepsilon)/19 - 0.013 \cdot (2\varepsilon)^2 + ... \quad (5.233)$$

The equations of critical scaling in the H_0 model have the standard form (5.108). From the general expressions (5.109) and (5.110) taking into account (5.221) and the data of Table 5.6, we find the critical dimensions Δ_F for the frequency ω and all the fields:

$$z \equiv \Delta_\omega = 4 - \gamma_\lambda^*, \quad \Delta_\psi = d_\psi + \gamma_\psi^*,$$
$$\Delta_{\psi'} = d_{\psi'} - \gamma_\psi^*, \quad \Delta_v = \Delta_{v'} = -1 + \Delta_\omega, \quad (5.234)$$

where $d_\psi = d/2 - 1$, $d_{\psi'} = d/2 + 1$ are the canonical dimensions of these fields, and $\gamma_\psi^* = \eta/2$ is the ordinary static exponent of the simple ψ^4 model. In the notation of [171], $2u_2 \equiv f$, $\gamma_\lambda^* = x_\lambda + \eta$, and $-\gamma_{g_2}^* = 4 - d = x_{\overline{\eta}} + x_\lambda + \eta$. The coefficients of the contributions $\sim u_2^2$ in (5.231) are obtained from the ε expansions of f_*, x_λ, and $x_{\overline{\eta}}$ given in [171]. The results of the two-loop calculation for model H are also given in [158], but they contain obvious misprints (even in the signs of the one-loop contributions) and do not agree with the results of [171]. Nevertheless, in analyzing the experimental data it is usual to use the values of the two correction exponents ω_f and ω_w given in [158], because they were not calculated in [171].

In conclusion, let us study the critical dimensions of the "effective" heat conduction and viscosity coefficients λ_{eff} and η_{eff}. These are important physical characteristics of the system [153] that can be determined from the renormalized response functions (5.227) of the H_0 model with $a = 0$ using the expressions

$$\lambda_{\text{eff}}(\tau) = -\chi_\psi \cdot p^{-2} \Gamma_{\psi'\psi}^R \big|_{\omega=0, p\to 0}, \quad \eta_{\text{eff}}(\tau, \omega) = -a^{-1} p^{-2} \Gamma_{v'v}^R \big|_{p\to 0},$$
$$(5.235)$$

where χ_ψ is the static susceptibility of the simple ψ^4 model, and $\Gamma^{\mathrm{R}}_{v'v}$ is understood as a scalar factor, the coefficient of the transverse projector in the indices.

The definition of λ_{eff} in (5.235) is equivalent to the approximate representation $\Gamma^{\mathrm{R}}_{\psi'\psi} \simeq i\omega - \lambda_{\mathrm{eff}}(\tau)p^2\chi_\psi^{-1}(\tau)$ at small (comparable to τ) p and ω. To obtain (5.235) for η_{eff} it is necessary to take into account the corrections to the H_0 model linear in a. They give, first of all, an addition $ia\omega$ to the function $\Gamma^{\mathrm{R}}_{v'v}$, and, second, corrections to the latter, which are multiples of ap^2 and which can be neglected as they are IR-irrelevant. The expression $ia\omega + \Gamma^{\mathrm{R}}_{v'v}$ at small (comparable to τ and ω) wave vectors p is approximately replaced by $a[i\omega - p^2\eta_{\mathrm{eff}}(\tau,\omega)]$, which leads to (5.235). We stress the fact that (5.235) involves the renormalized functions of the H_0 model with $a = 0$. Its function $\Gamma^{\mathrm{R}}_{v'v}$, in contrast to $\Gamma^{\mathrm{R}}_{\psi'\psi}$, does not contain a contribution of the type $i\omega$ and is proportional to p^2. This makes it possible to determine η_{eff} for $\omega \neq 0$, which is not natural for λ_{eff}.

The quantities (5.235) have definite critical dimensions that are easily found from the following arguments. The renormalized response functions $\langle\varphi\varphi\rangle$ of the H_0 model in coordinate space have dimensions $\Delta_\varphi + \Delta_{\varphi'}$, to which is added $-d - \Delta_\omega$ upon transformation to ω, p space and an overall minus sign upon passage to 1-irreducible functions of the type (5.227). Therefore,

$$\Delta[\Gamma^{\mathrm{R}}_{\varphi'\varphi}(\omega, p)] = d + \Delta_\omega - \Delta_\varphi - \Delta_{\varphi'}. \qquad (5.236)$$

From this, using (5.234) and $\Delta[\chi_\psi] = \eta - 2$ for the static susceptibility we find the critical dimensions of the quantities in (5.235):

$$\Delta[\lambda_{\mathrm{eff}}] = z + \eta - 4, \quad \Delta[\eta_{\mathrm{eff}}] = d - z, \quad (z \equiv \Delta_\omega), \qquad (5.237)$$

(in the notation of [171], $\Delta[\lambda_{\mathrm{eff}}] = -x_\lambda$ and $\Delta[\eta_{\mathrm{eff}}] = -x_{\overline{\eta}}$). Therefore, in the critical regime up to unimportant coefficients we have

$$\lambda_{\mathrm{eff}}(\tau) \sim \xi^{4-z-\eta}, \quad \eta_{\mathrm{eff}}(\tau,\omega) \sim \xi^{z-d}f(\omega\xi^z), \qquad (5.238)$$

where $\xi \sim \tau^{-\nu}$ with $1/\nu \equiv \Delta_\tau$ is the correlation length and f is a scaling function of an argument whose critical dimension is zero.

With this we conclude our discussion of model H. More details can be found in [153], [158], and [171].

5.26   Sound propagation near the critical point

In the simplest approximation, valid far from the critical point for sufficiently small frequencies (compared to molecular ones), sound waves are described by the hydrodynamical equations of an ideal fluid [172]:

$$\partial_t F + \partial_i(Fv_i) = 0, \quad F = S, \rho, \quad \rho[\partial_t v_i + (v\partial)v_i] + \partial_i p = 0, \qquad (5.239)$$

in which S and ρ are the entropy and mass (volume) densities and p is the pressure. The quantities $F = S, \rho, p$ are written as $F_0 + \delta F$, and all the δF and v are assumed small. Equations (5.239) are linearized in these quantities, which gives

$$\partial_t \delta F + F_0 \partial_i v_i = 0, \quad F = S, \rho, \quad \rho_0 \partial_t v_i + \partial_i \delta p = 0. \tag{5.240}$$

The solutions for δF and v are sought in the form of plane waves $\sim \exp[-i\omega t + ik\mathbf{x}]$, and so from (5.240) it follows that $v_\perp = 0$ and $\delta S' = 0$, where $S' \equiv S/\rho$ is the entropy per unit mass [from (5.240) we obtain $\partial_t \delta S' = 0$, which leads to $\delta S' = 0$ for the wave]. The linear coupling between $\delta \rho$ and δp is found from the relations of equilibrium thermodynamics: $\delta \rho(x) = \alpha \delta p(x)$, where $\alpha \equiv d\rho/dp$ for $S' = $ const. Then (5.240) reduces to the equations

$$\alpha \partial_t \delta p + \rho_0 \partial_i v_i = 0, \quad \rho_0 \partial_t v_i + \partial_i \delta p = 0. \tag{5.241}$$

This is a linear system of equations of the type $L\theta = 0$ for the pair of fields $\theta \equiv \delta p$, $v_\parallel$ with the 2×2 matrix

$$L = \begin{vmatrix} \alpha \partial_t, & \rho_0 \partial \\ \partial, & \rho_0 \partial_t \end{vmatrix} \sim \begin{vmatrix} -i\alpha\omega, & i\rho_0 k \\ ik, & -i\rho_0\omega \end{vmatrix} \tag{5.242}$$

(the second expression is the momentum–frequency representation). The condition $\det L = 0$ for the operator (5.242) gives the dispersion law (the relation between ω and k) for sound waves:

$$\det L = 0 \Rightarrow \omega = ck, \quad c^{-2} = \alpha = (d\rho/dp)\big|_{S'=\text{const}}, \tag{5.243}$$

where c is the speed of sound, defined by the last equation in (5.243). The derivative entering into this expression is proportional to the adiabatic compressibility $K_{S'} = \rho^{-1}(d\rho/dp)\big|_{S'=\text{const}}$ [see (1.24)] and can be expressed in terms of the specific heat C_p, $C_\rho \equiv C_V$ and the isothermal compressibility $K_T = \rho^{-1}(d\rho/dp)\big|_{T=\text{const}}$ using (1.27):

$$c^{-2} = \rho K_{S'} = \rho K_T C_\rho C_p^{-1}. \tag{5.244}$$

The theory described briefly above works well far from the critical point. It can be improved by including the effect of heat conduction and viscosity leading to the appearance of weak damping of the sound waves. Near the critical point the interaction of sound with critical fluctuations becomes important, i.e., we arrive at the problem of wave propagation in a medium with random parameters, "noise." It follows from the analysis of Sec. 5.24 that the random fields ψ and $v_\perp$ play the role of this noise, because the fluctuations of the fields m and $v_\parallel$ are IR-irrelevant. More precisely, the field m contains a "fluctuation part" that is expressed in terms of ψ by the first equation in (5.207) and, therefore, is eliminated. In analyzing the problem of sound propagation using the full model (5.202) or the stochastic equations (5.55) equivalent to it, we must therefore assume that

$$m = \overline{m} + \delta m, \quad \overline{m} \equiv g_{20}\psi^2/2, \tag{5.245}$$

where $\overline{m}$ is the fluctuation part defined by the first equation in (5.207) and δm is an additional sound component, the analog of δp in (5.241).

The effect of fluctuations on the sound wave is manifested explicitly in (5.244), because the quantities on its right-hand side contain critical singularities. Let us assume that an experiment is performed at $T > T_c$ for a definite amount of matter contained in a fixed volume. Then T_c is approached along a critical isochore [which justifies, in particular, the use of the functional (5.208) without the contribution $h\psi$], and the distance to T_c is determined by the parameter $\tau = T - T_c$. It is known (Sec. 1.8) that the quantities K_T and C_p in (5.244) have the same critical singularity $\sim \tau^{-\gamma}$ (like the magnetic susceptibility), while C_ρ has a weaker singularity $\sim \tau^{-\alpha}$. The singularities of K_T and C_p in (5.244) cancel each other out, and so

$$c^{-2} \sim C_\rho \sim \tau^{-\alpha}, \tag{5.246}$$

where α is the critical exponent of the specific heat for the simple ψ^4 model. It follows from (5.246) that the speed of sound tends to zero as $\tau^{\alpha/2}$ for $\tau \to 0$. The singularity (5.246) is the result of the interaction of the sound wave with critical fluctuations and must be explained by theory.

A simple modification of the simplified version (5.206) of the full fluctuation model (5.202) can be used as the basis of this theory. The modification consists of keeping the contributions involving $\partial_t m$ and $\partial_t v_\parallel$ in the coefficients of m' and $v'_\parallel$. These were discarded previously as they were IR-irrelevant for the critical dynamics of the fields ψ and $v_\perp$, but they are needed to describe sound. Then the integration over m' and $v'_\parallel$ leads to the sound equations instead of the simple relations (5.207):

$$m_0^{-1}\partial_t m + m_0^{-1}\lambda_{22}\partial^2 H_m + \partial v_\parallel = 0,$$
$$m_0^{-1}a\partial_t v_\parallel - \partial H_m = 0, \tag{5.247}$$

in which $m = \overline{m} + \delta m$ and $H_m = -m + \overline{m} = -\delta m$ according to (5.207) and (5.245) [the coefficients of $\partial_t m$ and $\partial_t v_\parallel$ have appeared owing to the dilatation (5.203) that occurs in passing from the functional (5.202) to (5.206)].

The term containing λ_{22} in (5.247) determines the wave damping owing to heat conduction effects (in Sec. 5.24 it was shown that the parameters λ_{11}, λ_{12}, and λ_{22} are proportional to λ_{SS} for the entropy, i.e., the heat conduction coefficient). These effects, like the effect of the term containing the longitudinal viscosity already discarded in (5.206), will not be included in what follows, because they are unimportant compared to the effects of interaction between the wave and the critical fluctuations.

For definiteness, we shall assume that the sound wave is generated by external harmonic [$\sim \exp(-i\omega t)$] sources $J \equiv J_m, J_v$ of low intensity added to the equations for the fields m and $v_{\|}$. These sources generate a sound wave of frequency $\omega = \omega_s$ representing the entire field $v_{\|}$ and the "sound component" δm of the field m in (5.245). Adding these sources to the right-hand side of (5.247), substituting into them m from (5.245) and $H_m = -\delta m$, and discarding the damping contribution involving λ_{22}, we find

$$m_0^{-1}\partial_t(\delta m + g_{20}\psi^2/2) + \partial v_{\|} = J_m,$$
$$m_0^{-1}a\partial_t v_{\|} + \partial(\delta m) = J_v. \tag{5.248}$$

The interaction of the sound wave with the critical fluctuations is determined by the contribution $\sim \psi^2$. Without it, (5.248) become a linear system analogous to (5.241):

$$L_0\theta = J, \quad \theta \equiv \delta m, \, v_{\|}, \quad L_0 = \begin{vmatrix} m_0^{-1}\partial_t, & \partial \\ \partial, & m_0^{-1}a\partial_t \end{vmatrix}. \tag{5.249}$$

The condition $\det L_0 = 0$ leads to the standard dispersion law:

$$\omega = c_0 k, \quad c_0^{-2} = am_0^{-2}, \tag{5.250}$$

where the constant c_0 is the speed of sound in this approximation.

Now we need to take into account the effect of the perturbation $\sim \psi^2$ in (5.248). In estimating this effect it is necessary to bear in mind the very important qualitative difference between the sound ($\theta \equiv \delta m, v_{\|}$) and fluctuation ($\varphi \equiv \psi, v_\perp$) fields. For the latter, the characteristic length for $\tau \to 0$ is the correlation length $\xi \sim \tau^{-\nu}$, the characteristic momentum (wave vector) is $k_{\mathrm{fl}} = \xi^{-1} \sim \tau^\nu$, and the characteristic frequency is $\omega_{\mathrm{fl}} \sim k_{\mathrm{fl}}^z \sim \tau^{z\nu}$, where $z \equiv \Delta_\omega$ is the critical frequency dimension in model H. Qualitative estimates can be made simply by using the canonical dimensions, neglecting anomalies. Then $\nu = 1/2$, $z = 4$, and for $\tau \to 0$ we obtain $k_{\mathrm{fl}} \sim \tau^{1/2}$ and $\omega_{\mathrm{fl}} \sim k_{\mathrm{fl}}^4 \sim \tau^2$. For sound fields $\theta \equiv \delta m, v_{\|}$ the frequency ω_s can be chosen arbitrarily if it is determined by an external source. From the experimental point of view the broad frequency range in which ω_s can be both much smaller and much larger than the characteristic fluctuation frequency ω_{fl} is of interest [173]. Therefore, for a qualitative estimate these can be assumed to be quantities of the same order: $\omega_s \sim \omega_{\mathrm{fl}} \sim \tau^2$. From this for the characteristic wave vectors of the sound fields we obtain $k_s = c_0^{-1}\omega_s \sim \tau^2$, which for $\tau \to 0$ leads to the inequality $k_s \ll k_{\mathrm{fl}}$.

Therefore, for qualitative estimates it can be assumed that

$$\omega_s \sim \omega_{\mathrm{fl}} \sim \tau^2, \quad k_s \sim \tau^2 \ll k_{\mathrm{fl}} \sim \tau^{1/2} \tag{5.251}$$

for $\tau \to 0$. Here $k_s = \lambda_s^{-1}$ is interpreted as the inverse wavelength of the sound, and $k_{\mathrm{fl}} = \xi^{-1}$ is the inverse correlation length, so that the inequality

(5.251) is equivalent to $\lambda_s \gg \xi$. This means that the wavelength of sound is much larger than the characteristic size of the fluctuations ξ. Therefore, in estimating the back effect of the sound fields θ on the fluctuation fields of model H, the fields θ can be treated as being nearly spatially uniform (although their rate of variation is comparable to the analogous quantity for the fluctuation fields). The inequality $\lambda_s \gg \xi$ is satisfied throughout the experimentally attainable region of ω_s and τ, and even when ω_s exceeds ω_{fl} by three or four orders of magnitude [173].

Let us return to (5.248). The quantity ψ^2 entering into these equations should be viewed as the corresponding random field of model H (5.210), lightly perturbed by the sound fields. This perturbation is weak (because the sound fields are determined by the sources in (5.248) and therefore can be assumed arbitrarily small), but it is impossible to completely neglect it. Actually, if we had proceeded in this manner, we would have had to assume that ψ in (5.248) is the random field of the unperturbed model H. It does not depend on the sound fields, and so the contribution involving ψ^2 in (5.248) would then be only a random addition to the external source J_m. Then only the sources would be changed in (5.249), and the form of the operator L_0 and the dispersion law (5.250) would remain the same.

Therefore, the effect of the sound fields θ on model H (5.210) needs to be taken into account. The corresponding perturbation is easily found by again examining the procedure of passing from model (5.206) to (5.210) taking into account the changes now introduced into (5.206). They reduce to the replacement of the simple constraints (5.207) by the sound equations (5.248). Their solutions are the quantities $\theta \equiv \delta m, v_\parallel$ formerly assumed to be zero. We therefore must now include the contribution $v_\parallel \neq 0$ and the addition δm to the field m in the terms of (5.206) corresponding to model H (5.210). This leads, first, to an addition from $v_\parallel$ in $\partial(v\psi)$, second, to the replacement of $H_m = 0$ by $H_m = -\delta m$, and, third, to the appearance of an addition of order δm to the variational derivative H_ψ of the action (5.197). We see from (5.197) that H_ψ contains the contribution $g_{20}m\psi$, and so the addition to H_ψ is $g_{20}\psi\delta m$. Taking into account all these additional terms, for the action functional of the perturbed model H we obtain

$$S'(\phi) = S(\phi) + \delta S(\phi), \tag{5.252}$$

where $S(\phi)$ is the unperturbed action and

$$\delta S(\phi) = \psi'[-\lambda_0 g_{20}\partial^2(\psi\delta m) - \partial(v_\parallel\psi)] + v'_\perp[g_{20}\psi\partial(\psi\delta m) - g_{20}\psi^2\partial(\delta m)/2] \tag{5.253}$$

is the addition due to the sound fields. To avoid misunderstandings, we note that in this section we use g_{20} to denote the second charge in (5.197). In model H (5.208), (5.210) this parameter does not appear at all, because in the absence of sound fields the interaction between ψ and m in (5.197) led only to the renormalization $g_{10} \to g_{10} - 3g_{20}^2$ (Sec. 5.24) and this entire combination was denoted by g_{10} in (5.208). The parameter g_{20} in (5.210) is the charge of the intermode interaction and has no relation to g_{20} from (5.197). We have used the notation (5.210) for the coefficient λ in (5.253) [$\lambda_0 \equiv \lambda'_{11}$ in (5.206)].

In (5.253) we have given the complete expression for $\delta S(\phi)$, but it is actually only the first term in it that is important. In fact, we are interested only in the regular harmonic component of the sound fields θ, which is generated by the contribution of the harmonic external sources in (5.248). Moreover, the solution of these equations also contains a fluctuation addition generated by the contribution $\sim \psi^2$ of the unperturbed model H, which plays the role of an additional (nonharmonic) random source. This extra fluctuation addition in θ is not of interest to us (Sec. 5.24) and can simply be eliminated by subtracting the zeroth approximation, i.e., the same quantity in the unperturbed model H, from the exact random quantity ψ^2 in (5.248).

Having done this, we can assume that $v_\parallel$ and δm in (5.253) are just small harmonic components of these fields. Owing to the inequality $\lambda_s \gg \xi$ (see above), these quantities can be treated in model H as being nearly spatially uniform, so that their gradients can be neglected. In this approximation the last two terms on the right-hand side of (5.253) vanish, and in the first expression $\partial^2(\psi \delta m)$ can be replaced by $\partial^2 \psi \cdot \delta m$. This addition to the action of model H corresponds to the temperature shift $\tau \to \tau - g_{20}\delta m$ neglecting corrections from the gradient δm. Since τ is small near the critical point, small additions $\sim \delta m$ to it can be important. This distinguishes the given term of (5.253) from the contribution $\sim \partial(v_\parallel \psi)$, which when the gradient $\partial v_\parallel$ is neglected is only a small correction $v_\parallel \partial \psi$ to the quantity $v_\perp \partial \psi$, which is not small in model H.

It then follows that in this problem the exact expression (5.253) can be replaced by

$$\delta S(\phi) = -\lambda_0 g_{20} \psi' \partial^2 \psi \cdot \delta m. \tag{5.254}$$

Here we have taken into account only the effect of the shift of τ corresponding to harmonic oscillations of the temperature in the sound wave due to adiabatic expansion and compression [172]. It is the main mechanism by which the wave interacts with critical fluctuations.

The effect of the small addition (5.254) to the action of model H on the random quantity ψ^2 in (5.248) is very simple to estimate. The unperturbed quantity ψ^2 corresponds to the functional average with weight $\exp S(\phi)$, and the exact quantity to the average with weight $\exp[S(\phi) + \delta S(\phi)] \cong [1 + \delta S(\phi)] \exp S(\phi)$, and so

$$\psi^2(x)\big|_{\text{exact}} \cong \psi^2(x) + \psi^2(x)\delta S(\phi) \tag{5.255}$$

with the fields of the unperturbed model H on the right-hand side.

Substituting (5.255) with $\delta S(\phi)$ from (5.254) into (5.248) and discarding the unimportant (in our case; see above) fluctuation contribution $\sim \psi^2$ to the source J_m, we obtain the system of equations

$$m_0^{-1}\partial_t[\delta m(x)-\lambda_0 A'\psi^2(x)\int dx'\,\psi'(x')\partial^2\psi(x')\cdot\delta m(x')]+\partial v_\|(x) = J_m(x),$$

$$m_0^{-1}a\partial_t v_\|(x) + \partial\delta m(x) = J_v(x), \qquad A' \equiv g_{20}^2/2.$$

$$(5.256)$$

Here we have explicitly indicated the integration over $x \equiv t, \mathbf{x}$, which was left implicit in expressions like (5.254).

We have arrived at the well-known problem of the propagation of a wave in a medium with random parameters (noise). The system of equations (5.256) for the sound fields $\theta \equiv \delta m, v_\|$ can be written compactly as

$$L\theta = J, \qquad L = L_0 - V, \tag{5.257}$$

where L_0 is the unperturbed operator (5.249) and V is a perturbation containing the random fields ψ, ψ' with distribution function specified by model H (5.210). The problem (5.257) is further defined by the natural retardation condition. We are interested in the solution $\theta = L^{-1}J$ averaged over the noise distribution, i.e., the quantity $\langle\theta\rangle = \langle G\rangle J$, $G \equiv L^{-1}$. Here $\langle G\rangle = \langle L^{-1}\rangle$ is interpreted as the full propagator of the wave. It can be calculated from perturbation theory by averaging the quantity $G = L^{-1} = (L_0 - V)^{-1} = G_0 + G_0 V G_0 + G_0 V G_0 V G_0 + ...$ with $G_0 \equiv L_0^{-1}$ over the noise distribution. The result can be represented by Feynman-like graphs, and the full propagator will satisfy the Dyson equation:

$$\langle G\rangle^{-1} = L_0 - \Sigma,$$

$$\Sigma = \langle V\rangle + \langle V G_0 V\rangle - \langle V\rangle G_0 \langle V\rangle + ..., \tag{5.258}$$

in which Σ is the sum of 1-irreducible self-energy graphs. The desired dispersion law is found by solving the equation

$$\det\langle G\rangle^{-1} = \det[L_0 - \Sigma] = 0 \tag{5.259}$$

in the momentum–frequency representation.

We note that it is not difficult to reformulate the problem as a quantum field model, representing the desired quantity $\langle G\rangle$ as a functional integral:

$$\langle G(x,x')\rangle \sim \int D\phi\, D\theta\, D\theta'\, \theta(x)\theta'(x')\exp[-\theta'L\theta + S(\phi)]. \tag{5.260}$$

Gaussian integration over the independent fields θ, θ' gives $L^{-1}(x,x') \equiv G(x,x')$, and subsequent integration over the noise fields ϕ with given weight $\exp S(\phi)$ [where here $S(\phi)$ is the action of model H (5.210)] corresponds to averaging over a given noise distribution. The representation (5.260) allows the desired quantity $\langle G\rangle$ to be identified as the Green function $\langle\theta(x)\theta'(x')\rangle$ of the quantum field model involving the set of fields θ, θ', ϕ, the action functional of which is the exponent in (5.260). Therefore, the standard Feynman diagram technique (Ch. 2) can be used to calculate $\langle G\rangle$ and Σ.

In our case things are easier, because in (5.258) we can limit ourselves to the simplest approximation $\Sigma = \langle V \rangle$. This contribution is determined by model H, and all the other terms give only IR-irrelevant corrections from the interaction of the H-model fields with the sound modes (Sec. 5.24). This can be checked explicitly by estimating the contributions of second order in V in (5.258) for Σ, representing them by the corresponding Feynman graphs. The UV-divergent integrals (natural for IR-irrelevant insertions) over the momenta of the internal sound line in these graphs can be cut off at fluctuation scales $k_{fl} \sim \tau^{1/2}$, and then the IR-irrelevance of these contributions becomes obvious.

Therefore, in our problem we have, up to IR-irrelevant corrections,

$$\langle G \rangle^{-1} = \langle G^{-1} \rangle = \langle L \rangle, \tag{5.261}$$

i.e., the operator L itself in (5.256), rather than $G = L^{-1}$, can be averaged over the noise distribution.

After the averaging $\langle ... \rangle$ the contribution of the perturbation V in (5.256) can be written as

$$2m_0^{-1} A' \partial_t \int dx' \overline{R}(x, x') \delta m(x'), \tag{5.262}$$

where $\overline{R}(x, x')$ is the correlator of the two composite operators of model H:

$$\overline{R}(x, x') \equiv \langle F(x) F'(x') \rangle, \tag{5.263}$$

$$F(x) \equiv [\psi^2(x) - \langle \psi^2(x) \rangle]/2, \quad F'(x) \equiv -\lambda_0 \psi'(x) \partial^2 \psi(x).$$

Subtraction of the constant $\langle \psi^2(x) \rangle$ from the operator $\psi^2(x)$ is allowed because it enters into (5.256) in the form $\partial_t \psi^2(x)$. The coefficients in the operators F and F' have been introduced for convenience.

In momentum–frequency space the convolution in (5.262) becomes the product $\overline{R}(\omega, k) \delta m(\omega, k)$, where ω and k are the parameters of the sound wave. Since the correlator $\overline{R}(\omega, k)$ is determined by model H, the natural scales for its arguments ω and k are the corresponding quantities ω_{fl} and k_{fl} from (5.251). Therefore, the inequality (5.251) for the wave vectors allows us to neglect the k dependence of the function $\overline{R}(\omega, k)$ and set $k = 0$ in it. In this approximation the integral in (5.262) is replaced by the expression

$$\int dt' R(t, t') \delta m(t', \mathbf{x}), \quad \text{where} \quad R(t, t') \equiv \int d\mathbf{x}' \overline{R}(x, x'). \tag{5.264}$$

The equation (5.262) in the approximation (5.264) in ω, k space takes the form $-2i\omega m_0^{-1} A' R(\omega) \delta m(\omega, k)$, and so for the operator L in (5.256) we have

$$\langle G \rangle^{-1} = \langle L \rangle = \begin{vmatrix} -i\omega m_0^{-1}[1 + 2A'R(\omega)] & ik \\ & \\ ik & -i\omega a m_0^{-1} \end{vmatrix}. \tag{5.265}$$

From this we find the dispersion law according to (5.259):

$$k^2 = \omega^2[c_0^{-2} + A_0 R(\omega)], \quad A_0 \equiv 2A' c_0^{-2}, \tag{5.266}$$

with the constants c_0^{-2} from (5.250) and A' from (5.256).

For given frequency ω the solution of the dispersion equation (5.266) for $k \equiv |k|$ is sought in the form $k = \omega c^{-1} + i\alpha$, from which we find the positive parameters c (the speed of sound) and α (the damping coefficient). It is usual to use not α but the dimensionless quantity $\alpha_\lambda \equiv \alpha \lambda_s = 2\pi c\alpha/\omega$, the damping per wavelength λ_s. Under realistic conditions $\alpha \ll \omega/c$, and then from (5.266) we have

$$c^{-2}(\omega) = c_0^{-2} + A_0 \mathrm{Re} R(\omega), \qquad \alpha_\lambda(\omega) c^{-2}(\omega) = \pi A_0 \mathrm{Im} R(\omega). \qquad (5.267)$$

We recall that $R(\omega)$ is the Fourier transform of the correlator $R(t,t')$ in (5.264). This correlator is interpreted physically as the function describing the response of the quantity $\langle F(x) \rangle$ to a temperature variation depending only on time $\delta\tau(t')$ in model H (up to a sign). The function $R(t,t')$ does not have a δ-function singularity for $t = t'$, and so its Fourier transform $R(\omega)$ vanishes for $\omega \to \infty$ and from (5.267) we have

$$R(\infty) = 0, \quad c(\infty) = c_0, \quad \alpha_\lambda(\infty) = 0. \qquad (5.268)$$

Therefore, the constant c_0 is interpreted as the limiting speed of sound for $\omega \to \infty$. We note that this quantity is independent of τ, in contrast to $c(0)$. From the viewpoint of the theory of critical behavior, the parameters c_0 and A_0 in (5.267) are nonuniversal constants independent of τ. The parameter c_0 does not enter into the right-hand side of the second equation in (5.267), and it can be eliminated from the first equation by subtracting the quantity $c^{-2}(\infty) - c_0^{-2}$.

The response function $R(t,t')$ can be expressed in terms of the correlator

$$D(t,t') \equiv \int d\mathbf{x}' \langle F(x) F(x') \rangle \qquad (5.269)$$

of two F operators from (5.263) using the FD theorem (5.77) for model H (5.210). Here it must be remembered that the formulation (5.77) corresponds to the standard notation for the action (5.61) with unit coefficients of all the $\partial_t \varphi$, whereas in (5.210) we have introduced the coefficient a in front of $\partial_t v$. Removing it by rescaling $v' \to a^{-1} v'$, for the coefficients α and β in (5.77) we obtain $\alpha_{\psi\psi} = -\lambda_0 \partial^2$, $\alpha_{vv} = -\eta_0 \partial^2$, and $\beta_{\psi v} = -\beta_{v\psi} = a^{-1} \partial\psi$. Taking the operator F from (5.263) as M in (5.77) and the vector $(\psi, 0)$ as N, we obtain

$$\overline{\varphi} N \equiv \varphi'(\alpha + \beta) N = \psi' \alpha_{\psi\psi} \psi + v' \beta_{v\psi} \psi = -\lambda_0 \psi' \partial^2 \psi - a^{-1} v' \psi \partial\psi.$$

The last contribution vanishes inside the integral $\int d\mathbf{x}'...$, and so the FD theorem (5.77) leads to the expression

$$R(t,t') = \theta(t-t') \int d\mathbf{x}' \langle F(x)\psi(x')\partial_{t'}\psi(x') \rangle = \theta(t-t')\partial_{t'} D(t,t') \qquad (5.270)$$

relating the function R from (5.264) to the correlator (5.269).

The equation (5.270) can be rewritten as

$$R(t, t') = \delta(t - t') D^{\text{stat}} + \partial_{t'} D^{\text{ret}}(t, t'),$$

$$D^{\text{stat}} \equiv D(t, t) = C, \qquad D^{\text{ret}}(t, t') \equiv \theta(t - t') D(t, t'). \tag{5.271}$$

According to the general principles governing the correspondence between dynamics and statics (Sec. 5.8), the correlator $D(t, t)$ is independent of t and coincides with the corresponding static object, in this case interpreted as the specific heat $C = C_V$ in the simple ψ^4 model. In frequency space (5.271) takes the form

$$R(\omega) = D^{\text{stat}} + i\omega D^{\text{ret}}(\omega), \qquad D^{\text{stat}} = C. \tag{5.272}$$

Equations (5.267), (5.268), and (5.272) are the final results for the dispersion law in this problem. The functions R and D entering into them are calculated in model H (5.210) and depend on all its parameters.

We are interested in the critical region, i.e., the IR asymptote $\tau \to 0$, $\omega \sim \tau^2 \to 0$ of these expressions. It can be studied using the renormalized H_0 model (Sec. 5.25), because the passage from unrenormalized to renormalized quantities in (5.267) leads only to a change of the nonuniversal parameters A_0 and c_0. The operator F in statics defined in (5.263) is renormalized multiplicatively (i.e., $F = Z_F F_R$) with $Z_F = Z_\tau^{-1}$ (Sec. 4.9), because the subtraction of $\langle \psi^2 \rangle$ in (5.263) eliminates the contribution of the constant in the renormalization. It follows from the general principles governing the correspondence between dynamics and statics that F is renormalized in exactly the same way in the dynamical H_0 model. The same arguments regarding renormalization hold for the operator $F' \sim \psi' \partial^2 \psi$ inside the integral $\int d\mathbf{x}...$ as for the operator $v' \partial_t v$ studied in Sec. 5.25. It follows from them that the operator F' in the H_0 model is also renormalized multiplicatively (it has nothing to mix with), and then from (5.271) for $t \neq t'$ we find that the renormalization constants of F and F' are equal:

$$Z_{F'} = Z_F = Z_\tau^{-1}. \tag{5.273}$$

We note that the constant $Z_{F'}$ for the operator F' in (5.263) is the product of Z_λ and the renormalization constant of the monomial $\psi' \partial^2 \psi$ itself.

Green functions with two composite operators are renormalized following the rule (3.159), and, in general, along with the replacement $F \to F_R$ it is necessary to introduce an additive counterterm for the operator pair (Sec. 3.25). From dimensional arguments [in generalizing (3.152) to dynamics, $d_{\varphi, F}$ should be viewed as the total dimensions and d should be replaced by $d + d_\omega$ everywhere] it is easily established that the pair counterterm is not required for the correlator (5.269) in dynamics, so that $D_R(\omega) = Z_F^{-2} D(\omega)$ and analogously for $D^{\text{ret}}(\omega)$ in (5.272). However, for the analogous static object $D^{\text{stat}} = C$ the pair counterterm is needed to generate the inhomogeneity in the RG equation for the renormalized specific heat C_R (Sec. 4.5). It then follows from (5.272) and (5.273) that the pair counterterm is needed also for the response function $R(\omega)$ and that it is exactly the same as for the specific heat C, so that the inhomogeneity in the RG equation for the renormalized response function $R_R(\omega)$ will also be exactly the same.

This can be summarized as follows:

$$C_R = Z_F^{-2} C + \Delta C, \quad R_R = Z_F^{-2} R + \Delta C, \quad D_R^{\mathrm{ret}} = Z_F^{-2} D^{\mathrm{ret}}, \qquad (5.274)$$

where ΔC is the pair counterterm in the specific heat, a constant independent of ω and τ and containing poles in ε. The second equation in (5.274) can be used to rewrite (5.267) in terms of renormalized quantities, which leads only to replacement of the nonuniversal parameters $c_0, A_0 \to c_R, A_R$:

$$c^{-2}(\omega) = c_R^{-2} + A_R \mathrm{Re} R_R(\omega), \quad \alpha_\lambda(\omega) c^{-2}(\omega) = \pi A_R \mathrm{Im} R_R(\omega),$$
$$\qquad (5.275)$$
$$c_R^{-2} = c_0^{-2} - A_0 Z_F^2 \Delta C, \quad A_R = A_0 Z_F^2.$$

It follows from (5.274) that the renormalized response function R_R satisfies the same RG equation as the specific heat C_R, the only difference being the particular form of the operator $\widetilde{\mathcal{D}}_\mu = \mathcal{D}_{\mathrm{RG}}$. If in (5.274) there were no contributions ΔC, this equation for any of the three X_R in (5.274) would have the form $[\mathcal{D}_{\mathrm{RG}} + 2\gamma_F] X_R = 0$ with $\gamma_F = -\gamma_\tau$ according to (5.273). The addition ΔC leads to the appearance of an inhomogeneous term $[\mathcal{D}_{\mathrm{RG}} + 2\gamma_F] \Delta C$ on the right-hand side of the RG equation, identical for R_R and C_R. This addition can be found by applying the operator $-\partial_\tau^2$ to the right-hand side of the RG equation (4.23) for the free energy $\overline{\mathcal{F}}'_R$, because $C_R = -\partial_\tau^2 \overline{\mathcal{F}}'_R$ (Sec. 4.5). In the end we obtain $[\mathcal{D}_{\mathrm{RG}} + 2\gamma_F] \Delta C = \mu^{-2\varepsilon} \gamma_0(g)$, where γ_0 is the vacuum RG function defined by (4.19) and $g \equiv g_1$ in the notation of model H.

Therefore, the RG equation for the renormalized response function $R_R = R_R(\omega, \tau, \lambda, g, \mu)$ in the H_0 model has the form

$$[\mathcal{D}_{\mathrm{RG}} - 2\gamma_\tau] R_R = \mu^{-2\varepsilon} \gamma_0, \quad \mathcal{D}_{\mathrm{RG}} = \mathcal{D}_\mu + \beta \partial_g - \gamma_\tau \mathcal{D}_\tau - \gamma_\lambda \mathcal{D}_\lambda \qquad (5.276)$$

with summation over the two charges $g = g_{1,2}$ in the term $\beta \partial_g$. We note that the RG functions β_1, γ_τ, and γ_0 depend only on the static charge g_1, and β_2 and γ_λ depend on the two charges g. Moreover, the counterterm ΔC has the form $\Delta C = \mu^{-2\varepsilon} b(g_1)$ and is the particular solution of the RG equation (5.276) (see above), from which for the function $b(g_1)$ we obtain

$$[-2\varepsilon - 2\gamma_\tau + \beta_1 \partial_{g_1}] b(g_1) = \gamma_0(g_1). \qquad (5.277)$$

The general solution of (5.276) can be studied exactly like that of the free energy in Sec. 4.5. First it is necessary to transform to a function of canonically dimensionless variables, taking, for example,

$$R_R = \tau^{-\varepsilon} \phi(s, w, g), \quad s \equiv \tau/\mu^2, \quad w \equiv \omega/2\lambda\mu^4 \qquad (5.278)$$

(the extra coefficient of $1/2$ in the definition of w is convenient for the calculations). Substituting (5.278) into (5.276), we obtain the equation

$$[-(2+\gamma_\tau)\mathcal{D}_s + \beta\partial_g - (4-\gamma_\lambda)\mathcal{D}_w - (2-\varepsilon)\gamma_\tau]\phi(s,w,g) = s^\varepsilon\gamma_0. \quad (5.279)$$

Its solution is analogous to (4.26), but with two invariant charges $\bar{g} \equiv \bar{g}_1, \bar{g}_2$ and the additional argument $\bar{w}$ of the function ϕ:

$$\phi(s,w,g) = s^\varepsilon b(g_1) + [\phi(1,\bar{w},\bar{g}) - b(\bar{g}_1)]\exp\left\{-(2-\varepsilon)\int_{g_1}^{\bar{g}_1} dx \frac{\gamma_\tau(x)}{\beta_1(x)}\right\}.$$
$$(5.280)$$

We have taken the function $s^\varepsilon b(g_1)$ with $b(g_1)$ from (5.277), which is proportional to the counterterm $\Delta C = \mu^{-2\varepsilon}b(g_1)$, as the particular solution of the inhomogeneous equation (5.279). The invariant variables $\bar{g}, \bar{w}$ in (5.280) are defined as the solutions of the corresponding Cauchy problem of the type (1.140) [note that (5.279) must be divided by $2+\gamma_\tau$ to bring it to the standard form], in particular,

$$\mathcal{D}_s\bar{w} = -\bar{w}[4-\gamma_\lambda(\bar{g})]/[2+\gamma_\tau(\bar{g})], \quad \bar{w}|_{s=1} = w. \quad (5.281)$$

In the critical regime $\bar{g} \to g_*$ and the exponent in (5.280) becomes $c_1 s^a$, while the solution of (5.281) takes the asymptotic form $\bar{w} = c_2 w s^b$, where $c_{1,2}$ are nonuniversal amplitude factors (Sec. 1.33), $a \equiv -(2-\varepsilon)\gamma_\tau^*/\Delta_\tau$, $b \equiv -\Delta_w/\Delta_\tau$, and $\Delta_\tau = 2 + \gamma_\tau^*$, $\Delta_w = 4 - \gamma_\lambda^*$ are the critical dimensions of the variables τ and w. Therefore, (5.278) and (5.280) lead to the following representation for the desired IR asymptote of the function $R_R(\omega)$ (the other arguments are not written out explicitly):

$$R_R(\omega)\big|_{\text{IR}} = \mu^{-2\varepsilon}b(g_1) + \mu^{-2\varepsilon}s^{-\alpha}c_1'f'(c_2\bar{w}), \quad \bar{w} \equiv ws^{-z\nu}, \quad (5.282)$$

in which $1/\nu \equiv \Delta_\tau$, $z \equiv \Delta_w$, $\alpha = 2 - d\nu = 2(\varepsilon+\gamma_\tau^*)/(2+\gamma_\tau^*)$ is the critical exponent of the specific heat in the static ψ^4 model, and the arguments s and w are defined in (5.278). The constant c_1' and the scaling function $f'(w)$ in (5.282) are defined by the relations $c_1' = -b_*c_1$, $f'(w) = 1 - b_*^{-1}\phi(1,w,g_*)$, i.e.,

$$f'(w) = 1 - b_*^{-1}\mu^{2\varepsilon}R_R(\omega,\tau,\lambda,g,\mu)\big|_{g=g_*,\tau=\mu^2,w=2w\lambda\mu^4}. \quad (5.283)$$

The constant $b_* \equiv b(g_{1*})$ appearing here can be found from (5.277), setting $g_1 = g_{1*}$ in it:

$$b_* = -\gamma_0^*/2(\varepsilon + \gamma_\tau^*) = -\nu\gamma_0^*/\alpha. \quad (5.284)$$

In the terminology of Sec. 1.33, $f'(w)$ in (5.282) is the reduced scaling function [we wish to preserve the notation $f(w)$ for the normalized scaling function, which will be introduced below]. The argument $c_2\bar{w}$ of the scaling function in (5.282) with zero critical dimension is proportional to $\omega/\omega_{\text{fl}}$ with $\omega_{\text{fl}} \sim \lambda\mu^4(\tau/\mu^2)^{z\nu}$. The correlation length ξ in this notation is proportional to $(\tau/\mu^2)^{-\nu}$, and so $\omega_{\text{fl}} \sim \xi^{-z}$.

It follows from (5.268), (5.272), and (5.274) that $R_R(0) = C_R$, $R_R(\infty) = \Delta C$, and so from (5.275) we have $c^{-2}(\infty) = c_0^{-2}$ in agreement with (5.268), while for $c^{-2}(0)$ we obtain the expected singularity (5.246). We note that the constant $R_R(\infty) = \Delta C$ coincides with the first term on the right-hand side of (5.282), and so the second contribution must vanish in the limit $\overline{w} \sim \omega \to \infty$.

In comparing with experiment it is more convenient to use the variable $t \equiv (T - T_c)/T_c$ instead of $\tau \equiv T - T_c$. Nonuniversal additive constants can be eliminated by subtracting the quantity $c^{-2}(\infty) = c_0^{-2}$ from the first equation of (5.275). Then from (5.275) and (5.282) using $\Delta C = \mu^{-2\varepsilon} b(g_1)$ we obtain

$$F_1(\omega, t) \equiv [c^{-2}(\omega) - c^{-2}(\infty)]t^\alpha = \mathrm{Re}\overline{f}'(\overline{w}),$$

$$F_2(\omega, t) \equiv \alpha_\lambda(\omega)c^{-2}(\omega)t^\alpha = \pi\mathrm{Im}\overline{f}'(\overline{w}), \tag{5.285}$$

$$\overline{f}'(w) \equiv C_1 f'(C_2 w), \quad \overline{w} \equiv \omega t^{-z\nu}, \quad t \equiv (T - T_c)/T_c$$

with the universal function $f'(w)$ from (5.283) and nonuniversal amplitude factors $C_{1,2}$.

It follows from the representation (5.285) that the experimentally measured quantities $F_{1,2}(\omega, t)$ actually depend not on the two variables ω, t, but only on their combination $\overline{w} \equiv \omega t^{-z\nu}$ up to IR-irrelevant corrections. This is the main statement of the theory of dynamical scaling as applied to this problem, and it is confirmed by the experimental data [173]. In experiments with different ω and t for a given material the constants $C_{1,2}$ in (5.285) are fixed, but they can be different [in contrast to the function $f'(w)$] for a different material. This is what we mean when we say they are nonuniversal.

Let us now discuss the properties of the scaling function $f'(w)$ defined by (5.283). Analyzing the graphs of the renormalized response function $R_R(\omega)$ in the H_0 model (Sec. 5.25), it is easily shown that this quantity is real for purely imaginary values of the frequency ω on the upper semiaxis ($\omega = ix$, $x > 0$) and that $\mathrm{Im}R_R(\omega) > 0$ for $\omega > 0$. Moreover, it is easily checked that for the function $R_R(\omega)$ for $\tau > 0$ the first two terms of the Taylor expansion in ω about the origin exist (i.e., the contributions of order 1 and ω), while for $\omega > 0$ the first term of the expansion in τ about the origin exists (the higher terms in these expansions do not exist owing to IR divergences in the coefficients). Taking into account the relation (5.282) between $R_R(\omega)$ and the scaling function $f'(w)$, it follows that for $f'(w)$ the first two terms of the expansion in w about the origin exist, and that this function has asymptote $\sim w^{-\alpha/z\nu}$ for $w \to \infty$. Taking into account also the above-noted properties of reality and positivity, we conclude that for this function

$$f'(w) \underset{w \to 0}{=} B' + iC'w + \dots, \quad f'(w) \underset{w \to \infty}{=} A'(-iw)^{-\sigma} + \dots, \quad \sigma \equiv \alpha/z\nu,$$

$$\tag{5.286}$$

where A', B', and C' are positive coefficients, and the ellipsis stands for corrections of order $w^{2+O(\varepsilon)}$ for small w and $w^{-1/2+O(\varepsilon)}$ for large w [in the ε expansion corrections $O(\varepsilon)$ in exponents become logarithms of w].

Owing to the uncertainty in the amplitude factors $C_{1,2}$ in (5.285), the scaling function $f'(w)$ entering into them can be replaced by any other function $f(w) = b_1 f'(b_2 w)$ with arbitrary positive constants $b_{1,2}$. This arbitrariness can be used to replace $f'(w)$ by the normalized scaling function $f(w)$, for which two of the three coefficients A, B, and C are fixed at the asymptote analogous to (5.286) [note that a felicitous choice of normalization significantly simplifies the form of $f(w)$ in actual calculations]. When calculating within the MS scheme it is convenient [174] to choose the normalization $B = 1$, $C = \alpha/z\nu \equiv \sigma$ at the asymptote of $f(w)$ analogous to (5.286). The third coefficient A remains unknown and must be calculated.

Therefore, for the normalized scaling function $f(w)$ we have

$$f(w) = b_1 f'(b_2 w), \tag{5.287a}$$

$$f(w) \underset{w\to 0}{=} 1 + i\sigma w + ..., \quad f(w) \underset{w\to\infty}{=} A(-iw)^{-\sigma} + ..., \quad \sigma \equiv \alpha/z\nu. \tag{5.287b}$$

The first of the equations in (5.287b) determines the coefficients $b_{1,2}$ in (5.287a) for known function $f'(w)$.

Starting from the definitions (5.283) and (5.287), the normalized scaling function $f(w)$ can be calculated in the form of an ε expansion from the graphs of the renormalized response function $R_{\mathrm{R}}(\omega)$:

$$f(w) = 1 + 2\varepsilon f_1(w) + (2\varepsilon)^2 f_2(w) + ... \tag{5.288}$$

for $d = 4 - 2\varepsilon$. The coefficient $f_1(w)$ is determined by the one-loop graph of the response function, which in the basic theory (5.211) corresponds to the expression

$$R_{\mathrm{B}}(\omega) = \int \frac{dk}{(2\pi)^d} \frac{\lambda k^2}{[k^2 + \tau]\,[-i\omega + 2\lambda k^2(k^2 + \tau)]}. \tag{5.289}$$

Calculating this integral and canceling out the pole in ε in it by adding an expression of the type $\mathrm{const}\cdot\mu^{-2\varepsilon}/\varepsilon$ (the pair counterterm), we can then pass to the limit $\varepsilon = 0$, which gives the one-loop approximation for the renormalized function $R_{\mathrm{R}}(\omega)$ for $\varepsilon = 0$. This is sufficient for using (5.283) and (5.287) to calculate the quantity $f_1(w)$ in the expansion (5.288) (in lowest order, $\alpha = 2\varepsilon/6$, $\nu = 1/2$, $\gamma_0^* = 1/16\pi^2$, $b_*^{-1} = -2\varepsilon \cdot 16\pi^2/3$). The calculation gives

$$f_1(w) = \frac{1}{6}\left\{1 + \frac{1-x}{2x}\ln(x) + \frac{1-3x}{2x\Delta}\ln\left[\frac{1+\Delta}{1-\Delta}\right]\right\}, \tag{5.290}$$

where $x \equiv -iw$, $\Delta \equiv (1 - 4x)^{1/2}$, and the phases of the logarithms are determined by the standard rule $\ln z = \ln|z| + i\arg z$, $-\pi < \arg z < \pi$, $\arg z = 0$

for $z > 0$. The function (5.290) has the asymptote $f_1(w) = -x/12 + ...$ for $w \to 0$ and $f_1(w) = (2 - \ln x)/12 + ...$ for $w \to \infty$. This agrees with the exact asymptotic formulas (5.287b) in the given order of the ε expansion and gives $A = 1 + 2\varepsilon/6 + ...$ (in the first two orders, $\sigma \equiv \alpha/z\nu = (2\varepsilon/12)[1 - 2\varepsilon \cdot 479/1026 + ...]$).

The one-loop calculation of the scaling function was performed in [175] and the two-loop one in [174]. The expression obtained in [174] for the function $f_2(w)$ in the expansion (5.288) contains several single integrals of complex functions. Here we shall give only the expression for the coefficient A in (5.287b) through two loops [174]:

$$A = 1 + \frac{2\varepsilon}{6} + \frac{(2\varepsilon)^2}{4}\left[\frac{89}{456}\pi^2 - \frac{7267}{3078}\right] + O(\varepsilon^3) \qquad (5.291)$$

[numerically, $A = 1 + 0.1667 \cdot (2\varepsilon) - 0.1086 \cdot (2\varepsilon)^2 + ...$]. We note that the passage to the normalized function (5.287a) completely eliminates the numerous contributions to $f'(w)$ involving the Euler constant and $\ln(4\pi)$, which considerably simplifies the form of $f(w)$.

When comparing with experiment, a subtraction at zero frequency is usually done in the first expression in (5.285), because it is easier to extract $c(0)$ from the experimental data than the constant $c(\infty) = c_0$. Replacing $f'(w)$ in (5.285) by the normalized function $f(w)$ (which changes only the unknown coefficients $C_{1,2}$), we obtain

$$[c^{-2}(0, t) - c^{-2}(\omega, t)]t^\alpha = C_1 \text{Re}[1 - f(C_2\bar{\omega})],$$

$$\alpha_\lambda(\omega, t)c^{-2}(\omega, t)t^\alpha = C_1 \pi \text{Im} f(C_2\bar{\omega}). \qquad (5.292)$$

The notation is the same as in (5.285), except that now for clarity we have explicitly indicated the dependence of the speed of sound and the damping on the temperature $t \equiv (T - T_c)/T_c$.

The left-hand sides $F_{1,2}$ of (5.292) are known from the experimental data. These data are most conveniently represented as points in the plane where $\ln \bar{\omega}$ is plotted along the X axis and $\ln F_i$ along the Y axis. It follows from (5.292) that the points obtained in this manner at different ω and t for the same material must lie on the same curve, and that the possible change of the coefficients $C_{1,2}$ in going to a different material must lead only to a parallel transport of this curve along the axes, this transport being identical for F_1 and F_2. It then follows that the shape of these curves for the two quantities $F_{1,2}$, which does not change in the parallel transport, is determined only by the form of the scaling function $f(w)$. This makes it possible to compare theory and experiment.

However, it is clear *a priori* that when an approximate function $f(w)$ is substituted into (5.292) we can hope to obtain agreement with experiment only when this approximate function gives a qualitatively correct reproduction of the asymptote (5.287b). It is impossible to use simply the initial segment of the ε expansion (5.288) as the approximate expression for f, because the asymptote for $w \to \infty$ is incorrectly reproduced in it: the fractional power of the variable w is replaced by logarithms. It is therefore necessary to perform some additional exponentiation of logarithms that allows the initial segment of the ε expansion of $f(w)$ to be rearranged into another approximate expression with qualitatively correct asymptotes.

This can be done in various ways. Here we shall describe the method used in [174] and [175]. This [175] consists of isolating from the function $f(w)$ a factor corresponding to some simple function with the correct asymptotes and regular behavior in the intermediate region. In particular, the authors of [174] and [175] use the representation

$$f(w) = (1 - iw)^{-\sigma} h(w) \tag{5.293}$$

with the exponent σ from (5.287b). From (5.293) and (5.287b) we have

$$h(w) \underset{w \to 0}{=} 1 + 0 \cdot iw + ..., \quad h(w) \underset{w \to \infty}{=} A + ... \tag{5.294}$$

with the constant A from (5.287b) known from (5.291) through two loops. The notation $0 \cdot iw$ in (5.294) emphasizes the absence of a contribution to $h(w)$ linear in w for $w \to 0$.

The ε expansion of $f(w)$ of known accuracy can be used to construct the ε expansion analogous to (5.288) of $h(w)$, defined in (5.293), with the same accuracy. In [174] and [175] the function (5.293) with known initial segment of the ε expansion of $h(w)$ was used as the approximate expression for $f(w)$ in (5.292). In comparing with experiment the exponents α in (5.292) and σ in (5.293) are taken from the experimental data for the three-dimensional problem ($\alpha \simeq 0.11$, $\nu \simeq 0.63$, $z \simeq 3.07$, $\sigma = \alpha/z\nu \simeq 0.057$). However, only the calculated first few terms of the ε expansion can be considered reliably known for the function $h(w)$. Analysis shows [174] that in the one-loop approximation for h there is a significant disagreement with experiment in the intermediate region. The inclusion of the two-loop correction significicantly improves the situation and makes the agreement quite satisfactory. We note that in [175] satisfactory agreement with experiment was obtained already in the one-loop approximation. However, this was achieved only by illegitimately exceeding the accuracy in the ε expansion of the function h, $h = 1 + 2\varepsilon h_1 + ...$: in [175] the coefficient 2ε of the one-loop contribution to h was replaced by the expression $3\alpha/\nu = 2\varepsilon + ...$ with substitution of the experimental value $3\alpha/\nu \simeq 0.5$ instead of $2\varepsilon = 1$ for the three-dimensional problem. It turns out that this replacement approximately reproduces the effect of the two-loop correction in the exact calculation, improving the agreement with experiment [174]. However, it is clear that exceeding the accuracy of the ε expansion in the function h is incorrect, and that the inclusion of two-loop corrections to the approximation [175] would lead to a further change that would make the results worse.

Here we conclude our discussion of the problem of sound propagation near the critical point of the liquid–gas transition. Additional information about this problem and its analogs in other models of critical dynamics can be found in [176] (phenomenology), [177], [178], [179] (model H and others), [180] (model F, He$_4$), [181] (models C and F), and [182] (model A).

In conclusion, we warn the reader that the presentation in Secs. 5.23 to 5.26 differs markedly from the traditional one. This is true already for the initial problem of choosing the dynamical variables of model H (Sec. 5.23), the role of the contributions involving $\partial_t v$ (Sec. 5.24), the technique of proving the multiplicative renormalizability (Sec. 5.25), and the derivation of the basic equations (5.292) for the sound wave. We note that this derivation is not completely correct in [175]. Our exposition is closer to that of [177], but with significant differences. All these differences concern only the ideology and the techniques for proving statements of a general nature and do not affect the results of particular calculations of universal quantities such as the critical exponents and normalized scaling functions.

Chapter 6

Stochastic Theory of Turbulence

In this chapter we present the main results obtained in the stochastic theory of turbulence and related problems using the RG technique. More detailed treatment of this subject can be found in the cited literature and the book [183].

6.1 The phenomenon of turbulence

The phenomenon of turbulence is well known to everyone. A typical example is the flow of a liquid through a pipe with a given pressure difference Δp between its ends. For small Δp the flow is smooth ("laminar"), and as Δp increases through a threshold Δp_{thr} the laminar flow becomes unstable and chaotic vortices appear whose intensity increases with increasing Δp. At the same time the fluid flow becomes more complicated: the characteristic scale of the eddies that first appear near threshold corresponds to the external scale of the system $L_{\max}$ (in our example, the pipe diameter), and as Δp grows these primary large-scale eddies break up into smaller and smaller ones. The regime of developed turbulence corresponds to $\Delta p \gg \Delta p_{\text{thr}}$. Then the system contains turbulent eddies of all possible sizes simultaneously from the external scale $L_{\max}$ to the dissipation length $L_{\min}$, at which eddy damping due to viscous friction becomes important. In the stationary regime all the energy that flows into the system from the external source creating the pressure gradient is finally transformed into heat owing to energy dissipation by small-scale eddies.

In general, the dimensionless parameter $\Delta p / \Delta p_{\text{thr}}$ corresponds to the Reynolds number $\text{Re} = u L_{\max} / \nu$, where u is the typical average flow velocity, $L_{\max}$ is the external scale, and ν is the kinematical viscosity of the medium.

When turbulence is present the total velocity field $v(x)$ can be written as the sum $v(x) = u(x) + \widehat{\varphi}(x)$, where $u(x)$ is the smooth laminar component and $\widehat{\varphi}(x)$ is the relatively small stochastic (fluctuating) component. In the theory of turbulence we study the statistical characteristics of the random field $\widehat{\varphi}(x)$, i.e., its correlation functions and the various response functions (Sec. 5.1). Near threshold, when the Reynolds number Re only slightly exceeds the threshold value Re_{thr}, the structure of the turbulent eddies that first appear with characteristic scale $L_{\max}$ is determined by the full geometry of the problem (the theory of this is presented in Chapter 3 of [172]), i.e., here the

581

turbulence still "remembers" the details of the global structure of the system. Problems of this type can only be solved individually for each specific system.

Essential simplifications can be expected to occur only in the case of developed turbulence, when Re $\gg$ Re$_{thr}$ and $L_{max} \gg L_{min}$. Then there exists a clearly expressed *inertial range* of lengths $L_{max} \gg L \gg L_{min}$, and one can speak of correlation functions of the field $\widehat{\varphi}(x)$ at these scales. In our example of flow through a pipe, we shall for clarity imagine a volume of liquid of size $L \ll L_{max}$, inside which are sensors that continuously record the instantaneous value of the flow velocity at several points. Since the characteristic scale for the laminar component of the velocity $u(x)$ is L_{max}, inside the region of dimension $L \ll L_{max}$ we can assume $u(x) = $ const. Therefore, when studying the structure of turbulent flow at these scales it is possible to ignore the nontrivial global structure and assume that the average velocity $u(x)$ is a constant. This constant can be eliminated by transforming to a comoving reference frame by means of a suitable Galilean transformation. This leads to the problem of uniform, isotropic, developed turbulence, in which the entire velocity field is identified with its stochastic (fluctuating) component $\widehat{\varphi}(x)$ and the primary large-scale eddies are assumed to be the energy source.

In going to a comoving system $v(x) \equiv v(t, \mathbf{x}) \rightarrow v(t, \mathbf{x} - ut) - u = \widehat{\varphi}(t, \mathbf{x} - ut)$, i.e., from the viewpoint of the uniform and isotropic problem, the sensor located at a given point $\mathbf{x}$ of the moving flow records the instantaneous values of the random quantity $\widehat{\varphi}(t, \mathbf{x} - ut)$. In the experimental analysis it is assumed that the explicit dependence of this quantity on t can be neglected compared to the dependence through the argument $\mathbf{x} - ut$, because the drift velocity u is much larger than the characteristic fluctuation velocity. Therefore, measurement of the correlator of the values of $\widehat{\varphi}$ for a single sensor at times t and $t + \Delta t$ (with averaging over t) is interpreted as measurement of the static (equal-time) correlator for homogeneous and isotropic turbulence for the separation $r = u\Delta t$.

Developed turbulence is observed experimentally for both liquids and gases (atmospheric turbulence, turbulence in wind tunnels) and obeys the same general laws. Since the typical velocity of turbulent fluctuations in real situations is much smaller than the speed of sound, in describing turbulence it is possible to neglect the compressibility of the medium, i.e., the vector velocity field can be assumed transverse.

6.2 The stochastic Navier–Stokes equation. The Kolmogorov hypotheses

It is widely recognized [154] that the stochastic Navier–Stokes equation (5.6) can be used to microscopically model the uniform, isotropic, developed turbulence of an incompressible liquid (gas):

$$\nabla_t \varphi_i(x) = \nu_0 \partial^2 \varphi_i(x) - \partial_i p(x) + \eta_i(x), \quad \nabla_t \equiv \partial_t + (\varphi \partial), \quad (6.1)$$

where φ is the transverse (owing to the incompressibility) vector velocity field ($\partial \varphi \equiv \partial_k \varphi_k = 0$), ν_0 is the kinematical viscosity coefficient, $p(x)$ and $\eta(x)$ are the pressure and transverse external random force per unit mass, and ∇_t is the Galilean-covariant derivative. We assume that η obeys a Gaussian distribution with zero mean and given correlator $\langle \widehat{\eta}_i(x)\widehat{\eta}_s(x')\rangle \equiv D_{is}(x,x')$ of the form

$$D_{is}(x,x') = \delta(t-t')(2\pi)^{-d}\int dk\, P_{is}(k)N(k)\exp ik(\mathbf{x}-\mathbf{x}'), \qquad (6.2)$$

where $P_{is}(k) = \delta_{is} - k_i k_s/k^2$ is the transverse projector, d is the dimension of $\mathbf{x}$ (and k) space, and $N(k)$ is a "pumping" function depending on $|k|$ and the model parameters.

We are therefore dealing with a special case of the problem (5.1), and we shall always study it in the standard formulation (Sec. 5.1), i.e., on the entire time axis with the retardation condition for (6.1) and zero boundary conditions on φ at infinity. The field φ in (6.1) is the fluctuating component of the velocity, because we are describing homogeneous, isotropic turbulence.

The random force η in (6.1) phenomenologically models stochasticity (which under realistic conditions must arise spontaneously as a consequence of the instability of laminar flow) and at the same time the pumping of energy into the system from the interaction with large-scale eddies. The average power of the energy pumping W (the amount of energy input to unit mass per unit time) is related to the function $N(k)$ in (6.2) as

$$W = [(d-1)/2(2\pi)^d]\int dk\, N(k) \qquad (6.3)$$

(see Sec. 6.8 for the proof).

The physical picture of uniform, isotropic, developed turbulence is the following. Energy from an external source [modeled by the random force η in (6.1)] flows into the system from large-scale motions (eddies) of characteristic size $L_{\max} \equiv 1/m$ (the parameter m defined by this equation will be referred to as the mass). It is then distributed throughout the spectrum (eddy fractionation) owing to the nonlinearity in (6.1), and begins to actively dissipate at scales $L_{\min} \equiv 1/\Lambda$ (the dissipation length) where viscosity begins to play an important role. The parameters W, ν_0, and $m \equiv L_{\max}^{-1}$ can be regarded as independent, and all the other parameters are expressed in terms of them using dimensional arguments ($m \sim L^{-1}$, $W \sim L^2 T^{-3}$, $\nu_0 \sim L^2 T^{-1}$, where L is length and T is time), for the lower limits in terms of the pair W and m, and for the upper ones in terms of the pair W and ν_0, in particular, $\Lambda = W^{1/4}\nu_0^{-3/4}$. Developed turbulence is characterized by large (of order 10^4–10^6) Reynolds number $\mathrm{Re} = (\Lambda/m)^{4/3}$ and, consequently, the presence of a wide inertial range determined by the inequalities $m \ll k \ll \Lambda$ for momenta (= wave numbers) and $\omega_{\min} \equiv W^{1/3}m^{2/3} \ll \omega \ll \omega_{\max} \equiv \nu_0\Lambda^2$ for frequencies.

The main premises of the phenomenological Kolmogorov–Obukhov theory were formulated [184] as two hypotheses. The first later had to be weakened.

We shall first give the original version (see Sec. 21 of [154]), referred to as hypothesis 1′, and then after some explanation we give the more accurate current version, hypothesis 1.

Hypothesis 1′ [154]: In the region $k \gg m$, $\omega \gg \omega_{\min} = W^{1/3}m^{2/3}$ the distribution of the Fourier components $\varphi(\omega, k)$ of the random velocity $\varphi(x) \equiv \varphi(t, \mathbf{x})$ depends on the total pumping power W, but is independent of the details of its structure, including the specific value of $m = L_{\max}^{-1}$.

Hypothesis 2 [154]: In the region $k \ll \Lambda$, $\omega \ll \omega_{\max} = \nu_0 \Lambda^2$ this distribution is independent of the viscosity coefficient ν_0.

It follows, in particular, from hypothesis 2 that in its region of applicability the dynamical correlator $\langle \widehat{\varphi}\widehat{\varphi} \rangle$ in ω, k-space for the d-dimensional problem has the form

$$\langle \widehat{\varphi}_i \widehat{\varphi}_s \rangle = P_{is} W^{1/3} k^{-d-4/3} f(W k^2/\omega^3, m/k), \tag{6.4}$$

where P_{is} is the transverse projector and f is an unknown scaling function of two independent dimensionless arguments.

In the inertial range, where the conditions of both hypotheses are satisfied, the second argument of the function f in (6.4) is small, and the first satisfies the inequalities $(k/\Lambda)^2 \ll W k^2/\omega^3 \ll (k/m)^2$, i.e., it is actually arbitrary. According to hypothesis 1′, the m dependence must vanish in this region, in other words, the function f in (6.4) must have a finite limit when its second argument m/k tends to zero. However, it has long been known that this is incorrect. Owing to the kinematical effect of the transport of turbulent eddies as a whole by large-scale motions with $k \sim m$ [185], the limit $m/k \to 0$ does not exist in general in dynamical objects. The following statement is better justified.

Hypothesis 1: In the region $k \gg m$ a finite limit for $m/k \to 0$ exists for the equal-time distribution function of spatial Fourier components $\varphi(t, k)$ of the random velocity field $\varphi(t, \mathbf{x})$.

For the static pair correlator, hypothesis 2 in its region of applicability leads to the representation

$$\langle \widehat{\varphi}_i(t, k)\widehat{\varphi}_s(t, k') \rangle = (2\pi)^d \delta(k + k') P_{is}(k) D^{\text{stat}}(k),$$

$$D^{\text{stat}}(k) = k^{-d}(W/k)^{2/3} f(m/k) \tag{6.5}$$

[in (6.4) and (6.5) and below we use the same letter f to denote different scaling functions].

It follows from hypothesis 1 that the function $f(m/k)$ in (6.5) has a finite limit $f(0)$ for $m/k \to 0$. The limit $f(0)$ is simply related to the known Kolmogorov constant (Sec. 6.11). The representation (6.5) is valid for all $k \ll \Lambda$, i.e., in both the inertial range, where the function $f(m/k)$ can be replaced by the constant $f(0)$, and in the "energy-containing" pumping region $k \lesssim m$, where the function $f(m/k)$ must be nontrivial (see below). In the dissipative region $k \gtrsim \Lambda$ the function f in (6.5) acquires the additional argument k/Λ, and the dependence on m/k must then vanish according to hypothesis 1.

Calculating the Fourier transform of the function D^{stat} from (6.5), we find that

$$D^{\text{stat}}(r) = (Wr)^{2/3} f(mr) \tag{6.6}$$

with region of applicability $r \gg 1/\Lambda$.

From the viewpoint of hypothesis 2, the representations (6.5) and (6.6) are completely equivalent, but this is not so regarding hypothesis 1. For the coordinate representation (6.6), in contrast to the momentum one, it is no longer possible to state that the function $f(mr)$ has a finite limit for $m \to 0$, i.e., that it reduces to a constant in the inertial range. The point is that owing to the strong singularity at the origin of the power of k isolated in (6.5), the existence of the Fourier transform (which is guaranteed physically) can be assured only when the power-law singularity in (6.5) is suppressed by the function $f(m/k)$. It then follows that for small k this function must be nontrivial, and its dependence on the argument m/k cannot be neglected in calculating the Fourier transform. This introduces an essential m dependence into $D^{\text{stat}}(r)$ for any r, including values in the inertial range $\Lambda^{-1} \ll r \ll m^{-1}$. However, here it can be manifested only as an additive constant $D^{\text{stat}}(r = 0)$, because for the difference $D^{\text{stat}}(r) - D^{\text{stat}}(r = 0)$ the power-law singularity of the function (6.5) is suppressed [with the natural assumption that the function $f(m/k)$ is analytic in k near $k = 0$], so that the limit $m \to 0$ of this difference must exist in the inertial range.

It then follows that up to corrections vanishing for $m \to 0$, (6.6) in the inertial range $\Lambda^{-1} \ll r \ll m^{-1}$ has the form

$$D^{\text{stat}}(r) = C_1 (W/m)^{2/3} - C_2 (Wr)^{2/3}, \tag{6.7}$$

where the first term is $D^{\text{stat}}(r = 0)$, the second is the difference $D^{\text{stat}}(r) - D^{\text{stat}}(r = 0)$, and $C_{1,2}$ are dimensionless positive constants.

Representations like (6.4) and (6.5) can be written down for any correlation functions of the field φ. They follow from only hypothesis 2 and altogether imply the presence of IR scaling (because the conditions of hypothesis 2 do not include any lower bounds on the frequency and momentum) with completely defined "Kolmogorov" dimensions $\Delta_F \equiv \Delta[F]$ of all IR-relevant quantities $F = \{\varphi \equiv \varphi(x), m, t \sim \omega^{-1}, r \sim k^{-1}\}$ for the nonessential parameters W and ν_0 (which are fixed in dilatations):

$$\Delta_\varphi = -1/3, \quad \Delta_t = -\Delta_\omega = -2/3, \quad \Delta_k = -\Delta_r = \Delta_m = 1. \tag{6.8}$$

The Kolmogorov hypotheses are phenomenological and are confirmed experimentally (see Sec. 6.12 for details). They must be explained by a microscopic theory, and this so far has not been completely accomplished. Some progress has been made only in justifying hypothesis 2, i.e., IR scaling in turbulence. In this regard there is an obvious analogy with the theory of critical behavior, and so IR scaling can be justified using the standard RG technique with particular assumptions about the form of the random-force correlator in

(6.1). In RG language, hypothesis 1 is a statement about the behavior for $m \to 0$ of the scaling functions entering into the RG representations for static objects. In the theory of turbulence, as in the theory of critical phenomena, such questions do not pertain directly to the competency of the RG method, because the explicit form of the scaling functions is not determined by the RG equations themselves (see Sec. 1.33).

The analog of m in the theory of critical behavior is the parameter $\tau = T - T_c$, and the standard method of studying the singularities of the scaling functions for $\tau \to 0$ (see Sec. 4.14) is the SDE technique, the Wilson operator expansion. This technique can also be used in the RG theory of turbulence. However, first the difference between these two problems should be emphasized. In the theory of critical behavior we are *a priori* certain that the massless $(T = T_c)$ theory with $\tau = 0$ exists because of the experimental evidence, and so we assume that expansions like (4.105) contain only a constant and corrections with positive powers of τ. Of course, this is guaranteed within the ε expansion, but, strictly speaking, for the realistic value $\varepsilon = 4 - d = 1$ we do not know the signs of the critical dimensions Δ_F of all the operators because we cannot calculate them exactly. In fact, the problem of the signs of the Δ_F is simply ignored. The problem would become important if someone could come up with at least one "dangerous" operator F with exactly known, strictly negative dimension $\Delta_F < 0$ for the real $\varepsilon = 1$. The contribution of this operator to the expansion (4.105) would contain a negative power of τ, and then the question would arise of how this is compatible with the existence of a finite limit for $\tau \to 0$. This problem is academic in the theory of critical behavior, because the dimensions Δ_F can in practice be calculated only as the first few terms of ε expansions, which allows no reliable judgement of their sign for $\varepsilon = 1$.

However, in the RG theory of turbulence this problem is crucial, as here the critical dimensions can be calculated exactly for some operators, and among these operators are dangerous ones with $\Delta_F < 0$ that generate contributions in the scaling function with negative powers of the parameter m. In this situation a finite limit for $m \to 0$ can be assured only when there are an infinite number of such contributions and some procedure for summing them. However, so far only the first few steps in this direction have been taken. These questions will be discussed in detail below in Secs. 6.9 and 6.10. Here we wish only to emphasize that there is a very important difference between Kolmogorov hypotheses 1 and 2 from the viewpoint of the RG technique.

6.3 Choice of the random-force correlator

In critical dynamics the form of the random-force correlator in Langevin equations like (5.4) is uniquely determined by the requirement that the dynamics and statics be mutually consistent. The stochastic Navier–Stokes equation (6.1) does not belong to this class, and so there is no unique rule for choosing this correlator. In the RG theory of turbulence it is chosen on the basis of both physical and technical arguments. The physical arguments are that realistic pumping for this problem must be infrared, i.e., the dominant contribution to the integral (6.3) must come from small momenta $k \sim m$ (energy pumping by large-scale eddies). On the other hand, in order to use the standard quantum field RG technique it is important that the pumping function $N(k)$ in (6.2) have a power-law asymptote at large k. This last condition is satisfied, in particular, by the function used in [186] and [187]:

$$N(k) = D_0 k^{4-d}(k^2 + m^2)^{-\varepsilon}, \tag{6.9}$$

where $\varepsilon \geq 0$ is an independent parameter of the model. In the RG theory of turbulence this parameter, which is completely unrelated to the spatial dimension d, plays the same role as the quantity $d_* - d$ in models of the theory of critical behavior (see Sec. 1.19). In other words, the quantity ε in (6.9) characterizes the degree of deviation from logarithmicity. The model becomes logarithmic for $\varepsilon = 0$ (see Sec. 6.4 for details), and realistic (i.e., infrared) pumping (6.9) occurs only for $\varepsilon > 2$. In the range $0 < \varepsilon < 2$ the pumping (6.9) is ultraviolet. The integral (6.3) diverges at large k for UV pumping. It is assumed to be cut off at $k \leq \Lambda$, most of the contribution to the integral coming from scales $k \sim \Lambda$. Then $W \sim \Lambda^{4-2\varepsilon}$, in contrast to $W \sim m^{4-2\varepsilon}$ for $\varepsilon > 2$.

Most studies on the RG theory of turbulence use a simpler, purely power-law pumping function:

$$N(k) = D_0 k^{4-d-2\varepsilon}, \tag{6.10}$$

which corresponds to $m = 0$ in (6.9). This choice is suitable if we are interested only in justifying IR scaling and the corresponding critical dimensions (which for any kind of pumping must be independent of m), and other objects like scaling functions are calculated from graphs only as ε expansions. Then the passage to the problem with $m = 0$ is straightforward, because the coefficients of the ε expansions of graphs have finite limits for $m \to 0$. However this, of course, cannot be considered a proof of Kolmogorov hypothesis 1, because for finite ε the $m \to 0$ limit may not exist in the region of realistic IR pumping $\varepsilon \geq 2$. (A simple example is the function $m^{2-\varepsilon}$: for $m \to 0$ the coefficients of its ε expansion vanish, and it itself diverges for $\varepsilon > 2$.) Therefore, hypothesis 1 can be discussed only in models like (6.9) involving the parameter m and necessarily outside the context of the ε expansion.

However, if we are not interested in these problems we can use the maximally simplified model (6.10). The "realistic" value for it is $\varepsilon_{\mathrm{r}} = 2$, corresponding to the boundary of the IR-pumping region in (6.9): for $\varepsilon > 2$ the integral in (6.3) with pumping function (6.10) does not exist owing to the IR divergence, and for $\varepsilon < 2$ the pumping is ultraviolet. We note that for $\varepsilon = 2$ the parameter D_0 in (6.10) acquires the dimension of W. Moreover, idealized pumping by infinitely large eddies corresponds to $N(k) \sim \delta(k)$, and the function Ak^{-d} with suitably chosen amplitude coefficient A can be considered as a power-law model of the d-dimensional δ function (see Sec. 6.11 for details).

Of course, a more realistic choice is the pumping model (6.9) or its generalization,

$$N(k) = D_0 k^{4-d-2\varepsilon} h(m/k), \qquad h(0) = 1 \tag{6.11}$$

with arbitrary "sufficiently well-behaved" (in particular, analytic in m^2 near
the origin) function h ensuring the convergence of the integral (6.3) for small k
and normalized to unity in the region $k \gg m$, where (6.11) becomes (6.10).
The function (6.9) is a special case of (6.11) with a particular choice of the
function h.

The assumption of power-law behavior of $N(k)$ at large k is actually the
only premise of the RG theory of turbulence vulnerable to criticism. In
its favor it can be said that physical considerations require that the energy
pumping be infrared, but nothing prevents it from having a tail in the region
$k \gg m$. It is only important that this tail not contribute noticeably to the
integral (6.3), which occurs for the function (6.9) for any $\varepsilon > 2$. A power of k
with arbitrary exponent is a completely natural model of the tail for a positive
function that falls off with increasing k. Any other hypothesis, for example,
finiteness or exponential falloff of the function $N(k)$, needs as much justification as the hypothesis of a power-law tail. In the exact theory the energy
pumping must be generated by the interaction between the fluctuating and
smooth components of the velocity, and so its characteristics for a particular
problem (for example, flow in a pipe with given pressure drop between its
ends) must, in principle, be calculable. However, as yet there is no complete
theory of this type, and within the stochastic problem (6.1), which is only a
simplified phenomenological version of this (hypothetical) exact theory, the
particular choice of pumping function $N(k)$ can be justified only by general
arguments and the results. It should be noted that in the RG theory of critical behavior widely used at present, the possibility of replacing the original
exact microscopic model (of, for example, a liquid) by a simplified fluctuation
model amenable to RG analysis is also simply a postulate and is justified only
by the results.

Most of the results in the RG theory of turbulence have been obtained
using the ε expansion for the model (6.10) with $m = 0$. The model (6.9) was
used in some studies, and their results generalize directly to the model (6.11)
with arbitrary function h specifying the global pumping structure. The results
pertaining to the inertial range are completely independent of the choice of
specific function h in (6.11).

For models like (6.11), in contrast to (6.10), it is possible to state and
attempt to solve the interesting problem of "freezing." Since the choice
of a particular value of ε in the range $\varepsilon > 2$ can be considered a struc-
tural detail of the IR pumping, it is desirable to show that the results pertaining to the inertial range are independent of this choice, in accordance
with Kolmogorov hypothesis 1. It is possible to do this for the critical
dimensions [187]. The critical dimensions, which depend on ε in the region
$0 < \varepsilon < 2$, reach the Kolmogorov values for $\varepsilon = 2$ and do not change as ε is

increased further (see Sec. 6.6 for details). This freezing of the dimensions in the region $\varepsilon \geq 2$ is consistent with Kolmogorov hypothesis 1, although, of course, it does not constitute a proof of it in the total volume, because the difficult problem of determining the dependence of the correlation functions on the parameter m in the inertial range remains unsolved.

6.4 UV divergences, renormalization, and RG equations of the quantum field model

According to the general rules of Sec. 5.3, the stochastic problem (6.1) is equivalent to the quantum field model for two transverse vector fields $\phi = \varphi, \varphi'$ with action functional (5.25), in this case

$$S(\phi) = \varphi' D \varphi'/2 + \varphi'[-\partial_t \varphi + \nu_0 \partial^2 \varphi - (\varphi \partial) \varphi], \tag{6.12}$$

where D is the random-force correlator (6.2). The contribution involving ∂p from (6.1) vanishes in (6.12) owing to the transversality of the field φ'.

The diagrammatic technique corresponding to the action functional (6.12) is given by the general rules in Sec. 5.3. The lines in the graphs correspond to the bare propagators (5.30). For the model (6.12) in ω, k space we have

$$\langle \widehat{\varphi} \widehat{\varphi}' \rangle_0 = \langle \widehat{\varphi}' \widehat{\varphi} \rangle_0^{\top} = P_\perp (-i\omega + \nu_0 k^2)^{-1},$$

$$\langle \widehat{\varphi} \widehat{\varphi} \rangle_0 = P_\perp N(k)/|-i\omega + \nu_0 k^2|^2, \qquad \langle \widehat{\varphi}' \widehat{\varphi}' \rangle_0 = 0, \tag{6.13}$$

where $P_\perp$ is the transverse projector in the vector indices of the fields and $N(k)$ is the pumping function from (6.2). For the interaction vertex in (6.12) the derivative inside the integral $\int dx...$ can be transferred to the auxiliary field φ':

$$V(\phi) = -\varphi'(\varphi \partial)\varphi = -\varphi'_i \varphi_k \partial_k \varphi_i = \partial_k \varphi'_i \cdot \varphi_k \varphi_i. \tag{6.14}$$

Rewriting this functional in the form $\varphi'_i v_{iks} \varphi_k \varphi_s/2$, we find the corresponding vertex factor

$$v_{iks} = -i[p_k \delta_{is} + p_s \delta_{ik}], \tag{6.15}$$

where p is the momentum flowing into the vertex through the field φ'.

When substituting a pumping function of the type (6.10) or (6.11) into (6.12) it is convenient to explicitly isolate the amplitude factor D_0, introducing the notation

$$N(k) = D_0 n(k), \qquad D_0 = g_0 \nu_0^3, \tag{6.16}$$

where $n(k) = k^{4-d-2\varepsilon}$ for the pumping function (6.10) or $k^{4-d-2\varepsilon} h(m/k)$ for (6.11) and its particular variant (6.9). The second equation in (6.16) determines the new parameter g_0, which in this problem plays the role of the bare charge.

Substituting (6.16) into (6.12), we obtain the functional

$$S(\phi) = g_0 \nu_0^3 \varphi' n \varphi'/2 + \varphi'[-\partial_t \varphi + \nu_0 \partial^2 \varphi - (\varphi \partial) \varphi], \tag{6.17}$$

which is interpreted as the unrenormalized action of this model.

Let us now discuss its renormalization. As usual, UV divergences are manifested in this model as poles in ε, but now the role of ε is played by the parameter (unrelated to the spatial dimension d) entering into the function $n(k) \sim k^{4-d-2\varepsilon}$. The structure of the UV divergences and the counterterms needed to eliminate them is determined by the general rules of Secs. 5.14 and 5.15 in terms of the canonical dimensions of the 1-irreducible Green functions Γ. We see from (6.17) that in this case we have $d_\omega = 2$ in (5.87). The canonical dimensions of all the fields and parameters of the model (6.17) along with their renormalized analogs (without the subscript 0) are given in Table 6.1.

Table 6.1 Canonical dimensions of the fields and parameters of the model (6.17) with pumping function $n(k) \sim k^{4-d-2\varepsilon}$ for arbitrary spatial dimension d.

F	$\varphi(x)$	$\varphi'(x)$	Λ, m, μ	ν, ν_0	W	g_0	g
d_F^p	-1	$d+1$	1	-2	-2	2ε	0
d_F^ω	1	-1	0	1	3	0	0
d_F	1	$d-1$	1	0	4	2ε	0

It follows from the positivity of the dimension $d[g_0] = 2\varepsilon$ of the bare charge that we are dealing with an IR problem like (1.105) for any $\varepsilon > 0$, including the region $2 > \varepsilon > 0$, where the pumping is ultraviolet (Sec. 6.3).

The basic action corresponding to the functional (6.17) has the form

$$S_B(\phi) = g\nu^3 \mu^{2\varepsilon} \varphi' n \varphi'/2 + \varphi'[-\partial_t \varphi + \nu \partial^2 \varphi - (\varphi \partial)\varphi], \qquad (6.18)$$

where g and ν are renormalized parameters and μ is the renormalization mass. As usual, we shall assume that the standard MS scheme, which preserves the symmetry (in particular, the Galilean invariance) of the basic theory, is used to renormalize the model (6.18).

The formal exponent of divergence of the graphs d_Γ^* is determined by (5.88) and (5.89) with $d_\omega = 2$ in this case. We see from Table 6.1 that in this model d_Γ is independent of ε, and so

$$d_\Gamma^* = d_\Gamma = d + 2 - n_\varphi - (d-1)n_{\varphi'}, \qquad (6.19)$$

where n_ϕ is the number of fields $\phi = \varphi, \varphi'$ in the given function Γ. However, the real exponent of divergence δ_Γ is smaller than the formal one, because it follows from the form of the vertex (6.14) that there must be a ∂ on each external φ' line in the functions Γ (and on the fields φ' in the corresponding counterterms), so that

$$\delta_\Gamma = d^*_\Gamma - n_{\varphi'}. \tag{6.20}$$

It is easily verified from the exponents (6.19) and (6.20) and the general rules of Sec. 5.15 that for $d > 2$ the counterterms in this model can be generated only by UV divergences of 1-irreducible functions of the type $\langle \varphi' \varphi \rangle_{1-\text{ir}}$ ($d^*_\Gamma = 2$, $\delta_\Gamma = 1$) and $\langle \varphi' \varphi \varphi \rangle_{1-\text{ir}}$ ($d^*_\Gamma = 1$, $\delta_\Gamma = 0$), and these counterterms must contain at least one ∂. It then follows that the graphs of $\langle \varphi' \varphi \rangle_{1-\text{ir}}$ can generate a counterterm $\sim \varphi' \partial^2 \varphi$, but do not generate counterterms $\sim \varphi' \partial_t \varphi$ of the same dimension, while the vertex graphs of $\langle \varphi' \varphi \varphi \rangle_{1-\text{ir}}$ can generate only counterterms $\varphi'(\varphi \partial)\varphi$ that are multiples of the vertex (6.14) when the transversality of all the fields is taken into account. However, these vertex counterterms allowed by the dimensions are in fact absent owing to the Galilean invariance of the model (6.18), which requires that the operator ∂_t enter into the counterterms only through the covariant derivative $\nabla_t = \partial_t + (\varphi \partial)$. Therefore, the absence of the counterterm $\sim \varphi'(\varphi \partial)\varphi$ follows from the absence of the counterterm $\sim \varphi' \partial_t \varphi$.

All the above pertains to the case $d > 2$. In the special case $d = 2$ an additional superficial divergence appears in the function $\langle \varphi' \varphi' \rangle_{1-\text{ir}}$ ($d^*_\Gamma = 2$, $\delta_\Gamma = 0$) which generates a local [in contrast to the corresponding contribution in (6.18)] counterterm $\sim \varphi' \partial^2 \varphi'$. The two-dimensional problem will be studied separately in Sec. 6.18, and for now we shall always assume that $d > 2$.

In this case we need only a single counterterm $\sim \varphi' \partial^2 \varphi$. Adding it to the functional (6.18), we obtain the renormalized action

$$S_R(\phi) = g\nu^3 \mu^{2\varepsilon} \varphi' n \varphi' / 2 + \varphi'[-\partial_t \varphi + Z_\nu \nu \partial^2 \varphi - (\varphi \partial)\varphi], \tag{6.21}$$

where Z_ν is a completely dimensionless renormalization constant, which can depend only on a single completely dimensionless parameter g (dependence on d and ε is always implicit here).

The renormalized action (6.21) is obtained from the unrenormalized one (6.17) by the standard procedure of multiplicative renormalization, in this case even without renormalization of the fields ϕ and the parameter m in a pumping function of the type (6.11):

$$Z_\phi = 1, \quad m_0 = m Z_m, \quad \nu_0 = \nu Z_\nu, \quad g_0 = g\mu^{2\varepsilon} Z_g, \quad Z_m = 1, \quad Z_g = Z_\nu^{-3}. \tag{6.22}$$

The equation $Z_m = 1$ and the relation between Z_g and Z_ν in (6.22) follow from the absence of renormalization of the nonlocal contribution of the random-force correlator in (6.21). It can be stated that in this model there are "anomalously few" divergences: the only independent renormalization constant is Z_ν.

The anomalous dimensions γ_a of the various quantities a and the β function of the charge g are determined by the general relations (1.113). In this case from (6.22) and (1.114) we have

$$\gamma_\phi = 0, \ \gamma_m = 0, \ \gamma_g = -3\gamma_\nu, \ \beta = g(-2\varepsilon + 3\gamma_\nu). \tag{6.23}$$

The RG equations for the connected Green functions have the standard form (1.107) with RG operator

$$\mathcal{D}_{\mathrm{RG}} = \mathcal{D}_\mu + \beta\partial_g - \gamma_\nu\mathcal{D}_\nu. \tag{6.24}$$

The one-loop calculation in the MS scheme gives [159][*]

$$Z_\nu = 1 - ag/2\varepsilon + O(g^2), \quad a = C_d/4(d+2), \tag{6.25}$$

where $C_d = (d-1)S_d/(2\pi)^d$ and $S_d = 2\pi^{d/2}/\|d/2\|$ is the d-dimension angular integral (1.87a). For the realistic 3-dimensional problem $(d = 3)$ we have $a = 1/20\pi^2$.

Using the constant (6.25) to calculate the corresponding RG function (1.113) γ_ν [which is most easily done using (1.120)] and taking into account the last equation in (6.23), we find

$$\gamma_\nu(g) = ag + ..., \quad \beta(g) = -2\varepsilon g + 3ag^2 + ..., \tag{6.26}$$

where the dots denote corrections of higher order in g. The positivity of the coefficient a in (6.26) ensures the existence of an IR-attractive fixed point in the physical region $g > 0$ and, consequently, the corresponding IR scaling. We shall discuss this regime in more detail in the following section.

6.5 General solution of the RG equations. IR scaling for fixed parameters g_0 and ν_0

For definiteness we shall consider the solution of the RG equation (1.107) in the model (6.21) with pumping of the type (6.11) for the example of the pair velocity correlator $D = \langle \widehat{\varphi}\widehat{\varphi} \rangle$ for $d > 2$. Owing to the absence of field renormalization (Sec. 6.4), the connected renormalized Green functions $W_{n\mathrm{R}}$ coincide with the unrenormalized functions W_n. The only difference is in the choice of variables and the form of the perturbation theory (in g or g_0). In the renormalized variables the dynamical correlator $D = W_2$ depends on k, ω, g, ν, m, and μ. From analysis of the canonical dimensions we have (always omitting the transverse projector in the vector indices)

$$D = \nu k^{-d} R(s, g, z, u), \quad s \equiv k/\mu, \ z \equiv \omega/\nu k^2, \ u \equiv m/k, \tag{6.27}$$

where R is a function of completely dimensionless arguments. The correlator $D = W_2$ satisfies the RG equation (1.107) with $\gamma_\phi = 0$, according to (6.23), and with the operator $\mathcal{D}_{\mathrm{RG}}$ from (6.24):

$$\mathcal{D}_{\mathrm{RG}}D = 0, \ \mathcal{D}_{\mathrm{RG}} = \mathcal{D}_\mu + \beta\partial_g - \gamma_\nu\mathcal{D}_\nu. \tag{6.28}$$

---

[*]The two-loop calculation is presented in [243].

According to the general rule (1.138), the solution of (6.28) has the form

$$D = \bar{\nu}k^{-d}R(1,\bar{g},\bar{z},\bar{u}), \quad \bar{z} = \omega/\bar{\nu}k^2, \quad \bar{u} = u = m/k, \tag{6.29}$$

where $\bar{g}$, $\bar{\nu}$, and $\bar{u}$ are the invariant variables corresponding to g, ν, and u (Sec. 1.29), i.e., the first integrals, depending on the scale parameter $s = k/\mu$, of a homogeneous equation of the type (1.132) with operator $L = \mathcal{D}_{\mathrm{RG}}$ from (6.28), normalized at $s = 1$ to g, ν, and u, respectively. The equation $\bar{u} = u = m/k$ follows from the absence of the contribution $\mathcal{D}_m$ in the RG operator (6.28), which follows from the fact that the parameter m in (6.22) is not renormalized. To prove (6.29) it is sufficient to note that its right-hand side certainly satisfies the RG equation (6.28) owing to the definition of the invariant variables (Sec. 1.29) and coincides with D at $s = 1$ because of their normalization conditions. It is also useful to recall that the dependence of the function $R(1,\bar{g},\bar{z},\bar{u})$ on the invariant variables is not determined by the RG equation (6.28) itself. The function R in (6.27) can in practice be calculated only using the graphs of the correlator D in the form of a series in g.

The analog of (6.29) for the static (equal-time) correlator

$$D^{\mathrm{stat}} = (2\pi)^{-1}\int d\omega\, D = \nu^2 k^{2-d} R(s,g,u) \tag{6.30}$$

is the representation (we immediately use $\bar{u} = u$)

$$D^{\mathrm{stat}} = \bar{\nu}^2 k^{2-d} R(1,\bar{g},u). \tag{6.31}$$

Substitution of (6.27) into (6.28) leads to a standard RG equation of the type (1.142) for the function R, which makes it possible to use the general relations (1.144), (1.147) to find the invariant variables $\bar{g} = \bar{g}(s,g)$ and $\bar{\nu} = \bar{\nu}(s,g,\nu)$ in (6.29): $\bar{g}$ is determined by (1.144), and for $\bar{\nu}$ we find

$$\bar{\nu} = \nu\exp\left[-\int\limits_g^{\bar{g}} dx\,\frac{\gamma_\nu(x)}{\beta(x)}\right] = (g\nu^3/\bar{g}s^{2\varepsilon})^{1/3} = (g_0\nu_0^3/\bar{g}k^{2\varepsilon})^{1/3}. \tag{6.32}$$

The second equation here follows from the first and from the relations (6.23) between the RG functions, and the third follows from the second and the renormalization formulas for the parameters (6.22). The invariant charge $\bar{g}$ in the one-loop approximation (6.26) has the form (1.154) with $g_* = 2\varepsilon/3a$.

Equations (1.144) and (6.32) express the invariant variables $\bar{e} \equiv \{\bar{g},\bar{\nu}\}$ in terms of the renormalized parameters $e \equiv \{g,\nu\}$ and the scale variable $s = k/\mu$: $\bar{e} = \bar{e}(e,s)$ and, conversely, $e = e(\bar{e},s)$. It is sometimes convenient to express $\bar{e}$ in terms of the momentum k and the bare parameters $e_0 \equiv \{g_0,\nu_0\}$ (the possibility of doing this was justified in Sec. 1.32). In our case, taking into account the nature of the dependences in (6.22), (1.144), (6.32) and dimensionality considerations, it is clear that $g_0 k^{-2\varepsilon} = F_1(\bar{g})$ and $\nu_0 = \bar{\nu}F_2(\bar{g})$ with as yet unknown functions $F_{1,2}$. Substituting (6.22) for g_0

into the first of these expressions, we obtain $g_0 k^{-2\varepsilon} = F_1(\bar{g}) = s^{-2\varepsilon} g Z_g(g)$, from which taking into account the normalization $\bar{g} = g$ for $s = 1$ we find that $F_1(g) = g Z_g(g)$. Similarly, from $\nu_0 = \bar{\nu} F_2(\bar{g})$ and the renormalization formulas for ν_0 in (6.22) we obtain $F_2(g) = Z_\nu(g)$. Therefore, the desired relations between the invariant and bare variables have the form

$$g_0 k^{-2\varepsilon} = \bar{g} Z_g(\bar{g}), \quad \nu_0 = \bar{\nu} Z_\nu(\bar{g}). \tag{6.33}$$

The constant Z_g can be expressed in terms of the β function (6.23) by (1.127), and Z_ν is expressed in terms of Z_g by the last equation in (6.22). The behavior of the function $g Z_g(g)$ is shown graphically in Fig. 1.2, and the accompanying discussion is the same as for the theory of critical behavior (Sec. 1.27). In the one-loop approximation $Z_g(g) = g_*/(g_* - g)$ according to (1.128), and from this using (6.33) we easily find the following one-loop representations for the invariant variables $\bar{e}$:

$$\bar{g} = g_*[1 + g_*/g_0 k^{-2\varepsilon}]^{-1}, \quad \bar{\nu} = \nu_0[1 + g_0 k^{-2\varepsilon}/g_*]^{1/3}, \tag{6.34}$$

which do not involve the renormalization mass μ.

Let us now consider the IR asymptote $k \to 0$, $\omega \to 0$, $m \to 0$ of the renormalized Green functions for fixed parameters g, ν, μ [this is equivalent to fixing g_0 and ν_0 taking into account the renormalization formulas (6.22)]. According to the general rules of Sec. 1.33, to obtain the leading terms of the desired asymptote it is necessary to simply replace all the invariant variables $\bar{e}$ in RG representations like (6.29) and (6.31) by their asymptotic expressions, which we shall denote as $\bar{e}_*$. In our case $\bar{g}_* = g_*$ =const is the coordinate of the IR-attractive fixed point and $\bar{\nu}_*$ is a nontrivial function easily found from (6.32):

$$\bar{\nu}_* = [g_0 \nu_0^3/g_* k^{2\varepsilon}]^{1/3} = (D_0/g_*)^{1/3} k^{-2\varepsilon/3}. \tag{6.35}$$

The parameters g_0 and ν_0 in this expression are grouped together to form the total amplitude factor $D_0 = g_0 \nu_0^3$ of the pumping function (6.16).

For the correlators D and D^{stat} in this regime the RG representations (6.29) and (6.31) give

$$D = \bar{\nu}_* k^{-d} f(\bar{z}_*, u), \quad \bar{z}_* \equiv \omega/\bar{\nu}_* k^2, \quad f(z, u) \equiv R(1, g_*, z, u), \tag{6.36a}$$

$$D^{\text{stat}} = \bar{\nu}_*^2 k^{2-d} f(u), \quad f(u) \equiv R(1, g_*, u). \tag{6.36b}$$

These are the usual expressions for the critical scaling (Sec. 1.33), with ε-dependent critical dimensions Δ_F, of quantities that are IR-relevant in this regime, i.e., the fields and variables k, ω, and m. The parameters g, ν, and μ are irrelevant (fixed in dilatations).

The "reduced scaling functions" f in (6.36) [terminology of Sec. 1.33] are expressed in terms of the corresponding functions R, which in practice can be calculated only as series of the type (1.172) in graphs of the renormalized correlator (6.27) or (6.30). The substitution of these series into the definitions (6.36) of the functions f leads to ε expansions of the type (1.173):

$$f(z, u) = \sum_{n=1}^{\infty} \varepsilon^n f_n(z, u), \quad f(u) = \sum_{n=1}^{\infty} \varepsilon^n f_n(u). \tag{6.37}$$

In lowest (first) order of renormalized perturbation theory for the model (6.11) we have

$$D = \frac{g\nu^3 \mu^{2\varepsilon} k^{4-d-2\varepsilon} h(u)}{|-i\omega + \nu k^2|^2}, \quad D^{\text{stat}} = \frac{g\nu^2 \mu^{2\varepsilon} k^{2-d-2\varepsilon} h(u)}{2}. \tag{6.38}$$

The first of these expressions was obtained by substituting the function (6.11) into the corresponding bare correlator (6.13) followed by replacement of the bare parameters by the basic ones ($\nu_0 \to \nu$, $g_0 \to g\mu^{2\varepsilon}$), and the second was obtained by integrating the first over ω following (6.30). From (6.38) we can easily find the first terms of the series (6.37). We note that all the coefficients f_n of these series turn out to be finite in the limit $u \equiv m/k \to 0$ in accordance with Kolmogorov hypothesis 1$'$ (but this does not constitute a proof of that hypothesis; see the discussion in Sec. 6.3).

In the theory of turbulence it is common to study not the static correlator (6.30), but the fluctuation energy spectrum $E(k)$, which is simply related to the function $D^{\text{stat}}(k)$ as

$$E(k) = C_d k^{d-1} D^{\text{stat}}(k)/2 \tag{6.39}$$

with constant C_d from (6.25). Its RG representation is obtained automatically from (6.31). In lowest order using (6.38) we find that

$$E(k) = C_d \bar{g} \, \bar{\nu}^2 k h(u)/4. \tag{6.40}$$

From this using $g_* = 2\varepsilon/3a$ in the approximation (6.26), for the IR asymptote of the function (6.40) we obtain

$$E(k) \simeq (C_d D_0)^{2/3} [\varepsilon(d+2)/24]^{1/3} k^{1-4\varepsilon/3} h(u). \tag{6.41}$$

When restricted to the inertial range $u \equiv m/k \ll 1$ the function $h(u)$ in all these expressions becomes $h(0) = 1$ according to the condition in (6.11).

IR representations analogous to (6.36) can be written down for any Green function. A systematic method of obtaining the equations of critical scaling and determining the corresponding critical dimensions in dynamics was described in Sec. 5.16 [see the text between (5.106) and (5.110)]. The only difference is that now ν plays the role of the parameter λ. The general equations of critical scaling have the form (5.108) with $a = m$ in this model, the frequency critical dimension is given by (5.109) with $\lambda \to \nu$ (which gives $\Delta_\omega = 2 - \gamma_\nu^*$), and the dimensions of the fields and parameter m are given by (5.110). These expressions involve the quantity $\gamma_\nu^* \equiv \gamma_\nu(g_*)$, which is found exactly from the form of the β function (6.23) without calculating graphs [this is a consequence of the relation between the constants Z_g and Z_ν in (6.22)]:

$$\gamma_\nu^* = 2\varepsilon/3. \tag{6.42}$$

Using (6.42) and the data of Table 6.1, from (5.110) and (5.109) with $\lambda \to \nu$ we obtain the following exact (without corrections $\sim \varepsilon^2$ and higher) expressions for the critical dimensions:

$$\Delta_\varphi = 1 - 2\varepsilon/3, \quad \Delta_{\varphi'} = d - \Delta_\varphi, \quad \Delta_\omega = -\Delta_t = 2 - 2\varepsilon/3, \quad \Delta_m = 1, \tag{6.43}$$

with the usual normalization $\Delta_k = 1$. In comparing (6.43) and (6.36), it must be remembered that in coordinate space the critical dimension of the correlator is equal to the simple sum of the dimensions of the fields entering into it, while in momentum space

$$\Delta[D(\omega, k)] = 2\Delta_\varphi - \Delta_\omega - d, \quad \Delta[D^{\mathrm{stat}}(k)] = 2\Delta_\varphi - d. \tag{6.44}$$

Using these expressions, it is easily checked that (6.36) correspond to IR scaling with the critical dimensions (6.43).

We again note that (6.43) are exact expressions, i.e., they do not have any corrections of order ε^2 and higher. For $\varepsilon = 2$ the critical dimensions (6.43) coincide with the Kolmogorov dimensions (6.8). This is considered one of the main achievements of this model. This result was first obtained in [186] and was later reproduced by many authors.

For massless pumping (6.10), $\varepsilon = 2$ is the only possible realistic value (Scc. 6.3). However, for the more realistic model (6.11) or its special case (6.9) the entire region $\varepsilon \geq 2$ corresponds to IR pumping (see the discussion in Sec. 6.3). For $\varepsilon > 2$ the dimensions (6.43) do not coincide with the Kolmogorov ones, which may appear to indicate complete failure of the RG theory of turbulence when it is used to try to at least partially (without hypothesis 1) explain Kolmogorov scaling. However, this is in fact not so, and the solution of the problem is described in the following section.

6.6   IR scaling at fixed parameters W and ν_0. Viscosity independence and the freezing of dimensions for $\varepsilon \geq 2$

The expressions (6.43) obtained for the critical dimensions correspond to IR scaling at fixed g_0 and ν_0, or g, ν, and μ in renormalized variables. This is how the problem is always formulated for models in the theory of critical behavior, and so expressions like (6.43) are the final results in those cases (but there the series in ε usually to not terminate). The problem is formulated differently in the Kolmogorov–Obukhov theory (Sec. 6.2), because here we deal with scaling at fixed values of the parameters W and ν_0. The pumping power W is related to the parameter $D_0 = g_0\nu_0^3 = g\nu^3\mu^{2\varepsilon}$ in (6.11) and (6.16) by (6.3), and to obtain the final results in terms of the Kolmogorov–Obukhov theory it is necessary to express g_0 in terms of W.

To do this, we shall take for definiteness the pumping model (6.9) and calculate the integral (6.3) with the cutoff $\Lambda = (W/\nu_0^3)^{1/4}$. This gives [187]

$$W = \frac{D_0 C_d}{4(2 - \varepsilon)} \left\{ (\Lambda^2 + m^2)^{2-\varepsilon} + \frac{m^{4-2\varepsilon}}{1 - \varepsilon} - \frac{(2 - \varepsilon)m^2(\Lambda^2 + m^2)^{1-\varepsilon}}{1 - \varepsilon} \right\} \qquad (6.45)$$

with constant C_d from (6.25). The Reynolds number $\mathrm{Re} = (\Lambda/m)^{4/3}$ is very large for developed turbulence, and so from (6.45) it follows that

$$D_0 = g_0 \nu_0^3 \equiv W B(\Lambda, m, \varepsilon) \simeq \begin{cases} c_1 W \Lambda^{2\varepsilon - 4} & \text{for } 2 > \varepsilon > 0, \\ c_2 W m^{2\varepsilon - 4} & \text{for } \varepsilon > 2, \end{cases} \qquad (6.46)$$

where $c_1 = 4(2 - \varepsilon)/C_d$, $c_2 = c_1(1 - \varepsilon)$. The definition of the exact function $B(\Lambda, m, \varepsilon)$ in (6.46) is obvious from comparison with (6.45), in particular, $B(\Lambda, m, 2) = 2[C_d \ln(\Lambda/m)]^{-1}$ for $\varepsilon = 2$. The simple approximations (6.46) are applicable outside the transition region near $\varepsilon = 2$. This region is narrow (of order $1/\ln \mathrm{Re}$), and so we shall assume below that the approximations (6.46) are valid right up to $\varepsilon = 2$. We have obtained them for the particular model (6.9), but the representation (6.46) is obviously valid for any model like (6.11): only the values of the coefficients $c_{1,2}$ in (6.46), which are not important for us, and the explicit form of the function $B(\Lambda, m, \varepsilon)$ in the narrow transition region near $\varepsilon = 2$ depend on the choice of the function h in (6.11).

As already mentioned, in the Kolmogorov–Obukhov theory the parameters ν_0 and W are assumed fixed (and in this sense have zero critical dimension). Therefore, the parameter $\Lambda = (W/\nu_0^3)^{1/4}$ is also fixed, while m is a dimensional parameter with $\Delta_m = 1$. We see from the representation (6.46) that for $2 > \varepsilon > 0$ the parameters D_0 and g_0 have zero critical dimension in the Kolmogorov regime, while for $\varepsilon > 2$ they acquire the critical dimension $2\varepsilon - 4$ owing to the relation $D_0 \sim g_0 \sim W m^{2\varepsilon - 4}$. Therefore, in the Kolmogorov regime of fixed ν_0 and W we have

$$\Delta[D_0] = \Delta[g_0] = \begin{cases} 0 & \text{for } 2 > \varepsilon > 0, \\ 2\varepsilon - 4 & \text{for } \varepsilon > 2. \end{cases} \qquad (6.47)$$

Equations (6.43) for the critical dimensions Δ_F were obtained assuming that the parameters $g_0 \sim D_0$ are dimensionless, and so for the Kolmogorov regime they are valid only in the region $2 > \varepsilon > 0$. For $\varepsilon > 2$ these parameters acquire the nontrivial dimension (6.47), and using it in (6.35) and (6.36a) leads to a change $\Delta_F \to \Delta'_F$ of the critical dimensions of the frequency and the field φ:

$$\Delta'_\omega = \Delta_\omega + \Delta[D_0]/3, \qquad \Delta'_\varphi = \Delta_\varphi + \Delta[D_0]/3. \qquad (6.48)$$

The change of the frequency dimension is quite important, as it causes all the other dimensions to change according to the general rule (5.110). It follows that in passing to the Kolmogorov regime the relation (6.43) between the dimensions of the fields φ and φ' is preserved ($\Delta_\varphi + \Delta_{\varphi'} = \Delta'_\varphi + \Delta'_{\varphi'} = d$), while the dimension of the parameter m remains the canonical one. The dimension Δ'_ω is determined from the form of the argument $\bar{z}_*$ in (6.36a) using (6.35) and (6.47).

Substituting (6.43) and (6.47) into (6.48), it is easily verified that for $\varepsilon \geq 2$ the dimensions (6.48) are independent of ε and coincide with their values for $\varepsilon = 2$, i.e., with the Kolmogorov dimensions (6.8). This is the meaning of the statement that the dimensions freeze in the Kolmogorov regime [187]. It is consistent with Kolmogorov hypothesis 1 in the sense that the dimensions characterize the behavior of the Green functions also in the inertial range, where there should not be any dependence on the detailed structure of the pumping (Sec. 6.2). The choice of specific value $\varepsilon > 2$ in the region of IR pumping can be considered one such detail.

The representation (6.46) allows yet another important statement to be proved, namely, the statement that the IR asymptotes of the Green functions in the Kolmogorov regime are independent of the viscosity coefficient ν_0 in the region of IR pumping $\varepsilon \geq 2$. In fact, it follows from expressions like (6.36) that the parameters g_0 and ν_0 enter into the IR asymptote only through $\bar{\nu}_*$ from (6.35) and are grouped together in it to form the combination $D_0 = g_0 \nu_0^3$. According to (6.46), in the Kolmogorov regime for $\varepsilon > 2$ we have $D_0 \sim W m^{2\varepsilon-4}$, i.e., in this case there is no ν_0 dependence in D_0. This proves the above statement regarding Kolmogorov hypothesis 2 in the region of IR pumping $\varepsilon \geq 2$ [187]. We note that this hypothesis is certainly invalid in the region of UV pumping $2 > \varepsilon > 0$, because the dependence on the viscosity coefficient ν_0 remains through the parameter $\Lambda = (W/\nu_0^3)^{1/4}$ in (6.46). We also note that hypothesis 2 automatically implies the presence of IR scaling with the Kolmogorov dimensions (6.8). The above derivation of (6.48) is therefore needed only to explain the mechanism by which the critical dimensions freeze.

Therefore, the standard RG analysis in the theory of turbulence can be used to prove Kolmogorov hypothesis 2 and, as a consequence, the presence of IR scaling with Kolmogorov dimensions (6.8) for the entire region of (realistic) IR pumping $\varepsilon \geq 2$. The fundamental unsolved problem is the justification of Kolmogorov hypothesis 1 or, stated broadly, the explanation of the dependence of scaling functions in representations like (6.36) on the argument $u = m/k$. The standard method of studying these problems in the theory of critical behavior is the SDE technique (Sec. 4.14) based on the theory of the renormalization of composite operators. It is natural to try to use a similar technique also in the RG theory of turbulence.

6.7 Renormalization of composite operators. Use of the Schwinger equations and Galilean invariance

The general theory of the renormalization of composite operators in critical statics was described in detail in Ch. 3. All its basic statements remain valid also for dynamical models if the canonical dimension d_F in all the expressions of Ch. 3 is now understood to be the total canonical dimension (5.87), and the spatial dimension d in (3.152) is replaced by $d + d_\omega$ with $d_\omega = 2$ in our model:

$$d_\Gamma = d + 2 - \sum_\phi d_\phi - \sum_F (d + 2 - d_F). \qquad (6.49)$$

The dimension $d_F \equiv d[F]$ of any local monomial F constructed from the fields $\phi = \varphi, \varphi'$ and their derivatives is equal to the sum of the dimensions of the fields and derivatives entering into it, i.e., of

$$d_\varphi = 1, \quad d_{\varphi'} = d - 1, \quad d[\partial] = 1, \quad d[\partial_t] = 2 \qquad (6.50)$$

according to the data of Table 6.1 for our model. The parameter ε does not appear in (6.50), and so in this model the values of d_F are independent of ε and (6.49) coincides with the formal exponent of divergence (5.89) for the 1-irreducible function Γ with any number of fields ϕ and composite operators F. We recall (Sec. 3.23) that superficial divergences are present only in those functions Γ for which the real exponent of divergence (3.153) (and, therefore, d_Γ) is a nonnegative integer.

In discussing the renormalization of composite operators, we shall always have in mind the formulation of the problem with $g_0 = \text{const}$, which is standard for the theory of critical behavior (rather than that with $W = \text{const}$ in the Kolmogorov regime) and the standard MS subtraction scheme. In this scheme the elements of the mixing matrix Q in (3.176) can contain only integer powers of the parameter m^2 (the analog of τ). In the massless model (6.10) only monomials F_i with the same value of d_F can mix in renormalization, while in models like (6.11) junior monomials with dimension $d_F - 2$, $d_F - 4$, and so on can also mix with these (Sec. 3.28). For the d-dimensional problem with arbitrary (for example, irrational) value of d, only monomials with the same number of fields φ' can mix in renormalization owing to the exclusivity of the value $d_{\varphi'} = d - 1$ in (6.50). From the viewpoint of physics, the most interesting operators are those without φ'. They always have integer dimension $d_F = 0, 1, 2, \ldots$. In the integer dimension $d = 3$ (we leave the special case of $d = 2$ for Sec. 6.18) we have $d_{\varphi'} = 2$, and so, in principle, it is possible for monomials with different numbers of fields φ' to mix. However, also in this case operators containing φ' usually do not mix with operators not containing φ' because each external φ' line in a graph of Γ decreases the real exponent of divergence (3.153) by one unit, in accordance with (6.14).

The unrenormalized ($\widehat{F}$) and renormalized ($\widehat{F}_\mathrm{R}$) composite operators corresponding to a given classical monomial F are determined by (3.184) with $\varphi \to \phi$ and $Z_\phi = 1$ in our model. The renormalization matrix Z_F for the closed (Sec. 3.28) family of operators $F = \{F_i\}$ and the corresponding matrix of anomalous dimensions γ_F are determined by (3.181), (3.183), and (4.31):

$$\widehat{F}_i = \sum_k Z_F^{ik} \widehat{F}_{kR}, \qquad \gamma_F = Z_F^{-1} \widetilde{\mathcal{D}}_\mu Z_F. \qquad (6.51)$$

Owing to the absence of field renormalization in this model, it follows from (3.181) that $Z_F^{-1} = Q$ is the mixing matrix.

The equation of critical scaling (Sec. 4.7) for the closed family of renormalized operators $\widehat{F}_R = \{\widehat{F}_{iR}\}$ is the analog of (4.42) in the dynamical version (5.108), i.e., in our case

$$[-\mathcal{D}_{\mathbf{x}} + \Delta_t \mathcal{D}_t + \Delta_m \mathcal{D}_m] \widehat{F}_R(x) = \Delta_F \widehat{F}_R(x), \qquad (6.52)$$

where $\Delta_{t,m}$ are the critical dimensions (6.43) and Δ_F is the matrix of critical dimensions, analogous to (5.110), of the system of operators $\widehat{F}_R = \{\widehat{F}_{iR}\}$:

$$\Delta_F = d_F^p + \Delta_\omega d_F^\omega + \gamma_F^* = d_F - \gamma_\nu^* d_F^\omega + \gamma_F^*, \text{ with } \Delta_\omega = 2 - \gamma_\nu^*. \qquad (6.53)$$

The equation $\Delta_\omega = 2 - \gamma_\nu^*$ was obtained from (5.109) with the replacement $\lambda \to \nu$. The diagonal elements of the matrix (6.53) neglecting γ_F^* are interpreted as the sum of the critical dimensions of the fields and derivatives making up a given operator ($\Delta[\partial] = 1$, $\Delta[\partial_t] = \Delta_\omega$), and $\gamma_F^* \equiv \gamma_F(g_*)$ is the addition from the operator renormalization. We also note that the rule (6.53) for Δ_F is valid not only for composite operators, but also for simple fields and parameters.

Those linear combinations (4.44) of renormalized operators

$$\widehat{F}'_{iR} = \sum_k U_{ik} \widehat{F}_{kR} \qquad (6.54)$$

for which the matrix (4.45) (in this case with $e = m$) is diagonal have definite critical dimensions. The diagonal elements of this matrix are the critical dimensions $\Delta[\widehat{F}'_{iR}]$ of the basis operators (6.54). In the massless model these dimensions are simply the eigenvalues of the matrix Δ_F (see Sec. 4.7 for more details).

In what follows we shall assume that all the diagonal elements U_{ii} of the matrix U in (6.54) are equal to unity. This auxiliary condition eliminates the arbitrariness (Sec. 4.7) in the definition of the matrix U in the absence of accidental degeneracy. Then the correspondence $F_i \to \widehat{F}'_{iR}$ is unique, which allows us to introduce the concept of the *basis operator* $\widehat{F}'_{iR}$ *associated with a given* F_i and the corresponding *associated critical dimension*:

$$\Delta^{\text{ass}}[F_i] \equiv \Delta[\widehat{F}'_{iR}]. \qquad (6.55)$$

This convenient term will frequently be used below.

Information about the renormalization of composite operators can sometimes be obtained without calculating graphs, by using the Schwinger equations (Sec. 3.29) and the Ward identities for Galilean transformations.

In particular, from (3.187) and (3.188) we find

$$\langle\langle \delta S_R(\phi)/\delta\varphi_i'(x)\rangle\rangle = -A_{\varphi'}^i(x), \tag{6.56a}$$

$$\langle\langle \varphi_i(x)\delta S_R(\phi)/\delta\varphi_i'(x)\rangle\rangle = -A_{\varphi'}^i(x)\delta W_R(A)/\delta A_\varphi^i(x), \tag{6.56b}$$

where $S_R(\phi)$ is the renormalized action (6.21) and $A(x)$ are the corresponding sources in linear form $A\phi \equiv A_\varphi^i\varphi_i + A_{\varphi'}^i\varphi_i'$. We recall [see the text following (5.26) in Sec. 5.3] that the source $A_{\varphi'}$ can be interpreted as a nonrandom external force.

The right-hand sides of (6.56) are UV-finite (Sec. 3.29) and have definite critical dimensions (Sec. 4.11). Therefore, the composite operators inside the average $\langle\langle ...\rangle\rangle$ on the left-hand sides of (6.56) must possess the same properties: they must be UV-finite operators with definite, known (from the form of the right-hand sides) critical dimensions. From this we can also obtain information about the renormalization matrices Z_F of the local monomials contained in these operators (see Sec. 6.8 for more details). We note that when shadow relations like (1.46) and (4.51) are generalized to dynamics, it is necessary to make the replacement $d \to d + d_\omega$ for the canonical dimensions and $d \to d + \Delta_\omega$ for the critical ones.

Useful information about the renormalization of composite operators can also be obtained using the Ward identities for Galilean transformations $\phi(x) \to \phi_v(x)$ with arbitrary velocity variable $v \equiv \{v_i(t)\}$ that falls off sufficiently quickly for $|t| \to \infty$:

$$\varphi_v(x) = \varphi(x_v) - v(t), \qquad \varphi_v'(x) = \varphi'(x_v),$$
$$x \equiv (t, \mathbf{x}), \quad x_v \equiv (t, \mathbf{x} + u(t)), \quad u(t) \equiv \int\limits_{-\infty}^{t} dt'v(t'). \tag{6.57}$$

We shall refer to equations of the type $H(\phi) = H(\phi_v)$ for functionals and $F(x; \phi_v) = F(x_v; \phi)$ for composite operators as *strict Galilean invariance* if they are satisfied for an arbitrary transformation (6.57), and simply as *Galilean invariance* if they hold only for ordinary transformations with $v = $ const, $u = vt$. For example, the functional (6.21) is invariant but not strictly so, because for it $S_R(\phi_v) = S_R(\phi) + \varphi'\partial_t v$. Invariance in the case of composite operators implies that the field transformation (6.57) leads only to a change of the argument x without the appearance of additional terms proportional to v, $\partial_t v$, and so on. Only operators constructed from the strictly invariant cofactors φ', $\partial\varphi$ and their covariant derivatives (∂ and ∇_t), for example, $\partial\varphi \cdot \partial\varphi$, $\varphi'\partial^2\varphi$, and $\varphi'\nabla_t(\partial\varphi)$, are strictly invariant. The cofactor $\nabla_t\varphi$ is invariant but not strictly so, and the cofactors φ and ∂_t are noninvariant.

In compact notation $F(x; \phi) \equiv F(x)$, $F(x; \phi_v) \equiv F_v(x)$ we have in general

$$F_v(x) = F(x_v) + \sum_{k\geq 1} v^k F_k(x_v) + ..., \tag{6.58}$$

where the dots denote possible contributions involving $\partial_t v$. The additional terms in (6.58) spoil the Galilean invariance and are polynomials in the velocity $v(t)$ and its derivatives, and the coefficients F_k are local operators of lower (compared to F) dimension that implicitly carry vector indices that are contracted with the indices on the factors of v. For any particular operator F it is easy to write down the complete expression (6.58), which will always contain a finite number of terms.

In [188] the Ward identities were used to show that the renormalization $F \to [F]_R$ (3.155) using the standard MS scheme can be permuted with the transformation (6.57):

$$([F(x)]_R)_v = [F_v(x)]_R = [F(x_v)]_R + \sum_{k \geq 1} v^k [F_k(x_v)]_R + ..., \qquad (6.59a)$$

$$(\text{counterterms of } F(x))_v = \text{counterterms of } (F_v(x)). \qquad (6.59b)$$

The second equation (6.59b) follows from (6.59a) and the definition (3.155) of the renormalized operator: $[F]_R = F + \text{counterterms of } F$.

Several useful consequences follow immediately from (6.59).

1. For any (strictly) Galilean-invariant operator F the corresponding renormalized operator $[F]_R$ and the sum of all the counterterms are also (strictly) invariant. Therefore, the critical dimensions (6.55) associated with Galilean-invariant operators are completely determined by the mixing of the invariant operators only among themselves — noninvariant operators cannot admix into invariant ones in renormalization.

2. If F is noninvariant but the contributions to F_v which violate Galilean invariance are UV-finite, then all its counterterms are invariant, i.e., $[F]_R = F + \text{invariant counterterms}$ (an example is $F = \varphi \partial^2 \varphi$, for which $F_v = F - v \partial^2 \varphi$ and the addition $-v \partial^2 \varphi$ is UV-finite).

3. In renormalization an operator of the type φ^n (with free vector indices or any contractions of them) cannot admix either into itself, or into any other operator F of the same dimension $d_F = n$. In fact, if one of the counterterms of F is φ^n, the left-hand side of (6.59b) will necessarily contain a contribution $\sim v^n$. However, the right-hand side of this equation cannot contain such a contribution: for $d_F = n$ the operator F contains no more than n fields φ, and if there are fewer than n such fields then F_v certainly does not contain a contribution $\sim v^n$, while if $F = \varphi^n$ there is a contribution $\sim v^n$ in F_v, but it vanishes upon selection of the counterterms owing to UV finiteness. The resulting contradiction proves the statement. It follows that the critical dimension (6.55) associated with an operator of the type φ^n does not contain corrections from γ_F^*, i.e., it is the simple sum of the critical dimensions of the cofactors [187]:

$$\Delta^{\text{ass}}[\varphi^n] = n\Delta_\varphi = n(1 - 2\varepsilon/3), \qquad (6.60)$$

according to (6.43).

It is clear from the derivation of (6.60) that the latter is due to the "exclusivity" of the monomial φ^n in the system of all monomials with a given $d_F = n$: the transformation (6.58) for $F = \varphi^n$ contains a UV-finite contribution $\sim v^n$ that cannot be generated by any other monomial F of the same dimension $d_F = n$. Therefore, the proof generalizes directly also to any other monomial F that is exclusive in the same sense: if F_v for a given F contains a UV-finite contribution that cannot be generated by any other monomial of the same dimension d_F, then this monomial F cannot be a counterterm, i.e., it cannot admix into any monomial of the same d_F, including with itself, in renormalization. If a given monomial F is chosen as one of the elements of the unrenormalized basis, the corresponding matrices Z_F and Δ_F will be block-triangular with $Z_{FF} = 1$ on the diagonal for the given operator. Therefore, the critical dimension (6.55) associated with it will be equal to the simple sum of the critical dimensions of the cofactors.

An example of such an exclusive operator is any monomial F constructed from p symbols ∂_t and n factors of φ (symbolically, $F = \partial_t^p \varphi^n$) with a specific set of vector indices (free or with contractions). Then F_v contains an exclusive contribution $\sim \partial_t^p v^n$, and so [189]

$$\Delta^{\mathrm{ass}}[\partial_t^p \varphi^n] = n\Delta_\varphi + p\Delta_\omega = n(1 - 2\varepsilon/3) + p(2 - 2\varepsilon/3). \qquad (6.61)$$

Instead of $\partial_t^p \varphi^n$, we can equally well choose the elements of the unrenormalized basis to be the polynomials obtained by replacing ∂_t by the covariant derivative ∇_t (which is more convenient from Galilean symmetry considerations). The rule (6.61) remains valid for such operators. The critical dimensions (6.43) of the two terms of the operator $\nabla_t = \partial_t + (\varphi\partial)$ are identical and equal to Δ_ω.

It then follows that (6.60) and (6.61) are special cases of the general, relation,

$$\Delta^{\mathrm{ass}}[\text{exclusive } F] = \sum \Delta[\text{cofactors of } F], \qquad (6.62)$$

following from (6.59). The general rule (6.62) also has other applications, for example, for the operators $F = \varphi'(\nabla_t\varphi)^n$ discussed in [186], which are also exclusive.

We conclude by noting that, in general, by substituting (6.51) into (6.59) we can obtain relations between the matrices Z_F of the original system of operators F and the analogous matrices for the simpler system of operators of lower dimension F_k in (6.58).

6.8 Renormalization of composite operators in the energy and momentum conservation laws

The equations (6.56), rewritten in the form (2.109), are equivalent to the following equations expressing the laws of energy and momentum conservation for composite operators (random quantities) [190]:

$$\partial_t \widehat{\varphi}_i + \partial_s \widehat{\Pi}_{is} = D_{is}\widehat{\varphi}'_s + \widehat{A}^i_{\varphi'}, \qquad (6.63a)$$

$$\partial_t \widehat{E} + \partial_i \widehat{S}_i = \widehat{W}_{\mathrm{dis}} + \widehat{\varphi}_i D_{is} \widehat{\varphi}'_s + \widehat{\varphi}_i \widehat{A}^i_{\varphi'}, \tag{6.63b}$$

where D is the correlator (6.2) for the action (6.21) and

$$\Pi_{is} = p\delta_{is} + \varphi_i \varphi_s - \nu_0(\partial_i \varphi_s + \partial_s \varphi_i), \tag{6.64a}$$

$$S_i = p\varphi_i - \nu_0 \varphi_s(\partial_i \varphi_s + \partial_s \varphi_i) + \varphi_i \varphi^2/2, \tag{6.64b}$$

$$W_{\mathrm{dis}} = -\nu_0(\partial_i \varphi_s \cdot \partial_i \varphi_s + \partial_i \varphi_s \cdot \partial_s \varphi_i) \tag{6.64c}$$

with $\nu_0 = \nu Z_\nu$ according to (6.22) [here $g\nu^3 \mu^{2\varepsilon} = g_0 \nu_0^3 \equiv D_0$ in (6.21)]. The contribution involving the nonlocal composite operator

$$p = -\partial^{-2}[\partial_i \partial_s(\varphi_i \varphi_s)], \tag{6.65}$$

interpreted as the pressure, arises in (6.63) because, owing to the transverse nature of the field φ', the result of the formal differentiation of the functional (6.21) with respect to φ' in (6.56) must be contracted with the transverse projector $P_{is}^\perp = \delta_{is} - \partial_i \partial_s/\partial^2$.

The equations (6.63) express the energy and momentum conservation laws (all quantities are per unit mass): φ_i is the momentum density, $E \equiv \varphi^2/2$ is the energy density, Π_{is} is the stress tensor, S_i is the energy flux density vector, W_{dis} is the energy dissipation rate, and the right-hand sides of (6.63) involve the contribution of the nonrandom external force $A_{\varphi'}$ and the contribution of the random force with its correlator (6.2). We stress the fact that (6.63) are equations for the random quantities themselves (as indicated explicitly by the hat), and not simply their expectation values.

The expression given in (6.3) for the energy pumping power W is obtained by averaging the corresponding contribution in (6.63b):

$$W = \langle \widehat{\varphi} D \widehat{\varphi}' \rangle \equiv \int dx' D_{is}(x, x') \langle \widehat{\varphi}_i(x) \widehat{\varphi}'_s(x') \rangle. \tag{6.66}$$

Since the correlator (6.2) is a δ function in time and is symmetric with respect to the permutation of t and t', the definition of the response function $\langle \widehat{\varphi} \widehat{\varphi}' \rangle$ in (6.66) must be made precise at $t = t'$ by using the half-sum of the limits (5.31) (with an additional transverse projector in the vector indices), which when substituted into (6.66) leads to (6.3).

The correlation functions of composite operators can, in principle, be measured experimentally, and objects like (6.64) constructed from unrenormalized monomials and bare parameters are real observables. In particular, experimental data are available on the static pair correlator of the dissipation operator (6.64c) that indicate the presence of IR scaling with critical dimension $\Delta[\widehat{W}_{\mathrm{dis}}] = 0$ in the Kolmogorov regime (see Sec. 6.12 for more details). This dimension differs greatly from the canonical value $d[\widehat{W}_{\mathrm{dis}}] = 4$, which must be the result of a renormalization-generated anomaly like order-ε contributions in (6.43) from the anomalous dimension γ_ν^*.

The first question that arises in the theoretical study of operators like (6.64) is whether a given operator $\widehat{F}$ has definite critical dimension $\Delta[\widehat{F}]$ at all. This question is nontrivial, because only basis operators (6.54) possess definite dimensions, and unrenormalized operators $\widehat{F}$ and their renormalized analogs $\widehat{F}_R$ are in general linear combinations of operators (6.54) with various critical dimensions. Therefore, any particular operator $\widehat{F}$, with rare exceptions, cannot be treated separately: in order to expand it in the basis (6.54) the latter must first be constructed, and this requires analysis of the renormalization of the entire closed system of operators to which the original $\widehat{F}$ belongs.

This analysis was performed for all the monomials entering into (6.64) in [190] using the MS scheme for the massless model (6.10). We note that the masslessness eliminates the admixture of junior operators in senior ones, but does not distort the critical dimensions of the latter, because they are uniquely determined by the senior–senior block (Sec. 4.7).

It was shown in [190] that all the operators in (6.63) are UV-finite and possess definite critical dimensions equal to

$$\Delta[\partial_t \widehat{\varphi}_i] = \Delta[\partial_s \widehat{\Pi}_{is}] = \Delta[\widehat{A}^i_{\varphi'}] = \Delta_\varphi + \Delta_\omega = 3 - 4\varepsilon/3 \qquad (6.67a)$$

for the terms in (6.63a) and

$$\Delta[\partial_t \widehat{E}] = \Delta[\partial_i \widehat{S}_i] = \Delta[\widehat{W}_{\mathrm{dis}}] = 2\Delta_\varphi + \Delta_\omega = 4 - 2\varepsilon \qquad (6.67b)$$

for the terms in (6.63b). Here the operators $E = \varphi^2/2$ and Π_{is} themselves are UV-finite and have dimensions

$$\Delta[\widehat{E}] = \Delta[\widehat{\Pi}_{is}] = 2\Delta_\varphi = 2 - 4\varepsilon/3, \qquad (6.68)$$

and the energy flux density vector S_i defined by (6.64b) is the sum of the UV-finite contribution with dimension $3 - 2\varepsilon$ and the addition $a\partial^2 \varphi_i$ with the different dimension $2 + \Delta_\varphi = 3 - 2\varepsilon/3$ and UV-divergent coefficient a. Since this addition is transverse it does not contribute to (6.63b) and can simply be discarded, which leads to a physically reasonable redefinition of the energy flux density S_i.

Let us use this example to explain the technique of working with composite operators. We begin with the scalar $F(x) = \varphi^2(x) = 2E(x)$ with $d_F = 2$. There are no other scalars with $d_F = 2$ in the massless model and this operator cannot admix into itself [consequence 3 of (6.59)], so that it is not renormalized [i.e., in (3.155) $[F]_R = F$ and in (3.184) $\widehat{F} = \widehat{F}_R$ owing to the absence of field renormalization] and it has definite critical dimension $\Delta[F] = 2\Delta_\varphi$ according to the rule (6.60).

Let us now consider (6.63a). All the contributions except $\partial_s \Pi_{is}$ are certainly UV-finite and therefore not renormalized, and so this contribution must also be UV-finite. Since any operator of the type ∂F is renormalized by the renormalization of F itself (Sec. 3.24), it is sufficient to study the 3×3

renormalization matrix (6.51) for the three symmetric tensor operators with $d_F = 2$ appearing in (6.64a):

$$F_1 \equiv \partial_i \varphi_s + \partial_s \varphi_i, \quad F_2 \equiv \varphi_i \varphi_s, \quad F_3 \equiv p \delta_{is}. \tag{6.69}$$

The last one is nonlocal but needed because it enters into (6.64a). According to the general rule of locality of counterterms (Ch. 3), nonlocal operators like F_3 in (6.69) cannot be counterterms, i.e., in renormalization they cannot admix into any other operators, including into themselves (but local operators can admix into nonlocal ones). The operator $\varphi^2 \delta_{is}$ is also allowed in (6.69) by dimensions and index structure, but it is not renormalized (see above) and does not admix into the other operators (6.69) [consequence 3 of (6.59)], and so it is completely decoupled.

In the system (6.69) the operator F_1, which is a multiple of the simple field φ, is not renormalized [i.e., $\widehat{F}_1 = \widehat{F}_{1R}$, which leads to $Z_{11} = 1$, $Z_{12} = Z_{13} = 0$ for the elements $Z_{ik} \equiv Z_F^{ik}$ of the matrix Z_F in (6.51)], and for the operators $F_{2,3}$ we have $\widehat{F}_2 = Z_{21} \widehat{F}_{1R} + Z_{22} \widehat{F}_{2R}$ and $\widehat{F}_3 = Z_{31} \widehat{F}_{1R} + Z_{32} \widehat{F}_{2R} + \widehat{F}_{3R}$ (the nonlocal operator F_3 does not enter into the counterterms). The constants Z in these expressions are related to each other, because from (6.65) and (6.69) we have $F_3 = -P^{\|} F_2$ up to δ_{is}, which leads to $\widehat{F}_{3R} = -P^{\|} \widehat{F}_{2R}$, because external factors like the longitudinal projector $P_{is}^{\|} = \partial_i \partial_s / \partial^2$ do not affect the renormalization (Sec. 3.24). From this we find that

$$\widehat{F}_3 = Z_{31} \widehat{F}_{1R} + Z_{32} \widehat{F}_{2R} + \widehat{F}_{3R} = -P^{\|} \widehat{F}_2 =$$
$$= -P^{\|} [Z_{21} \widehat{F}_{1R} + Z_{22} \widehat{F}_{2R}] = -Z_{22} P^{\|} \widehat{F}_{2R} = Z_{22} \widehat{F}_{3R},$$

because the contribution $\widehat{F}_{1R}$ operated on by $P^{\|}$ vanishes since the field φ is transverse. Equating the coefficients of the operators $\widehat{F}_{iR}$ in this expression, we find $Z_{31} = Z_{32} = 0$ and $Z_{22} = 1$. The remaining unknown element Z_{21} can be found from requiring that the divergence

$$\partial \widehat{\Pi} = \partial [\widehat{F}_3 + \widehat{F}_2 - \nu Z_\nu \widehat{F}_1] = \partial [\widehat{F}_{3R} + Z_{21} \widehat{F}_{1R} + \widehat{F}_{2R} - \nu Z_\nu \widehat{F}_{1R}]$$

be UV-finite (see above). The three operators $\partial \widehat{F}_{iR}$ are independent, and so the coefficient of each must be UV-finite, i.e., it must be the same as its UV-finite part (Sec. 4.7). For the coefficient of $\partial \widehat{F}_{1R}$ we then obtain $Z_{21} - \nu Z_\nu = -\nu$, i.e., $Z_{21} = \nu[Z_\nu - 1]$. We have used the fact that in the MS scheme constants like Z_ν and the diagonal elements of any matrix Z_F have the form (1.104), and so their UV-finite part is equal to unity, while all the nondiagonal elements of Z_F consist only of poles in ε, and so they do not have any UV-finite part.

Therefore, for the system (6.69) the matrices Z_F and γ_F in (6.51) can be found without calculating any graphs:

$$Z_F = \begin{pmatrix} 1 & 0 & 0 \\ \nu(Z_\nu - 1) & 1 & 0 \\ 0 & 0 & 1 \end{pmatrix}, \quad \gamma_F = \begin{pmatrix} 0 & 0 & 0 \\ \nu\gamma_\nu & 0 & 0 \\ 0 & 0 & 0 \end{pmatrix} \tag{6.70}$$

with RG function γ_ν from (6.23). We can then use the known (Table 6.1 in Sec. 6.4) canonical dimensions $d_F = 2, 2, 2$ and $d_F^\omega = 1, 2, 2$ of the operators (6.69) to construct the matrix of critical dimensions (6.53) and the matrix U in (6.54) which diagonalizes it:

$$\Delta_F = \begin{pmatrix} 2 - \gamma_\nu^* & 0 & 0 \\ \nu\gamma_\nu^* & 2 - 2\gamma_\nu^* & 0 \\ 0 & 0 & 2 - 2\gamma_\nu^* \end{pmatrix}, \quad U = \begin{pmatrix} 1 & 0 & 0 \\ -\nu & 1 & 0 \\ 0 & 0 & 1 \end{pmatrix}, \quad (6.71)$$

where $\gamma_\nu^* = 2\varepsilon/3$ according to (6.42). Since the matrix Δ_F is triangular, in this case its diagonal elements are the desired critical dimensions, and the corresponding basis operators (6.54), i.e., $\widehat{F}_{1R}' = \widehat{F}_{1R}$, $\widehat{F}_{2R}' = \widehat{F}_{2R} - \nu\widehat{F}_{1R}$, and $\widehat{F}_{3R}' = \widehat{F}_{3R}$ in our case, possess these dimensions. The tensor (6.64a) originally represented in terms of unrenormalized operators (6.69) can be expressed in terms of renormalized operators using (6.51) with matrix Z_F known from (6.70), and then in terms of the basis operators (6.54):

$$\widehat{\Pi} = \widehat{F}_3 + \widehat{F}_2 - \nu Z_\nu\widehat{F}_1 = \widehat{F}_{3R} + \nu(Z_\nu - 1)\widehat{F}_{1R} + \widehat{F}_{2R} - \nu Z_\nu\widehat{F}_{1R} = \widehat{F}_{2R}' + \widehat{F}_{3R}'.$$

The result involves two operators $\widehat{F}_{iR}'$ with the same (see above) critical dimension $\Delta[F] = 2 - 2\gamma_\nu^* = 2 - 4\varepsilon/3$.

We have presented the detailed analysis of the renormalization of the relatively simple system (6.69) as an example; more complicated systems are treated similarly [190]. To analyze the vector (6.64b) we need to study the closed system of the following six vector operators with $d_F = 3$:

$$F = \{\partial_t\varphi_i,\ \partial^2\varphi_i,\ \partial_i\varphi^2,\ \partial_s(\varphi_i\varphi_s),\ \varphi^2\varphi_i,\ p\varphi_i\}. \quad (6.72)$$

The analogous system for scalars in (6.63b) consists of seven operators with $d_F = 4$:

$$F = \{\partial_t\varphi^2,\ \partial^2\varphi^2,\ \partial_i\partial_s(\varphi_i\varphi_s),\ \partial_i(\varphi_i\varphi^2),\ \varphi\partial^2\varphi,\ \varphi^4,\ \partial_i(p\varphi_i)\}. \quad (6.73)$$

On the basis of dimensions, we might expect that the system (6.73) could also include the nonlocal operator $\varphi D\varphi'$ from (6.63b) and, in the special case $d = 3$, also the operator $\varphi'\varphi'$, which has the same dimension $d_F = 4$ for $d = 3$. However, these two operators are not renormalized and do not mix with the operators of the system (6.73) [159], and so we can neglect them.

It has been shown [190] that the Schwinger equations and the Ward identities asociated with Galilean transformations can be used to find the 6×6 renormalization matrix Z_F for the system (6.72) up to two unknown elements, and the analogous 7×7 matrix for the system (6.73) up to three unknown elements. All the other nontrivial elements of these matrices are expressed in terms of the constant Z_ν. It is important that the remaining unknown elements of Z_F do not affect the critical dimensions and are unimportant for analyzing the contributions in (6.63b) in terms of renormalized operators.

The desired critical dimensions of the basis operators (6.54) for the system (6.72) are $3-\gamma_\nu^*$, $3-2\gamma_\nu^*$ (triply degenerate), and $3-3\gamma_\nu^*$ (doubly degenerate), while for the family (6.73) we obtain $4-2\gamma_\nu^*$ (doubly degenerate), $4-3\gamma_\nu^*$ (four-fold degenerate), and $4-4\gamma_\nu^*$ with γ_ν^* from (6.42) [190].

From the viewpoint of experiment (see Sec. 6.12 for details), it is the dissipation operator (6.64c) that is of greatest interest in (6.63). It is strictly Galilean-invariant and is represented as a linear combination of unrenormalized operators $\widehat{F}_i$ for the system (6.73) [with the F_i numbered in the order they occur in (6.73)], and after renormalization reduces to one of the basis operators (6.54) for this system [190]:

$$\widehat{W}_{\mathrm{dis}} = \nu_0[\widehat{F}_5 - \widehat{F}_3 - \widehat{F}_2/2] = \nu[\widehat{F}_{5\mathrm{R}} - \widehat{F}_{3\mathrm{R}} - \widehat{F}_{2\mathrm{R}}/2] = \nu\widehat{F}'_{5\mathrm{R}},$$

$$\Delta[\widehat{W}_{\mathrm{dis}}] = 4 - 2\varepsilon. \tag{6.74}$$

For the realistic value $\varepsilon = 2$ in the massless model (Sec. 6.3) we have $\Delta[\widehat{W}_{\mathrm{dis}}] = 0$, in agreement with the experimental data for the static pair correlator of the dissipation operator [see (6.100) and the accompanying discussion in Sec. 6.12].

6.9   Study of the $m \to 0$ asymptote of the scaling functions of the pair velocity correlator using the SDE

The RG representations (6.36) with $\overline{\nu}_*$ from (6.35) describe the leading terms of the IR asymptote $k \to 0$, $m \sim k \to 0$, $\omega \sim k^{\Delta_\omega} \to 0$ of the (dynamical and static) pair velocity correlator of the model (6.21) for arbitrary fixed value of the parameter $u \equiv m/k$. The inertial range is specified by the additional condition $u \equiv m/k \ll 1$, and we are interested in the m dependence of the correlators in precisely this region for the Kolmogorov regime with $W = \mathrm{const}$ (Sec. 6.6).

According to Kolmogorov hypothesis 1 (Sec. 6.2), the static correlator in this region should not depend on m, i.e., in this regime (6.36b) must have a finite limit for $m \to 0$. It then follows from (6.35) and (6.36b) that the m dependence in the combination $D_0^{2/3} f(u)$ with D_0 from (6.46) must vanish for $u = m/k \to 0$, so that the scaling function $f(u)$ of the static correlator (6.36b) must have the following asymptotes:

$$f(u) \underset{u \to 0}{=} \begin{cases} \mathrm{const} = f(0) & \text{for } 2 \ge \varepsilon > 0, \\ \mathrm{const} \cdot u^{4(2-\varepsilon)/3} & \text{for } \varepsilon \ge 2. \end{cases} \tag{6.75}$$

Strictly speaking, the entire Kolmogorov theory pertains only to the case of realistic IR pumping $\varepsilon \ge 2$. The expression (6.75) for $2 \ge \varepsilon > 0$ should be viewed as a natural generalization of hypothesis 1 to the region of UV pumping [187].

It should be stressed that the explicit form of the scaling functions $f(z, u)$ and $f(u)$ defined in terms of R in (6.36) is independent of the choice of asymptotic regime (the Kolmogorov regime with $W = \text{const}$ or the "ordinary" critical regime with $g_0 = \text{const}$). The dependence on this choice enters into (6.36) only through the overall factors $\overline{\nu}_*$ and the form of the argument $\overline{z}_*$ in (6.36a) owing to the changed meaning of the parameter D_0 in (6.35): for the ordinary critical regime $D_0 = \text{const}$, and for the Kolmogorov regime D_0 is given by (6.46). Therefore, the statement (6.75) about the asymptote of $f(u)$ following from hypothesis 1 must be an objective property (to be proved) which is independent of the regime, so that the latter can be chosen arbitrarily when studying the function $f(u)$. Always in what follows when studying these problems we shall choose the ordinary critical regime with $g_0 = \text{const}$, because it is more convenient when using the SDE technique (see below). In this regime the critical dimensions of the fundamental quantities are determined by (6.43) for any $\varepsilon > 0$. We also note that the entire discussion of the technique of renormalizing composite operators in Secs. 6.7 and 6.8 corresponds to this regime with $g_0 = \text{const}$, which is standard for the theory of critical behavior. In principle, it is easy to check how the critical dimensions of composite operators change in going from the ordinary regime ($g_0 = \text{const}$) to the Kolmogorov regime ($W = \text{const}$). However, we do not need to do this and just refer the reader to the more detailed discussion in [191] and the book [183].

As already mentioned, in the theory of critical behavior the standard method of studying the asymptote of the scaling functions for $\tau \to 0$ (τ is the analog of m^2) is the SDE technique; this was discussed in detail in Sec. 4.14. The analog of the SDE relation (3.217) in dynamics is the following formal representation (with infinite summation and neglecting the residual term; see the remark at the end of Sec. 3.33):

$$\widehat{\varphi}(x_1)\widehat{\varphi}(x_2) = \sum_F B_F(x, m^2) \widehat{F}_{\mathrm{R}}'(x'), \qquad (6.76)$$

in which $x = x_1 - x_2$, $x' = (x_1 + x_2)/2$. The subscript R on the fields $\widehat{\varphi}$ on the left-hand side of (6.76) has been omitted owing to the absence of field renormalization in our model. We recall (Sec. 3.33) that relations like (6.76) describe the $x \to 0$ asymptote for fixed x', and in dynamics $x \equiv (t, \mathbf{x})$ and $x' \equiv (t', \mathbf{x}')$. The summation in (6.76) runs over the various local composite operators F (more details follow), $\widehat{F}_{\mathrm{R}}'$ are the basis operators (6.54) associated with F and having definite critical dimensions Δ_F, and B_F are the corresponding UV-finite Wilson coefficients. For the ordinary critical regime ($g_0 = \text{const}$) with fixed values of the renormalized variables g, ν, and μ, the B_F in the MS scheme are represented as series in g with coefficients (Sec. 3.33) regular in m^2 (the analog of τ). It is this property of regularity in m^2 that is the main advantage of choosing the ordinary critical regime; in the Kolmogorov regime ($W = \text{const}$) for $\varepsilon > 2$ a nontrivial, nonanalytic m dependence would appear in the coefficients B_F through the parameters $g(m)$ and $\nu(m)$ owing to (6.22) and (6.46).

According to the general principles of constructing expansions like (6.76) (Sec. 3.33), in general the right-hand side involves all operators F that can mix in renormalization with the local monomials present in the formal Taylor expansion of the left-hand side of (6.76) in x. These monomials are constructed only from the fields φ and their derivatives, and so in the general case of arbitrary dimension $d > 2$ operators involving fields φ' cannot mix with them (Sec. 6.7). It can therefore be assumed that the SDE representation (6.76) involves only operators F constructed from fields φ and their derivatives without the fields φ'.

The dynamical correlator is obtained by averaging (6.76):

$$\langle \widehat{\varphi}(x_1)\widehat{\varphi}(x_2)\rangle = \sum_F B_F(x, m^2)\langle \widehat{F}'_{\mathrm{R}}(x')\rangle. \tag{6.77}$$

The expectation values of local operators $\langle \widehat{F}'_{\mathrm{R}}(x')\rangle$ are independent of x' owing to translational invariance. Their leading contributions to the IR asymptote $m \to 0$ are given by an expression analogous to (4.104):

$$\langle \widehat{F}'_{\mathrm{R}}(x')\rangle \underset{m\to 0}{=} \mathrm{const} \cdot m^{\Delta_F}, \tag{6.78}$$

where Δ_F is the critical dimension of a given operator [in (6.78) we have used $\Delta_m = 1$]. Substitution of (6.78) into (6.77) leads to a relation analogous to (4.105)

$$\langle \widehat{\varphi}(x_1)\widehat{\varphi}(x_2)\rangle = \sum_{n=0}^{\infty} \sum_F c_{nF} m^{2n+\Delta_F} \tag{6.79}$$

determining the dependence of the dynamical correlator on the parameter m.

In the ordinary critical regime, which we are considering at present, the m dependence enters into the correlators (6.36) only through the argument $u = m/k$. Therefore, (6.79) determines the $u \to 0$ asymptote of the scaling functions f in (6.36):

$$f(u, ...) \underset{u\to 0}{=} \sum_{n=0}^{\infty} \sum_F c_{nF}(...) u^{2n+\Delta_F} \tag{6.80}$$

with changed coefficients c_{nF}. The ellipsis in (6.80) denotes the other arguments of the scaling function, for example, the argument z for the dynamical correlator.

The expansion (6.76) involves all possible local operators constructed from the fields φ and their derivatives, including tensor operators. Then $\widehat{F}'_{\mathrm{R}}$ and B_F are understood to carry the appropriate indices, which are contracted with each other. In addition, (6.76) involves the two free vector indices of the fields φ. Since the structure of the correlators (6.36) in these indices is known (transverse projector), these indices can be contracted in calculating the scaling functions, and then all the contributions on the right-hand side of (6.76) become scalar ones.

Only those operators $\widehat{F}'_{\mathrm{R}}$ that have nonzero expectation value (6.78) contribute to the correlator (6.77). All operators of the type ∂F with external derivatives ∂ are thereby immediately excluded, as their expectation values are zero owing to translational invariance. All nonscalar irreducible tensor operators are also excluded for the same reason [owing to translational invariance, the expectation values (6.78) can depend on the vector indices only through the δ_{ik} and therefore cannot be irreducible].

Therefore, in terms of irreducible representations only scalar operators F without external derivatives ∂ can contribute to (6.80) for the scaling functions of the correlators (6.36).

Even more stringent restrictions on the operators F in (6.76) arise in going to the static correlator (6.36b). It was shown in [192] that the expression $\widehat{\varphi}_i(x_1)\widehat{\varphi}_i(x_2) - [\widehat{\varphi}^2(x_1)]_R/2 - [\widehat{\varphi}^2(x_2)]_R/2$ contracted in the indices for $t_1 = t_2$ is a strictly Galilean-invariant object and that its SDE representation (6.76) contains contributions from only strictly Galilean-invariant operators F. After the averaging $\langle...\rangle$ the subtracted terms $\sim \langle[\widehat{\varphi}^2(x)]_R\rangle$ become coordinate-independent constants representing the first term on the right-hand side of (6.7):

$$\langle\widehat{\varphi}(x_1)\widehat{\varphi}(x_2)\rangle\,|_{t_1=t_2} = \text{const} + \sum_{\substack{\text{strictly}\\ \text{G.-inv.}F}} B_F(\mathbf{x}, m^2)\langle\widehat{F}'_R(x')\rangle. \tag{6.81}$$

In passing to momentum space (6.36b) the constant on the right-hand side of (6.81) can be discarded [formally it corresponds to the contribution $\sim \delta(k)$ that vanishes for $k \neq 0$]. Therefore, from (6.80) and (6.81) for the scaling function $f(u)$ of the static correlator (6.36b) we find that

$$f(u)\underset{u\to 0}{=} \sum_{n=0}^{\infty} \sum_{\substack{\text{strictly}\\ \text{G.-inv.}F}} c_{nF}u^{2n+\Delta_F} \tag{6.82}$$

with summation over all strictly Galilean-invariant scalar operators with nonzero expectation values (6.78).

Within the ε expansion $\Delta_F = d_F + O(\varepsilon)$, and so the operators F are naturally classified according to the value of d_F. For scalar operators without φ' we have (Sec. 6.7) $d_F = 0, 2, 4, ...$ (scalars with odd d_F do not exist). The simplest of them is $F = 1$ with $d_F = 0$, then operators with $d_F = 2$, $d_F = 4$, and so on.

The contributions of only strictly Galilean-invariant scalars without external derivatives remain when we go to the SDE representation (6.82) of the scaling function of the static correlator. In the lowest dimensions there are only a few such operators. They are the trivial operator $F = 1$ with $d_F = 0$; invariant scalars with $d_F = 2$ do not exist at all, and in the class of scalars with $d_F = 4$ the only strictly Galilean-invariant operator is the dissipation operator (6.64c) with dimension $\Delta_F = 4 - 2\varepsilon$ known exactly from (6.67b).

In the class of Galilean-invariant scalars without external derivatives with $d_F = 6$ there are only two independent operators, namely,

$$F_1 = \partial_i \varphi_k \cdot \partial_i \varphi_s \cdot \partial_k \varphi_s, \qquad F_2 = \partial^2 \varphi_i \cdot \partial^2 \varphi_i. \qquad (6.83)$$

All the other invariant scalars with the same dimension $d_F = 6$ reduce to linear combinations of the operators (6.83) up to contributions involving external derivatives ∂, which are unimportant in (6.82). The critical dimensions Δ_i associated with the operators (6.83) were calculated in [189]:

$$\Delta_1 = 6 - 2\varepsilon, \qquad \Delta_2 = 6 - 8\varepsilon/7 + O(\varepsilon^2). \qquad (6.84)$$

The first of these is the exact value and the second is the one-loop approximation. It is important for what follows that both dimensions in (6.84) are positive at the boundary $\varepsilon = 2$ of the IR-pumping region.

The above discussion for the scaling function of the static correlator leads to a representation analogous to (4.107):

$$f(u) = C_1 + C_2 u^2 + C_3 u^4 + C_4 u^{4-2\varepsilon} + C_5 u^6 + C_6 u^{\Delta_1} + C_7 u^{\Delta_2} + ..., \quad (6.85)$$

where the Δ_i are the dimensions (6.84) and the ellipsis denotes contributions of order $u^{8+O(\varepsilon)}$ and higher. The coefficients C_i in (6.85) can be calculated as ε expansions exactly as in Sec. 4.14 [we note that in our model (6.21) all ε expansions C_i begin with contributions of order ε or higher].

In the theory of critical behavior (Sec. 4.14) the study of the analogous asymptote $\tau \to 0$ ends at this point, because, first, there is no reason to doubt that all the exponents in relations like (6.80) and (6.82) are positive, and, second, there is no possibility of reliably judging their signs when in practice only the first few terms of their (infinite) ε expansions are known.

The situation is different in the theory of turbulence, in particular because here the critical dimensions of some composite operators can be found exactly and it is clear that they become negative with increasing ε. Operators with dimensions $\Delta_F < 0$ and their contributions to the SDE will be termed *dangerous*. In (6.80) and (6.82) the contributions with negative powers of the parameter $u = m/k$, which diverge for $m \to 0$, correspond to such operators.

It follows from (6.60) and (6.61) that all operators of the type φ^n become dangerous for $\varepsilon > 3/2$ and operators of the type $\partial_t^p \varphi^n$ become dangerous for $\varepsilon > 3(2p+n)/2(p+n) \equiv \varepsilon_0$, with the boundary ε_0 for $2p < n$ also still located in the region of UV pumping $\varepsilon < 2$. All of these operators are not strictly Galilean-invariant even after the replacement $\partial_t \to \nabla_t$ (Sec. 6.7), and so they cannot contribute to the representation (6.82) for the static correlator. The first nontrivial contribution to (6.82) is generated by the strictly invariant dissipation operator (6.64c) with critical dimension $\Delta_F = 4-2\varepsilon$ known exactly from (6.67b), which becomes dangerous in passing through the boundary of the region of IR pumping $\varepsilon = 2$. As ε increases new dangerous operators will appear, for example, F_1 from (6.83) with Δ_1 from (6.84) becomes dangerous for $\varepsilon > 3$. It was assumed, but not rigorously proved, in [193] that an equation like (6.60) is satisfied also by the dissipation operator (6.64c), and then all its powers are dangerous operators for $\varepsilon > 2$. However, this assumption is poorly justified judging from the results of [194], where among other things it was shown that the squared dissipation operator must mix in renormalization with other operators of the same canonical dimension.

According to (6.75), in the region of UV pumping $2 > \varepsilon > 0$ the scaling function of the static correlator $f(u)$ must have a finite limit for $u = m/k \to 0$, and this is consistent with the fact that no one has yet found a strictly Galilean-invariant scalar operator with exactly known critical dimension that would become dangerous for $\varepsilon < 2$. Of course, this still does not prove the Kolmogorov hypothesis 1 for this region, but also it does not contradict it. If such an operator were found, we would be faced with a difficult problem (which for now we can simply ignore, as is actually also done in the theory of critical behavior).

However, the analogous problem is quite important in the IR pumping region, because there we know of a strictly invariant scalar dissipation operator (6.74) with $\Delta_F = 4 - 2\varepsilon$ that becomes dangerous right after crossing the boundary of the IR pumping region $\varepsilon = 2$. Since the exponent $4(2 - \varepsilon)/3$ of the power of u isolated explicitly in (6.75) cannot be the critical dimension Δ_F of any operator (because they all must be integers for $\varepsilon = 0$), we can hope to justify the representation (6.75) equivalent to Kolmogorov hypothesis 1 in the IR pumping region $\varepsilon > 2$ only using some procedure for summing the infinite number of contributions from the various dangerous operators. An example of this procedure is given in the following section.

In conclusion, we add that in going from the ordinary critical regime to the Kolmogorov one ($W = \mathrm{const}$) the dimensions Δ_F in (6.78) would change, while additional common factors $m^{\cdots}$ with fractional powers of m would appear at the IR asymptote of the Wilson coefficients B_F. These two effects cancel each other out, so that the common power of m for any term on the right-hand side of (6.77) remains unchanged. The final form of the SDE representations (6.80), (6.82) with the ordinary (not Kolmogorov) critical dimensions Δ_F is not changed. This is natural, because the form of the scaling functions does not depend at all on the choice of regime (see the beginning of this section).

6.10 Summation of the contributions of dangerous operators φ^n and $\partial_t^p \varphi^n$ in the dynamical velocity correlator

The contributions of all operators of the type φ^n were summed in [187] using one version of infrared perturbation theory (IRPT; see, for example, [23]). The formulation of IRPT used in [187] is based on splitting the fields in the functional integral into soft (long-wavelength) and hard (short-wavelength) components [23] and then neglecting the space-time nonuniformity of the soft field. For the dynamical correlator in the inertial range the summation of the contributions of all operators of the type φ^n in (6.79) leads to the following expression (we present the results of [187] in the later, improved formulation of [196]):

$$\langle \widehat{\varphi}(x_1)\widehat{\varphi}(x_2)\rangle = \sum_{n=0}^{\infty} \frac{t^n}{n!}\langle((\widehat{\varphi}\partial)^n\rangle B_1(x) + ...,$$ (6.86)

where B_1 is the Wilson coefficient of the first operator $\widehat{F}'_R = 1$ in (6.76), $x \equiv (t,\mathbf{x})$, $x' \equiv (t',\mathbf{x}')$, and $\varphi\partial \equiv \varphi_i(x')\partial_i$ with $\partial_i \equiv \partial/\partial x_i$. The operators ∂_i in (6.86) act only on the function $B_1(x)$ and can be taken out of the average $\langle ... \rangle$, which then contains only the product of n fields $\widehat{\varphi}(x')$ with free vector indices. This product is an unrenormalized monomial and contains the contribution of the corresponding associated operator with unit coefficient [a consequence of (6.59)]. It is only this dangerous contribution with dimension (6.60) that interests us now, because admixtures of other renormalized operators are unimportant due to their dimensions [187].

Equation (6.86) can be rewritten compactly as

$$\langle \widehat{\varphi}(x_1)\widehat{\varphi}(x_2)\rangle = \langle\langle B_1(t, \mathbf{x} + \widehat{v}t)\rangle\rangle,$$ (6.87)

where $\langle\langle ... \rangle\rangle$ denotes the average over the distribution of the x-independent random quantity (not a field) $\widehat{v}$, given by

$$\langle\langle \widehat{v}...\widehat{v}\rangle\rangle = \langle \widehat{\varphi}(x')...\widehat{\varphi}(x')\rangle = \text{const} \cdot m^{n\Delta_\varphi} + ...$$ (6.88)

with free vector indices on the n factors of $\widehat{v}$ and $\widehat{\varphi}$. The moments of the random quantity $\widehat{v}$ defined by the first equation in (6.88) are nonzero only when the number of factors n is even, and in their vector indices they are multiples of the symmetrized product of δ symbols [the index structure has been omitted in (6.88)]. The second equation in (6.88) represents the leading term of the IR asymptote $m \to 0$ generated by the dangerous operator of dimension (6.60) associated with φ^n, and the ellipsis denotes irrelevant (less singular) corrections from possible admixtures of other operators involving ∂ symbols and fewer fields φ.

A representation like (6.87) for the dynamical correlator $\langle\widehat{\varphi}\widehat{\varphi}\rangle$ and its analog for the response function $\langle\widehat{\varphi}\widehat{\varphi}'\rangle$ were obtained in [187] neglecting the interaction of the hard components of the fields. This corresponds to lowest-order perturbation theory for the function $B_1(t,\mathbf{x})$ in (6.87) and its analog for $\langle\widehat{\varphi}\widehat{\varphi}'\rangle$. In this approximation these functions coincide with the corresponding bare propagators (6.13) with $\nu_0 \simeq \nu$, so that in ω, k space we have [187]

$$\langle\widehat{\varphi}\widehat{\varphi}'\rangle = \langle\langle\frac{1}{[-i(\omega - \widehat{v}k) + \nu k^2]}\rangle\rangle,$$ (6.89a)

$$\langle\widehat{\varphi}\widehat{\varphi}\rangle = \langle\langle\frac{1}{[-i(\omega - \widehat{v}k) + \nu k^2]}N(k)\frac{1}{[i(\omega - \widehat{v}k) + \nu k^2]}\rangle\rangle.$$ (6.89b)

When the leading singularities are chosen the analytic m^2 dependence of a pumping function like (6.11) can be neglected in (6.89b), which corresponds to going to the massless model (6.10). We note that the representation (6.89a) for the response function $\langle \widehat{\varphi}\widehat{\varphi}' \rangle$ was obtained by a different method in [195].

The expressions (6.87) and (6.89) are the result of summing the contributions of all dangerous operators of the type φ^n in the SDE representations. They contain [through the moments (6.88) of the random quantity $\widehat{v}$] a nontrivial dependence on the parameter m. However, this dependence vanishes when we pass to the static correlator $\langle \widehat{\varphi}\widehat{\varphi} \rangle$ with $t_1 = t_2$: in the language of the representation (6.87), this case corresponds to $t = 0$, while in the language of (6.89b) it corresponds to integration over the frequency ω, causing the dependence on its shift by $\widehat{v}k$ to vanish. Therefore, dangerous operators like φ^n do not contribute to the static correlator $\langle \widehat{\varphi}\widehat{\varphi} \rangle$, as must be the case according to general arguments (Sec. 6.9).

The generalization (6.87) of the simplest approximation (6.89b) with exact function B_1 was obtained in [196]. There it was shown that relations like (6.86) to (6.88) can also be obtained without IRPT, by using the operator expansion (SDE) for the product $\widehat{\varphi}\widehat{\varphi}$ itself and not only for its expectation value $\langle \widehat{\varphi}\widehat{\varphi} \rangle$. One important advantage of this approach is that we can avoid using the artificial IRPT parameter—the momentum k_* at which the regions of soft ($k < k_*$) and hard ($k > k_*$) momenta meet. We also note that the SDE technique (Ch. 3) contains a constructive recipe for calculating the Wilson coefficients, in particular, the function B_1 in (6.87).

An important generalization of (6.86) to (6.88) was obtained in [189], where the authors managed to sum not only the contributions of all operators of the type φ^n, as in (6.87), but also the contributions of all operators of the type $\partial_t^p \varphi^n$, which can also be dangerous even in the region of UV pumping $\varepsilon < 2$ (Sec. 6.9). Here the quantity $\widehat{v}t$ with $t = t_1 - t_2$ in (6.87) is replaced by the integral over the range $[t_1, t_2]$ of a time-dependent "random velocity" $\widehat{v}(t)$ with known distribution of the type (6.88) (different times t of the fields φ for the same spatial coordinate $\mathbf{x}$). It is important that this integral vanishes at $t_1 = t_2$, as does the expression $\widehat{v}t$ in (6.87), i.e., new dangerous operators of the type $\partial_t^p \varphi^n$ also do not contribute to the static correlator.

However, the nontrivial m dependence remains in dynamical objects through the distribution of the random quantity $\widehat{v}$ or $\widehat{v}(t)$, i.e., the Kolmogorov hypothesis 1 does not generalize to dynamical objects. We also note that representations like (6.89) are well known in the technique of self-consistency equations for the Green functions of a given model (see, for example, [195], [197], [198]) and have a simple physical interpretation [185]: they describe the kinematical effect of the transport of turbulent eddies as a whole by the large-scale field $\widehat{v}$. The new feature in the approach of [187] and subsequent studies in the same area is the combination of this simple physical idea (IRPT) with the elements of the RG technique and the SDE.

On the basis of the results of analyzing operators of lower dimension it can be assumed (with no hope of a rigorous proof) that in the region of UV pumping $2 > \varepsilon > 0$ there are no dangerous operators other than φ^n and $\partial_t^p \varphi^n$. If this is true, then the results of Secs. 6.9 and 6.10 suggest that the Kolmogorov hypothesis 1 (6.75) is valid for the static correlator in this region. A new dangerous dissipation operator (6.64c) appears in going to the realistic region of IR pumping $\varepsilon \geq 2$ and no clearly expressed infinite series of dangerous operators that one could try to sum is evident. Therefore, the problem of justifying the Kolmogorov hypothesis (6.75) for this region remains unsolved.

However, it must be said that a purely theoretical solution of this problem within the framework of the RG technique and the SDE is impossible in principle because it is impossible to study all composite operators, and experiment will obviously have the last word. Usually only the dependence on the distance or momentum (see Sec. 6.12) is measured in experiments on turbulence, whereas to check hypothesis 1 it is desirable to have direct experimental data on the dependence of various quantities on the external scale $L_{\max} = m^{-1}$ with the other parameters fixed. Such data could provide answers to many questions that are inaccessible to a purely theoretical analysis.

6.11   The problem of singularities for $\varepsilon \to 2$ in the massless model.
ε-expansion of the Kolmogorov constant

If we are not concerned with the difficult problem of determining the dependence of the scaling functions on the parameter m discussed in Secs. 6.9 and 6.10, we can use the simplified massless pumping model (6.10) suitable for calculating m-independent quantities like the critical dimensions. Most of the studies on the RG theory of turbulence and generalizations of it have been performed for just this model.

As explained in Sec. 6.3, this model can be studied only in the region $\varepsilon \leq 2$, and realistic IR pumping corresponds to the $\varepsilon \to 2$ limit taken from this region. Several particular problems arise in this approach, as we now explain.

It follows from (6.45) with $m = 0$ that for $\varepsilon \to 2 - 0$

$$D_0 = 4(2 - \varepsilon)C_d^{-1}W\Lambda^{2\varepsilon - 4} \tag{6.90}$$

with the constant C_d from (6.25). Owing to the appearance in (6.90) of the factor $2 - \varepsilon$, the pumping function (6.10) with D_0 from (6.90) in the limit $\varepsilon \to 2 - 0$ for fixed W and $\Lambda = (W\nu_0^{-3})^{1/4}$ vanishes for any $k \neq 0$, although the integral (6.3) for it remains finite. This means that in this limit (6.10) actually reduces to a d-dimensional momentum δ function, namely [taking into account the explicit form of C_d in (6.25)],

$$N(k) \xrightarrow[\varepsilon \to 2]{} \delta(k) \cdot 2(2\pi)^d/(d - 1), \tag{6.91}$$

with the well-known power-law representation of the δ function:

$$\delta(k) = \lim_{\Delta \to +0} (2\pi)^{-d} \int dx (\Lambda x)^{-\Delta} \exp(ikx) =$$

$$= S_d^{-1} k^{-d} \lim_{\Delta \to +0} [\Delta (k/\Lambda)^{\Delta}]$$
(6.92)

with d-dimensional arguments k, x. We note that it is the function $N(k) \sim \delta(k)$ that is usually considered a natural model of pumping by infinitely large eddies ($m = 0$) corresponding to the Kolmogorov theory for uniform, isotropic, developed turbulence. It follows from the above that the function (6.10) with amplitude (6.90) in the limit $\varepsilon \to 2 - 0$ can be assumed to be the power-law equivalent of δ-like pumping.

As already mentioned in Sec. 6.6, the parameter D_0 enters into asymptotic representations like (6.36) only through the expression (6.35) for the asymptote of the invariant viscosity $\bar{\nu}_* \sim D_0^{1/3}$. Therefore, in static objects F like (6.36b) the entire D_0 dependence is isolated as an overall factor $\bar{\nu}_*^n \sim D_0^{n/3}$, where $n = d_F^\omega$ is the frequency dimension of F (static scaling functions and their arguments do not depend on D_0). Since $D_0 \sim (2-\varepsilon)$ according to (6.90), for a finite limit to exist for $\varepsilon \to 2 - 0$ the corresponding static scaling functions f must have singularities $\sim (2 - \varepsilon)^{-n/3}$ to cancel the zeros $\sim (2 - \varepsilon)^{n/3}$ of the overall amplitude factor $D_0^{n/3}$.

The rigorous proof of the presence of such singularities is a complicated unsolved theoretical problem that apparently bears some relation to the problem of justifying the Kolmogorov hypothesis 1 for the massive model. Starting from this hypothesis, it can be assumed that a finite limit for $\varepsilon \to 2 - 0$ in the massless model must exist for all strictly Galilean-invariant static objects, because the limit function (6.91) corresponds to pumping by infinitely large eddies ($m = 0$). We also note that the problem of singularities of the scaling functions for $\varepsilon \to 2-0$ occurs only in the massless model (6.10). When we go to a massive generalization like (6.9) the quantity D_0 becomes nonzero in the limit $\varepsilon \to 2 - 0$ [see the text following (6.46)].

This is related to the problem of calculating the Kolmogorov constant within the ε expansion. This constant is determined from the observed fluctuation energy spectrum (6.39). According to the experimental data, this spectrum in the inertial range has the form

$$E(k) = C_K W^{2/3} k^{-5/3},$$
(6.93)

where the dimensionless coefficient C_K is the Kolmogorov constant. Different experiments obtain values ranging from 1.3 to 2.7 for C_K (a review of the experimental data and references can be found in [199]; see also [200]). There is no generally accepted, convincing explanation of why this spread in values is so large. There have been attempts to relate it to the possible influence of anisotropy effects, which are always present to some degree in an actual experiment. However, it is not impossible that this large spread is direct evidence of the invalidity of Kolmogorov hypothesis 1 in the sense that even for the static correlator in the inertial range the dependence on the shape of the pumping function is preserved, for example, dependence on the parameter ε in (6.9) in the region $\varepsilon \geq 2$. We recall that the shape of the pumping function and its scale m are only "phenomenological equivalents" of the parameters of the real problem stated in terms of an abstract uniform, isotropic, developed turbulence, and so these two characteristics can be different in different experiments.

Therefore, the constant C_K in the law (6.93) can be calculated only for a specific choice of pumping function and only in the region where the law (6.93) itself can be justified. The latter is valid only for the realistic problem with IR pumping, i.e., for the region $\varepsilon \geq 2$ with pumping of the type (6.9). However, we cannot justify (6.93) in this region owing to the unsolved problem of determining the m dependence (Secs. 6.9 and 6.10). The existence of a finite limit for $m \to 0$ for the spectrum (6.39) can be assumed sufficiently well justified (Sec. 6.10) only in the region of UV pumping $0 < \varepsilon \leq 2$, so that the real constant C_K in (6.93) can be calculated only using the $\varepsilon \to 2$ limit from this region. It includes small ε, and so the constant C_K can be calculated as an ε expansion. This calculation was performed in the lowest order of the ε expansion in [187] and will be given below.

Let us begin with a few general remarks. First, it should be noted that in the theory of critical behavior amplitude factors like C_K in (6.93) are considered nonuniversal constants (in particular, because the normalization of the order-parameter field cannot be determined) and therefore are, in general, never calculated. We are always interested only in universal amplitude ratios independent of the normalization. The situation is different in the theory of turbulence, because here the field φ has a direct physical meaning and is not renormalized, so that constants like C_K in (6.93) can be measured experimentally.

There is yet another important distinction between these two classes of problems. Ordinary fluctuation models in the theory of critical behavior are certainly approximate, because when they are constructed all IR-irrelevant contributions in the action functional are discarded (Sec. 1.16). Such contributions play a dual role. First, they generate corrections to scaling with the corresponding correction exponents ω_i (Sec. 4.8), and, second, their inclusion changes the coefficients of the IR-relevant contributions, i.e., the numerical parameters of the basic model. This is another reason why the calculation of nonuniversal constants in terms of the parameters of the basic model is meaningless in the theory of critical behavior.

In the stochastic theory of turbulence (6.1) the original Navier–Stokes equation is essentially also phenomenological, and it must also be assumed that correction terms can be present on its right-hand side. Of course, these corrections must not spoil the symmetry of the problem (Galilean invariance) and they must be IR-irrelevant in the canonical dimension, i.e., they must contain a larger (compared to the basic contributions) number of factors φ and (or) ∂.

Formally, here there is complete analogy with the theory of critical behavior, but in reality there is a very important difference: in the theory of critical behavior there is only one characteristic UV scale Λ (and it is the cutoff parameter in the integrals), so that the bare coefficients of all the contributions in the action are made dimensionless by the same quantity Λ. Therefore, the effect of IR-irrelevant corrections on the coefficients of the IR-relevant contributions (i.e., on the parameters of the basic model) can be very significant, and only results independent of the particular numerical values of these parameters are of objective value. Examples of such objective results are the universal characteristics of critical behavior: the exponents and normalized scaling functions.

In the stochastic theory of turbulence there is a UV scale $\Lambda = (W\nu_0^{-3})^{1/4}$, the inverse dissipation length (and it is also the cutoff parameter in UV-divergent integrals). However, this parameter Λ cannot be the characteristic scale for the coefficients of the correction terms in the Navier–Stokes equation itself. These corrections are not related to the intensity of the energy pumping and must obviously be determined only by the microscopic structure of the medium. Therefore, the characteristic parameter for them must be the microscopic scale $\Lambda_0^{-1} = r_0$, the average intermolecular separation. Here we thus have two UV scales Λ and Λ_0, and physically $\Lambda_0 \gg \Lambda$, because the size of even the smallest turbulent eddies must be much larger than the intermolecular separation. Owing to this the indirect effect of possible corrections on the coefficients of the basic contributions is suppressed in the theory of turbulence. In this sense the model (6.1) can be considered "more microscopic" than the usual fluctuation models of the theory of critical behavior. The coefficient ν_0 appearing in (6.1) can be identified with the experimentally measured kinematical viscosity, and the field φ can be identified with the observed fluctuation component of the velocity.

Let us make a final remark regarding corrections. Their effect is usually estimated from the canonical dimensions, but in reality it is determined by the corresponding exact critical dimensions (Sec. 4.8). Therefore, in principle, it is possible that as ε increases a correction operator that is IR-irrelevant in its canonical dimension somehow can become IR-relevant in the exact dimension even before reaching the realistic value $\varepsilon_{\rm r}$. This would imply that the original model neglecting this correction does not admit extrapolation from small ε up to the required value $\varepsilon_{\rm r}$. In the theory of critical behavior the possibility of such an extrapolation is actually just postulated, because the exact dimensions are unknown, and the predictions based on this postulate of the theory usually agree with experiment.

In the theory of turbulence the critical dimensions of several composite operators can be found exactly, so that this problem (already pointed out in [186]) is more important. In this regard we can say only the following. Up to now, no one has found a Galilean-invariant correction operator with exactly known critical dimension that would be dangerous in the above sense, i.e., which would become IR-relevant even before reaching the boundary of the region of realistic IR pumping $\varepsilon = 2$. We stress the fact that here we can speak only of operators with exactly known critical dimension; any arguments at the level of the first order in ε are certainly incorrect when there are unknown corrections of order ε^2 and higher.

Let us return to the calculation of the Kolmogorov constant [187] for the massless model (6.10). From (6.35), (6.36b), (6.39), and (6.90) we have

$$E(k) = C_K(\varepsilon)W^{2/3}k^{-5/3}(\Lambda/k)^{4(\varepsilon-2)/3}, \tag{6.94}$$

where $C_K(\varepsilon)$ is the analog of the constant C_K in (6.93) for arbitrary $\varepsilon \leq 2$:

$$C_K(\varepsilon) = (2C_d/g_*^2)^{1/3}(2-\varepsilon)^{2/3}f(\varepsilon), \tag{6.95}$$

with constant C_d from (6.25) and the function $f(\varepsilon) \equiv f(u = 0, \varepsilon)$ from (6.36b) (now we also indicate the ε dependence explicitly). Accepting the hypothesis that the product $(2-\varepsilon)^{2/3}f(\varepsilon)$ is finite for $\varepsilon \to 2$ (see the beginning of the section) and expanding it in a series in ε, we arrive at the following ε expansion of the constant $C_K(\varepsilon)$ in (6.94):

$$C_K(\varepsilon) = \varepsilon^{1/3}\sum_{n=0}^{\infty}c_n\varepsilon^n. \tag{6.96}$$

In lowest order, from (6.36b) and (6.38) we have $f(\varepsilon) = g_*/2$, where $g_* = 2\varepsilon/3a$ with the constant a from (6.26). Therefore, in this approximation from (6.95) we find that

$$C_K(\varepsilon) = 2[(d+2)\varepsilon/3]^{1/3}, \tag{6.97}$$

which for $d = 3$ and $\varepsilon = 2$ gives numerically $C_K \simeq 3$ [187].

The expression (6.97) is the first term of the ε expansion (6.96). The fact that it disagrees with experiment is not surprising, since there are no obvious reasons for the corrections to (6.97) in (6.96) to be numerically small at the realistic value $\varepsilon = 2$.

Another problem arises in connection with the ε expansion (6.96). In deriving it we used (6.90), obtained by computing the integral (6.3) with cutoff $k \leq \Lambda$ for massless pumping (6.10). For the integral (6.3) which is UV-divergent at small ε, a cutoff at momenta of order Λ is natural, but here we are speaking, of course, only of the order of magnitude and not of the exact value. Therefore, nothing hinders us from making the replacement $\Lambda \to b\Lambda$ in (6.90) and, consequently, in (6.94), where b is an arbitrary coefficient of order unity. This leads to the appearance of an additional factor $b^{4(\varepsilon-2)/3}$ on the right-hand side of (6.95). This factor does not change the physical value of the Kolmogorov constant C_K, defined as the limit of $C_K(\varepsilon)$ for $\varepsilon \to 2$, but it does significantly affect the coefficients of the ε expansion (6.96).

It therefore follows that there is no unique determination of the ε expansion of the Kolmogorov constant C_K, and in any calculation of this type it is necessary to first clearly fix the definition of $C_K(\varepsilon)$, the continuation of the physical constant C_K to the entire region $0 < \varepsilon \le 2$. In our case this definition is (6.94), which leads to the result (6.97) in the lowest order of the ε expansion. A value of C_K closer to experiment was obtained in lowest-order perturbation theory in [201]–[203]; a detailed discussion of this can be found in [183].

6.12 Deviations from Kolmogorov scaling for composite operators

There are many studies devoted to the discussion of the experimental data for the following static Green functions involving composite operators:

$$\langle [\widehat{\varphi}_\parallel(x_1) - \widehat{\varphi}_\parallel(x_2)]^n \rangle |_{t_1=t_2} \equiv D_n, \qquad (6.98a)$$

$$\langle \widehat{W}_{\mathrm{dis}}(x_1)\widehat{W}_{\mathrm{dis}}(x_2)\rangle |_{t_1=t_2} \equiv D_W, \qquad (6.98b)$$

where $W_{\mathrm{dis}}(x)$ is the dissipation operator (6.64c), and $\varphi_\parallel$ in (6.98a) is the longitudinal component of the velocity field along the direction of the vector $\mathbf{x}_1 - \mathbf{x}_2$ [in real experiments $\varphi_\parallel = (\varphi n)$ and $\mathbf{x}_1 = \mathbf{x}_2 + nr$, where $n = u/|u|$ is the unit vector in the direction of the average flux velocity u and $r = |\mathbf{x}_1 - \mathbf{x}_2|$]. We note that the quantities (6.98) are Galilean-invariant and that in (6.98a) we have excluded contributions like the first term on the right-hand side of (6.7) by construction.

The quantities (6.98) can depend only on the distance $r = |\mathbf{x}_1 - \mathbf{x}_2|$ and on the model parameters. If in the spirit of the Kolmogorov theory (Sec. 6.2) we assume that only the dependence on r and on the total pumping power W remains in the inertial range, then for this region from the known canonical dimensions of the quantities (6.98) we obtain

$$D_n = \mathrm{const} \cdot (Wr)^{n/3}, \qquad D_W = \mathrm{const} \cdot W^2. \qquad (6.99)$$

Experiment confirms the power-law r dependence of (6.98) in the inertial range, but the measured exponents of the powers of r differ slightly from those predicted by (6.99).

In relation to this it should be stressed that in practice only the exponent of r is measured, and any difference between it and the predictions of (6.99), which is correct from the viewpoint of the canonical dimensions, leads to an important problem: what parameter (m or Λ) makes the experimentally observed additional powers of r dimensionless? Of course, the best way to solve this problem would be by direct experimental study. Unfortunately, the experimental determination of the dependence on parameters like m or Λ in the theory of turbulence is a very difficult technical problem, and there are no such data (as far as the author knows). It is therefore impossible to give any reliable, unique answer to this question. However, it is usually assumed on the basis of various qualitative considerations of a model nature (see, for example, [154]), that the parameter that makes the extra [compared to (6.99)] powers of r dimensionless must be the external scale m^{-1} rather than the dissipation length Λ^{-1}. Therefore, the results of experimental measurement of (6.98) in the inertial range are usually represented as

$$D_n = \text{const} \cdot (Wr)^{n/3}(mr)^{q_n}, \qquad D_W = \text{const} \cdot W^2(mr)^{-\mu}, \qquad (6.100)$$

where q_n and μ are exponents characterizing the degree of deviation from the laws (6.99). According to experiment, $\mu = 0.20 \pm 0.05$ ([204], where a graph of the n dependence of q_n for even n in the range from 2 to 18 is also given). The experimental value of q_2 is practically indistinguishable from zero, and the equation $q_3 = 0$ is considered exact on the basis of arguments based on the energy-balance equation [154]. For $n \geq 4$ the exponents q_n are negative and grow in absolute value with increasing n, and so the discrepancies between the laws (6.99) and (6.100) for D_n become more pronounced.

These discrepancies are usually referred to as deviations from Kolmogorov scaling, but this terminology is not completely correct. Strictly speaking, from Kolmogorov hypotheses 1 and 2 in the form in which they were stated in Sec. 6.2, it is completely impossible to make definite conclusions about the behavior of Green functions involving composite operators without additional assumptions.

Let us explain in more detail. These hypotheses deal with the properties of the distribution function of the Fourier components $\varphi(t, k)$ of the random velocity field $\varphi(t, \mathbf{x})$ in the region $k \gg m$ for hypothesis 1 (we shall refer these components as m-hard) and in the region $k \ll \Lambda$ for hypothesis 2 (Λ-soft). The components $\varphi(t, k)$ that are simultaneously Λ-soft and m-hard correspond to the inertial range. The random field $\varphi(x)$ as a whole contains contributions from both components. Therefore, any local composite operator of the type $\varphi^2(x)$, $\varphi(x)\partial^2\varphi(x)$, and so on constructed from $\varphi(x)$ contains all possible combinations of these two components. In this situation it is impossible to arrive at any conclusion about the properties of Green functions involving composite operators on the basis of the Kolmogorov hypotheses regarding the distribution function of the components $\varphi(t, k)$, and for mixed constructions neither of the two hypotheses is applicable. Arguments about Green functions involving composite operators based on the Kolmogorov hypotheses can be made only using additional considerations, for example, the statement that the main contribution to such-and-such an operator comes only from Λ-soft components, and so hypothesis 2 about independence from ν_0 is applicable to it. This is the situation regarding the Green functions (6.98a), because it is clear from their skeleton-graph representations with dressed lines that since the Kolmogorov dimension (6.8) of the field φ is negative, the main contribution to operators of the type φ^n comes from small momenta, where only the Λ-soft components are important. We note that this argument justifies the use of m as the parameter making the extra powers of r dimensionless in the first relation in (6.100). However, even in this relatively favorable situation we can speak of the applicability of only one of the two Kolmogorov hypotheses and not of both at the same time. The latter would become possible only when it could be strictly proved that only components $\varphi(t, k)$ from the inertial range give the main contribution to this operator. It is difficult to imagine how this can happen, and in any case it is obvious that it is impossible without additional considerations leading outside the framework of the Kolmogorov hypotheses regarding the distribution of the components $\varphi(t, k)$. We note that for the dissipation operator (6.64c) this (hypothetical) mechanism for justifying Kolmogorov scaling (6.99) is certainly impossible, because it would lead to the establishment of scaling for the operator construction itself in (6.64c) without the overall coefficient ν_0. This would directly contradict the hypothesis (6.99), from which it follows that for a given operator construction the correlator in the inertial range must have the form const $\times \nu_0^{-2} W^2$, i.e., it must explicitly depend on ν_0. We add that owing to the energy balance, the expectation value of the operator (6.64c) coincides with the pumping power W and is independent of ν_0, although this factor does occur in the operator (6.64c) itself.

The main conclusion is that the usual Kolmogorov hypotheses (Sec. 6.2) do not lead directly to representations like (6.99) for composite operators. Therefore, the expressions (6.99) must be understood not as a consequence but as a generalization (assumed, not proven) of the usual Kolmogorov hypotheses.

A more reliable theoretical basis for discussing the behavior of (6.98) in the region of the IR asymptote $\Lambda r \gg 1$ (including in the inertial range) is the RG technique for composite operators discussed earlier. It is known (Sec. 6.8) that $\widehat{W}_{\text{dis}}$ is an operator with definite critical dimension (6.74), so that $\Delta[D_W] = 2\Delta[\widehat{W}_{\text{dis}}] = 2(4 - 2\varepsilon)$ with the "Kolmogorov value" $\Delta[D_W] = 0$ for $\varepsilon = 2$. For D_n the quantity averaged over in (6.98a) is a linear combination of expressions like $\widehat{\varphi}^k(x_1)\widehat{\varphi}^{n-k}(x_2)$. Any unrenormalized monomial $\widehat{\varphi}^p(x)$ can be expanded in a complete set of corresponding basis operators (6.54), among which will be an operator associated with φ^p and having critical dimension $p\Delta_\varphi$ according to (6.60). This contribution is the leading one at the IR asymptote (it becomes dangerous for $\varepsilon > 3/2$) and all the others are IR-irrelevant corrections that can be neglected. It then follows that the leading term in D_n at the IR asymptote has critical dimension $\Delta[D_n] = n\Delta_\varphi = n(1-2\varepsilon/3)$, which reaches the Kolmogorov value $\Delta[D_n] = -n/3$ at $\varepsilon = 2$.

It can be shown [191] that in going from the ordinary critical regime to the Kolmogorov one (Sec. 6.6), the critical dimensions of the composite operators in question (the one associated with φ^n and the dissipation operator) freeze at their Kolmogorov values in the entire region of IR pumping $\varepsilon \geq 2$, and that in this region the ν_0 dependence vanishes [along with the dependence on the cutoff parameter $\Lambda = (W\nu_0^{-3})^{1/4}$]. It therefore follows that in this region the RG analysis leads to the following representations for the leading contributions of the IR asymptote of (6.98):

$$D_n = (Wr)^{n/3} f_n(mr), \qquad D_W = W^2 f_W(mr) \qquad (6.101)$$

with unknown scaling functions $f_{n,W}(mr)$.

It is useful to use (6.100) and (6.101) as an example to again explain the difference between the critical dimensions in the ordinary (ν_0 and g_0 fixed) and Kolmogorov (ν_0 and W fixed) critical regimes. For both regimes the variables r and m are dilated in scale transformations ($m \to \lambda m$, $r \to \lambda^{-1}r$), i.e., they have nonzero critical dimensions, while their product mr has zero critical dimension. The critical dimensions of the quantities (6.101) are therefore determined only by the explicitly isolated factors of $(Wr)^{n/3}$ for D_n and W^2 for D_W. In the Kolmogorov regime the parameter W is assumed fixed under dilatations of m and r, i.e., it is assumed to have zero critical dimension, and so in this regime $\Delta[D_n] = -n/3$ and $\Delta[D_W] = 0$. In the ordinary critical regime the parameter g_0 is assumed fixed, and then from (6.46) in the region of IR pumping we have $W \sim m^{4-2\varepsilon}$, i.e., in this regime W [and therefore $\Lambda = (W\nu_0^{-3})^{1/4}$] is a parameter with nonzero critical dimension. Taking into account the dimension of W in this regime, we have $\Delta[D_n] = n(1 - 2\varepsilon/3)$, $\Delta[D_W] = 2(4 - 2\varepsilon)$ in agreement with (6.60) and (6.74). We stress the fact that all the expressions given above for the critical dimensions are exact. The additional powers of mr in (6.100) do not affect the dimensions, because in any regime they have zero critical dimension (but if powers of Λr instead of mr appeared, they would contribute to the critical dimensions).

We recall that experiments determine not the dimensions themselves, but only the exponents of r. It is therefore possible to judge the dimensions from the experimental data only when we know the parameter (m or Λ) making some of the powers of r dimensionless. If it is m, as in (6.100), then the naive Kolmogorov dimensions are exact. For IR pumping ($\varepsilon \geq 2$) the RG analysis leads to the representations (6.101), i.e., it unambiguously indicates that it is m that makes the additional powers of r dimensionless.

The representations (6.101) are valid for the entire region of the IR asymptote $\Lambda r \gg 1$, and the auxiliary condition $mr \ll 1$ corresponds to the inertial range. Therefore, to obtain specific predictions for the inertial range from (6.101) it is necessary to know the leading terms of the $mr \to 0$ asymptote of the scaling functions $f_{n,W}(mr)$. The experimental data for the inertial range are represented by (6.100). They suggest that the leading terms of this asymptote have the form

$$f_n(mr) \underset{mr \to 0}{\simeq} \text{const} \cdot (mr)^{q_n}, \quad f_W(mr) \underset{mr \to 0}{\simeq} \text{const} \cdot (mr)^{-\mu}. \quad (6.102)$$

This asymptote was obtained from the experimental data. Like the Kolmogorov hypothesis 1, (6.102) is very difficult to justify theoretically. There are many studies in which the authors attempt to theoretically justify (6.100) and then calculate the exponents q_n and μ appearing there. A review and critical analysis of these studies can be found in [183]. We also mention [205] and [206], in which the relations between the various functions D_n are discussed and the experimental data are analyzed not only for the inertial range, but also for the dissipation region.

6.13 Turbulent mixing of a scalar passive admixture

In this problem we consider a pair of fields φ, θ, where φ is the usual transverse vector velocity field satisfying the stochastic equation (6.1), and θ is an addditional scalar admixture field satisfying the equation

$$\nabla_t \theta = \nu_0' \partial^2 \theta, \quad \nu_0' = u_0 \nu_0 \tag{6.103}$$

with the usual covariant derivative ∇_t from (6.1). In the simplest case that we shall consider here, no proper noise is introduced into the equation for θ. The impurity admixture is referred to as passive because the introduction of the additional field θ has no effect on the stochastic problem (6.1), (6.2) for the velocity field φ.

The field θ can have a different physical meaning, for example, it may be the concentration of impurity particles in a turbulent atmosphere, or the temperature field in heat transfer problems, and so on. In the first case the parameter ν_0' is the molecular diffusion coefficient, and in the second it is the heat conduction coefficient. For definiteness we shall use the terminology of the first problem. Instead of the pair of coefficients ν_0 from (6.1) and ν_0' from (6.103), it is more convenient to use the pair ν_0, u_0, where $u_0 = \nu_0'/\nu_0$ is the dimensionless parameter introduced in (6.103) (the inverse Prandtl number). In what follows it will play the role of the second bare charge in addition to g_0 from (6.17).

According to the general rules of Sec. 5.3, the stochastic problem (6.1), (6.2), (6.103) is equivalent to the quantum field model for the system of fields $\phi = \varphi, \varphi', \theta, \theta'$ with unrenormalized action

$$S(\phi) = S(\varphi, \varphi') + \theta'[-\partial_t \theta + u_0 \nu_0 \partial^2 \theta - (\varphi \partial) \theta], \tag{6.104}$$

in which $S(\varphi, \varphi')$ is the functional (6.17) for the problem without the admixture. The model (6.104) involves an additional triple vertex of the type $\theta' \varphi \theta$, and the impurity fields θ, θ' correspond to bare propagators analogous to (6.13):

$$\langle \widehat{\theta \theta'} \rangle_0 = \langle \widehat{\theta' \theta} \rangle_0^\top = [-i\omega + u_0 \nu_0 k^2]^{-1},$$
$$\langle \widehat{\theta \theta} \rangle_0 = \langle \widehat{\theta' \theta'} \rangle_0 = 0. \tag{6.105}$$

The equation $\langle \widehat{\theta' \theta'} \rangle_0 = 0$ follows from the general rule (5.30) and $\langle \widehat{\theta \theta} \rangle_0 = 0$ is a consequence of the absence of noise in (6.103). The graphs with the lines (6.105) of the new fields θ, θ' do not contribute to the Green functions of the old fields φ, φ', which remain the same as in the model without the admixture.

The basic action corresponding to (6.104) has the form

$$S_B(\phi) = S_B(\varphi, \varphi') + \theta'[-\partial_t \theta + u\nu \partial^2 \theta - (\varphi \partial) \theta], \tag{6.106}$$

where $S_B(\varphi, \varphi')$ is the functional (6.18) and u is the renormalized parameter corresponding to u_0.

The canonical dimensions of the fields and parameters involved in $S_B(\varphi, \varphi')$ are known from the data of Table 6.1. The new parameters u_0 and u for the problem with the admixture are completely dimensionless in the momentum and frequency, and for the new fields θ, θ' from (6.106) we can find only the canonical dimensions of the product $\theta\theta'$, and not of each field separately:

$$d^p[\theta\theta'] = d[\theta\theta'] = d, \quad d^\omega[\theta\theta'] = 0. \tag{6.107}$$

However, this is sufficient, because only Green functions with identical numbers of the fields θ and θ' are nonzero in this model. This is most easily verified by noting that the functional (6.106) is invariant under the transformation $\theta \to b\theta$, $\theta' \to b^{-1}\theta'$ with arbitrary constant b.

These Green functions have a simple physical interpretation. In the problem of the turbulent mixing of an impurity, the simplest response function $\langle\widehat{\theta}(x)\widehat{\theta}'(x')\rangle$ with $x = t, \mathbf{x}$ and $x' = t', \mathbf{x}'$ has the meaning of the function describing the distribution in the coordinate $\mathbf{x}$ at time t of an impurity particle, which at time $t' < t$ had coordinate $\mathbf{x}'$ (see the remark at the end of Sec. 5.4). The higher-order Green functions of this type are related to the analogous multiparticle distribution functions $p_n(x_1...x_n; x_1'...x_n')$ [154] as

$$\langle\widehat{\theta}(x_1)...\widehat{\theta}(x_n)\widehat{\theta}'(x_1')...\widehat{\theta}'(x_n')\rangle = \sum_{\text{perms.}} p_n(x_1...x_n; x_1'...x_n') \tag{6.108}$$

with summation over all $n!$ permutations of the arguments $x_1...x_n$ (or $x_1'...x_n'$, which is equivalent).

The UV divergences and counterterms for the model (6.106) are analyzed using the standard rules discussed in Sec. 5.15. For the formal exponent of divergence (5.89) of 1-irreducible graphs of Γ with identical numbers $n_\theta = n_{\theta'}$ of fields θ and θ' we find the following, using (6.107) and the data of Table 6.1:

$$d_\Gamma = d + 2 - n_\varphi - (d-1)n_{\varphi'} - dn_\theta. \tag{6.109}$$

However, as usual, the real exponent of divergence δ_Γ is smaller than the formal one $d_\Gamma = d_\Gamma^*$. This is because, first, in the basic vertex (6.14) the derivative ∂ can always be moved onto the field φ', which decreases the real exponent by $n_{\varphi'}$ according to (6.20). Second, in the impurity vertex $\theta'(\varphi\partial)\theta$ of the functional (6.106) the derivative ∂ can always be moved onto θ' because the field φ is transverse. It then follows that the ∂ can be moved to the external lines of both the fields θ and θ', decreasing the real exponent by $n_\theta + n_{\theta'}$. Then, using the equation $n_\theta = n_{\theta'}$, we have

$$\delta_\Gamma = d_\Gamma - n_{\varphi'} - 2n_\theta. \tag{6.110}$$

Using (6.109) and (6.110), it is easily verified that in the general case of arbitrary dimension $d > 2$, only counterterms of the type $\varphi'\partial^2\varphi$ and $\theta'\partial^2\theta$ are needed to eliminate the UV divergences. The first of these must be exactly the same as for the model without the admixture, and so

$$S_R(\phi) = S_R(\varphi, \varphi') + \theta'[-\partial_t\theta + Z_2 u\nu\partial^2\theta - (\varphi\partial)\theta], \qquad (6.111)$$

where $S_R(\varphi, \varphi')$ is the functional (6.21). The renormalized action (6.111) is obtained from the unrenormalized functional (6.104) by renormalizing the parameters (6.22) and also

$$u_0 = u Z_u, \qquad Z_2 = Z_u Z_\nu, \qquad (6.112)$$

without field renormalization ($Z_\phi = 1$), as in the model (6.17).

The one-loop calculation of Z_2 in the MS scheme gives [207]

$$Z_2 = 1 - ga\alpha/2\varepsilon u(u+1) + ..., \quad \alpha \equiv 2(d+2)/d \qquad (6.113)$$

with coefficient a from (6.25). The renormalization constants Z_ν and $Z_g = Z_\nu^{-3}$ remain the same as in the model without the admixture, because the admixture has no effect on the Green functions of the fields φ, φ' (see above).

The RG equation for our model has the usual form (5.106) with $\gamma_\phi = 0$ and $a = u, \nu$ in this case. Since the parameter u enters into the constant (6.113), it must also be considered to be a charge, and the contribution $-\gamma_u \mathcal{D}_u = -u\gamma_u\partial_u$ in an RG operator of the type (5.106) should be understood as $\beta_u\partial_u$:

$$\beta_u = -u\gamma_u, \qquad \gamma_u = \widetilde{\mathcal{D}}_\mu \ln Z_u. \qquad (6.114)$$

The one-loop expressions for the RG functions γ_ν and β_g are known from (6.26), and for β_u it follows from (6.113) and (6.114) that

$$\beta_u = ag[u - \alpha/(u+1)] + ... \qquad (6.115)$$

with constants a from (6.25) and α from (6.113).

Therefore, we are dealing with a two-charge model with charges $g \equiv g_1$ and $u \equiv g_2$. Its RG operator has the form

$$\mathcal{D}_{RG} = \mathcal{D}_\mu + \sum_g \beta_g\partial_g - \gamma_\nu\mathcal{D}_\nu \qquad (6.116)$$

with summation over these two charges in the term involving $\beta_g\partial_g$.

The possible critical regimes are determined by the fixed points of the system of β functions (6.26) and (6.115), and the type of point is determined by the general rules of Sec. 1.42. We are interested only in points from the region $g_1 \equiv g \geq 0$, $g_2 \equiv u \geq 0$ owing to their physical meaning. It is easily checked that within the ε expansion, the β functions (6.26) and (6.115) have only one IR-attractive fixed point in this region, with coordinates

$$g_{1*} \equiv g_* = 2\varepsilon/3a + ..., \quad g_{2*} \equiv u_* = [(1+4\alpha)^{1/2} - 1]/2 + ..., \qquad (6.117)$$

where g_* is the nontrivial fixed point of the β function (6.26) (the same as for the model without the admixture) and u_* is the positive root of the equation $u(u+1) = \alpha$. Substituting into (6.117) the expression for α known from (6.113), for $d = 3$ we obtain numerically $u_* \simeq 1.393$. It can be shown that all phase trajectories [solutions of the Cauchy problem (1.188)] with natural initial data leaving the region $0 < g < g_*$ (Sec. 1.27) and $u \geq 0$ end up at the fixed point (6.117) at the IR asymptote $s \equiv k/\mu \to 0$.

The correction exponents ω_i are defined as the eigenvalues of the matrix (1.189). In our case this matrix is triangular, and so the eigenvalues are simply its diagonal elements $\omega_{11} = d\beta_g(g)/dg$ and $\omega_{22} = \partial\beta_u(g,u)/\partial u$ at the fixed point. In the one-loop approximation, from (6.26) and (6.115) we have

$$\omega_{11} = 2\varepsilon, \qquad \omega_{22} = 2\varepsilon(2u_* + 1)/3(u_* + 1) \qquad (6.118)$$

with u_* from (6.117). The fact that the exponents (6.118) are positive indicates that the fixed point (6.117) is IR stable.

The general solutions of the RG equations are RG representations of the Green functions analogous to (6.36), but now with two charges g, u [the latter should not be confused with the variable $u \equiv m/k$ from (6.36)]. At the IR asymptote $k/\mu \to 0$ the invariant charge $\bar{u}$ tends to the fixed point (6.117), i.e., the "effective Prandtl number" $(\bar{u})^{-1}$ in the IR regime becomes a fully defined quantity independent of the bare value of this parameter.

The critical dimensions $\Delta_F = \Delta[F]$ of the various quantities F (fields and parameters) are determined by the general rule (6.53). The equations (6.42) and (6.43) remain unchanged, and for the new fields θ, θ', since they are not renormalized ($\gamma_\theta = \gamma_{\theta'} = 0$), from (6.53) and (6.107) we obtain

$$\Delta[\theta\theta'] = d[\theta\theta'] = d, \qquad (6.119)$$

i.e., the dimension of the product $\theta\theta'$ remains canonical.

A consequence of (6.119) is the well-known Richardson four-thirds law ([154], Sec. 24) characterizing the spreading rate of a cloud of impurity particles in a turbulent medium (for example, the atmosphere). If the field $\theta(t,\mathbf{x})$ is the concentration of impurity particles, the rms radius R at time $t > 0$ of a cloud of such particles which started at $t' = 0$ from the origin $\mathbf{x}' = 0$ is defined as

$$R^2 = \int d\mathbf{x}\, \mathbf{x}^2 \langle \widehat{\theta}(t,\mathbf{x})\widehat{\theta}'(0,0)\rangle. \qquad (6.120)$$

Using (6.119) and the normalization $\Delta[\mathbf{x}] = -1$, from (6.120) we obtain $\Delta[R^2] = -2$, so that $\Delta[dR^2/dt] = -2 - \Delta_t = -4/3$ at the Kolmogorov value (6.8) of the dimension $\Delta_t = -2/3$ ($\varepsilon = 2$). According to the dimensions it then follows that

$$dR^2/dt \sim R^{4/3}, \qquad (6.121)$$

which is Richardson's law.

This RG analysis of the model (6.103) was performed in [207], and then later reproduced and extended (in particular, by explicit expressions for the invariant charge $\bar{u}$ in the one-loop approximation) in [201] to [203]. The authors of [203] and [208] studied a generalization of the model (6.103) with nonzero correlator $\langle \widehat{\theta\theta} \rangle$ [in (6.103) this correlator is zero]. In [203] it was assumed that the correlator $\langle \widehat{\theta\theta} \rangle \neq 0$ appears owing to the nonstationarity of the problem, but formally it is easier to obtain $\langle \widehat{\theta\theta} \rangle \neq 0$ by introducing a new "noise" in (6.103) [183]. The static correlator $\langle \widehat{\theta\theta} \rangle$ and the corresponding spectrum of the type (6.39) involve an amplitude factor analogous to C_{K} in (6.93), called the Batchelor constant C_{Ba}. The ratio $C_{\mathrm{K}}/C_{\mathrm{Ba}}$ was calculated in [203] and [208], but the technique used by those authors differs from the standard quantum field RG technique described here. The calculation of the ratio $C_{\mathrm{K}}/C_{\mathrm{Ba}}$ using the latter is given in [183].

The generalization of the problem (6.103) to the case of a "chemically active" scalar admixture ([154], Sec. 14.5) was studied in [209]. A proper noise term and a nonlinearity $\sim \theta^n$ with integer $n \geq 1$ are introduced into the right-hand side of (6.103). It was shown in [209] that this model is multiplicatively renormalizable only for certain values of the spatial dimension d and exponent n. A discussion of the results of this study and a commentary can be found in [183].

6.14 Stochastic magnetic hydrodynamics (MHD)

Magnetic hydrodynamics is the study of a conducting, turbulent medium of charged particles that is electrically neutral as a whole. The motion of the particles generates currents and, consequently, a magnetic field, which in turn affects the motion of the charged particles. In the stochastic problem the corresponding system of equations has the form [210], [211]

$$\nabla_t \varphi_i = \nu_0 \partial^2 \varphi_i - \partial_i p + (\theta\partial)\theta_i + \eta_i^{\varphi}, \tag{6.122a}$$

$$\nabla_t \theta_i = u_0\nu_0 \partial^2 \theta_i + (\theta\partial)\varphi_i + \eta_i^{\theta} \tag{6.122b}$$

with covariant derivative ∇_t from (6.1). In (6.122) φ is the usual transverse vector velocity field, p is the pressure, $\theta_i = B_i/(4\pi\rho)^{1/2}$, and $u_0\nu_0 \equiv \nu_0' = c^2/4\pi\sigma$, where c is the speed of light, ρ and σ are the density and conductivity of the medium, u_0^{-1} is the magnetic Prandtl number, and B_i is the transverse (owing to Maxwell's equations) vector field of the magnetic induction. The expression (6.122b) without the contribution of the random force is obtained from Maxwell's equations $\partial B = 0$, $\mathrm{curl}\, E + c^{-1}\partial_t B = 0$, $\mathrm{curl}\, B = 4\pi j/c$ (neglecting the displacement current), and Ohm's law $j = \sigma(E + [\varphi \times B]/c)$ for a moving conducting medium [210]. The term $(\theta\partial)\theta_i$ on the right-hand side of (6.122a) is the contribution of the Lorentz force $\sim [j \times B] \sim [\mathrm{curl}\, B \times B] = (B\partial)B - \partial(B^2/2)$, and the last term is included in the pressure. In (6.122a) the transverse random force η_i from (6.1) is denoted as η_i^{φ}, and the analogous contribution η_i^{θ} on the right-hand side of (6.122b) is interpreted as the curl of the random current. The velocity field φ_i and the noise η_i^{φ} corresponding to it are true vectors, while θ_i and η_i^{θ} are pseudovectors.

The RG analysis of the stochastic problem (6.122) was first performed in [211] using the Wilson recursion relations (Sec. 1.1). Here we shall present the equivalent quantum field formulation [212].

According to the general rule (5.25), the stochastic problem (6.122) corresponds to the quantum field model for a system of four transverse vector fields $\phi = \varphi, \theta, \varphi', \theta'$ with unrenormalized action functional,

$$S(\phi) = \varphi' D^{\varphi\varphi} \varphi'/2 + \theta' D^{\theta\theta}\theta'/2 + \varphi' D^{\varphi\theta}\theta' + \varphi'[-\partial_t\varphi + \nu_0\partial^2\varphi-$$

$$-(\varphi\partial)\varphi + (\theta\partial)\theta] + \theta'[-\partial_t\theta + u_0\nu_0\partial^2\theta - (\varphi\partial)\theta + (\theta\partial)\varphi].$$

(6.123)

The new parameter u_0 in (6.123) is completely dimensionless, and the new fields θ and θ' have the same canonical dimensions as the fields φ and φ', respectively (see Table 6.1 in Sec. 6.4).

The massless model with pumping of the type (6.10) was studied in [211] and [212]. The matrix of the random-force correlators for the real 3-dimensional problem is then written as follows in the general case [212]:

$$D_{is}^{\varphi\varphi} = D_0^{\varphi} P_{is} k^{4-d-2\varepsilon}, \quad D_{is}^{\theta\theta} = D_0^{\theta} P_{is} k^{4-d-2\alpha\varepsilon},$$

$$D_{is}^{\varphi\theta} = D_0^{\varphi\theta} \varepsilon_{ism} k_m k^{3-d-\varepsilon-\alpha\varepsilon},$$

(6.124)

where $P_{is} = \delta_{is} - k_i k_s/k^2$ is the transverse projector, ε_{ism} is the completely antisymmetric pseudotensor with $\varepsilon_{123} = 1$, all the constants D_0 are amplitude factors, and the positive coefficient α in the exponents is an additional parameter of the theory. These exponents are chosen such that the interactions generated by the three correlators (6.124) become logarithmic simultaneously for $\varepsilon = 0$ (otherwise, some of the contributions might be discarded as being IR-irrelevant). The requirement that the interactions generated by the three correlators become logarithmic simultaneously determines the exponents of k in (6.124) up to corrections that vanish for $\varepsilon = 0$. The notation for (6.124) used in [211] and [212] corresponds to the simplest assumption that these corrections are linear in the parameter ε.

The index structure of the correlators (6.124) is determined by the requirements that all fields be transverse and that spatial parity be conserved. Since the noise η_i^φ is a true vector and η_i^θ is a pseudovector, the diagonal correlators $D^{\varphi\varphi}$ and $D^{\theta\theta}$ must be true tensors, and the mixed correlator $D^{\varphi\theta}$ must be a pseudotensor. We note that the form $\varepsilon_{ism} k_m$ is automatically transverse in the indices i, s.

In calculations performed in the spirit of dimensional regularization, the symbols δ_{is} and ε_{ism} can formally be used also for arbitrary spatial dimension d, but in the final results the symbol ε_{ism}, in contrast to δ_{is}, is meaningful only for the realistic dimension $d = 3$. In the other important case of integer dimension $d = 2$ there is no pseudotensor transverse in both indices, and so the mixed correlator $D^{\varphi\theta}$ must then be assumed zero. The mixed correlator was not introduced in [211], and the general case (6.124) was studied in [212].

The 2-dimensional problem was also treated in those studies, but both contain errors: in [212] the additional counterterm of the type $\varphi'\varphi'$ arising for $d = 2$ was erroneously assumed to be multiplicative, and in [211] it was ignored completely. Therefore, in what follows we shall limit ourselves to the case $d > 2$ with application to the realistic case of $d = 3$ in mind.

We shall use a representation analogous to (6.16) for the amplitude factors of the correlators (6.124):

$$D_0^\varphi = g_0\nu_0^3, \quad D_0^\theta = g_0'\nu_0^3, \quad D_0^{\varphi\theta} = \rho[D_0^\varphi D_0^\theta]^{1/2}, \quad |\rho| \le 1. \tag{6.125}$$

The two bare charges g_0, g_0' and the completely dimensionless parameter ρ are thereby determined. The inequality $|\rho| \le 1$ is a necessary condition for the correlator matrix (6.124) to be positive-definite.

The diagrammatic technique for the model (6.123) is determined by the standard rules of Sec. 5.3. The graphs will contain vertices of three types, $\varphi'\varphi\varphi$, $\varphi'\theta\theta$, $\theta'\varphi\theta$, and five nonzero lines (propagators) of the type $\varphi\varphi'$, $\theta\theta'$, $\varphi\varphi$, $\theta\theta$, $\varphi\theta$, where the last three are proportional to the correlators (6.124). The only pseudotensor among them will be the propagator $\langle\widehat{\varphi\theta}\rangle$.

Let us now turn to the analysis of the renormalization of the model (6.123). From the general rules of Sec. 5.14 we can find the canonical dimensions of the bare charges in the amplitudes (6.125):

$$d[g_0] = 2\varepsilon, \quad d[g_0'] = 2\alpha\varepsilon, \quad d^\omega[g_0] = d^\omega[g_0'] = 0. \tag{6.126}$$

Therefore, the basic action $S_B(\phi)$ is obtained from the unrenormalized functional (6.123) by the following replacement of all the bare parameters by basic ones: $\nu_0 \to \nu$, $u_0 \to u$, $g_0 \to g\mu^{2\varepsilon}$, $g_0' \to g'\mu^{2\alpha\varepsilon}$. The renormalized action $S_R(\phi)$ is obtained by adding to the basic action all the needed counterterms determined by the general rules of Sec. 5.15. For the formal exponent (5.89) in the model (6.123) we have (recalling that the canonical dimensions of the fields φ and θ are identical)

$$d_\Gamma = d_\Gamma^* = d + 2 - n_\varphi - (d-1)n_{\varphi'} - n_\theta - (d-1)n_{\theta'}. \tag{6.127}$$

However, the real exponent δ_Γ, as usual, is smaller than the formal one

$$\delta_\Gamma = d_\Gamma - n_{\varphi'} - n_{\theta'}, \tag{6.128}$$

because we can always extract a ∂ at any vertex of the action (6.123) for the fields φ', θ' since all the fields are transverse.

On the basis of the exponents (6.127) and (6.128) and the general rules of Sec. 5.15 we conclude that counterterms in the model (6.123) for $d > 2$ can be generated only by the 1-irreducible Green functions $\langle\widehat{\varphi'\varphi}\rangle_{1-ir}$, $\langle\widehat{\theta'\theta}\rangle_{1-ir}$, $\langle\widehat{\varphi'\theta}\rangle_{1-ir}$, $\langle\widehat{\theta'\varphi}\rangle_{1-ir}$ (for all $d_\Gamma = 2$, $\delta_\Gamma = 1$) and by the vertex functions $\langle\widehat{\varphi'\varphi\varphi}\rangle_{1-ir}$, $\langle\widehat{\varphi'\varphi\theta}\rangle_{1-ir}$, $\langle\widehat{\varphi'\theta\theta}\rangle_{1-ir}$, $\langle\widehat{\theta'\varphi\varphi}\rangle_{1-ir}$, $\langle\widehat{\theta'\varphi\theta}\rangle_{1-ir}$, $\langle\widehat{\theta'\theta\theta}\rangle_{1-ir}$ (for all $d_\Gamma = 1$, $\delta_\Gamma = 0$), where the auxiliary fields φ', θ' can enter into the counterterms only in the form $\partial\varphi'$ and $\partial\theta'$. Therefore, all self-energy counterterms must contain two symbols ∂ and all vertex counterterms must contain one symbol ∂, and the derivative ∂_t cannot enter into the counterterms. We note that additional counterterms from Green functions of the type $\langle \widehat{\varphi}' \widehat{\varphi}' \rangle_{1-\mathrm{ir}}$ and $\langle \widehat{\theta}' \widehat{\theta}' \rangle_{1-\mathrm{ir}}$ would appear for $d = 2$.

Many of the counterterms allowed by dimensions are actually absent, which ensures the multiplicative renormalizability of the model. To prove this statement, we split up all the Green functions into even and odd ones according to the number of pseudovector fields θ and θ' they contain. It is clear that all even functions must be true tensors and odd ones must be pseudotensors, and that the single odd term in the action (6.123) is the contribution of the mixed correlator $D^{\varphi\theta}$ generating the "odd" line $\varphi\theta$. Even Green functions can contain only an even number of lines $\varphi\theta$ and odd ones can only contain an odd number of such lines, and so in the absence of the mixed correlator $D^{\varphi\theta}$ all odd Green functions must be equal to zero. However, even when the correlator $D^{\varphi\theta}$ is present, odd 1-irreducible Green functions of the type $\varphi'\theta$, $\theta'\varphi$, $\varphi'\varphi\theta$, $\theta'\varphi\varphi$, $\theta'\theta\theta$ cannot generate counterterms when the formal scheme of dimensional regularization without the cutoff Λ is used for the calculations (Sec. 1.20), because in that approach all counterterms must be polynomials in the momentum (or ∂), and without Λ it is impossible to construct pseudotensor structures of the required dimension which are polynomials in the momentum.

Therefore, counterterms can be generated only by even 1-irreducible functions of the type $\varphi'\varphi$, $\theta'\theta$, $\varphi'\varphi\varphi$, $\varphi'\theta\theta$, $\theta'\varphi\theta$, i.e., objects of the same form as the terms of the functional (6.123). Since the fields φ', θ' can enter into the counterterms only along with ∂ (see above), 1-irreducible functions of the type $\varphi'\varphi$ and $\theta'\theta$ can generate only counterterms $\sim \varphi'\partial^2\varphi$ and $\theta'\partial^2\theta$, but cannot generate counterterms $\sim \varphi'\partial_t\varphi$ and $\theta'\partial_t\theta$.

Let us now consider the vertex counterterms. Since all the fields are transverse, it is easily verified that the counterterm of the $\varphi'\varphi\varphi$ vertex containing a single ∂ must have the form $\sim \varphi'(\varphi\partial)\varphi$, the counterterm of the $\varphi'\theta\theta$ vertex is proportional to $\varphi'(\theta\partial)\theta$, and for the $\theta'\varphi\theta$ vertex two counterterms are possible, namely, $\theta'(\varphi\partial)\theta$ and $\theta'(\theta\partial)\varphi$. Two of these four vertex counterterms are forbidden by Galilean invariance, from which it follows that the absence of counterterms $\sim \varphi'\partial_t\varphi$ and $\theta'\partial_t\theta$ (see above) means that counterterms $\sim \varphi'(\varphi\partial)\varphi$ and $\theta'(\varphi\partial)\theta$ in which $(\varphi\partial)$ is grouped together with ∂_t to form the covariant derivative ∇_t must also be absent.

Of the two remaining vertex counterterms $\sim \varphi'(\theta\partial)\theta$ and $\theta'(\theta\partial)\varphi$ the second is actually also forbidden [212]. It is generated by a vertex of the type $\theta'\varphi\theta$, which in the original functional (6.123) corresponds to the expression $\theta'(\theta\partial)\varphi - \theta'(\varphi\partial)\theta \equiv \theta'_i v_{ism}\theta_s\varphi_m$ with vertex factor $v_{ism} = \delta_{im}\partial_s - \delta_{is}\partial_m$ that is transverse in the index i of the field θ': $\partial_i v_{ism} = 0$. This transversality is preserved in any graph containing the $\theta'\varphi\theta$ vertex (the contraction of ∂ with a bare vertex adjacent to an external θ' line vanishes), and so it is also preserved in the divergent part of these graphs. Therefore, the counterterms of this vertex must be grouped together to form the combination $\theta'(\theta\partial)\varphi - \theta'(\varphi\partial)\theta$ which is a multiple of the bare contribution, so that the absence of one of these two terms (owing to Galilean invariance) implies the absence of the other.

The renormalization of the basic model corresponding to (6.123) then requires only three counterterms of the type $\varphi'\partial^2\varphi$, $\theta'\partial^2\theta$, and $\varphi'(\theta\partial)\theta$. The renormalized action functional then has the form

$$S_R(\phi) = \varphi' D^{\varphi\varphi}\varphi'/2 + \theta' D^{\theta\theta}\theta'/2 + \varphi' D^{\varphi\theta}\theta' + \varphi'[-\partial_t\varphi + Z_1\nu\partial^2\varphi -$$

$$- (\varphi\partial)\varphi + Z_3(\theta\partial)\theta] + \theta'[-\partial_t\theta + Z_2 u\nu\partial^2\theta - (\varphi\partial)\theta + (\theta\partial)\varphi]$$

$$(6.129)$$

with all the bare parameters in the amplitudes (6.125) replaced by their basic analogs: $D_0^\varphi = g_0\nu_0^3 \to g\mu^{2\varepsilon}\nu^3$, $D_0^\theta = g_0'\nu_0^3 \to g'\mu^{2\alpha\varepsilon}\nu^3$. The parameter ρ in the mixed correlator is not renormalized ($\rho_0 = \rho$).

The functional (6.129) is obtained from (6.123) by the standard multiplicative renormalization of the fields and parameters: $\phi \to Z_\phi\phi$ and

$$\nu_0 = \nu Z_\nu, \quad u_0 = u Z_u, \quad g_0 = g\mu^{2\varepsilon} Z_g, \quad g_0' = g'\mu^{2\alpha\varepsilon} Z_{g'},$$

$$(6.130)$$

$$Z_1 = Z_\nu, \quad Z_2 = Z_\nu Z_u, \quad Z_3 = Z_\theta^2, \quad Z_{\theta'} = Z_\theta^{-1}, \quad Z_\varphi = Z_{\varphi'} = 1.$$

The fields θ, θ' in this model are renormalized, in contrast to the fields φ, φ'. According to the general rule (Sec. 3.13), the nonlocal terms in the action corresponding to the correlators (6.124) are not renormalized, i.e., they have the same form in the basic and renormalized actions. It then follows that the amplitude factors D_0^φ and D_0^θ in (6.125), which transform into $g\mu^{2\varepsilon}\nu^3$ and $g'\mu^{2\alpha\varepsilon}\nu^3$, respectively, in the basic action, remain the same in the renormalized action (6.129), and so (using $Z_{\varphi'} = 1$)

$$g_0\nu_0^3 = g\mu^{2\varepsilon}\nu^3, \quad g_0'\nu_0^3 Z_{\theta'}^2 = g'\mu^{2\alpha\varepsilon}\nu^3. \tag{6.131}$$

The charge renormalization constants in (6.130) are thereby expressed in terms of the other constants Z as

$$Z_g = Z_\nu^{-3}, \quad Z_{g'} = Z_\nu^{-3} Z_{\theta'}^{-2} = Z_\nu^{-3} Z_\theta^2. \tag{6.132}$$

As already mentioned, the parameter ρ entering into the mixed correlator (6.125) is not renormalized (formally, $\rho_0 = \rho Z_\rho$, $Z_\rho = 1$).

It follows from (6.130) and (6.132) that the field and parameter renormalization constants are related to $Z_{1,2,3}$ in (6.129) as

$$Z_\nu = Z_1, \quad Z_u = Z_2 Z_1^{-1}, \quad Z_g = Z_1^{-3}, \quad Z_{g'} = Z_3 Z_1^{-3},$$

$$(6.133)$$

$$Z_\theta = Z_{\theta'}^{-1} = Z_3^{1/2}, \quad Z_\varphi = Z_{\varphi'} = 1.$$

From this for the corresponding RG functions $\gamma_a = \widetilde{\mathcal{D}}_\mu \ln Z_a$ we obtain

$$\gamma_\nu = \gamma_1, \quad \gamma_u = \gamma_2 - \gamma_1, \quad \gamma_g = -3\gamma_1, \quad \gamma_{g'} = \gamma_3 - 3\gamma_1,$$

$$(6.134)$$

$$\gamma_\theta = -\gamma_{\theta'} = \gamma_3/2, \quad \gamma_\varphi = \gamma_{\varphi'} = 0.$$

The renormalized charges in this model are the parameters g, g', and u, and they correspond to three β functions of the type (1.113):

$$\beta_g = g(-2\varepsilon + 3\gamma_1), \quad \beta_{g'} = g'(-2\alpha\varepsilon + 3\gamma_1 - \gamma_3), \quad \beta_u = u(\gamma_1 - \gamma_2). \quad (6.135)$$

The one-loop calculation of the constants $Z_{1,2,3}$ in (6.129) for arbitrary dimension $d > 2$ gives [212]

$$Z_1 = 1 - gd(d-1)/4B\varepsilon - g'(d^2+d-4)/4B\alpha\varepsilon u^2 + ...,$$

$$Z_2 = 1 - g(d+2)(d-1)/2B\varepsilon u(u+1) - g'(d+2)(d-3)/2B\alpha\varepsilon u^2(u+1) + ...,$$

$$Z_3 = 1 + g/B\varepsilon u - g'/B\alpha\varepsilon u^2 + ...$$

$$(6.136)$$

with constant $B \equiv 2d(d+2)(2\pi)^d/S_d = d(d+2)(4\pi)^{d/2}\|d/2\|$. The parameter ρ from (6.125) does not enter into (6.136), because the graphs of 1-irreducible functions of the type $\varphi'\varphi$, $\theta'\theta$, and $\varphi'\theta\theta$ defining these constants in the one-loop approximation do not contain pseudotensor lines $\varphi\theta$ [212]. However, a ρ dependence can appear in the constants (6.136) in higher orders of perturbation theory.

The ratios g/u and g'/u^2 enter into the constants (6.136) and the corresponding RG functions. They can be eliminated by going from g and g' to the new charges

$$g_1 = g/Bu, \quad g_2 = g'/Bu^2, \quad (6.137)$$

for which from (6.134) and (6.135) we have

$$\beta_{g_1} = g_1(-2\varepsilon + 2\gamma_1 + \gamma_2), \quad \beta_{g_2} = g_2(-2\alpha\varepsilon + \gamma_1 + 2\gamma_2 - \gamma_3), \quad \beta_u = u(\gamma_1 - \gamma_2).$$
$$(6.138)$$

From this point on we shall restrict ourselves to the case of realistic dimension $d = 3$. In the new variables (6.137) for $d = 3$ using the constants (6.136) we obtain the following expressions for the one-loop RG functions $\gamma_i = \widetilde{\mathcal{D}}_\mu \ln Z_i$ [212]:

$$\gamma_1 = 3g_1 u + 4g_2, \quad \gamma_2 = 10g_1/(u+1), \quad \gamma_3 = -2g_1 + 2g_2. \quad (6.139)$$

Substituting these expressions into (6.138), we obtain the one-loop β functions for $d = 3$:

$$\beta_{g_1} = g_1[-2\varepsilon + 6g_1 u + 8g_2 + 10g_1/(u+1)],$$

$$\beta_{g_2} = g_2[-2\alpha\varepsilon + 3g_1 u + 2g_1 + 2g_2 + 20g_1/(u+1)], \quad (6.140)$$

$$\beta_u = u[3g_1 u + 4g_2 - 10g_1/(u+1)].$$

The system of β functions has the following fixed points:

1) The line $g_{1*} = g_{2*} = 0$ with arbitrary u_*;

2) $g_{2*} = u_* = 0$, $\quad g_{1*} = \varepsilon/5$;

3) $g_{1*} = u_* = 0$, $\quad g_{2*} = \alpha\varepsilon$;

$$(6.141)$$

4) $g_{2*} = 0$, $\quad g_{1*} = 2\varepsilon/9u_*$, $\quad u_* = [(43/3)^{1/2} - 1]/2 \simeq 1.393$;

5) $u_* = 0$, $\quad g_{1*} = \varepsilon(4\alpha - 1)/39$, $\quad g_{2*} = \varepsilon(11 - 5\alpha)/39$;

6) $g_{1*} = \varepsilon(u_* + 1)/15$, $\quad g_{2*} = \varepsilon[10 - 3u_*(u_* + 1)]/60$,

where u_* for point No. 6 is the positive root of the equation $3u^2 + 7u + 54 = 60\alpha$ with parameter α from (6.124), which exists for $\alpha \geq 0.9$.

Calculating the matrix (1.189) using the β functions (6.140), it is easily checked that of the fixed points listed above, only points No. 3 and 4 can be IR-stable. In [211] these were referred to as the magnetic and kinetic fixed points, respectively. The magnetic point No. 3 is IR-stable for any $\alpha \geq 0.25$, and the kinetic point No. 4 is stable for any $\alpha \leq 1.16$. In the intermediate region $0.25 \leq \alpha \leq 1.16$ both these points are IR stable. In this situation the choice between the two possible critical regimes depends on the actual values of the initial data in equations of the type (1.188) for the system of β functions (6.140), and in this sense the critical behavior is nonuniversal. An attempt has been made to study the basin of attraction of the magnetic and kinetic fixed points for the realistic (see below) case $\alpha = 1$ in the correlators (6.124) [213]. However, the authors of that study ignored the need to renormalize the $\varphi'\theta\theta$ vertex (because they erroneously considered it to be an effect of higher order in the charges), and so the results in [213] cannot be viewed as fully reliable.

In conclusion, let us briefly discuss the problem of choosing the realistic value of the parameter α in the correlators (6.124). In this section we have studied, following [211] and [212], the massless problem with pumping of the type (6.10) and with arbitrary positive parameters ε and α in (6.124). Actually, the massless model is meaningful only in the region $0 \leq \varepsilon \leq 2$ with the realistic value $\varepsilon = 2$, which corresponds to δ-function pumping (6.91). However, then it is natural to require that the correlator $D^{\theta\theta}$ in (6.124) for $\varepsilon = 2$ also become δ-function IR pumping of the type (6.91), which is possible only for $\alpha = 1$. Therefore, in the massless model the natural realistic values of the parameters ε and α must be $\varepsilon = 2$ and $\alpha = 1$. We note that in a massive model of the type (6.9) the energy pumping by the correlator $D^{\theta\theta}$ would be infrared for any $\alpha > 0$ for sufficiently large $\varepsilon > 0$. However, models of this type have not been studied in stochastic magnetic hydrodynamics.

6.15   Critical dimensions in MHD

If the critical dimensions Δ_F of various quantities F are calculated following the standard rule (6.53), then from the known canonical dimensions taking into account (6.134) for $d = 3$ we obtain

$$\Delta_\omega = -\Delta_t = 2 - \gamma_1^*, \quad \Delta_\varphi = 1 - \gamma_1^*, \quad \Delta_\theta = 1 - \gamma_1^* + \gamma_3^*/2,$$

$$\Delta_\varphi + \Delta_{\varphi'} = \Delta_\theta + \Delta_{\theta'} = d = 3. \tag{6.142}$$

In the kinetic regime [point No. 4 in (6.141)] from (6.138) we have $\gamma_1^* = \gamma_2^* = 2\varepsilon/3$ (exact values), and for γ_3^* from (6.139) in the one-loop approximation we find that $\gamma_3^* = -4\varepsilon/9u_* + ...$ with $u_* = 1.393$. This gives the former results (6.43) for $F = \omega, \varphi, \varphi'$, and for the field θ from (6.142) we obtain

$$\Delta_\theta = 1 - 2\varepsilon/3 - 2\varepsilon/9u_* + ... \tag{6.143}$$

The ellipsis always denotes corrections of order ε^2 and higher.

At the magnetic fixed point [point No. 3 in the list (6.141)] in the one-loop approximation (6.139) we have

$$\gamma_1^* = 4\alpha\varepsilon + ..., \quad \gamma_2^* = 0 + ..., \quad \gamma_3^* = 2\alpha\varepsilon + ... \tag{6.144}$$

with the parameter α from (6.124). In the normal case the desired values of Δ_F for the magnetic regime must be obtained by substituting (6.144) into (6.142).

This is the conclusion arrived at in [212] regarding the critical dimensions in the two possible IR regimes. However, it is actually incorrect (as noted in [214]), because here we are dealing with the special case $g_* = 0$ for several charges. We note that we have already encountered this problem in Sec. 4.28 in discussing the Goldstone asymptote of the correlator D_σ.

Let us explain the problem in more detail for the model we are discussing. Let $F_R \equiv W_{nR}$ be an arbitrary renormalized connected Green function with n_ϕ fields $\phi = \varphi, \varphi', \theta, \theta'$ and $\gamma_F = \sum n_\phi \gamma_\phi$ be its anomalous dimension (we recall that in this model the fields θ and θ' are renormalized). The function F_R satisfies the standard RG equation (5.106), in this case with $a = \nu$ and with summation over the three charges $g = \{g_1, g_2, u \equiv g_3\}$ in the term $\beta_g \partial_g$. It is convenient to transform to canonically dimensionless variables, introducing a representation analogous to (6.27):

$$F_R = k^{d_F} \cdot \nu^{d_F^\omega} \cdot R(k/\mu, \omega/\nu k^2, g, ...), \quad k/\mu \equiv s, \tag{6.145}$$

in which d_F and d_F^ω are the corresponding canonical dimensions of the function F_R and g is the set of all dimensionless renormalized charges. In general, the function F_R depends on several momenta and frequencies. Any of them can be used as k and ω in (6.145), and then the function R will contain additional arguments of the type k_i/k_s and ω_i/ω_s. These are denoted by the ellipsis in (6.145) and are unimportant for what follows.

If F_R in the form (6.145) is substituted into an RG equation of the type (5.106), then from the general rules given in Sec. 1.29 we obtain the following RG representation for F_R:

$$F_R = k^{d_F} \cdot (\overline{\nu})^{d_F^{\omega}} \cdot R(1, \omega/\overline{\nu}k^2, \overline{g}, ...) \cdot R_{\gamma_F}, \qquad (6.146)$$

in which $\overline{\nu}$ and $\overline{g}$ are the corresponding invariant variables and R_{γ_F} is an additional factor generated by the field renormalization:

$$R_{\gamma_F} = \exp\left\{ \int_1^s \frac{ds'}{s'} \gamma_F(\overline{g}(s', g)) \right\}. \qquad (6.147)$$

Its form is determined by (1.141) with $F = \ln R$ and $\gamma = -\gamma_F$ in this case.

At the IR asymptote $s \equiv k/\mu \to 0$ we have $\overline{g} \to g_*$, $\overline{\nu} \to \overline{\nu}_*$ (see Sec. 6.5) and in the factor (6.147) $\gamma_F \to \gamma_F^*$, so that

$$R_{\gamma_F} \to \text{const} \cdot (k/\mu)^{\gamma_F^*}, \qquad \gamma_F^* \equiv \gamma_F(g_*). \qquad (6.148)$$

From this with the additional assumption that the scaling function R in (6.146) is finite (i.e., that its arguments and the function R as a whole are finite) we obtain the standard expressions (6.53) for the critical dimensions.

Since the scaling function is finite we can assume that it has zero critical dimension in the IR regime. This is the normal situation typical of most problems of this type (we recall that the requirement that the argument $\omega/\overline{\nu}k^2$ of the scaling function be finite determines the relative rate of falloff of the variables ω and k at the IR asymptote). However, for the IR regimes in the model (6.129) the assumption of finiteness of the scaling function R in (6.146) is not satisfied for several Green functions, which leads to violation of the rule (6.53). We shall consider the two possible IR regimes separately [183].

1. *The kinetic regime.* Here $g_{2*} = 0$, and the problem is that for Green functions with $n_\theta > n_{\theta'}$ which vanish for $g_2 = 0$ the function R in (6.146) contains overall factors of $\overline{g}_2 \to 0$. In particular, for the renormalized correlator $F_R = \langle \widehat{\theta\theta} \rangle_R$ a single factor of g_2 is isolated from the function R in (6.145). Introducing the notation $R = g_2 \widetilde{R}$ for it, for the function R in (6.146) at the IR asymptote $s \equiv k/\mu \to 0$ we obtain (we indicate explicitly only the charge dependence important for our analysis)

$$R = \overline{g}_2 \widetilde{R}(\overline{g}_1, \overline{g}_2, \overline{u}) \sim s^{\omega_2} \cdot \widetilde{R}(g_{1*}, 0, u_*), \qquad (6.149)$$

where ω_2 is the correction exponent (Sec. 1.42) characterizing the rate of falloff of the invariant charge for $s \to 0$: $\overline{g}_2 \to g_{2*} = 0$. Using the β functions (6.138) with the relations $\gamma_1^* = \gamma_2^* = 2\varepsilon/3$ for the kinetic fixed point, it is easy to show that in this case

$$\omega_2 = \partial_{g_2} \beta_{g_2} |_{g_*} = 2\varepsilon - 2\alpha\varepsilon - \gamma_3^*. \qquad (6.150)$$

Therefore, in the kinetic regime the scaling function R in (6.146) for the renormalized correlator $\langle \widehat{\theta\theta} \rangle_R$ turns out to be a quantity with critical dimension $\Delta[R] = w_2$, which corresponds to adding $w_2/2$ to the critical dimension of the field θ.

Using $\widetilde{\Delta}_F$ to denote the critical dimensions including this addition, from (6.142) and the above we find

$$\widetilde{\Delta}_\theta = \Delta_\theta + w_2/2 = 1 - 2\varepsilon/3 + \varepsilon(1-\alpha), \quad \widetilde{\Delta}_\theta + \widetilde{\Delta}_{\theta'} = d = 3. \quad (6.151)$$

We note that the results in (6.151) are exact, i.e., there are no order-ε^2 and higher corrections, because the contribution of γ_3^* in (6.151) cancels.

The dimension of the field θ in (6.151) is determined from the form of the correlator $\langle \widehat{\theta\theta} \rangle_R$. However, it can be shown that the result thus obtained is universal, i.e., all the Green functions in the kinetic regime display IR scaling with critical dimensions $\widetilde{\Delta}_F$ coinciding with the earlier dimensions (6.43) for $F = w, \varphi, \varphi'$ and with the new dimensions (6.151) for the fields θ and θ'. To prove this it is necessary to make the change of variable $\phi(x) \to \widetilde{\phi}(x)$ in the functional (6.129), transforming to the new fields $\widetilde{\theta}, \widetilde{\theta}'$ [214]:

$$\widetilde{\theta}(x) = \theta(x) \cdot g_2^{-1/2}, \quad \widetilde{\theta}'(x) = \theta'(x) \cdot g_2^{1/2} \quad (6.152)$$

without changing the fields φ and φ' ($\widetilde{\varphi} = \varphi$, $\widetilde{\varphi}' = \varphi'$). It is easily shown that for the new fields $\widetilde{\phi}$ the anomalous dimension $\gamma_F = \sum n_\phi \gamma_\phi$ in the RG equation (5.106) acquires the addition $(n_\theta - n_{\theta'})\beta_{g_2}/2g_2$, which in the IR regime in question is equivalent to adding $\pm w_2/2$ to the critical dimensions of the fields θ and θ'. Now the critical dimensions of all the basic quantities can be calculated using the usual rule (6.53), because the problem of the zero in the R functions disappears after the replacement (6.152). In fact, after this replacement the charge g_2 moves from the propagators to the $\varphi'\theta\theta$ vertex, so that this vertex can be simply neglected when calculating the IR asymptote in the regime $\bar{g}_2 \to 0$. It is important that all the Green functions of the new fields $\widetilde{\phi}$ remain finite, and the magnetic field then plays the role of a simple passive admixture. We also note that for the realistic values $\varepsilon = 2$, $\alpha = 1$ the corrected critical dimensions (6.151), in contrast to the original ones Δ_θ and $\Delta_{\theta'}$ in (6.142), coincide with the corresponding Kolmogorov dimensions (6.8) for the fields φ and φ'.

2. *The magnetic regime.* In this regime $u_* = 0$, $g_{1*} = 0$, $g_{2*} \neq 0$, and the invariant variables $\bar{u}$ and $\bar{g}_1$ fall off for $s \equiv k/\mu \to 0$ according to power laws:

$$\bar{u} \sim s^{w_u} \to 0, \quad \bar{g}_1 \sim s^{w_1} \to 0 \quad (6.153)$$

with exponents

$$w_u = \gamma_1^* - \gamma_2^* = 4\alpha\varepsilon + ..., w_1 = -2\varepsilon + 2\gamma_1^* - \gamma_2^* = -2\varepsilon + 8\alpha\varepsilon + ... \quad (6.154)$$

For $g_1 = 0$, $g_2 \neq 0$, $u \neq 0$ the Green functions with odd value of the sum $n_\theta + n_{\theta'}$ vanish (we note that they can be generated only by graphs involving the mixed correlator $\varphi\theta$), and all the other Green functions remain finite. Excluding the special case of odd $n_\theta + n_{\theta'}$ for now, we conclude that for the other Green functions the problem of the zero for $\overline{g}_1 \to g_{1*} = 0$ in the scaling functions does not exist. However, at the magnetic fixed point there remains a second problem, that of singularities for $\overline{u} \to u_* = 0$ (in contrast to the charges $g_{1,2}$, the parameter u can also occur in denominators, and so we must now speak not of zeros but of singularities of the scaling functions). For the dynamical Green functions in the variables (6.137) this problem was solved in [214] by a change of variable $\phi(x) \to \widetilde{\phi}(x)$ in the functional (6.129), analogous to (6.152):

$$\varphi(x) = u\widetilde{\varphi}(\widetilde{x}), \quad \varphi'(x) = \widetilde{\varphi}'(\widetilde{x}), \quad \theta(x) = u^{1/2}\widetilde{\theta}(\widetilde{x}), \quad \theta'(x) = u^{-1/2}\widetilde{\theta}'(\widetilde{x}),$$
(6.155)

where $x \equiv (t, \mathbf{x})$, $\widetilde{x} \equiv (ut, \mathbf{x})$. After substituting (6.155) into the functional (6.129), the parameter u enters only as the coefficient of the terms $\widetilde{\varphi}'\partial_t\widetilde{\varphi}$ and $\widetilde{\varphi}'(\widetilde{\varphi}\partial)\widetilde{\varphi}$. Therefore, at the asymptote (6.153) these contributions to the action vanish, but the remaining ones are sufficient to ensure the finiteness of the limit $\overline{u} = 0$, $\overline{g}_1 = 0$ for all dynamical Green functions with even $n_\theta + n_{\theta'}$. It then follows that for these objects the scaling functions in the RG representations remain finite, so that the critical dimensions can be found from the general scheme of Sec. 5.16, combining the RG equation with the scale equations. It is easily checked that for the Green functions of the new fields $\widetilde{\phi}$, the operator $\mathcal{D}_{\mathrm{RG}}$ acquires the addition $[\mathcal{D}_t + n_\varphi + n_\theta/2 - n_{\theta'}/2]\beta_u/u$ when the replacement (6.155) is made in an RG equation like (5.106). In this IR regime we have $\beta_u/u \to \omega_u = \gamma_1^* - \gamma_2^*$, and when the RG equation is combined with the scale equations (Sec. 5.16) we find the following new values of the critical dimensions [214] $(d = 3)$:

$$\widetilde{\Delta}_\omega = -\widetilde{\Delta}_t = 2 - \gamma_2^*, \quad \widetilde{\Delta}_\varphi = 1 - \gamma_2^*, \quad \widetilde{\Delta}_{\varphi'} = 2 + \gamma_1^*,$$

$$\widetilde{\Delta}_\theta = 1 - \alpha\varepsilon + \gamma_2^*/2, \quad \widetilde{\Delta}_\theta + \widetilde{\Delta}_{\theta'} = d = 3$$
(6.156)

with γ_i^* from (6.144). We note that the critical dimensions of the fields ϕ and $\widetilde{\phi}$ coupled by the transformation (6.155) coincide, because the factors of u in (6.155) have zero critical dimension.

Therefore, the dynamical Green functions of the fields $\widetilde{\phi}$ with even $n_\theta + n_{\theta'}$ in the magnetic regime display IR scaling with the critical dimensions (6.156). This automatically implies the analogous property also for the static correlation functions of the fields $\widetilde{\varphi}$ and $\widetilde{\theta}$, but only when these objects exist in the limit theory with $u = g_1 = 0$, i.e., when the integrals over frequencies of the corresponding dynamical functions in the limit theory with $u = g_1 = 0$ are convergent. In the model under consideration this condition is satisfied for all static objects except the correlation functions $\widetilde{\varphi}\widetilde{\varphi}$ and $\widetilde{\varphi}\widetilde{\varphi}\widetilde{\varphi}$. For them the integrals over frequencies diverge at $u = 0$, which implies the appearance of an additional singularity in u in the static scaling function.

This problem was studied in detail in [211] for the static correlator $\widetilde{\varphi}\widetilde{\varphi}$ and it was shown that

$$\langle\widehat{\widetilde{\varphi}\widetilde{\varphi}}\rangle_{\text{stat}} \sim \begin{cases} k^{-d+2-2\alpha\varepsilon+\cdots} & \text{for } \alpha > 1/3, \\ k^{-d+2-2\varepsilon+4\alpha\varepsilon+\cdots} & \text{for } \alpha < 1/3. \end{cases} \tag{6.157}$$

The most realistic value (see the remark at the end of Sec. 6.14) $\alpha = 1$ corresponds to the first variant in (6.157). We note that the authors of [211] used the old technique of Wilson recursion relations (Sec. 1.1) and studied only the pair correlators $\varphi\varphi$ and $\theta\theta$, and not IR scaling as a general property of all the Green functions. The derivation of the asymptote (6.157) in the language used in the present book can be found in [183].

In the normal situation when there is scaling with dimensions $\widetilde{\Delta}_\phi$ for the fields $\widetilde{\phi}$, their static correlators must behave as

$$\langle\widehat{\widetilde{\phi}\widetilde{\phi}}\rangle_{\text{stat}} \sim k^{-d+2\widetilde{\Delta}_\phi}. \tag{6.158}$$

In the regime in question the rule (6.158) holds for the $\widetilde{\theta}\widetilde{\theta}$ correlator but not for the $\widetilde{\varphi}\widetilde{\varphi}$ one, i.e., in this case there is no universal IR scaling with definite dimensions of the fields and parameters for all the Green functions.

This conclusion is also confirmed by analysis of the IR asymptote of the dynamical Green functions with odd $n_\theta + n_{\theta'}$, which so far we have not considered. Such functions vanish when the mixed correlator (6.124) vanishes, and so they all must contain at least one additional factor of $g_1^{1/2}$. In the IR regime under consideration this leads to the appearance of an extra small factor $\sim (\overline{g}_1)^{1/2} \sim s^{\omega_1/2}$ compared to the Green functions considered earlier, and this also violates universality.

This RG analysis of the IR asymptote for the magnetic fixed point is useful methodologically. It shows that in nontrivial models the standard RG equations at the IR asymptote may lead not only to the usual universal IR scaling, but also to much more complicated behavior of the Green functions. In this sense the RG technique is much richer than the simple idea of critical scaling.

6.16   The turbulent dynamo in gyrotropic MHD

The turbulent dynamo in MHD is [215] the effect of generation of a large-scale magnetic field by the energy of the turbulent motion of a medium. This effect can occur only when gyrotropy, i.e., nonconservation of spatial parity, is present.

In gyrotropic MHD the noise correlators in (6.122) are written as the sum of tensor and pseudotensor contributions [216], [217], in particular (d=3)

$$D_{is}^{\varphi\varphi} = D_0^\varphi k^{1-2\varepsilon}[P_{is} + i\rho\varepsilon_{ism}k_m/k] \tag{6.159}$$

instead of the purely tensor structure in (6.124). The real parameter ρ in (6.159) characterizes the magnitude of the gyrotropy, and the requirement that the correlator $D^{\varphi\varphi}$ be positive-definite imposes the constraint $|\rho| \leq 1$.

Gyrotropic stochastic hydrodynamics with correlator of the type (6.159) was first studied in [216] using the RG technique of Wilson recursion relations. There it was shown that in ordinary hydrodynamics (without a magnetic field) the introduction of gyrotropy does not spoil the IR stability of the critical regime and does not change the values of the critical dimensions (6.43) of the basic quantities, so that only the explicit form of the scaling functions is affected. The generation of "curl terms" (see below) in gyrotropic MHD leading to instability of the system was discovered [216] . It was therefore concluded [216] that the RG technique is inapplicable in that case. However, this is not really true. It was later shown [217] that the appearance of a large-scale magnetic field (i.e., the dynamo phenomenon) stabilizes the system and makes it possible to perform the standard RG analysis of the model in the new stable regime. The dynamo effect is explained [217] as just the usual mechanism of spontaneous symmetry breaking: the normal solution with $\langle \widehat{\theta}(x) \rangle = 0$ is unstable and is stabilized by the appearance of a spatially uniform average magnetic field $\langle \widehat{\theta} \rangle = C \neq 0$.

Let us now discuss the results of [217] in more detail. Those authors studied the model (6.122) without magnetic noise ($\eta^\theta = 0$) and with the correlator (6.159) for the noise η^φ. Analysis shows that the introduction of gyrotropy does not spoil the multiplicative renormalizability of the model in calculations using the formal scheme without the cutoff Λ (Sec. 1.20). It then follows that the renormalized action has the usual form (6.129), except now without the contributions involving $D^{\theta\theta}$ and $D^{\varphi\theta}$ since there is no magnetic noise:

$$S_R(\phi) = \varphi' D^{\varphi\varphi} \varphi'/2 + \varphi'[-\partial_t \varphi + Z_1 \nu \partial^2 \varphi - (\varphi\partial)\varphi +$$
$$+ Z_3 (\theta\partial)\theta] + \theta'[-\partial_t \theta + Z_2 u \nu \partial^2 \theta - (\varphi\partial)\theta + (\theta\partial)\varphi]$$

(6.160)

with the usual substitution $D_0^\varphi = g_0 \nu_0^3 = g\nu^3 \mu^{2\varepsilon}$ in the amplitude of the correlator (6.159). The gyrotropy parameter appearing here is not renormalized: $\rho_0 = \rho$.

For the normal solution, which is represented by the graphs of perturbation theory in the model (6.160), all correlators $\langle \widehat{\theta} \rangle$, $\langle \widehat{\theta\theta} \rangle$, $\langle \widehat{\theta\theta\theta} \rangle$, ... with any number of fields θ are equal to zero owing to the absence of the $\theta\theta$ propagator. However, this normal solution is actually unstable, as shown by analysis of the response functions $\langle \widehat{\varphi\varphi'} \rangle$ and $\langle \widehat{\theta\theta'} \rangle$ in the model (6.160).

We recall (Sec. 5.5) that in dynamical models the stability criterion is the requirement that all small perturbations be damped out, which is ensured by the correct behavior of the response functions: all their poles in ω must lie in the lower half-plane.

The response functions in the model (6.160) can be written as

$$\langle \widehat{\varphi}\widehat{\varphi}' \rangle = [(-i\omega + Z_1 \nu k^2)P - \Sigma^{\varphi'\varphi}(\omega, k)]^{-1},$$

$$\langle \widehat{\theta}\widehat{\theta}' \rangle = [(-i\omega + Z_2 u\nu k^2)P - \Sigma^{\theta'\theta}(\omega, k)]^{-1},$$

(6.161)

in which P is the transverse projector and the matrices $\Sigma^{\phi'\phi}$ for the fields $\phi = \varphi, \theta$ are the contributions of 1-irreducible self-energy graphs with the lines and vertices of the model (6.160). In the simplest one-loop approximation,

$$\Sigma^{\varphi'\varphi} = \underset{\varphi' \quad \varphi \quad \varphi \quad \varphi}{\overset{\varphi \quad\quad\quad \varphi'}{\frown}} \quad , \quad \Sigma^{\theta'\theta} = \underset{\theta' \quad \varphi \quad \varphi \quad \theta}{\overset{\theta \quad\quad\quad \theta'}{\frown}} . \tag{6.162}$$

The indices at the line ends specify the types of propagator and vertex, and the charge g enters as a factor in the $\varphi\varphi$ line. All the contributions in (6.161) are transverse in the vector indices, and the symbol -1 should be understood as the inverse on the subspace of the transverse vectors.

Stability in zeroth order is ensured by the correct sign for the bare terms $\sim \nu k^2$ in (6.161). The corrections from the graphs (6.162) contain the factor g, which is assumed to be a small parameter in perturbation theory, and, as shown by direct calculation, they do not have singularities in the variables ω, k which could compensate for the smallness of g. Therefore, it appears at first glance that the corrections from the graphs of $\Sigma^{\phi'\phi}$ cannot compete with the bare contributions at small g, and thus cannot lead to instability.

This is true, but with one exception. If the expansion of $\Sigma^{\phi'\phi}(\omega, k)$ in ω and k contains a term linear in k, then its contribution $\sim gk$ for $k \to 0$ will be more important than the contribution of the bare term $\sim \nu k^2$ independently of the value of g, and the stability can be violated for small k. This is obviously impossible when gyrotropy is absent, because there is no true tensor that is linear in k and transverse in the vector indices. However, when gyrotropy is introduced the requirement that there be a true tensor is lifted (nonconservation of spatial parity), and then a contribution $\sim i\varepsilon_{ism}k_m$ with real coefficient is allowed in the matrix Σ_{is}. In coordinate space such contributions correspond to additions $\sim \mathrm{curl}\phi$ to the right-hand side of dynamical equations like (6.122), and in what follows these will be referred to as curl terms.

The appearance of curl terms in $\Sigma^{\phi'\phi}$ certainly leads to instability at small k, because the matrix $i\varepsilon_{ism}k_m$ has eigenvalues $\pm|k|$, i.e., it does not have definite sign.

It is therefore important to know whether curl terms are present in $\Sigma^{\phi'\phi}$. The calculation [217] of the graphs (6.162) for the model (6.160) with correlator (6.159) shows that there are no curl terms in $\Sigma^{\varphi'\varphi}$, but they are present in $\Sigma^{\theta'\theta}$, namely,

$$\Sigma_{is}^{\theta'\theta}(\omega, k)\Big|_{curl} = \frac{\rho g \nu \Lambda (\mu/\Lambda)^{2\varepsilon}}{6\pi^2 (1+u)(1-2\varepsilon)} \cdot i\varepsilon_{ism} k_m, \qquad (6.163)$$

where ρ is the gyrotropy parameter from (6.159) and Λ is the UV cutoff. Since the renormalized charge g must be assumed to be of order ε (Sec 1.27), the ε dependence of the other factors in (6.163) in lowest order of the ε expansion can be neglected, as their inclusion at one loop would exceed the accuracy.

It should be stressed that the expression (6.163) in the terminology of Sec. 1.19 is a regular contribution and would be completely lost if the calculations were performed in the formal scheme without the cutoff Λ. As explained in detail in Sec. 1.20, regular contributions can be avoided when they lead only to the renormalization of insignificant parameters like T_c. However, in our case the situation is different, because there are no bare curl terms in the original equations (6.122). Therefore, their appearance when the interaction is switched on is an objective fact indicating instability of the normal solution with $\langle \widehat{\theta}(x) \rangle = 0$ with respect to fluctuations of the magnetic field θ.

This instability is manifested at small momenta k and is maximal for $k \to 0$, i.e., for spatially uniform fluctuations of the field θ. It is therefore natural to assume that the system is stabilized owing to the appearance of a nonzero spatially uniform mean field $\langle \widehat{\theta}_i(x) \rangle \equiv C_i \neq 0$, just as a ferromagnet below T_c is stabilized owing to the appearance of spontaneous magnetization.

To test this assumption, we make the shift $\theta(x) \to \theta(x) + C$ in (6.160) and then verify that the instability can be made to vanish by choosing a suitable value of C. This calculation was performed in the one-loop approximation in [217]. There it was shown that after the shift terms linear in the momentum k do not appear in $\Sigma^{\varphi'\varphi}$, while in $\Sigma^{\theta'\theta}$ such terms take the form

$$\Sigma_{is}^{\theta'\theta}(\omega, k) \simeq \frac{ig\rho u^{-1/2}}{16\pi(1+u)} \left\{ \left[\frac{8\Lambda\nu}{3\pi u^{-1/2}} - |C| \right] \varepsilon_{ism} k_m + |C|[k \times n]_i n_s \right\}, \qquad (6.164)$$

where $n_i \equiv C_i/|C|$ is the unit vector in the direction of the spontaneous field C_i. The answer (6.164) is given in lowest order of the ε expansion neglecting corrections in ε [see the text following (6.163)].

We see from (6.164) that the shift $\theta_i(x) \to \theta_i(x) + C_i$ generates along with the curl term a second contribution linear in k, which in [217] is referred to as an exotic term. It is also more important than the contribution $\sim \nu k^2$ at small k, and so it could also generate an instability. If this happened, stabilization by shifting θ would become impossible, because the two contributions to (6.164) cannot simultaneously be eliminated by the choice of C. However, in [217] it was shown that the exotic contribution does not generate an instability, so that in this regard only the curl terms are dangerous. They can be cancelled in (6.164) by the choice

$$|C| = 8\nu\Lambda u^{1/2}/3\pi \qquad (6.165)$$

(the most important thing here is that the sign of $|C|$ in (6.164) be the one needed to cancel the curl terms).

The expression (6.165) determines the absolute value $|C|$ of the spontaneous uniform magnetic field arising in the one-loop approximation (the dynamo effect). Its direction remains completely arbitrary, as must be the case in a problem involving spontaneous symmetry breaking. For a numerical estimate the parameters ν and u in (6.165) can be replaced by their bare values, and Λ by the inverse dissipation length. It should also be remembered that the vector $\langle \widehat{\theta}_i(x) \rangle \equiv C_i$ differs from the magnetic induction B_i by a known factor [see the explanation of (6.122)].

An important property of the solution (6.165) is the absence of the parameter ρ in it. The spontaneous field C remains finite for arbitrarily weak gyrotropy, even though without the latter the field could not appear at all. We can thus state that gyrotropy is needed only to switch on the mechanism of spontaneous symmetry breaking, and its numerical value plays no role. In this sense gyrotropy plays the same role in this problem as an external magnetic field for a ferromagnet. It can therefore be assumed, guided by the general idea of quasiaverages [218], that the dynamo effect in MHD must also occur in the absence of gyrotropy.

When the stability condition (6.165) is satisfied, only the exotic contribution linear in k remains in (6.164). The authors of [217] studied its effect on the structure of long-wavelength excitations of the linearized equations (6.122) in the dynamo regime, i.e., with $\theta_i \to \theta_i + C_i$. In the linear approximation neglecting viscosity and the exotic term the solutions of these equations are the two ordinary Alfvén wave modes (Sec. 69 of [210]). The inclusion of the exotic term causes one of these two modes (polarized orthogonally to the direction of the spontaneous field) to grow linearly with time. Formally this is also an instability, but a very weak one (compared to the usual exponential growth), and so it vanishes when the exponential damping $\sim \exp[-\nu k^2 t]$ owing to viscosity is taken into account.

6.17 Critical dimensions in the dynamo regime

Let us now consider the problem of combining the dynamo effect and the RG technique [183], [217].

The renormalization procedure for the unrenormalized model corresponding to (6.160) can be performed in two steps. The first is Λ renormalization (Sec. 1.20). In this case it reduces to the subtraction of regular contributions like (6.163) which are powers of Λ from all the unrenormalized graphs of $\Sigma^{\theta'\theta}$ (there are no regular contributions in the other graphs) and the replacement of all these contributions by the equivalent explicit addition of the curl term $\nu_0 h_0 \theta' \mathrm{curl}\theta$ to the unrenormalized action functional. In this manner a new bare parameter h_0 is calculated from the graphs. It has the dimension of momentum and in perturbation theory is represented by a series like (1.99):

$$h_0 = \Lambda \sum_{n=1}^{\infty} (g_0 \Lambda^{-2\varepsilon})^n c_n(u_0, \rho, \varepsilon), \qquad (6.166)$$

containing only bare parameters and the cutoff Λ. The first term of the series (6.166) is defined by (6.163): $c_1 = \rho/6\pi^2(1+u_0)(1-2\varepsilon)$.

In this way we arrive at the unrenormalized theory with the addition of the bare curl term $\nu_0 h_0 \theta' \mathrm{curl}\theta$ to the action, along with the subtraction of the regular contributions from all the graphs. The latter implies that now all the graphs can be calculated within the formal scheme without the cutoff Λ and then they will have poles in ε (Sec. 1.20). These poles are eliminated by the standard procedure of multiplicative ε renormalization, equivalent to the introduction of the three constants Z indicated in (6.160) and another new constant $Z_4 = Z_\nu Z_h$ for the curl term, in terms of which the renormalization constant of the new parameter h_0 is determined: $h_0 = h Z_h$. Here it is important that in a formal scheme of the MS type the dimensional parameter h cannot enter into the renormalization constants Z. Therefore, the addition of the curl term has no effect on the three constants Z in (6.160) calculated in the formal scheme without this term. This addition leads only to the appearance of a new renormalization constant Z_h, which is not really needed in the final RG equations (see below). The dimensionless gyrotropy parameter ρ can enter into the constants Z, but this does not occur at one loop.

The Green functions of the renormalized model obtained in this manner satisfy the usual RG equations of the type (5.106) with the additional term $-\gamma_h \mathcal{D}_h$ in the operator $\mathcal{D}_{\mathrm{RG}}$ from the new variable h. Here we are speaking of the renormalized Green functions for the normal solution with zero expectation values $\langle \widehat{\theta} \rangle$, $\langle \widehat{\theta}\widehat{\theta} \rangle, \ldots$ and with the instability-generating UV-finite (owing to renormalization) curl contribution in the response function $\langle \widehat{\theta}\widehat{\theta'} \rangle$. This contribution is eliminated by the dynamo effect, i.e., by a suitable shift $\theta \to \theta + C$ of the magnetic field θ. This procedure can be performed either before the ε renormalization, thereby leading to an expression $C_0 = C_0(h_0, g_0, \nu_0, u_0, \rho, \varepsilon)$ for the unrenormalized spontaneous field C_0, or after the ε renormalization, which gives an analogous equation $C = C(h, g, \nu, u, \rho, \mu, \varepsilon)$ for the renormalized parameters. We note that the cutoff Λ does not enter explicitly into these expressions [but it does enter implicitly via h_0, according to (6.166)], because all the calculations after the Λ renormalization are already performed within the formal scheme of Sec. 1.20. We also note that C_0 does not contain any dependence on the renormalization mass μ, i.e., the quantity C_0 is renormalization-invariant (as it must be in view of its physical interpretation), while in the renormalized version the relation between the parameters C and h is UV-finite (no poles in ε), because it is defined using only renormalized quantities.

The renormalized Green functions in the dynamo regime can be obtained by shifting the argument $\theta(x) \to \theta(x) + C$ in the generating functional $\Gamma_{\mathrm{R}}(\phi)$

of the 1-irreducible renormalized Green functions of the original theory without the shift. Since a shift by a UV-finite amount in a UV-finite functional $\Gamma_R(\phi)$ does not violate the UV-finiteness of the latter, it is clear that the shift parameter must be renormalized by the same constant Z as the field θ:

$$C_0 = C Z_C, \quad Z_C = Z_\theta. \tag{6.167}$$

After the shift, the renormalized Green functions involve also the quantity C, but it does not need to be regarded as a new variable, because it is expressed in terms of h and the other parameters of the model. It is thus clear that in choosing h as an independent variable, the passage to the dynamo regime by shifting θ need not at all change the form of the RG equations for the Green functions discussed above.

However, another possibility arises in the dynamo regime. Now it is possible to take C instead of h as the independent variable, which leads to the replacement $-\gamma_h \mathcal{D}_h \to -\gamma_C \mathcal{D}_C$ in the operator $\mathcal{D}_{\mathrm{RG}}$. The advantage of this choice is that now instead of the RG function $\gamma_h = \tilde{\mathcal{D}}_\mu \ln Z_h$, which requires the calculation of a new constant Z_h, the RG equation involves the RG function $\gamma_C = \gamma_\theta$ known from (6.167). Then all the quantities needed to write the RG equations are expressed in terms of the three renormalization constants Z of the model (6.160) with the correlator (6.159). Here the expressions (6.133)–(6.135) are preserved (but now $g' = 0$), and so in RG equations of the type (5.106) for renormalized connected Green functions in the dynamo regime we have

$$\mathcal{D}_{\mathrm{RG}} = \mathcal{D}_\mu + \beta_g \partial_g + \beta_u \partial_u - \gamma_\nu \mathcal{D}_\nu - \gamma_C \mathcal{D}_C,$$
$$\gamma_\varphi = \gamma_{\varphi'} = 0, \quad \gamma_\theta = -\gamma_{\theta'} = \gamma_3/2, \quad \gamma_C = \gamma_\theta \tag{6.168}$$

with β functions from (6.135).

The renormalization constants Z in (6.160) can also depend on the gyrotropy parameter ρ in (6.159). The ρ dependence does not yet appear in the one-loop approximation, and so the three constants in (6.160) coincide with the expressions (6.136) for $g' = 0$ of the model without gyrotropy. However, in any case owing to the absence of renormalization of the fields φ, φ' and the preservation of the relation $Z_g = Z_\nu^{-3}$ in (6.133), the expressions (6.43) will also remain valid for the critical dimensions of the quantities $\omega, \varphi, \varphi'$.

The RG equations (5.106) with $\mathcal{D}_{\mathrm{RG}}$ from (6.168) for the Green functions in the dynamo regime lead to ordinary RG representations of the type (6.146) with the additional invariant variable $\overline{C}$. In the model we are considering [217] without magnetic noise ($g' = 0$) only the kinetic IR regime is possible (Sec. 6.15). The problem of zeros $\sim g_2$ in the R functions discussed in Sec. 6.15 is now absent, so that the critical dimensions can be calculated using the standard rule (6.53). The point is that now we are dealing with the model with $g_2 = 0$ in the notation of Sec. 6.15. In the dynamo regime the factors of g_2 in the R functions are replaced by $|C|^2$, which from the start are assumed to have nonzero critical dimensions (in contrast to the charge g_2, which acquires critical dimensions only in the special case $g_{2*} = 0$). The critical dimensions in the dynamo regime are therefore determined by the usual relations (6.142) for the kinetic fixed point, to which we must now add the equation $\Delta_C = \Delta_\theta$ following from (6.167). As explained in Sec. 6.15, the equations (6.142) lead to the usual expressions (6.43) for the critical dimensions Δ_F of the quantities $F = \omega, \varphi, \varphi'$ and expression (6.143) for $\Delta_\theta = \Delta_C$. Comparing Δ_φ and Δ_θ (we recall that the canonical dimensions of the fields φ and θ are identical), we conclude that in the dynamo regime

$$\Delta_\theta = \Delta_C = \Delta_\varphi + \gamma_\theta^*, \quad \gamma_\theta^* = \gamma_3^*/2 = -2\varepsilon/9u_* + \dots \qquad (6.169)$$

with $u_* = 1.393$.

The main qualitative result of the above RG analysis [217] is that the magnetic quantities θ and C, in contrast to the velocity field φ, have critical dimensions for realistic $\varepsilon = 2$ different from the Kolmogorov value $\Delta_\varphi = -1/3$ owing to the additional term γ_θ^* from the nontrivial renormalization of the field θ. This possibility is usually not taken into account in studies devoted to generalizing the Kolmogorov phenomenology to the turbulent dynamo regime (see, for example, [219]).

6.18 Two-dimensional turbulence

Let us return to the basic stochastic problem (6.1) and, following [220], discuss the case of two-dimensional turbulence ($d = 2$), which so far we have ignored.

As explained in Sec. 6.4, for $d = 2$ an additional quadratic divergence appears in a 1-irreducible function of the type $\varphi'\varphi'$, which requires the introduction of an additional local counterterm $\sim \varphi'\partial^2\varphi'$. In some studies (for example, [212], [221]) there have been erroneous attempts to eliminate this additional divergence by introducing a new renormalization constant Z for nonlocal action term $\sim \varphi'D\varphi'$ (while in [211] it was completely ignored). However, this is incorrect. Even though it is actually possible to eliminate the divergence in this way in the simplest one-loop approximation, this does not work in higher orders, because this procedure contradicts the fundamental principal of locality of the counterterms on which the entire theory of UV renormalization is based (Ch. 3).

The authors of [220] studied a generalization of the two-dimensional problem with variable spatial dimension $d = 2 + 2\Delta$. The arbitrary parameter Δ is treated as a second regularizer in addition to ε from (6.10). The logarithmic theory corresponds to the values $\varepsilon = 0$, $\Delta = 0$ (i.e., $d = 2$), and the realistic theory is obtained by extrapolation from the logarithmic value to the realistic values $\varepsilon_r = 2$ and Δ_r given by the line $\Delta \sim \varepsilon$. The two-dimensional problem directly corresponds to the special case $\Delta = 0$, but by choosing a suitable proportionality constant a on the ray $\Delta = a\varepsilon$, for $\varepsilon_r = 2$ we can obtain any *a priori* specified value of the real dimension $d_r = 2 + 2a\varepsilon_r$, including $d_r = 3$.

This formulation of the problem was studied in [220] and [221], but in [221] the renormalization of the contribution $\sim \varphi'\varphi$ was erroneously assumed to be multiplicative. This error was corrected in [220], the results of which are discussed briefly below. A more detailed discussion of all these problems can be found in the book [183].

The authors of [220] studied a modification of the model (6.12) with massless pumping of the type (6.10) in dimension $d = 2 + 2\Delta$. The modification is that a local term proportional to k^2 is added to the usual pumping function (6.16) with $d = 2 + 2\Delta$:

$$N(k) = g_0 \nu_0^3 k^{2-2\Delta-2\varepsilon} + g_0' \nu_0^3 k^2, \quad d[g_0] = 2\varepsilon, \quad d[g_0'] = -2\Delta. \quad (6.170)$$

The addition $\sim k^2$ is needed to ensure that the renormalization is multiplicative, because this contribution appears as a counterterm. The coefficient g_0' plays the role of a second bare charge [its total canonical dimension is given in (6.170)], and its realistic value for the original model (6.10) in the theory of turbulence must be taken to be $g_0' = 0$ in order to have IR pumping at $\varepsilon = 2$.

The UV divergences of the basic theory are now poles in the set of regularizers ε and Δ, which should be treated formally as quantities of the same degree of smallness. In [220] it was shown that the renormalization of this model requires only counterterms of the type $\varphi'\partial^2\varphi$ and $\varphi'\partial^2\varphi'$, and so the renormalized action has the form

$$S_R(\phi) = \varphi'[g\nu^3\mu^{2\varepsilon}k^{2-2\Delta-2\varepsilon} + Z_2 g'\nu^3\mu^{-2\Delta}k^2]\varphi'/2+$$
$$+\varphi'[-\partial_t\varphi + Z_1\nu\partial^2\varphi - (\varphi\partial)\varphi]. \quad (6.171)$$

The kernel D of the quadratic form $\varphi'D\varphi'$ in (6.171) is written symbolically, and the exact meaning of the notation is clear from the context. The renormalization constants Z in (6.171) are functions of the two dimensionless renormalized charges g, g' and the two regularizers ε, Δ.

According to the theory of analytic renormalization with several regularizers [222], by making a suitable choice of the constants Z in (6.171) it is possible to eliminate all UV divergences, i.e., poles in the small parameters ε and Δ, from the Green functions. More precisely, this means that for these functions there exists a finite limit as ε and Δ tend to zero simultaneously, and that this limit is independent of the ratio Δ/ε, i.e., on how the regularizers tend to zero.

The action (6.171) is obtained from the unrenormalized functional (6.12) with the pumping (6.170) by multiplicative renormalization of the parameters without field renormalization:

$$g_0 = g\mu^{2\varepsilon}Z_g, \quad g_0' = g'\mu^{-2\Delta}Z_{g'}, \quad \nu_0 = \nu Z_\nu,$$

$$Z_\varphi = Z_{\varphi'} = 1, \quad Z_\nu = Z_1, \quad Z_g = Z_\nu^{-3}, \quad Z_{g'} = Z_2 Z_\nu^{-3}. \quad (6.172)$$

From this for the RG functions we obtain

$$\gamma_\nu = \gamma_1, \quad \gamma_g = -3\gamma_1, \quad \gamma_{g'} = \gamma_2 - 3\gamma_1, \quad \gamma_\varphi = \gamma_{\varphi'} = 0,$$

$$\beta_g = g[-2\varepsilon + 3\gamma_1], \quad \beta_{g'} = g'[2\Delta - \gamma_2 + 3\gamma_1]. \tag{6.173}$$

The one-loop calculation of the constants Z gives [220]

$$Z_1 = 1 - g/64\pi\varepsilon + g'/64\pi\Delta + ...,$$

$$Z_2 = 1 - g^2/64\pi g'(\Delta + 2\varepsilon) - g/32\pi\varepsilon + g'/64\pi\Delta + ... \tag{6.174}$$

The anomalous dimensions $\gamma_i = \widetilde{\mathcal{D}}_\mu \ln Z_i$ are calculated using these constants:

$$\gamma_1 = (g + g')/32\pi + ..., \quad \gamma_2 = (g + g')^2/32\pi g' + ... \tag{6.175}$$

Their substitution into (6.173) gives all the desired RG functions in one-loop order.

Analysis of the β functions thereby obtained shows [220] that they have the following three fixed points:

1. $g_* = g'_* = 0$;

2. $g_* = 0, \quad g'_*/32\pi = -\Delta;$ $\tag{6.176}$

3. $g_*/32\pi = 2\varepsilon(2\varepsilon + 3\Delta)/9(\varepsilon + \Delta), g'_*/32\pi = 2\varepsilon^2/9(\varepsilon + \Delta).$

In [220] point No. 2 is referred to as the thermal point, and No. 3 as the Kolmogorov point.

The standard analysis (Sec. 1.42) shows that point No. 1 is IR-attractive in the quadrant $\varepsilon < 0$, $\Delta > 0$, point No. 2 in the sector $\Delta < 0$, $2\varepsilon + 3\Delta < 0$, and point No. 3 in the sector $\varepsilon > 0$, $2\varepsilon + 3\Delta > 0$. The boundaries between these regions are crossover lines at which the IR-regime changes owing to the change of the fixed point governing a particular regime.

In the theory of turbulence, the only interesting region is $\varepsilon > 0$ and $\Delta \geq 0$ (i.e., $d \geq 2$), in which only the Kolmogorov point No. 3 is IR-stable. We see from (6.173) that the exact equation (6.42) remains valid at this point, and therefore equations (6.43) for the critical dimensions of the basic quantities also remain valid. This is the main result of [220]: the appearance of a new UV divergence for $d = 2$ does not affect the equations (6.43) for the critical dimensions [we recall that for $\varepsilon = 2$ they take the Kolmogorov values (6.8)].

6.19 Langmuir turbulence of a plasma

The Langmuir turbulence of a plasma as a physical phenomenon differs greatly from the ordinary developed turbulence of a liquid or a gas. However, this is an interesting stochastic case with a nontrivial IR asymptote (the dielectric constant has singularities near the Langmuir frequency for small wave vectors) which also can be studied using the standard RG technique.

Let us briefly state the problem. In the simplest model a plasma is treated as a mixture of two gases, electrons (e) and ions (i) with charges $e < 0$ for electrons and $-e > 0$ for ions, with masses $m_e \ll m_i$, equilibrium temperatures $T_e \gg T_i$, and equilibrium concentrations $n_e = n_i$. Two types of oscillation are possible in a plasma: low-frequency ion–sound waves with ordinary dispersion law $\omega(k) = ck$, where $c = (T_e/m_i)^{1/2}$ is the speed of sound, and high-frequency Langmuir oscillations with dispersion law $\omega^2(k) = \omega_e^2[1 + 3r_e^2k^2]$, where ω_e is the Langmuir frequency and r_e is the Debye radius ($\omega_e^2 = 4\pi e^2 n_e/m_e$, $r_e^2 = T_e/m_e\omega_e^2$). The interaction of these waves is described by the Zakharov equations [223] (see below for details), and most of the work on this topic is devoted to the study of specific solutions of these equations (solitons, collapse of Langmuir waves, and so on [223]–[225]).

Here we are interested in a different aspect of the problem [226], namely, the stochastic properties of the stationary regime. Under realistic conditions the system acquires energy from an external source (for example, energy pumping by an external resonance field). This energy is then redistributed among the various degrees of freedom via all their interaction mechanisms and in the end is dissipated as heat. As usual, one can attempt to model all these complicated processes by introducing a suitable random noise along with the dissipation (which is also absent in the original Zakharov equations), because in the stationary regime the gain and loss of energy by the system must cancel each other out. This formulation of the problem was first studied in [226]; here we present an improved version [227].

Let us begin with the basic equations. Wave processes correspond to small deviations of the densities from their equilibrium values: $\delta n_i \equiv \theta$, $\delta n_e \equiv \theta + \theta'$. The contribution θ represents ion–sound waves conserving electrical neutrality, and θ' represents Langmuir oscillations of the charge density. The latter are conveniently characterized by the longitudinal electric field they produce, $E_s = -\partial_s\psi$, with the potential ψ determined by the Poisson equation: $\partial^2\psi = -4\pi e\theta'$. The interaction of these waves is described by the Zakharov equations [223] with the addition [226] of the random noise E_s^{stoch}, which also must be assumed to be a longitudinal vector:

$$LE_s = \omega_e^2[n_e^{-1}\theta E_s + E_s^{\text{stoch}}], \quad L \equiv -\partial_t^2 - \omega_e^2(1 - 3r_e^2\partial^2), \tag{6.177}$$

$$[\partial_t^2 - c^2\partial^2]\theta = m_i^{-1}\partial^2[\overline{E^2}/8\pi]. \tag{6.178}$$

The contribution $\sim \overline{E^2}$ is the high-frequency pressure generated by the Langmuir oscillations, and the overline here and below denotes the average over rapid oscillations. When the noise and nonlinearities are neglected, (6.177) describes high-frequency Langmuir oscillations, and (6.178) describes low-frequency sound waves.

To isolate the rapid oscillations we write the solution of (6.177) as

$$E = \frac{1}{2} w_e m_e^{1/2} [\varphi \exp(-iw_e t) + \overset{+}{\varphi} \exp(iw_e t)] \tag{6.179}$$

and similarly for the random field E^{stoch}. The complex-conjugate vector amplitudes $\varphi, \overset{+}{\varphi}$ are assumed to vary slowly compared to the exponentials in (6.179). Therefore, when they are substituted into (6.177) their second derivatives with respect to time can be neglected and the coefficients of each of the two fast harmonics can be equated separately. The substitution $\overline{E^2} = m_e w_e^2 \overset{+}{\varphi} \varphi / 2$ is made in (6.178). After these transformations, (6.177) and (6.178) take the form

$$[i\partial_t + \nu_0(\partial^2 - \lambda_1 \theta)]\varphi = w_e \varphi^{\text{stoch}}/2, \tag{6.180}$$

$$[\partial_t^2 - c^2 \partial^2]\theta = \lambda_2 \partial^2(\overset{+}{\varphi}\varphi), \tag{6.181}$$

where $\nu_0 = 3w_e r_e^2/2$, $\lambda_1 = 1/3n_e r_e^2$, $\lambda_2 = m_e w_e^2/16\pi m_i$ are known constants, and φ^{stoch} is the corresponding coefficient in the representation of E^{stoch} analogous to (6.179). The solution of (6.180) is specified by the natural retardation condition, which is equivalent to introducing an infinitesimal imaginary term $-i0$ into the parameter ν_0.

The desired IR asymptote corresponds to small k and ω in (6.180), i.e., $k \simeq 0$ and $\omega^2 \simeq w_e^2$ in (6.177). In this region (6.181) can be simplified considerably by discarding the contribution $\partial_t^2 \sim \omega^2$, because (6.180) gives $\omega \sim k^2$ and so the term with $\partial_t^2 \sim \omega^2 \sim k^4$ in (6.181) is unimportant for $k \to 0$ compared to $\partial^2 \sim k^2$. From the viewpoint of physics this implies neglecting the retardation of the ion–sound waves, which is legitimate if their speed is much greater than the characteristic speed ω/k in (6.180). The dispersion law $\omega \sim k^2$ in this equation guarantees that ω/k is small for sufficiently small k.

Equation (6.181) can be solved exactly in this "static" approximation: $\theta = -c^{-2}\lambda_2 \overset{+}{\varphi}\varphi$. The substitution of this solution into (6.180) leads to a standard stochastic problem of the type (5.1):

$$[i\partial_t + \nu_0(\partial^2 + g_0'\overset{+}{\varphi}\varphi)]\varphi = \eta, \qquad \eta \equiv w_e \varphi^{\text{stoch}}/2 \tag{6.182}$$

with parameters $\nu_0 = 3w_e r_e^2/2$, $g_0' = 1/48\pi n_e r_e^4$.

To complete the formulation of the problem we still need to specify the correlator $D_{is}(x, x') \equiv \langle \hat{\eta}_i(x)\hat{\eta}_s(x') \rangle$ of the random force η in (6.182). It is chosen [226], [227] to be the usual white noise:

$$D_{is}(x, x') = c_0' P_{is}^{\|} \delta(x - x'), \tag{6.183}$$

where c_0' is a positive constant and $P_{is}^{\|} = \partial_i \partial_s/\partial^2$ is the longitudinal projector.

On the basis of physical considerations it would be more correct to assume that the noise η is concentrated in a certain frequency band near the origin, which in (6.177) corresponds to the neighborhood of the Langmuir frequency (resonance pumping). The IR singularities of interest are determined by precisely this region. Inside it the white noise (6.183) coincides with the correct distribution, and so it can be hoped that the replacement of the latter by the former does not affect the leading IR singularities.

When introducing energy pumping via noise it is also necessary to introduce a proportional dissipation. This is done [226], [227] by going to the complex parameter $\nu_0 = a_0(1 - iu_0)$ with the needed sign $u_0 > 0$. The coefficient c_0' in (6.183) determining the strength of the pumping must then be assumed to be proportional to the imaginary part of ν_0: $c_0' = 2a_0 u_0 Q$, where Q is a positive dimensionless coefficient. This coefficient can be completely eliminated by an appropriate dilatation of the quantities φ and η in (6.182). In such a dilatation $c_0' \to c_0'/Q \equiv c_0$ and $g_0' \to g_0' Q \equiv g_0$, so that we arrive at the problem with parameters

$$\nu_0 = a_0(1 - iu_0), \quad c_0 = -2\,\mathrm{Im}\,\nu_0 = 2a_0 u_0, \quad g_0 = g_0' c_0'/c_0. \tag{6.184}$$

According to the general rules of Sec. 5.3, the stochastic problem (6.182) modified in this manner corresponds to a quantum field model with unrenormalized action functional

$$S(\phi) = c_0 \overset{+}{\varphi}{}' \varphi' - \overset{+}{\varphi}{}'[i\partial_t + \nu_0(\partial^2 + g_0 \overset{+}{\varphi}\varphi)]\varphi + \text{H.c.} \tag{6.185}$$

Here and below +H.c. denotes the addition of the term that is the complex conjugate of the preceding expression. The model (6.185) describes a system of four complex fields $\phi = \varphi, \overset{+}{\varphi}, \varphi', \overset{+}{\varphi}{}'$, where all are longitudinal vectors.

The most interesting physical quantity in this problem is the longitudinal dielectric constant $\varepsilon_\|(\omega, k)$. It is expressed as

$$\varepsilon_\|^{-1}(\omega, k) - 1 = \omega_e[W(\omega - \omega_e, k) + W^*(-\omega - \omega_e, k)]/2 \tag{6.186}$$

in terms of the linear response function (Sec. 5.1) of the model (6.185) [the dilatations of φ and η made in going to (6.185) do not affect this function]:

$$\langle \widehat{\varphi}_i(x) \overset{+}{\varphi}{}'_s(x') \rangle = (2\pi)^{-d-1} \int d\omega \int dk\, P_{is}^\| W(\omega, k) \exp[ik(\mathbf{x} - \mathbf{x}') - i\omega(t - t')]. \tag{6.187}$$

The zeroth approximation W_0 for the response function coincides with the corresponding bare propagator for the action (6.185). It is easily found using the general rules discussed in Sec. 5.3:

$$W_0(\omega, k) = [\omega - \nu_0 k^2]^{-1}. \tag{6.188}$$

We see from this that at the IR asymptote $k \to 0$, $\omega \to 0$ the response function $W(\omega, k)$ diverges. It then follows from (6.186) that the quantity $\varepsilon_\|(\omega, 0)$ tends to zero for $\omega \to \pm\omega_e$, so that

$$\varepsilon_\|^{-1}(\omega + \omega_e, k) \simeq \omega_e W(\omega, k)/2 \tag{6.189}$$

in the vicinity of the Langmuir frequency ω_e [i.e., for small ω and k in (6.189)] and, similarly, $\varepsilon_{\parallel}^{-1}(\omega - \omega_e, k) \simeq \omega_e W^*(-\omega, k)/2$ near $-\omega_e$.

Therefore, the problem the theory must address is that of the IR asymptote of small ω and k for the response function $W(\omega, k)$ in (6.189). As usual, this problem can be solved by using the RG technique. The canonical dimensions of the fields and parameters in the functional (6.185) are easily found from the general rules of Sec. 5.14, and for the bare charge g_0 we obtain $d[g_0] = 4 - d$. Therefore, the model (6.185) is logarithmic for $d = 4$, so that it is natural to study it using dimensional regularization $d = 4 - 2\varepsilon$ with ordinary renormalization in the MS scheme.

The analysis of UV divergences performed in [227] using the standard rules of Sec. 5.15 showed that the model (6.185) is not multiplicatively renormalizable. In addition to the bare structure $(\overset{+}{\varphi}{}'\varphi)(\overset{+}{\varphi}\varphi) \equiv \overset{+}{\varphi}{}'_i \varphi_i \overset{+}{\varphi}_s \varphi_s$, a second vertex structure $(\overset{+}{\varphi}{}'\overset{+}{\varphi})(\varphi\varphi) \equiv \overset{+}{\varphi}{}'_i \overset{+}{\varphi}_i \varphi_s \varphi_s$ appears as a counterterm. Therefore, to ensure that the model is multiplicatively renormalizable (which is necessary in order to use the standard RG technique), we must add to the original interaction in (6.185) the new structure arising as a counterterm with an independent coefficient, which plays the role of a second bare charge. We note that the new vertex counterterm appears already in the simplest one-loop approximation. This was not taken into account in [226], where this stochastic problem was first studied.

Introducing the second interaction into (6.185), we find that

$$
\begin{aligned}
S(\phi) = c_0 \overset{+}{\varphi}{}'\varphi' - \overset{+}{\varphi}{}'[i\partial_t + \nu_0 \partial^2]\varphi + \text{H.c.} - \\
-\nu_0 [g_{10}(\overset{+}{\varphi}{}'\varphi)(\overset{+}{\varphi}\varphi) + g_{20}(\overset{+}{\varphi}{}'\overset{+}{\varphi})(\varphi\varphi)/2] + \text{H.c.}
\end{aligned}
\tag{6.190}
$$

with ν_0 and c_0 from (6.184) and two bare charges g_{i0} with identical canonical dimensions $d[g_{i0}] = 4 - d = 2\varepsilon$, $i = 1, 2$.

The renormalized analog of (6.190) containing all the needed counterterms is the functional [227]

$$
\begin{aligned}
S_{\text{R}}(\phi) = Z_1 c \overset{+}{\varphi}{}'\varphi' - \overset{+}{\varphi}{}'[Z_2 i\partial_t + Z_3 \nu \partial^2]\varphi + \text{H.c.} - \\
-\nu\mu^{2\varepsilon}[Z_4 g_1(\overset{+}{\varphi}{}'\varphi)(\overset{+}{\varphi}\varphi) + Z_5 g_2(\overset{+}{\varphi}{}'\overset{+}{\varphi})(\varphi\varphi)/2] + \text{H.c.}
\end{aligned}
\tag{6.191}
$$

with dimensionless renormalized charges $g_{1,2}$ and renormalized analogs of the parameters (6.184): $\nu = a(1 - iu)$, $c = -2\,\text{Im}\,\nu = 2au$.

Calculation shows that for complex ν the constant Z_1 in (6.191) is real and all the other constants Z are complex. Therefore, to ensure that the renormalization is multiplicative, in general it is also necessary to assume that the charges in (6.190) and (6.191) are complex. The following parameterization was used in [227]:

$$
\nu = a(1 - iu), \quad g_k = 16\pi^2(g'_k + iug''_k), \quad k = 1, 2
\tag{6.192}
$$

with real g'_k and g''_k.

It is obvious for complex g_k and ν that the action (6.191) is obtained from (6.190) by the standard procedure of multiplicative renormalization of the fields and parameters, and so the usual RG technique is applicable in its entirety to this model. In terms of real variables, the charges (i.e., the parameters on which the constants Z depend) are g'_k, g''_k, and u from (6.192), i.e., this model has five charges. The only independent parameter like a mass in the action (6.191) is $a = \mathrm{Re}\nu$ in (6.192) [we recall that the coefficient c in (6.191) is expressed in terms of a and u as $c = 2au$].

In [226] only the one-loop β functions were calculated neglecting the second interaction, and in [227] the one-loop β functions of all the above five charges were calculated, along with the two-loop anomalous dimensions γ_F of the parameter a and all the fields ϕ (the one-loop graphs do not contribute to these γ_F).

Analysis of the resulting β functions showed [227] that they have five fixed points. However, only one of these can be IR-stable, namely, the IR-attractive focus (Sec. 1.42) with coordinates $g'_{1*} = 0.458 \cdot (2\varepsilon)$, $g''_{1*} = 0.763 \cdot (2\varepsilon)$, $g'_{2*} = -1.431 \cdot (2\varepsilon)$, $g''_{2*} = -0.702 \cdot (2\varepsilon)$, $u_* = 1.737$ for $d = 4 - 2\varepsilon$. The anomalous dimensions γ_F^* of the parameter a and fields ϕ at this fixed point are found to be

$$\gamma_a^* = 0.0804(2\varepsilon)^2 + ..., \quad \gamma_\varphi^* = -0.3211(2\varepsilon)^2 + ...,$$

$$\gamma_{\varphi'}^* = [0.2874 - 0.0935i](2\varepsilon)^2 + ... \tag{6.193}$$

The analogous quantities for the fields $\overset{+}{\varphi}$, $\overset{+}{\varphi}{}'$ are obtained from the corresponding expressions (6.193) by complex conjugation.

Computer analysis of the equations (1.188) for this five-charge model has shown that when the initial data are chosen to conform with the original Zakharov equations ($\nu g_1 > 0$, $g_2 = 0$, $0 < u \ll 1$), the phase trajectory reaches the IR-attractive point, but for other initial data it can go off to infinity. Therefore, for realistic initial data the critical regime is actually reached. It is dissipative in the terminology of [223]–[225], and there cannot be any other regimes in this stochastic problem because there are no other IR-attractive points.

Using (6.189) it is easy to show that for this critical regime the function $\varepsilon_\parallel(\omega, k)$ has a scaling form at the IR asymptote $\omega \to \omega_e$, $k \to 0$:

$$\varepsilon_\parallel(\omega, k) = k^2 s^{-\gamma_a^* - \gamma_w^*} f(\xi), \quad \xi = (\omega - \omega_e) s^{\gamma_a^*}/ak^2, \tag{6.194}$$

where $s \equiv k/\mu$, $\gamma_w^* = \gamma_\varphi^* + \overline{\gamma_{\varphi'}^*}$ is the critical anomalous dimension of the response function (6.187) (the line over $\gamma_{\varphi'}^*$ denotes complex conjugation), and $f(\xi)$ is a scaling function of the argument given in (6.194). The equation $\xi = \mathrm{const}$ determines the dispersion law for Langmuir oscillations at the IR asymptote of small k:

$$\omega - \omega_e = \mathrm{const} \cdot k^{2-\gamma_a^*}. \tag{6.195}$$

For the three-dimensional problem with $2\varepsilon = 4 - d = 1$ in lowest order of the ε expansion from (6.193) we have

$$\gamma_a^* = 0.0804, \qquad \gamma_w^* = -0.0337 + 0.0935i. \qquad (6.196)$$

Judging from these numbers, the corrections from the anomalous dimensions are quantitatively small. However, the complex nature of γ_w^* in (6.194) must lead to a nontrivial qualitative effect: the appearance of oscillations of the dielectric constant $\varepsilon_\parallel$ at small k.

Relations (6.194) and (6.195) can be viewed as the main results of the RG analysis of this stochastic problem. It is discussed in more detail in [226] and [227].

Addendum

Here we briefly sketch some of the new results pertaining to the material in this book. They are arranged according to section.

Secs. 4.29, 4.39. The value of ω_2 in the nonlinear σ model (and thereby also in the O_n φ^4 model), i.e., the coefficient of $1/n^2$ in the $1/n$ expansion of the correction exponent ω, has recently been calculated [228].

Secs. 4.33, 4.34. In [229] relations like (4.273) and (4.285) were generalized to the case of second order in $1/n$. This made it possible to find the contributions $\sim 1/n^2$ to the critical dimensions (4.297) for the operator ϕ^l [229] and (4.300) for the operator ψ^l [230]. The latter result was obtained using the value of ω_2 already known [228] in the nonlinear σ model (the special case $l = 2$).

Sec. 4.36. Nontrivial new technical tricks for calculating multiloop massless graphs can be found in [228], [231], and [232].

Secs. 4.45, 4.49. The question of whether the GN model in the $2 + \varepsilon$ scheme is internally closed has now been answered, in the negative. By direct calculation it has been shown [233] that the GN interaction V_0 generates a new vertex counterterm of the type V_3 from (4.486) already at three-loop order (this was not noticed by the authors of the earlier studies [142], [144], in which the three-loop β function of the GN model was calculated). Therefore, the GN model with the single interaction V_0 in the $2 + \varepsilon$ scheme is not multiplicatively renormalizable and should be studied within the context of the complete four-fermion model [151], including all the vertices (4.486) (see Sec. 4.49). The critical regime of the GN model corresponds to one of the fixed points of this complete model [152], and so the concept of critical exponents of the GN model and their $2 + \varepsilon$ expansions remains meaningful, in contrast to the concept of RG functions of the GN model. We also note that the proof of critical conformal invariance presented in Sec. 4.47 carries over without change to the Green functions of the complete model at its GN fixed point, because all the additional interactions (4.486) are not dangerous operators in the sense of the definition in Sec. 4.41.

Sec. 5.25. The results of the two-loop calculation of the basic critical exponents of model H in [171] and the two correction exponents ω in [158], which were regarded as classical and were widely quoted, have been shown to contain numerical errors (we recall that in Sec. 5.25 we presented the results of [171] converted to the MS scheme; a non-MS scheme was used for the calculations in [171]). The accurate calculation of the two-loop contributions in model H using the MS scheme [234] leads to the following changes in the coefficients of the contributions $\sim u_2^2$ in the RG functions

(5.231) and, accordingly, of the contributions $\sim (2\varepsilon)^2$ in (5.232) and (5.233): $0.160 \to 0.007789$ in γ_λ, $-0.206 \to -0.05987$ in γ_{g_2}, $-0.0635 \to -0.02678$ in u_{2*}, and $-0.013 \to -0.018544$ in γ_λ^*. The most important is the correction to γ_λ^*, because this quantity is related by (5.234) to the subtraction-scheme-independent critical exponent $z \equiv \Delta_\omega$, in terms of which the traditional exponents x_λ and $x_{\overline{\eta}}$ [171] are expressed: $x_\lambda = 4 - z - \eta$, $x_{\overline{\eta}} = z - d$. These changes also affect the universal Kawasaki amplitude constant R [171], which was not studied in Sec. 5.25, and the two correction exponents ω_f and ω_W [158]. In the notation of Sec. 5.25 the exponent $\omega_f = \partial \beta_{g_2} / \partial g_2 |_{g*}$ is calculated from the known β function (5.231) of the charge g_2, and for the quantity ω_W introduced in [158] it has been shown [234] that this exponent is uniquely expressed in terms of z as $\omega_W = 2z - d - 2$. The coefficients of the two-loop contributions to these exponents obtained in this manner differ markedly from those given in [158]: for $d = 4 - 2\varepsilon$ from the data of [234] we have $\omega_f = 2\varepsilon + (2\varepsilon)^2 \cdot 0.005364$ and $\omega_W = 2 - 17(2\varepsilon)/19 + (2\varepsilon)^2 \cdot 0.0371$, whereas in [158] $0.005364 \to 0.121$ in ω_f and $0.0371 \to 0.163$ in ω_W. It was also shown in [234] that the two correction exponents [158] are insufficient for describing the corrections to scaling, because ω_W is only one of a series of correction exponents $\omega = 2 + O(\varepsilon)$ generated by the family of all those operators with which the IR-irrelevant operator $F_1 = v' \partial_t v$ in the functional (5.210) can mix in renormalization.

Secs. 6.9, 6.12, 6.13. The Kraichnan model [239] has recently received a great deal of attention in the theory of turbulence (see [235] to [238] and references therein). This model describes the mixing of a passive scalar admixture (the field θ) by a random velocity field v with a simple Gaussian distribution, the correlator of which is assumed to be a δ function of time. The model is defined by the usual stochastic equation (6.103) with the addition of self-noise η_θ with correlator $\langle \widehat{\eta}_\theta \widehat{\eta}_\theta \rangle = \delta(t - t') f(Mr)$ in $t, \mathbf{x}$ space [M^{-1} is the corresponding external scale and $f(0) = \text{const}$]. This differs from the problem studied in Sec. 6.13 in that in the latter the random velocity field v in the covariant derivative $\nabla_t = \partial_t + (v\partial)$ was assumed to be determined by the stochastic Navier–Stokes equation (6.1), whereas in the Kraichnan model it is assumed to be described by a simple Gaussian distribution with given correlator which is a time δ function: $\langle \widehat{v}_i \widehat{v}_s \rangle = \delta(t - t') P_{is} C_0 k^{-d-\varepsilon}$ (t, k space), where P_{is} is the transverse projector in the vector indices of the fields v, C_0 is the amplitude factor, and ε is an arbitrary parameter. Formally, its real value can be taken to be $\varepsilon_{\rm r} = 4/3$, which corresponds to $\varepsilon_{\rm r} = 2$ in the dynamical correlator (6.36a) for $\omega = 0$ (a δ function in time) for the massless model, assuming that it exists.

In this problem one studies correlators of the type (6.98) for the field θ and the dissipation operator $W_{\rm dis} = \partial \theta \cdot \partial \theta$, along with their generalizations: the static correlators $\langle F_n F_m \rangle$ for $F_n \equiv W_{\rm dis}^n$, i.e., powers of the dissipation operator. For all these objects we obtain representations like (6.100) with exponents q_n, μ and analogs of μ for the correlators $\langle F_n F_m \rangle$ which can be calculated. This is the main difference between this problem and the analogous one in the theory of turbulence. In the terminology of Sec. 6.9 this is related to the fact that in the Kraichnan model for quantities like (6.98) there exists within the ε expansion a single principal dangerous operator with minimum value $\Delta_F < 0$ that determines the leading contribution of the asymptote $Mr \to 0$ in (6.102), i.e., the difficult problem of summing the contributions of infinite series of dangerous operators is absent here. For the quantities $\langle [\theta(x_1) - \theta(x_2)]^{2n} \rangle$ the principal dangerous operator is F_n, while for $\langle F_n(x_1) F_m(x_2) \rangle$ it is the operator F_{n+m}, so that the exponents in relations like (6.100) for these objects are determined by the critical dimensions $\Delta_n = \Delta^{as}[F_n]$ of the operators associated with F_n [see (6.55)]. The quantities Δ_n were calculated in first order in ε in [235], and in first order in $1/d$ in [236]. This was all done without the RG apparatus, using tricks that work specifically for that model. The standard RG technique and the SDE (Sec. 6.9) were applied to this problem in [238], and the critical dimensions Δ_n were calculated through order ε^2 [238] and then ε^3 [240].

Bibliography

CHAPTER 1

[1] Andrews, T. (1869) *Philos. Trans. R. Soc. London* **159**, 575.

[2] Van der Waals, J. D. (1873) Ph. D. Thesis, University of Leiden.

[3] Weiss, P. (1907) *J. Phys. (Paris) (4)* **6**, 661.

[4] Landau, L. D. (1937) *Zh. Éksp. Teor. Fiz.* **7**, 19, 627, 1232 [in Russian].

[5] Onsager, L. (1944) *Phys. Rev.* **65**, 117.

[6] Guggenheim, E. A. (1943) *J. Chem. Phys.* **13**, 253.

[7] Brush, S. G. (1967) *Rev. Mod. Phys.* **39**, 883.

[8] Lee, T. D. and Yang, C. N. (1952) *Phys. Rev.* **87**, 410.

[9] Domb, C. (1960) *Adv. Phys.* **9**, 149.

[10] Lentz, W. (1920) *Z. Phys.* **21**, 613.

[11] Ising, E. (1925) *Z. Phys.* **31**, 253.

[12] Heisenberg, W. (1928) *Z. Phys.* **49**, 619.

[13] Peierls, R. (1936) *Proc. Cambridge Philos. Soc.* **32**, 477.

[14] Kramers, H. A. and Wannier, G. H. (1941) *Phys. Rev.* **60**, 263.

[15] Shedlorsky, T. and Montroll, E. (1963) *J. Math. Phys.* **4**, 145.

[16] Onsager, L. (1949) *Nuovo Cimento (Suppl.)* **6**, 261.

[17] Montroll, E. W., Potts, R. B., and Ward, J. C. (1963) *J. Math. Phys.* **4**, 308.

[18] Yang, C. N. (1952) *Phys. Rev.* **85**, 808.

[19] Kramers, H. A. and Wannier, G. H. (1941) *Phys. Rev.* **60**, 252.

[20] Montroll, E. (1941) *J. Chem. Phys.* **9**, 706.

[21] Kaufman, B. (1949) *Phys. Rev.* **76**, 1232.

[22] Kaufman, B. and Onsager, L. (1949) *Phys. Rev.* **76**, 1244.

[23] Popov, V. N. (1983) *Functional Integrals in Quantum Field Theory and Statistical Physics*. Reidel, Dordrecht [Russian ed. 1976].

[24] Fisher, M. (1965) *The Nature of Critical Points*. University of Colorado Press, Boulder.

[25] Domb, C. and Hunter, D. L. (1965) *Proc. Phys. Soc. London* **86**, 1147.

[26] Widom, B. (1965) *J. Chem. Phys.* **43**, 3892, 3898.

[27] Patashinskiĭ, A. Z. and Pokrovskiĭ, V. L. (1966) *Sov. Phys. JETP* **23**, 292.

[28] Kadanoff, L. P. (1966) *Physics* **2**, 263.

[29] Patashinskiĭ, A. Z. and Pokrovskiĭ, V. L. (1964) *Sov. Phys. JETP* **19**, 677.

[30] Stueckelberg, E. and Petermann, A. (1953) *Helv. Phys. Acta* **26**, 499.

[31] Gell-Mann, M. and Low, F. E. (1954) *Phys. Rev.* **95**, 1300.

[32] Landau, L. D., Abrikosov, A. A., and Khalatnikov, I. M. (1954) *Dokl. Akad. Nauk SSSR* **95**, 773, 1177; **96**, 261 [in Russian].

[33] Bogolyubov, N. N. and Shirkov, D. V. (1955) *Dokl. Akad. Nauk SSSR* **103**, 203, 391 [in Russian].

[34] Shirkov, D. V. (1955) *Dokl. Akad. Nauk SSSR* **105**, 972 [in Russian].

[35] Bogoliubov, N. N. and Shirkov, D. V. (1980) *Introduction to the Theory of Quantized Fields*, 3rd ed. Wiley, New York [Russian eds. 1957, 1973, 1976, 1984].

[36] Gross, D. J. and Wilczek, F. (1973) *Phys. Rev. Lett.* **30**, 1343; Politzer, H. D. (1973) *Phys. Rev. Lett.* **30**, 1346.

[37] Wilson, K. G. (1971) *Phys. Rev. B* **4**, 3174.

[38] Larkin, A. I. and Khmel'nitskiĭ, D. E. (1969) *Sov. Phys. JETP* **29**, 1123.

[39] Wilson, K. G. (1971) *Phys. Rev. B* **4**, 3184.

[40] Wilson, K. G. and Fisher, M. E. (1972) *Phys. Rev. Lett.* **28**, 240.

[41] Wilson, K. G. (1972) *Phys. Rev. Lett.* **28**, 548.

[42] Wilson, K. G. and Kogut, J. B. (1974) *Phys. Rep. C* **12**, 75.

[43] Brezin, E., Le Guillou, J. C., and Zinn-Justin, J. (1976) Field theoretical approach to critical phenomena. *Phase Transitions and Critical Phenomena*, Vol. 6 (ed. C. Domb and M. S. Green). Academic Press, London.

[44] Amit, D. J. (1978) *Field Theory, the Renormalization Group, and Critical Phenomena*. McGraw-Hill, New York.

[45] Zinn-Justin, J. (1989, 1993, 1996, 2002) *Quantum Field Theory and Critical Phenomena*. Clarendon Press, Oxford.

[46] Lipatov, L. N. (1977) *Sov. Phys. JETP* **45**, 216.

[47] Brezin, E., Le Guillou, J. C., and Zinn-Justin, J. (1977) *Phys. Rev. D* **15**, 1544.

[48] Ma, S. (1976) *Modern Theory of Critical Phenomena*. Benjamin, Reading, MA.

[49] Patashinskiĭ, A. Z. and Pokrovskiĭ, V. L. (1979) *Fluctuation Theory of Phase Transitions*. Pergamon Press, Oxford [Russian eds. 1973, 1982].

[50] Stanley, H. E. (1971) *Introduction to Phase Transitions and Critical Phenomena*. Clarendon Press, Oxford.

[51] Landau, L. D. and Lifshitz, E. M. (1980) *Statistical Physics*, Part 1, 3rd ed. Pergamon Press, Oxford.

[52] Anisimov, M. A. (1987) *Critical Phenomena in Liquids and Liquid Crystals* [in Russian]. Nauka, Moscow.

[53] Vasiliev, A. N. (1998) *Functional Methods in Quantum Field Theory and Statistical Physics*. Gordon and Breach, New York [Russian ed. 1976].

[54] Anisimov, M. A., Gorodetskiĭ, E. E., and Zaprudskiĭ, V. M. (1981) *Sov. Phys. Usp.* **24**, 57.

[55] Lawrie, I. D. and Sarbach, S. (1984) Theory of tricritical points. *Phase Transitions and Critical Phenomena*, Vol. 9 (ed. C. Domb and J. L. Lebowitz). Academic Press, London.

[56] Aharony, A. and Fisher, M. E. (1973) *Phys. Rev. B* **8**, 3323.

[57] De Gennes, P. G. (1972). *Phys. Lett. A* **38**, 339.

[58] Aharony, A. (1973) *Phys. Rev. B* **8**, 3363.

[59] Brezin, E. and Zinn-Justin, J. (1976) *Phys. Rev. B* **13**, 251.

[60] Collins, J. C. (1984) *Renormalization: an Introduction to Renormalization, the Renormalization Group, and the Operator–Product Expansion*. Cambridge University Press, Cambridge.

[61] 't Hooft, G. (1973) *Nucl. Phys. B* **61**, 455.

[62] Di Castro, C. (1972) *Lett. Nuovo Cimento* **5**, 69.

[63] Brezin, E., Le Guillou, J. C., Zinn-Justin, J., and Nickel, B. G. (1973) *Phys. Lett. A* **44**, 227.

[64] Vladimirov, A. A., Kazakov, D. I., and Tarasov, O. V. (1979) *Sov. Phys. JETP* **50**, 521.

[65] Chetyrkin, K. G., Kataev, A. L., and Tkachov, F. V. (1981) *Phys. Lett. B* **99**, 147; **101**, 457(E).

[66] Gorishny, S. S., Larin, S. A., and Tkachov, F. V. (1984) *Phys. Lett. A* **101**, 120.

[67] Kazakov, D. I. (1983) *Phys. Lett. B* **133**, 406.

[68] Kleinert, H., Neu, J., Shulte-Frohlinde, V., Chetyrkin, K. G., and Larin, S. A. (1991, 1993) *Phys. Lett. B* **272**, 39; **319**, 545(E).

[69] Le Guillou, J. C. and Zinn-Justin, J. (1977) *Phys. Rev. Lett.* **39**, 95.

[70] Le Guillou, J. C. and Zinn-Justin, J. (1985, 1987, 1989) *J. Phys. Lett. (Paris)* **46**, L137; *J. Phys. (Paris)* **48**, 19; **50**, 1365; Bervillier, C. (1986) *Phys. Rev. B* **34**, 8141.

[71] Mermin, N. D. and Wagner, H. (1966) *Phys. Rev. Lett.* **17**, 1133.

[72] Nienhuis, B. (1982) *Phys. Rev. Lett.* **49**, 1062.

[73] Baker, G. A., Nickel, B. G., and Meiron, D. I. (1978) *Phys. Rev. B* **17**, 1365.

[74] Yorke, J. and Yorke, E. (1979) *J. Stat. Phys.* **21**, 263.

CHAPTER 2

[75] Feynman, R. P. and Hibbs, A. R. (1965) *Quantum Mechanics and Path Integrals*. McGraw-Hill, New York.

[76] Berezin, F. A. (1971) *Theor. Math. Phys.* **6**, 141.

[77] Vasil'ev, A. N., Perekalin, M. M., and Pis'mak, Yu. M. (1983) *Theor. Math. Phys.* **55**, 529.

[78] Polyakov, A. M. (1970) *JETP Lett.* **12**, 381.

[79] Jevicki, A. (1977) *Nucl. Phys. B* **127**, 125; Lüscher, M. (1978) *Phys. Lett. B* **78**, 465; Perelomov, A. M. (1987) *Phys. Rep.* **146**, 135.

[80] Zakharov, V. E. and Mikhaĭlov, A. V. (1978) *Sov. Phys. JETP* **47**, 1017; Brezin, E., Itzykson, C., Zinn-Justin, J., and Zuber, J.-B. (1979) *Phys. Lett. B* **82**, 442.

[81] Emery, V. J. (1975) *Phys. Rev. B* **11**, 3397; Edwards, S. F. and Anderson, P. W. (1975) *J. Phys. F* **5**, 965.

CHAPTER 3

[82] Speer, E. R. (1968) *J. Math. Phys.* **9**, 1404.

[83] 't Hooft, G. and Veltman, M. (1972) *Nucl. Phys. B* **44**, 189.

[84] Zav'yalov, O. I. (1979) *Renormalized Feynman Diagrams* [in Russian]. Nauka, Moscow.

[85] Hepp, K. (1969) *Théorie de la Renormalisation*. Springer-Verlag, Berlin.

[86] Vladimirov, A. A. (1980) *Theor. Math. Phys.* **43**, 417.

[87] Vasil'ev, A. N. (1989) *Theor. Math. Phys.* **81**, 1244.

[88] Chetyrkin, K. G. and Tkachov, F. V. (1982) *Phys. Lett. B* **114**, 340; Chetyrkin, K. G. and Smirnov, V. A. (1984) *Phys. Lett. B* **144**, 419; Smirnov, V. A. and Chetyrkin, K. G. (1985) *Theor. Math. Phys.* **63**, 462.

[89] Vasil'ev, A. N., Pis'mak, Yu. M., and Honkonen, Yu. R. (1981) *Theor. Math. Phys.* **47**, 465.

[90] Chetyrkin, K. G. and Tkachev, F. V. (1979) A new approach to calculating multiloop Feynman integrals. Preprint P-0118. Institute for Nuclear Research, USSR Academy of Sciences, Moscow [in Russian]; *see also* Chetyrkin, K. G., Kataev, A. L., and Tkachov, F. V. (1980) *Nucl. Phys. B* **174**, 345.

[91] Wilson, K. G. (1969, 1971) *Phys. Rev.* **179**, 1499; *Phys. Rev. D* **3**, 1818.

[92] Zimmermann, W. (1973) *Ann. Phys. (N.Y.)* **77**, 536, 570.

CHAPTER 4

[93] Prudnikov, A. P., Brychkov, Yu. A., and Marichev, O. I. (1986) *Integrals and Series*, Vol. 2: Special Functions. Gordon and Breach, New York [Russian ed. 1983].

[94] Fisher, M. E. and Aharony, A. (1974) *Phys. Rev. B* **10**, 2818.

[95] Fisher, M. E. and Langer, J. S. (1968) *Phys. Rev. Lett.* **20**, 665.

[96] Bray, A. J. (1976) *Phys. Rev. B* **14**, 1428.

[97] Antonov, N. V., Vasil'ev, A. N., and Stepanenko, A. S. (1991) *Theor. Math. Phys.* **88**, 779.

[98] Chang, R. F., Burstyn, H., and Sengers, J. V. (1979) *Phys. Rev. A* **19**, 866.

[99] Brezin, E. and Wallace, D. J. (1973) *Phys. Rev. B* **7**, 1967.

[100] Wegner, F. (1979) *Z. Phys. B* **35**, 207; Schafer, L. and Wegner, F. (1980) *Z. Phys. B* **38**, 113; Hikami, S. (1985) *Prog. Theor. Phys. Suppl.* **84**, 120.

[101] Lee, P. A. and Ramakrishnan, T. V. (1985) *Rev. Mod. Phys.* **57**, 287.

[102] Efetov, K. B., Larkin, A. I., and Khmel'nitskiĭ, D. E. (1980) *Sov. Phys. JETP* **52**, 568; Kravtsov, V. E. and Lerner, I. V. (1985) *Sov. Phys. JETP* **61**, 758.

[103] Patashinskiĭ, A. Z. and Pokrovskiĭ, V. L. (1973) *Sov. Phys. JETP* **37**, 733.

[104] Nalimov, M. Yu. (1989) *Theor. Math. Phys.* **80**, 819.

[105] Harris, A. B. and Lubensky, T. C. (1974) *Phys. Rev. Lett.* **33**, 1540; Lubensky, T. C. (1975) *Phys. Rev. B* **9**, 3573.

[106] Khmel'nitskiĭ, D. E. (1975) *Sov. Phys. JETP* **41**, 981.

[107] Shalaev, B. N. (1977) *Sov. Phys. JETP* **26**, 1204; Jayaprakash, C. and Katz, H. J. (1977) *Phys. Rev. B* **16**, 3987.

[108] Sokolov, A. I. and Shalaev, B. N. (1981) *Sov. Phys. Solid State* **23**, 1200.

[109] Borin, V. F., Vasil'ev, A. N., and Nalimov, M. Yu. (1992) *Theor. Math. Phys.* **91**, 446.

[110] Lewis, A. L. and Adams, F. M. (1978) *Phys. Rev. B* **18**, 5099.

[111] Lawrie, T. D. (1985) *Nucl. Phys. B* **257** [FS14], 29.

[112] Macfarlane, A. J. and Woo, G. (1974, 1975) *Nucl. Phys. B* **77**, 91; **86**, 548(E).

[113] Wu, F. Y. (1982) *Rev. Mod. Phys.* **54**, 235.

[114] Brezin, E. and Zinn-Justin, J. (1976) *Phys. Rev. B* **14**, 3110.

[115] Brezin, E., Le Guillou, J. C., and Zinn-Justin, J. (1976) *Phys. Rev. D* **14**, 2615.

[116] Hikami, S. and Brezin, E. (1978) *J. Phys. A* **11**, 1141.

[117] Hikami, S. (1983) *Nucl. Phys. B* **215** [FS7], 555; Bernreuther, W. and Wegner, F. J. (1986) *Phys. Rev. Lett.* **57**, 1383; Wegner, F. (1989) *Nucl. Phys. B* **316**, 663.

[118] Brezin, E., Le Guillou, J. C., and Zinn-Justin, J. (1976) *Phys. Rev. B* **14**, 4976.

[119] Ma, S. K. (1973, 1974) *Phys. Rev. A* **7**, 2172; **10**, 1818.

[120] Abe, R. (1973) *Prog. Theor. Phys.* **49**, 1877.

[121] Kondor, I. and Temesvari, T. (1978) *J. Phys. Lett. (Paris)* **39**, L99; Kondor, I., Temesvari, T., and Herenyi, L. (1980) *Phys. Rev. B* **22**, 1451.

[122] Okabe, Y. and Oku, M. (1978) *Prog. Theor. Phys.* **60**, 1277.

[123] Vasil'ev, A. N., Pis'mak, Yu. M., and Honkonen, Yu. R. (1982) *Theor. Math. Phys.* **50**, 127.

[124] Fisher, M. E. and Aharony, A. (1973) *Phys. Rev. Lett.* **31**, 1238; Aharony, A. (1974) *Phys. Rev. B* **10**, 2834; Abe, R. and Hikami, S. (1974) *Prog. Theor. Phys.* **51**, 1041.

[125] Ma, S. K. (1973) *Rev. Mod. Phys.* **45**, 589; Wilson, K. G. (1973) *Phys. Rev. D* **7**, 2911.

[126] Vasil'ev, A. N. and Nalimov, M. Yu. (1983) *Theor. Math. Phys.* **55**, 423.

[127] Ruhl, W. and Lang, K. (1991) *Z. Phys. C* **51**, 127.

[128] Vasil'ev, A. N. and Stepanenko, A. S. (1993) *Theor. Math. Phys.* **95**, 471.

[129] Hikami, S. (1979, 1980) *Prog. Theor. Phys.* **62**, 226; **64**, 1425.

[130] Vasil'ev, A. N. and Nalimov, M. Yu. (1983) *Theor. Math. Phys.* **56**, 643; Vasil'ev, A. N., Nalimov, M. Yu., and Honkonen, Yu. R. (1984) *Theor. Math. Phys.* **58**, 111.

[131] D'Eramo, M., Peliti, L., and Parisi, G. (1971) *Lett. Nuovo Cimento* **26**, 878.

[132] Broadhurst, D. J. (1986) *Z. Phys. C* **32**, 249.

[133] Gorishnyĭ, S. G. and Isaev, A. P. (1985) *Theor. Math. Phys.* **62**, 232.

[134] Kazakov, D. I. (1985) *Theor. Math. Phys.* **62**, 84.

[135] Gracey, J. A. (1991) *Int. J. Mod. Phys. A* **6**, 395, 2775(E).

[136] Vasil'ev, A. N., Derkachev, S. É., Kivel', N. A., and Stepanenko, A. S. (1993) *Theor. Math. Phys.* **94**, 127.

[137] Gracey, J. A. (1994) *Int. J. Mod. Phys. A* **9**, 727.

[138] Parisi, G. (1972) *Lett. Nuovo Cimento* **4**, 777.

[139] Gross, D. and Neveu, A. (1974) *Phys. Rev. D* **10**, 3235.

[140] Zinn-Justin, J. (1991) *Nucl. Phys. B* **367**, 105.

[141] Wetzel, W. (1985) *Phys. Lett. B* **153**, 297.

[142] Luperini, C. and Rossi, P. (1991) *Ann. Phys. (N.Y.)* **212**, 371.

[143] Gracey, J. A. (1990) *Nucl. Phys. B* **341**, 403.

[144] Gracey, J. A. (1991) *Nucl. Phys. B* **367**, 657.

[145] Vasil'ev, A. N., Derkachev, S. É., Kivel', N. A., and Stepanenko, A. S. (1992) *Theor. Math. Phys.* **92**, 1047.

[146] Vasil'ev, A. N. and Stepanenko, A. S. (1993) *Theor. Math. Phys.* **97**, 1349.

[147] Gracey, J. A. (1994) *Int. J. Mod. Phys. A* **9**, 567.

[148] Kivel, N. A., Stepanenko, A. S., and Vasil'ev, A. N. (1994) *Nucl. Phys. B* **424** [FS], 619.

[149] Bernreuther, W. and Wegner, F. (1986) *Phys. Rev. Lett.* **57**, 1383; Gracey, J. A. (1991) *Phys. Lett. B* **262**, 49.

[150] Derkachov, S. E., Honkonen, J., and Pis'mak, Yu. M. (1990) *J. Phys. A* **23**, 5563.

[151] Bondi, A., Curci, G., Paffuti, G., and Rossi, P. (1990) *Ann. Phys. (N.Y.)* **199**, 268.

[152] Vasil'ev, A. N., Vyazovskiĭ, M. I., Derkachev, S. É., and Kivel', N. A. (1996) *Theor. Math. Phys.* **107**, 441, 710.

CHAPTER 5

[153] Hohenberg, P. C. and Halperin, B. I. (1977) *Rev. Mod. Phys.* **49**, 435.

[154] Monin, A. S. and Yaglom, A. M. (1971, 1975) *Statistical Fluid Mechanics*, Vols. 1 and 2. MIT Press, Cambridge, MA. [Russian eds. 1967, 1996].

[155] Wyld, H. W. (1961) *Ann. Phys. (N.Y.)* **14**, 143.

[156] Martin, P. C., Siggia, E. D., and Rose, H. A. (1973) *Phys. Rev. A* **8**, 423.

[157] Janssen, H. K. (1976) *Z. Phys. B* **23**, 377.

[158] De Dominicis, C. and Peliti, L. (1978) *Phys. Rev. B* **18**, 353.

[159] Adzhemyan, L. Ts., Vasil'ev, A. N., and Pis'mak, Yu. M. (1983) *Teor. Math. Phys.* **57**, 1131.

[160] Dohm, V. (1987) *J. Low Temp. Phys.* **69**, 51.

[161] Fixman, M. (1962) *J. Chem. Phys.* **36**, 310.

[162] Kadanoff, L. P. and Swift, J. (1968) *Phys. Rev.* **166**, 89.

[163] Kawasaki, K. (1967, 1970) *J. Phys. Chem. Solids* **28**, 1277; *Ann. Phys. (N.Y.)* **61**, 1.

[164] Bausch, R., Janssen, H. K., and Wagner, H. (1976) *Z. Phys.* B **24**, 113.

[165] Antonov, N. V. and Vasil'ev, A. N. (1984) *Theor. Math. Phys.* **60**, 671.

[166] Halperin, B. I., Hohenberg, P. C., and Ma, S. (1972) *Phys. Rev. Lett.* **29**, 1548.

[167] De Dominicis, C., Brezin, E., and Zinn-Justin, J. (1975) *Phys. Rev. B* **12**, 4945.

[168] De Dominicis, C., Ma, S., and Peliti, L. (1977) *Phys. Rev. B* **15**, 4313.

[169] Halperin, B. I., Hohenberg, P. C., and Ma, S. (1976) *Phys. Rev. B* **13**, 4119.

[170] Dohm, V. (1991) *Phys. Rev. B* **44**, 2697.

[171] Siggia, E. D., Halperin, B. I., and Hohenberg, P. C. (1976) *Phys. Rev. B* **13**, 2110.

[172] Landau, L. D. and Lifshitz, E. M. (1987) *Fluid Mechanics*, 2nd ed. Pergamon Press, Oxford [Russian eds. 1944, 1953, 1986].

[173] Roe, D., Wallace, B., and Meyer, H. (1974) *J. Low Temp. Phys.* **16**, 51; Roe, D. and Meyer, H. (1978) *J. Low Temp. Phys.* **30**, 91.

[174] Adzhemyan, L. Ts., Vasiljev, A. N., and Serdukov, A. V. (1998) *Int. J. Mod. Phys. B* **12**, 1255; *Sov. Phys. JETP* **87**, 934.

[175] Kroll, D. M. and Ruhland, J. M. (1980) *Phys. Lett. A* **80**, 45.

[176] Ferrel, R. A. and Bhattacharjee, J. K. (1981, 1982, 1985) *Phys. Lett. A* **86**, 109; **88**, 77; *Phys. Rev. A* **31**, 1788; Ferrel, R. A., Mirhoshen, B., and Bhattacharjee, J. K. (1987) *Phys. Rev. B* **35**, 4662.

[177] Dengler, R. and Schwabl, F. (1987) *Europhys. Lett.* **4**, 1233.

[178] Dengler, R. and Schwabl, F. (1987) *Z. Phys. B* **69**, 327.

[179] Folk, R. and Moser, G. (1995) *Phys. Rev. Lett.* **75**, 2706.

[180] Pankert, J. and Dohm, V. (1989) *Phys. Rev. B* **40**, 10842, 10856.

[181] Drossel, B. and Schwabl, F. (1993, 1994) *Z. Phys. B* **91**, 93; **95**, 141(E).

[182] Schorgg, A. M. and Schwabl, F. (1994) *Phys. Rev. B* **49**, 11682.

CHAPTER 6

[183] Adzhemyan, L. Ts., Antonov, N. V., and Vasiliev, A. N. (1999) *The Field Theoretic Renormalization Group in Fully Developed Turbulence.* Gordon and Breach, New York.

[184] Kolmogorov, A. N. (1941) *Dokl. Akad. Nauk SSSR* **30**, 299; Obukhov, A. M. (1941) *Dokl. Akad. Nauk SSSR* **32**, 22 [in Russian].

[185] Kraichnan, R. H. (1964, 1965, 1966) *Phys. Fluids* **7**, 1723; **8**, 575; **9**, 1728, 1884.

[186] De Dominicis, C. and Martin, P. C. (1979) *Phys. Rev. A* **19**, 419.

[187] Adzhemyan, L. Ts., Antonov, N. V., and Vasil'ev, A. N. (1989) *Sov. Phys. JETP* **68**, 733.

[188] Antonov, N. V. (1992) *Vestnik SPburg. Gos. Univ., Ser. Fiz. Khim.* No. 4, 6 [in Russian].

[189] Adzhemyan, L. Ts., Antonov, N. V., and Kim, T. L. (1994) *Theor. Math. Phys.* **100**, 1086.

[190] Adzhemyan, L. Ts., Vasil'ev, A. N., and Gnatich, M. (1988) *Theor. Math. Phys.* **74**, 115.

[191] Antonov, N. V. and Vasil'ev, A. N. (1997) *Theor. Math. Phys.* **110**, 97.

[192] Antonov, N. V. (1992) *Vestnik SPburg. Gos. Univ., Ser. Fiz. Khim.* No. 3, 3 [in Russian].

[193] Yakhot, V., She, Z.-S., and Orszag, S. A. (1989) *Phys. Fluids A* **1**, 289.

[194] Antonov, N. V., Borisenok, S. V., and Girina, V. I. (1996) *Theor. Math. Phys.* **106**, 75.

[195] Belinicher, V. I. and L'vov, V. S. (1987) *Sov. Phys. JETP* **66**, 303.

[196] Antonov, N. V. (1988, 1991) *Zap. Nauchn. Seminarov LOMI* **169**, 18; **189**, 15 [in Russian].

[197] L'vov, V. S. (1991) *Phys. Rep.* **207**, 1.

[198] Moiseev, S. S., Tur, A. V., and Yanovskiĭ, V. V. (1984) *Sov. Phys. Dokl.* **29**, 926.

[199] Yaglom, A. M. (1981) *Izv. Akad. Nauk SSSR, Fiz. Atmosf. Okeana* **17**, 1235 [in Russian].

[200] Kuznetsov, V. R., Praskovskiĭ, A. A., and Sabel'nikov, V. A. (1988) *Izv. Akad. Nauk SSSR, Ser. Mek. Zhidk. Gaza* **6**, 51 [in Russian].

[201] Yakhot, V. and Orszag, S. A. (1986) *Phys. Rev. Lett.* **57**, 1722.

[202] Yakhot, V. and Orszag, S. A. (1986) *J. Sci. Comput.* **1**, 3.

[203] Dannevik, W. P., Yakhot, V., and Orszag, S. A. (1987) *Phys. Fluids* **30**, 2021.

[204] Anselmet, F., Gagne, Y., Hopfinger, E., and Antonia, R. A. (1984) *J. Fluid Mech.* **140**, 63.

[205] Benzi, R., Ciliberto, S., Tripiccione, R., Baudet, C., Massaioli, F., and Succi, S. (1993) *Phys. Rev. E* **48**, 29.

[206] Stolovitzky, G. and Sreenivasan, K. R. (1993) *Phys. Rev. E* **48**, 33.

[207] Adzhemyan, L. Ts., Vasil'ev, A. N., and Gnatich, M. (1984) *Theor. Math. Phys.* **58**, 47.

[208] Yakhot, V. and Orszag, S. A. (1987) *Phys. Fluids* **30**, 3.

[209] Gnatich, M. (1990) *Theor. Math. Phys.* **83**, 601.

[210] Landau, L. D. and Lifshitz, E. M. (1960) *Electrodynamics of Continuous Media*, transl. of 1st Russ. ed. Pergamon Press, Oxford [Russian eds. 1956, 1982].

[211] Fournier, J.-D., Sulem, P.-L., and Pouquet, A. (1982) *J. Phys. A* **15**, 1393.

[212] Adzhemyan, L. Ts., Vasil'ev, A. N., and Gnatich, M. (1985) *Theor. Math. Phys.* **64**, 777.

[213] Camargo, S. J. and Tasso, H. (1992) *Phys. Fluids B* **4**, 1199.

[214] Adzhemyan, L. Ts., Hnatich, M., Horvath, D., and Stehlik, M. (1995) *J. Mod. Phys. B* **9**, 3401.

[215] Vaĭnshtein, S. I., Zel'dovich, Ya. B., and Ruzmaĭkin, A. A. (1980) *The Turbulent Dynamo in Astrophysics*. Nauka, Moscow [in Russian]; Vaĭnshtein, S. I. (1983) *Magnetic Fields in the Cosmos*. Nauka, Moscow [in Russian]; Moffat, H. K. (1978) *Magnetic Field Generation in Electrically Conducting Fluids*. Cambridge University Press, Cambridge.

[216] Pouquet, A., Fournier, J.-D., and Sulem, P.-L. (1978) *J. Phys. Lett. (Paris)* **39**, 199.

[217] Adzhemyan, L. Ts., Vasil'ev, A. N., and Gnatich, M. (1987) *Theor. Math. Phys.* **72**, 940.

[218] Bogolyubov, N. N. (1961) Quasiaverages in statistical mechanics problems. Preprint D-788. JINR, Dubna [in Russian].

[219] Kraichnan, R. H. (1965) *Phys. Fluids* **8**, 1385.

[220] Honkonen, J. and Nalimov, M. Yu. (1996) *Z. Phys. B* **99**, 297.

[221] Ronis, D. (1987) *Phys. Rev. A* **36**, 3322.

[222] Speer, E. R. (1969) *Annals Math. Studies*, Vol. 62.

[223] Zakharov, V. E. (1972) *Sov. Phys. JETP* **35**, 908.

[224] Zakharov, V. E. (1984) Collapse and self-focusing of Langmuir waves. *Fundamentals of Plasma Physics*, Vol. 2. Énergoatomizdat, Moscow [in Russian].

[225] D'yachenko, A. I., Zakharov, V. E., Rubenchik, A. I., Sagdeev, R. Z., and Shvets, V. F. (1986) *JETP Letters* **44**, 648.

[226] Pelletier, G. (1980) *J. Plasma Phys.* **24**, 421.

[227] Adzhemyan, L. Ts., Vasil'ev, A. N., Gnatich, M., and Pis'mak, Yu. M. (1989) *Theor. Math. Phys.* **78**, 260.

ADDENDUM

[228] Broadhurst, D. J., Gracey, J. A., and Kreimer, D. (1997) *Z. Phys. C* **75**, 559.

[229] Derkachov, S. E. and Manashov, A. N. (1998) *Nucl. Phys. B* **522**, 301.

[230] Derkachov, S. E. and Manashov, A. N. (1997) *Phys. Rev. Lett.* **79**, 1423.

[231] Broadhurst, D. J. and Kreimer, D. (1995) *Int. J. Mod. Phys. C* **6**, 519.

[232] Kreimer, D. (1995) *Phys. Lett. B* **354**, 117.

[233] Vasil'ev, A. N. and Vyazovskiĭ, M. I. (1997) *Theor. Math. Phys.* **113**, 1277.

[234] Adzhemyan, L. Ts., Vasil'ev, A. N., Kabrits, Yu. S., and Kompaniets, M. V. (1999) *Theor. Math. Phys.* **119**, 454.

[235] Gawedski, K. and Kupiainen, A. (1995) *Phys. Rev. Lett.* **75**, 3834; Bernard, D., Gawedski, K., and Kupiainen, A. (1996) *Phys. Rev. E* **54**, 2564.

[236] Chertkov, M., Falkovich, G., and Lebedev, V. (1995) *Phys. Rev. E* **52**, 4924; Chertkov, M. and Falkovich, G. (1996) *Phys. Rev. Lett.* **76**, 2706.

[237] Kraichnan, R. H. (1994, 1997) *Phys. Rev. Lett.* **72**, 1016; **78**, 4922.

[238] Adzhemyan, L. Ts., Antonov, N. V., and Vasiliev, A. N. (1998) *Phys. Rev. E* **58**, 1823.

[239] Kraichnan, R. H. (1968) *Phys. Fluids* **11**, 945.

[240] Adzhemyan, L. Ts., Antonov, N. V., Barinov, V. A., Kabrits, Yu. S., and Vasil'ev, A. N. (2001) *Phys. Rev. E* **63**, 025303 (R); **64**, 056306.

[241] Adzhemyan, L.Ts. and Vasil'ev, A.N. (1998) *Theor. Math. Phys.* **117**, 1223.

[242] Vasil'ev, A.N. (2000) *Theor. Math. Phys.* **122**, 323.

[243] Adzhemyan, L. Ts., Antonov, N. V., Kompaniets, M. V., and Vasil'ev, A. N. (2003) *Int. J. Mod. Phys. B* **17**, 2137.

Subject Index

Printed and bound by CPI Group (UK) Ltd, Croydon, CR0 4YY

24/10/2024

01778277-0018